B and T Cells in Immune Recognition

B and T Cells in Immune Recognition

Edited by
F. Loor and G. E. Roelants
Basel Institute for Immunology
Switzerland

A Wiley–Interscience Publication

JOHN WILEY & SONS
London · New York · Sydney · Toronto

Library of Congress Cataloging in Publication Data:
Main entry under title:

B and T cells in immune recognition.

'A Wiley–Interscience publication.'
Includes bibliographies and index.
1. Lymphocytes. 2. Cellular recognition.
3. Immune response. I. Loor, F. II. Roelants, G. E.
[DNLM: 1. Immunity, Cellular. 2. B-lymphocytes.
3. T-Lymphocytes. WH200 B11]
QR185.8.L9B18 599'.02'9 76-26913

ISBN 0 471 99438 3

Set on Linotron and printed in Great Britain
by J. W. Arrowsmith Ltd., Bristol.

Contributing Authors

Erika R. Abney — *Division of Immunology and Experimental Biology, National Institute for Medical Research, Mill Hill, London, NW7 1AA, England.* Present address: *Unitad de Biologia Experimental, Facultad de Medicina U.N.A.M., C.U., Apdo Postal 70343, Mexico, 20.D.F. Mexico.*

Peter C. L. Beverley — *ICRF Tumour Immunology Unit, Department of Zoology, University College London, Gower Street, London, WC1E 6BT, England.*

Harry G. Bluestein — *University of California, San Diego, School of Medicine, La Jolla, California 92037, USA.*

Peter A. Bretscher — *The John Curtin School of Medical Research, Department of Microbiology, P.O. Box 334, Canberra City, ACT 2601, Australia.*

K. Theodor Brunner — *The Lausanne Unit of Human Cancer Immunology, Ludwig Institute for Cancer Research,* and *Department of Immunology, Swiss Institute for Experimental Cancer Research, CH1011 Lausanne, Switzerland.*

Jean-Charles Cerottini — *The Lausanne Unit of Human Cancer Immunology, Ludwig Institute for Cancer Research,* and *Department of Immunology, Swiss Institute for Experimental Cancer Research, CH1011 Lausanne, Switzerland.*

Max D. Cooper — *Department of Pediatrics and Microbiology and the Comprehensive Cancer Center, University of Alabama in Birmingham, University station, Birmingham, Alabama 35294, USA.*

Jean Dausset — *Université de Paris VII, Institut de Recherches sur les Maladies du Sang, Hôpital Saint-Louis, 2 Place du Docteur-Fournier, F 75475 Paris Cedex 10, France.*

Didier Fradelizi — *Université de Paris VII, Institut de Recherches sur les Maladies du Sang, Hôpital Saint-Louis, 2 Place du Docteur-Fournier, F 75475 Paris Cedex 10, France.*

Kenneth H. Fye *Clinical Immunology Section, Veterans Administration Hospital, Department of Medicine, University of California, San Francisco, 4150 Clement Street, San Francisco, California 94121, USA.*

Joel W. Goodman *Department of Microbiology, University of California, San Francisco, Medical Center, San Francisco, California 94143, USA.*

Gerry G. B. Klaus *Division of Immunology, National Institute for Medical Research, Mill Hill, London, NW7 1AA, England.*

Abraham I. Kook *Department of Cell Biology, The Weizmann Institute of Science, Rehovot 76100, Israel.*

Nicole M. Le Douarin *Institut d'Embryologie du CNRS et du Collège de France, 49 bis Avenue de la Belle Gabrielle, F 94130 Nogent-sur-Marne, France.*

Francis Loor *Basel Institute for Immunology, 487 Grenzacherstrasse, CH Postfach 4005, Basel 5, Switzerland.*

Vincenzo C. Miggiano *Basel Institute for Immunology, 487 Grenzacherstrasse, CH Postfach 4005, Basel 5, Switzerland.*
Present address: *Universitá degli Studi di Calabria, Dipartimento di Biologia Cellulare, Arcavacata, Cosenza, Italy.*

N. Avron Mitchison *Tumor Immunology Unit, Department of Zoology, University College London, Gower Street, London WC1E 6BT, England.*

Haralampos Moutsopoulos *Clinical Immunology Section, Veterans Administration Hospital, Department of Medicine, University of California, San Francisco, 4150 Clement Street, San Francisco, California 94121, USA.*

Markus Nabholz *Basel Institute for Immunology, 487 Grenzacherstrasse, CH Postfach 4005, Basel 5, Switzerland.*
Present address: *Department of Immunology, Swiss Institute for Experimental Cancer Research, CH1011 Lausanne, Switzerland.*

Albert A. Nordin *Basel Institute for Immunology, 487 Grenzacherstrasse, CH Postfach 4005, Basel 5, Switzerland.*
Present address: *Gerontology Research Center, Baltimore City Hospitals, Baltimore, Maryland 21224, USA.*

John J. T. Owen *Department of Anatomy, The Medical School, The University of Newcastle-upon-Tyne, Newcastle-upon-Tyne NE1 7RU, England.*

R. Michael E. Parkhouse *Division of Immunology and Experimental Biology, National Institute for Medical Research, Mill Hill, London NW7 1AA, England.*

Georges E. Roelants *Basel Institute for Immunology, 487 Grenzacherstrasse, CH Postfach 4005, Basel 5, Switzerland.*
Present address: *International Laboratory for Research on Animal Diseases (ILRAD), P.O. Box 30709, Nairobi, Kenya.*

Lars P. Ryder *Tissue Typing Laboratory, University Hospital (Righospitalet), Blegdamsvej, 9, DK-2100 Copenhagen Ø, Denmark.*

Max H. Schreier *Basel Institute for Immunology, 487 Grenzacherstrasse, CH Postfach 4005, Basel 5, Switzerland.*

M. Seligmann *Laboratoire d'Immunochimie et d'Immunopathologie, (INSERM U.108), Institut de Recherches sur les Maladies du Sang, Hôpital Saint-Louis, 2 Place du Docteur-Fournier, F 75475 Paris Cedex 10, France.*

Myra Small *Department of Cell Biology, The Weizmann Institute of Science, Rehovot 76100, Israel.*

Jonathan Sprent *Tumour Immunology Unit, Department of Zoology, University College London, Gower Street, London WC1E 6BT, England.*
Present address: *Immunobiology Research Unit, Department of Pathology, University of Pennsylvania, School of Medicine, Philadelphia, Pennsylvania 19174, USA.*

A. Svejgaard *Tissue Typing Laboratory, University Hospital (Righospitalet), Blegdamsvej 9, DK-2100, Copenhagen Ø, Denmark.*

Norman Talal *Immunology and Arthritis Section (151T), Veterans Administration Hospital, Department of Medicine, University of California, San Francisco, 4150 Clement Street, San Francisco, California 94121, USA.*

Nathan Trainin *Department of Cell Biology, The Weizmann Institute of Science, Rehovot 76100, Israel.*

Contents

Foreword

NIELS K. JERNE

Immunology is probably the most rapidly advancing branch of vertebrate biology. This is the result partly of the continuing exponential growth of the number of immunological research workers, and partly of the spectacular advances achieved during the 1960s. In 1960, the primary structure of the antibody molecule was yet unknown. It was not even the target of much investigation, because the generally accepted theory of antibody formation by antigen instruction suggested that the primary structure of antibody polypeptide chains would be less interesting than their three-dimensional folding. This changed when it was demonstrated that the primary structure (the amino acid sequence) determines the specificity of an antibody molecule. As amino acid sequence is the expression of nucleotide sequence in DNA, it became clear that every antibody specificity is encoded in the genome of the cell that synthesizes this antibody. Instruction by antigen had to be given up; and the selection by antigen of cells already synthesizing fitting antibodies before the arrival of antigen became the central notion of immunological theory.

Also, in 1960, the involvement of the small lymphocyte in antibody formation was by no means generally accepted. The plasma cell was known to secrete immunoglobulin, but not until the early 1960s was it demonstrated that the small lymphocyte is a plasma cell precursor. It may sound strange that fifteen years ago immunologists were still searching for the cells in which the phenomena they had been studying for seventy years originate.

Thirdly, though graft rejection was of course known in 1960, it was then not evident that this phenomenon would turn out to be more intimately involved with the behaviour of the immune system than the induction by any other 'foreign' antigen of cellular immunity, or of antibody formation. This insight resulted mainly from genetic studies which showed that genes belonging to the major histocompatibility complex have a determining effect on the range of responsiveness of the immune system to foreign antigens.

Towards the end of that decade, the thymus and the bone marrow (and the bursa of Fabricius in birds) emerged as primary organs of lymphocyte differentiation; the fundamental distinction between T and B lymphocytes became established. All these advances in knowledge were accompanied by refinements in biochemical and biological techniques; lymphocyte cultures were shown to respond to antigen *in vitro*, both with antibody formation and with the development of cellular immunity.

It should be clear from this brief account of some of the outstanding advances, that a textbook written in 1960 would not have been of much use in 1970. The problems of providing a useful text for teachers and students of basic immunology has since remained with us. The rate at which new research data appear in the journals, and the impossibility for a single scientist of keeping abreast along the entire advancing edge of immunology make it impossible for one author to cover the entire field; instead, the cooperation of many authors, each dealing with the subject of his or her expertise has become mandatory. The present book is the result of the most recent endeavour along these lines. The authors chosen are all scientists of high repute, who have responded to the invitation of the editors with remarkably clear and succinct accounts of the present status in their fields of research.

The insight that the immune system consists of the total population of lymphocytes—that the lymphocytes *are* the immune system—is reflected in the first chapters which deal with various aspects of this cell. The lymphocyte is probably the most interesting vertebrate cell. Its enormous heterogeneity—resulting from the commitment of each cell to the expression of only one antigen-recognizing specificity—is not matched by any other cell type. This heterogeneity of synthetic potentialities, the diversity of the immune system, is the basis of its astonishing functional capabilities.

In studying the origin of this diversity, we have to go back to the early ontogeny of the lymphocytes and of the organs in which they differentiate under the stimulatory influence of specific differentiation antigens, or hormones. We also have to consider the migration patterns, and the lifespan and turnover of these cells. Particular interest is focused on the surface membranes, which carry the molecules that express the recognizing specificity of the cells as well as other structures that enable the cells to respond to signals from their environment.

Considerable space is rightly devoted to the behaviour of lymphocytes when cultured *in vitro,* in their response both to antigens and to allogeneic lymphocytes, and in their maturation to antibody secreting cells or to cells that can kill appropriate target cells. In all these responses we are confronted with phenomena of mutual cooperation and suppression among lymphocytes. Many of these phenomena appear to depend on signals and receptors that are determined by genes in the major histocompatibility region of the genome—a region that is distinct from the chromosomal regions that determine antibody specificity. The clarification of the relationship between these two sets of structural genes lies at the centre of current attempts towards basic immunological understanding since both appear to determine and restrict the recognition repertoire that is available to the immune system of an individual.

The presentations of the various physiological aspects of the immune system are followed by a number of reviews that examine the role of lymphocytes in pathological situations: cancer, autoimmunity, lymphocytic disorders, and the association of histocompatibility genes and antigens with transplantation and with histocompatibility associated diseases.

A final chapter reveals the difficulty of encompassing the multitude of immunological phenomena within a simplifying theoretical framework.

A basic immunologist would have liked to see the inclusion of a chapter devoted to the molecular genetics of the immune system: the analysis of the origin of lymphocyte diversity at the DNA level; also of a chapter on the emerging techniques for studying lymphocyte clones derived from single cells. The time is not yet ripe, however, for reviews of these important and rapidly advancing approaches to the solution of basic issues. Classical immunology has always had dual roots, in biology and in medicine. This dual motivation has led immunology to the central position it now occupies in modern biology. The contributions assembled here span the area of common interest to basic and clinical immunologists and the book thus maintains this tradition. As the understanding of an expert within a restricted area is naturally more advanced than the general understanding, these texts should retain their basic validity for both teachers and students for a number of years to come.

Basel Institute for Immunology, August 1976

Preface

During the last few years the field of cellular immunology has become extremely complex. Newcomers find it increasingly difficult to make their way through the abundant, often dispersed, literature while the specialists may lose touch with one or the other facet of the field.

This book attempts to collect our present knowledge of various topics chosen for their more immediate interest. It does not constitute a textbook nor does it pretend to cover all aspects of cellular immunology. Specialists of each subject were asked to treat their field *ex professo* and were left free to concentrate on the most salient points rather than to present a systematic coverage of the literature. They were also encouraged to interpret the phenomena in a personal way—which may not always be the one of the editors.

The immune system is dynamic in both time and space: during the animal lifetime it differentiates and it reorganizes itself continuously. The book starts with chapters entirely devoted to these 'accessory' problems of immune recognition, because the recognition of an antigen by the immune system is far more complex than the antigen–antibody reaction and is not a phenomenon limited to the binding of antigens to lymphocyte membrane receptors. Thus it recalls how primary lymphoid organs emerge and what the ancestors of the lymphocytes are (Chapters 1 and 2), that there is a definite heterogeneity among these lymphocytes (Chapter 3), that lymphocytes have definite pathways of migration, localization, and circulation throughout the body (Chapter 4), that the function of the whole lymphoid system is regulated by its own hormones and also influenced by hormones produced by other organs (Chapter 5) and that macrophages play a complex regulatory role in lymphocyte activation (Chapter 6).

Until recently most of these 'accessory' events were considered as only perturbing the understanding of basic phenomena; still, we have now to deal with these further complications, for—whether we like it or not—the immune response is a phenomenon which happens *in vivo,* and all the aforementioned complexity plays a role in its development. Should it be surprising, therefore, that immune responses obtained *in vitro* do not always reflect the responses shown by whole animals? A critical evaluation of one of the *in vitro* models most used in immunology, is presented in Chapter 7. This is the only chapter which we requested should consist mainly of actual original data. The very minimum conclusion that should be drawn from these data is that utmost caution is required when trying to understand *in vivo* phenomena with *in vitro* assay systems.

The book continues with the analysis of various aspects of the lymphocyte surface which is the site where most important steps of immune recognition take place. The physiology of the lymphocyte membrane is first surveyed (Chapter 8). Two contributors (Chapters 9 and 10), who are neither over-committed nor hypersensitive to the two current, irreconciliable concepts regarding the repertoire and the nature of recognition structures on B and T cells, then treat the major problems of what the actual recognition structures on the membranes of various lymphocyte types are.

The book then turns to the problem of what stimulates or represses the immune system. Besides the problem of B cell stimulation and tolerance (Chapter 11) we have chosen to develop two other topics which we consider especially attractive and important: what activates T cells (Chapters 12 and 13) and how T and K cells kill their targets (Chapter 14).

The final part of the book deals with some medical aspects of immune recognition. A few topics in which advances have been made recently were selected: the problem of involvement of B and T cells in cancer (Chapter 15), in autoimmunity (Chapter 16), in immunodeficiency and in lymphoproliferative diseases (Chapter 17); the problems of organ transplantation (Chapter 18) and the recently discovered association of certain HLA with specific diseases (Chapter 19). This does not mean that these topics are the most important from a humanistic point of view; for other problems, like resistance to parasites or virus infections, were they to be taken up and solved by modern immunology, would alleviate much more suffering than, let us say, heart transplants.

The final chapter of this book aims at an integration of the immune system, in terms of B and T cell activation (the problem of generation of diversity is not considered here). Clearly, it is one among many, for lymphocyte activation is a field especially flourishing in theories, hypotheses and models, and interpretation often primes or even precedes the mere production of solid data. Therefore, why among so many speculations did we select this activation and regulation model? Mainly, because it still has the favour of most immunologists while its basic principles will soon be ten years old. This long-lived characteristic makes it different from the many spectacular, but kite-like theories (as defined by M. Sela: 'they often fly high, but usually not for very long') which appear daily.

Within this scope, and given these limitations, we hope to contribute to clarifying current knowledge and focusing attention on crucial, unanswered questions.

April, 1976

Francis Loor, Basel
Georges E. Roelants, Nairobi

Chapter 1

Ontogeny of Primary Lymphoid Organs

N. M. LE DOUARIN

Abbreviations

CAM: chorio-allantoic membrane

1. INTRODUCTION

The cells responsible for the immune competence in the mature organism are found in a number of locations: blood and lymphatic vessels, lymph nodes, bone marrow and other tissues. Active research, especially that carried out on birds and mammals during the last fifteen years, has led to a new picture of how the immune system differentiates and to a considerable insight into the ontogeny of the immune function. Two families of lymphoid cells have been distinguished. One depends on the thymus—T lymphocytes—the other is non-thymodependent—B lymphocytes— and, in birds, differentiates within a cloacal lymphoepithelial organ, the bursa of Fabricius.[1] The crucial period, during which thymus and bursa influence the immunological development, occurs during embryonic and early postnatal life, when the lymphoid tissue is forming and immunological capacity is maturing. The thymus, the bursa of Fabricius and a still conjectural bursa equivalent in non-avian vertebrates, are the primary lymphoid organs in which lymphocyte precursor cells receive an inductive influence and then migrate into the widely deployed peripheral lymphoid system composed of lymph nodes, spleen, Peyer's patches, etc.

Experiments involving removal of the thymus and of the bursa of Fabricius at an early time of life made it possible to understand their respective functions.

By removal of the thymus from mice on the day of birth, Miller found that a correct and full development of immunological potential failed to occur.[2] In the decade following Miller's original discovery, an enormous amount of work on the immunological functions of the thymus was performed. It appeared that the thymus is the site of differentiation of a population of lymphocytes responsible for cell-mediated immunity and is essential to the development of lymphocytes involved in delayed hypersensitivity reactions, in allograft rejection and in the initiation of graft-versus-host reactions. By contrast neonatally bursectomized chicken fail to form humoral antibodies but maintain their power to reject skin allografts. The bursa dependent lymphocytes—B lymphocytes—were thus considered as responsible for humoral immunity.[3–6]

The developmental processes through which epithelio-mesenchymal organs like the thymus and the bursa of Fabricius undergo lymphoid differentiation have been the subject of a long debate in embryological literature. Both the origin of lymphocytes and the respective contribution of endoderm, ectoderm and mesenchyme to thymus and bursa histogenesis were the subject of various interpretations.

This chapter will briefly review the embryological origin of the thymic primordium in the Vertebrate series and report recent investigations on the histogenetic processes which occur during the ontogeny of the thymus and the bursa of Fabricius.

2. THE THYMUS

The presumed origin of the thymic primordium in the various vertebrates, which will be first considered, is essentially inferred from histological analysis during their embryonic development. Experimental analysis of the origin of the cells constituting the thymus will be presented in a second section.

It should still be pointed out that the actual immunological function of the thymus is well documented only for Aves and Mammals; only little information has been produced concerning the thymus-equivalent organs found in the other vertebrates.

2.1 Origin of the thymic primordium in vertebrates

2.1.1 *Cyclostomes*

Müller[7] and Stannius[8] were among the first to describe a 'thymus' in the Atlantic hagfish, *Mixina glutinosa*, which, in fact, was later shown to be the pronephros! Cole described a lymphoid organ in the pharyngeal velum[9] which was considered by Kampmeier as a lymph propulsor rather than as a thymus.[10] Finally it has been generally accepted that the hagfish has no thymus. When challenged with appropriate antigens, the Pacific hagfish is, however, capable of a variety of immunological responses.[11–13] The search for a thymus in young specimens of *Eptatetrus stontii* resulted in the discovery of a phagocytic and antigen receptive cell population associated with the pharyngeal velar muscle complex which probably contains a prothymus or a precursor of the thymus of higher vertebrates.[14]

In larvae of *Lampetra* a primitive thymus is formed by seven median placodes located on the epipharyngeal fold.[15] However, no thymus is present in the lampreys.[16,17]

2.1.2 *Selacians*

In the *Rajidae* the thymus originates from the first six branchial pouches which proliferate and give rise to thickenings except for the derivatives of the first and sixth pouches, these become separated from the pharynx and fuse in a quadrilobulated thymus. A similar developmental process is observed in *Squalus* and *Spinax* for instance. In certain species like *Chimaera* and *Scyliorhinus* the thymus has only three lobes, owing to the early regression of the primordium originating from the fourth pouch. The thymus degenerates in Selacians when sexual maturity is reached.[18,19]

2.1.3 *Chondrosteans*

In this group the thymus arises from the first four branchial pouches. In *Acipenser ruthenus* the primordium of the first pair degenerates. The thymus is still present and well developed in 3-year-old animals of this species.[20]

2.1.4 *Holosteans*

In *Amia* the thymic rudiment is formed by the second, third and fourth pharyngeal pouches which give rise to a paired thymus located above the branchial cavity.[21]

2.1.5 *Teleosts*

The thymus remains attached to the pharynx and the thymic tissue is in continuity with the endodermal epithelium. It first appears as two symmetrical masses on the internal side of the branchial cavity and starts to involute when the animal reaches sexual maturity.

2.1.6 *Dipneusts*

The thymus of *Lepidosiren* originates from epithelial buds arising from the dorsal side of the second, third and fourth branchial pouches. The fourth buds degenerate and the definitive organ results only from the fusion of the second and third rudiments.[22]

2.1.7 *Amphibians*

Thymic histogenesis has been studied in various species.

In Urodels and Anurans the thymus arises from the pharyngeal epithelium of the second branchial pouch.

In *Xenopus laevis* larvae the thymic primordium becomes separated from the endoderm lining the pharyngeal cavity at stage 45 of Nieuwkoop and Faber.[23] At stage 49, considerable growth and differentiation of the organ has occurred; cortex and medulla are easily distinguishable.[24]

In *Rana pipiens*, by the end of the first week after fertilization, when the animals have reached the last embryonic stage (stage 25, of Shumway),[25] the two thymus rudiments are small clusters of epithelial cells on both sides of the head in the vicinity of the developing auditory vesicles. During the next three weeks, i.e. the first four larval stages of Taylor and Kollros,[26] the thymus undergoes major changes and becomes lymphoid.[27]

2.1.8 *Reptiles*

In Reptiles the thymus is derived from outgrowths of the dorsal side of the branchial pouches while the parathyroids are formed ventrally. The level of origin of the thymic buds varies according to the species considered. For instance the thymus originates from the pouches II and III in the Lizzard, IV and V in Ophidians and III in Turtles (according to Grassé).[28]

2.1.9 *Aves*

Verdun[29] first reported that the thymic rudiment in birds originates from pharyngeal pouches III and IV. This view has been confirmed by more recent observations.[30–34]

The thymic epithelial rudiment arises from the dorsal wall of pouches III and IV with major contribution of pouch III. The thymus is first visible as a rudiment in the 4-day chick and quail[34] and in the 3-day duck (*Anas platyrhynchos*) embryos.[35]

In chick and quail embryos the thymus primordium is at first a mass of epithelial cells which separates from the pharynx at 5 days and elongates to form a cord extending along the jugular vein.

The endodermal primordium is surrounded by a thin mesenchymal capsule, later responsible for the lobulation of the organ concomitant with its vascularization.

Lymphoid differentiation takes place at 10 days in the duck[31] and the quail and at 12 days in the chick.[34]

2.1.10 *Mammals*

The thymus was considered as a mesodermal derivative before Kölliker observed in early rabbit embryo that the thymic rudiment was actually made up of a tubular epithelial structure associated with the branchial pouches.[36]

Thymus histogenesis has been described only in a few forms of mammals and appears to originate from different levels of the pharynx according to the species considered.

2.1.10.1 *Eutherians*

Schematically, the various observations reported on *eutherian mammals* can be summarized as follows:

(1) In some species like rat,[37] mouse[38–40] and man[41] the thymus arises exclusively from a ventral outgrowth of the pharyngeal pouch III. In the rat, the fourth pouch regresses early and neither thymus IV nor parathyroid IV rudiments develop.
(2) In most higher Mammals, the primordia arising from the third pouches are *by far* the most important contributors to the thymus but a small amount of thymic tissues arises also in some species from pouch IV. Such is the situation for instance in the pig (see Reference 42), the cat, the dog and the calf.[43]

Although in most cases, an exclusively endodermal origin of the thymus has been recognized, the participation of ectoderm of *sinus cervicalis* to thymic histogenesis has also been suggested in certain species;[44–46] this point is highly controversial (see Reference 37) and more thorough studies of the fate of the *sinus cervicalis* are required in various species of Mammals.

In all the species studied, the first *Anlage* of the thymus appears as an epithelial tube surrounded by a thin capsule of mesenchyme. The thymic epithelium detaches early from the pharynx and migrates in a cranio–caudal direction. During its caudal movement the thymic rudiment can be fragmented in a cervical and a thoracic lobe (for instance in *Ovis aries*).[47]

In the mouse embryo, on the tenth day of gestation, the thymic rudiment is still in relation with the pharynx at the level of the third branchial pouch. At this stage, the thymic primordium is composed by only two cell types—pharyngeal endoderm separated from the mesenchyme by a basement membrane. At the end of the eleventh day, thymic epithelium appears as masses of pale-staining epithelial cells in close relation to the third branchial arch artery and ganglion nodosum. From the fourteenth day, lymphoid differentiation begins.[48]

2.1.10.2 *Marsupials*

In Marsupials the position of thymic bodies may be cervical and thoracic or purely cervical as is the case in eutherian mammals. According to Fraser and Hill the source of the epithelial components undergoing thymic transformation is variable in the different species observed.[49] In *Trichosurus* the thymus is considered as derived from ectoderm of the cervical vesicle and endoderm from both pharyngeal pouch III and IV producing respectively cervical and thoracic bodies. In *Phascolomys* and *Phascolarctus* the thymus arises from the cervical vesicle and related endoderm but only in the cervical location. On the other hand, the cervical vesicle of *Parameles* does not participate in thymic histogenesis which entirely depends on the development of endodermal pouches III and IV. A similar observation has been reported for the Opossum.[50] In this species, however, an accessory thymus body frequently develops from the cervical vesicle.

2.2 Experimental analysis of thymic development

2.2.1 *Embryonic origin of the epithelium and mesenchyme of the thymic primordium*

Although the epithelio-mesenchymal structure of the primary thymic rudiment has been recognized early, the precise embryonic origin of both epithelium and mesenchyme of the thymus was not clearly established. On the other hand, owing to the lymphoid differentiation of the organ, the evolution of its two components is difficult to follow and thus is not well understood.

According to descriptive studies of thymic development in various forms of vertebrates, the epithelium is considered either endodermal or ectodermal or dual in origin. Experimental studies were carried out to clarify this question in the avian embryo. From excision and marking experiments, Hammond suggested that in the chick the thymus is a derivative of the branchial ectoderm located in the dorsal region of the pharynx corresponding to pouches III and IV.[33] According to this author, endoderm–ectoderm interaction is necessary for the development of branchial ectoderm into epithelium of the thymic *Anlage*.

Although a certain contribution of the pharyngeal ectoderm to thymic histogenesis cannot be completely excluded, Le Douarin and her collaborators showed that the precursor of thymic reticular cells is actually the pharyngeal endoderm of the third and fourth branchial pouches.[51] The ventro-lateral part of the pharyngeal endoderm can be separated by trypsinization from the underlying mesenchyme between 6- and 30-somite stages (Figure 1).[52] If taken from embryos at the 15-somite stage onwards, pharyngeal

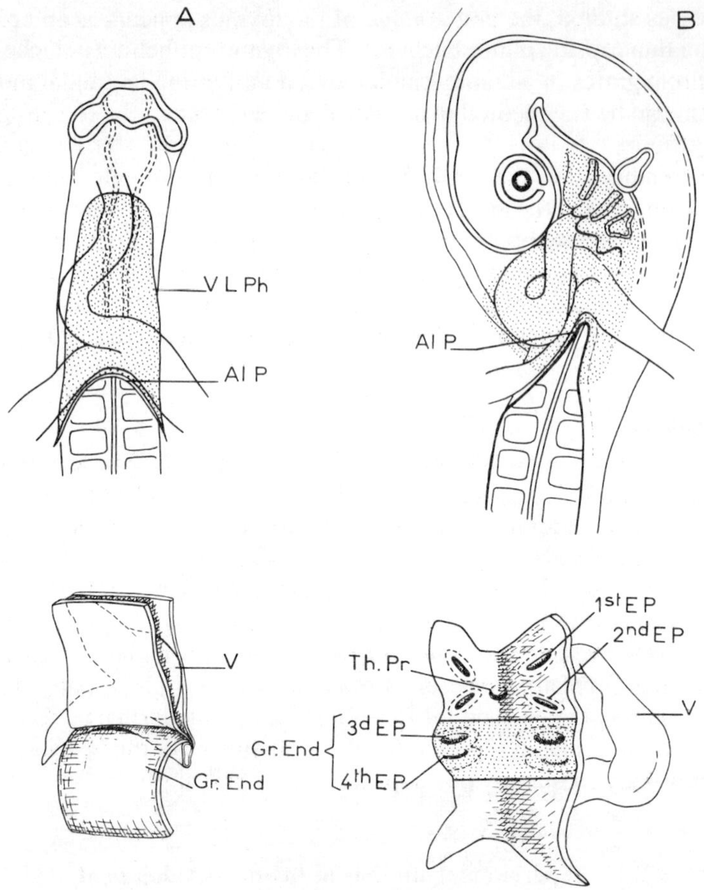

Figure 1. Schematic representation of the endodermal area taken from quail embryos at various stages of development involving the presumptive thymic epithelium. (A) In 15- to 22-somite embryos the ventrolateral wall of the pharynx (V L Ph) is cut and the pharyngeal endoderm (Gr. End) is isolated by trypsinization from the cardiac fold and grafted into the somatopleure of a chick embryo. (B) In older embryos a piece of the ventrolateral endoderm including the third and fourth branchial pouches (3d, 4th EP) is selected for the graft. AIP: anterior intestinal portal; V: ventricle; Th. Pr: thyroid primordium

endoderm can differentiate into a thymus when associated with appropriate mesenchyme, proving that no intervention of ectoderm is required for thymic differentiation to occur.[53]

The embryonic origin of the thymic mesenchyme was investigated in birds[54] using the quail-chick marker system. This cell marking technique[55,56] is based on structural differences in the interphase nucleus of the chick and quail. In the Japanese quail (*Coturnix coturnix japonica*) a large amount of heterochromatic DNA is conspicuous in all embryonic and adult cells. This feature is unusual in birds or mammals and especially does not exist in the chick. In this species, the chromatin is evenly dispersed in the nucleoplasm with some small chromocentres and does not participate much in nucleolar structure. Owing to the different disposition of the chromatin, quail and chick cells experimentally associated in chimeras, can be easily distinguished (see Figure 2 (Plate I)).

We have studied the embryogenesis of the pharyngeal region in a series of experiments involving transplantation of quail tissues into chick or the reciprocal.[54,57] Our attention was essentially focused on the contribution of the mesectoderm (i.e. the mesenchyme derived from the ectoderm via the cephalic neural crest) to the head and the neck in higher vertebrates.[58] Results of heterospecific grafts of the neural rudiment between quail and chick embryos (Figure 3) lead to the conclusion that the whole mesenchyme of the

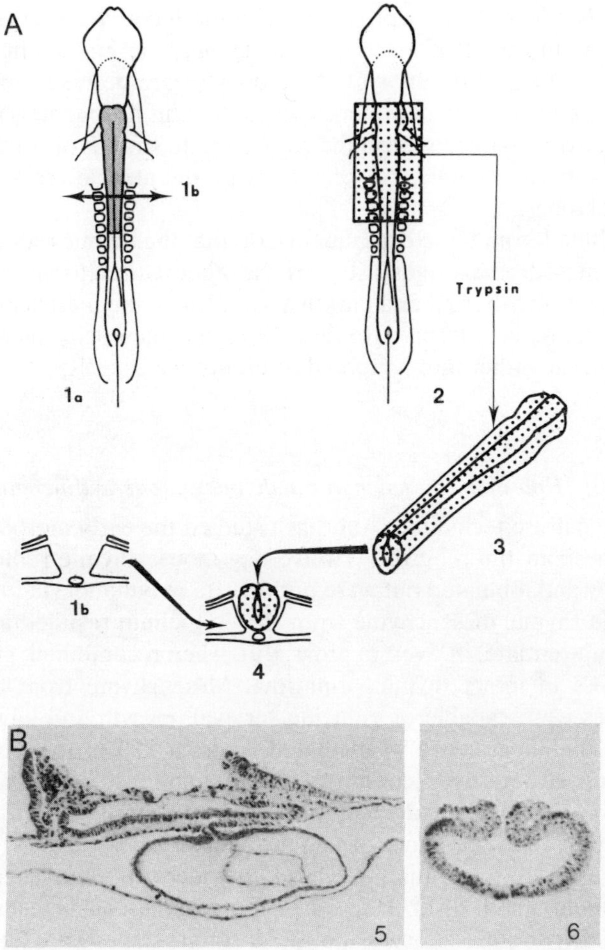

Figure 3. Isotopic and isochronic graft of a quail rhombencephalon into a chick embryo at 6- to 9-somite stages. (A) Schematic drawing of the experimental procedure. (1) The rhombencephalon (hatched area) is surgically removed from the chick host embryo. (a) Dorsal view; (b) transverse section at the level of the excision. (2) The corresponding area is taken in a quail embryo at the same developmental stage and submitted to a solution of trypsin. (3) The quail rhombencephalon is isolated by trypsinization and in (4) grafted orthotopically into the chick host. (B) Transverse sections of the chick embryo (5) and of the isolated quail neural tube (6) before the graft. Feulgen-Rossenbeck. (5) × 80; (6) × 180

branchial arches is derived from the mesencephalic and the rhombencephalic neural crest except for the muscle plate which is formed by a central core of mesodermal cells.[5,7]

If isotopic and isochronic grafts of quail rhombencephalic primordium are carried out into chick embryos at the 6- to 9-somite stage the third and fourth arches of the chick host are invaded by quail cells and the thymic rudiment is chimeric from its early development. The host endodermal pouch and later on the thymic epithelial cord are surrounded by quail mesenchymal cells (see Figure 4 (Plate II). When thymus lobulation occurs the mesenchymal cells penetrate the epithelium with the blood vessels (see Figure 5 (Plate III)). Interestingly, in this kind of chimerical thymus, the endothelium of the thymus capillaries is always of host origin while the pericytes are derived from the ectodermal mesenchyme. The contribution of mesenchyme to thymus histogenesis appears restricted to the interlobular connective tissue in the cortex and to strands of connective cells lining the blood vessels in the medulla. No cells showing the nuclear marker are seen in the lymphoid cortical lobes.

It can be concluded from these experiments: (a) that the thymic rudiment is invaded by capillary buds of mesodermal origin just as are the other tissues (for instance, the dermis of the face and neck or the thyroid and parathyroid glands), the mesenchyme of which also originates from the mesectoderm;[57] (b) that the thymic mesenchymal component has no ability to differentiate either into lymphoid or into reticular cells.

2.2.2 *Role of the mesenchyme in thymic endoderm differentiation*

Using an *in vitro* culture technique, Auerbach studied the early morphological events of thymus histogenesis in the mouse.[59] Twelve-day mouse thymic rudiments cultured in plasma clots grew and lobulated but were not the site of lymphocyte formation. Trypsin-separation of the thymic mesenchyme from the epithelium resulted in failure of either component to differentiate, or even to grow. But, when recombined, growth and lobulation occurred like in intact thymic primordia. Mesenchyme from a wide variety of embryonic tissues were capable of inducing survival, growth and lobulation of thymus epithelium and this effect could be mediated across a 20 μm thick millipore diffusion membrane. Using an improved culture method, lymphoid development in embryonic thymus explants could be obtained.[60] Lymphoid cells appeared within the epithelial part of the *Anlage* but only in the presence of mesenchyme.[61,62]

The developing potencies of the pharyngeal endoderm were investigated in the avian embryo by Le Douarin and coworkers. The pharyngeal mesenchyme and the endodermal rudiment from chick or quail embryos at 6- to 30-somite stages were separated by trypsin. Then, the endoderm was associated, either in culture or by grafting, with various types of mesenchymes belonging to 3–5-day-old embryos. Thymic potencies first appeared in the endoderm around the 15-somite stage[53] and they developed only in contact with mesenchyme of lateral plate origin. Neither somitic nor limb bud mesoderm could stimulate endoderm growth and subsequent lymphoid differentiation of the *Anlage*. On the contrary, a thymus developed when the endoderm was grafted into the somatopleure or in the splanchnopleure of 3-day-old embryos. Thymic differentiation was also obtained when pulmonary mesenchyme from 5-day embryos were associated *in vitro* with the pharyngeal endoderm.

Similar observations were made concerning various other endodermal derivatives. Liver and pancreatic endoderm for instance could not differentiate when associated with mesenchyme of dorsal origin, while their evolution was normal with any kind of lateral plate mesodermal derivatives.[51,52]

2.2.3 *Origin of thymic lymphocyte stem cells*

There have been several theories to explain the appearance of lymphocytes in the thymus *Anlage*. As early as 1879, Kölliker considered that thymus lymphocytes and reticular cells were entirely derived by *transformation* of epithelial cells of the early primordium.[36] This view was subsequently supported by several authors who claimed to demonstrate transitional forms between thymic epithelial and lymphoid cells.[48,63–76] The *transformation* theory was modified by de Winiwarter[77] and Deanesly.[78] The latter considered that while the thymic lymphocytes were of epithelial origin the reticular cells were derived from mesenchymal cells invading the epithelial rudiment.[78] De Winiwarter proposed a dual origin for thymic lymphocytes: in the early stages of thymus development they would arise from epithelial cell transformation but later on they would derive from prethymic mesenchymal cells invading the epithelium.[77]

According to the *substitution* theory, proposed by Hammar,[79–82] lymphocytes arise exclusively from thymic mesenchymal cells invading the epithelial *Anlage* which would be responsible for the formation of reticular cells and Hassal's corpuscles. The studies of Maximow[83,84] showed that lymphocyte precursors—the so-called *Wanderzellen*—originated from undifferentiated mesenchyme cells which first became large amoeboid basophilic cells and then migrated into the thymic epithelium. The *substitution* theory was further supported by a number of studies carried out in several species: pigs,[46] guinea pigs,[85,86] humans,[87] and birds.[31,88,89]

All these conclusions were based on simple histological observations of developing thymuses only. Experimental studies of thymic histogenesis, using tissue culture and grafting techniques, also provided conflicting results. The first attempts to culture thymic rudiments *in vitro* failed to demonstrate lymphopoiesis.[90–92] When grafted either in the mouse anterior eye chamber or on the chick chorio-allantoic membrane (CAM), prelymphoid mouse thymus also failed to show lymphopoiesis.[93,94] However, lymphoid development was obtained in 12–12$\frac{1}{2}$-day mouse thymus in culture either *in vitro*[60] or on the CAM.[59,61] By grafting on the CAM chimeric combinations of prelymphoid mouse thymic epithelium and chick mesenchyme, Auerbach further concluded that the epithelial component produced lymphocytes while the mesenchyme generated connective tissue and glandular stroma. This experimental approach of thymic histogenesis provided powerful support for the transformation theory, i.e. the epithelial origin of thymic lymphocytes.

Recently, Turpen, Volpe and Cohen[95] investigated the origin of thymic lymphocytes in *Rana pipiens*; they concluded that in amphibians the lymphocytes are derived from prelymphoid cells which originate in the thymic rudiment itself, although it is unclear whether they come from the endoderm or from the mesenchyme. In Anurans the primary *Anlage* of the thymus is derived from the epithelium of the second pharyngeal pouch of the gill area. The gill arch region was removed in 72-hour-old embryos and reciprocally exchanged between diploid and triploid specimens prior to thymus differentiation and the appearance of blood islands and vascular system. The relative percentages of host and donor cells which persisted in the differentiated transplant and which populated the peripheral lymphoid tissue of the host were analysed in post-metamorphic chimerical animals. It was concluded that the vast majority of lymphoid cells of the frog thymus were not derived from circulating embryonic mesenchymal cells but differentiated from elements belonging to the thymic rudiment itself. The differentiated thymic lymphocytes subsequently migrated into the peripheral lymphoid organs, the kidney, spleen and bone marrow. In early post-metamorphic life, the thymus lymphopoiesis was self-sustaining since during that period no sign of an afferent stream of cells entering the thymus could be found.

Completely different results were obtained in *Pleurodeles waltlii* by Deparis and Jaylet.[96] When the liver rudiment was transplanted between diploid and tetraploid embryos at several developmental stages, all haemopoietic organs of the host—including the thymus—were subsequently colonized by cells originated from the grafted granulopoietic tissue. In addition, 18 months after transplanting the thymus rudiment between diploid and tetraploid larvae, the lymphoid population of the grafted thymus was entirely of host type. Therefore, thymic lymphocytes did not actually have an intrinsic origin but were rather derived from the granulopoietic liver tissue. Embryonic aspects of avian thymic lymphopoiesis were also studied with cell tracer techniques. Moore and Owen[97] used a sex-chromosome marker system in paired chick embryos, which were joined by vascular anastomosis of chorio-allantoic or yolk-sac blood vessels. Chromosome analysis following yolk-sac anastomosis at 4–5 days of incubation revealed high levels of chimerism in the thymus (44%–70%). When the anastomosis was established later in development, only low levels of thymic chimerism were found. This suggests that an inflow of blood-borne stem cells is responsible for lymphoid differentiation in the chick thymus; the correlation between the time of parabiosis establishment and the extent of chimerism further showed that lymphoid precursor cells enter the thymus rudiment at early developmental stages. The haematogenous theory of thymic lymphocyte origin both in the mouse and in the chick is supported by the experiments reported above and also by the results obtained by Owen and Ritter:[98] when mouse or chick thymic rudiments of different developmental stages were cultured in cell-impermeable diffusion chambers on the chicken chorio-allantois, 10-day mouse or 7-day chick thymic rudiments failed to become lymphoid while thymuses were able to sustain lymphopoiesis.

Since the chromosome marker system gives information only for dividing cells, a possible contribution of endodermal or mesodermal thymic components could not be completely excluded. The quail-chick marker system was thus applied to this problem of the origin of thymic lymphocytes, since it allows evaluation of the composition of the whole cell population of thymic tissue at a given time through simple histological observations.[54,99] Three-day-old chick embryos were grafted into the somatopleure with quail grafts either of the presumptive thymic endoderm or of whole thymic rudiments taken at two different periods of development: either 15-somite stage to 4 days, or 4 days to 10 days. The thymus was then removed at 14 days of total age (i.e. age at the time of the graft plus duration of the graft), fixed, and stained by the Feulgen-Rossenbeck procedure.

When only quail endoderm was grafted, a chimerical thymus developed: its reticular cells showed the nuclear marker characteristic of the quail graft while the connective tissue, the endothelium of the blood vessels and the whole lymphoid population were of host (chick) type (see Figure 6 (Plate IV)). Therefore the endoderm of the thymic rudiment did not differentiate into lymphocytes. Whether the thymic mesenchyme itself could participate to some extent in lymphopoiesis, when transplanted in association with the endoderm, was then tested. When complete thymic rudiments (endoderm plus pharyngeal mesenchyme) from 15-somite to 4-day quail embryos were transplanted, all thymic lymphocytes were again of host origin, just as when the thymic endoderm was grafted alone. However, most of the connective elements were now of graft type, though some contribution of the somatopleural host mesenchyme was still observed. The reticular cells were again of graft type. Thus the thymic mesenchyme is also unable to differentiate into lymphocytes, confirming the result obtained by selective labelling of the thymic mesenchymal component in the experiment reported above.

Since the whole lymphoid population of the avian embryo thymus is of extrinsic origin, similar experiments were carried out to find out at which stage thymic rudiments are colonized by blood-borne stem cells. The thymus *Anlage* taken from 4–10-day-old quail and chick embryos were used as heterospecific grafts and examined at 14 days of total age. This is schematically represented in Figure 7. The inflow of lymphoid stem cells in the thymus rudiment of the quail takes place for about 24 hours during the sixth day of development. In the chick embryo it starts on the second half of the seventh day of development and lasts 36 hours. Indeed, when the thymus *Anlage* to be heterospecifically grafted is taken at earlier times than day 5 (quail) and day $6\frac{1}{2}$ (chick) the lymphocytes found in the thymus at total age of 14 days are exclusively of host type. When, on the other hand, the donor of the thymus graft is older than 6 days (quail) and 8 days (chick), the thymus

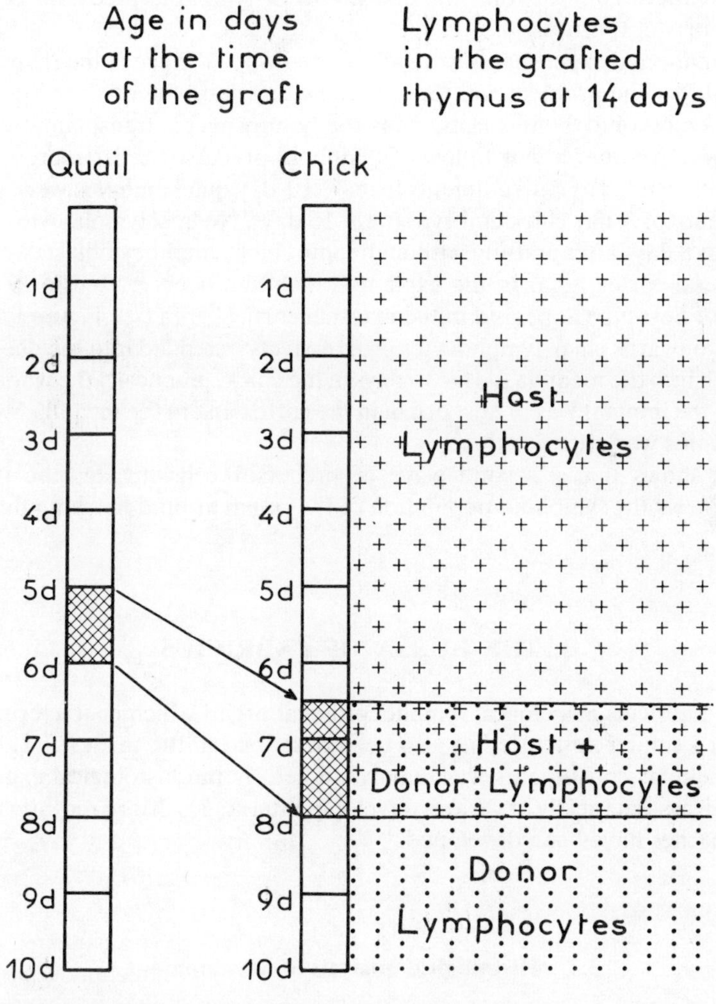

Figure 7. Schematic representation of the timing for the colonization of the thymus rudiment by blood-borne lymphoid stem cells. See text for details

lymphocytes at day 14 total age are of donor type, showing a very slow—if not completely non-existent—penetration of stem cells in the grafted thymus. A clear mixture of host and donor lymphocytes is obtained only for grafts which are made during the critical period mentioned above.

The presence in the embryo of thymic lymphocyte stem cells before the stage when they normally participate in thymus histogenesis, was demonstrated as follows: thymic rudiments were dissected from 4-day quail embryos, grafted for 2 days in 3-day chick somatopleure and then retransplanted into a 3-day quail for 8 days (schematically shown in Figure 8). At this time the quail thymus was thus 14 days old; its lymphoid population was of chick type, while its reticular and connective cells exhibited the quail nuclear marker. This observation shows that haemocytoblasts able to colonize the thymus rudiment are available in the embryo as early as the fourth or the fifth day of incubation, the quail thymus being invaded by host stem cells at this stage. On the other hand, the fact that none, or very few, quail cells were found in the thymus after the second transplantation in quail embryo confirms that the first inflow of lymphoid precursor cells into the thymus is a short-term process.

The aforementioned experiments were all stopped when the age of the thymic rudiment graft was 14 days. The problem was then to know how long the non- or low attractivity period of the embryonic thymus lasts. Thus the heterospecific transplantation time was prolonged to see whether a new inflow of lymphoid precursor cells entered the thymus during embryonic life. Thymic rudiments from 7–10-day quail embryos were grafted into the somatopleure of 3-day chick embryos for 8–15 days. No host lymphocytes were found in quail thymus 8 days after grafting, although some chick lymphocytoblasts were detected in the peripheral cortex area. In the explants which had been grafted for 9 days, large lymphocytes of host type appeared in the external cortical area (see Figure 9 (Plate IV)). From 10 days onwards chick lymphocytes progressively extended into the deeper cortical layers and then into the medulla. After 15 days in the chick, practically the whole lymphoid population of the implant was of host origin in the cortex and in the medulla, whatever the age of the thymus was after grafting.

This result shows that a second wave of precursor cells invades the thymus and completely renews the lymphoid population of the organ around hatching time.

3. THE BURSA OF FABRICIUS

The bursa of Fabricius is a cloacal lymphoepithelial organ which characterizes the class Aves, though a similar structure has also been described in the turtle.[100] The ontogenic origin of its cellular constituents was mainly studied by the histological examination of embryos taken systematically at various developmental stages. More recently new experimental approaches have been developed.

3.1 Histological analysis of development

The primordium of the bursa develops in relation with the proctodeum and the anal plate. In the chick embryo it first appears by 4–5 days of incubation as a median lamina of endodermal epithelium proliferated dorsally and caudally from the anal plate between the

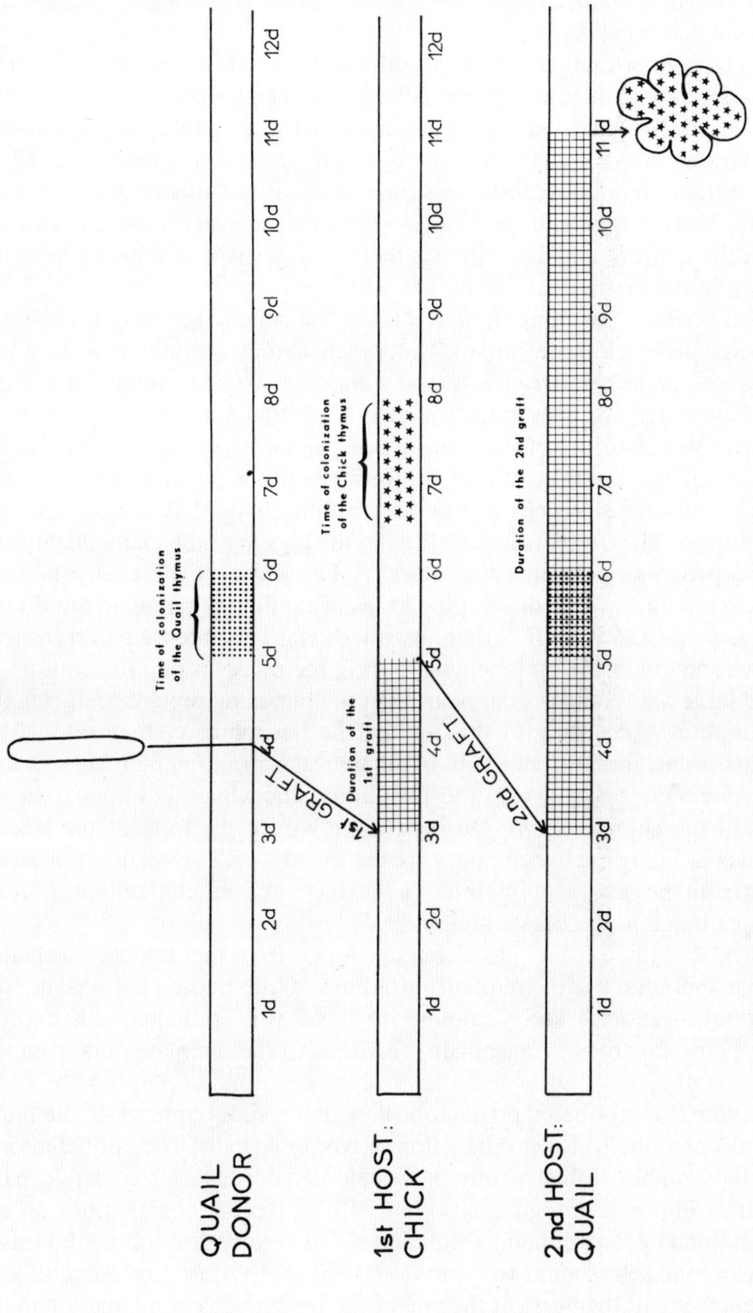

Figure 8. Diagram of the experimental procedure devised to test the availability of haemopoietic stem cells in the chick embryo before the normal time of thymic colonization. A quail thymic rudiment from a 4-day-old embryo is grafted into a 3-day-old chick for 2 days and then transplanted into a 3-day quail host embryo. When the total age of the grafted thymus reached 14 days it is observed histologically and shows a lymphoid cell population entirely of chick origin. This demonstrates that the stem cell inflow into the quail thymus occurred exclusively during the first transplantation in the chick host

latter and the caudal intestine.[101–103] Recent work suggested a possible participation of the proctodeal ectoderm which is in contact with the endoderm in the anal plate.[104,105]

At 6 days, the epithelial rudiment of the bursa projects from the dorso-caudal corner of the urodeal membrane as a slightly rounded protuberance with a convex caudal margin. At this stage, the epithelium of both the urodeal membrane and the bursa begins to vacuolate as a result of epithelial cell breakdown.

At 7 days, the bursal primordium develops laterally and extends further caudally. The proctodeum is now curved with its fundus turned caudad.[102] The vacuolated spaces of the bursal epithelium coalesce to form a lumen lying in line with the proctodeal sinus. The two cavities remain separated for a short time by an ecto-endodermal membrane which ruptures on the seventh or eighth day. From this time onwards the bursa rotates forward around the transverse body axis and on the ninth day becomes oriented vertically and lies almost parallel with the coprodeum. The bursa is then oval in shape and grows out on a cylindrical stalk open in the proctodeal cavity.

By the tenth to the eleventh day, longitudinal folds of the cuboidal or cylindrical lining epithelium develop and project into the lumen. The mesenchymal component of the organ differentiates into a thin *serosa,* a *muscularis* and a *mucosa* which develops as a loose vascular network of mesenchymal cells in the folds of the epithelium. The ventral surface of the bursa lies on the dorsal surface of the coprodeum and the bursal serosa is fused with the serosa of the coprodeum. Dorsally, it is in continuity with the peritoneum.

At 12 days, groups of epithelial cells proliferate forming buds that project into the underlying tunica propria. The bud formation starts in the region of the gland distal from the proctodeum and progresses towards the latter.[106] The adjacent mesenchymal cells progressively encircle the buds, while developing loops of capillaries surround but do not penetrate them. Ackerman and Knouff[107] demonstrated relatively intense and transitory alkaline phosphatase activity of the mesenchyme during the phase of bud formation.

Jolly[102] described large amoeboid cells appearing in the bursal mesenchyme during the twelfth day of incubation. According to this author the basophilic cells develop by a transformation of the stellate mesenchyme cells of the tunica propria and then migrate into the epithelial bud; there they give rise to the lymphoid cells which first appear in the follicles at 16 days in the chick embryo. During the growth of the follicles the closely associated epithelial cells lining the bursal cavity are the site of a cytological differentiation and become distinct from the general epithelium. In the duck the epithelial cells proliferate and enlarge forming a bulge in the bursal lumen.[102]

Electron microscopy study of the chick bursa shows that the follicle-associated epithelium has dense cytoplasm with frequent infoldings of the plasma membrane and numerous apical tubules, vesicles and vacuoles. At this level, epithelial cells express pinocytotic activity, being capable of transporting ferritin and Indian ink tracers from the lumen.

It is suggested that the transepithelial pinocytotic flow of intestinal contents in the bursa may provide a significant stimulus for proliferation of lymphoid cells. The epithelial cells not associated with the lymphoid follicles contain flocculent material localized between the nucleus and the bursal lumen (see Figure 10 (Plate V)).[108] Recent observations on the bursal primordium stained by Pappenheim's technique[109] at various developmental stages have shown that basophilic cells similar to those described by Jolly are first detectable in the peripheral mesenchyme of the bursa at the end of the seventh day of incubation in the chick embryo and at $6\frac{1}{2}$ days in the quail embryo. Later on, their number increases significantly in the mesenchyme. Some of them remain in the tunica propria where they differentiate into granulocytes, while others migrate towards the endoderm which they actively penetrate from the ninth day of incubation in the quail and eleventh day in the chick.

3.2 Experimental analysis of development

The first attempt to analyse the developmental mechanisms which take place during bursa organogenesis is due to Moore and Owen.[110] They used the same method as for the analysis of thymus development, i.e. the production of parabiotic union with vascular anastomosis between 6–11-day chicken embryos, which could be distinguished by sex chromosome markers. After several days of incubation, the embryos were treated with colcemid and the bursas were prepared for chromosome analysis. In embryos parabiosed at 6–11 days of incubation and sampled at 17–20 days, a cell chimerism was observed: between 26% and 50% of the dividing cells scored in the bursas of the parabionts were derived from the opposite partner. Since the majority of mitoses at this stage were confined to the follicles, the authors concluded that the chimerism actually involved the lymphoid cell population implying that an afferent stream of cells enters and proliferates in the bursa during embryogenesis and that the majority of bursal lymphocytes are derived from blood-borne stem cells. When transplanted to the chorio-allantoic membrane, 10-day bursal rudiments became lymphoid in most cases. When the transplant and the host were of opposite sex, the chromosome analysis of those bursal grafts showed 80% to 90% host metaphases when sampled 7–9 days later.

Finally, when testosterone-treated hosts were used, the host bursa always failed to develop and the bursal transplants showed a wide range of histological development when sampled 9 days later. In the grafted bursas, about 50% of the dividing cells were of host type.

Recently, bursal histogenesis was investigated using the quail-chick marker system.[106] Epithelio-mesenchymal rudiments of the bursa of Fabricius were taken from 5–11-day quail embryos and grafted into 3-day-old chick somatopleure. At the time of fixation, the total age of the bursa was 18 or 19 days. The species specificity of the grafted bursal lymphoid population studied after staining by Feulgen-Rossenbeck's reaction, appeared to depend on the stage at which the organ was transferred from donor to host embryo. In bursas transplanted before the end of the seventh day of embryonic development the whole lymphoid population was of host origin (see Figure 11 (Plate VI)). Grafts of 8–11 days bursal *Anlage* contained a mixture of host and donor lymphocytes. When taken from quail embryos older than 11 days, the bursas contained only lymphoid cells of donor type (see Figure 12 (Plate VI)).

This observation confirms that an immigration of lymphoid cells occurs during bursal development and shows that haemopoietic differentiation depends entirely on the seeding of stem cells of extrinsic origin. The immigration process which results in differentiation of the first lymphocytes takes place from 8 to 11 days post-fertilization in the quail embryo.

The chimerism observed in the bursas grafted during this period of time varied from one follicle to another. In some areas of the graft the follicles contained lymphocytes exclusively of one type (host or donor) while in others both quail and chick lymphoid cells co-existed. No data are available to allow an estimate of the number of stem cells seeding into each developing bursal follicle. However, the observation of follicles with chimerical lymphoid populations indicates that more than one stem cell often seeds into individual follicles. Most of the rapid subsequent increase in lymphoid cell numbers within the follicles can be accounted for by a cell-doubling time of about 9 hours in the 15-day chick bursa.[111] At hatching time, the lymphoid follicles contain roughly 10^3–10^4 lymphocytes. This number is influenced by the exit of B lymphocytes from the bursa, a process which starts around the sixteenth day.[112,113]

Reverse grafts of chick bursal primordium into quail embryos were also carried out. Like in the quail, the haemopoietic differentiation in the chick entirely depends on blood-borne stem cells. The latter begin to invade the organ at the end of the eighth day of incubation.

4. LYMPHOID STEM CELL IDENTIFICATION IN THE DEVELOPING THYMUS AND BURSA OF FABRICIUS

Numerous observations support the idea that the basophilic cells, first described by Maximow[83,84] which appear at a precise stage of development in both thymus and bursa are the precursors of the lymphocytes. However, no undisputable demonstration of this fact has been given.

By grafting chick thymic and bursal rudiments into quail embryos before the time of colonization by lymphoid stem cells we have clearly identified the basophilic cells as the lymphoid precursors.[54,106] Six-day chick thymus and 7-day chick bursas were grafted into 3-day quail embryos for 3–7 days. The panoptic technique first applied to the grafts made it possible to detect the basophilic cells. Subsequently, the sections were treated for DNA staining. It appeared that all the cells with a strong cytoplasmic basophilia showed the quail nuclear marker and thus had invaded the chick thymic and bursal rudiments during the implantation period (see Figure 13 (Plate VII)). Lymphoid differentiation of the quail cells which colonize the chick tissue could be followed in explants which were grafted for a longer period of time into the quail host.

At the electron microscope level the stem cells could also be recognized by their large DNA-rich nucleolus and their cytoplasm with a high ribosomal particles content. The epithelial cells have a clear cytoplasm and, since they are of chick origin, a small nucleolus. In the thymus, they show many cytoplasmic processes with conspicuous desmosomes and sometimes rudimentary ciliary structures (see Figure 14 (Plate VIII)).

5. CONCLUSIONS

The primary lymphoid organs, thymus and bursa of Fabricius, are both epithelio-mesenchymal derivatives in which the epithelium is predominantly derived from the endoderm. Although a small contribution of ectoderm may occur in the histogenesis of these two lymphoid tissues, the reticular cells are derived beyond any doubt from endoderm. The mesenchymal component originates from different germ layers in the two organs: bursa mesenchyme arises from mesoderm while thymus mesenchyme arises from ectomesenchyme, that is ectoderm. Anyhow, the thymic mesenchyme, though necessary for the evolution of the presumptive thymic endoderm, does not exert a specific effect. It can be replaced in this function by various types of mesenchymes originating from the lateral mesoderm.

Neither thymic nor bursal epithelium and mesenchyme have the ability to give rise to lymphocytes. The latter are entirely derived from blood-borne stem cells which are available in the chick and quail embryos as early as the fourth day of incubation, i.e. several days before either thymus or bursal rudiments are differentiated enough to be colonized.

Our results do not support the assumption that bursal stem cells originate from the bone marrow while thymic stem cells come from the yolk sac.[114] Our findings show that stem cells present in the blood prior to differentiation of the bone marrow are able to colonize the bursal as well as the thymic rudiments and to differentiate into lymphocytes.

The lymphoid precursor cells which invade the bursal and thymic primordia are in close contact with the endodermal epithelium during the whole process of lymphocytic differentiation which quite likely is the result of cell interactions occurring between endoderm and lymphoid stem cells. For this reason several endomesodermal derivatives have been proposed as bursa-equivalent in mammals. It has been suggested that the gut-associated lymphoid tissue such as appendix,[115] tonsils,[116] Peyer's patches[4] and foetal liver[117] exert bursa-like functions in the control of humoral immune response. However, this view has not gained universal acceptance particularly because gut-associated lymphoid tissue does not have the fundamental characteristics of the primary lymphoid organs which are: (a) the absence of antibody formation *in situ* and (b) the independence of the mitosis rate from antigenic stimulation among primitive lymphoid cells. Peyer's patches, for instance, are the site of antibody production and are very poorly developed in germ-free animals.[118,119]

6. REFERENCES

1. Szenberg, A. and Warner, N. L. (1963). *Nature,* **198**, 1012–1013.
2. Miller, J. F. A. P. (1961). *Lancet,* **2**, 748–749.
3. Perey, D. Y., Cooper, M. D. and Good, R. A. (1967). *Transplantation,* **5**, 615–623.
4. Perey, D. Y., Cooper, M. D. and Good, R. A. (1968). *Science,* **161**, 265–266.
5. Perey, D. Y. and Good, R. A. (1968). *Lab. Invest.,* **18**, 15–26.
6. Sutherland, D. E. R., Archer, O. K. and Good, R. A. (1964). *Proc. Søc. exp. Biol. Med.,* **115**, 673–676.
7. Müller, J. (1845). *Abh. Ak. Berlin,* **31**, 109–170.
8. Stannius, F. (1854). General anatomy of myxinoids. *Zootomie d. Fische.*
9. Cole, F. J. (1926). *Trans. Roy. Soc. Edinburgh,* **54**, 309–342.
10. Kampmeier, O. F. (1969). *Evolution and Comparative Morphology of the Lymphatic System.* Charles C. Thomas, Springfield, Illinois.
11. Hildemann, W. H. and Thoenes, G. H. (1969). *Transplantation,* **7**, 506–521.
12. Hildemann, W. H. and Thoenes, G. H. (1969). In *Developmental Aspects of Antibody Formation and Structure* (J. Sterzl and M. Riha, Eds.), Czech. Acad. Sci., Prague, pp. 170–212.
13. Acton, R. T., Weinheimer, P. F., Hildemann, W. H. and Evans, E. E. (1971). *Infec. Immunity,* **4**, 160–166.
14. Holliston, B. R., Cooper, E. L., Reddy, A. L. and Hildemann, W. H. (1975). *Amer. Zool.,* **15**, 39–49.
15. Wallin, I. E. (1917). *Amer. J. Anat.,* **22**, 127–158.
16. Keibel, F. (1927). *Zeitschr. f. mikr. Anat. Forsch.,* **8**, 408–476.
17. Sterba, G. (1953). *Wissench. Zeit. Fried. Schiller Univ. Jena, Mathem. Naturwiss. Reihe,* **H.2**, 239–298.
18. Hammar, J. A. (1912). *Zool Jahrb., Abt. Anat.,* **33**, 135–188.
19. Grassé, P. P. (1958). *Traité de Zoologie,* Vol. 13 (3). Masson, Paris, pp. 2634–2637.
20. Sokoloff, A. J. (1912). *Mém. Sci. Univ. Imp. Kazan.,* **79**, 1–33.
21. Hill, B. H. (1935). *J. Morphol.,* **57**, 61–89.
22. Bryce, T. H. (1906). *J. Anat. Physiol.,* **40**, 91–99.
23. Nieuwkoop, P. D. and Faber, J. (1967). *Normal Table of Xenopus laevis,* 2nd ed. North-Holland, Amsterdam.
24. Manning, M. J. and Horton, J. D. (1969). *J. Embryol. exp. Morphol.,* **22**, 265–277.
25. Shumway, W. (1940). *Anat. Rec.,* **78**, 139–147.
26. Taylor, A. C. and Kollros, J. J. (1946). *Anat. Rec.,* **94**, 7–24.
27. Curtis, S. K., Volpe, E. P. and Cowden, R. R. (1972). *Zeitschr. Zellforsch.,* **127**, 323–346.

28. Grassé, P. P. (1970). *Traité de Zoologie*, Vol. 14(3). Masson, Paris, p. 946.
29. Verdun, P. (1898). *C. R. Soc. Biol.*, **5**, 243–244.
30. Venzke, W. G. (1942). *Iowa State Coll. J. Sci.*, **17**, 145–148.
31. Venzke, W. G. (1952). *Amer. J. Vet. Research*, **13**, 395–404.
32. Schrier, J. E. and Hamilton, H. L. (1952). *J. exp. Zool.*, **119**, 165–187.
33. Hammond, W. S. (1954). *J. Morphol.*, **95**, 501–522.
34. Jotereau, F. (1975). *Arch. Biol.*, **86**, 139–161.
35. Hamilton, B. (1913). *Anat. Anz.*, **44**, 417–439.
36. Kölliker, A. (1879). *Entwicklungsgeschichte des Menschen und der höheren Tiere*. 2nd ed. Wilhelm Engelmann, Leipzig, pp. 815–880.
37. Rogers, W. M. (1929). *Amer. J. Anat.*, **44**, 283–329.
38. Crisan, C. (1935). *Z. Anat. Entwickl. Gesch. Dtsch.*, **104**, 327–358.
39. Bargmann, W. (1939). In *Möllendorff's Handbuch der mikroskopischen Anatomie des Menschen*, Vol. 6(2). Springer Verlag, Berlin, pp. 137–196.
40. Treilhou-Lahille, F. (1969). *Bull. Soc. Zool. Fr.*, **94**, 101–110.
41. Delmas, A. (1943). *Bull. Acad. Méd.*, **127**, 99.
42. Patten, B. M. (1948). *Embryology of the pig*, 3rd ed. McGraw-Hill, New York.
43. Kingsbury, B. F. (1936). *Amer. J. Anat.*, **60**, 149–183.
44. Roud, A. (1900). *Bull. Soc. Vaud Sci. Nat.*, **36**, 239–300.
45. Zottermann, A. (1911). *Anat. Anz.*, **38**, 514–530.
46. Badertscher, J. A. (1915). *Amer. J. Anat.*, **17**, 317–337.
47. Luckhaus, G. (1966). *Berlin. Munch. tierärztl. Wochen.*, **79**, 183–188.
48. Sanel, F. T. (1967). *Zeitschr. Zellforsch.*, **83**, 8–29.
49. Fraser, E. A. and Hill, J. P. (1915). *Phil. Trans. Roy. Soc. London, B.* **207**, 1–85.
50. Kingsbury, B. F. (1940). *Amer. J. Anat.*, **67**, 393–435.
51. Le Douarin, N., Bussonnet, C. and Chaumont, F. (1968). *Ann. Embryol. Morphog.*, **1**, 29–40.
52. Le Douarin, N. (1964). *Bull. Biol. Fr. Belg.*, **98**, 543–676.
53. Le Douarin, N. (1967). *C. R. Soc. Biol.*, **161**, 431–434.
54. Le Douarin, N. and Jotereau, F. (1975). *J. exp. Med.*, **142**, 17–40.
55. Le Douarin, N. (1969). *Bull. Biol. Fr. Belg.*, **103**, 435–452.
56. Le Douarin, N. (1973). *Develop. Biol.*, **30**, 217–222.
57. Le Lièvre, C. and Le Douarin, N. (1975). *J. Embryol. exp. Morphol.*, **34**, 124–154.
58. Le Douarin, N. (1974). *Med. Biol.*, **52**, 281–319.
59. Auerbach, R. (1960). *Develop. Biol.*, **2**, 271–284.
60. Ball, W. D. and Auerbach, R. (1960). *Exp. Cell Res.*, **20**, 245–247.
61. Auerbach, R. (1961). *Develop. Biol.*, **3**, 336–354.
62. Auerbach, R. (1964). In *The Thymus*, Wistar Inst. Symp. Monograph No. 2 (V. Defendi and D. Metcalf, Eds.), Wistar Inst. Press, Philadelphia, pp. 1–8.
63. Prenant, A. (1894). *Cellule*, **10**, 85–184.
64. Beard, T. (1899). *Lancet*, **1**, 144–193.
65. Beard, T. (1902). *Zool. Jahrb.*, **17**, 403–480.
66. Bell, E. T. (1906). *Amer. J. Anat.*, **5**. 29–62.
67. Dustin, A. P. (1920). *Arch. Biol.*, **30**, 601–693.
68. Stohr, P. (1906). *Anat. Hefte*, **31**, 409–458.
69. Cheval, M. (1908). *Bibliograph. Anat.*, **17**, 189–201.
70. Baillif, R. N. (1949). *Amer. J. Anat,*, **84**, 457–510.
71. Ackerman, G. A. and Knouff, R. A. (1964). *Anat. Rec.*, **149**, 191–216.
72. Ackerman, G. A. and Knouff, R. A. (1965). *Anat. Rec.*, **152**, 35–54.
73. Ackerman, G. A. (1967). *Anat. Rec.*, **158**, 387–400.
74. King, J., Ackerman, G. A. and Knouff, R. A. (1964). *Anat. Rec.*, **148**, 300–301.
75. Weakley, B. S., Patt, D. J. and Shepro, S. (1964). *J. Morphol.*, **115**, 319–354.
76. Tachibana, F., Imai, Y. and Kojima, M. (1974). *J. Reticuloendoth. Soc.*, **15**, 475–496.
77. de Winiwarter, H. (1924). *Bull. Histol.*, **1**, 1–15.
78. Deanesly, R. (1927). *Quart. J. Microsc. Sci.*, **71**, 113–146.
79. Hammar, J. A. (1905). *Anat. Anz.*, **27**, 23–30.
80. Hammar, J. A. (1908). *Arch. f. mikr. Anat.*, **73**, 1–68.
81. Hammar, J. A. (1910). *Ergebn. Anat. u. Entwickl.*, **19**, 1–274.
82. Hammar, J. A. (1911). *Anat. Hefte*, **43**, 201–242.

83. Maximow, A. (1909). *Arch. f. mikr. Anat.*, **74**, 525–621.
84. Maximov, A. (1912). *Arch. f. mikr. Anat.*, **79**, 560–611.
85. Grégoire, C. (1932). *Arch. Intern. Med. exper.*, **7**, 513–629.
86. Klapper, C. E. (1946). *Amer. J. Anat.*, **78**, 139–179.
87. Norris, E. (1938). *Contrib. to Embryol.*, Carnegie Inst., Washington, **27**, 191–207.
88. Dantchakoff, V. (1908). *Arch. mikr. Anat.*, **73**, 117–181.
89. Dantchakoff, V. (1916). *J. exp. Med.*, **24**, 87–105.
90. Pappenheimer, A. (1913). *Amer. J. Anat.*, **14**, 299–332.
91. Choi, M. H. (1931). *Folia Anat. Jap.*, **9**, 495–503.
92. Emmart, E. W. (1936). *Anat. Rec.*, **66**, 59–73.
93. Grégoire, C. (1935). *Arch. Biol.*, **46**, 717–820.
94. Grégoire, C. (1958). *Quart. J. Microsc. Sci.*, **99**, 511–515.
95. Turpen, J. B., Volpe, E. P. and Cohen, N. (1975). *Amer. Zool.*, **15**, 51–61.
96. Deparis, P. and Jaylet, A. (1975). *J. Embryol. exp. Morphol.*, **33**, 665–683.
97. Moore, M. A. S. and Owen, J. J. T. (1967). *J. exp. Med.*, **126**, 715–726.
98. Owen, J. J. T. and Ritter, M. A. (1969). *J. exp. Med.*, **129**, 431–437.
99. Le Douarin, N. and Jotereau, F. (1973). *Nature* (*New Biol.*), **246**, 25–27.
100. Sidky, Y. A. E. and Auerbach, R. (1968). *J. exp. Zool.*, **167**, 187–196.
101. Minot, C. S. (1900). *J. Boston Soc. med. Sci.*, **4**, 153–164.
102. Jolly, J. (1915). *Arch. Anat. Microsc. Morphol. Exp.*, **16**, 363–547.
103. Boyden, E. A. (1922). *Amer. J. Anat.*, **30**, 163–201.
104. Ruth, R. F. (1960). *Fed. Proc.*, **19**, 579–585.
105. Ruth, R. F., Allen, C. P. and Wolfe, H. R. (1964). In *Thymus in Immunobiology* (Good, R. A. and Gabrielsen, A. D., Eds.), Hoeber, New York, pp. 183–206.
106. Le Douarin, N., Houssaint, E., Jotereau, F. and Belo, M. (1975). *Proc. Nat. Acad. Sci.*, **72**, 2701–2705.
107. Ackerman, G. A. and Knouff, R. A. (1963). *Anat. Rec.*, **146**, 23–27.
108. Bockman, D. E. and Cooper, M. D. (1973). *Amer. J. Anat.*, **136**, 455–478.
109. Pappenheim, A. (1910–1911). In *Techniques histologiques* (Gabe, M., Ed.), Masson, Paris, pp. 802–804.
110. Moore, M. A. S. and Owen, J. J. T. (1966). *Develop. Biol.* **14**, 40–51.
111. Rubin, E., Cooper, M. D. and Kraus, F. W. (1971). *Bacteriol. Proc.*, **71**, 67.
112. Cooper, M. D., Cain, W. A., Van Alten, P. J. and Good, R. A. (1969). *Int. Arch. Allergy Appl. Immunol.*, **35**, 242–252.
113. Kincade, P. W., Self, K. S. and Cooper, M. D. (1973). *Cell. Immunol.*, **8**, 93–102.
114. Leene, W., Duyzings, M. J. M. and van Steeg, C. (1973). *Zeitschr. Zellforsch.*, **136**, 521–533.
115. Archer, O. K., Sutherland, D. E. R. and Good, R. A. (1963). *Nature*, **200**, 337–339.
116. Peterson, R. D., Cooper, M. D. and Good, R. A. (1965). *Amer. J. Med.*, **38**, 579–604.
117. Owen, J. J. T., Cooper, M. D. and Raff, M. C. (1974). *Nature*, **249**, 361–363.
118. Cooper, G. N. and Turner, K. (1967). *Aust. J. exp. Biol. Med. Sci.*, **45**, 363–378.
119. Cooper, G. N. and Turner, K. (1968). *Aust. J. exp. Biol. Med. Sci.*, **46**, 415–424.

Chapter 2

Ontogenesis of Lymphocytes

J. J. T. OWEN

Abbreviations

Ig: immunoglobulin
MLR: mixed lymphocyte reaction
PHA: phytohaemagglutinin
θ: Thy-1 determined antigen

1. INTRODUCTION

The origin of lymphocytes in embryonic life has been of considerable interest to scientists since the beginning of this century. In many ways it is surprising that interest goes back so far, since the importance of lymphocytes in immune responses was not fully appreciated until the early 1960s. However, over fifty years ago, great controversy raged about the sites of origin of lymphocytes in embryogenesis and, in particular, whether the lymphocytes of major embryonic lymphoid organs, such as the thymus, are derived by the differentiation of cells intrinsic to the organ primordium or from migrant cells originating elsewhere in the embryo. Renewed impetus for the resolution of these problems can with the demonstration of the role of lymphocytes as 'immunocompetent' cells. It could then be seen that knowledge of the origins of lymphocytes during embryonic life was basic to many fundamental questions of immunology: questions, for example, such as the acquisition of immunocompetence and the basis for tolerance to self antigens. In addition, there is the important question of the nature of immune deficiency defects, any rational approach to treatment of which must be based upon a knowledge of underlying defects of development.

The introduction of cell marker methods has led to clarification of some of the problems posed above; the results of these studies have been discussed in the previous chapter and elsewhere.[1] Suffice it to say here that these studies have led to the notion that the lymphoid system, and indeed the haemopoietic system as a whole, is in a dynamic state of change during embryogenesis with active stem cell migration as well as cell production and maturation being prominent features.[2]

The first blood-forming organ of mammals, and, indeed, also of birds, is the yolk sac. During mammalian development, haemopoietic activity in yolk sac is succeeded and replaced by haemopoietic activity in foetal liver and in foetal spleen. Finally, bone marrow becomes the major haemopoietic organ of adult mammals, although the potentiality for haemopoiesis is retained by spleen and, to some extent, liver. The interrelationships between these various haemopoietic organs has been discussed extensively by Metcalf and Moore.[2] In particular, the notion that yolk sac is the primordial site of haemopoietic stem cell differentiation and that blood formation in all other sites is dependent upon migration of stem cells from yolk sac has been supported by experiments in which whole mouse embryos have been cultured with or without their yolk sacs. Without the yolk sac, intra-embryonic haemopoiesis fails to proceed.[3] However, these experiments do not prove that active cell migration is involved in the establishment of liver haemopoietic stem cells and further work is required to resolve this matter.

In the following sections, I will discuss the ontogeny of lymphocytes in various embryonic organs from both morphological and functional standpoints. I will then outline current knowledge of the maturation pathways of T and B lymphocytes during embryogenesis and try to relate these pathways to problems such as the generation of tolerance to self antigens. I will not attempt to provide a comprehensive review of the literature, but, nonetheless, I hope to present those aspects of the subject in which advances are being made, together with the experimental approaches which are proving most fruitful.

2. LYMPHOCYTE PRODUCTION IN THE YOLK SAC

2.1 Methods of investigation

As mentioned in the previous section, yolk sac is the first haemopoietic organ of the mammalian embryo and although its activity is largely limited to the production of erythrocytes, multipotential haemopoietic stem cells have been demonstrated within it.[3]

The presence of multipotential cells indicates a capability for production of a variety of blood cell types. However, the extent to which yolk sac does produce a broad range of blood cells is a debatable matter. Certainly, during the terminal stages of its activity, the presence of mature blood cell types within it cannot be used as definitive evidence that yolk sac produces these cells, since by this stage other haemopoietic organs are active and hence may be the source of such cells.

Furthermore, distinction must be made between two broad lines of experimental evidence concerning the potentialities of yolk sac. First of all, there are those observations which have been made on the capacity of yolk sac to produce blood cells *in situ*—for example, by studying the generation of cells in yolk sac cultured *in vitro*. On the other hand, there are experiments in which yolk sac cells have been transplanted from the embryo to irradiated adult hosts. The potentiality of yolk sac cells to become functional T and B cells has then been tested[4,5] in the irradiated recipient. These latter experiments demonstrate that yolk sac is a source of stem cells, but they do not prove that yolk sac *in situ* produces the variety of cell types which can be shown to develop in irradiated hosts. Clearly, important host factors may contribute to the development of these cells.

2.2 Morphological studies

Morphological studies on yolk sac can produce at best only tentative evidence for the presence of lymphocytes. Haemopoietic tissues contain very heterogeneous populations of cells ranging in size from large 'blast-like' cells to small mature erythrocytes, granulocytes etc. Within these populations, cells of lymphocyte morphology are found. However, morphological study alone cannot determine whether these are 'immunologically competent' lymphocytes or precursors of these cells. Even with this qualification, there are very few, if any, cells in yolk sac which are of lymphoid morphology. In mouse embryo yolk sac, the majority of cells are developing 'first generation' erythrocytes (see Figure 1). However, 'blast' cells with large nuclei and basophilic cytoplasm can be found and these cells show no obvious morphological signs of differentiation into any particular blood cell type (i.e. their cytoplasm does not exhibit the characteristic acidophilia associated with haemoglobin, nor are there granules indicative of granulocyte maturation (see Figure 2 (Plate X)). This population might contain lymphocyte progenitors as well as progenitors of other cell types.

2.3 Cell markers and functional studies

In recent years, the introduction of cell markers for identification of T and B lymphocytes has been enormously valuable for many purposes, but especially for study of cells within the developing T and B cell series. Assays for those functions generally associated with T and B lymphocytes are also likely to be of value in this context. With regard to yolk sac, there are no reports of the presence of cells bearing T markers. Certainly in the mouse embryo, θ (Thy-1) positive cells in yolk sac have not been demonstrated. There is, however, an intriguing report that yolk sac cells of the mouse embryo are capable of mounting a graft versus host reaction *in vitro*.[6] This immunological reaction is generally accepted as a parameter of T cell function and the result is therefore rather surprising in that it suggests that function normally attributable to mature T cells is present in yolk sac prior to development of thymus. The result is a provocative one, but it must await further confirmation and identification of the reactive cell type before firm conclusions can be drawn.

The question as to whether yolk sac is a site of B lymphocyte maturation is also uncertain. In the mouse embryo, cells bearing surface immunoglobulin are not found in yolk sac at any stage.[7] Moreover, if yolk sac is isolated in organ culture and maintained for up to 7 days, cells bearing surface immunoglobulin do not develop.[7] However, these experiments do not rule out the possibility that pre-B cells, i.e. cells synthesizing immunoglobulin within their cytoplasm but not expressing it on their surfaces, may be present (see Section 9). Recently it has been claimed that cells in yolk sac of 10-day mouse embryos are capable of binding several protein antigens.[8] However, the immunological significance of this antigen binding and the nature of the cell involved requires further investigation.

In the chick embryo there is evidence for immunoglobulin synthesis within cells of yolk sac prior to the appearance of immunoglobulin synthesizing cells within the bursa of Fabricius.[9] However, the number of immunoglobulin positive cells within yolk sac is low and the results require further confirmation and analysis.

In summary, evidence to date indicates the yolk sac is a site of proliferation of haemopoietic stem cells and that these cells are capable of full maturation into a variety of blood cell types, including lymphocytes, when transferred to irradiated hosts. However, the extent of maturation of lymphoid cells *in situ* within yolk sac is a more debatable matter and most evidence at the moment suggests that, if there is lymphocyte maturation within the yolk sac of mammals, it occurs to only a very early stage in the series of maturation steps which are required to obtain fully immunocompetent T and B lymphocytes. This is an interesting area for future work.

3. LYMPHOCYTE PRODUCTION IN FOETAL LIVER AND SPLEEN

3.1 Morphological Studies

The foetal liver is a very prominent haemopoietic organ during a major part of embryogenesis in mammals. However, foetal liver is not a haemopoietic organ in birds. Foetal spleen is not such a large organ as liver and not such a major haemopoietic site. However, it is concerned with blood cell production in a comparable way to liver and, therefore, I have included both organs in this section.

As in yolk sac, early blood cell production in foetal liver and spleen is largely directed toward the maturation of erythroid cells (but of the 'second generation' type). However, with increasing gestation time, various leukocytes, such as granulocytes and macrophages become evident (see Figure 3 (Plate XI)). Again, as in yolk sac, liver is a site of multiplication of multipotential haemopoietic stem cells which are fully capable of maturation into a variety of cell types upon transfer to irradiated hosts.[4,5] However, this result does not answer the question as to whether mature lymphoid cells are generated within foetal liver or foetal spleen *in situ*.

From a morphological point of view, it is possible to discern cells with characteristic small lymphocyte morphology within mouse embryo foetal liver from about the seventeenth day of gestation onwards. At this time, there are also considerable numbers of monocytic cells as well as granular leukocytes, although the majority of cells are erythroid in type. 'Blast'-like cells are present in foetal liver as soon as haemopoiesis in initiated. Large 'lymphoid' cells can be seen at all stages of liver and spleen haemopoiesis, but without further methods of cell identification (see below), it is impossible to make judgements as to their functional significance.

3.2 Cell markers and functional studies

3.2.1 *T cells*

Cells bearing the T cell marker θ (Thy-1) can be detected in foetal liver of mouse embryos a few days before birth. However, θ positive lymphocytes do not develop in 14-day foetal liver maintained in organ culture for 7 days[7] and so it seems likely that these cells are derived from thymus which contains large numbers of θ positive cells by this stage. Recently, agents which elevate cellular cyclic AMP levels have been shown to induce the expression of θ antigen on 14-day mouse embryo liver cells.[10] Although the full significance of this result awaits further clarification, the cells which are induced to express θ may be thymic stem cells (see Section 8). However, it is important to note that expression of θ does not necessarily denote functional capacity.

In the human foetus, the capacity to mount a mixed lymphocyte reaction (MLR), which is usually accepted as a function of T cells, has been reported as early as 7·5 weeks gestation in human liver cells.[11] Thus, foetal liver cells have been shown to respond by proliferation to histo-incompatible adult lymphocytes and the direction of the response has been confirmed by demonstrating the foetal karyotype in the proliferating cells. This result is intriguing for a number of reasons; firstly, because reactivity is found in foetal liver 5 weeks before the appearance of comparable reactivity in thymus and, secondly, because reactivity to the mitogen phytohaemagglutinin (PHA) was not found in liver in parallel tests. This dissociation between PHA and MLR reactivity may denote the fact that separate subsets of T cells are involved in these two responses and this would fit with other evidence for lymphocyte subsets (see P. C. L. Beverley, herein). However, the reasons why T cell activity should be found in human foetal liver prior to the development of thymus are obscure. The immunological specificity of this reaction and the nature of the responsive cell require further investigation.

In spleen, the situation with respect to the presence of θ positive cells in the mouse embryo is comparable to that in liver, namely that positive cells are only present after their appearance in thymus. It seems likely that these cells are thymus-derived, but it has not been fully proved that they all are (see Section 8). Spleen cells of neonatal mice can respond in a mixed lymphocyte reaction but killer cell activity does not arise until 7 days of age.[12]

3.2.2 *B cells*

B lymphocytes can be detected in foetal liver of a variety of mammalian species from a fairly early stage of gestation. For example, in the mouse embryo, lymphocytes bearing surface immunoglobulin can be detected at about the fifteenth to seventeenth days of gestation depending upon strain.[7,13] The question then arises as to whether these B cells are generated within foetal liver or whether they are migrants from other sites of B cell differentiation. A comparable question can be asked with respect to the B cells which are found in spleen at about the same time.

These questions have been investigated by isolating foetal liver and foetal spleen *in vitro* prior to the appearance of cells bearing surface immunoglobulin *in vivo*. The ability of the isolated explants to generate B cells over various culture periods has then been tested.[7,14] The results indicate that B lymphocytes develop independently in foetal liver and in foetal spleen; they do not develop in cultures of yolk sac and thymus. In further experiments of this type, we have shown (Owen, J. J. T., Raff, M. C. and Cooper, M. D., unpublished observations) that mouse foetal bone marrow is also a site of B cell maturation. In conclusion, it is likely that B lymphocyte maturation in mammals is multifocal and is not dependent upon gastrointestinal influences as suggested previously.[15]

In more recent studies,[16] cells have been found as early as the twelfth day of gestation in mouse foetal liver which exhibit cytoplasmic immunoglobulin but not surface immunoglobulin. The significance of intracellular immunoglobulin within these pre-B cells well before immunoglobulin is inserted into the cell surface is uncertain (see Section 9). There is general agreement between various studies that in the mouse embryo the major, if not only, immunoglobulin class synthesized is IgM. In the human embryo, the earliest expression of immunoglobulin synthesis have been found in 7·5 week liver in which some large lymphocytes stain for cytoplasmic IgM but not for surface Ig.[17] In the same study, surface IgM has first been detected on cells at 9 weeks in liver and 11 weeks in spleen. Surface IgD and surface IgG have been found on separate subpopulations of IgM positive cells at 10–11 weeks.

4. LYMPHOCYTE PRODUCTION IN FOETAL BONE MARROW

There is considerable evidence that bone marrow is the major source of haemopoietic stem cells in late foetal life and throughout adult life. Not only have multipotential stem cells been demonstrated within this organ, but other categories of unipotential cells have also been found (reviewed in Reference 2). Although bone marrow in adult animals is known to contain T lymphocytes, it is assumed that these cells are derived from the thymus. Certainly, no experiments have been performed to demonstrate that mature T lymphocytes can develop *in situ* within bone marrow. However, cells exhibiting T cell markers have been demonstrated in bone marrow of athymic 'nude' mice and it has been suggested that these cells are thymic precursor cells.[18] In the absence of a normal thymus, they presumably are unable to differentiate further along the T cell maturation pathway (see Section 8).

With regard to B lymphocytes, there is increasing evidence that a considerable number of these cells are generated within adult bone marrow.[19,20] The situation with regard to the bone marrow of the embryo is less clear, but recently, by culturing whole limb bones from mouse embryos, we have been able to show that marrow can generate B lymphocytes in organ culture (see Section 3). Thus, in addition to foetal liver and foetal spleen, embryonic bone marrow is a site of differentiation of immunoglobulin-bearing B lymphocytes.

5. LYMPHOCYTE PRODUCTION IN FOETAL THYMUS

The thymus is a major site of lymphocyte production in all mammals. It is an organ which initially is derived from pharyngeal pouch epithelium, but it is soon invaded by stem cells of blood origin (see N. M. Le Douarin, herein). These stem cells differentiate within thymus to become small lymphocytes which subsequently migrate to other lymphoid organs. Thus, thymus is the major site, and probably the exclusive site, of T lymphocyte production. *In vitro* experiments in which embryonic mouse thymus has been placed in organ culture for periods of up to 7 days, have shown that all of the lymphocytes produced are θ bearing and no cells with readily-detectable surface immunoglobulin (i.e. B cells) are generated.[14] Thus, thymus is a site of production of T cells but not of B cells.

Many of the developing lymphoid cells within thymus do not have functional capabilities. Indeed, the first morphological appearance of lymphocytes and the expression of θ antigen on their surface does not coincide with functional capacity in the mouse embryo. For example, small lymphocytes can be seen within the thymus as early as the fifteenth day of gestation in the mouse and many of these cells are θ positive.[21] However, the ability to respond to mitogens is acquired at a later stage of gestation and reactivity in the MLR test is first present at, or just before, birth.[22] Hence, there is a considerable delay between the initial maturation of T cells in thymus and their full functional differentiation.

A number of reports deal with the responsiveness of human foetal thymocytes to mitogens and to allogeneic cells *in vitro*.[23,24] Phytohaemagglutinin reactivity has been detected in thymus at 10 weeks' gestation and mixed lymphocyte reactivity at 12·5 weeks. In each instance, this is some weeks after the initiation of thymus lymphopoiesis, suggesting that there is a delay between the development of morphologically distinct lymphocytes and their functional capability as in the mouse embryo.

6. LYMPHOCYTE PRODUCTION IN THE BURSA OF FABRICIUS

Since the early work of Glick *et al.*[25] on the effects of bursectomy, the bursa of Fabricius has been looked upon as an extremely important organ in the maturation of B lymphocytes in birds. Early bursectomy has been found to have a marked effect on the development of the B cell series[26] and certainly the bursa of Fabricius is a very early site of lymphocyte maturation in the chick embryo. Moreover, many of the cells within the bursa during these early stages of development are synthesizing immunoglobulin which is inserted into their cell membranes.[27] In view of the evidence that some of these B lymphocytes migrate to other sites of the developing embryo[28,29] it seems reasonable to look upon the bursa as a crucial organ for the production of B lymphocytes, rather comparable to the way in which the thymus is responsible for the production of T lymphocytes.

B lymphocytes produced within the bursa are derived from stem cells which migrate into the organ from extra-bursal sites.[30] In view of the tentative evidence that immunoglobulin synthesis may be initiated within the yolk sac (see Section 2), it is possible that stem cells migrating into the bursa may derive from yolk sac and may have already initiated immunoglobulin synthesis before they arrive within bursal tissue. Furthermore, it has been claimed that bursectomy at very early stages of gestation causes only a moderate depletion of IgM-containing cells.[31] The question arises as to whether the bursa is the sole site of early B lymphocyte maturation in birds. Whatever the answer to this question, most evidence supports the notion that the bursa is the major site of differentiation of stem cells into B lymphocytes.

The bursa is also an important organ for the generation of different immunoglobulin classes on B cells. Thus, Kincade and Cooper[27] have shown that IgM synthesis is initiated within bursal lymphocytes prior to the synthesis of IgG. Further, on the basis of data collected from birds injected as embryos with anti-μ sera in order to suppress IgM synthesis, it has been suggested that there is a switch within a single cell line from IgM to IgG, since IgG synthesis is also suppressed.[32] The appearance of B cells in the bursa and the switch from IgM to IgG is probably independent of antigen stimulation.[27] Of course, the subsequent activation of B cells to become high-rate antibody secreting cells is antigen dependent and the possibility of an antigen-driven switch in immunoglobulin class at this stage has not been excluded, although the fact that some bursectomized chicks which produce IgM following antigen stimulation do not produce IgG at all argues against it.

7. SITES OF LYMPHOCYTE PRODUCTION: SUMMARY

From the foregoing account, it can be seen that lymphocytes are produced in diverse organs during embryogenesis. The thymus is a major site of production of T lymphocytes in both birds and mammals, and in birds B lymphocytes are produced within the bursa of Fabricius. However, in mammals, B lymphocytes are produced in a variety of major haemopoietic organs including foetal liver, foetal spleen and embryonic bone marrow. Although early lymphoid cells can be detected in these sites by morphological means and surface marker techniques, these cells may still not be fully mature in a functional sense. Indeed, there is evidence to the contrary, namely that there is a considerable period of time during which cells undergo maturation before they reach their final functional capability. The question as to whether cells may be more susceptible to tolerance induction during these formative stages is an extremely interesting one and this notion will be considered in the following sections.

8. THE DIFFERENTIATION PATHWAYS OF EMBRYONIC T LYMPHOCYTES

8.1 Are thymic stem cells precommitted to T differentiation before migration to the thymus?

On the basis of chromosome marker and tissue culture studies, it has been argued[33,34] that the migrant stem cells that enter the epithelial thymic are large 'basophilic' cells (see Figure 4 (Plate XII)). This has been confirmed in the avian embryo thymus by using the structural differences between quail and chick nuclei as a marker in interspecific chimeras (Reference 35, and N. M. Le Douarin, herein). In 13–14-day mouse embryos, these basophilic cells do not express T cell alloantigens,[21] but during the course of the following day or so *in vivo* and *in vitro* T cell antigens such as θ appear on the surfaces of maturing thymus lymphocytes.[21] It has been concluded, therefore, that one of the first maturational steps which takes place after stem cells have entered the thymus is marked by the appearance of cell surface alloantigens.[36]

Further evidence that thymic stem cells do not express T cell antigens before entry to the thymus has been obtained by showing that migration of stem cells to the irradiated thymus is unaffected by treatment of donor cell suspensions with anti-T sera and complement.[37] However, low amounts of T antigens have been detected on spleen and bone marrow cells of athymic (nude) mice by immunofluorescence techniques.[18] It has been suggested that these cells are T cell precursors which are unable to differentiate fully in the absence of a normal thymus.[18] Hence, these observations support the notion that thymic stem cells may already express some T cell antigens, albeit in low amounts, prior to migration to the thymus. In this sense, thymic stem cells are already precommitted to T differentiation in sites of general haemopoiesis.

There are other observations which support the concept of precommitted T stem cells. Thus, a variety of agents (including thymus hormones) which elevate cellular cyclic AMP levels can induce the expression of T surface antigens on cells of foetal liver and adult spleen *in vitro*.[10] Furthermore, pretreatment of bone marrow suspensions with thymopoietin (a putative thymus hormone) renders stem cells present in the suspensions which are capable of populating the thymus, susceptible to the action of anti-T cell sera and complement.[37] This experiment suggests that thymic stem cells can be induced to express (or increase their expressions of) T surface antigens prior to migration to the thymus.

Other types of data indicate that differences exist between multipotential haemopoietic stem cells and T precursor cells in haemopoietic tissues. For example, there are density differences between the two types of cell.[38] However, it must be admitted that the matter is not finally resolved since adequate assays for identifying and enumerating thymic stem cells are not available. Also, expression of T surface antigens is not necessarily a good parameter of immunological function (see below) and too much weight should not be placed on the expression of surface antigens on precursor cells at this stage.

The idea of precommitment of T stem cells has been taken a step further recently with the suggestion that there are pathways of T cell maturation which do not require migration through the thymus.[39] The major evidence quoted in support of this idea is the observation that thymectomy of foetal sheep at early stages of gestation is only partially successful in preventing the ontogeny of T cell function.[40,41] However, the thymus is fully lymphoid and presumably capable of export of T cells some time before thymectomy was performed in these experiments (R. K. Jordan, unpublished observations). Furthermore, the sheep thymus is an elongated structure with multiple lobes and there must be some doubt about the completeness of the operative procedures.

In summary, the balance of evidence at the moment favours the idea that stem cells destined for T cell maturation undergo some differentiation along the T cell pathway prior to their migration to thymus. However, the latter step is vital for the generation of full T cell function *in vivo* and in this context, alternative pathways of maturation have not been convincingly demonstrated.

8.2 Maturation of T lymphocytes within the embryonic thymus

8.2.1 In vivo *and* in vitro *studies*

Morphological studies have suggested that the initial inflow of stem cells into the epithelial thymic rudiment occurs at about the eleventh day of gestation in the mouse embryo.[34] Cells expressing the θ antigen can be detected by immunofluorescence at 13 days' gestation and shortly afterwards by cytotoxicity tests. By the fifteenth day of gestation, large numbers of small lymphoid cells are present within the thymus and the majority of these express T cell alloantigens.[21] However, these cells are still functionally immature as tested by responsiveness to mitogens and alloantigens *in vitro*.[22] Indeed, mixed lymphocyte reactivity is not manifest in the mouse thymus until shortly before birth.

The sequence of events which takes place in thymus *in vivo*, in terms of acquisition of surface alloantigens and also functional capability, can be examined *in vitro* using an organ culture system.[21,42] Thus, if 14-day mouse embryo thymus is cultured for periods of 7 to 28 days, considerable numbers of θ positive cells are generated and reactivity to mitogens such as phytohaemagglutinin, concanavalin A and pokeweed mitogen can be demonstrated.[42] The main advantages of this *in vitro* approach are that (a) maturation of T cells can be observed without complexities due to cell migration to or from the thymus arising and that (b) the conditions under which cell maturation proceeds can be controlled. It has been shown that a five-fold greater magnitude of mitogen responsiveness can be obtained in organ culture as compared to thymus *in vivo*, indicating that mature cells are selectively concentrated *in vitro*.[42] Furthermore, recent studies have demonstrated the generation of mixed lymphocyte reactivity in the same *in vitro* system (J. Robinson, unpublished observations).

It is clear, therefore, that maturation of stem cells within thymus may lead to the generation of cell populations exhibiting a variety of functions without the participation of other organs. There is considerable evidence for functional T cell heterogeneity in peripheral lymphoid organs (see P. C. L. Beverley, herein) and, at least in part, this heterogeneity is thought to arise in the thymus,[43,44] However, the question as to whether there are separate lines of differentiation involved or whether the various functional capabilities are a reflection of the stage of maturation reached by various cells has not been resolved. The *in vitro* model may be an ideal one for investigating this problem as well as others such as the induction of tolerance and the functional capabilities of T cells in the absence of B cells (B lymphocytes are not generated in embryonic thymus *in vitro*).

8.2.2 *Microenvironmental and hormonal influences*

The nature of the interaction between the epithelial primordium of the thymus and the stem cells within it is still largely unresolved. Of course, there is evidence that humoral thymic factors (thymic hormones) play a part in T cell maturation (see N. Trainin *et al.*, herein). However, the importance of 'microenvironments' (influences extending only over short distances) is undecided. The fact that animals with profound T cell deficiencies can only be fully restored to immunological competence by thymus grafts as opposed to thymus hormones or thymus in diffusion chambers[45,46] argues for the importance of influences operating within the thymus.

The nature of such influences remains obscure. However, observations that cells (? thymic precursors) within peripheral organs can be influenced to express T cell alloantigens by agents which elevate levels of intracellular cyclic AMP have been applied to stem cells within the 14-day mouse embryo thymus. It has been shown that the expression of θ antigen on 14-day thymus cells can be induced by treatment with cAMP, DB-cAMP or prostaglandin E,[47] hormones such as glucagon and prolactin, and agents such as isoproterenol and histamine.[48] The effects of the latter two can be blocked by β adrenergic and H2 antagonists respectively.[48] Although it is not certain whether the initiation of θ synthesis is involved in these experiments or whether an increased rate of θ synthesis and expression is induced, the system is of potential value for examining the factors involved in stem cell maturation in the thymus.

The 'nude' mouse offers a model for studying disordered thymus development since in this instance an epithelial thymic remnant develops but lymphocyte formation does not occur within it,[49] despite the fact that these mice have thymic stem cells.[50,51] The basis of the defect is not known but it seems reasonable to suppose that, in part at least, it is due to the failure of the epithelial component of the thymus to support stem cell proliferation and maturation.

In summary, intrathymic maturation of T lymphocytes is an important and probably an essential step in T cell maturation. T cell alloantigens such as θ appear on maturing thymus cells some time before functional capability is expressed. Hence, expression of θ is not a reliable parameter of T cell function in the embryo. The significance of the long period of T cell maturation which precedes the emergence of 'competent' lymphocytes is obscure (see also B cell maturation—Section 9). Likewise the manner in which functionally heterogeneous populations are generated requires more investigation before anything useful can be concluded. It is hoped that the *in vitro* approach described above may be valuable in this context.

9. THE DIFFERENTIATION PATHWAYS OF EMBRYONIC B LYMPHOCYTES

9.1 Pre-B cells

As mentioned in Section 3, surface immunoglobulin-bearing B lymphocytes are first detected by immunofluorescence in mouse foetal liver and spleen at 17 days' gestation in Balb/c mouse embryos.[7] However, cells which have intracellular IgM but no surface Ig can be detected in mouse foetal liver as early as 12 days' gestation.[16] Intracellular IgM positive/surface Ig negative cells persist in liver up to birth. They are also found in foetal spleen and adult bone marrow, but are not found in adult lymph node or spleen.[16] This distribution and the fact that these cells are synthesizing Ig suggests that they are B lymphocyte precursors (pre-B cells). IgM-containing plasma cells and other IgM-secreting lymphoid cells in adult lymphoid tissues are readily distinguished from pre-B cells because secreting cells always stain more intensely for IgM and they always have readily detectable IgM on their surface. Intracellular positive/surface IgM negative cells have also been found in human foetal liver.[17]

The pre-B cells in adult bone marrow may be the same as the cells described in other studies which lack surface Ig but can differentiate into surface Ig-bearing cells *in vitro* and *in vivo*.[19,20] It also seems likely that pre-B cells in foetal liver may be the 'type 1' B cells which incorporate radioactive leucine into rapidly turning over 7–8 s IgM.[52]

The absence of surface Ig on pre-B cells has been confirmed in experiments in which foetal liver fragments were maintained in organ culture in the presence or absence of anti-μ antibodies (20 μg/ml). Whilst the presence of anti-μ antibodies induces the disappearance and prevents re-expression of cell surface Ig (see below), the expression of intracellular Ig in pre-B cells is unaffected.[16] Thus, it seems most unlikely that pre-B cells have Ig available at the cell surface.

9.2 B cells in foetal tissues and adult bone marrow—'immature' B cells

Organ culture techniques have been used to demonstrate the generation of B lymphocytes in mouse foetal liver and spleen.[14] The presence of purified anti-μ antibodies in the culture medium completely inhibits the appearance of Ig-bearing cells in 14-day foetal liver cultures and causes the disappearance of Ig-bearing cells already present in late foetal and newborn liver.[53] Moreover, this suppression is irreversible and can be obtained with low concentrations of anti-Ig antibody. However, cultures of adult lymph node and spleen can only be reversibly suppressed and require much higher concentrations of antibody for suppression.[53] It is interesting to note that B cells in adult bone marrow can be irreversibly suppressed in the same manner as foetal liver by low concentrations of anti-Ig antibody.

These results suggest a fundamental difference between newly formed and more mature B cells which may have important implications for B cell tolerance to self antigens. They lend support to the notion that B cells pass through a stage in their normal differentiation when they are highly susceptible to tolerance induction.[54] The high susceptibility of immature B cells to Ig suppression, which has also been observed by Sidman and Unanue,[55] may also explain the observations that anti-idiotype antibody induces prolonged clonal deletion when injected into neonatal mice, but when injected into adults reversible receptor blockade is produced.[56] Clearly, this is an important area for future study.

Other parameters of B cell maturation have recently been described. Changes in generation time,[57] cell density and size,[52,58] Ig receptor turnover,[52] response to B cell mitogens[52,59] and acquisition of complement receptors[60] take place in ontogenetic sequence.

As mentioned in Section 3, the predominant and probably only class of Ig expressed in B cells in the mouse embryo is IgM, although with the recent evidence for the existence of an IgD class in the mouse (References 61 and 62, and R. M. E. Parkhouse and E. R. Abney, herein) the situation is not entirely resolved. Certainly, in human development there is some evidence that IgD is present on lymphocyte surfaces at an early stage.[63] The question of a switch in immunoglobulin class during the development of B lymphocytes in the mouse embryo has not been examined to the same extent as the situation in the avian bursa of Fabricius (see Section 6). However, Kearney and Lawton[64] have found that cultures of mouse foetal liver and spleen, stimulated with bacterial lipopolysaccharide, give rise to plasma cells staining for IgM, IgG_1 and IgA and in view of the great preponderance of IgM-bearing cells in the foetal period, it seems likely that many of the precursors of cells synthesizing various classes of Ig are IgM positive.

The generation of diversity of specificity of Ig on B cells in the mouse embryo has been examined in recent studies. D'Eustachio and Edelman[65] found that although the numbers of antigen-binding cells in foetuses and young mice are smaller than in adults, there is no restriction in the variety of specificities expressed in the foetus, either with respect to the kinds of antigens bound or to the range of avidities of binding. They suggest that their data is consistent with models of diversity in which genes coding for the full repertoire of antibodies are generated somatically from a small number of germ line genes in the absence of any strong selection with respect to antigenic specificity. On the other hand, there is evidence for sequential development of clonal diversity among responsive B cells in the mouse.[66,67] Further work is required to resolve these differences which may reflect the fact that although clonal diversity of mouse B cells is fully developed by the time of surface Ig expression, further differentiation is required before all classes can be activated by antigen and auxiliary cells. Indeed, the biological significance of the long period between the synthesis of Ig in pre-B cells and the emergence of reactive B cells is enigmatic. It seems most likely that this Ig is destined for the plasma membrane rather than for secretion, but then it is not clear why there is such a long period between synthesis and incorporation into the cell surface.

The question of 'microenvironmental' or 'humoral' influences on B cell maturation has scarcely been examined. Work described earlier indicates that B cells mature in haemopoietic tissues such as foetal liver, spleen and adult bone marrow. However, bone marrow appears not to be essential for B cell maturation.[68,69] Recently, agents which increase levels of intracellular cyclic AMP have been shown to induce phenotypic conversion of Ia negative to Ia positive B lymphocytes.[70] This work may mark a beginning to the analysis of the nature of the events involved.

10. GENERAL SUMMARY

I hope that it will be clear from this discussion of the ontogeny of lymphocyte populations that many basic questions of immunology are essentially problems of lymphocyte differentiation. In the last few years an increasing level of interest has been shown in ontogenetic aspects of immunity and *in vitro* systems are now available for study of the critical stages of lymphocyte maturation.

I think that it is clear that although humoral factors have been implicated in lymphocyte maturation, T and B lymphocytes differentiate in special anatomical sites and it is hard to escape the conclusion that there are local factors which are important in determining this, just as local factors have been shown to be of importance in haemopoiesis in general.[71]

Lymphocytes undergo maturation over a period of time and the stages of maturation which take place are beginning to be identified. With the development of techniques for isolating cell populations of various stages of maturity the way is open for an analysis of problems of diversity of responsiveness and tolerance.

Acknowledgements

I wish to acknowledge the excellent assistance of Mr Layfield and Mr McFarlane in the preparation of the photomicrographs and Ms C. Grainger in the preparation of the manuscript.

11. REFERENCES

1. Owen, J. J. T. (1970). In *Immune Reactions, Handbuch der Allemeinen Pathologie*, Bd VII/3 (Studer, A. and Cottier, H., Eds.), Springer-Verlag, Berlin, pp. 129–181.
2. Metcalf, D. and Moore, M. A. S. (1971). Haemopoietic cells: their origin, migration and differentiation. In *Frontier of Biology*, Vol. 24, North-Holland, Amsterdam.
3. Moore, M. A. S. and Metcalf, D. (1970). *Brit. J. Haematol.*, **18**, 279–296.
4. Tyan, M. L. (1968). *J. Immunol.*, **100**, 535–542.
5. Tyan, M. L., Cole, L. J. and Herzenberg, L. A. (1967). *Proc. Soc. exp. Biol.*, **124**, 1161–1163.
6. Hofman, F. and Globerson, A. (1973). *Eur. J. Immunol.*, **3**, 197–181.
7. Owen, J. J. T., Cooper, M. D. and Raff, M. C. (1974). *Nature*, **249**, 361–363.
8. Decker, J. M., Clarke, J., Bradley, L. M. Miller, A. and Sercarz. E. E. (1974). *J. Immunol.*, **113**, 1823–1833.
9. Albini, B. and Wick, G. (1975). *Int. Arch. Allergy Appl. Immunol.*, **48**, 513–529.
10. Scheid, M. P., Hoffmann, M. K., Komuro, K., Hammerling, U., Abbott, J., Boyse, E. A., Cohen, G. H., Hooper, J. A., Schulof, R. S. and Goldstein, A. L. (1973). *J. Exp. Med.*, **138**, 1027–1032.
11. Stites, D. P., Carr, M. C. and Fudenberg, H. H. (1974). *Cell. Immunol.*, **11**, 257–271.
12. Wu, S., Bach, F. H. and Auerbach, R. (1975). *J. Exp. Med.*, **142**, 1301–1305.
13. Nossal, G. J. V. and Pike, B. (1973). *Immunology*, **25**, 33–45.
14. Owen, J. J. T., Raff, M. C. and Cooper, M. D. (1975). *Eur. J. Immunol.*, **5**, 468–473.
15. Perey, D. Y. E., Cooper, M. D. and Good, R. A. (1968). *Science*, **161**, 265–266.
16. Raff, M. C., Megson, M., Owen, J. J. T. and Cooper, M. D. (1976). *Nature*, **259**, 224–226.
17. Gathings, W. E., Cooper, M. D., Lawton, A. R. and Alford, C. A. (1976). *Fed. Proc.* (*Abstr.*), **35**, 276.
18. Roelants, G. E., Loor, F., von Boehmer, H., Sprent, J., Hass, L., Mayor, K. S. and Ryden, A. (1975). *Eur. J. Immunol.*, **5**, 127–131.
19. Osmond, D. G. and Nossal, G. J. V. (1974). *Cell. Immunol.*, **13**, 132–145.
20. Ryser, J. E. and Vassalli, P. (1974). *J. Immunol.*, **113**, 719–728.
21. Owen, J. J. T. and Raff, M. C. (1970). *J. exp. Med.*, **132**, 1216–1232.
22. Mosier, D. E. (1974). *J. Immunol.*, **112**, 305–340.
23. Hayward, A. R. and Ezer, G. (1974). *Clin. exp. Immunol.*, **17**, 169–179.
24. Carr, M. C., Stites, D. P. and Fudenberg, H. H. (1975). *Transplantation*, **20**, 410–413.
25. Glick, B., Chang, T. G. and Jaap, R. G. (1956). *Poultry Sci.*, **35**, 224–225.
26. Cooper, M. D., Cain, W. A., van Alten, P. J. and Good, R. A. (1969). *Int. Arch. Allergy Appl. Immunol.*, **35**, 242–252.
27. Kincade, P. W. and Cooper, M. D. (1971). *J. Immunol.*, **106**, 371–382.
28. Ivani, K., Murgatroyd, L. B. and Lydyard, P. M. (1972). *Immunology*, **23**, 107–111.
29. Kincade, P. W., Self, K. S. and Cooper, M. D. (1973). *Cell. Immunol.*, **8**, 93–102.
30. Moore, M. A. S. and Owen, J. J. T. (1966). *Develop. Biol.*, **14**, 40–51.

31. Jankovic, B. D., Knezevic, Z., Isakovic, K., Mitrovic, K., Markovic, B. M. and Rajcevic, M. (1975. *Eur. J. Immunol.*, **5**, 656–659.
32. Kincade, P. W., Lawton, A. R., Bockman, D. E. and Cooper, M. D. (1970). *Proc. Natl. Acad. Sci.*, **69**, 1918–1925.
33. Moore, M. A. S. and Owen, J. J. T. (1967). *J. exp. Med.*, **126**, 715–726.
34. Owen, J. J. T. and Ritter, M. A. (1969). *J. exp. Med.*, **129**, 431–442.
35. Le Douarin, N. M. and Jotereau, F. V. (1975). *J. exp. Med.*, **142**, 17–40.
36. Owen, J. J. T. (1972). In *Ontogeny of Acquired Immunity* (Porter, R. and Knight, J., Eds.), Associated Scientific Publishers, Amsterdam, pp. 35–54.
37. Komuro, K., Goldstein, G. and Boyse, E. A. (1975). *J. Immunol.*, **115**, 195–198.
38. El-Arini, M. O. and Osoba, D. (1973). *J. exp. Med.*, **137**, 821–837.
39. Bryant, B. J. (1974). In *Progress in Immunology* II, Vol. 3, (Brent, L. and Holborrow, J., Eds.), North-Holland, Amsterdam, pp. 5–14.
40. Silverstein, A. M. and Prendergast, R. A. (1970), In *Developmental Aspects of Antibody Formation and Structure* (Sterzl, J. and Riha, I., Eds.), Academia, Prague, pp. 69–77.
41. Cole, G. J. and Morris, B. (1971), *Aust. J. exp. Biol. Med. Sci.*, **49**, 33–53.
42. Robinson, J. H. and Owen, J. J. T. (1976). *Clin. exp. Immunol.*, **23**, 347–354.
43. Zeiller, K., Pascher, G., Wagner, G., Leibich, H. G., Holzberg, E. and Hannig, K. (1974). *Immunology*, **26**, 995–1012.
44. Droege, W. and Zucker, R. (1975). *Transplant.* Rev., **25**, 3–25.
45. Stutman, O., Yunis, E. J. and Good, R. A. (1969). *J. exp. Med.*, **130**, 809–819.
46. Pierpaoli, W. and Besedovsky, H. O. (1975). *Brit. J. exp. Path.*, **56**, 180–182.
47. Singh, U. and Owen, J. J. T. (1975). *Eur. J. Immunol.*, **5**, 286–288.
48. Singh, U. and Owen, J. J. T. (1976). *Eur. J. Immunol.*, **6**, 59–62.
49. Owen, J. J. T., Jordan, R. K. and Raff, M. C. (1975). *Eur. J. Immunol.*, **5**, 653–655.
50. Wortis, H. H., Nehlsen, S. and Owen, J. J. T. (1971). *J. exp. Med.*, **134**, 681–692.
51. Loor, F. and Kindred, B. (1973). *J. exp. Med.*, **138**, 1044–1055.
52. Melchers, F., von Boehmer, H. and Phillips, R. A. (1975). *Transplant. Rev.*, **25**, 26–58.
53. Raff, M. C., Owen, J. J. T., Cooper, M. D., Lawton, A. R., Megson, M. and Gathings, W. E. (1975). *J. exp. Med.*, **142**, 1052–1064.
54. Nossal, G. J. V. and Pike, B. L. (1975). *J. exp. Med.*, **141**, 904–917.
55. Sidman, C. L. and Unanue, E. R. (1975). *Nature*, **257**, 149–151.
56. Strayer, D. S., Lee, W. M. F., Rowley, D. A. and Kohler, H. (1975). *J. Immunol.*, **114**, 728–733.
57. Strober, S. (1975). *J. Immunol.*, **114**, 877–885.
58. Lafleur, L., Miller, R. G. and Phillips, R. A. (1973). *J. exp. Med.*, **187**, 954–966.
59. Gronowicz, E. and Coutinho, A. (1975). *Scand. J. Immunol.*, **41**, 429–437.
60. Gelfand, M. C., Elfenbem, G. J., Frank, M. M. and Paul, W. E. (1974). *J. exp. Med.*, **139**, 1125–1153.
61. Abney, E. R. and Parkhouse, R. M. E. (1974). *Nature*, **252**, 600–602.
62. Vitetta, E. S., Melcher, U., McWilliams, M., Lamm, M., Phillips-Quagliata, J. M. and Uhr, J. W., (1975). *J. exp. Med.*, **141**, 206–215.
63. Rowe, D. S., Hug, K., Faulk, W. P., McCormick, J. N. and Gerber, H. (1973). *Nature*, (*New Biol.*), **242**, 155–157.
64. Kearney, J. F. and Lawton, A. R. (1975). *J. Immunol.*, **115**, 677–681.
65. D'Eustachio, P. and Edelman, G. M. (1975). *J. exp. Med.*, **142**, 1078–1091.
66. Press, J. L. and Klinman, N. R. (1973). *J. Immunol.*, **111**, 829–885.
67. Press, J. L. and Klinman, N. R. (1974). *Eur. J. Immunol.*, **4**, 155–159.
68. Kincade, P. W., Moore, M. A. S., Schlegel, R. A. and Pye, J. (1975). *J. Immunol.*, **115**, 1217–1222.
69. Phillips, R. A. and Miller, R. G. (1974). *Nature*, **251**, 444–446.
70. Hammerling, U., Chin, A. F., Abbott, J. and Scheid, M. P. (1975). *J. Immunol.*, **115**, 1425–1431.
71. Trentin, J. J. (1970). In *Regulation of Haematopoiesis*, Vol. 1 (Gordon, A. S., Ed.), 161–186 Appleton-Century-Crofts, New York, pp. 161–186.

Chapter 3

Lymphocyte Heterogeneity

P. C. L. BEVERLEY

Abbreviations

ALS: anti-lymphocytic serum
ATx: adult thymectomized
CML: cell-mediated lympholysis
Con A: concanavalin A
CRBC: chicken red blood cell
CRT: cortisone resistant thymus
DHR: delayed hypersensitivity
FcR: receptor for Fc of Ig
GvH(R): graft-versus-host (response)
Ia: I associated antigen
K: killer
LD: lymphocyte defined

LPS: lipopolysaccharide
MLC: mixed leukocyte culture
MLR: mixed leukocyte reaction
PBL: peripheral blood lymphocytes
PHA: phytohaemagglutinin
PPD: purified protein derivative
PWM: pokeweed mitogen
RFc: rosette forming cells
SD: serologically defined
SRBC: sheep red blood cell
TDL: thoracic duct lymphocyte
T_E: early T cells
T_H: helper T cells
$T_{C,S}$: cytotoxic, suppressor T cells

For markers (θ, Tla, Ly, . . .), see tables.

1. INTRODUCTION

Lymphocytes were first defined by solely morphological criteria and compelling evidence of their involvement in immune responses was not presented until the late 1950s.[1] In the short period since then, extensive studies of the biology of lymphocytes have been performed and have revealed a bewildering array of functions. The question then arose as to whether all these functions could be attributed to one cell type, the small lymphocyte, or whether the apparent homogeneity of these cells concealed a heterogeneity of physical and chemical properties to match the functional diversity.

In the last few years it has become clear that among the small lymphocytes there is indeed great diversity of physico-chemical properties such as charge, density and size, of surface antigenic structure and biological behaviour. In this chapter an attempt will be made to describe this heterogeneity, and to relate distinct lymphocyte phenotypes to particular functions. No attempt will be made to review the subject exhaustively, but those phenotypic properties which allow clear separation of lymphocytes into functionally distinct subpopulations will be emphasized. Because the most powerful technique for identifying lymphocyte subpopulations has been the use of alloantisera in inbred mouse strains, most of what follows will be information obtained from murine studies. Less complete information in other species points to at least broadly similar heterogeneity.

The best established division of lymphocytes is into two major categories (reviewed in Reference 2), thymus-derived, or T, and bursa-equivalent-derived, or B, lymphocytes. These two sets of cells have been shown to mediate different functions. The principal function of B cells is undoubtedly to synthesize and secrete antibody while T cells perform functions which may be grouped under the heading of cell-mediated immune responses. The evidence for this classification came first from studies of neonatally thymectomized animals[3–6] or bursectomized birds[7–9] which were complemented by clinical observations on patients with immune deficiency diseases.[10] Although no 'central' lymphoid organ exactly equivalent to the avian bursa has been identified in mammals, experiments on the origin of antibody forming cells suggest that these are derived from the bone marrow.[11]

Since the division into T and B lymphocytes is now firmly established, it will be only briefly considered and the major part of this review will be devoted to more recent studies concerning heterogeneity within the T or B populations.

2. TECHNICAL CONSIDERATIONS

2.1 Introduction

Procedures used for separation of lymphocytes into subpopulations have two main aims: (a) the separation itself may be the objective, thereby demonstrating physico-chemical heterogeneity in the population examined and (b) the separated populations may be assayed for functional activity. Many methods have been used to achieve these objectives, and we shall outline some of the difficulties inherent in the more important.

2.2 Physical methods

Size,[12] density,[13] and charge[14] have all been used to separate lymphocytes. Methods utilizing these properties allow general statements to be made on the properties of resting lymphocytes, for example that T cells are largeı,[15] denser[16] and more negatively charged[14] than B cells. There is unfortunately, however, considerable overlap even between T and B cells so that resolving subpopulations of T or B cells presents considerable difficulties. These methods do, however, often allow considerable enrichment of particular cell populations[17] and can be used for separation of very large numbers of cells.[18] Charges in physical properties of cells after, for example, antigenic stimulation can also conveniently be studied by these techniques.[19] Adherence to glass, plastic or particularly nylon wool columns[20] have also been widely used to separate B from T cells, which are less adherent. This method suffers from the drawback that there is considerable loss of T cells and evidence that this may be selective.[21]

2.3 Immunosuppressive agents

A variety of drugs and physical or biological agents have been shown, mainly *in vivo*, to alter the proportions of lymphocyte populations. Irradiation by γ- or X-rays has been widely used. Data on the sensitivity of various lymphocyte populations suggest that, in general, humoral are more sensitive than cell-mediated immune responses[22] and that this is because antibody forming cell precursors, B cells, are more radiosensitive than at least some T effector precursors.[23,24] It is also clear that after antigen or mitogen stimulation lymphocytes may become less radiosensitive.[25] There is evidence also that T cells may vary in radiosensitivity.[26]

Corticosteroids have been shown to preferentially deplete peripheral lymphoid tissues of B cells.[27] Some of the observed effects, however, may be due to alterations in migratory properties[28] rather than actual elimination of cells. In the thymus the immunologically competent medullary lymphocytes are steroid resistant.[29]

A variety of other drugs, particularly cytotoxic agents, have preferential effects on different lymphoid cell compartments. For example cyclophosphamide preferentially depletes thymus-independent areas of spleen and lymph nodes[30] and azathioprine (*in vitro*) preferentially inhibits T rosette forming cells.[31]

All these agents suffer, as tools for defining lymphocyte subpopulations, from problems associated with their often ill-understood multiple effects both *in vivo* and *in vitro* and in particular from the fact that cells in the same lineage differ widely in sensitivity, according to their physiological state (see above).

2.4 Antibodies

Antisera of two types have been used to define surface antigens of lymphocytes, those raised in another species (heteroantisera) and those raised in another individual or inbred strain of the same species (alloantisera). Both types of serum may identify antigens present on all cells (transplantation antigens) or antigens restricted to particular sets of cells (differentiation antigens). This section concerns mainly sera recognizing differentiation antigens since these have been most useful in studying lymphocyte functions; however, even sera recognizing antigens present on all cells may demonstrate heterogeneity because the quantity of antigens on different cells varies.[32] All antisera suffer from common problems as tools for defining subsets. They commonly contain contaminating antibodies against undesired specificities.[33,34] The number of cells identified by a particular serum also depends on the sensitivity of the technique used and in cytotoxic tests the source of complement may completely alter the result. With guinea pig complement, for example, the alloantigen PC-1 appears to be restricted to IgG producing plasma cells, but with rabbit complement both IgM and IgG antibody forming cells are lysed.[35] Table 1 lists some antigens which have been useful in studies of murine lymphocyte heterogeneity.

In a number of cases, particularly the Ly series of antigens (discussed more fully in Section 3), attempts have been made to overcome the problems of contaminating antibodies and autoantibody by using congenic mouse stocks differing at a single Ly locus. Interestingly, it is often not possible to raise significant titres of anti-Ly antibody by immunization between congenic partners.[36] The Ly antisera are therefore usually made in F_1 animals by immunization with lymphocytes of one congenic partner and can be absorbed to remove contaminating specificities with cells of the opposite congenic partner. A further advantage is that the appropriately absorbed serum can be tested in both congenic partners, or better still, pairs of sera can be tested in a criss-cross fashion on both congenic strains. Even in such a defined experimental sytsem as this, it is now clear that anomalies may occur and reciprocal sera tested on congenic partners do not always have the same effects in abrogating functional properties.[37] The reasons for this are not clear, but may reside in residual undetected genetic disparities between congenic stocks, or differences in titre of sera (Reference 38, and Shen *et al.*, submitted for publication).

In summary, the use of antisera to cell surface components, particularly combined with breeding of congenic stocks, allows approaches to two questions. The first is: whether presence or absence of an antigen, defined by an antiserum in a particular strain, marks a particular functional subset of lymphocytes. The second is: what is the genetic basis for expression of that particular antigen? We are mainly concerned here with the first question, but antisera have been used to great effect in studies of genes controlling expression of antigens related to murine leukaemia viruses.[39]

2.5 Miscellaneous markers

Subpopulations of lymphocytes may also be defined by the presence or absence of surface structures not defined by antisera. Among the most important of these are Fc receptors

Table 1. Mouse Lymphocyte cell surface antigens

Antigens defined by:	Linkage group	Thymocytes	T lymphocytes	B lymphocytes	Plasma cells	References
Heteroantisera						
MSLA	–	++	+	–	–	40
MBLA	–	–	–	+	+	41, 42
MSPCA	–	–	–	–	++	43
G IX[a]	IX[a]	++	(+)[b]	(+)[b]	(+)[b]	44, 45
Alloantisera						
TL	IX	++	–	–	–	39, 46
Thy-1	II	++	+	–	–	47, 48, 49
Ly-1	XII	++	+	–	–	36, 50
Ly-2	XI	++	+	–	–	36, 50, 51
Ly-3	XI	++	+	–	–	36, 51
Ly-4	?	–	–	+	+	52
Ly-5	?	+	+	–	–	53
PC-1	?	–	–	–	++	35
ALa-1	?		(+)[c]	(+)[c]	+	(Feeney *et al.*, submitted)
Ia	IX	±	±	+	+	54, 55, 56
H-2	IX	+	++	++	+	39

[a] Strains of mice are G IX positive or negative. Positive strains show different levels of antigen expression and there is evidence for interaction of several genes in the control of antigen expression.
[b] Antigen is expressed in peripheral cells during MuLV infection.
[c] Antigen appears on activated T and B cells only.

(FcR),[57,58] complement receptors,[59] receptors for sheep red blood cells (SRBC) found on human T cells,[60] for human red cells on activated human T cells,[61] for viruses[62,63] and for cholera toxin.[64] These are particularly useful in human studies (e.g. References 65, 66) because it is difficult (and sometimes unethical) to raise specific alloantisera.

3. T LYMPHOCYTE HETEROGENEITY

3.1 The T_1–T_2 hypothesis

The division of the lymphoid system into two major compartments, B and T, has already been described. Compelling evidence suggesting heterogeneity within the T lymphocytes came from studies of the graft-versus-host response (GvHR). Cantor and Asofsky[67] showed that a small dose of anti-lymphocyte serum (ALS) reduced the GvH activity of spleen cells, but that addition of small numbers of peripheral blood lymphocytes (PBL), which alone gave no GvHR, restored the GvH activity of the spleen cells. Similar synergy was demonstrated between thymus and PBL and between spleen cells from ALS treated and adult thymectomized (ATx) animals.[68,69] This data together with evidence of heterogeneity within the T cells forming specific rosettes with sheep red blood cells (SRBC)[70,71] lead to the formulation of the hypothesis that there might be two types of T cells.[72] T_1 cells, found mainly in thymus and spleen, are high in surface Thy-1, short-lived and are lost preferentially after ATx. In contrast, T_2 cells are found mainly in lymph nodes and peripheral blood, have less Thy-1, are resistant to ATx but sensitive to ALS. It was proposed that T cell responses seen after immunization are mediated by T_2 cells, while T_1 cells on contact with antigen become T_2 (memory) cells.

This hypothesis is in accord with data showing, for example, that ATx does not affect primary humoral immune responses but that secondary responses are decreased[69] and with a variety of data suggesting heterogeneity of peripheral T cells (e.g. References 26, 69, 70, 73). It fails to account satisfactorily for the synergistic response of T_1 and T_2 cells in GvH or mixed lymphocyte responses if the only response of T_1 cells is to become T_2 cells, especially as the synergy is seen most clearly in responses to histocompatibility antigens in which it is proposed that the T_2 pool is already large.[72] Nor does this scheme deal with the crucial question of whether functionally distinct effector T cells, for example, killer and helper cells, are distinct cell types or not, and whether their precursors, prior to antigen stimulation, are also separate. Possible schemes for T cell differentiation are illustrated in Figure 1. In the following subsections we shall consider data which strongly suggests that scheme A most nearly approximates the path of T cell differentiation in the mouse.

3.2 Precursor T cell heterogeneity

In this discussion precursors will be considered to be cells taken from animals which have not been deliberately immunized. The possible influence of cross-reacting environmental antigens will not be considered. If T precursors can be demonstrated, which upon antigenic stimulation can only mediate one effector function, but not another, then it is implicit that differentiation to an antigen sensitive precursor, committed to a particular cell lineage, is antigen independent (scheme A, Figure 1). In contrast, if a single precursor can be shown to carry out more than one effector function, then the terminal stages of T cell differentiation are antigen driven (schemes B, C and D, Figure 1).

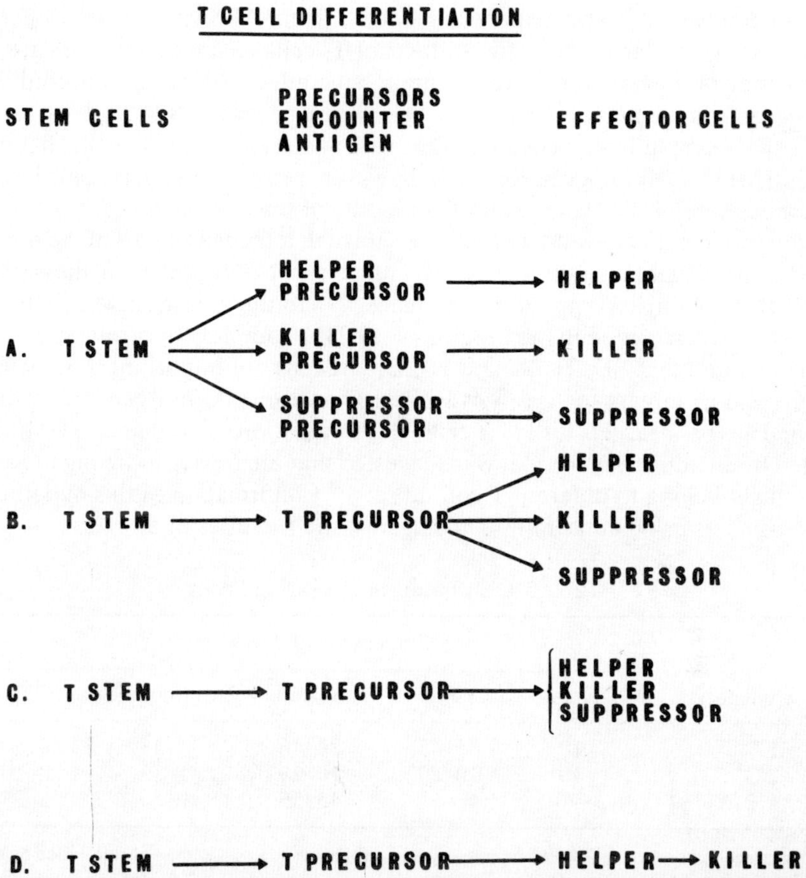

Figure 1. It is not suggested that helpers, killers and suppressors are the only distinct T cell subsets, they are merely used as examples to illustrate possible schemes for T cell differentiation

Since non-immune B cells may be readily distinguished from non-immune T cells (e.g. References 41, 49, 52), it is clear that precursor heterogeneity exists. The experiments on GvHR, discussed in the preceding section, suggest heterogeneity within the T compartment. Furthermore, several authors have demonstrated synergy between subsets of T cells in the *in vitro* generation of killer cells during mixed leukocyte culture (MLC), a model which may be an *in vitro* correlate of the GvHR. The two cells involved have been characterized as a short-lived, high Thy-1, ATx sensitive cell (T_1) and a long-lived, low Thy-1, ALS sensitive cell. (T_2).[69,73–76] It has also been demonstrated that both populations can be separated from the normal spleen T population, using the Fluorescence Activated Cell Sorter. This suggests that synergy may operate as a physiological mechanism *in vivo.*[77]

Although the evidence cited above clearly demonstrates synergy between T cell subpopulations in GvHR and MLC/CML (cell-mediated lympholysis), a number of problems require clarification. In the studies of GvHR,[67,68] experiments using parent into F_1 strain combinations identified the precursor of the GvHR effector cell as a T_1 type cell, while long-lived recirculating cells (T_2) acted as amplifiers. In contrast, in MLC/CML, the weight of evidence suggests strongly that the precursors of the cytotoxic cells are T_2 cells[69,75–77] although a contrary report has been published.[78] Furthermore, although it is possible to demonstrate synergy between T_1 cells (from ALS spleens) and T_2 cells (from ATx spleens), particularly when low numbers of T_2 cells are used, if T_1 cells are added to larger number of T_2 cells they have a suppressive effect.[69] This is in accord with data showing that some immune responses are increased after ATx.[73,75,79,80] It is clear therefore that a simple T_1–T_2 scheme does not adequately account for all the findings in the GvHR and MLC/CML models. Recently, however, two sets of experiments have thrown light on the nature of the synergy and the identity of the cells involved.

Experiments in both man and mouse have shown that the major part of the proliferative response in MLC/CML is stimulated by antigens not identical with the serologically defined (SD) major histocompatibility antigens.[81,82] In the mouse these LD (lymphocyte defined) antigens are coded in the I region of the H–2 complex. In further experiments, it was demonstrated that SD, H-2K and H-2D, antigens are important in sensitization of killer cells and as targets for cytolysis while LD antigens do not generate strong killer activity and are poor targets for killer cells. LD stimulation can enhance a CML response stimulated by an SD difference.[83] It was suggested that the cells responding to SD and LD antigens might belong to different T cell subsets.[84] Confirmation of this hypothesis came from experiments using antisera directed against the Ly series of antigens.

Table 2. Distribution of Ly bearing T cells

	Percentage of T cells in:[a]				
Ly phenotype[b]	Thymus	CRT	Spleen	Lymph node	TDL
1^{+c}	>10	35–45	30–40	35–45	40–60
2^+3^+	>10	10–20	10–20	5–15	2–10
$1^+2^+3^+$	>90	40–55	45–55	40–50	35–55

[a] Based on numbers of cells killed by anti-Ly sera expressed as a percentage of cells killed by anti-Thy-1.
[b] Data for C57Bl/6, except for TDL in CBA/CA (Sprent and Beverley, unpublished) in which only antisera to Ly-1 and Ly-3 antigens have been used.
[c] Number of Ly-1^+ cells obtained by subtracting number killed by anti-2/3 antiserum from 100%. Similar calculations for 2^+3^+ and $1^+2^+3^+$ cells. Serial cytotoxicity experiments give lower figures for Ly-1^+ and 2^+3^+ cells.[86]

Each antigen has two alleles 1 and 2 and the genes for Ly-2 and Ly-3 are closely linked but separate from Ly-1 (Table 1). The antigens are expressed only on T cells.[50] Table 2 summarizes data on the Ly phenotypes of T cells in different sites (References 85, 86, and unpublished data). Anti-Ly sera have now been used to investigate the phenotype of the cells involved in MLC/CML.

The results show that T cells of phenotype Ly-2^+3^+ are able to generate killer activity while those of phenotype Ly-1^+ are not. On the other hand, pretreatment of responder cells with either anti-Ly-1 or anti-Ly-2 or 3 considerably reduced proliferation in the MLC (against the whole H-2 complex). If the incompatibility was restricted to the I region, only anti-Ly-1 reduced proliferation while anti-Ly-2 or 3 gave increased thymidine incorporation.[86] The implication is that Ly-1^+ cells proliferate in response to I region determinants while Ly-2^+3^+ cells proliferate and differentiate to killer cells in response to H-2K and H-2D region determinants.*

* It is already clear that this is an oversimplification since recent reports show that I region determinants can act as targets for cytotoxic T cells; see, for example, Wagner, M., Gotze, D., Ptschelinzew, L. and Rollinghoff, M. (1975). *J. exp. Med.*, **142**, 1477–1487.

Experiments using mixtures of Ly-1^+ and Ly-2^+3^+ cells also showed synergy between the two cell types in the generation of Ly-2^+3^+ killer cells.[87] Furthermore, it was shown that this was not due to recruitment of cells from the Ly-1^+ subpopulations and also that removal of cells bearing I region antigens from the stimulating population abolished the synergy.[87]

Preliminary studies on the GvHR show that pretreatment of spleen cells with either anti-Ly-1 or anti-Ly-2 abolishes their ability to cause spleen enlargement in neonatal F_i hosts.[85] This could mean that GvHR is mediated by Ly-$1^+2^+3^+$ cells, or more likely, in view of the MLC/CML data cited above, that synergy between Ly-1^+ and Ly-2^+3^+ cells is a prerequisite for a positive GvHR.

In summary it appears clear that at least the majority of killer cell precursors are Ly-2^+3^+ and that these respond predominantly to H-2K and H-2D region determinants. Ly-1^+ cells on the other hand do not generate killer activity but respond by proliferation to I region antigens. It should be emphasized, however, that these observations have been made only on cells from C57Bl/6 and Balb/c mice both with phenotype Ly-1·2, 2·2 and 3·2 (see Section 2.4).

Another function which has been shown to be mediated by T cells is cooperation with B cells in the humoral response.[88] Studies of the effects of adult thymectomy on the primary antibody response to SRBC have shown that T helper cells involved are long-lived cells resistant to ATx[73,89] but sensitive to small doses of ALS (T_2). Helper cells also show different dose requirements for priming compared to cytotoxic or suppressor cells.[89,91] Direct evidence that the precursors of helper and killer cells differ, has come from studies with anti-Ly sera. Both *in vivo*[86] and *in vitro*[92] it has been shown that they are Ly-1^+.

There is now also *in vitro* evidence suggesting that just as in the generation of mature killer cells there is an interaction of two T cells, so also are two T cells required for production of mature helper cells.[93] Both *in vivo* and *in vitro* the precursor of the mature helper cell appears to be an ATX resistant cell,[89,93] but *in vitro* a second ALS resistant T cell is also required for its maturation.[93] The reason for the difference between the *in vivo* and *in vitro* data is as yet unclear.

A further activity of T cells which also reveals heterogeneity, is the response to mitogens. Cortical (cortisone sensitive) thymocytes respond well only to Con A,[95] while cortisone resistant thymus (CRT) cells respond both to Con A and PHA. In the peripheral lymphoid tissues, two types of lymphocyte have been distinguished on the basis of their mitogen responses. The first responds equally well to both PHA and Con A, and has high density of Thy-1 and is found in both lymph nodes and spleen. The second responds better to Con A and is found mainly in the spleen. It has a lower density of surface Thy-1.[96] Subsets of T cells responding differentially to mitogens can also be separated using physical means.[97] Studies using Ly antiserum pretreatment of spleen or CRT have suggested that cells responding to Con A are Ly-1^+ while the PHA responding cells have the phenotype Ly-1^+2^+.[92] However, in some experiments Ly-2^+3^+ cells are able to respond to Con A (Cantor, personal communication), a finding which is in accord with the two observations that suppressor activity is easily induced by Con A stimulation[98] and that suppressor cells have the phenotype Ly-2^+3^+ or Ly-$1^+2^+3^+$(see Section 3.3).

3.3 Effector T cell heterogeneity

Several experimental approaches have been used to show that primed helper cells and cytotoxic T lymphocytes (killer cells) are distinct cell types. In one series of experiments it was demonstrated that differences in antigen dose favoured development of helper activity or killer cells.[91] Similarly, it has been shown that spleen cells from skin grafted mice appear to have mainly helper activity 13–28 days after grafting, while spleen cells immunized *in vitro* to the same alloantigens show strong cytotoxicity but not helper activity.[90] Another approach depends upon physical separation of cells and it appears that under some circumstances primed populations may contain separable effector subpopulations.[99,100]

More definitive separation of helper and killer T cells has been achieved using antisera to Ly antigens. Treatment of primed helper cells with anti-Ly sera and complement shows that helper cells primed to SRBC[85,86] or keyhole limpet haemocyanin (KLH) (Feldmann *et al.*, submitted for publication), have the phenotype Ly-1^+. Experiments on alloimmune peritoneal exudate lymphocytes[37] and alloimmune killer cells[86] generated *in vitro* show that at least the majority of killer cells in the C57Bl/6 mouse) have the phenotype Ly-2^+3^+. The same is true in Balb/c mice (also Ly-1·2, 2·2, 3·2),[86] but experiments using sera directed against the alternative alleles (Ly-1·1, 2·1, 3·1) in congenic mice[37] have not given such clear-cut results since anti-Ly-1·1 reduces cytotoxicity approximately 50% and anti-2·1 or anti-3·1 fail to abolish cytotoxicity. The reasons for this are not clear (see Section 2.4 and Reference 37).

A further function which has been investigated with these sera is suppressor T cell activity. In CBA mice (Ly-1·1, 2·1, 3·2) specific suppressor cells generated *in vitro* in response to high antigen dose[93] have the phenotype Ly-2^+.[94] A similar finding has been obtained in studies of suppressor cells found in allotype suppressed mice (Reference 101 and Herzenberg *et al.*, submitted for publication). Data from another model in which specific T suppressor cells can be demonstrated[102] suggests that suppressor cells may carry a histamine receptor,[103] a characteristic which they share with T killer cells (Shearer *et al.*, submitted for publication). Helper T cells in contrast have no histamine receptor.[103,104]

A number of authors have described non-specific suppressive effects of T cells which may be induced by culture of cells in the absence[21,105] or presence of antigen or mitogen.[106,107] Suppressor activity can be assayed *in vitro* or *in vivo* and may affect primary or secondary antibody responses, mitogen responses,[108,109] GvHR,[110] mixed lymphocyte reaction (MLR),[98] or CML.[107] In two experimental series the non-specific suppressor activity has been characterized as due to a T cell capable of adhering to nylon wool.[21,108] Treatment of suppressor cells, generated by culture in the absence of antigen and capable of suppressing SRBC antibody responses, with anti-Ly sera and complement, shows that this type of activity is partially sensitive to both anti-Ly-1 and anti-Ly-2. A mixture of cells treated with either antisera does not reconstitute the suppressor activity. This suggests that the suppressor activity is not due to an interaction of Ly-1^+ and Ly-2^+, but to cells of phenotype Ly-1^+2^+.[92] It is unclear at present how these observations fit with data showing that this type of suppression can be generated from both ALS treated and ATx spleen cells[105] and that ATx spleens are depleted of Ly-$1^+2^+3^+$ cells,[86] though it is likely that several different types of suppressive activity are detected in different experimental models.

The foregoing discussion shows that there is considerable evidence that at least two well-characterized effector cell types, helper cells and killer cells, are separate cell types. Furthermore, specific helper and suppressor cells are also distinguishable. Killer cells and specific suppressor cells, which have the same Ly phenotype have not yet been distinguished. The evidence also suggests (in the case of helper and killer cells) that precursor cells of a particular Ly phenotype retain the same phenotype as they differentiate to effectors.

3.4 Further evidence of T cell heterogeneity

In the preceding sections an attempt has been made to identify, by physico-chemical and antigenic properties, functionally distinct T cell subsets. In this section experimental systems in which T cells play a part, but in which they have been less fully characterized will be mentioned. Similarly, we shall discuss data on surface markers which are present on only a proportion of T cells, when the functional properties of that subset are as yet unclear.

Clinical studies of immunodeficient patients[10] and thymectomized animals[4,5] suggest that both contact hypersensitivity and delayed hypersensitivity (DHR) are T dependent functions. Experimentally in mice, both effects can be abrogated by treatment with anti-Thy-1 and complement in a cell transfer system.[111,112] Similarly, in guinea pigs, antigen induced cell proliferation *in vitro*, which correlates well in timing[113] and specificity[114] with DHR, can be abrogated by heterologous anti-T serum.[115] There is evidence also that T lymphocytes may also be responsible for production of at least some lymphokines.[116]

Although the cells mediating these various functions have not been clearly identified, several observations suggest that more than one subset of T cells are involved. It is clear that cells mediating DHR have a different anatomical distribution from those capable of helping a humoral response.[117] Furthermore, DHR cells can be separated from helper cells by velocity sedimentation.[118] Similarly, it has been recently shown that cells capable of mediating contact sensitivity and cytotoxicity directed to the same hapten, have very different antigen priming requirements.[119]

Two markers present on B lymphocytes (Section 4.3) have been demonstrated also on some T cells. These are Fc receptors (FcR) and antigens coded in the I region of the H-2 complex. The presence of receptors capable of binding Ig on normal T cells is controversial. Some studies have failed to demonstrate them on normal thymus cells[120] while others show large numbers of Fc binding cells in the same organ.[121] Several lymphomas of apparently T cell origin (bearing T markers such as Thy-1, Tl, etc.) have been shown to bind Fc[122] and a number of studies have also demonstrated that T cells activated *in vivo* to H-2 determinants also carry FcR.[123] Recently, it has been demonstrated that donor cells in the spleens of irradiated recipients of allogeneic T cells are more than 40% FcR positive, but intriguingly, donor T cells recovered from the thoracic duct of the same animals are FcR negative. These activated TDL (thoracic duct lymphocyte) can, however, generate FcR positive cells in the spleens of a second host.[124] In spite of the absence of demonstrable FcR, B cell derived Ig can be detected on activated TDL.[125] Whether these discrepancies indicate differences in affinity or number of FcR on T subsets, or the existence of more than one type of FcR is not clear. Recent data suggests at least that helper T cells have few or no FcR,[126] although it appears that all Ly separable subsets may have both FcR^+ and FcR^- cells within them (Herzenberg, personal communication).

The magnitude of a number of immune responses has been shown to be controlled by genes mapped to the I region of the H-2 complex (e.g. References 127–129). It was clearly of great interest, when sera were produced which detected antigens (I associated, Ia) also coded in this region, to know which lymphocytes expressed them. As in the question of FcR, it is clear that most, or all, B cells express Ia antigens (Section 4.3) but the position with regard to T cells is more obscure. Early reports suggested that Ia antigens were present on T cells.[130] In later experiments using purified T and B cell populations, other authors concluded that the antigens were present exclusively on B cells.[131] A resolution of these differences is suggested by the finding that some anti-Ia sera consistently give a biphasic titration curve in cytotoxic tests on lymph node or spleen but not CRT cells, suggesting two antibody populations or that T cells differ from B cells in amount of antigen expressed.[132] Failure to detect the T cell reactivity might in some cases be due to low antibody titre[131] or differences in the specificity of the antisera used.

Studies of the Ia phenotype of T cells activated by various means have now also been reported. In one such report 20%–30% of Con A and 5%–10% of PHA activated blast cells appeared to be positive although specificity controls for the antiserum were lacking.[133] Another study showed that the majority of H-2 activated donors cells in both spleen and TDL of F_1 recipients were Ia positive.[124] The latter cells were FcR negative (see above), a finding of interest in view of the reported close association of Ia and FcR.[134]

In conclusion, it should be said that there is little to suggest which functional subsets of T cells bear Ia antigens. From the data on activated TDL it seems likely that at least some cells involved in MLC/CML carry Ia specificities. Preliminary data from this laboratory (Feldmann, unpublished) suggests also that specific suppressor T cells may be Ia positive, a finding in agreement with other data suggesting that specific suppressor factor is Ia related.[135]

3.5 A new T cell classification?

The T_1–T_2 scheme led to numerous experiments revealing T cell heterogeneity, however (Section 3.1), it failed to account adequately for synergy between subsets of T cells, nor did it deals with the question of whether different T cell functions are mediated by one or more subsets of cells. Recent data, particularly that obtained with antisera to Ly antigens (discussed in the preceding sections) indicates very clearly that not only are the effectors of various T cell functions clearly separate but so also are the precursors of at least some of these effectors. This implies that differentiation of a presumed T stem cell into precursors of the various T subsets is antigen independent[86] (scheme A, Figure 1).

It is possible that even in the absence of exogenous antigen, the majority of T stem cells are driven to the precursor stage by contact with cross-reacting antigens of the external or internal environment. This possibility is supported by data suggesting that most antigen sensitive precursors have properties very similar to memory cells.[136] A resolution of this question might be achieved by indentification of the T stem cell and studies of its properties.

A possible candidate is the post-thymic T cell described by Stutman (reviewed in Reference 137). This cell is Thy-1^+, but immuno-incompetent as judged by ability to mediate GvHR, MLR and PHA response. Under the influence of the thymus it becomes immuno-competent. In many other properties such as sensitivity to ATx, tissue distribution and migration pattern, it is identical to the T_1 cell of Raff and Cantor.[72] Since the ability of this cell to synergize with T_2 cells has not been tested it may indeed be, that the two cells are indentical.

An alternative T stem cell is the Ly-$1^+2^+3^+$ cell which appears in the peripheral lymphoid tissues early in ontogeny, and is also sensitive to adult thymectomy.[86] The tissue distribution of this cell differs from both the T_1 and the post-thymic cell in that it is found in lymph nodes[85] and TDL (Sprent and Beverley, unpublished) as well as the spleen. As yet, nothing is known of the functional properties of this cell in the adult because of technical problems in isolating it. Neonatal mouse spleen cells, however, which are mainly Ly-$1^+2^+3^+$[86] give a very poor MLR.[138] Whether Ly-$1^+2^+3^+$ cells are capable of synergizing with T_2 cells is not known.

If the identity of the precursor of T cells, capable of differentiating into effector cells on contact with antigen is unclear, so also is the nature of the cells which interact with these precursors to amplify their response. In the studies of GvHR and MLC/CML previously discussed (Sections 2.1–2.3), it appeared that cells with properties of T_1 and T_2 were able to synergize in production of effector cells. Current data derived from use of Ly antisera demonstrates synergy between Ly-1^+ amplifying cells and Ly-2^+3^+ precursors in generation of effectors.[86,87] It is clear, however, that the Ly-1^+cell is not equivalent to the T_1 cell since it differs in tissue distribution (Table 1), sensitivity to ATx[86] and, most importantly, in ability to respond in MLC.[86,139] A partial resolution of these discrepancies may be deduced from two observations. First, it is clear that T_1 cells are not a clearly defined cell population, and in some experiments they can show both amplyifying and suppressive effects in MLC/CML.[73] Secondly, the only function ascribed to Ly-$1^+2^+3^+$ cells so far is suppressive.[92] It may be therefore that most T_1 populations contain a mixture of Ly-1^+ and Ly-$1^+2^+3^+$ cells. The Ly-1^+ cells involved would, however, have to be relatively spared by small doses of ALS *in vivo* and relatively sensitive to adult thymectomy. Alternatively, there may be heterogeneity as yet undetected within the Ly-$1^+2^+3^+$ population, with some cells able to mediate amplifier function while others are stem cells.

There are similar problems in interpreting older data on the differentiation of precursors into effectors. Studies of cells forming rosettes (RFc) with SRBC[70] were interpreted as indicating two population of T-RFc. Both subsets were considered to be T helpers at different stages of differentiation, however, since functional evidence of helper activity in both subsets has not been obtained, and it is clear now that antigen stimulation can lead to differentiation of T cells with quite distinct functional and antigenic phenotypes, this interpretation may well be an oversimplification. A further observation which lends weight to this view is the finding that in previously primed animals ALS treatment and thoracic duct drainage fail to affect the secondary humoral response,[140] suggesting that not all memory can be ascribed to a long-lived recirculating (T_2) cell.

In spite of these and other ambiguities (reviewed in Reference 141) in our present understanding of T lymphocyte differentiation and heterogeneity, it may be worthwhile to attempt a classification of T cells in order to try to bring together some of the older and more recent data. Such a scheme has been proposed[142] and is reproduced in Table 3. It is immediately clear that this classification requires several qualifications. First of all, it should be made clear that the T_H class includes T helpers for antibody as well as cells which give a good MLR and synergize with T killer precursors in the generation of T killers, but it is not implied that these are necessarily the same cell. The same disclaimer should be made for the cytotoxic and suppressor cells in the $T_{C,S}$ class.

With respect to T_E cells, although they undoubtedly appear early in ontogeny and the majority are sensitive to ATx, the finding that they are present in TDL would imply that at least some are long-lived cells since all TDL T cells have been shown to be so.[143] T_E therefore does not necessarily imply a precursor relationship to the T_H and $T_{C,S}$ subsets.

In summary therefore, this classification, which it is assumed will be of a very temporary nature, brings together some recent data on T cells and while it does not attempt to explain T cell differentiation, it may at least help to identify some areas of ignorance.

4. B LYMPHOCYTE HETEROGENEITY

4.1 Introduction

The problems of B cell heterogeneity appear at first sight to differ from those of the T cell compartment. In contrast to the functional heterogeneity of T cells, B cells have only one

Table 3. Classification of T cells

Property	Subset T_H[a]	$T_{C,S}$[a]	T_E[a]	References 142
Ly phenotype[b]	1^+	23^+	123^+	35, 85–87
Ontogeny	late	late	early	86
ATx	resistant	resistant	sensitive	86
Recirculation[c]	+	+	+	(Sprent and Beverley, unpublished)
Histamine receptor		+		103, 104
Ia phenotype	?+	?+		133, Feldmann (unpublished)
PHA response	+	+	?	92
Con A	+	?	?	92
Cytotoxic precursor		+		86,87
Cytotoxic effector		+		37, 86, 87
Proliferative response to: 'LD'	+		?	86, 87
'SD'		+	?	86, 87
Specific T helper	+			85, 86
Specific T suppressor		+		(94, (Herzenberg *et al.*, submitted)
Non-specific T suppressor			?+	92

[a] T_H is a T helper, $T_{C,S}$ is T cytotoxic, suppressor and T_E is T early.
[b] Ly phenotype defined in C57Bl/6 mice Ly-1·2, 2·2, 3·2; not necessarily applicable in other strains, see Reference 37 and Section 2.4.
[c] Recirculation = presence in TDL.
Reproduced by permission of *Nature* from Reference 142.

important function, that of antibody secretion. However, this difference may be more apparent than real since different antibody classes have very different functional properties and an analogy may therefore be drawn between cells secreting antibody of different classes and the functionally distinct T effector cells described in preceding sections.

If this analogy is accepted then the same questions may be asked in relation to B cell heterogeneity as in the case of T cells. In particular, which stages of B cell differentiation are antigen driven and when do cells destined to produce particular antibody classes become programmed to do so? Although as yet, answers to these questions are by no means definitive, there is ample evidence of heterogeneity within the B compartment.

4.2 Surface immunoglobulin

Although other aspects of lymphocyte surface immunoglobulin are considered in this book, it seems appropriate here, to consider them as markers of B cell heterogeneity. In all species studied the majority of B cells bear the μ heavy chain and a light chain. In the mouse this is mainly of the κ type. A number of observations strongly suggest that IgM is the earliest immunoglobulin to appear.[144] Furthermore, studies on the immunosuppressive activity of anti-μ sera have shown both *in vivo*[145] and *in vitro*[146] that not only IgM synthesis but also IgG and IgA is inhibited. This implies an IgM to IgG (or IgA) switch during maturation of B cells. Also in support of this concept is data on rosette inhibition showing that non-immune B cell SRBC rosettes can only be inhibited by anti-μ sera while at the peak of an anti-SRBC response they can be inhibited by anti-γ also.[147] A number of direct labelling studies have also demonstrated cells with both IgM and IgG on their surface[148] or surface IgM and intracytoplasmic IgG.[149] There is also evidence for the presence of more than one IgG class on some mouse lymphocytes.[150]

In contrast to the apparent freedom in expression of heavy chain class in B cells, the majority of authors (e.g. References 147, 151) find that only one light chain type is produced by a single B cell. The second restriction in immunoglobulin expression is in the allotype. Although earlier data suggested exclusion of one parental allotype in antibody secreted by heterozygous cells[152] more recently, evidence has accumulated suggesting that, at least early in the primary response, both allotypes may be detected on single cells.[150]

The foregoing discussion would suggest that the B cell receptor for antigen binding is an IgM molecule, and that IgM bearing cells may represent a relatively uncommitted cell which can, under appropriate stimulus, differentiate to a cell producing another immunoglobulin class. Experiments with anti-immunoglobulin coated columns[153] and on antigen binding cells during the course of the response to SRBC[147] suggest that cells become restricted to one antibody class only after exposure to antigen.

The simplified view of B cell surface immunoglobulin given above has been complicated by results which suggest that in primates, IgD may also function as a B cell receptor for antigen. IgD is found in very low concentration in human serum but is present on the majority of B cells.[154,155] In the mouse, no IgD, or IgD homologue, has been detected in the serum, but lacto-peroxidase surface labelling followed by immunoprecipitation and SDS polyacrylamide gel electrophoresis, has identified a new heavy chain class on mouse B cells.[156,157] It is likely that this is murine IgD.

In both man and mouse, IgM can be detected earlier in ontogeny than IgD[158,159] and interestingly there appear to be more IgD bearing cells in lymph nodes than in spleen.[158] In the spleen, cells bearing IgM only, IgD only, or IgM+IgD have been detected.[158] It is suggested therefore that IgD^+ B cells are more mature and may be memory B cells.[158] An alternative proposal is that interaction of antigen with IgM^+ cells induces tolerance, interaction of antigen and T cells with IgM^+IgD^+ cells leads to IgM production and IgM memory, and interaction of antigen with IgD^+ cells leads to IgG production.[160] This concept is supported by the finding that treatment of monkeys with anti-IgD serum leads to production of large amounts of IgG.[161] In summary, the data support the proposition that IgD is a B cell receptor for antigen, but as yet it is unclear what functional role it plays in triggering the cell.

4.3 Other B cell surface markers

Table 4 lists a number of surface antigens and other markers present on B cells. A number of these will not be discussed further since we are concerned here with heterogeneity within the B cell lineage and these markers appear to be present on all B cells. These include MBLA, Ly-4, Ia and H-2. A further group of the antigens distinguishes plasma cells from their precursors. These markers, MSPCA, PC-1 and ALa-1 are evidence of the major reorganization of B cell structure which takes place as the cell differentiates into a high-rate antibody secreting cell. Equally striking is the loss from the surface of plasma cells of detectable surface Ig and Fc receptors.[58]

Of particular interest among these markers is Th-B which appears to be present on the majority of spleen B lymphocytes and on 50% of thymocytes. Although it can be detected in bone marrow by absorption it fails to prevent development of PFC in a transfer system.[162] This suggests that there is antigenic heterogeneity among the precursors of

Table 4. B lymphocyte surface markers

Antigens defined by:	Presence on B lymphocytes	Presence on Plasma cells	Presence on T lymphocytes	References
Heteroantisera				
MBLA	+	+	−	41, 42
Th-B	+	+	+	162
MSPCA	−	+	−	43
Alloantisera				
PC-1	−	+	−	35
Ly-4	+	+	−	52
ALa-1	−	+	+	(Feeney *et al.*, submitted)
Ia	+	+	?	54–56
H-2	++	+	+	39
Other markers				
Fc receptor	+	−	+	58
Complement receptor	+	−	−	59

PFC, an observation which is in accord with functional studies demonstrating differences between spleen and bone marrow B cells (Section 4.4).

4.4 Functional approaches to B cell heterogeneity

Because polyclonal activators or mitogens activate relatively large numbers of cells when compared to even complex antigens such as SRBC they have been used to attempt to define functionally distinct subsets of B cells,[163] and to develop concepts of B cell activation. On the basis of responses to dextran sulphate, lipopolysaccharide (LPS) and PPD measured as either numbers of Ig producing cells or synthesis of DNA, three types of B cell have been proposed[164] differing in ontogeny and distribution in the adult mouse. Other studies using mainly pokeweed mitogen (PWM) and LPS[163] show that even under optimal culture conditions in which IgM and IgG synthesis can be observed, no more than 50% of B cells respond to either mitogen. The same study showed that some cells, particularly in lymph nodes and Peyer's patches respond by DNA synthesis while the remainder synthesize DNA and go on to develop extensive endoplasmic reticulum and high-rate antibody secretion. The latter population predominates in the spleen and large numbers of these IgM secreting plasmablasts are also seen in 'nude' spleen suggesting that maturation of these spleen IgM cells is T independent. Nude lymph nodes, on the other hand, have fewer than normal IgM blasts, suggesting that this may be a T dependent response.[165] It is suggested that three classes of cells can be distinguished by their response to LPS and PWM, those that fail to respond, those which respond by proliferation, and not Ig production, and those that both proliferate and become high-rate antibody synthesizing cells. The fact that stimulation with a given mitogen also leads to appearance of significant numbers of IgG containing and secreting cells[163] suggests in addition that each of the three B cell types contains cells pre–programmed to express a certain class of immunoglobulin when appropriately stimulated, irrespective of the class of the surface receptor of the resting cell. It is suggested that these three types may represent an ontogenetic sequence or maturation steps in the adult animals. The triggering requirements for the three cell types are different and it is suggested that the 'early' types may require potentiating factors, e.g.

T or macrophage help to express Ig synthesis, while the 'mature' type does not. This suggestion is supported by the observation that thymus dependent antibody responses arise earlier in ontogeny than thymus independent ones.[166]

Studies of the requirement of B cells for T helper cells also reveal heterogeneity. On the basis of experiments examining synergy between thymus cells and spleen or bone marrow, two classes of B lymphocyte were proposed: B_1, T independent and present in both bone marrow and spleen, and B_2, T dependent and present mainly in spleen.[167] Somewhat similar experiments suggest that these two categories of B cell may be distinguished with regard to both antibody synthesis and tolerance induction.[168] The interpretation of these experiments is now complicated by more recent data demonstrating T–T interactions[87] and it may be that the B cell containing populations (BM or spleen), vary in their content of T cells able to synergize with the T population (thymus cells) provided.

It is clear from the foregoing discussion that functional heterogeneity among B cells can be demonstrated. Attempts to define the cells in these functionally distinct categories have depended mainly on physical methods because of the lack of antigenic markers distinguishing B cell subsets. The heterogeneity of immunoglobulin bearing cells in spleen and bone marrow has been studied by sedimentation and cell electrophoresis.[169] Sedimentation profiles of bone marrow Ig bearing cells show a peak at 2·6 mm/hr with a faster sedimenting shoulder. The sedimentation velocity of this peak is identical to that of spleen Ig bearing cells and in functional assays contains antibody forming cell precursors. On the other hand, the cells sedimenting at a faster rate are able to transfer to irradiated hosts the ability to respond after a delay.[170] It is not clear, however, whether pre-B cells correspond to the Ig bearing cells in this fraction, although anti-immunoglobulin antiserum suppresses the transfer of pre-B activity.

The same techniques have also been used to distinguish precursors for IgM and IgG responses to T dependent or independent antigens. In unprimed mice, no clear differences in sedimentation profile were found between the precursors of IgM antibody producing cells responding to T dependent and some T independent antigens. For some T independent antigens there were higher numbers of large precursor cells.[171] Using the same technique IgM and IgG precursors for a secondary response to a thymus dependent antigen could be separated but not those for a secondary response to a T-independent antigen. Electrophoretic separation of unprimed IgM and memory IgG B cells (for a T dependent antigen) has also been achieved.[172] The data suggests therefore that the most readily separable subsets of B cells are unprimed IgM precursors and the primed precursors of IgG cells. These two populations of B cells have also been shown in the rat to differ in mechanism of stimulation, migration pattern, tissue distribution and turnover rate.[136]

5. NULL CELLS

It is clear that in a number of experimental systems, antibody coated target cells may be killed by effector cells which do not have T cell markers (e.g. Reference 173). The identity of these effector cells is the subject of this section. The cells will be called K (killer) cells as distinct from the T killer cell discussed above.

A number of authors have demonstrated that removal of actively phagocytic cells (macrophages) does not remove K cell activity from mouse spleen cell suspensions.[174] However, other reports show that some activity also resides among cells which can be removed by treatment with carbonyl iron.[175] Other experiments showing that K cell killing can be inhibited by anti-Ig sera led to the suggestion that the effector might be a B cell[174] but later data showed that removal of cells bearing surface Ig did not deplete K cell activity and furthermore that the effect of anti-Ig sera might well be a non-specific effect.[173] It was suggested therefore that K cell activity might be due to a null cell bearing neither surface Ig nor T cell markers.[173] That K cells carry a receptor for Fc and that the recognition of antibody–target cell complexes required for cytolysis is via this receptor, has been clearly demonstrated.[176]

Recent attempts to clarify some of the contradictions apparent in the earlier studies quoted above have revealed heterogeneity within the K cell population. In studies of the cytotoxicity of non-immune mouse spleen cells for antibody coated chicken red blood cells (CRBC) at least two populations of K cells were defined, a 'myeloid' cell which is not removed by carbonyl iron and does not carry B or T cell markers (except for Fc receptors) but adheres readily to glass bead columns,[177] and a 'lymphoid' K cell which is present in low concentration and is distinguished by weak adherence to glass, large size and complement receptors.[178] A further population capable of lysis of CRBC has been defined in the guinea pig and mouse. These cells are phagocytic and have all the properties of macrophages.[179]

Further evidence for K cell heterogeneity comes from studies utilizing a variety of antibody coated, non-erythrocyte target cells. While phagocytic effector cells can kill coated CRBC in mice, guinea pigs and man, but not Chang liver cells in human or rat, or lymphoma lines in the mouse, non-phagocytic effector cells can kill both CRBC and all the other non-erythrocyte target cells.[180] This suggests heterogeneity in the mechanism of cytolysis.

A functional approach to distinguishing subsets of killing cells has recently been developed which appears to indicate heterogeneity both within the target cell population and the effectors. It was found that four different types of killing by human peripheral blood cells differed in their dependence on the presence of Ca^{2+} or Mg^{2+} ions in the medium. T cell mediated killing required only Ca^{2+}, non-T killing of antibody coated SRBC requires Mg^{2+} or no cations, while non-T killing of antibody coated Chang cells requires Ca^{2+}. Particularly interesting is the additional finding that 'spontaneous' killing of Chang cells in the absence of antibody requires both Ca^{2+} and Mg^{2+} for full expression.[181] Data on T and non-T killing in the mouse also reveals differences in ion dependence.[182] Preliminary data also suggests that the differences in cation requirements represent differences in the nature of the killer cells.[182]

It is clear from the foregoing discussion that K cell activity is heterogeneous and it is probable that clear definition of K subsets will depend on a better understanding of mechanisms of killing combined with refinement of physical methods of separating cells or the discovery of antigenic markers for K cells.[183]

6. SIGNIFICANCE OF HETEROGENEITY

In the preceding sections is has been argued that, by the stage of differentiation at which a lymphocyte becomes an antigen sensitive precursor, commitment to a particular cell lineage and effector function has already occurred. This commitment is indicated by the presence of phenotypic markers detected on the cell surface. The part which these markers play in the subsequent behaviour of the cell is far from clear.

In the case of T cells, however, some data suggests that the surface phenotype may play a role in determining the migration pattern of the cell. It is known that T lymphocytes of differing origin show different distribution patterns on injection.[184] Enzymatic treatment of the cell surface interferes with these patterns, although the specificity of this effect is unclear.[185]

One may speculate that the surface phenotype might play a role in cell to cell interactions. Since Ly antigens are markers for T subsets which can synergize, it might be expected that they would play a part in these interactions; however, antisera to Ly antigens (without complement) do not interfere with T lymphocyte-mediated cytolysis[37] and it has also been demonstrated that T cells with differing Ly genotype are able to synergize in MLC/CML.[87] More surprisingly, anti-H-2 sera directed to T killer cells are in general unable to block cytotoxicity in the absence of complement,[186] although a contrary result has been reported.[187] Recent data, however, suggests that anti-H-2, but not anti-Ia, may inhibit antigen binding of (T, G-A–L).[188] In summary, the evidence derived from inhibition of antigen binding, that H-2 or I region coded surface structures play a part in recognition of antigen by T cells, is weak.

On the other hand, genetic evidence has established that genes in the I region of the H-2 complex control the level of response to many antigens.[127–129] Furthermore, in one experimental system, a T cell produced helper factor, which has an antigen binding site, can be absorbed by antibody to IA subregion coded antigens.[189] This implies that the IA coded structure may be the T cell receptor. Failure to detect I coded recognition structures directly may be due to heterogeneity of T cell I region products, analogous to Ig classes of B cells. Detection of a particular receptor would then depend on the specificity of the antiserum used and the rate of synthesis and secretion of the receptor. Surface Ig is difficult to detect on high-rate antibody secreting plasma cells and there is evidence for a rapid turnover rate for T cell recognition structures.[190]

In contrast to problems of detection for T cells, and identity (IgM or IgD) for B cells, of the specific receptor for antigen, the majority of B cells have non-specific Fc and complement receptors and some T cells also bind antigen–antibody complexes (Sections 3.4 and 4.3). The physiological role of these non-specific receptors is not yet clear but it is clear that K cell cytotoxicity is mediated via Fc receptor (Section 5). Furthermore, injected cells carrying bound antigen–antibody complexes give these up to the surrounding white pulp in the spleen (Reference 2, p. 54). Other studies have suggested an important role of cell bound complexes in regulation of immune responses,[191] a view which is reinforced by the data on immune response genes discussed above and the reported close association of Ia antigens and Fc receptors on the cell surface.[134]

In conclusion, studies of lymphocyte heterogeneity have begun to reveal the differentiation pathways of lymphocytes and support the concept of an early commitment of precursors to a particular effector function (scheme A, Figure 1). The already large number of distinct effector cell subsets, some with amplifying or suppressive functions, lend support to the view of the immune system as a network[192] with positive and negative feedback controls operating at many levels.

7. REFERENCES

1. Gowans, J. L. and McGregor, D. D. (1965). *Progr. Allergy*, **9**, 1–37.
2. Greaves, M. F., Owen, J. J. T. and Raff, M. C. (1973). *T and B lymphocytes*, Excerpta Medica, Amsterdam.
3. Jankovic, B. D. and Isvaneski, M. (1963). *Int. Arch. Allergy Appl. Immunol.*, **23**, 188–201.
4. Cooper, M. D., Peterson, R. D. A., South, M. A. and Good, R. A. (1966). *J. exp. Med.*, **123**, 75–118.

5. Miller, J. F. A. P. (1962). *Proc. Roy. Soc. B*, **156**, 415–428.
6. Waksman, B. H., Arnason, B. G. and Jankovic, B. D. (1962). *J. exp. Med.*, **116**, 187–207.
7. Mueller, A. P., Wolfe, H. R. and Meyer, R. K. (1960). *J. Immunol.*, **85**, 172–179.
8. Warner, N. L., Szensberg, A. and Burnet, F. M. (1962). *Aust. J. exp. Biol. Med. Sci.*, **40**, 373–387.
9. Cooper, M. D., Perey, D. Y., McKneally, M. F., Gabrielson, A. E., Sutherland, D. E. and Good, R. A. (1966). *Lancet*, **1**, 1388–1391.
10. Good, R. A., Biggars, W. D. and Park, B. H. (1971). In *Progress in Immunology* (Amos, B., Ed.), Academic, New York, pp. 699–722.
11. Davies, A. J. S., Leuchars, E., Walles, V., Marchant, R. and Elliott, E. V. (1967). *Transplantation*, **5**, 222–231.
12. Milier, R.G. and Phillips, R. A. (1969). *J. Cell Physiol.*, **73**, 191–201.
13. Shortman, K. (1968). *Aust. J. exp. Biol. Med. Sci.*, **46**, 375–396.
14. Nordling, S., Andersson, L. and Hayry, P. (1972). *Eur. J. Immunol.*, **2**, 405–410.
15. Osoba, D. (1970). *J. exp. Med.*, **132**, 368–383.
16. Shortman, K., Cerottini, J.-C. and Brunner, K. T. (1972). *Eur. J. Immunol.*, **2**, 313–319.
17. Bianco, C., Patrick, R. and Nussenzweig, V. (1970). *J. exp. Med.*, **132**, 702–720.
18. Phillips, R. A. and Cowan, D. H. (1972). *Med. Clin. North Am.*, **56**, 1334–1355.
19. Gorczynski, R. M. and Norbury, C. (1974). *Brit. J. Cancer*, **30**, 118–128.
20. Julius, M. H., Simpson, E. and Herzenberg, L. A. (1973). *Eur. J. Immunol.*, **3**, 645–649.
21. Hodes, R. J. and Hathcock, K. S. (1976). *J. Immunol.*, **116**, 167–177.
22. Uhr, J. W. and Scharff, M. (1960). *J. exp. Med.*, **112**, 65–76.
23. Cunningham, A. J. and Sercarz, E. E. (1971). *Eur. J. Immunol.*, **1**, 413–421.
24. Mitchison, N. A. (1971). *Eur. J. Immunol.*, **1**, 18–27.
25. Schrek, R. and Stefani, S. (1964). *J. Nat. Cancer Inst.*, **32**, 507–522.
26. Stobo, J. D. and Paul, W. E. (1973). *J. Immunol.*, **110**, 362–375.
27. Cohen, J. J. and Claman, H. N. (1971). *J. exp. Med.*, **133**, 1026–1034.
28. Cohen, J. J. (1972). In *Cell Interactions* (Silvestri, L., Ed.), North-Holland, Amsterdam, pp. 162–163.
29. Blomgren, H. and Andersson, B. (1969). *Exp. Cell Res.*, **57**, 185–192.
30. Turk, J. L. and Poulter, L. W. (1972). *Clin. exp. Immunol.*, **10**, 285–296.
31. Bach, J. F., Dardenne, M. and Fournier, C. (1969). *Nature*, **222**, 998–999.
32. Boyse, E. A., Old, L. J., Stockert, E. (1968). *Proc. Nat. Acad. Sci.*, **60**, 886–893.
33. Greaves, M. F. and Raff, M. C. (1971). *Nature* (*New Biol.*), **233**, 239–241.
34. Boyse, E. A., Bressler, E., Iritani, C. and Lardis, M. (1970). *Transplantation*, **9**, 339–341.
35. Takahashi, T., Old, L. J. and Boyse, E. A. (1970). *J. exp. Med.*, **131**, 1325–1341.
36. Scheid, M., Boyse, E. A., Carswell, E. A. and Old, L. J. (1972). *J. exp. Med.*, **135**, 938–955.
37. Shiku, H., Kisielow, P., Bean, M. A., Takahashi, T., Boyse, E. A., Oettgen, H. F. and Old, L. J. (1975). *J. exp. Med.*, **141**, 227–241.
38. Flaherty, L. and Bennett, D. (1973). *Transplantation*, **16**, 505–514.
39. Boyse, E. A. and Old, L. J. (1968). *Annu. Rev. Genet.*, **3**, 269–290.
40. Shigeno, N., Arpels, C., Hammerling, U., Boyse, E. A. and Old, L. J. (1968). *Lancet*, **2**, 320–323.
41. Raff, M. C., Nase, S. and Mitchison, N. A. (1971). *Nature*, **230**, 50–51.
42. Niederhuber, J. E. and Moller, E. (1972). *Cell. Immunol.*, **3**, 559–568.
43. Takahashi, T., Old, L. J., Chen-Jung, N. and Boyse, E. A. (1972). *Eur. J. Immunol.*, **1**, 478–482.
44. Stockert, E., Old, L. J. and Boyse, E. A. (1971). *J. exp. Med.*, **133**, 1334–1355.
45. Ikeda, H., Stockert, E., Rowe, W. P., Boyse, E. A., Lilly, F., Sato, H., Jacobs, S. and Old, L. J. (1973). *J. exp. Med.*, **137**, 1103–1107.
46. Old, L. J., Boyse, E. A. and Stockert, E. (1963). *J. Nat. Cancer Inst.*, **31**, 977–986.
47. Reif, A. E. and Allen, J. M. V. (1963). *Nature*, **200**, 1332–1333.
48. Sclesinger, M. and Yron, I. (1969). *Science*, **164**, 1412–1413.
49. Raff, M. C., and Wortis, H. H. (1970). *Immunology*, **18**, 931–942.
50. Boyse, E. A., Miyazawa, M., Aoki, T. and Old, L. J. (1968). *Proc. Roy. Soc. B*, **178**, 175–193.
51. Boyse, E. A., Itakura, K., Stockert, E., Iritani, C. A. and Miura, M. (1971). *Transplantation*, **11**, 351–353.
52. Snell, G. D., Cherry, M., McKenzie, I. F. C. and Bailey, D. W. (1973). *Proc. Nat. Acad. Sci.*, **70**, 1108–1111.
53. Komuro, K., Itakura, K., Boyse, E. A. and John, M. (1975). *Immunogenetics*, **1**, 452–456.

54. Klein, J. (1975). *Contemp. Topics in Immunobiol.* (in press).
55. Sachs, D. H. and Cone, J. L. (1973). *J. exp. Med.*, **138**, 1289–1304.
56. Shreffler, D. C. and David, C. S. (1972). *Tissue Antigens*, **2**, 232–240.
57. Uhr, J. W. and Phillips, J. M. (1966). *Ann. N.Y. Acad. Sci.*, **129**, 792–808.
58. Basten, A., Miller, J. F. A. P., Sprent, J. and Pye, J. (1972). *J. exp. Med.*, **135**, 610–626.
59. Dukor, P., Bianco, C. and Nussenzweig, V. (1971). *Eur. J. Immunol.*, **1**, 491–493.
60. Jondal, M., Holm, G. and Wigzell, H. (1972). *J. exp. Med.*, **136**, 207–215.
61. Sheldon, P. J. and Holborrow, E. J. (1975). *J. Immunol. Method.*, **7**, 379–386.
62. Valdimarsson, H., Agnarsdottir, G. and Lachmann, P. J. (1975). *Nature*, **255**, 554–556.
63. Jondal, M. and Klein, G. (1973). *J. exp. Med.*, **138**, 1365–1378.
64. Revesz, T. and Greaves, M. F. (1975). *Nature*, **257**, 103–106.
65. Jondal, M. and Pross, H. (1975). *Int. J. Cancer*, **15**, 596–605.
66. Jondal, M., Svedmyr, E., Klein, E. and Singh, S. (1975). *Nature*, **255**, 405–407.
67. Cantor, H. and Asofsky, R. (1972). *J. exp. Med.*, **135**, 764–779.
68. Cantor, H. and Asofsky, R. (1970). *J. exp. Med.*, **131**, 235–246.
69. Simpson, E. and Cantor, H. (1975). *Eur. J. Immunol.*, **1**, 337–343.
70. Bach, J. F. and Dardenne, M. (1973). *Immunology*, **25**, 353–366.
71. Greaves, M. F. and Moller, E. (1970). *Cell. Immunol.*, **1**, 372–385.
72. Raff, M. C. and Cantor, H. (1971). In *Progress in Immunology* (Amos, B., Ed.), Academic, New York, pp. 83–93.
73. Cantor, H. and Simpson, E. (1975). *Eur. J. Immunol.*, **5**, 330–336.
74. Cohen, L. and Howe, M. L. (1973). *Proc. Nat. Acad. Sci.*, **70**, 2707–2710.
75. Stobo, J. D., Paul, W. E. and Henney, C. S. (1973). *J. Immunol.*, **110**, 652–660.
76. Wagner, H. (1973). *J. exp. Med.*, **138**, 1379–1397.
77. Cantor, H., Simpson, E., Sato, V. L., Fathman, C. G. and Herzenberg, L. A. (1975). *Cell. Immunol.*, **15**, 180–196.
78. Howe, M. L. and Cohen, L. (1973). In *Proc. 8th Leukocyte Cult. Conf.* (Lindahl-Kiessling, K., Ed.), Academic, New York, pp. 523–527.
79. Mosier, D. and Cantor, H. (1971). *Eur. J. Immunol.*, **1**, 459–461.
80. Andersson, B. and Blomgren, H. (1971). *Cell. Immunol.*, **2**, 411–424.
81. Eijsvoogel, V. P., du Bois, R. S., Melief, C. J. M., Zeylemaker, W. P., Raat-Koning, L. and de Groot-Kooy, L. (1973). *Transplant. Proc.*, **5**, 415–420.
82. Schendel, D. J., Alter, B. J. and Bach, F. H. (1973). *Transplant. Proc.*, **5**, 1651–1655.
83. Schendel, D. J. and Bach, F. H. (1974). *J. exp. Med.*, **140**, 1534–1546.
84. Bach, F. H., Segall, M., Zier, K. S. *et al.* (1973). *Science*, **180**, 403–406.
85. Kisielow, P., Hirst, J. A., Shiku, H., Beverley, P. C. L., Hoffmann, M. K., Boyse, E. A. and Oettgen, H. F. (1975). *Nature*, **253**, 219–220.
86. Cantor, H. and Boyse, E. A. (1975). *J. exp. Med.*, **141**, 1376–1389.
87. Cantor, H. and Boyse, E. A. (1975). *J. exp. Med.*, **141**, 1389–1399.
88. Claman, H. N., Chaperon, E. A. and Triplett, R. F. (1966). *Proc. Soc. exp. Biol. Med.*, **122**, 1167–1183.
89. Kappler, J. W., Hunter, P. C., Jacobs, D. and Lord, E. (1974). *J. Immunol.*, **113**, 27–39.
90. Janeway, C. A., Sharrow, S. O. and Simpson, E. (1975). *Nature*, **253**, 544–546.
91. Dennert, G. (1974). *Nature*, **249**, 358–360.
92. Hirst, J. A., Beverley, P. C. L., Kisielow, P., Hoffmann, M. K. and Oettgen, H. F. (1975). *J. Immunol.*, **115**, 1555–1557.
93. Feldmann, M., Erb, P., Kontiainen, S. and Dunkley, M. (1975). In *Membrane Receptors of Lymphocytes* (Seligmann, M., Preud'homme, J. L. and Kourilsky, F. M., Eds.), North-Holland, Amsterdam, pp. 305–310.
94. Feldmann, M., Beverley, P. C. L., Dunkley, M. and Kontiainen, S. (1975). *Nature*, **258**, 114–115.
95. Blomgren, H. and Svedmyr, E. (1971). *Cell. Immunol.*, **2**, 285–299.
96. Stobo, J. D. and Paul, W. E. (1973). *J. Immunol.*, **110**, 362–375.
97. Shortman, K., Byrd, W., Cerottini, J. C. and Brunner, K. T. (1973). *Cell. Immunol.*, **6**, 25–40.
98. Rich, R. and Rich, S. (1975). *J. Immunol.*, **114**, 112–119.
99. Elliott, B. E., Haskill, J. S. and Axeland, M. A. (1975). *J. exp. Med.*, **141**, 584–599.
100. Elliott, B. E. and Haskill, J. S. (1975). *J. exp. Med.*, **141**, 600–607.
101. Herzenberg, L. A. and Herzenberg, L. A. (1974). In *Contempory Topics in Immunobiology*, Vol. 3 (Cooper, M. D. and Warner, N. L., Eds.), Plenum Press, New York, pp. 41–75.
102. Eichmann, K. (1974). *Eur. J. Immunol.*, **4**, 296–301.

103. Eichmann, K. (1975). *Eur. J. Immunol.*, **5**, 511–517.
104. Shearer, G. M., Mehmon, M. I., Weinstein, Y. and Sela, M. (1972). *J. exp. Med.*, **136**, 1302–1307.
105. Dorr Burns, F., Marrack, P. C., Kappler, J. W. and Janeway, C. A. (1975). *J. Immunol.*, **114**, 1345–1347.
106. Folch, H. and Waksman, B. H. (1974). *J. Immunol.*, **113**, 140–144.
107. Peavy, D. L. and Pierce, C. W. (1974). *J. exp. Med.*, **140**, 356–369.
108. Folch, H. and Waksman, B. H. (1973). *Cell. Immunol.*, **9**, 12–24.
109. Gershon, R. K., Gery, I. and Waksman, B. H. (1974). *J. Immunol.*, **112**, 215–221.
110. Hardin, J. A., Chused, T. M. and Steinberg, A. D. (1973). *J. Immunol.*, **111**, 650–655.
111. Asherson, G. L. and Zembala, M. (1974). *Eur. J. Immunol.*, **4**, 804–807.
112. Youdim, S., Stutman, O. and Good, R. A. (1973). *Cell. Immunol.*, **6**, 98–109.
113. Oppenheim, J. J. (1968). *Fed. Proc.*, **27**, 21–28.
114. Schlossman, S. F. (1972). *Transplant. Rev.*, **10**, 97–111.
115. Shevach, E. M., Green, I., Ellman, L. and Maillard, J. (1972). *Nature* (*New Biol.*), **235**, 19–21.
116. Oates, C. M., Bisenden, J. F., Maini, R. N., Payne, L. N. and Dumonde, D. C. (1972). *Nature* (*New Biol.*), **239**, 137–139.
117. Bloom, B. R. (1971). *Advan. Immunol.*, **13**, 102–140.
118. Elliott, B. E. and Haskill, J. S. (1974). *Nature*, **252**, 607–608.
119. Dennert, G. and Harlen, L. E. (1975). *Nature*, **257**, 486–488.
120. Basten, A., Warner, N. L. and Mandel, T. (1972). *J. exp. Med.*, **135**, 627–642.
121. Andersson, C. L. and Grey, H. M. (1974). *J. exp. Med.*, **139**, 1175–1188.
122. Harris, A. W., Bankhurst, A. D., Mason, S. and Warner, N. L. (1973). *J. Immunol.*, **110**, 431–438.
123. Yoshida, T. O. and Andersson, B. (1972). *Scand. J. Immunol.*, **1**, 401–415.
124. Kramer, P. H., Hudson, L. and Sprent, J. (1975). *J. exp. Med.*, **142**, 1403–1415.
125. Hudson, L., Sprent, J., Miller, J. F. A. P. and Playfair, J. H. L. (1974). *Nature*, **251**, 60–62.
126. Stout, R. D. and Herzenberg, L. A. (1975). *J. exp. Med.*, **141**, 611–621.
127. McDevitt, H. O. and Sela, M. (1967). *J. exp. Med.*, **122**, 969–981.
128. Benaccerraf, B. and McDevitt, H. O. (1972). *Science*, **175**, 273–279.
129. Munro, A. J. and Taussig, M. J. (1975). *Nature*, **256**, 103–106.
130. David, C. S., Shreffler, D. C. and Frelinger, J. A. (1973). *Proc. Nat. Acad. Sci.*, **70**, 2509–2514.
131. Hammerling, G. J., Mauve, G., Goldberg, E. and McDevitt, H. O. (1975). *Immunogenetics*, **1**, 428–437.
132. Frelinger, J. A., Niederhuber, J. E., David, C. S. and Schreffler, D. C. (1974). *J. exp. Med.*, **149**, 1273–1284.
133. Wagner, H., Hammerling, G. J. and Rollinghoff, M. (1975). *Immunogenetics*, **2**, 257–268.
134. Dickler, H. B., Arbeit, R. D. and Sachs, D. H. (1975). In *Membrane Receptors of Lymphocytes* (Seligmann, M., Preud'homme, J. L. and Kourilsky, F. M., Eds.), North-Holland, Amsterdam, pp. 259–266.
135. Okumara, K. and Tada, T. (1974). *J. Immunol.*, **112**, 783–791.
136. Strober, S. (1975). *Transplant. Rev.*, **24**, 84–112.
137. Stutman, O. (1975). In *The Biological Activity of Thymic Hormones* (van Bekkum, D. W., Ed.), Kooyker, Rotterdam, pp. 87–94.
138. Spear, P. G., Wang, A. L., Rutishauser, U. and Edelman, G. M. (1973). *J. exp. Med.*, **138**, 557–573.
139. Bach, M. A. and Bach, J. F. (1972). *Transplant. Proc.*, **4**, 165–167.
140. Lance, E. M. (1970). *Clin. exp. Immunol.*, **6**, 789–802.
141. Bach, J. F., Cantor, H., Roelants, G. and Stutman, O. (1975). In *Biological Activity of Thymic Hormones* (van Bekkum, D. W., Ed.), Kooyker, Rotterdam, pp. 157–168.
142. Medawar, P. B. and Simpson, E. (1975). *Nature*, **258**, 106–108.
143. Sprent, J. and Basten, A. (1973). *Cell. Immunol.*, **7**, 40–59.
144. Kincade, P. W. and Cooper, M. D. (1971). *J. Immunol.*, **106**, 371–382.
145. Lawton, A. R., Asofsky, R. A., Hylton, M. B. and Cooper, M. D. (1972). *J. exp. Med.*, **135**, 277–297.
146. Pierce, C. W., Solliday, S. M. and Asofsky, R. A. (1972). *J. exp. Med.*, **135**, 675–697.
147. Greaves, M. F. and Hogg, N. M. (1971). In *Progress in Immunology* (Amos, B., Ed.), Academic, New York, pp. 111–126.
148. Nossal, G. J. V., Warner, N. L., Lewis, H. and Sprent, J. (1972). *J. exp. Med.*, **135**, 405–428.

149. Pernis, B. (1971). In *Immunologic Intervention* (Uhr, J. W. and Landy, M., Eds.), Academic, New York, pp. 149–152.
150. Anderson, H. R. (1972). *Eur. J. Immunol.*, **2**, 11–18.
151. Rabellino, E., Colon, S., Grey, H. M. and Unanue, E. R. (1971). *J. exp. Med.*, **113**, 156–167.
152. Pernis, B., Forni, L. and Amante, L. (1970). *J. exp. Med.*, **132**, 1001–1018.
153. Wigzell, H. (1970). *Transplant. Rev.*, **5**, 76–97.
154. Fu, S. M., Winchester, R. J. and Kunkel, H. G. (1974). *J. exp. Med.*, **139**, 451–456.
155. Rowe, D. S., Hug, K., Forni, L. and Pernis, B. (1973). *J. exp. Med.*, **138**, 965–972.
156. Abney, E. R. and Parkhouse, R. M. E. (1974). *Nature*, **252**, 600–602.
157. Melcher, U., Vitetta, E. S., McWilliams, M., Lamm, M. E., Philips-Quagliata, J. M. and Uhr, J. W. (1974). *J. exp. Med.*, **140**, 1427–1431.
158. Parkhouse, R. M. E. and Abney, E. R. (1975). In *Membrane Receptors of Lymphocytes* (Seligmann, M., Preud'homme, J. L. and Kourilsky, F. M., Eds.), North-Holland, Amsterdam, pp. 51–64.
159. Vossen, J. M. and Hijmans, W. (1975). *Ann. N.Y. Acad. Sci.*, **254**, 262–279.
160. Vitetta, E. S. and Uhr, J. W. (1975). In *Membrane Receptors of Lymphocytes* (Seligmann, M., Preud'homme, J. L. and Kourilsky, F. M., Eds.), North-Holland, Amsterdam, pp. 27–37.
161. Pernis, B. (1975). In *Membrane Receptors of Lymphocytes* (Seligmann, M., Preud'homme, J. L. and Kourilsky, F. M., Eds.), North-Holland, Amsterdam, pp. 25–26.
162. Yutoku, M., Grossberg, A. L. and Pressman, D. (1975). *J. Immunol.*, **195**, 69–74.
163. Greaves, M. F. and Janossy, G. (1975). *Transplant. Rev.*, **24**, 177–236.
164. Gronowicz, E. and Coutinho, A. (1975). *Transplant. Rev.*, **24**, 3–40.
165. Andersson, J., Sjoberg, O. and Moller, G. (1972). *Transplant. Rev.*, **11**, 131–157.
166. Andersson, B. and Blomgren, H. (1975). *Nature*, **253**, 476–477.
167. Playfair, J. H. L. and Purves, E. C. (1971). *Nature (New Biol.)*, **231**, 149–151.
168. Gershon, R. K. and Kondo, K. (1970). *J. Immunol.*, **18**, 723–738.
169. Osmond, D. G., Miller, R. G. and von Boehmer, H. (1975). *J. Immunol.*, **114**, 1230–1236.
170. Lafleur, L., Miller, R. G. and Phillips, R. A. (1972). *J. exp. Med.*, **135**, 1363–1374.
171. Gorczynski, R. M. and Feldmann, M. (1975). *Cell. Immunol.*, **18**, 88–97.
172. Schlegel, R. A., von Boehmer, H. and Shortman, K. (1974). *Cell. Immunol.*, **16**, 203–217.
173. Greenberg, A. H., Hudson, L., Shen, L. and Roitt, I. M. (1973). *Nature (New Biol.)*, **242**, 111–113.
174. Van Boxel, J. A., Stobo, J. D., Paul, W. E. and Green, I. (1972). *Science*, **175**, 194–196.
175. Dennert, G. and Lennox, E. S. (1973). *J. Immunol.*, **111**, 1844–1854.
176. Moller, G. and Svehag, S. E. (1972). *Cell. Immunol.*, **4**, 1–19.
177. Greenberg, A. H., Shen, L. and Roitt, L. M. (1973). *Clin. exp. Immunol.*, **15**, 251–259.
178. Greenberg, A. H., Shen, L., Walker, L., Arnaiz-Villena, A. and Roitt, I. M. (1975). *Eur. J. Immunol.*, **5**, 474–480.
179. Temple, A., Loewi, G., Davies, P. and Howard, A. (1973). *Immunology*, **24**, 655–699.
180. Greenberg, A. H., Shen, L. and Medley, G. (1975). *Immunology*, **29**, 719–729.
181. Golstein, P. and Fewtrell, C. (1975). *Nature*, **255**, 491–492.
182. Golstein, P. and Smith, E. T. (1975). *Eur. J. Immunol.*, **6**, 31–37.
183. Greaves, M. F. (1975). *Progr. Haematol.* IX, 255–303.
184. Zatz, M. M. and Lance, E. M. (1970). *Cell. Immunol.*, **1**, 3–17.
185. Woodruff, J. J. and Gesher, B. M. (1969). *J. exp. Med.*, **129**, 551–567.
186. Cerottini, J.-C. and Brunner, K. T. (1974). *Advan. Immunol.*, **18**, 67–132.
187. Wekerle, H., Lonai, P. and Feldmann, M. (1973). *Transplant. Proc.*, **5**, 133–138.
188. Hammerling, G. J., Lonai, P. and McDevitt, H. O. (1975). in *Membrane Receptors of Lymphocytes* (Seligmann, M., Preud'homme, J. L. and Kourilsky, F. M., Eds.), North-Holland, Amsterdam, pp. 121–126.
189. Taussig, M. J., Munro, A. J., David, C. S. and Staines, N. A. (1975). *J. exp. Med.*, **142**, 694–700.
190. Ramseier, H. (1975). *Eur. J. Immunol.*, **5**, 589–593.
191. Gorczynski, R., Kontiainen, S., Mitchison, N. A. and Tigelaar, R. (1974). In *Cellular Selection and Regulation in the Immune Response* (Edelman, G., Ed.), Raven Press, New York, pp. 143–154.
192. Jerne, N. K. (1974). In *Cellular Selection and Regulation in the Immune Response* (Edelman, G., Ed.), Raven Press, New York, pp. 1–11.

Chapter 4

Migration and Lifespan of Lymphocytes

J. SPRENT

Abbreviations

ASRL: antigen-specific selective recruitment of recirculating lymphocytes
'B' mice: mice deprived of T lymphocytes
^{3}HTdR: tritiated thymidine
NTx: neonatally thymectomized
PALS: periarteriolar lymphocyte sheath
PCV: post-capillary venule
RLP: recirculating lymphocyte pool
θ: Thy-1 locus determined antigen
TDL: thoracic duct lymphocytes
TL: thymus leukaemia determined antigen
T.TDL: H_2-activated T cells derived from thoracic duct lymph
TxBM: adult thymectomized, irradiated and bone-marrow-protected

1. INTRODUCTION

Much of our current knowledge of lymphocyte behaviour and function has been derived from techniques for culturing lymphocytes successfully *in vitro*. Invaluable though these techniques have proved, *in vitro* systems by their nature provide little information on the crucial question of how lymphocytes behave and interact in their natural environment. The purpose of this chapter is to consider in brief the ontogeny, migratory streams and longevity of the two main classes of lymphocytes, viz. thymus-derived (T) lymphocytes and bursa- or bursa-equivalent-derived (B) lymphocytes. Many of the topics covered in this chapter have been reviewed in detail elsewhere.[1–4]

2. ORIGIN AND DISTRIBUTION

2.1 Primary lymphoid organs

The primary lymphoid organs are generally considered to include the thymus, present in all vertebrates, and the bursa of Fabricius, found only in birds;[5,6] the lymphoid component of the bone marrow also probably falls within this category. Primary lymphoid organs differ from secondary lymphoid organs, e.g. the spleen and lymph nodes, in several respects. Firstly, primary lymphoid organs contain cells derived not only from mesoderm (lymphocytes) but also from ectoderm (epithelial cells). Secondly, lymphopoiesis in these organs is intense, independent of antigenic stimulation and prominent before birth. Thirdly, primary lymphoid organs rarely manifest cytological hallmarks of an immune response.*

A considerable body of evidence suggests that the main function of the primary lymphoid organs is to act as a centre for lymphocyte formation and dissemination. Ultimately, all lymphocytes are believed to arise from pluripotential haemopoietic stem cells.[4–6,8] These cells appear initially in the yolk sac and later in the foetal liver;[8–10] in adults, stem cells are found predominantly in the bone marrow but are also present in small numbers in the spleen and blood. As considered in more detail by J. J. T. Owen, herein, pluripotential stem cells, or possibly partly differentiated 'lymphoid stem cells', enter the primary lymphoid organs at a stage early in embryonic life when these organs are alymphoidal. The differentiation of these immigrant stem cells into lymphocytes, though poorly understood, is probably controlled by the epithelial cell component (Reference 11 and *vide infra*).

2.1.1 *Thymus*

Histologically, the thymus is divided into two regions: the cortex, which contains the vast majority of cells (85% or more), and the medulla.[8,12] Cortical thymus cells have restricted immunocompetence† and consist mostly of small lymphocytes with a rapid turnover.[13,14] In mice they are characterized by a high content of θ antigen, low content of H2 antigen and are cortisone-sensitive and (in TL-positive strains) express the TL antigen.[15–22] The cells of the medulla have a slower turnover, express less θ antigen but more H2 antigen, are

* Although germinal centre formation in the thymus and marrow is rare, these organs may nevertheless contain considerable numbers of antibody-forming cells, particularly in hyperimmune states.[7] Whether these cells differentiate *in situ* or are recruited from other regions is not clear.

† In young mice and chickens the cortex is reported to contain a discrete subpopulation of cells which exert a suppressive effect on a variety of immune responses.[12]

cortisone-resistant and TL-negative. Cortical and medullary thymus cells also show subtle differences in size and electrophoretic mobility.[12,23] The interrelationship between these two cell types is controversial. Some workers claim that medullary cells are derived from the cortex[24–27] while others maintain the two populations develop independently.[14,18] Both cell types appear to arise by sequential division from large lymphocytes.[14]

The most direct evidence that lymphocytes migrate from the thymus has come from studies involving intrathymic injection of radioactive cell-markers such as tritiated thymidine (^{3}HTdR). Experiments with guinea pigs,[28,29] hamsters,[30] calves,[31] rabbits,[32] mice[33] and rats[34] have all shown that, shortly after injection of the isotope, labelled cells appear in the spleen, lymph nodes and intestines. Seeding of labelled cells is particularly prominent in neonatal animals.

Several studies have shown that lymphocytes migrate from chromosomally-marked thymus grafts placed in thymus-deprived mice, e.g. mice thymectomized at birth (NTx mice) (reviewed in References 5, 35 and 36). Within the first two weeks of grafting most of the dividing cells in the grafts are of donor (i.e. graft) origin. Soon afterwards cells carrying the graft marker appear in the peripheral blood and reach high levels after 1–2 months. Subsequently the proportion of donor-derived lymphocytes in the blood decreases. This is considered to reflect gradual replacement of the graft with lymphocytes of host origin, i.e. cells presumably derived from immigrant host stem cells.

Other less direct evidence favouring efflux of cells from the thymus comes from reports that absolute and differential lymphocyte counts in blood from thymic veins of the rat[37] and the guinea pig[38] are higher than in the thymic artery or in venous blood from other regions. In the calf it has been reported that the thymic veins and lymphatics, but not vessels in others regions, contain large numbers of cells carrying thymocyte-specific markers.[39]

A large proportion of lymphocytes in the secondary lymphoid organs are known to be thymus(T)-dependent. For example, in certain species in which the young are embryologically immature at birth, e.g. mice and rats, neonatal thymectomy causes a profound depletion of lymphocytes from certain (T-dependent) areas of the secondary lymphoid organs (reviewed in Reference 5). Lymphoid depletion in such species is much less obvious when thymectomy is delayed until later life. Nevertheless, if thymectomy in adulthood is followed by heavy irradiation (which destroys nearly all lymphocytes) and protection with stem-cell-containing populations, e.g. bone marrow or foetal liver cells (TxBM mice), the degree of lymphoid depletion is as profound as that induced by neonatal thymectomy. In some species, e.g. the sheep, the thymus develops well before birth and lymphoid depletion following thymectomy is only prominent when the thymus is extirpated *in utero*.[40] Profound lymphoid depletion is also a feature of conditions associated with congenital thymic aplasia, e.g. the nude (*nu nu*) mouse mutant.[41]

Neonatal thymectomy or congenital thymic aplasia (particularly in mice) leads to a characteristic runting syndrome associated with a marked deficiency of certain immune functions (reviewed in Reference 5). Such animals are unable to reject allogeneic or xenogeneic skin grafts,[42,43] express delayed-type hypersensitivity[44] or give more than minimal responses to certain (T-dependent) antigens;[5] lymphocytes from thymus-deprived mice are also unable to evoke graft-versus-host reactions.[45] These various immune responses are thus considered to be mediated by T-dependent lymphocytes.

In mice, lymphocytes in the secondary lymphoid organs are divided into two broad categories, viz. (a) cells which express the θ antigen but not easily-detectable surface immunoglobulin (Ig) and (b) θ-negative cells which carry a high density of surface Ig.[46–52] Most lymphocytes expressing the θ antigen are T-dependent, i.e. there is a specific deficiency of these cells in NTx, TxBM and nude mice (henceforth referred to as 'B' mice).[53–55] While it is generally assumed that the vast majority of -positive cells are indeed thymus-derived, i.e. rather than merely thymus-dependent, this notion requires qualification.

On the basis of the amount of θ antigen they express, *θ-positive lymphocytes* can be divided into three subgroups:

(1) *High θ content.* These cells are largely restricted to the thymic cortex[51] but are also found in small numbers in the spleen;[56] the latter may represent recent immigrants from the thymus.
(2) *Intermediate θ content.* These cells are considered to arise in the thymus and account for the vast majority of θ-positive cells in the secondary lymphoid organs. They are very rare in the marrow, comprise approximately 30% of spleen cells, 15% of Peyer's patch cells and occur at high concentrations in lymph nodes and the central lymph (70%–90%).
(3) *Low θ content.* Although only about 1% of spleen cells from nude mice stain with anti-θ sera raised in mice (which detects cells displaying a high or intermediate concentration of θ antigen), up to 20% stain weakly with antisera raised in rabbits against mouse brain (some brain cells being θ-positive).[57] These 'θ^{+weak}' cells—which are also found in TxBM mice—have a very rapid turnover, occur predominantly in the spleen, are rare in lymph nodes and absent from thoracic duct lymph.[58] These cells are also found in the bone marrow, which suggests that some may originate in this region.[59] Their relationship to other θ-positive cells is not clear. The observation that θ^{+weak} cells are rare in normal mice and disappear rapidly in thymus-deprived mice implanted with a thymus graft[59] suggests they may represent, or be derived from, prethymic stem cells.

Precisely how prethymic stem cells differentiate into immunocompetent T cells is not clear. In recent years much interest has centred on the role of factors released from thymic epithelial cells. Many workers have observed that implantation of NTx mice with thymus grafts contained in cell-impermeable diffusion chambers causes a limited restoration of T-dependent functions, e.g. rejections of skin allografts (reviewed in Reference 5). There are also numerous reports that thymus-deprived mice partially regain immunocompetence when treated with purified extracts of thymus, e.g. 'thymosin'[60] and 'thymus humoral factor (THF)',[61] or with 'thymic factor (TF)' derived from serum[62] (reviewed by N. Trainin *et al.*, herein). More recently it has been found that the partial restoration of T-dependent functions under these conditions does not occur with (a) NTx mice aged 40 days or more prior to grafting,[63] or (b) nude mice.[64] From this and other evidence, Stutman[65] has postulated that certain cells are functionally immature on leaving the thymus but subsequently attain immunocompetence in the periphery under thymic hormonal control; such 'immature post-thymic T cells' are considered to survive for several weeks after neonatal thymectomy (owing to peripheralization before birth) but are absent in athymic (nude) mice.

There is also evidence that thymic factors act on prethymic cells. Thus, murine bone marrow or foetal liver cells incubated with thymic factors *in vitro* rapidly assume some of the surface membrane characteristics of thymic cortical cells, i.e. there is expression of the θ and TL antigens.[66] Similar results have been obtained with thymus cells removed from 14-day embryos, i.e. at a stage when the cells are θ-negative.[67] Two comments should be made concerning these data. Firstly, the capacity to induce expression of θ and TL antigens is not a property unique to thymic factors. Indeed it seems to be a general property of agents which stimulate adenyl cyclase and thereby increase levels of intracellular cyclic AMP.[66,67] Secondly, there is as yet no evidence that thymic factors can induce differentiation of TL-positive cells (which are immunoincompetent) to an immunocompetent TL-negative state. A clear understanding of the role of thymic factors in T cell differentiation will only emerge through further investigation.

Finally, mention should be made of recent evidence that, in mice, T cells can be subdivided into three groups on the basis of their Ly 1, 2 and 3 surface antigens. As considered in details by P. C. L. Beverley, herein, T lymphocytes comprise a mixture of cells: (a) Ly $1^+ 2^- 3^-$ cells, (b) Ly $1^- 2^+ 3^+$ cells, and (c) Ly $1^+ 2^+ 3^+$ cells. Growing evidence suggests that these three cell classes are functionally distinct.

2.1.2 *Bursa of Fabricius and the mammalian bursal equivalent*

Chickens bursectomized in embryonic life, e.g. at 17 days' incubation,* have reduced levels of serum Ig[69] and few plasma cells,[69] and show a profound depletion of small lymphocytes bearing surface Ig.[70] *A priori*, lymphocytes with these characteristics (B lymphocytes) could either differentiate peripherally under bursal control or originate within the bursa. The second possibility is favoured by evidence that injection of radioactive cell markers into the bursa rapidly leads to the appearance of labelled cells in organs such as the spleen.[71–73]

The identity of the bursal equivalent in mammals is much debated. Various organs have been accorded this status, particularly the gut-associated lymphoid tissues or parts thereof.[74,75] Direct evidence that lymphocytes with the characteristics of B cells are produced in these tissues, however, is lacking.

In the foetus there is strong evidence that B lymphocytes arise in the liver. In the mouse embryo, Ig-synthesizing cells have been detected as early as the fourteenth day of gestation (F. Melchers, personal communication). The cells at this time are probably lymphoid precursors since they are larger than normal B lymphocytes, release Ig at a rapid rate and do not express detectable Ig on their surface membrane. Their precursor state is evident from the fact that if 14-day foetal liver cells are cultured *in vitro* they differentiate into typical small lymphocytes expressing surface-bound Ig.[76] Lymphocytes with easily-detectable surface Ig first appear in the spleen at about 16 days' gestation but are not found in bone marrow until the time of birth.[77]

In the adult, it is probable, though yet to be proved conclusively, that most B cells arise in bone marrow. The proportion of small lymphocytes in marrow varies considerably from species to species but in rodents comprises approximately 25% of the nucleated cells.[78] Marrow small lymphocytes have a rapid turnover, over 80% being labelled after repeated intraperitoneal injections of ^{3}HTdR for 3–4 days.[79,80] Most of these cells seem to be B lymphocytes at various stages of differentiation since over 50% show varying amounts of surface Ig and very few, at least in mice, carry T-cell markers.[80] Studies on the turnover of Ig-positive cells in murine marrow indicated that within 36 hours of giving repeated injections of ^{3}HTdR 40% of total marrow small lymphocytes, but only 1% of Ig-positive cells, were labelled.[80] By 72 hours, however, labelling of Ig-positive cells reached 60%. These data clearly suggest that small lymphocytes in marrow are initially Ig-negative (by surface staining) but then rapidly express increasing amounts of surface Ig. This notion is supported by evidence that Ig-negative marrow small lymphocytes labelled with ^{3}HTdR *in vivo* become Ig-positive soon after intravenous injection (i.e. when recovered from the spleen) or when cultured *in vitro*.[81]

* Paradoxically, removal of the bursa at a much earlier stage, e.g. at 3 days' incubation, has little effect on B cell differentiation.[68]

Definite evidence that at least a proportion of lymphocytes in the marrow migrate to other organs has come from studies of Brahim and Osmond[82,83] in the guinea pig. These workers observed that after marrow cells were labelled *in situ*, i.e. by giving intramyeloid injection of ^{3}HTdR, labelled lymphocytes appeared shortly afterwards in the spleen and, to a lesser extent, the lymph nodes. Linna and Liden[84] have reported similar findings. Quantitative estimations of the extent of lymphocyte migration from the marrow, however, are not yet available.

At present there is insufficient evidence to regard the bone marrow in the adult as the bursal equivalent, i.e. the sole source of B lymphocytes. Indeed this is unlikely since destruction of the marrow, i.e. by treating mice with the bone-seeking isotope ^{89}Sr, has little demonstrable effect on B cell differentiation.[85] It should be added, however, that in view of the well-known phenomenon of extra-medullary haemopoiesis, e.g. in haemolytic anaemia, these data cannot be construed as evidence against the notion that, under *normal physiological conditions*, most B cells arise in the marrow. It would nevertheless seem probable that some degree of B cell formation does occur in the spleen since this organ, like the marrow but unlike lymph nodes, contains (a) haemopoietic stem cells,[8] (b) large lymphocytes releasing 7s IgM at a rapid rate* (i.e. B cell precursors)[86] and (c) 'pre-B cells'—immature cells which take longer than typical small B lymphocytes to differentiate into antibody-forming cells in the presence of antigen.[87,88] Whether or not B cells are formed in the gut-associated lymphoid tissues has yet to be determined.

B lymphocytes are usually defined as cells which express easily-detectable membrane-bound Ig. While this definition is convenient and in most instances has proved valid, it needs qualification. Firstly, evidence cited above suggests that Ig is difficult to detect on B cells at early stages of differentiation. Secondly, it cannot be assumed necessarily that all Ig-bearing cells are B cells. In this respect, some workers,[89] though not others,[90] report that with certain techniques, i.e. surface iodination in the presence of lactoperoxidase, T (θ-positive) lymphocytes express large amounts of surface Ig. It has also been found that certain populations of activated T cells carry Ig demonstrable by surface immunofluorescence;[91] it should be added that this material is probably not synthesized by the T cells but passively adsorbed from B cells.

2.2 Secondary lymphoid organs and the recirculating lymphocyte pool

The secondary lymphoid organs have four main distinguishing features.[5,6] Firstly, they contain a mixture of T and B lymphocytes. Secondly, they usually show prominent cytological evidence of immune responses, e.g. germinal centre foundation. Thirdly, they are reduced in size in animals raised in a gnotobiotic (germ-free) environment, i.e. where contact with antigen is reduced. Fourthly, lymphopoiesis in these organs is relatively slow.

Secondary lymphoid tissue occupies the bulk of the spleen and lymph nodes and is also found in certain regions of the alimentary tract, e.g. tonsil, Peyer's patches and appendix. The proportions of T and B lymphocytes in these tissues varies considerably. In rodents, most cells in the spleen and Peyer's patches are B cells (50%–60%) whereas T cells predominate in lymph nodes (70%–80%).

The majority of cells in the secondary lymphoid organs are not sessile but migrate continuously from one region to another via the bloodstream and lymph (*vide infra*). This is evident from the fact that procedures such as treatment with antilymphocyte serum,[92]

* The release of monomeric (7s) IgM from small lymphocytes, by contrast, is very slow (half-time release = 20 hr).[86] Cells releasing 7s IgM at a rapid rate are absent from lymph nodes and TDL.

extracorporeal irradiation of the blood[93] and attachment of ^{32}P-impregnated strips to the spleen[94] induce not only a marked lymphocytopenia but also a profound cellular depletion of the secondary lymphoid tissues.

Marked depletion of the secondary lymphoid tissues also follows chronic thoracic duct drainage.[1,55,95–99] Maximal outputs of thoracic duct lymphocytes (TDL) occur within the first 24 hours of establishing the fistulas. Thereafter, cell numbers decline rapidly and reach a low plateau (10–20-fold below normal) after 1–2 weeks.

Gowans[100–102] made the crucial discovery that the decline in TDL outputs from rats could be prevented by reinjecting the mobilized cells intravenously. By radio-labelling the injected TDL he proved that the lymph-borne cells were derived at least in part from the bloodstream. Most of the injected cells entered the lymph within 48 hours and nearly all were collected by 5 days. These data provided the first unequivocal evidence that lymphocytes recirculate continuously between blood and lymph. Lymphocyte recirculation has since been demonstrated in a number of different species including the mouse,[55] sheep[103–106] and man;[107] it appears to be very limited in the pig.[108]

It is now considered that the bulk of small lymphocytes in the secondary lymphoid tissues reside within the 'recirculating lymphocyte pool' (RLP),[102] i.e. can be mobilized through a lymphatic fistula. Based on the degree of lymphoid depletion which follows chronic thoracic duct drainage, recirculating lymphocytes probably comprise >60% of lymphoid cells in the spleen and blood and >90% in lymph nodes. Evidence that nearly all recirculating lymphocytes are long-lived cells will be considered in a later section.

Evidence from a number of sources indicates that recirculating lymphocytes comprise two distinct subpopulations. This was first apparent from demonstrations that extracorporeal irradiation of the blood or thoracic duct lymph of calves causes the cellularity of the lymph to decline in a biphasic fashion.[109–111] From this approach, Cronkite and Chanana[111] calculated that one component of small lymphocytes in the central lymph can be mobilized very rapidly (half-time = 1·2 days) whereas mobilization of the second component is very slow (half-time > 25 days).

Studies in mice suggest that these two populations represent T and B lymphocytes, respectively. Of TDL collected from mice within the first 24 hours of cannulation, approximately 80% of the small lymphocytes are T (θ-positive) cells and the remainder B (Ig-positive, Fc-receptor-bearing cells).[55] Thereafter, the proportions of B and T lymphocytes rise and fall, respectively, and stabilize at 40%–45% after 5 days. In absolute terms, total outputs of T lymphocytes decrease by 10-fold during this time, a plateau being reached at 4–5 days. Outputs of B lymphocytes, by contrast, decrease only very slowly. Thus even after continuous drainage for 10 days, total outputs of B cells are only 2–3-fold less than at 1 day. Slow mobilization of B cells is further supported by the findings that TDL outputs from 'B' mice, though initially much lower than from normal mice, decrease only very slowly during prolonged thoracic duct drainage.[55] It should be mentioned that the vast majority of B lymphocytes collected during prolonged drainage are long-lived cells, i.e. they fail to incorporate label during repeated administration of ^{3}HTdR.[112]

These data clearly imply that T and B lymphocytes recirculate at different rates. This has been verified by injecting radio-labelled populations of purified T and B cells intravenously and monitoring their appearance in the lymph by autoradiography.[112,113] Whereas T cells (TDL) enter the lymph in large numbers soon after injection, B cells appear much more slowly, particularly in mice. In this species less than 10% of B cells were recovered in the lymph within 72 hours of injection (compared with >50% of T cells).[112] Less striking differences are reported for the rat; Howard[113] obtained recoveries of 20% and 33% for B and T cells, respectively, over a drainage period of 48 hours.

The size of the RLP is difficult to measure precisely because each of the two components have to be considered separately. In mice, approximately 2×10^8 long-lived small T lymphocytes can be mobilized through a thoracic duct fistula before a plateau is reached at 5 days.[55] This provides a rough estimate for the size of the recirculating T cell pool. In calves, it is estimated that the easily-mobilizable lymphocyte component—presumably equivalent to the T cell pool in mice—is $4\cdot8\times10^9$ small lymphocytes per kilogram body weight (compared with approximately 8×10^9 per kilogram for mice).

With respect to the B cell pool, thoracic duct drainage of normal or nude mice for 10 days has been found to yield a total of 1–$1\cdot5\times10^8$ long-lived small B lymphocytes.[55,112] Calculating the size of the B cell pool from these data is difficult for two reasons. Firstly, long-lived B cells were still entering the lymph in appreciable numbers when the fistulas were arbitrarily closed at 10 days. Secondly, it is difficult to exclude the possibility that cells which enter the lymph during late stages of drainage do not normally recirculate, the cells being mobilized only as a consequence of depletion of the lymphoid organs. It can be calculated nevertheless that the number of *potentially* mobilizable small B lymphocytes in mice is of the order of $1\cdot5$–2×10^8 cells (6–8×10^9 cells/kg), i.e. similar to the number of recirculating T cells. This seems to be much higher than in calves where the total number of slowly-mobilizable lymphocytes (? B cells) collected during 30 days' drainage amounted to only $1\cdot2\times10^8$ lymphocytes per kilogram.[111]

3. LIFESPAN

In the light of present knowledge, it is salutary to consider that as recently as 20 years ago small lymphocytes were regarded as short-lived cells of unknown function.[114] Studies of Otteson[115] in 1954 in man provided the first clear indication that some small lymphocytes have a long lifespan. He injected patients with ^{32}P and at various intervals studied the radioactivity of DNA extracted from peripheral blood lymphocytes. With this approach he calculated that about 20% of blood lymphocytes had an average lifespan of 2–3 days whereas the remainder retained their radioactivity (remained in interphase) for 6 months or more. These findings were subsequently confirmed by other workers who observed radiation-induced 'lethal' chromosome damage in mitogen-stimulated lymphocytes from patients given therapeutic irradiation up to 10 years previously.[116,117]

Quantitative information on the lifespan of lymphocytes in rodents has arisen largely from studies on the rate at which cells become labelled during repeated injection of ^{3}HTdR or, conversely, lose their label after cessation of the injection. The characteristic rapid turnover of lymphocytes in the primary lymphoid organs has already been mentioned. Nearly all small lymphocytes in the thymus and bone marrow incorporate label within 5 days of continuous administration of ^{3}HTdR and then rapidly lose their label when the injections are interrupted.[13,27,80,118]

The turnover of lymphocytes in the secondary lymphoid tissues varies considerably depending on the tissue concerned. Everett and his colleagues observed that lymphocytes in certain regions, e.g. the spleen and blood, labelled in a biphasic fashion.[13,79] These workers arbitrarily defined lymphocytes which became labelled within 5 days of a course of ^{3}HTdR injections as 'short-lived' cells and the remainder as 'long-lived'. On this basis, 'short-lived' cells comprised about 50% of lymphocytes in the spleen and blood and 20% in lymph nodes. Though useful, this classification is not entirely satisfactory for two reasons. Firstly, it cannot be assumed that lymphocytes which label rapidly with ^{3}HTdR necessarily have a short lifespan. For example, to account for the progressive increase in mass of the lymphoid tissues in young growing animals, the rate of lymphocyte formation must clearly exceed destruction. *A priori*, rapid labelling of lymphocytes in such animals is therefore likely to indicate not extensive cell destruction but rapid formation of cells destined to become long-lived lymphocytes. When using young growing animals it is therefore essential to take into consideration the overall increase in weight of the lymphoid tissues during the term of the experiments—a factor neglected by a number of investigators. This of course is less of a problem when adult animals are used.

Secondly, finding a clear-cut biphasic rate of lymphocyte labelling has in most hands proved the exception rather than the rule. For example, the labelling pattern of small lymphocytes in the blood is usually paraboloid rather than biphasic,[124,125] thus implying that the cells in this situation turn over at not two but several different rates. With other cell populations, e.g. small lymphocytes in thoracic duct lymph, the rate of labelling (or loss of label after ceasing administration of the isotope) is slow and close to being linear for several weeks (*vide infra*). Such populations thus appear to consist almost entirely of long-lived lymphocytes.

3.1 Lymphocytes with a slow turnover

The longevity of most lymphocytes in blood was established by the remarkable experiments of Robinson *et al.*[119] who infused rats intravenously with ^{3}HTdR for many months. Although approximately 30% of blood lymphocytes were labelled after 5 days, 5%–8% remained unlabelled even after continuous administration of isotope for over 9 months. Caffrey *et al.*[13,121] demonstrated long-lived lymphocytes in various tissues of rats by giving repeated injection of ^{3}HTdR for 16 days and then studying the subsequent decline in levels of labelled cells. At 8 weeks after terminating the injections, 5%–10% of lymphocytes in blood, spleen, lymph nodes and TDL remained heavily labelled; some cells retained their label for up to 1 year. Similar results were obtained in mice by Claësson *et al.*[122]

Small lymphocytes in lymph nodes[122,123] and the liver[124] appear to consist almost entirely of long-lived cells. Lymphocytes with a slow turnover also comprise about 50% of lymphoid cells in the epidermis.[124] Long-lived lymphocytes are generally very rare in the bone marrow, although they do exist in small numbers in certain strains of mice[125,126] and rats[127,128] and probably also in other species. These cells—which are more common in older mice[129]—are probably derived from the bloodstream since they can be mobilized by prolonged thoracic duct drainage.[127,128]

With regard to the lifespan of recirculating lymphocytes, Sprent and Basten[112] observed that approximately 40% of TDL from adult mice were labelled after repeated administration of ^{3}HTdR for 2 months. If the pattern of labelling had continued at the same rate, 100% of the cells would therefore have been labelled after 4–6 months. Since TDL consist of approximately 80% T cells and 20% B cells, this would imply that both classes of lymphocytes in the central lymph have long lifespan. Direct evidence on the turnover of lymph-borne T and B cells was obtained by pulsing mice repeatedly with ^{3}HTdR for 6 weeks and then examining the identity of the labelled cells.[112] By using a double-labelling technique it was established that 38% of the T cells and 80% of the B cells were labelled at this time. These data suggested that the average lifespan (strictly speaking the duration of G_0) was less for B cells than T cells. This was confirmed by comparing the turnover of TDL in normal and 'B' mice. Whereas only 10% of TDL from normal mice were labelled after administration of ^{3}HTdR for 2 weeks, labelling of TDL from nude and TxBM mice was of the order of 20%–30%; labelling throughout this period was linear. From these and the above data it was calculated that the average lifespan of recirculating lymphocytes was in the range of 4–6 months for T cells and 5–7 weeks for B cells.

Several other studies have provided direct evidence that a large proportion of both T and B lymphocytes in the secondary lymphoid tissues are long-lived cells. Parrott and de Sousa[130] examined the homing patterns of long-lived lymphocytes from the spleen and lymph nodes of mice given repeated injections of ^{3}HTdR for 4 weeks. When cells were harvested 3 days after the last injection of isotope (to allow disappearance of the rapidly-dividing cells) autoradiography showed that, on adoptive transfer, the labelled cells homed to both the T and B dependent areas of the lymphoid tissues. Bianco *et al.*[131] used a similar approach to label lymph node cells and observed that the long-lived (labelled) cells comprised a mixture of T and B lymphocytes. With regard to B cells, close to 100% of lymph node lymphocytes in nude mice appear to be long-lived cells. Thus, Röpke and Hougen[123] demonstrated that labelling of nude lymph node cells during repeated administration of ^{3}HTdR was slow, only 5% being labelled at 5 days and 30% at 30 days. Termination of the injections after 30 days caused a slow linear decline in the levels of labelled cells. A slow rate of turnover of B lymphocytes from thoracic duct lymph of TxBM rats has been reported by Howard.[113]

The above data clearly suggest that the vast majority of recirculating T and B lymphocytes—and therefore the bulk of lymphocytes in the secondary lymphoid tissues—are long-lived cells.

3.2 Lymphocytes with a rapid turnover

The fast turnover of a proportion of lymphocytes in the spleen and blood has already been mentioned. The size of this pool is difficult to define but is probably of the order of 10%–30% of the lymphocytes present. Quantitative information on the surface membrane characteristics of these cells is not yet available—hence their precise identity is unknown. In view of evidence considered earlier, many probably represent recent immigrants from the thymus and bone marrow. Others could be newly-formed B cells produced *in situ*. Yet others are likely to represent newly-formed or long-lived lymphocytes responding to specific antigen.

Recently-divided lymphocytes comprise a large proportion of lymphoid cells residing within the epithelial lining of the gut. In mice, approximately 50% of lymphocytes in the epithelium overlying Peyer's patches and lining the jejunum become labelled within 3–5 days during repeated administration of ^{3}HTdR.[120,124] A similar proportion of lymphocytes in the epidermis label within this time.[124] In the intestines most of the recently-divided lymphocytes are probably T cells since intraepithelial lymphocytes are rare in 'B' mice,[132,133] and in normal mice stain with heteroantisera specific for T cells.[134] The identity of intraepithelial lymphocytes in the skin is not known.

Rapidly-dividing lymphocytes are rare in lymph nodes and thoracic duct lymph. Contrary to the claims of some workers,[13] most studies on the rate at which lymph node lymphocytes or TDL become labelled during long-term administration of ^{3}HTdR (or the rate at which the label is subsequently lost) have failed to detect a discrete subpopulation of small lymphocytes with a rapid turnover.[112,123] Nevertheless, recent evidence of Strober[135,136] suggests that some recently-formed small (B) lymphocytes do exist in thoracic duct lymph. He observed that when rats were pulsed repeatedly with very high ('suicidal') doses of ^{3}HTdR for 48 hours, the capacity of their TDL to mount an adoptive immune response to ferritin was abolished. These rapidly-dividing cells presumably account for only a very small proportion of TDL.

Unlike small lymphocytes, nearly all large lymphocytes have a very rapid turnover. This applies to plasma cells, cells found in germinal centres and the large lymphocytes in thoracic duct lymph.[102,112,137,138] Most of these cells are probably the descendants of small T and B lymphocytes engaged in immune responses to various environmental antigens. While some of these stimulated large lymphocytes may be destined to differentiate into long-lived memory cells, the majority probably have a short lifespan. Their fate will be considered in a later section.

4. MIGRATION

4.1 Migratory properties of normal lymphocytes

It has already been mentioned that T and B lymphocytes recirculate at different rates and are distributed in unequal proportions in the secondary lymphoid tissues. Evidence will now be considered that T and B cells segregate in quite separate regions of the lymphoid tissues.

4.1.1 *Long-lived recirculating lymphocytes*

4.1.1.1 *T lymphocytes**

Procedures which selectively deplete the easily-mobilizable component of the RLP, e.g. short-term thoracic duct drainage, cause a characteristic depletion of lymphocytes from certain regions of the secondary lymphoid tissues.[55,95–99] These regions include the periarteriolar lymphocyte sheath (PALS) of the spleen, the lymph node paracortex and the interfollicular areas of Peyer's patches. These areas are thymus-dependent since they show marked hypocellularity in 'B' mice.[41,139] Direct evidence that long-lived recirculating T cells traverse the T-dependent areas has been obtained by preparing autoradiographs from recipients of normal TDL or lymph node cells labelled with such isotopes as ^{3}H-uridine;[55,97,102,130,140,141] as considered in the previous section, approximately 80% of TDL and lymph node cells are long-lived recirculating T cells.

In the spleen, labelled T cells appear within a few minutes of intravenous injection in the marginal zone and the red pulp venous sinusoids. From there the cells migrate to the PALS of the white pulp. The precise route by which T cells leave the spleen from the white pulp is not clear although it may be via 'bridging channels (pseudolymphatics)' connecting the PALS with the red pulp sinusoids.[140] The lymphatic supply to the spleen is very limited in most species and the majority of cells probably leave via the splenic vein.

Whereas all blood-borne cells are carried non-specifically to the red pulp of the spleen (though not to the white pulp), the capacity to migrate to lymph nodes is restricted to lymphocytes. Entry to lymph nodes is governed by specialized venules with a high endothelium, termed post-capillary venules (PCV), which are located predominantly in the paracortical region.[102] T cells appear in the walls of the PCV soon after injection and from there pass between[142] or through[143] the endothelial cells to enter the paracortex. By some as yet undefined route the cells then migrate to the medullary sinus, thus gaining access to the efferent lymphatics.

Although information on T cell migration to Peyer's patches is scanty, most cells probably enter the T-dependent areas via PCV.[144] They presumably leave these areas via lymphatics although this has yet to be proved.

* The discovery that T cells (including TDL) show a variety of Ly phenotypes may necessitate re-evaluating the current dogma that all recirculating T cells display uniform migratory patterns.

Autoradiographs provide little indication of the *extent* of lymphocyte migration through the lymphoid tissues. This can be studied conveniently by following the distribution of radioactivity in whole organs removed from recipients of lymphocytes labelled with the γ-emitting isotope, ^{51}Cr.[145] Within 1–4 hours of injecting ^{51}Cr-labelled T cells (TDL) intravenously, a high proportion (30%–40%) of the injected counts localize in the spleen.[55,145,146] Thereafter counts in the spleen decrease steadily and reach a plateau of 15%–20% after 24 hours. Counts in lymph nodes, by contrast, are low initially but increase progressively to reach a maximum after about 24 hours. At this stage, lymph node counts are about 50% higher than at 4 hours and exceed the counts remaining in the spleen.

Interpreting these data requires consideration of the relative speeds at which lymphocytes traverse the spleen and lymph nodes. Studies of Ford[147] on isolated spleens perfused *in vitro* indicate that the mean transit time of T cells (TDL) through the spleen is 5–7 hours. The mean transit time through lymph nodes, however, is probably of the order of 18 hours.[55,102,113,148]

In the above experiments with ^{51}Cr-labelled cells it would thus appear that the slow increase in lymph node counts observed between 4 and 24 hours reflects a gradual increase in the lymph nodes of cells which initially homed to and then migrated through the spleen. Lymph node homing therefore probably consists of 2 components: (a) *primary migration*—rapid accumulation of cells which enter lymph nodes directly from the circulation soon after injection; and (b) *secondary migration*—gradual accumulation of cells which reach lymph nodes only after temporary sequestration in other regions, e.g. the spleen.

4.1.1.2 *B lymphocytes*

Certain regions of the secondary lymphoid tissues are of normal cellularity in 'B' mice.[41,139] These 'thymus-independent' areas include (a) the peripheral white pulp and primary follicles of the spleen, (b) the superficial cortex, follicles and medulla of lymph nodes and (c) the follicles of Peyer's patches; germinal centres in these regions also fall within this category. Although the T-independent areas show minimal loss of cellularity during short-term thoracic duct drainage, prolonged drainage, e.g. for 7–10 days, does cause extensive depletion.[55] These areas thus harbour the slow-moving component of the RLP, i.e. long-lived recirculating B cells. In this respect, autoradiography has demonstrated that radio-labelled cells consisting predominantly of long-lived B cells, e.g. TDL, lymph node or spleen cells from 'B' mice[55,130,140] or rats,[141] localize selectively in these areas after intravenous injection.

In the spleen, labelled B cells, like T cells, are first detected in the marginal zone and the red pulp venous sinusoids. From there B cells migrate to the peripheral region of the white pulp, possibly by way of the PALS. As with T cells they then presumably leave the spleen via the splenic vein.

In lymph nodes, B cells, again like T cells, are first seen in the walls of the PCV. The cells then migrate selectively to the primary follicles and the superficial cortex; whether B cells reach these areas by penetrating the walls of the PCV where these vessels approach the follicles or, conversely, leave the PCV as the latter traverse the paracortex is not clear. At later periods labelled cells reach the medullary cords and by 24 hours are seen in the medullary sinus.

B cells probably gain access to the follicles of Peyer's patches via PCV.[144] Like T cells they appear to leave these structures through fine lymphatics.

Studies with ^{51}Cr-labelled lymphocytes show that a large proportion (30%–40%) of B cells localize in the spleen soon after injection.[55] Whereas in recipients of T cells counts in the spleen fall rapidly soon after injection, this is much less evident with B cells. The propensity of B cells to remain in the spleen is associated with a restricted capacity to localize in lymph nodes.[55,149] Counts in the nodes are very low within 4 hours of injection and increase to only a limited extent between 4 and 24 hours. Throughout this time counts in lymph nodes are 3–4-fold less in recipients of B cells than of T cells.

These findings can be interpreted in two ways. Firstly, it can be argued that B cells migrate through the spleen as rapidly as T cells, their limited localization in lymph nodes reflecting an impaired capacity to penetrate the walls of PCV. B cells migrating through the spleen would therefore tend to re-enter the spleen rather than lymph nodes. The alternative possibility is that B cells are not restricted in their capacity to penetrate PCV but accumulate in the spleen because their mean transit time through the latter is very slow. Without direct evidence on the speed of B cell migration through the spleen it is difficult to choose between these two possibilities (see References 2 and 3 for discussion).

4.1.2 *Short-lived non-recirculating lymphocytes*

Since the thymus and bone marrow rarely sustain discernible immune responses and lymphopoiesis in these tissues is active in germ-free animals, it seems likely that lymphocytes leaving the primary lymphoid organs are virgin cells which have yet to encounter foreign antigen. What is the relationship between these cells and long-lived recirculating lymphocytes in the secondary lymphoid organs? There are two main possibilities:

(1) Cells leaving the primary lymphoid organs feed directly into the RLP, their rate of release from these organs paralleling the rate at which long-lived lymphocytes enter the RLP.
(2) Emigrants from the primary lymphoid organs fail to enter the RLP without first undergoing further differentiation in the secondary lymphoid tissues.

In the case of T cells, the first possibility is in line with the popular view that, since the properties of thymic medullary cells (cortisone-resistant thymus cells) and most peripheral T cells are virtually identical, the cells leaving the thymus must therefore arise in the thymic medulla rather than the cortex. A variety of evidence (considered in detail elsewhere)[2] is against this view. Firstly, the rate of cell release from the thymus is probably much greater than the limited rate of cell entry into the RLP (10^6 T cells per day in mice).[2] Secondly, adult thymectomy causes a rapid depletion from the T-dependent areas of the spleen[150,151] but takes many months to reduce the size of the RLP.[152] Thirdly, circumstantial evidence suggests that most cortisone-resistant thymus cells do *not* leave the thymus.[153] Fourthly, and most importantly, labelled lymphocytes leaving the thymus after intrathymic injection of isotope migrate to the spleen in large numbers but do not enter thoracic duct lymph (Reference 154 and I. Weissman, personal communication).

The same objection holds for the B cell lineage. Thus, whereas large numbers of lymphocytes are considered to leave the marrow, the rate at which B cells enter the RLP is very low (10^6 cells per day in mice).[2] Moreover, purified populations of small lymphocytes from marrow have only a very limited capacity to recirculate into thoracic duct lymph after intravenous injection.[155]

The second possibility thus appears the more attractive. To account for the various data cited above one is led to the conclusion that, if most cells migrating from the primary lymphoid organs do not join the RLP, they must have a restricted lifespan; if not, the secondary lymphoid tissues would rapidly become hypertrophic. The question then arises of what drives the few surviving cells to enter the RLP. The most likely explanation—developed in detail elsewhere[2] and originally postulated by Miller and Mitchell[152]—is that such differentiation is a sequel to contact with specific antigen. According to this view, the vast majority of (virgin) lymphocytes migrating from the primary lymphoid organs do not encounter specific antigen and as a result soon die. Those that do meet specific antigen mount a primary response to the antigen, this encounter directing a proportion of the responding cells or their progeny to become long-lived recirculating memory cells.

If correct, this theory implies that all recirculating lymphocytes, i.e. the vast majority of lymphocytes outside the primary lymphoid organs, are memory cells primed to environmental antigens. Two pieces of evidence are at least consistent with this viewpoint. Firstly, the size of the RLP is greatly reduced in germ-free mice (Sprent and Miller, unpublished). Secondly, in the case of B cells, Strober and his colleagues[135,136,155,156] have produced impressive evidence that priming with certain antigens, e.g. ferritin, induces marked changes in the properties of the responding lymphocytes. These workers observed that the cells from unprimed rats which gave a primary adoptive response to ferritin had a rapid turnover, a reduced density of surface Ig and were unable to recirculate. With primed rats, by contrast, the cells controlling the response to ferritin had the properties of typical long-lived recirculating lymphocytes. These data are supported by evidence that in the case of the response of mice to certain haptenic determinants, e.g. NIP, the cells giving a primary response to the hapten differ in size, density, adherence properties and electrophoretic mobility from cells controlling the secondary response.[157–159] It is important to stress that, by the parameters used in these and the above studies, the properties of cells responding to some antigens, e.g. sheep erythrocytes, are not discernibly affected by antigen priming, i.e. both the primary and secondary responses appear to be controlled by long-lived recirculating lymphocytes. To explain this discrepancy it has been postulated that sheep erythrocytes cross-react with environmental antigens, the 'primary' response measured in the laboratory thus in effect representing a secondary response.

Although the evidence cited above is clearly consistent with the notion that most lymphocytes seeding from the primary lymphoid organs have a limited lifespan, this has yet to be proved. Moreover, there is as yet no direct information on whether the distinct differences in the properties of 'virgin' and 'memory' B cells also apply to the T cell lineage. In this respect, however, it may be germane to mention that two types of T lymphocytes—'T_1' and 'T_2' cells—are reported to collaborate synergistically to induce graft-versus-host reactions[160] and to differentiate into cytotoxic lymphocytes.[161] 'T_1' cells are considered to be short-lived, non-recirculating lymphocytes of recent thymic origin and 'T_2' cells as typical long-lived recirculating lymphocytes. Without further information it would be premature to conclude that this synergistic effect exerted by 'T_1' cells reflects a true primary response of virgin T cells to antigen.

Obtaining direct information on the migratory properties of recent emigrants from the thymus and marrow is difficult because of the problem of obtaining these cells in a purified form. Since the proportion of cells which become labelled during short-term parenteral administration of ^{3}HTdR is high in the spleen but low in lymph nodes (and TDL), it would appear that most cells leaving the thymus and marrow preferentially home to (and die in) the spleen. In the case of T cells, studies on the distribution of labelled cells leaving the thymus after intrathymic injection of ^{3}HTdR have shown nevertheless that the cells localize not only in the spleen but also in lymph nodes and Peyer's patches.[33,34] It is important to emphasize, however, that for most of these studies newborn animals were used, the microarchitecture of the secondary lymphoid organs in the neonate being quite different from that in the adult, at least in mice.[162] In this species, T cells are virtually absent in the spleen at birth but abundant in mesenteric lymph nodes and particularly prominent in Peyer's patches. Curiously, however, adult T cells transferred to neonatal mice localize well in the spleen. These data defy simple explanation. One possibility is that to counter the intense concentration of antigens entering the gut soon after birth, the lymphoid architecture is so arranged at this stage that newly-formed (virgin) lymphocytes are directed *en masse* to the intestinal region.

4.2 Factors controlling lymphocyte migration

A number of studies have shown that lymphocyte migration is radically altered following treatment of cells *in vitro* with neuraminidase,[163] sodium periodate,[164] concanavalin A[165–168] or phytohaemagglutinin.[167] Such treatment reduces homing to both spleen and lymph nodes but increases localization in the liver and lungs. The above agents all attach to carbohydrate-containing moieties on the cell surface and, at least in the case of neuraminidase and sodium periodate, cleave terminal sialic acid residues. These data have been widely interpreted as signifying that cell surface carbohydrates play a specific role in controlling lymphocyte migration. This notion is difficult to accept. Perhaps a more plausible explanation is that the reduced localization of lymphocytes in the spleen and lymph nodes after treatment with the above agents is simply the result of the cells being preferentially trapped in the liver. In this respect it is known that liver cells express receptors which non-specifically bind glycoproteins stripped of terminal sialic acid.[169,170]

In contrast to the apparently non-specific inhibitory effects of neuraminidase-like agents, treatment of lymphocytes with the proteolytic enzyme, trypsin, has a highly specific effect on cell migration. Woodruff and Gesner and their coworkers,[146,171,172] have shown that if TDL are incubated with low concentrations of trypsin before injection, the cells home in above normal numbers to the white pulp of the spleen but are totally unable to localize in lymph nodes or to recirculate; significantly, liver localization—an index of cell damage—is unaltered. This effect is of only short-term duration and after several hours the cells begin to enter lymph nodes and TDL in large numbers, presumably as the result of mobilization from the spleen. Trypsin-treated lymphocytes also regain their lymph-node-seeking ability if they are maintained *in vitro* for 6–12 hours before transfer. This reversal of their homing defect is temperature-dependent and requires protein synthesis. These data strongly suggest that lymphocytes carry and synthesize trypsin-sensitive surface 'receptors' which control the ability of the cells to penetrate the walls of post-capillary venules (see also F. Loor, herein).

Since entry to the spleen is not impaired by trypsin treatment, receptors for post-capillary venules presumably play no role in spleen-homing. The nature of the receptors which control entry to the splenic white pulp is obscure. (That such receptors do exist is evident from the fact that lymphocytes alone are able to traverse this region.)

Although there is no convincing evidence that sialic acid residues *per se* play a specific role in controlling lymphocyte migration (*vide supra*), it is clear nevertheless that migration cannot proceed normally unless these moieties are intact. In this respect it is of interest that certain viruses which cleave sialic acid from the cell surface, e.g. Newcastle disease virus, cause a marked alteration in lymphocyte migration. Woodruff and Woodruff[173–176] reported that rats infected with Newcastle disease virus develop a profound lymphocytopenia and depletion of lymphocytes from the lymph node paracortex, PALS of the spleen and thoracic duct lymph. Within three days of infection the cellularity of the blood, lymphoid tissues and lymph returns to normal. Since most of the lymphocytes which reappear are not newly-formed, their prior disappearance presumably reflects temporary sequestration rather than destruction. Where sequestration occurs is unknown, although the neuraminidase activity of the virus would suggest the liver as a likely candidate.

A marked alteration in lymphocyte migration is also a feature of infection with *Bordetella pertussis.*[177–180] As with Newcastle disease virus infections, there is a profound though temporary depletion of lymphocytes from the lymphoid tissues and thoracic duct lymph which is not associated with cell destruction. Curiously, however, this depletion is associated with a marked lymphocytosis rather than lymphocytopenia.

The above data are of some interest because they clearly suggest that the transient elevation or depression of blood lymphocyte counts which characterize a wide variety of disease states result simply from an altered pattern of lymphocyte migration rather than a change in the rate of cell formation. It goes without saying that a direct effect on lymphocyte migration might interfere with the initiation of immune responses and thus contribute substantially to the pathogenicity of the micro-organisms concerned. Elucidating precisely how micro-organisms influence lymphocyte traffic would seem an area worthy of future study.

Nothing has been said so far of why T and B lymphocytes segregate in different regions of the lymphoid tissues, nor why short-lived and long-lived lymphocytes have different migratory properties. In fact virtually nothing is known of this important subject. Conceivably, lymphocytes may possess a variety of surface receptors (possibly related to their surface markers), each facet of migration being controlled by a separate receptor. Assessing this or other possibilities without further information is clearly difficult.

4.3 Lymphocyte traffic during antigenic stimulation

4.3.1 *Selective recruitment of recirculating lymphocytes*

Evidence was considered in the preceding section suggesting that the RLP is formed as the result of constant exposure of non-recirculating virgin lymphocytes to environmental antigens. This leads us to consider how recirculating lymphocytes themselves behave on encountering specific antigen.

Studies in mice[181,182] and rats[183] have shown that if antigens such as heterologous erythrocytes or hapten–protein conjugates are injected intravenously, the reactivity of TDL towards the injected antigens disappears within 24 hours. This disappearance (a) is antigen-specific, (b) affects both T and B lymphocytes, (c) applies to IgG as well as to IgM antibody production, and (d) is demonstrable whether responses are measured *in vivo* or *in vitro.*[182,212] The degree of unresponsiveness observed depends on the dose of antigen injected; it is minimal with low doses and complete with high doses.[182] The unresponsiveness of TDL lasts only 1–2 days; specifically-reactive lymphocytes reappear in the lymph after 3 days and by 5 days give enhanced responses.

A similar phenomenon affects recirculating lymphocytes reactive to major transplantation antigens. Thus TDL from mice injected intravenously 1 day previously with large numbers of allogeneic lymphoid cells are totally and specifically depleted of (T) lymphocytes capable of producing a graft-versus-host reaction, cell-mediated lympholysis or skin allograft rejection towards the injected alloantigens.[181,184] (Curiously, there is only a partial removal of lymphocytes reactive in mixed-lymphocyte culture.) Again, the period of unresponsiveness lasts for only 1–2 days and by 3–4 days the lymph contains above normal numbers of antigen-reactive cells.

These data have been interpreted in terms of antigen-induced selective recruitment of recirculating lymphocytes (ASRL);[181] in other words, that contact with antigen leads to selective withdrawal of specifically-reactive lymphocytes from the RLP to the lymphoid tissues. Here the cells are temporarily immobilized and engage in immune responses to the injected antigen. After 2–3 days the activated lymphocytes or their progeny then re-enter the circulation in expanded numbers.

Three lines of evidence support this concept:

(1) In marked contrast to TDL, spleens from mice given antigen, e.g. sheep erythrocytes, 1 day previously contain above normal numbers of specifically-reactive lymphocytes.[212] Cells from lymph nodes, by contrast, are, like TDL, totally unresponsive to the injected antigen.[182] In this respect it should be mentioned that when heterologous erythrocytes are injected intravenously they localize predominantly in the spleen (and liver) but fail to reach lymph nodes.[182] These findings imply that when large amounts of antigen localize in the spleen and remain confined to this region, the spleen selectively filters out virtually all specifically-reactive lymphocytes from the RLP, i.e. including those present in lymph nodes. *A priori*, this is not difficult to imagine since most recirculating lymphocytes probably traverse the spleen at least once a day (see Section 4.1).
(2) If isolated spleens are perfused *in vitro* and antigen is added to the perfusate, lymphocytes reactive to the antigen cease to circulate and become specifically trapped in the spleen.[185]
(3) When parental strain T cells are transferred intravenously to irradiated F_1 hybrid recipients, cells of donor origin present in thoracic duct lymph 24 hours later are almost devoid of lymphocytes reactive to host-type histocompatibility determinants.[186] By radio-labelling the injected cells it has been demonstrated that most of the host-reactive lymphocytes localize in the spleen and there undergo blast transformation.[187,188]

The above data suggest that when antigen is given intravenously, ASRL takes place predominantly in the spleen. It would follow that if antigen localized in lymph nodes, e.g. after subcutaneous injection, ASRL would occur preferentially to this region. On this point, studies of Hay *et al.*[189] are of particular relevance. These workers injected allogeneic lymphocytes subcutaneously into the drainage area of the popliteal lymph node of sheep and studied the capacity of cells recovered from the efferent lymphatic vessels to respond in mixed-lymphocyte culture. The reactivity of these cells to lymphocytes of donor origin was high within the first 2 days of injection but at later times completely disappeared; this disappearance was specific since the cells responded well to other antigens. Specifically-reactive lymphocytes reappeared in the lymph in large numbers after 5–7 days.

Of significance is the fact that when cells emerging from the injected node were diverted from the body, the reactivity against donor determinants also disappeared from the lymph-borne cells of the uninjected contralateral node. Furthermore, this disappearance continued indefinitely. Analogous findings have recently been reported for sheep given repeated local injections of hapten–protein conjugates.[190] These data are of considerable importance because they raise the possibility that if antigen is confined to a single lymph node, the latter might be able to screen the entire RLP for antigen-reactive lymphocytes. If so, removal of the node at an appropriate time should leave the animal depleted of specifically-reactive lymphocytes. There are of course less attractive explanations which have yet to be ruled out, e.g. production of blocking factors, suppressor T cells etc. Nevertheless the obvious need for devising methods of selective immunosuppression e.g.

for organ transplants, should make it imperative to elucidate the mechanism of unresponsiveness in this model.

Mention should also be made of two other studies suggestive of ASRL to lymph nodes. Firstly, specifically-reactive lymphocytes labelled with radioisotopes show a marked tendency to localize in lymph nodes stimulated with the sensitizing antigen; localization in nodes stimulated with other antigens is much less.[191,192] Secondly, the immune response of lymph nodes to a particular antigen can be abolished ('pre-empted') by a prior intraperitoneal injection of the antigen.[193] This phenomenon is at least partly non-specific, however, since third-party antigens have a similar though less profound effect.

The factors which initiate and control ASRL are poorly understood. One possibility suggested[194] is that when lymphocytes encounter specific antigen the surface membrane components which govern their migratory properties undergo a series of progressive changes, these changes leading the cells to seek and then transiently remain in regions of the lymphoid tissues conducible for initiating immune responses. This hypothesis was advanced to explain the curious observation that if spleens from mice given antigen intravenously 1 day previously are transferred *in vivo* together with more antigen, there is a lag period of several days (compared to normal spleen cells) before the cells begin to make specific antibody.[194] Conversely, if the cells are cultured *in vitro*, production of specific antibody occurs rapidly (*vide supra*). It is thus argued that when lymphocytes are removed during ASRL and introduced into the bloodstream, the alteration in their cell membrane configuration is such that the cells are temporarily unable to migrate to the antibody-forming sites of the lymphoid tissues in their new host. This problem of course would not arise during culture *in vitro*.

The foregoing discussion has concerned *specific* alterations in cell migration following confrontation with antigen. It should be mentioned that antigen injection also produces certain non-specific effects. Thus, when normal unstimulated lymphocytes are transferred to antigen-injected animals, the cells tend to home in above normal numbers to regions in which the antigen is concentrated.[195] This phenomenon has been interpreted as indicated that sequestration of antigen induces non-specific 'trapping' of lymphocytes, possibly as the result of local release of lymphokines.

4.3.2 *Activated lymphocytes*

ASRL at 1–2 days after antigen injection is followed by the appearance of large numbers of blast cells in the central lymph.[186,189,196,197,200] These cells have a rapid turnover, express a variety of effector functions and most are probably the progeny of lymphocytes responding to the injected antigen.* Large blast cells are also found in the lymph of normal animals, presumably reflecting the constant exposure of the host to environmental antigens. Most of these cells—which comprise 2%–5% of lymph-borne cells—seem to belong to the B cell lineage since a large proportion show varying amounts of endoplasmic reticulum;[143,201] moreover, in mice only about 10% are θ-positive.[112] Of the small proportion of lymph-borne cells which are in S phase (0·5% of TDL), however, the majority (75%) appear to be T cells, i.e. they stain with an anti-T cell heteroantiserum.[134] The unique migratory properties of T and B blasts are considered below.

* This may be an oversimplification since studies with blast cells derived from locally stimulated lymph nodes of sheep have shown that most of the Ig secreted by the cells does not have specificity for the injected antigen (B. Morris, personal communication). This may reflect that many antigens are non-specifically mitogenic.

4.3.2.1 *T blasts*

Information on the migration of activated T cells has come from two approaches: (1) By pulsing blast cells *in vitro* with ^{3}HTdR and, by autoradiography, following the migratory properties of the T cell component with an anti-T cell heteroantiserum.[134] (2) By using a variety of radioisotopes to study the fate of purified populations of H2-activated T cells;[199,202] these cells (T.TDL) can be collected in large numbers from the lymph of irradiated F_1 hybrid mice injected 3–4 days previously with parental strain T cells.[199]

Both approaches have given essentially similar results. The cells home predominantly to the spleen soon after injection and localize in the T-dependent areas of the white pulp. Many of the cells then leave the spleen but, unlike normal small T lymphocytes, they become redistributed to the gut rather than to lymph nodes. In the small intestine the cells localize in high concentrations in Peyer's patches but are also found in the adjacent lamina propria and between the surface epithelial cells. A few cells appear to penetrate the epithelial lining and enter the gut lumen. Appreciable numbers of T blasts also migrate to the large intestine; localization here occurs more slowly than to the small intestine.

The limited capacity of activated T cells to migrate to lymph nodes has led some workers to conclude that T blasts do not recirculate.[203] In fact, T blasts (T.TDL) enter thoracic duct lymph after intravenous injection (the only direct test for recirculation) just as readily as normal T cells, recirculation occurring while the cells are still in a transformed state.[199] How do T blasts gain access to the lymphatics? This is not clear, although their gut-seeking ability suggests that recirculation might occur predominantly via the gut-associated lymphoid tissues.

Most T blasts differentiate into typical small lymphocytes soon after transfer. Studies with T.TDL labelled with ^{51}Cr or ^{3}HTdR suggest that the vast majority of the cells die in the lymphoid tissues within 1–2 weeks of injection.[202] A small proportion, however, differentiate into long-lived recirculating lymphocytes.[204] These lymphocytes appear to represent memory cells since their antigenic specificity is restricted to the H2 determinants to which the progenitors of the cells were originally activated.[205]

4.3.2.2 *B blasts*

A number of studies have shown that most large lymphocytes from the central lymph of normal or antigen-stimulated mice[134,202] or rats[103,198,206–210] home to the gut where they soon differentiate into plasma cells. Both T and B blasts thus share a propensity for migrating to the intestines. The homing patterns of these two cell types nevertheless display subtle differences. Firstly, B blasts, unlike T blasts, localize predominantly in the lamina propria of the small intestine; very few cells reach Peyer's patches or the surface epithelium. Secondly, homing to the large intestine is very restricted with B blasts. Thirdly, B blasts have only a limited capacity to recirculate.

Like small B lymphocytes, B blasts differ with respect to the class of surface Ig they express.[134] For example, IgA-containing blasts account for 85% of total Ig-positive blasts in TDL, 50% in mesenteric lymph nodes but only 5% in peripheral lymph nodes. Conversely, blasts bearing either IgG or IgM are common in peripheral lymph nodes (40%–50%) but rare in TDL (3%–14%). The class of Ig borne by the cells seems to have a radical effect on their migratory properties. Thus, whereas most IgA-bearing blasts localize in the lamina propria, blasts carrying IgM or IgG fail to reach the gut and home predominantly to the spleen and lymph nodes.

Although these data suggest a close correlation between gut-homing and the expression of surface IgA, it should be mentioned that certain IgA-bearing blasts, i.e. those present in Peyer's patches, have virtually no capacity to migrate to the lamina propria. To account for this anomaly, Guy-Grand *et al.*[134] postulate that IgA on the cell surface must be expressed in a polymeric form before cells can home to the gut. They base this suggestion on the fact that IgA-bearing blasts in Peyer's patches, unlike those in TDL or mesenteric lymph nodes, contain virtually no intracellular IgA. They propose that Peyer's patch blasts reach maturity by migrating to the mesenteric lymph nodes where they synthesize intracellular polymeric IgA; this material then replaces monomeric IgA on the cell surface and thus imbues the cells with the capacity to migrate to the gut. They consider that the secretory component of IgA—which is synthesized by the intestinal epithelium and binds to polymeric but not monomeric IgA—might play a crucial role in directing the cells to the lamina propria.

Another possibility that has to be considered is that blast cells are attracted to the intestines by exogeneous antigens. This seems unlikely in view of evidence that blast cells home to the small intestine of unsuckled neonates delivered by caesarian section[211] and also to pieces of foetal gut implanted in adults,[134,144,209] i.e. in situations where the antigen content of the gut would presumably be very low.

5. REFERENCES

1. Gowans, J. L. and McGregor, D. D. (1965). *Progr. Allergy*, **9**, 1–78.
2. Sprent, J. (1975). In *The Lymphocyte: Structure and Function* (Marchalonis, J. J., Ed.) in press.
3. Ford, W. L. (1975). *Progr. Allergy*, **19**, 1–59.
4. Greaves, M. F., Owen, J. J. T. and Raff, M. C. (1973). *T and B Lymphocytes*, Excerpta Medica, Elsevier, New York.
5. Miller, J. F. A. P. and Osoba, D. (1967). *Physiol. Rev.*, **47**, 437–520.
6. Miller, J. F. A. P. and Davies, A. J. S. (1964). *Annu. Rev. Med.*, **15**, 23–36.
7. Benner, R., Meima, F., van der Meulen, G. M. and van der Muiswinkel, W. B. (1974). *Immunology*, **26**, 247–255.
8. Metcalf, D. and Moore, M. A. S. (1971). *Haemopoietic Cells*, North-Holland, Amsterdam.
9. Moore, M. A. S. and Owen, J. J. T. (1967). *J. exp. Med.*, **126**, 715–726.
10. Owen, J. J. T. and Ritter, M. A. (1969). *J. exp. Med.*, **129**, 431–442.
11. Moore, M. A. S. and Owen, J. J. T. (1967). *Lancet*, **2**, 658–659.
12. Droege, W. and Zucker, R. (1975). *Transplant. Rev.*, **25**, 3–25.
13. Everett, N. B. and Tyler, R. W. (1967). *Int. Rev. Cytol.*, **22**, 205–237.
14. Shortman, K. and Jackson, H. (1974). *Cell. Immunol.*, **12**, 230–246.
15. Cerottini, J.-C. and Brunner, K. T. (1967). *Immunology*, **13**, 395–403.
16. Aoki, T., Hämmerling, U., de Harven, E., Boyse, E. A. and Old, L. J. (1969). *J. exp. Med.*, **130**, 979–1002.
17. Lance, E. M., Cooper, S. and Boyse, E. A. (1970). *Cell. Immunol.*, **1**, 536–544.
18. Schlesinger, M. (1972). *Progr. Allergy*, **16**, 214–299.
19. Raff, M. C. (1971). *Nature (New Biol.)*, **229**, 182–184.
20. Blomgren, H. and Andersson, B. (1969). *Exp. Cell Res.*, **57**, 185–192.
21. Blomgren, H. and Andersson, B. (1971). *Cell. Immunol.*, **1**, 545–560.
22. Leckband, E. and Boyse, E. A. (1971). *Science*, **172**, 1258–1260.
23. Zeiller, K., Pascher, G., Wagner, G., Liebich, H. G., Holzberg, E. and Hannig, K. (1974). *Immunology*, **26**, 995–1012.
24. Sainte-Marie, G. and Leblond, C. P. (1958). *Proc. Soc. exp. Biol. Med.*, **98**, 909–915.
25. Sainte-Marie, G. (1973). In *Contemporary Topics in Immunobiology*, Vol. II (Davies, A. J. S. and Carter, R. L., Eds.), Plenum Press, New York, pp. 111–117.
26. Weissman, I. L. (1973). *J. exp. Med.*, **137**, 504–510.
27. Fathman, C. G., Small, M., Herzenberg, L. A. and Weissman, I. L. (1975). *Cell. Immunol.*, **15**, 109–128.
28. Nossal, G. J. V. (1964). *Ann. N.Y. Acad. Sci.*, **120**, 171–181.
29. Murray, R. G. and Woods, P. A. (1964). *Anat. Rec.*, **150**, 113–128.
30. Linna, T. J. (1968). *Blood*, **31**, 727–746.

31. Iorio, R. J., Chanana, A. D., Cronkite, E. P. and Joel, D. D. (1970). *Cell Tissue Kinet.*, **3**, 161–173.
32. Linna, T. J. (1967). *Int. Arch. Allergy Appl. Immunol.*, **31**, 313–337.
33. Joel, D. D., Hess, M. W. and Cottier, H. (1972). *J. exp. Med.*, **135**, 907–923.
34. Weissman, I. L. (1967). *J. exp. Med.*, **126**, 291–304.
35. Davies, A. J. S., Leuchars, E., Wallis, V. and Doenhoff, M. J. (1971). *Proc. Roy. Soc. Lond. B.*, **176**, 369–384.
36. Davies, A. J. S. and Carter, R. L. (1972). In *Contemporary Topics in Immunobiology*, Vol. 1 (Hanna, M. G., Ed.), Plenum Press, New York, pp. 1–31.
37. Sainte-Marie, G. and Leblond, C. P. (1964). *Blood*, **23**, 275–299.
38. Ernström, U., Gyllensten, L. and Larsson, B. (1965). *Nature*, **207**, 540–541.
39. Williams, R. M., Chanana, A. D., Cronkite, E. P. and Waksman, B. H. (1971). *J. Immunol.*, **106**, 1143–1146.
40. Cole, G. J. and Morris, B. (1971). *Aust. J. exp. Biol. Med. Sci.*, **49**, 33–53.
41. de Sousa, M. A. B., Parrott, D. M. V. and Pantelouris, E. M. (1969). *Clin. exp. Immunol.*, **4**, 637–644.
42. Miller, J. F. A. P. (1962). *Ann. N.Y. Acad. Sci.*, **99**, 340–354.
43. Manning, D. D., Reed, N. D. and Shaffer, C. F. (1973). *J. exp. Med.*, **138**, 488–494.
44. Arnason, B. G., Janković, B. D., Waksman, B. H. and Wennersten, C. (1962). *J. exp. Med.*, **116**, 177–186.
45. Miller, J. F. A. P., Mitchell, G. F. and Weiss, N. S. (1967). *Nature*, **214**, 992–997.
46. Reif, A. E. and Allen, J. M. V. (1964). *J. exp. Med.*, **120**, 413–433.
47. Raff, M. C. (1969). *Nature*, **224**, 378–379.
48. Raff, M. C. (1970). *Immunology*, **19**, 637–650.
49. Miller, J. F. A. P. and Sprent, J. (1971). *Nature* (*New Biol.*), **230**, 267–270.
50. Rabellino, E., Colon, S., Grey, H. M. and Unanue, E. R. (1971). *J. exp. Med.*, **133**, 156–167.
51. Raff, M. C. (1971). *Transplant. Rev.*, **6**, 52–80.
52. Nossal, G. J. V., Warner, N. L., Lewis, H. and Sprent, J. (1972). *J. exp. Med.*, **135**, 405–428.
53. Raff, M. C. and Wortis, H. H. (1970). *Immunology*, **18**, 931–942.
54. Schlesinger, M. and Yron, I. (1970). *J. Immunol.*, **104**, 798–804.
55. Sprent, J. (1972). *Cell. Immunol.*, **7**, 10–39.
56. Olsson, L. and Claësson, M. H. (1973). *Nature* (*New Biol.*), **244**, 50–51.
57. Loor, F. and Roelants, G. E. (1974). *Nature*, **251**, 229–230.
58. Roelants, G. E., Loor, F., von Boehmer, H., Sprent, J., Hägg, L.-B., Mayor, K. S. and Rydén, A. (1975). *Eur. J. Immunol.*, **5**, 127–131.
59. Roelants, G. E., Mayor, K. S., Hägg, L.-B. and Loor, F. (1976). *Eur. J. Immunol.*, **6**, 75–81.
60. Goldstein, A. L., Thurman, G. B., Cohen, G. H. and Hooper, J. A. (1975). In *Biological Activity of Thymic Hormones* (van Bekkum, D. W., Ed.), Kooyker Scientific Publications, Rotterdam, pp. 173–197.
61. Trainin, N., Small, M., Zipori, D., Umiel, T., Kook, A. I. and Rotter, V. (1975). In *Biological Activity of Thymic Hormones* (van Bekkum, D. W., Ed.), Kookyer Scientific Publications, Rotterdam, pp. 117–144.
62. Bach, J.-F., Bach, M.-A., Charriere, J., Dardenne, M., Fournier, C., Papiernik, M. and Pleau, J.-M. (1975). In *Biological Activity of Thymic Hormones* (van Bekkum, D. W., Ed.), Kooyker Scientific Publications, Rotterdam, 145–158.
63. Stutman, O., Yunis, E. J. and Good, R. A. (1969). *J. exp. Med.*, **130**, 809–819.
64. Stutman, O. (1974). *Fed. Proc.*, **33**, 736 (Abstract).
65. Stutman, O. (1975). In *Biological Activity of Thymic Hormones* (van Bekkum, D. W., Ed.), Kooyker Scientific Publications, Rotterdam, pp. 87–94.
66. Scheid, M. P., Hoffman, M. K., Komuro, K., Hammerling, U., Abborr, J., Boyse, E. A., Cohen, G. H., Hooper, J. A., Schulof, R. S. and Goldstein, A. L. (1973). *J. exp. Med.*, **138**, 1027–1032.
67. Singh, U. (1975). In *Biological Activity of Thymic Hormones* (van Bekkum, D. W., Ed.), Kooyker Scientific Publications, Rotterdam, pp. 29–30.
68. Janković, B. D., Knežević, Z., Isaković, K., Mitrović, K., Marković, B. M. and Rajčević, M. (1975). *Eur. J. Immunol.*, **5**, 656–659.
69. Cooper, M. D., Cain, W. A., van Alter, P. J. and Good, R. A. (1969). *Int. Arch. Allergy Appl. Immunol.*, **35**, 242–252.
70. Kincade, P. W., Self, K. S. and Cooper, M. D. (1973). *Cell. Immunol.*, **8**, 93–102.

71. Linna, T. J., Bäch, R., and Hemmingsson, E. (1971). In *Morphological and Functional Aspects of Immunity* (Lindahl-Kiessling, K., Alm, G. and Hanna, M. G., Jr., Eds.), Plenum Press, New York, pp. 149–159.
72. Hemmingsson, E. (1972). *Int. Arch. Allergy Appl. Immunol.*, **42**, 764–774.
73. Bäck, R. and Linna, T. J. (1973). *Eur. J. Immunol.*, **3**, 147–152.
74. Peterson, R. D., Cooper, M. D. and Good, R. A. (1965). *Amer. J. Med.*, **38**, 579–604.
75. Perey, D. Y. E., Cooper, M. D. and Good, R. A. (1968). *Science*, **161**, 265–266.
76. Owen, J. J. T., Cooper, M. D. and Raff, M. C. (1974). *Nature*, **249**, 361–363.
77. Nossal, G. J. V. and Pike (1973). *Immunology*, **25**, 33–46.
78. Osmond, D. G. and Nossal, G. J. V. (1974). *Cell. Immunol.*, **13**, 117–131.
79. Everett, N. B., Caffrey, R. W. and Rieke, W. O. (1964). *Ann. N.Y. Acad. Sci.*, **113**, 887–897.
80. Osmond, D. G. and Nossal, G. J. V. (1974). *Cell. Immunol.*, **13**, 132–145.
81. Ryser, J.-E. and Vassalli, P. (1974). *J. Immunol.*, **113**, 719–728.
82. Brahim, F. and Osmond, D. G. (1970). *Anat. Rec.*, **168**, 139–159.
83. Brahim, F. and Osmond, D. G. (1973). *Anat. Rec.*, **175**, 737–746.
84. Linna, T. J. and Liden, S. (1969). *Int. Arch. Allergy Appl. Immunol.*, **35**, 35–46.
85. Miller, R. A. and Phillips, R. G. (1974). *Nature*, **251**, 444–446.
86. Melchers, F., Cone, R., von Boehmer, H. and Sprent, J. (1975). *Eur. J. Immunol.*, **5**, 382–387.
87. Lafleur, L., Underdown, B. J., Miller, R. G. and Phillips, R. A. (1972). *Ser. Haemat.*, **5**, 50–63.
88. Lafleur, L., Miller, R. G. and Phillips, R. A. (1973). *J. exp. Med.*, **137**, 954–966.
89. Marchalonis, J. J., Cone, R. E. and Atwell, J. (1972). *J. exp. Med.*, **135**, 956–971.
90. Vitetta, E. A., Bianco, C., Nussenzweig, J. W. and Uhr, J. W. (1972). *J. exp. Med.*, **136**, 81–93.
91. Hudson, L. and Sprent, J. (1976). *J. exp. Med.*, **143**, 444–449.
92. Parrott, D. M. V. (1967). *J. Clin. Pathol. Suppl.*, **20**, 456–465.
93. Cottier, H., Cronkite, E. P., Jansen, C. R., Rai, K. R., Singer, S. and Sipe, C. R. (1964). *Blood*, **24**, 241–253.
94. Ford, W. L. (1968). *Brit. J. exp. Pathol.*, **49**, 502–510.
95. McGregor, D. D. and Gowans, J. L. (1963). *J. exp. Med.*, **117**, 303–320.
96. McGregor, D. D. (1966). *Fed. Proc.*, **25**, 1713–1719.
97. Goldschneider, I. and McGregor, D. D. (1968). *J. exp. Med.*, **127**, 155–168.
98. Fish, J. C., Mattingly, A. T., Ritzmann, S. E., Sarles, H. E. and Remmers, A. R., Jr. (1969). *Arch. Surg.*, **99**, 664–668.
99. Dineen, J. K. and Adams, D. B. (1970). *Immunology*, **19**, 11–30.
100. Gowans, J. L. (1957). *Brit. J. exp. Pathol.*, **38**, 67–78.
101. Gowans, J. L. (1959). *J. Physiol. (London)*, **146**, 54–69.
102. Gowans, J. L. and Knight, E. J. (1964). *Proc. Roy. Soc. Lond. B*, **159**, 257–282.
103. Hall, J. G. and Morris, B. (1964). *Lancet*, **1**, 1077–1078.
104. Hall, J. G. and Morris, B. (1966). *J. exp. Med.*, **121**, 901–910.
105. Hall, J. G. (1967). *Quart. J. exp. Physiol.*, **52**, 76–85.
106. Frost, H., Cahill, R. N. P. and Trnka, Z. (1975). *Eur. J. Immunol.*, **5**, 839–843.
107. Perry, S., Irvin, G. L. and Whang, J. (1967). In *The Lymphocyte in Immunology and Haemopoiesis* (Yoffey, J. M., Ed.), Edward Arnold, London, p. 99.
108. Binns, R. M. and Hall, J. G. (1966). *Brit. J. exp. Pathol.*, **47**, 275–280.
109. Ruchti, C., Cottier, H., Cronkite, E. P., Jansen, C. R. and Rai, K. R. (1970). *Cell Tissue Kinet.*, **3**, 301–315.
110. Schnappauf, H. P. and Schnappauf, U. (1968). *Blut.*, **16**, 209–220.
111. Cronkite, E. P. and Chanana, A. D. (1970). In *Formation and Destruction of Blood Cells* (Greenwalt, T. J. and Jamieson, G. A., Eds.), Lippincott, Philadelphia, pp. 284–303.
112. Sprent, J. and Basten, A. (1973). *Cell. Immunol.*, **7**, 40–59.
113. Howard, J. C. (1972). *J. exp. Med.*, **135**, 185–199.
114. Yoffey, J. M. and Courtice, F. C. (1956). *Lymphatics, Lymph and Lymphoid Tissue*, Harvard University Press, Cambridge, Mass.
115. Otteson, J. (1954). *Acta Physiol. Scand.*, **32**, 75–93.
116. Buckton, K. E., Jacobs, P., Court-Brown, W. M. and Doll, R. (1962). *Lancet*, **2**, 676–682.
117. Norman, A., Sasaki, M. S., Ottoman, R. E. and Fingerhut, A. G. (1965). *Science*, **147**, 745.
118. Matsuyama, M., Wiadrowski, M. N. and Metcalf, D. (1966). *J. exp. Med.*, **123**, 559–576.
119. Robinson, S. H., Brecher, G., Lowrie, I. S. and Haley, J. E. (1965). *Blood*, **26**, 281–295.
120. Cottier, H., Schindler, R., Bürki, H., Sordat, B., Joel, D. D. and Hess, M. W. (1971). *Int. Arch. Allergy Appl. Immunol.*, **41**, 4–12.

121. Caffrey, R. W., Rieke, W. O. and Everett, N. B. (1962). *Acta Haematol.*, **28**, 145–154.
122. Claësson, M. H., Röpke, C. and Hougen, H. P. (1974). *Scand. J. Immunol.*, **3**, 597–604.
123. Röpke, C. and Hougen, H. P. (1974). In *Proceedings of the First International Workshop on Nude Mice* (Rygaard, J. and Povlsen, C. O., Eds.), Gustav Fischer Verlag, Stuttgart, pp. 51–60.
124. Lemmel, E.-M. and Fichtelius, K. E. (1971). *Int. Arch. Allergy Appl. Immunol.*, 716–728.
125. Röpke, C., Hougen, H. P. and Everett, N. B. (1975). *Cell. Immunol.*, **15**, 82–93.
126. Haas, R. J., Bohne, F. and Fliedner, T. M. (1969). *Blood*, **34**, 791–805.
127. McGregor, D. D. (1966). *J. exp. Med.*, **127**, 953–966.
128. Howard, J. C. and Scott, D. W. (1972). *Cell. Immunol.*, **3**, 421–429.
129. Miller, S. C. and Osmond, D. G. (1975). *Cell Tissue Kinet.*, **8**, 97–110.
130. Parrott, D. M. V. and de Sousa, M. A. B. (1971). *Clin. exp. Immunol.*, **8**, 663–684.
131. Bianco, C., Patrick, R. and Nussenzweig, V. (1970). *J. exp. Med.*, **132**, 702–720.
132. Ferguson, A. and Parrott, D. M. V. (1972). *Clin. exp. Immunol.*, **12**, 477–488.
133. Parrott, D. M. V. and de Sousa, M. A. B. (1974). In *Proceedings of The First International Workshop on Nude Mice* (Rygaard, J. and Povlsen, C. O., Eds.), Gustav Fischer Verlag, Stuttgart, pp. 61–70.
134. Guy-Grand, D., Griscelli, C. and Vassalli, P. (1974). *Eur. J. Immunol.*, **4**, 435–443.
135. Strober, S. (1972). *J. exp. Med.*, **136**, 851–871.
136. Strober, S. (1975). *Transplant. Rev.*, **24**, 84–112.
137. Fliedner, T. M., Kesse, M., Cronkite, E. P. and Robertson, J. S. (1964). *Ann. N.Y. Acad. Sci.*, **113**, 578–594.
138. Rieke, W. O., Caffrey, R. W. and Everett, N. B. (1963). *Blood*, **22**, 674–689.
139. Parrott, D. M. V., de Sousa, M. A. B. and East, J. (1966). J. exp. Med., **123**, 191–204.
140. Mitchell, J. (1972). *Immunology*, **22**, 231–245.
141. Howard, J. C., Hunt, S. V. and Gowans, J. L. (1972). *J. exp. Med.*, **135**, 200–219.
142. Schoefl, G. L. (1972). *J. exp. Med.*, **136**, 568–588.
143. Marchesi, V. T. and Gowans, J. L. (1964). *Proc. Roy. Soc. Lond. B*, **159**, 283–290.
144. Parrott, D. M. V. and Ferguson, A. (1974). *Immunology*, 26, 571–588.
145. Zatz, M. M. and Lance, E. M. (1970). *Cell. Immunol.*, **1**, 3–17.
146. Woodruff, J. and Gesner, B. M. (1968). *Science*, **161**, 176–178.
147. Ford, W. L. (1969). *Cell Tissue Kinet.*, **2**, 171–191.
148. Ford, W. L. and Simmonds, S. J. (1972). *Cell Tissue Kinet.*, **5**, 175–189.
149. Taub, R. N. and Lance, E. M. (1971). *Transplantation*, **11**, 536–542.
150. Waksal, S. D., Weprin, L. and St. Pierre, R. L. (1973). *J. Reticuloendothel. Soc.*, **13**, 343–344 (Abstract).
151. Cantor, H. and Boyse, E. A. (1975). In *Biological Activity of Thymic Hormones* (van Bekkum, D. W., Ed.), Kooyker Scientific Publications, Rotterdam, pp. 77–82.
152. Miller, J. F. A. P. and Mitchell, G. F. (1969). *Transplant. Rev.*, **1**, 3–42.
153. Elliot, E. V. (1973). *Nature* (*New Biol.*), **242**, 150–152.
154. de Sousa, M. A. B. (1973). In *Contemporary Topics in Immunobiology*, Vol. 2 (Davies, A. J. S. and Carter, R. L., Eds.), Plenum Press, New York, pp. 119–136.
155. Strober, S. and Dilley, J. (1973). *J. exp. Med.*, **138**, 1331–1344.
156. Strober, S. and Dilley, J. (1973). *J. exp. Med.*, **137**, 1275–1292.
157. Schlegel, R. A., von Boehmer, H. and Shortman, K. (1975). *Cell. Immunol.*, **16**, 203–217.
158. Schlegel, R. A. and Shortman, K. (1975). *J. Immunol.*, **115**, 94–99.
159. Schrader, J. W. (1974). *Cell. Immunol.*, **10**, 380–393.
160. Raff, M. C. and Cantor, H. (1971). *Progr. Immunol.*, **1**, 83–93.
161. Wagner, H. (1973). *J. exp. Med.*, **138**, 1379–1397.
162. Friedberg, S. H. and Weissman, I. L. (1974). *J. Immunol.*, **113**, 1477–1492.
163. Woodruff, J. J. and Gesner, B. M. (1969). *J. exp. Med.*, **129**, 551–567.
164. Zatz, M., Goldstein, A. L., Blumenfeld, O. O. and White, A. (1972). *Nature* (*New Biol.*), **240**, 252–255.
165. Gillette, R. W., McKenzie, G. O. and Swanson, M. H. (1973). *J. Immunol.*, **111**, 1902–1905.
166. Taub, R. N. (1974). *Cell. Immunol.*, **12**, 263–270.
167. Schlesinger, M. and Israel, E. (1974). *Cell. Immunol.*, **14**, 66–79.
168. Rodriguez, B. A., Rich, R. R. and Rossen, R. D. (1975). *J. Immunol.*, **115**, 771–776.
169. Pricer, W. E., Jr. and Ashwell, G. (1971). *J. Biol. Chem.*, **246**, 4825–4833.
170. Van Lenten, L. and Ashwell, G. (1972). *J. Biol. Chem.*, **247**, 4633–4640.
171. Gesner, B. M., Woodruff, J. J. and McCluskey, R. T. (1969). *Amer. J. Pathol.*, **57**, 215–230.

172. Woodruff, J. J. (1974). *Cell. Immunol.*, **13**, 378–384.
173. Woodruff, J. F. and Woodruff, J. J. (1970). *Cell. Immunol.*, **1**, 333–354.
174. Woodruff, J. F. and Woodruff, J. J. (1972). *Cell. Immunol.*, **5**, 296–306.
175. Woodruff, J. J. and Woodruff, J. F. (1972). *Cell. Immunol.*, **5**, 307–317.
176. Woodruff, J. J. and Woodruff, J. F. (1974). *Cell. Immunol.*, **10**, 78–85.
177. Morse, S. I. (1965). *J. exp. Med.*, **121**, 49–68.
178. Morse, S. I. and Riester, S. K. (1967). *J. exp. Med.*, **125**, 401–408.
179. Morse, S. I. and Riester, S. K. (1967). *J. exp. Med.*, **125**, 619–628.
180. Taub, R. N., Rosett, W., Adler, A. and Morse, S. I. (1972). *J. exp. Med.*, **136**, 1581–1593.
181. Sprent, J., Miller, J. F. A. P. and Mitchell, G. F. (1971). *Cell. Immunol.*, **2**, 171–181.
182. Sprent, J. and Miller, J. F. A. P. (1974). *J. exp. Med.*, **139**, 1–12.
183. Rowley, D. A., Gowans, J. L., Atkins, R. C., Ford, W. L. and Smith, M. E. (1972). *J. exp. Med.*, **136**, 499–513.
184. Sprent, J. and Miller, J. F. A. P. (1976). *J. exp. Med.*, **143**, 585–600.
185. Ford, W. L. (1972). *Clin. exp. Immunol.*, **12**, 243–254.
186. Ford, W. L. and Atkins, R. C. (1972). *Nature* (*New Biol.*), **234**, 178–180.
187. Atkins, R. C. and Ford, W. L. (1975). *J. exp. Med.*, **141**, 664–680.
188. Ford, W. L., Simmonds, S. J. and Atkins, R. C. (1975). *J. exp. Med.*, **141**, 681–696.
189. Hay, J. B., Cahill, R. N. P. and Trnka, A. (1974). *Cell. Immunol.*, **10**, 145–153.
190. McConnell, I., Lachmann, P. J. and Hobart, M. J. (1974), *Nature* (*New Biol.*), **250**, 113–116.
191. Thursh, D. R. and Emeson, E. E. (1972). *J. exp. Med.*, **135**, 754–763.
192. Thursh, D. R. and Emeson, E. E. (1973). *J. exp. Med.*, **138**, 659–671.
193. O'Toole, C. M. and Davies, A. J. S. (1971). *Nature*, **230**, 187–189.
194. Sprent, J. and Miller, J. F. A. P. (1973). *J. exp. Med.*, **138**, 143–162.
195. Zatz, M. M. and Lance, E. M. (1971). *J. exp. Med.*, **134**, 224–241.
196. Hall, J. G. and Morris, B. (1963). *Quart. J. exp. Physiol.*, **48**, 235–247.
197. Cunningham, A. J., Smith, J. B. and Mercer, E. H. (1966). *J. exp. Med.*, **124**, 701–714.
198. Delorme, E. J., Hodgett, J., Hall, J. G. and Alexander, P. (1969). *Proc. Roy. Soc. Lond. B*, **174**, 229–236.
199. Sprent, J. and Miller, J. F. A. P. (1972). *Cell. Immunol.*, **3**, 385–404.
200. Sprent, J. and Miller, J. F. A. P. (1972). *Cell. Immunol.*, **3**, 213–230.
201. Basten, A., Warner, N. L. and Mandel, T. (1972). *J. exp. Med.*, **135**, 627–642.
202. Sprent, J. (1976). *Cell. Immunol.*, **21**, 278–302.
203. Jacobsson, H. and Blomgren, H. (1973). *Clin. exp. Immunol.*, **13**, 439–453.
204. Sprent, J. and Miller, J. F. A. P. (1976). *Cell Immunol.*, **21**, 303–313.
205. Sprent, J. and Miller, J. F. A. P. (1976). *Cell. Immunol.*, **21**, 314–326.
206. Griscelli, C., Vassalli, P. and McCluskey, R. J. (1969). *J. exp. Med.*, **130**, 1427–1451.
207. Hall, J. G. and Smith, M. E. (1970). *Nature*, **226**, 262–263.
208. Hall, J. G., Parry, D. M. and Smith, M. E. (1972). *Cell Tissue Kinet.*, **5**, 269–281.
209. Moore, A. R. and Hall, J. G. (1972). *Nature*, **239**, 161–162.
210. Moore, A. R. and Hall, J. G. (1973). *Cell. Immunol.*, **8**, 112–119.
211. Halstead, T. E. and Hall, J. G. (1972). *Transplantation*, **14**, 339–346.
212. Sprent, J. and Lefkovits, I. (1976). *J. exp. Med.*, **143**, 1289–1298.

Chapter 5

The Role of Thymic Hormones in Regulation of the Lymphoid System

N. Trainin
M. Small
A. I. Kook

Abbreviations

ALS: anti-lymphocytic serum
BSA: bovine serum albumin
CFU-S: colony forming units
Con A: concanavalin A
GvH: graft-versus-host
LPS: *E. coli* lipopolysaccharide
MIF: macrophage inhibiting factor
MLC: mixed lymphocyte culture

PHA: phytohaemagglutinin
PPD: purified protein derivative of tuberculin
PVP: polyvinylpyrrolidone
SRBC: sheep red blood cell
STH: somatotropic growth hormone
TF: J.-F. Bach's thymus factor
THF: N. Trainin's thymus factor

1. INTRODUCTORY REMARKS

During recent years an increasing body of evidence has accumulated suggesting that the functions of the thymus are in part mediated by a hormone-like mechanism. Experiments performed in mammals at different levels of the phylogenetic scale have shown that humoral products derived from this organ are capable of influencing development and activity of the lymphoid system. In the scope of this chapter we intend to place some more recent developments concerning thymic humoral function against a background of experiments which investigate the relation between the thymus and the endocrine system. For more detailed results obtained in our and other laboratories supporting the concept of a thymic hormone, the reader is referred to publications of symposia held during the past year (*Ann. N.Y. Acad. Sci.*, Vol. 249, 1975, and *The Biological Activity of Thymic Hormones*, ed. D. W. van Bekkum, Kooyker Scientific Publications, Rotterdam, 1975).

2. HORMONAL CONTROL OF THE LYMPHOID SYSTEM

Early investigations presented by Dougherty[1,2] and the work of Comsa[3] suggested that hormones of the anterior pituitary, thyroid, adrenals and gonads can exert regulatory functions on the immune system. Since such hormonal regulation could theoretically affect establishment of the lymphoid organs, could regulate the functions of such organs to maintain homeostasis under changing demands, or could act on the cellular level to trigger individual lymphoid cells into activity, evaluation in terms of immune responsiveness only begins to elucidate these complexities. Without being exhaustive, the following examples indicate some of the diversified effects of hormones on different levels of the lymphoid system.

A lymphoid–hypophyseal axis has been investigated in some detail and the studies point to a role of growth hormone (STH) rather than other pituitary hormones in this relationship. In Snell-Bagg dwarf mice, hereditary hypopituitary function was shown to be accompanied by underdeveloped lymphoid structures and decreased T-cell reactivity.[4] Normalization of lymphoid histology and enhanced immunoreactivity followed administration of STH to such mice.[5,6] In normal animals, the deficiency syndrome could be reproduced by hypophysectomy[7] or by injection of anti-STH serum.[8] Furthermore, STH has been found to bring about recovery of lymphoid tissue after stress-induced involution[9,10] and to affect metabolism[7] and immune reactivity[11] of normal thymocytes. This last effect appeared restricted to immature thymocytes and was not detected in cortisone-resistant thymocytes, spleen cells or lymph node cells. More recently, this increase of thymocyte reactivity was shown to occur after direct *in vitro* contact with STH and with physiological concentrations of the hormone. Moreover, the development of T cells in newborn mice appeared to be influenced by STH treatment (S. Arrenbrecht, unpublished results).

Lytic effects on the lymphoid system have been seen after administration of adrenal glucocorticoids. Cortisone-induced wasting of newborn mice[12] and alterations in lymphocyte migration[13] have been described, and cytolysis of a certain sector of the lymphoid system was exerted *in vivo* or by direct contact *in* vitro.[2,14–19] Mature T cells appeared to be less sensitive to the lytic effects of cortisone than were immature T cells.[20,21] On the other hand, some suggestion of stimulatory effects of corticoids on lymphoid cells have also been reported to occur at lower concentrations,[22] raising the question of which effects occur under physiologic conditions.

Although investigated to a lesser extent, several other hormones have been found to contribute in different degree to the homeostasis of the lymphoid system. Thyroxin has been found to influence lymphocyte development[5,23] and thymocyte migratory behaviour[24] and also to stimulate epithelial cords in the newborn rat thymus.[25] This last activity was also exerted by estrogens which, in addition, were shown to affect proliferation of lymphoid tissues.[26] Insulin was also found to stimulate thymocyte proliferation,[7] while surgical removal of the rat pancreas was followed by reduction in thymus weight.[27] Parathyroid hormone also stimulated proliferation of rat thymocytes *in vitro*[28] by a pathway which appeared to depend on the availability of calcium.[29]

Receptors for many hormones have been found on lymphocytes[30–32] and cyclic AMP has been implicated as a mediator of hormonal influences on the lymphoid system.[33–35]

3. EVIDENCE FOR THE ENDOCRINE NATURE OF THYMIC FUNCTION

Indications of interaction between the endocrine and the lymphoid systems include effects exerted by the thymus on other endocrine organs. Removal of the thymus early after birth is followed by dramatic changes in the hypophysis,[8,36,37] such as degranulation of STH producing cells. Effects of the thymus on the adrenal cortex and the thyroid have been reported,[38,39] but since these experiments were performed in nude mice the decreased thyroxin levels in the blood and enlargement of the zona reticularis of the adrenal cortex may not necessarily be the exclusive consequence of thymic deprivation. Derangement of gonadal functions seems to follow removal of the thymus. Neonatal thymectomy of females sometimes resulted in lack of ovarian follicles and corpora lutea and hyperplasia of interstitial cells.[40–42] More recent work confirms an influence of the thymus on maturation of female sexual function in conventional and germ-free mice.[43] This was reflected in neonatally thymectomized mice by a delay of puberty and slower maturation of the ovaries and uterus. Nude mice, which exhibited similar defects, were not restored by injection of thymocytes while immune reactivity was increased by such treatment.

Classical evidence of endocrine activity requires loss of function by removal of the organ under question and restoration by the hormones produced therein. Surgical removal of the thymus results in varied signs of immunodeficiency (Section 4), which can be reversed by cell-free substances derived from the thymus, either diffusing through Millipore chambers (Section 5.1), or administered in the form of thymic extracts (Sections 5.2 and 6). Another method of repair was observed during pregnancy when immune functions of thymectomized mothers were restored by thymus factors of the embryos diffusing through the placenta.[44] A circulating thymic factor in the serum of mice and men has been described.[45] This factor disappears soon after thymectomy and is reported to decrease with ageing. Further analysis of the mechanism of action of this serum factor and of thymic factors extracted from thymic tissue (see Section 8) reveals that the biochemical mechanism involved is common to the pathway followed by many hormonal stimuli.

4. DEFECTS IN THE IMMUNE SYSTEM FOLLOWING THYMUS DEPRIVATION

It has been demonstrated repeatedly (reviewed in Reference 46) in various species that the capacity of mammalian organisms to react in response to many types of immune challenge is dependent upon activity of the thymus. Neonatal thymectomy or adult thymectomy combined with irradiation are followed by characteristic depletion of particular areas of lymph node and spleen tissue, by decreased numbers of peripheral blood lymphocytes, and by reduction in the numbers of circulating lymphocytes obtained by cannulation of the thoracic duct. These structural defects are accompanied by impairment in lymphocyte activity which could be expressed in the whole gamut of immune functions. A wasting syndrome often develops following neonatal thymectomy of rodents. *Vis-à-vis* homograft responses, removal of the mouse thymus at birth results in a marked impairment of the ability to reject skin grafts and tumours of allogeneic or xenogeneic origin, and to mount delayed hypersensitivity reactions. Most antibody responses are decreased following neonatal thymectomy because of loss of activity of thymus-dependent helper cells while the actual antibody forming cells remain apparently unimpaired.

In addition to these effects in the total animal, the defects following neonatal thymectomy have also been traced *vis-à-vis* populations of lymphocytes which are distributed in various compartments of the lymphoid system. Thymectomized mice are deficient in cells distinguished by surface entigens, such as θ and Ly (References 47–49, and P. C. L. Beverley, herein). This is reflected in reduced *in vitro* reactivity of these populations in mixed lymphocyte cultures,[50] and in equivalent populations of rats in decreased responsiveness to phytohaemagglutinin (PHA).[51] The ability of spleen, thoracic duct, or lymph node cells to induce a graft-versus-host type reaction after inoculation into appropriate recipients is also impaired after neonatal thymectomy of mice or rats.[46]

Similar defects were found when adult thymectomized mice were given lethal irradiation to deplete the existing lymphocyte complement and repopulated with haemopoietic stem cells which differentiated in the absence of the thymus.[46,52–56] These findings have been reproduced in a variety of species and lead to the solidly based conclusion that the thymus is essential for establishment of lymphoid structure and function during early life. In addition, a continuing role of the thymus during adulthood has also been demonstrated. Although thymectomy of adult rodents did not always result in the same generally defective homograft response seen after neonatal thymectomy,[57] removal of the thymus from young adult mice was sometimes followed by prolonged survival of allogeneic skin or tumour grafts, of the same H-2 type,[58] or when severe antigenic challenge depleted the resources of the host.[59] At extended time intervals after removal of the thymus, conventional antigenic challenges also elicited reduced responses,[60–62] possibly reflecting turnover of the lymphocytes involved.

Recent investigations have revealed the existence of a short-lived population of thymus-derived lymphocytes, and the continuing function of the thymus appears essential either for the production or for the activation of these cells. When differential distribution of Ly antigens was studied, one of the subpopulations of spleen cells was reduced by 50% within three weeks.[63] One function of such short-lived T cells appears to involve regulation of the activity of other T cell subpopulations. Loss of suppressor T cells soon after adult thymectomy has been detected by increased responses to certain antigens such as polyvinylpyrrolidone (PVP),[64,65] and by increased cytotoxicity[66] and mixed lymphocyte culture (MLC) reactivity.[67] At somewhat longer intervals after adult thymectomy, a decrease, at least in cytotoxicity[68] has been found. A loss has also been reported in a population of T cells which appears to be involved in enhanced syngeneic tumour growth.[69] In other experiments, a progressive decrease in those spleen T-cells which form rosettes with sheep erythrocytes (SRBC) began within a week after adult thymectomy. This function was found to be dependent upon the level of a circulating serum factor (TF) of thymic origin.[45] This thymic hormone which is lacking after adult thymectomy will be described more fully below (Section 7.3).

An interesting new development indicates that the thymus also plays a role in the regulation of haemopoiesis.[70–73] This was measured in terms of colony-forming units (CFU-S), which were found to be reduced after neonatal thymectomy. Thus, when lethally irradiated mice were injected with bone marrow cells from neonatally thymectomized syngeneic donors, the number of haemopoietic colonies appearing in the spleen was significantly diminished in comparison to the number resulting from administration of normal bone marrow. In addition, bone marrow stem cells from such thymectomized mice were impaired in the capacity to exert a radio-protective effect upon injection into irradiated recipients. This defect was further characterized as a proliferative impairment as it was found associated with a low rate of stem cell cycling. Thus, in the absence of the thymus, these cells were insensitive to the *in vivo* cytotoxic effect of chlorambucil and to *in vitro* killing by tritiated thymidine in high doses.[73]

5. WAYS OF REPAIR OF THE IMMUNE RESPONSE AFTER THYMUS DEPRIVATION

5.1 Effects of diffusion chambers containing thymic tissue[74,75]

Following implantation of thymic tissue subcutaneously or under the kidney capsule, structural and functional immunological defects usually apparent in thymus deprived mice (neonatally thymectomized or adult thymectomized-irradiated) are either prevented or reversed. In order to determine whether this restoration is of a hormonal nature, thymus grafts were enclosed in cell-impermeable diffusion chambers and repair of structure and function was studied. It was observed that those changes characteristic of thymus deprivation did not occur with the intensity and degree manifested in thymectomized controls. Thus, body weights increased steadily, though remaining below normal, and most of the stigmata associated with the wasting syndrome were absent. The decline in peripheral blood lymphocyte levels was prevented or arrested and the depleted areas of the spleen, lymph nodes, and Peyer's patches were partially repopulated. T cell helper function in the SRBC response was regained as measured by serum haemolysins and haemagglutinins and by the number of plaque-forming cells in the spleen. The level of anti-human γ-globulin and anti-BSA (bovine serum albumin) was restored in other experiments. Cell-mediated responses restored by thymus tissue in diffusion chambers include: rejection of skin allografts, ability to resist xenogeneic tumour grafts, delayed hypersensitivity responses to SRBC, BSA and tuberculin and the capacity of lymphoid cells to induce a graft-versus-host (GvH) response. Responses to virus were restored as measured by the prevention of polyoma oncogenesis. In addition, the susceptibility of mice to the fatal effects of lymphocytic choriomeningitis virus infection which was diminished by thymectomy returned when such mice were implanted with thymus tissue in diffusion chambers. These changes were brought about by normal syngeneic thymus tissue of young, newborn or embryonic animals, by allogeneic thymus or by a non-lymphoid functional thymoma. No restorative effects were apparent when control tissues from spleen or lymph node were placed within the diffusion chambers. This fact lends strength to the claim that the diffusion chambers were indeed impermeable to passage of lymphocytes, a point which has been supported by other experimental evidence. Most of these experiments were performed in mice, and similar repair was apparent when thymic tissues in diffusion chambers were tested in rats, golden hamsters and rabbits.

5.3 Thymic humoral factors[74–78]

The results obtained with Millipore chamber experiments indicate that the thymus produces a non-cellular agent which can diffuse out from cell-tight chambers and prevent most of the deficits that follow thymectomy. Further analysis of this problem was performed using cell-free extracts prepared from thymic tissue and injected into either intact or thymectomized animals.

Repair and stimulation of lymphoid structure and increase in the lymphocyte complement of the peripheral blood were considered indicative of activity in early investigations when crude preparations of thymic tissue were injected. Enlargement of spleen lymphoid follicles, increase in the number of mitotic figures in the depleted areas of spleens of thymectomized mice, increased lymph node weight and incorporation of tritiated thymidine into lymph nodes were noted. These extracts were also reported to prevent the weight loss and death which are characteristic of the wasting syndrome that can follow neonatal thymectomy.

As the role of the thymus in immune responses was gradually clarified, emphasis shifted from evaluation of thymus preparations by morphological criteria to functional assay. Since thymus derived cells are particularly involved in cell-mediated reactions, the influence of thymus extracts was assayed in such responses. Functions such as delayed hypersensitivity to BSA, rejection of allogeneic first and second set skin grafts and rejection of allogeneic tumours were restored by repeated injections of crude thymic extracts to neonatally thymectomized mice. Injections of thymic extracts also restored the capacity of spleen cells from such mice to induce a graft-versus-host response when transferred to appropriate hosts. Both the splenomegalic response (Simonsen assay) and the lethal runting syndrome were thus repaired.

These results indicate that some stage in the homograft rejection mechanism is dependent on a thymic humoral factor. Regarding the T cell helper function in antibody responses to SRBC, the thymic preparation from one laboratory was found to cause a slight increase in the haemolysin response of neonatally thymectomized mice. A similar effect was exerted in neonatal animals by one fraction extracted from thymus tissue although an inhibitory effect followed administration of a second electrophoretic fraction of this material.

The early preparations from various species (mice, rats, rabbits, sheep and calves) were obtained according to several different procedures as followed in various laboratories. The results indicated that an agent found in the different preparations was controlling some property involved in the reactivity of T cells. Since immunocompetence was thus conferred upon animals which were lacking such reactivity, it seemed possible that the acquisition of competence by the lymphocytes was associated with development to maturity of cells which at some stage are dependent on a thymic humoral influence. Subsequent implementation of *in vitro* assays facilitated further investigation of this hypothesis. As will be discussed in Sections 6 and 8, such techniques have permitted elucidation of the types of cell affected, the nature of the cellular changes brought about under control of a thymic hormone and the mechanism by which such changes occurred. In addition, the development of additional assays has facilitated testing of a range of responses known to require T cell activity and all have been found to be under the influence of thymic humoral factors.

6. TARGET CELLS FOR THYMIC HORMONES AND TYPES OF IMMUNE REACTIVITY CONFERRED

As described above, administration of thymic extracts had repaired certain defects of the lymphoid system and partially restored several immunological functions damaged by thymectomy. In order to clarify the nature of the relationship between the thymus factor and the cells which are ultimately reactive in an immune response, it was necessary to investigate the possibility of a direct interaction between thymus extracts and suspensions of lymphocytes *in vitro*. We thus designed experiments[79] to test the competence induced in spleen cells from neonatally thymectomized mice by direct contact with THF, the thymic extract prepared in our own laboratory. Defined requirements of *in vitro* techniques permitted the use of syngeneic extracts discarding any antigenic stimulation which xenogeneic preparations might carry. Immunocompetence was evaluated by determining the ability of lymphocytes to induce a graft-versus-host response which was measured according to an *in vitro* procedure[80] shown to reflect an immune process and to depend upon the presence of mature T-cells. While a positive response is consistently induced by spleen cells from intact parental mice, inocula containing the same number of spleen cells from neonatally thymectomized donors fail to induce a reaction of splenomegaly. Syngeneic thymic extract, added for the duration of the assay, conferred reactivity upon the deficient cell populations. Control extracts of syngeneic lymph node or spleen prepared and tested in parallel had no such effect. The activity of thymic extracts was thus shown to be exerted directly upon isolated spleen cells without requiring anatomical integrity of the lymphoid tissue. The closed system permitted the conclusions that extrasplenic processing of thymic hormones was unnecessary and that migration of cells from other organs was not required for the attainment of competence in a GvH reaction. In further experiments the capacity of thymic preparations to confer immunological reactivity was shown to be exerted prior to antigenic stimulation of the cells.[79] Contact between the lymphoid cells and thymic extracts for one hour was sufficient to enable reactivity of these lymphocytes suggesting a rapid change in cells already present in the spleen of neonatally thymectomized mice.

Using the same *in vitro* GvH assay as a measure of immunocompetence, target cells were found within the bone marrow population of intact mice.[81] However, in addition to contact of these cells with thymic extract, a complementary process occurring in the environment of the peripheral lymphoid organs appeared to be essential in the differentiation of these cells to immune reactivity. This was demonstrated by the fact that neither bone marrow cells assayed directly after incubation in THF nor bone marrow cells exposed to thymus extract and cultured with non-responding thymic tissue were reactive. In contrast, aliquots of the same bone marrow cells exposed to THF and cultured *in vitro* with non-responding spleen tissue or injected into non-responding mice and recovered from the spleen exhibited immunocompetence when assayed for GvH activity.

Target cells for thymic hormones could theoretically be either prethymic cells and presumably devoid of θ antigen or a post-thymic element already possessing the θ antigen and having passed a certain degree of thymus processing but still lacking complete immunological reactivity. To clarify this point lymphocytes were treated with THF after exposure to anti-θ serum and complement, and submitted to test in the GvH assay.[82] While spleen cells from neonatally thymectomized mice exposed to calf thymus extract acquired immunoreactivity in the *in vitro* GvH assay, acquisition of competence did not occur in those aliquots of cells pretreated with anti-θ serum prior to THF. When spleen cells from intact mice were treated with anti-θ serum no restorative effect of THF was seen in terms of GvH reactivity. In contrast, when intact mice were injected with low doses of anti-lymphocyte serum (ALS) presumed to decrease the circulating mature lymphocyte compartment without depleting immature T cells, the impaired GvH reactivity of the spleen cells was restored by THF treatment. No increase was detected in the number of θ-positive cells of spleen, lymph node or bone marrow after incubation in THF, or in lymph node cells after injection of THF to neonatally thymectomized mice. With the reservations attached to the use of anti-θ serum of non-congenic origin, these experiments raised the possibility that the target cell for THF activity is already committed to the T cell pathway and characterized by the presence of θ antigen.

Conflicting results have been obtained with other thymic factors. Using an assay based on the θ sensitivity of rosette forming cells, TF, the thymic factor extracted from serum by the group of J.-F. Bach, was found to induce the appearance or increase the expression of θ antigen in cells from bone marrow, spleen of adult thymectomized mice, or spleen of nude mice.[45,83] Using a cytotoxicity assay after cell fractionation, thymopoietin a thymic factor isolated by the group of Gideon Goldstein was found to induce appearance of both θ and TL antigens,[84,85] and θ was similarly induced[86,87] by treatment with thymosin, the thymic preparation of Allan Goldstein's laboratory.

The sensitivity of T-rosette-forming cells to azathioprine is the basis of the standard assay for TF activity. With this test, target cells for TF appeared in the spleen within a week following adult thymectomy. Using higher concentrations of TF target cells in this assay have also been detected in normal bone marrow and in spleen of nude mice.[45,83]

As a result of treatment with thymic factors, lymphocytes acquire properties that enable them to react in capacities characteristic of mature T cells. One such reactivity is the mitogenic response to lectins Con A and PHA.[88] When spleen cells or thymocytes from intact mice were incubated for 24 hours in medium containing THF and then stimulated by PHA or Con A, a significant increase in the mitogenic reactivity of both types of cells was observed. In contrast, THF did not modify the response of spleen cells to LPS (*E. coli* lipopolysaccharide), a B cell mitogen. It has been suggested that T cell maturation is reflected by the ratio of the mitogenic response to PHA and Con A and the mature T cells have a relatively higher response to PHA.[89] By testing increasing concentrations of THF it was possible to establish a differential effect upon target cells since low doses of THF were relatively more effective on Con A reactivity while higher doses increased the PHA response. These results suggest that progressive maturation by THF increases the number of cells responding to PHA by permitting these cells to reach higher levels of maturation. When spleen cells from nude mice were tested in parallel no effect of THF on the T mitogen responses was detected (V. Rotter, unpublished results). While pretreatment of normal lymphocytes with THF produced the increased mitogenic response described, addition of the same mitogens in the presence of THF resulted in decreased responses to both PHA and Con A. On the other hand, responsiveness to T lectins in nude mice was reported both with thymosin[90] and TF.[83] An *in vitro* effect of thymosin on bone marrow cells and an *in vivo* effect on lymph node cells has also been reported *vis-à-vis* T mitogen responsiveness.[90] Also, thymopoietin was found to increase Con A reactivity in spleen cells and in fractionated bone marrow cells.[85]

The mixed lymphocyte reaction (MLC) combines the advantages of the mitogen response and the GvH assay since results are quantitative (as in the former) and reflect an immune response (as in the latter). The proliferation measured in MLC is considered to represent the recognition phase of the response against allogeneic tissue (Reference 91, and M. Nabholz and V. C. Miggiano, herein) and requires the reactivity of mature T cells. The one-way MLC response of parental spleen cells from neonatally thymectomized mice against F_1 spleen cells is reduced from that obtained with cells from normal mice. THF restored this response to normal values after 1 hour's preincubation. Moreover, THF treatment of parental spleen cells from intact donors also resulted in an increased response, though of smaller increment. The proliferative response of thymocytes was dramatically increased by incubation of the cells in THF before assay. In contrast, no effect of THF was evident when bone marrow derived spleen cells were assayed in the MLC response. Finally, the T cells found in the lymph nodes, and in the thymus after cortisone treatment, exhibited an initially strong response which was not further increased by THF treatment.[92] Thymosin preparations were also reported to increase MLC reactivity of thymocytes while depressing a similar response in spleen, lymph node or bone marrow cells.[90]

Since an effector function of cell-mediated immunity was not measured in this assay, it was of interest to determine the effect of THF on the cell-mediated lysis (CML) which can develop later in the course of MLC. Addition of THF to the cultures of spleen cells from adult thymectomized mice restored the CML activity of these cells to the level observed with normal cells. Cytolytic activity of normal spleen cells was also increased by addition of THF during the sensitization phase of the reaction.[93] These results indicate that the lymphocytes involved in CML are under the control of THF and that target cells for THF are also found in the spleens of adult thymectomized mice.

Recent work by Cantor and Boyse indicates that killer cells such as those active in CML, develop as a T cell line which is distinct from the T cell subpopulation involved in a helper capacity in the SRBC response.[63] Our data indicate that lymphocytes can also be activated by THF to manifest helper function in a response against SRBC.[94] When thymocytes were incubated in THF and transferred to thymectomized irradiated mice which received B cells as well, the number of direct plaques was double that seen with control thymocytes not reactivated with THF. A similar effect was observed when THF was injected into the thymectomized irradiated hosts together with injection of thymocytes. Thus, the helper function in an anti-SRBC IgM response appears to be under the control of THF and target cells for this activity are found within the thymocyte population. This effect of THF was shown to involve metabolic activation of the lymphocytes since the effect seen at 37 °C was not detected at 4 °C.

The question of which cells within the thymus are sensitive to THF was approached by velocity-sedimentation fractionation of the thymocyte population and assay of PHA reactivity with and without treatment of each fraction with THF. In parallel, the distribution of cortisone-resistant thymocytes was assayed and the fractions activated by THF appeared to be cortisone-sensitive.[95] Also, as described above, cortisone-resistant thymocytes were unaffected by THF in the MLC assay. When the survival of thymocytes was evaluated after exposure to hydrocortisone succinate *in vitro*, it was found that THF treatment of the thymocytes increased their resistance to the cytolytic effect of cortisone.[96] It thus appears that THF increased the cortisone-resistant compartment of the thymus by transforming cortisone-sensitive cells into resistant ones. In addition, TF has been reported to reduce the density of steroid receptors on thymocytes.[83]

The accumulated data indicate that as a result of processing by thymic hormones, developing T cells acquire characteristics which enable reactivity in a variety of T cell responses. From the range of functions affected, it seems that different sublines of T cells may undergo a similar differentiational event under the influence of thymic factors permitting reactivity of each subpopulation in the response for which it is equipped. Alternatively, the thymic hormone-induced change could occur in a common T cell precursor of the lymphocytes that are subsequently committed to parallel pathways.

Taken together, the results presented here are compatible with the hypothesis that THF is controlling a step in the maturation of T cells from an immature stage corresponding to Cantor's T_1[97] to a mature stage (perhaps T_2). A similar conclusion was reached by Stutman regarding the target cells for a thymic humoral factor released by syngeneic tissue in diffusion chambers.[98] Such target cells were tested in a GvH assay and were found in adult bone marrow and spleen and in the liver and spleen of newborn mice. These cells were characterized as spleen-seeking, non-recirculating, immunoincompetent, short-lived, resistant to short terms ALS treatment, rapidly dividing, relatively large and light in density and adherent to nylon wool. The target cells in Stutman's model were TL negative and lacking immunoglobulins. Activation of those target cells was abolished when the cells were treated with anti-θ serum and complement and thus they are characterized as post-thymic T cells. On the other hand, as mentioned previously, thymosin, TF and thymopoietin appeared to induce θ antigens in target lymphocytes. Finally, in the light of recent experiments by Roelants and Loor,[99,100] the possibility of prethymic θ-bearing lymphocytes cannot be ignored. Thus, the question of whether thymic hormones influence T cells already released from the thymus or function during processing of cells not previously exposed to the thymic environment cannot yet be unequivocally answered.

The properties conferred by THF upon lymphocytes also find an expression in the ability of T cells to react against tumour cells both *in vitro* and *in vivo*. Following early demonstrations that injections of THF restored the reactivity of neonatally thymectomized mice against allogeneic tumour grafts,[101] experiments were initiated to measure the influence of THF in the anti-tumour reactivity of lymphocytes from intact mice *vis-à-vis* syngeneic tumours. In order to distinguish between a possible effect of THF on the sensitization of lymphocytes against tumour cells or on the subsequent effector phase of the reaction, sensitization was carried out *in vitro* and reactivity measured in terms of cytotoxicity against identical target cells with and without addition of THF to either or both phases of the reaction.[102,103] Addition of THF during the process of sensitization of spleen cells or thymus cells resulted in an increased capacity of these lymphocytes to destroy tumour cells in a cytotoxic test found to depend upon T cells. In contrast, THF added only during the cytotoxicity assay (the effector phase) was devoid of influence.

When the *in vivo* behaviour of these tumour-sensitized cells was studied, tumour enhancement was encountered, but addition of THF to the cultures during sensitization reduced the degree of enhancement caused by these cells *in vivo*: Under conditions in which enhancement was bypassed, lymphocytes sensitized in the presence of THF exhibited increased anti-tumour reactivity *in vivo* as well as *in vitro* and significantly inhibited the growth of syngeneic tumours in two strains of mice; however, total abrogation of the tumours has not yet been achieved.[103] These experiments and others, carried out in our laboratory,[69,104] suggest that the anti-tumour reactivity of T cells which is under the control of THF may be obscured *in vivo* by other subpopulations of T cells with opposing activity. Elimination of the counteracting enhancing cells has resulted in greatly strengthened anti-tumour reactivity by those subpopulations which are capable of reacting against the tumour cells. The aim of future experiments is thus to find a means of functional elimination of the enhancing cells which arise during the natural development of a tumour to enable full reactivity of those lymphocytes which acquired anti-tumour properties under the influence of THF.

To the extent that the influence of thymosin[90] and TF[83] on anti-tumour reactivity have been investigated these thymic preparations manifested increased lymphocyte reactivity against tumours.

While the maturation of lymphocytes by THF leads to acquisition of reactivity *vis-à-vis* all the immune responses described above, a simultaneous loss of activity has also been found in lymphocytes treated with THF: loss of autoreactivity. The first suggestion that THF may play a role in control of sensitization against self antigens came from experiments in which either THF or serum from normal (but not thymectomized) mice added during the process of *in vitro* autosensitization of lymphocytes dramatically decreased their cytotoxicity against tumour cells of syngeneic origin.[105]

Further clarification of this role was achieved by using normal syngeneic tissues as the target of autoreactivity.[106,107] The *in vitro* graft-versus-host assay described previously for detecting allogeneic reactivity was modified for this purpose. Spleen cells sensitized on syngeneic (but not allogeneic) monolayers induced a GvH reaction against syngeneic tissue that was clearly inhibited by addition of THF to the sensitizing cultures. By testing the effect of THF on autosensitized spleen cells from neonatally thymectomized mice it was possible to evaluate the simultaneous action of THF on lymphoid cells participating in autoreactivity or alloreactivity: the same cell suspensions which lost anti-self reactivity after interaction with THF gained reactivity against allogeneic tissue after THF processing. This suggested that the process of T cell maturation involves acquisition of immunocompetence against allogeneic tissue and concomitant loss of autoreactivity as a result of THF directed maturation of lymphocytes.

The possibility that lymphocytes of thymectomized mice might undergo a spontaneous autosensitization process *in vivo* was submitted to test and the results indicated that spleen cells from mice deprived of the thymus had indeed developed reactivity against components of self *in vivo*. This autoreactivity was also prevented by exposure of the cells to THF. When the entire process of self-sensitization, administration of THF and testing of GvH reactivity was reproduced *in vivo*, a syngeneic GvH response followed injection of spleen cells from neonatally thymectomized mice to syngeneic recipients and was clearly reduced by injection of THF to the donor mice. In addition an inhibitory effect of THF on the proliferation of autosensitized lymphocytes in a syngeneic MLC suggests that the effect may occur in the recognition phase of the anti-self reactivity. Similar effects of THF were exerted on thymocytes while cortisone-resistant thymocytes did not manifest detectable autoreactivity. A possible role of suppressor cells in autoimmune processes and the effect of THF on such a population requires further clarification. At the present time, the picture emerging from these experiments suggests the involvement of immature lymphocytes in autoreactivity and the maturation of these cells to a non-self reactive state. Thymic involvement in the processes controlling autoreactivity appears to be at least partially mediated by THF.

Parallel findings have emerged from studies of the effect of TF on the formation of auto-rosettes.[83] Formation of autologous rosettes was found to be a property of thymocytes and of spleen cells from aged mice or especially from adult thymectomized animals. In this last case, the auto-rosette level could be normalized by TF treatment either *in vivo* or *in vitro*.

As indicated above, thymus deprivation is followed by a defect in the bone marrow stem cell population, which is expressed by reduced capacity of these cells to repopulate syngeneic lethally irradiated recipients.[70–72] This impairment was accompanied by a striking decrease in the proportion of cycling cells in the bone marrow manifested by the inability of high doses of tritiated thymidine to kill CFU-S in this population. Repair of humoral thymus function resulted in restoration of the proliferative capacity of bone marrow CFU-S from neonatally thymectomized mice.[73] This was achieved either by reimplantation of thymus tissue in dialysis bags or by *in vitro* THF treatment of the bone marrow cells. On the other hand, THF treatment had no influence on the bone marrow cycling of cells from normal mice. It is therefore probable that the thymus controls this activity of stem cells via a hormonal mechanism. Such cells were thus induced to synthesize DNA while, as will be described in Section 8, THF influences the acquisition of immune competence in later stages of T cell maturation by inhibition of cell division and induction of protein synthesis. Thus, it could be hypothesized that THF is a lymphopoietic hormone which controls the differentiation of the lymphoid cell line by inducing stem cells to divide and more mature T cells to reduce their rate of proliferative activity concomitant with the acquisition of immunocompetent properties. In this respect, THF may resemble erythropoietin, another acidic protein, which regulates the differentiation of the erythroid line by inducing stem cells to divide and by controlling protein synthesis as these cells mature.[108]

Finally, it should be mentioned that in a simultaneous coded test in which Trainin's THF, A. L. Goldstein's thymosin, and Bach's TF were compared, similar activities were revealed in terms of numerous immune and biochemical assays.[109]

7. ISOLATION AND CHEMICAL CHARACTERIZATION OF THYMIC HORMONES

7.1 THF[111,112]

During isolation and investigation of the nature of the thymic hormone, the biological activity of THF was assayed by its ability to induce competence in spleen cells from neonatally thymectomized mice to react in the *in vitro* GvH test and to augment their response to antigenic stimulation in the MLC assay.

The standard procedure used in our laboratory for preparation of THF is as follows. All steps are carried out in the cold. Fresh calf thymus is homogenized in 2 × volumes of 0·005 M Na phosphate buffer pH 7·4. The homogenate is centrifuged at 2500 rev/min for 20 min and the supernatant further centrifuged at 105·000*g* for 5 h. This supernatant is then dialysed against a 20 times larger volume of water for 60 h. THF passes through Union Carbide dialysis sacs No. 27/32 or 23/32, suggesting that the mol. wt. of the active material is roughly 6000 or less.[114] This dialysate which is lyophylized and diluted to a standard protein concentration contains all the humoral activity of the thymus, and is devoid of substances of mol. wt. above 10·000. Keeping strict measures of sterility during the preparation process, the preparation is endotoxin and pyrogen free and can be kept frozen for long periods of time, but loses its activity after 48 h at room temperature. The activity of the preparation is abolished by pronase treatment but not with DNase or RNase. It is with this preparation of THF that most of our experimental and clinical work has been done.

Further purification is achieved along the following steps: the lyophylized dialysate is fractionated by gel filtration through a G-10 Sephadex column, where the activity is

PLATE I

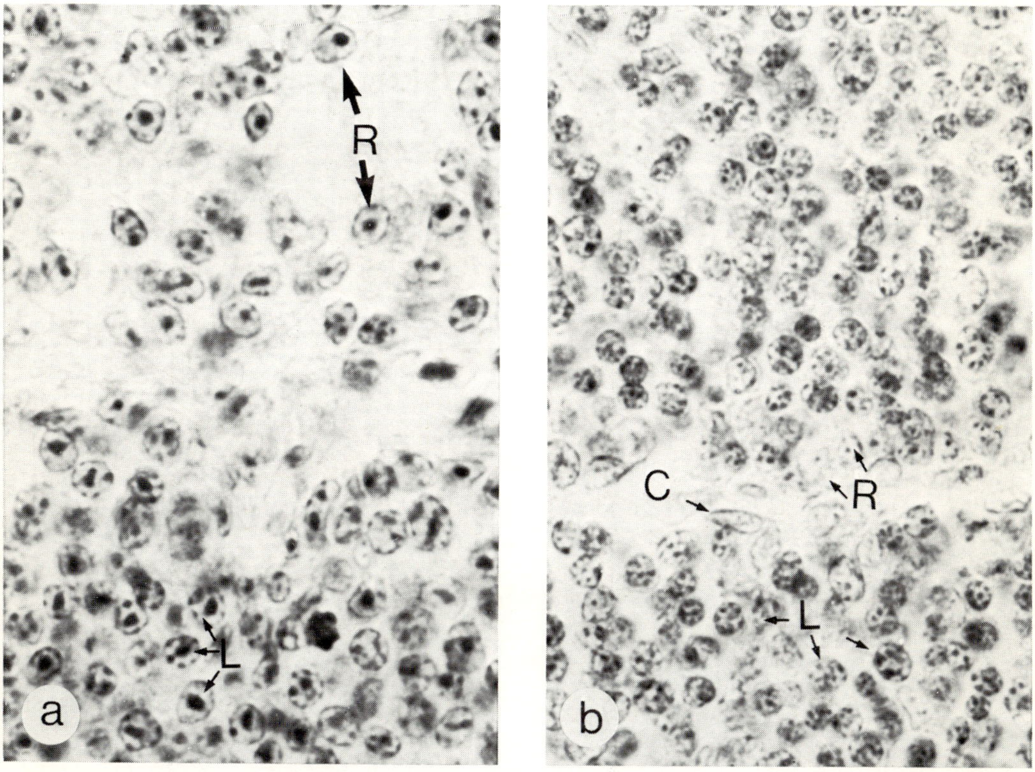

Figure 2. Cell recognition in quail and chick lymphoid tissue. Thymus of a 16-day quail (a) and 16-day chick (b) embryos stained according to the Feulgen-Rossenbeck procedure (R: reticular cells; C: connective cells; L: lymphocytes). In the quail the lymphocyte nuclei contain one large heterochromatic mass and several smaller ones attached to the nuclear membrane. In chick lymphocytes, several small chromocentres are dispersed in the nucleoplasm. Quail reticular and connective cell nuclei show usually one single centronuclear mass of heterochromatin while in the same cell types of the chick the nucleus contains a fine network of evenly distributed chromatin. The same disposition of the chromatin material is observed in the corresponding cell types of the differentiated bursa of Fabricius.

PLATE II

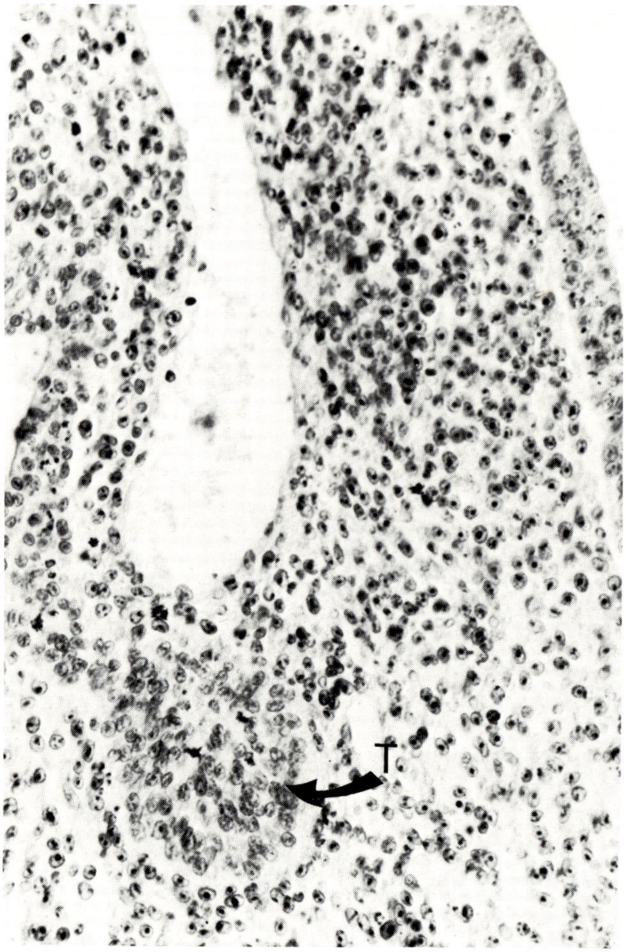

Figure 4. Transverse section in the branchial arch III of a 4, 5-day chick embryo which has received at 7-somite stage the isochronic graft of the rhombencephalon of a quail. The branchial arch mesenchyme is made up of quail cells originating from the grafted neural crest. The epithelia thymic primordium (T) of chick origin is surrounded by quail mesenchymal cells. Feulgen-Rossenbeck's staining. ×400

PLATE III

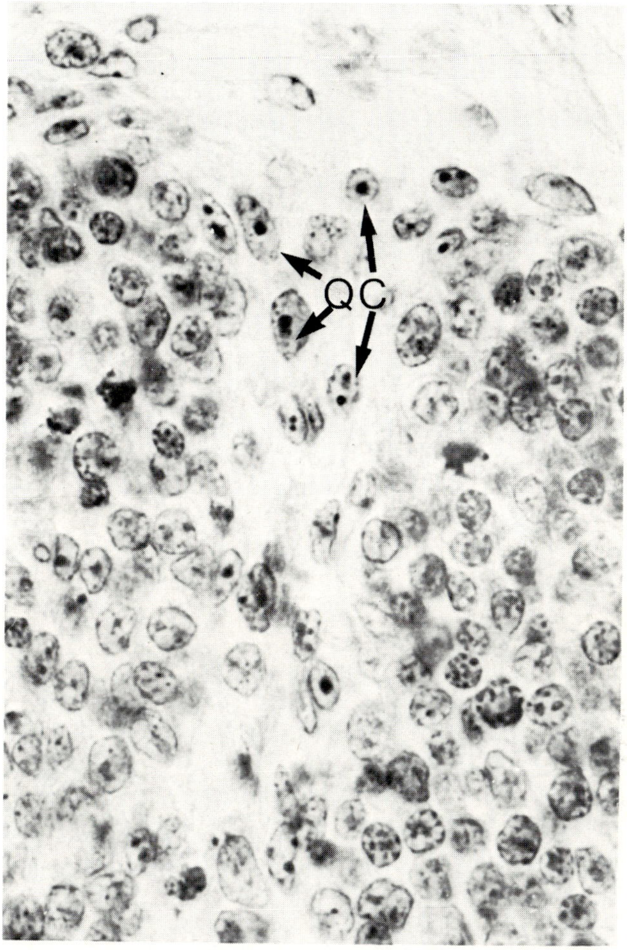

Figure 5. The same experiment as described in Figure 4 (Plate II). The thymus of the grafted embryo is observed at 13 days of incubation. Quail mesenchymal (QC) cells originating from the rhombencephalic neural crest penetrate into the interlobular space along the blood vessels. No quail cells are found in the cortical areas. ×1500

PLATE IV

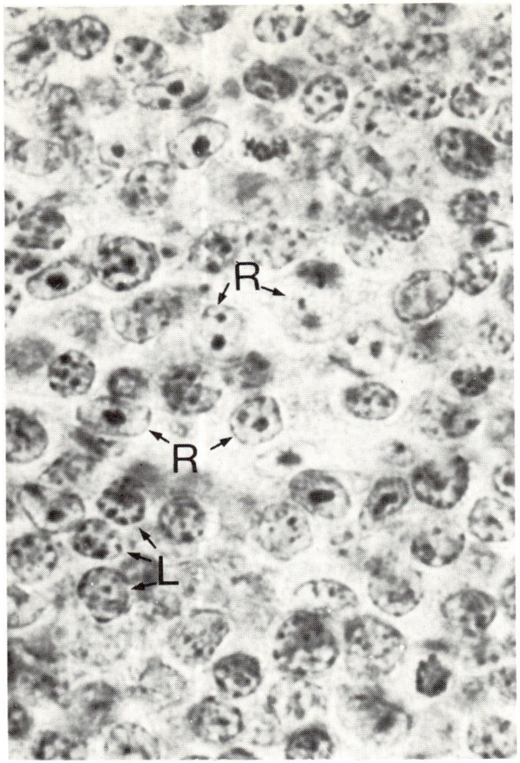

Figure 6. Chimerical thymus resulting from the graft of a thymic endodermal rudiment of a 16-somite quail embryo into the somatopleure of a 3-day-old chick. 13 days after grafting. The lymphocytes (L) belong to the chick species and the reticular cells (R) to quail. Feulgen-Rossenbeck's staining. ×1800

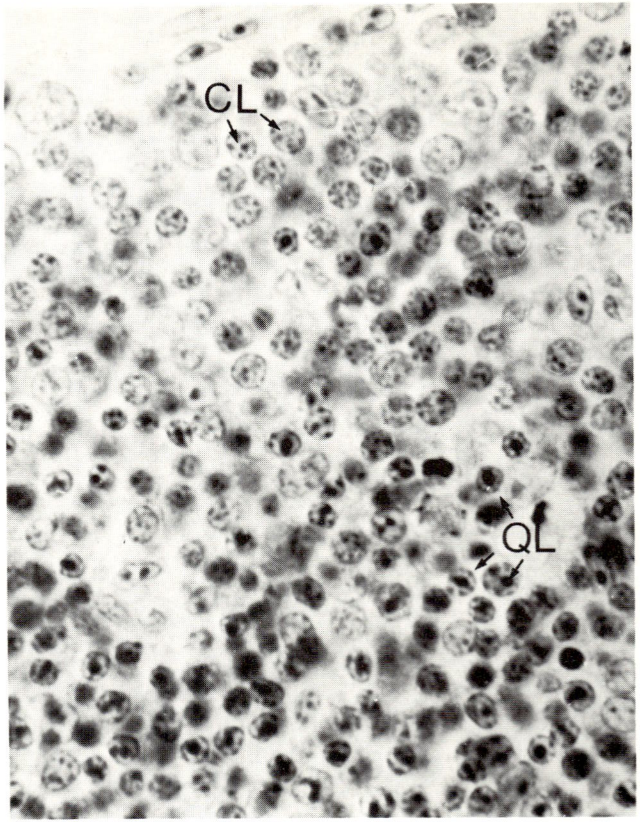

Figure 9. Thymus of a 8-day-old quail embryo grafted for 9 days into a chick host. In the external region of the cortex quail lymphocytes have been replaced by host ones (CL). Quail lymphocytes are present in the internal region of the cortex (QL). Feulgen-Rossenbeck. ×1000

PLATE V

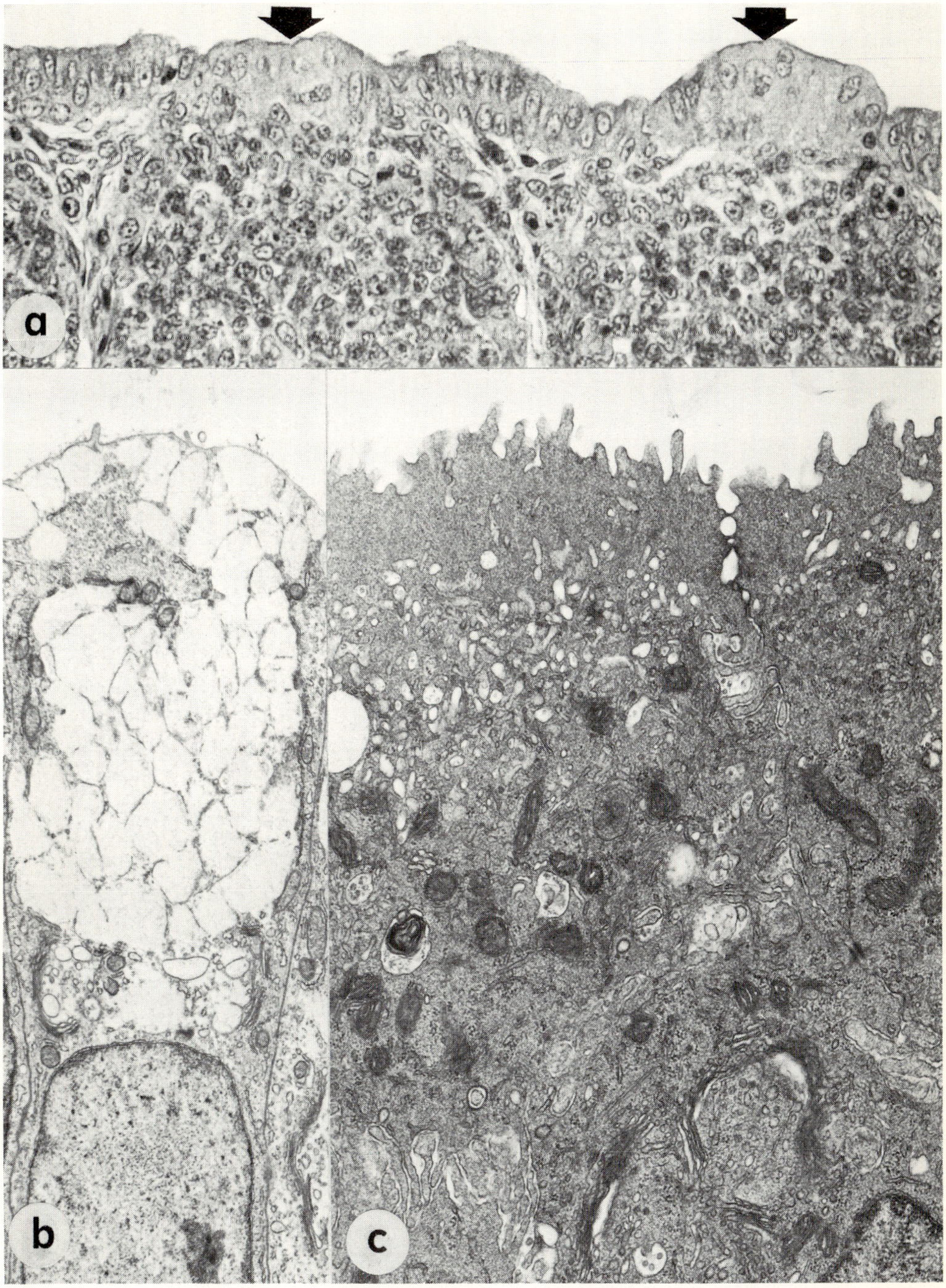

Figure 10. (a) Light micrograph of the bursa of Fabricius from a newly hatched chick. Follicle-associated epithelium (arrows) at the apex of the lymphoid follicle is clearly distinguishable from the intervening epithelium. Paraffin section, haematoxyline-eosine. ×625. (b) Chick bursa. An electron micrograph of an epithelial cell unassociated with a lymphoid follicle. Light, flocculent material occupies the area between nucleus (below) and bursal lumen (top). ×9000. (c) Chick bursa. An electron micrograph of a follicle-associated epithelium (FAE). Apical portions of these cells contain numerous vesicles and tubular profiles. Many irregular microvilli project into the bursal lumen (top). ×17 000 (from Bockman, D. E. and Cooper, M. D., *Amer. J. Anat.* (1973), **136**, 455–477; reproduced by permission of The Wistar Press)

PLATE VI

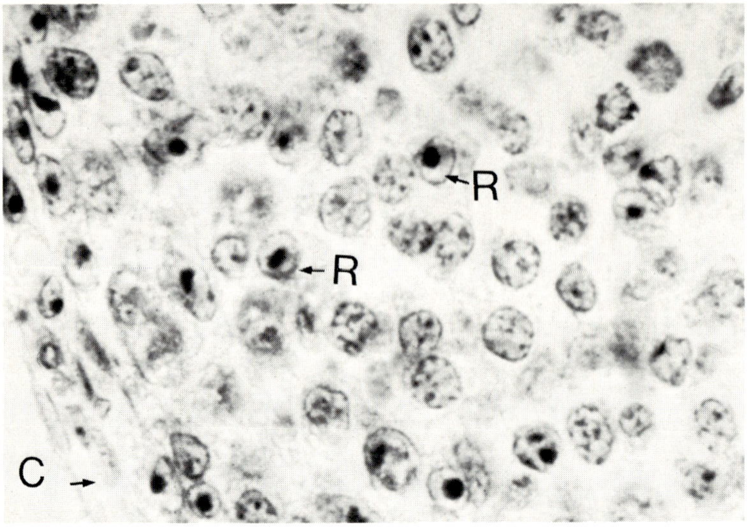

Figure 11. Quail bursal rudiment taken at 5 days of incubation from the donor and grafted for 13 days into a 3-day-old chick host. The perifollicular connective tissue (C) and the reticular cells (R) of the follicles belong to the quail donor while the lymphoid cells are of chick host type. Feulgen-Rossenbeck's staining. ×1500

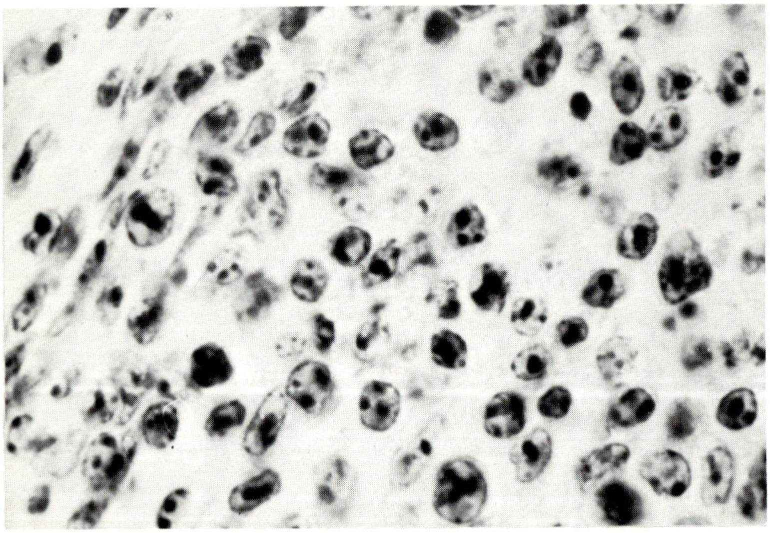

Figure 12. Quail bursal rudiment taken at 11 days of incubation and grafted for 8 days in a 3-day chick host. The lymphoid cells are of donor type. Feulgen-Rossenbeck's staining. ×1500

PLATE VII

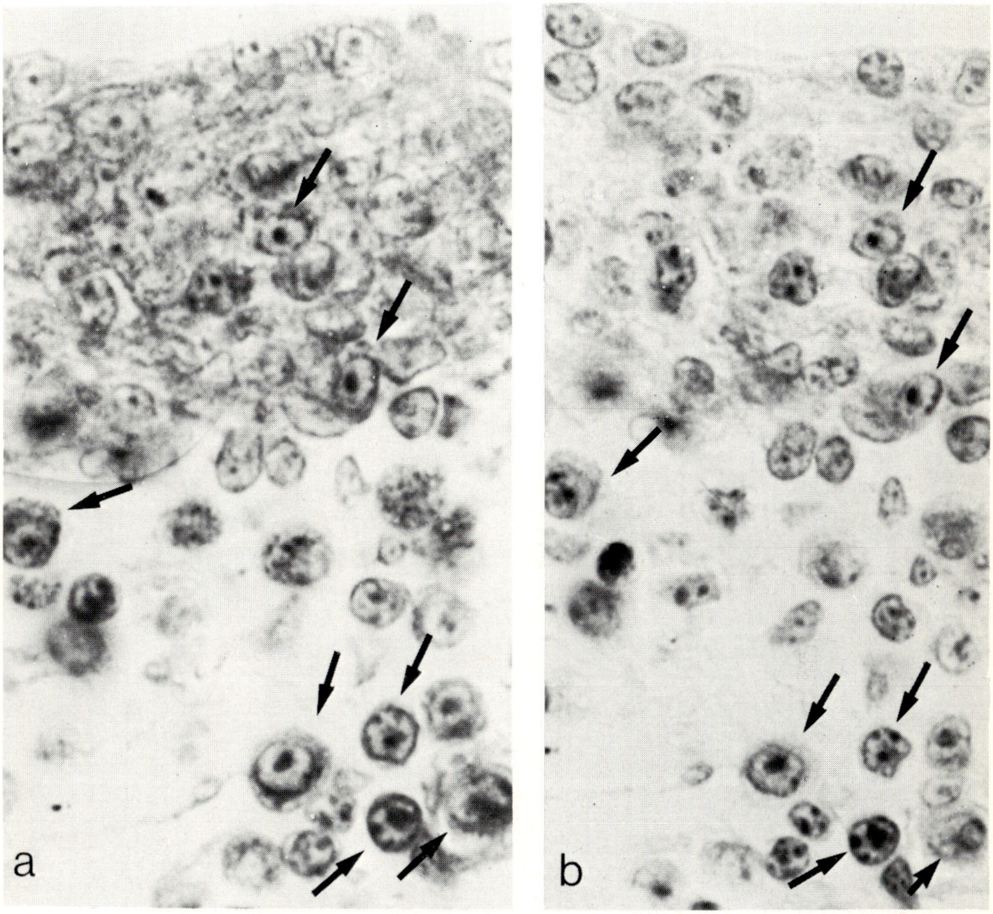

Figure 13. 7-day chick bursal rudiment grafted for 7 days into a quail embryo. (a) Pappenheim's staining showing basophilic cells in both epithelium and mesenchyme. (b) Feulgen-Rossenbeck's staining of the same section. The basophilic cells show the quail nuclear marker and thus are of host origin. ×1500

PLATE VIII

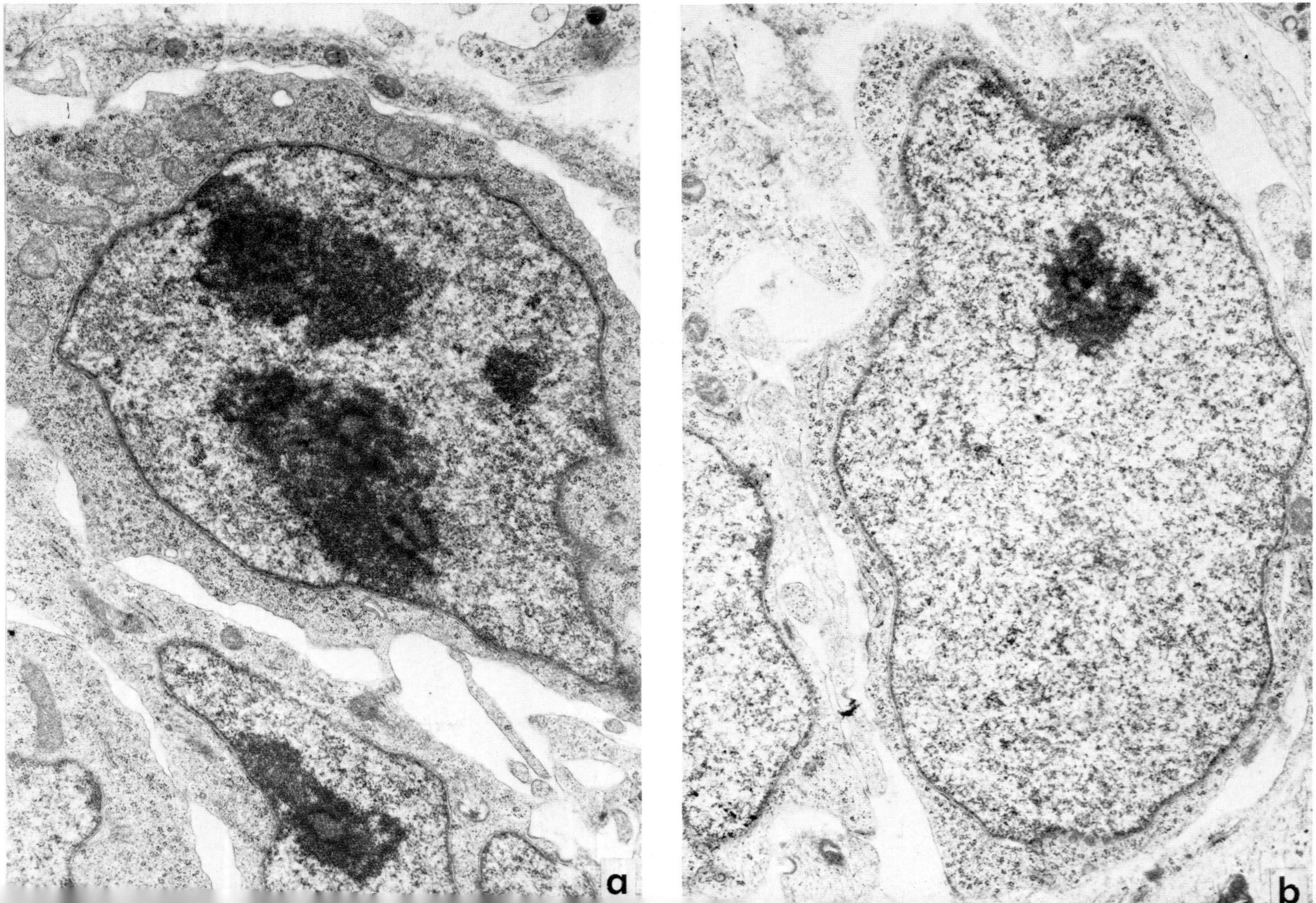

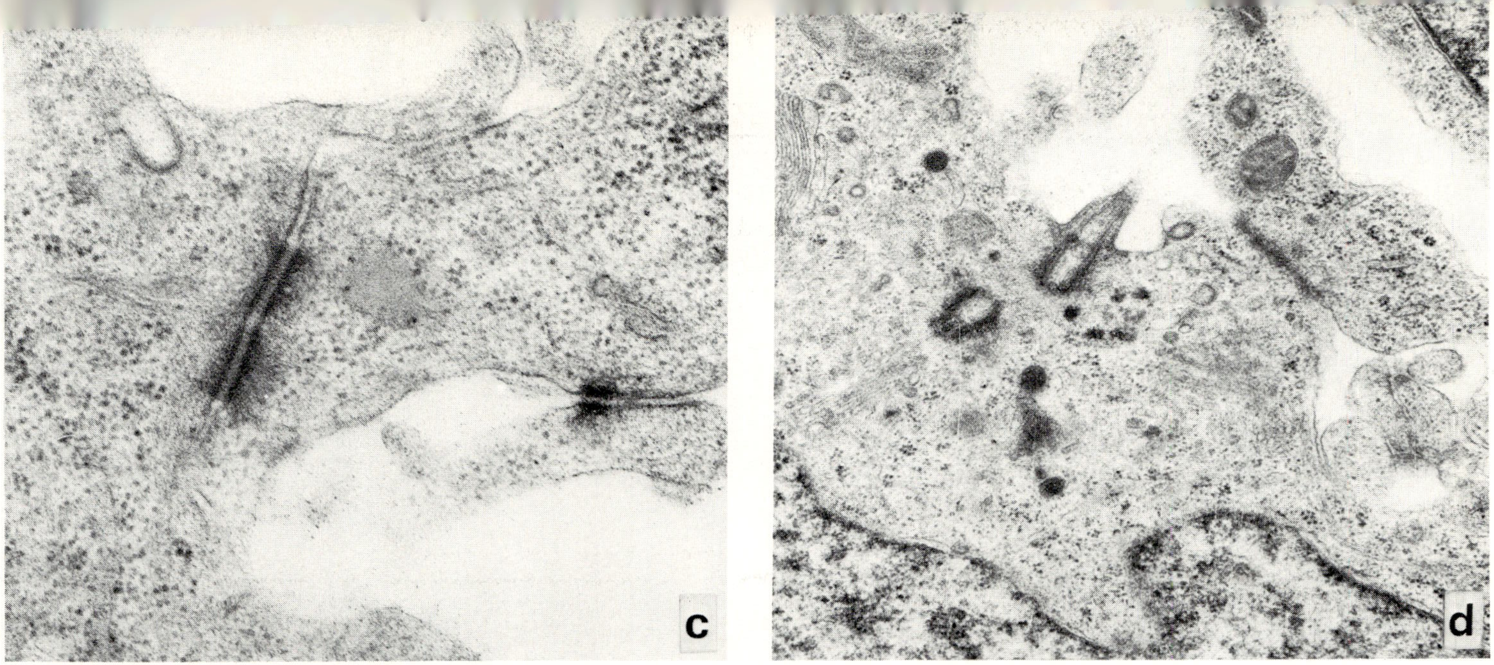

Figure 14. 6-day chick thymic rudiment grafted for 4 days into a 3-day quail embryo. Glutaraldehyde–osmium tetroxide fixation; uranyl acetate–lead citrate staining. (a) Quail lymphoid stem cell with a large nucleolar structure. ×11 600. (b) Chick epithelial cells: small nucleolus. ×11 6000. (c) Cell processes of differentiating reticular cells with desmosomes. ×40 000. (d) Ciliary structures in differentiating reticular cells. ×40 000

PLATE IX

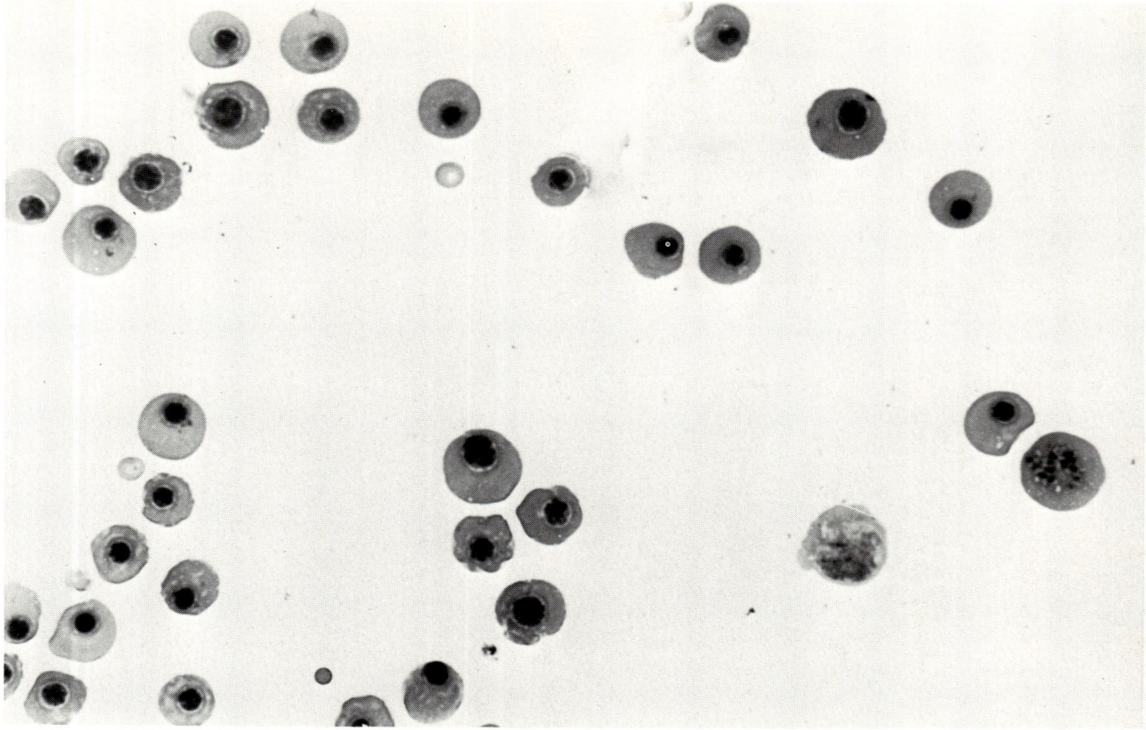

Figure 1. 'Cytocentrifuge' preparation of 10-day mouse yolk sac cells. Note the predominance of 'first generation' nucleated erythrocytes. McNeil's stain. ×700

PLATE X

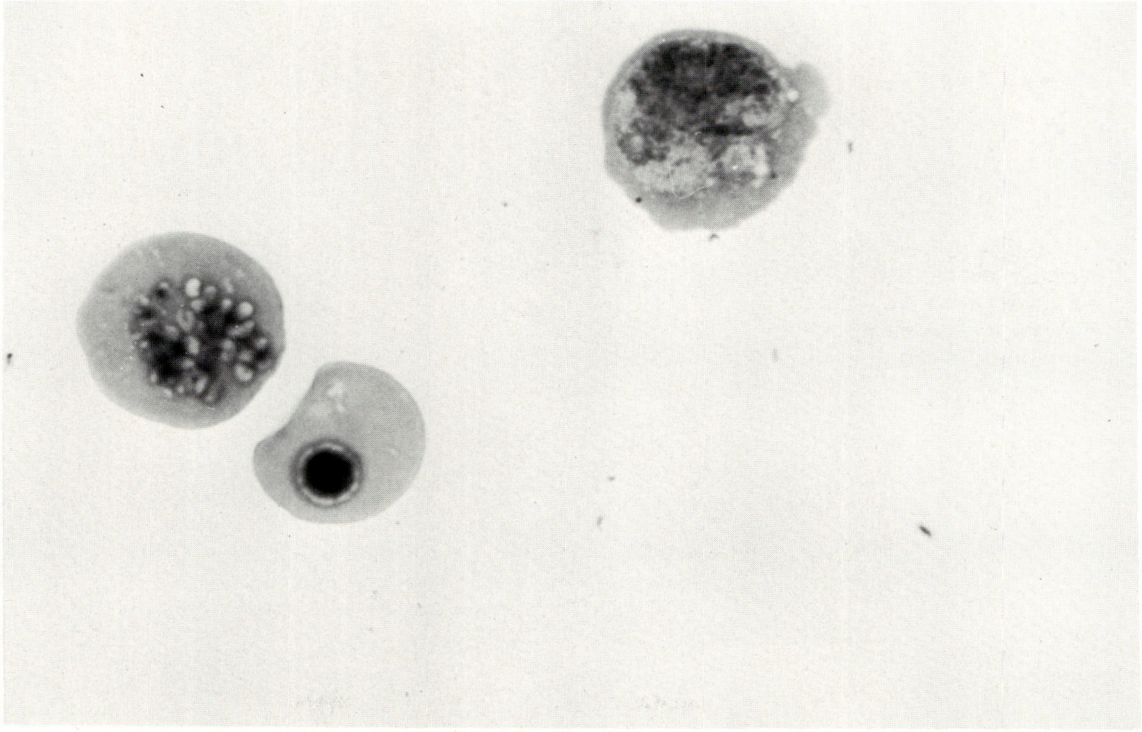

Figure 2. 'Cytocentrifuge' preparation of 10-day mouse yolk sac showing an example of the large undifferentiated 'blast' cells which are present at this time. A nucleated red cell and a mitotic figure can also be seen. McNeil's stain. ×1750

PLATE XI

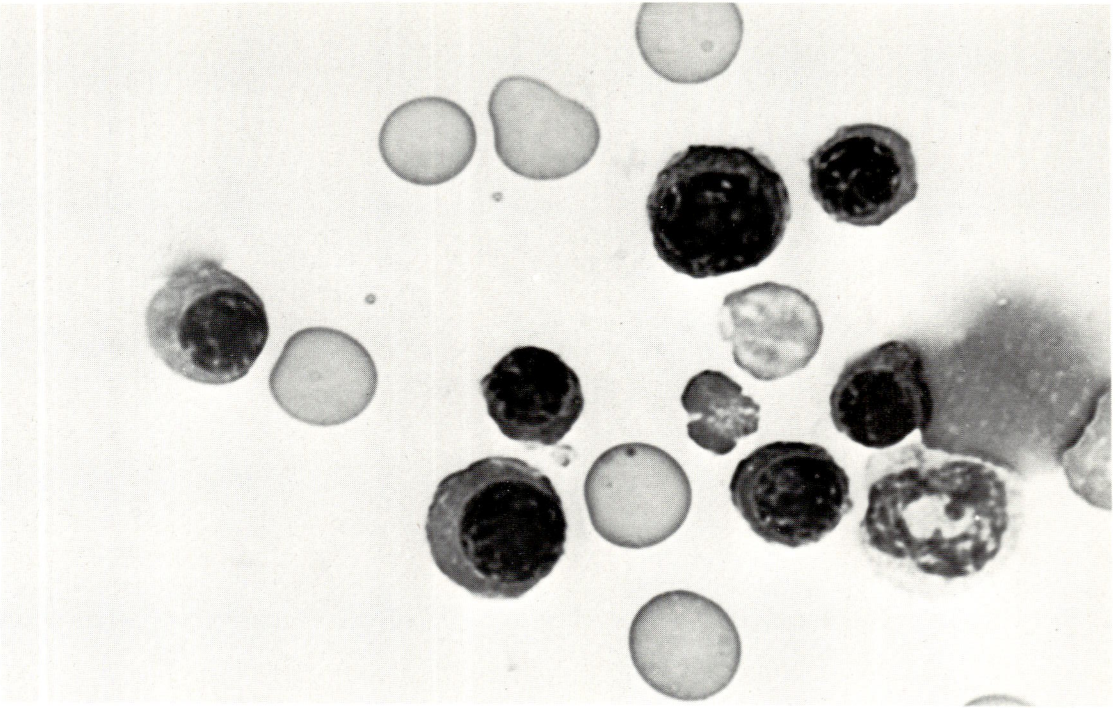

Figure 3. 'Cytocentrifuge' preparation of 16-day foetal liver. A granulocyte (ring-shaped nucleus) can be seen among developing erythrocytes. McNeil's stain. ×1750

PLATE XII

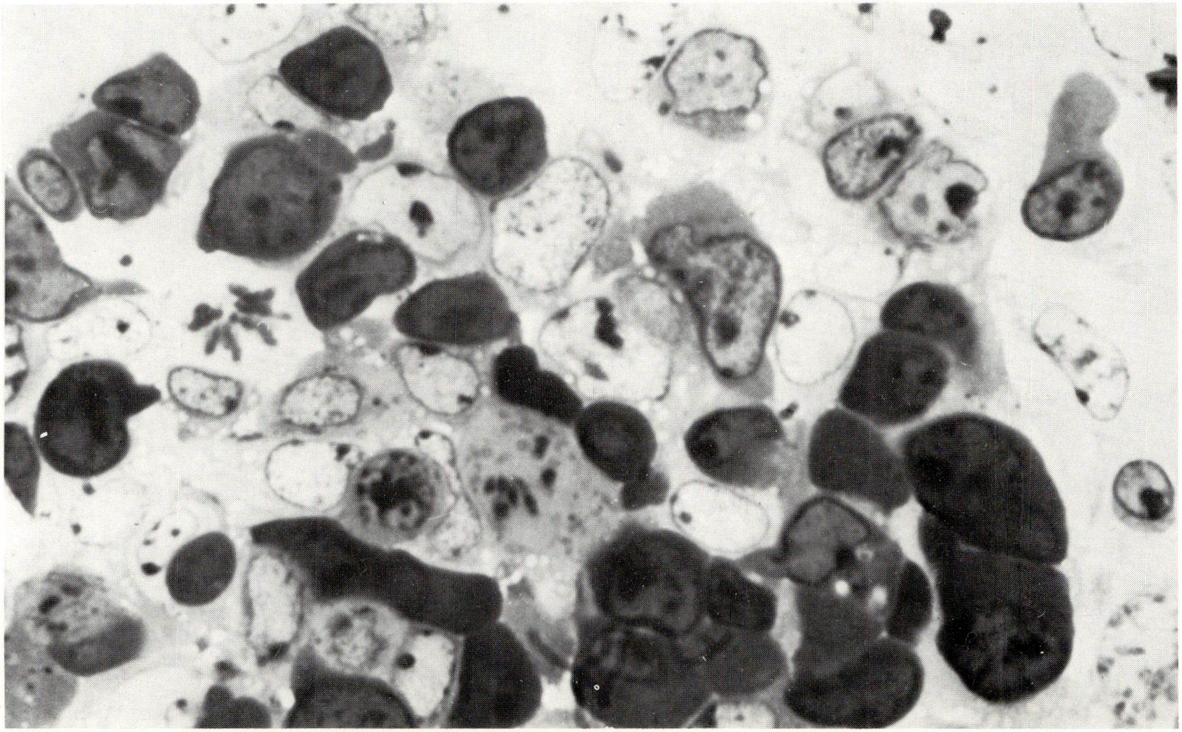

Figure 4. Section of 14-day mouse embryo thymus showing the presence of darkly-staining 'blast' cells. Epithelial cells show two levels of staining, some are very pale, others have slightly darker cytoplasm. Araldite embedded 1 micron section, toluidine blue stain. ×1750

PLATE XIII

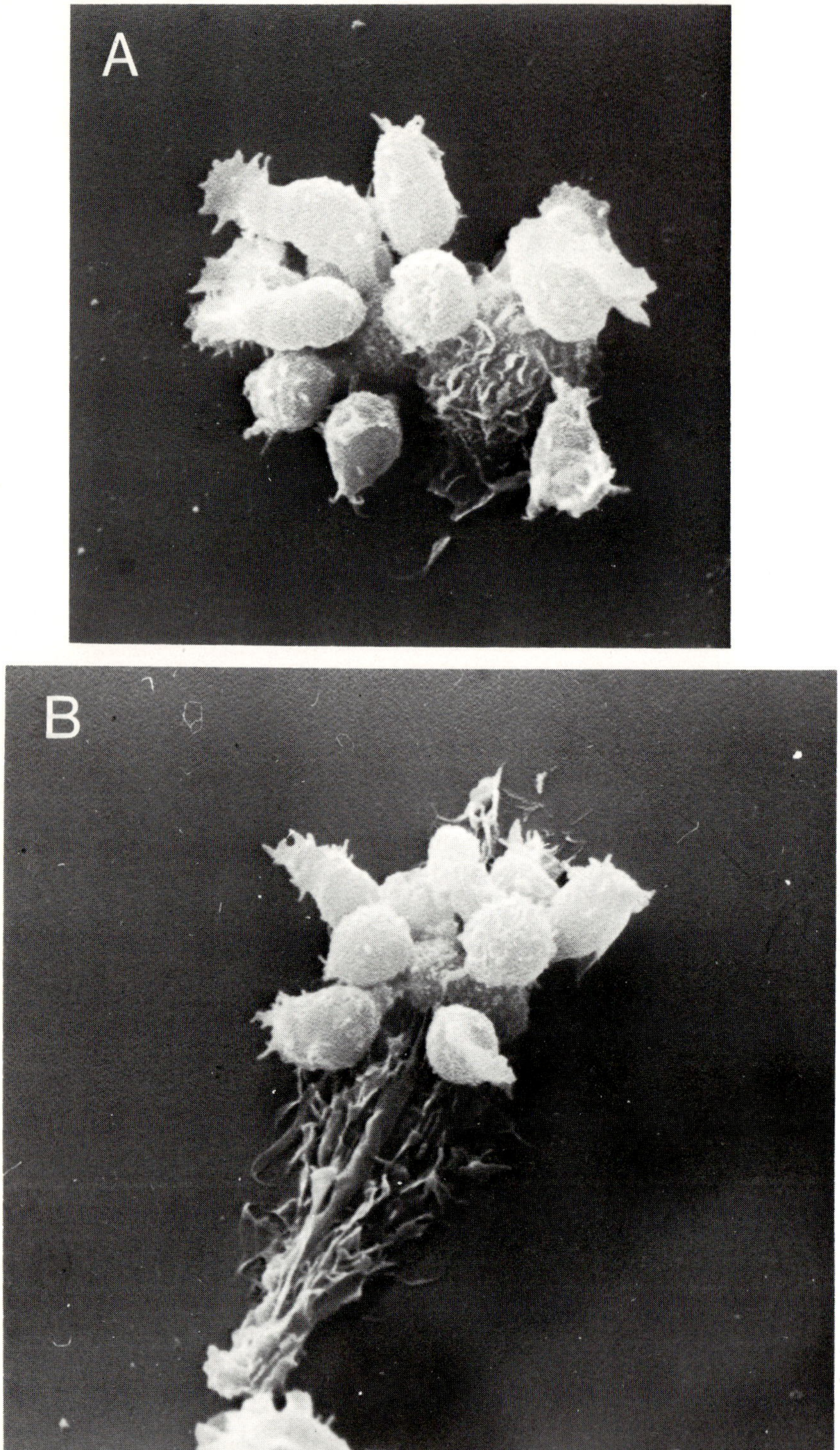

PLATE XIII (cont.)

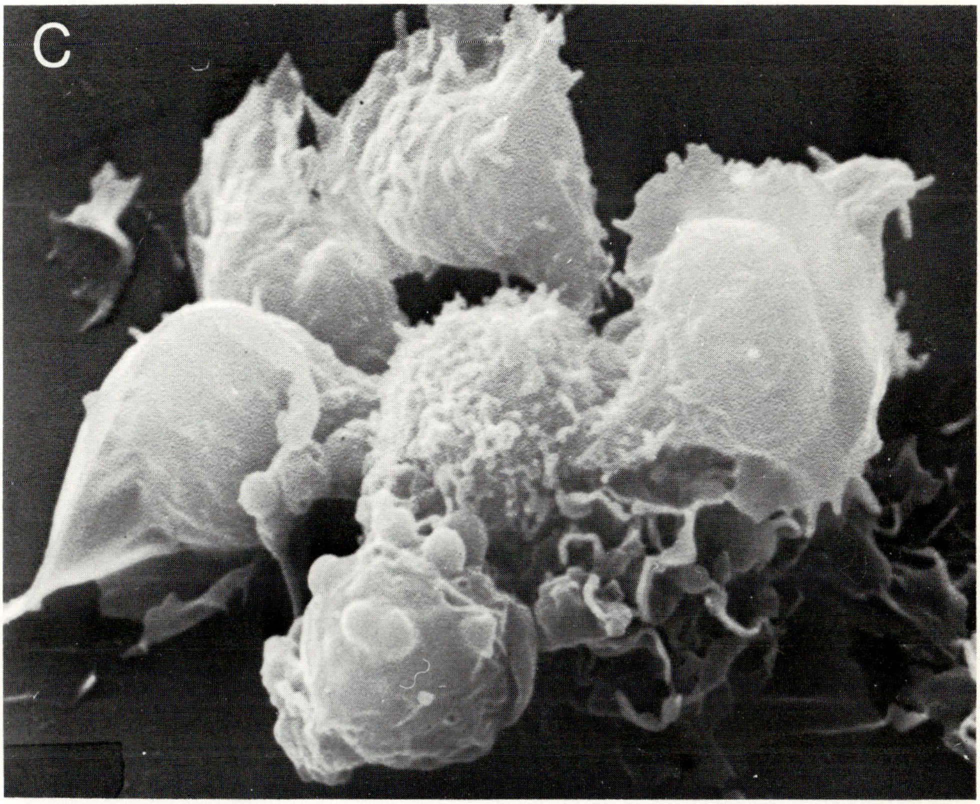

Figure 1. Scanning electron micrograph of lymphocyte-macrophage clusters. Peripheral lymphocytes are attached to a central lymphocyte which itself rests on the surface of a macrophage. The smooth surface of the peripheral (T) lymphocytes form a sharp contrast to the microvillous surface of the central (B) lymphocyte. (A, B ×2000, C ×4300); reproduced by courtesy of Nielsen, M. H., Jensen, H., Braendstrup, O. and Werdelin, O.,[127] with permission of the Journal of Experimental Medicine (Vol. 140, p. 1260 (1974))

recovered in the void volume. This activity is further fractionated by gel filtration on a G-25 Sephadex column and the peak containing the active material chromatographed on DEAE Sephadex A-25 anion exchanger developed with a linear concentration gradient of NaCl. The activity concentrated in one peak eluted at 0·15 N NaCl (between 300–500 thousandfold purification), is represented by one active protein species having a pI of 5·7–5·9 as analysed by isoelectric focusing on polyacrylamide gels. The activity of this preparation was again abolished by pronase treatment but not by DNase or RNase. No glycopeptide was detected on the gels. Analysis of the amino acid composition of THF revealed the presence of 31 residues of 10 amino acids. The mol. wt. based on the amino acid composition is 3220. Some purified preparations of THF also reveal the presence of one residue of cysteine.[115]

7.2 Thymosin[90,110,116]

Thymosin is usually isolated from calf thymus by a series of steps involving heating, acetone, as well as ammonium sulphate precipitation followed by gel filtration on Sephadex G-25, anion exchange chromatography on DEAE cellulose, gel filtration on Sephadex G-50 and polyacrylamide gel electrophoresis. Thymosin is a polypeptide having a mol. wt. of approximately 12 200 and is composed of 108 amino acid residues.

The acidic nature of thymosin and THF is expressed by the presence in both of a large proportion of glutamic and aspartic residues (over 50%), and the similar immunological and biochemical activities suggest that the two preparations contain at least the same active site. THF may represent an active subunit of thymosin. Indeed, it has been recently reported that reduction and S-carboxymethylation of thymosin followed by gel filtration on Sephadex G-50 columns equilibrated with 6 M guanidimium chloride results in the detection of smaller subunits having mol. wt. of 2400 and 3200.

7.3 TF[45,83]

TF has been found in the serum of several species including man. It is usually extracted from plasma of pigs and has been isolated by defibrination followed by dialysis, concentration on a UM 2 Amicon membrane, filtration on Sephadex G-25, CM cellulose chromatography, thin layer chromatography and electrophoresis. TF has been shown to be released into the circulation from the thymus. It is a polypeptide of mol. wt. about 1000. At this time, the relationship of serum TF with other thymic hormones is still a matter of speculation. It should be taken into account, however, that variations in molecular weights and in some of the biological activities studied might possibly reflect the complexity of immunoreactive forms of peptide hormones in plasma and tissue extracts.

7.4 Thymopoietin[113]

Two closely related polypeptides called thymopoietin I and II have been isolated from the thymus. These factors appear to have an effect on neuromuscular transmission as well as inducing in lymphocytes antigenic and function properties which are characteristic of T cells. Thymopoietin I and II were isolated to homogeneity from calf thymus by saline extraction and heating, by molecular sizing on membranes and by molecular sieve chromatography, adsorption chromatography on hydroxyl-apatite and ion-exchange chromatography on QAE-Sephadex. The two polypeptides showed a line of identity in Ouchterlony immunodiffusion, behaved similarly in functional tests, and might represent variant forms of a single hormone. Molecular weight has been determined to be 7000 and current evidence suggests that the hormone is produced by thymic epithelial cells and not by liver, kidney, spleen, lymph node, muscle or thyroid.

8. MECHANISM OF ACTION OF THYMIC HUMORAL FACTORS

8.1 The hormonal nature of thymic preparations[117]

It was found that THF induces a rapid activation of membranal adenyl cyclase and an increase in cellular levels of cAMP in thymocytes and spleen cells from neonatally thymectomized mice. This induction of adenyl cyclase activity is not dependent on protein synthesis,[118] and is probably due to structural changes induced in the cell membrane by THF. (The THF preparation itself does not contain any detectable amounts of cAMP.) The level of adenyl cyclase activity was found to be low in spleen cells from neonatally thymectomized mice and was restored to that of spleen cell populations from intact mice, following exposure of the cells to THF. The activation of adenyl cyclase and the increase in cellular levels of cAMP were found to be directly related to the process of induction of immune competence in spleen cells of neonatally thymectomized mice by THF. Thus, substances which were found to increase intracellular levels of cAMP such as DBcAMP, theophylline (an inhibitor of cyclic nucleotide phosphodiesterase known to increase levels of cAMP,[119] or prostaglandin E_2(PGE_2), were found to mimic the effect of THF and confer reactivity to spleen cells from neonatally thymectomized mice in the *in vitro* GvH assay. These substances exhibited a dose response curve typical of effects mediated via cAMP. Moreover, it was found that the increase in cellular cAMP levels is an obligatory event in the process of induction of immune competence by THF. Thus, the ability of spleen cells from neonatally thymectomized mice to respond to allogeneic stimulation was abolished when the increase of cellular cAMP was prevented, while the immune competence of spleen cells from intact mice was not impaired under this condition.[119]

These results taken together with the observed physiological disorders which appear following thymus deprivation and with the demonstration that a THF-like substance secreted by the thymus is found in the circulation,[45] are consistent with the postulates established by Sutherland to define hormonal action mediated by cAMP,[120] and permit the classification of THF as a thymus hormone. In addition, it was recently found in our laboratory (A. I. Kook, unpublished observations) that propranolol inhibits the inductive effect of THF when measured in terms of both MLC or GvH reactivity of lymphoid cells. Although the characteristics of the cellular receptors for THF are still undefined, these results suggest that they are of β-adrenergic nature.

8.2 Intracellular events involved in the induction of immune competence in lymphoid cells by thymic hormones[118,121,122]

Results of previous experiments[79,117] demonstrated that THF itself does not provide antigenic stimulation to immature spleen cells from neonatally thymectomized mice. Moreover, it was possible to show that exposure of immature spleen cells to THF, DBcAMP or theophylline for 1 h before their stimulation by alloantigens restored the ability of the cells to react in the *in vitro* GvH assay.[117] Thus, differentiation and/or maturation of T cells is initiated by a rise in cellular cAMP levels. The observation that immune maturation takes place in the absence of antigenic stimulation allowed the investigation of biochemical events which lead to the acquisition of immunocompetence and which follow the rise in cellular levels of cAMP induced by THF. Indeed, it was possible to demonstrate, using cycloheximide as a reversible inhibitor of protein synthesis that a protein synthesis step is initiated following the rise in cellular cAMP which is a prerequisite for spleen cells from neonatally thymectomized mice to acquire immune maturity. The inhibition of protein synthesis prevents these cells from responding in the *in vitro* GvH assay, despite activation of adenyl cyclase by THF. It was thus concluded that, in agreement with established patterns of differentiation,[123–125] a protein synthesis step is involved in the chain of biochemical events induced by THF via cellular cAMP, which lead to maturation of lymphoid cells and acquisition of cell-mediated immune capacity.

There is ample evidence that cAMP exerts a basic regulatory function on cell growth.[126,127] Thus, the addition of DBcAMP to cultures of human lymphoid cells resulted in a delay of mitotic activity of these cells.[128] In light of the above information, it was of interest to investigate the effect of THF on DNA synthesis in lymphoid cells. It was found that both thymocytes and spleen cells from neonatally thymectomized mice exposed to THF exhibited increased levels of cAMP and adenyl cyclase activity which persisted for the duration of the pre-incubation period and were accompanied by a marked reduction in DNA synthesis and the proliferative activity of these cells. These results suggest that, by preventing cell proliferation, THF via elevated levels of cAMP stimulates protein synthesis and induces the cells to acquire immunocompetence allowing the otherwise immature spleen cells from neonatally thymectomized mice to respond in the *in vitro* GVH assay.

In addition to thymocytes and spleen cells from neonatally thymectomized mice, THF also induces adenyl cyclase activity and suppresses tritiated thymidine incorporation into DNA of spleen cell populations obtained from intact mice. Young and relatively undifferentiated thymus derived cells are known to be present in normal spleen cell populations,[129] and it was postulated that these cells mature under the influence of THF and increase the proportion of competent cells in the population. Therefore, the control exerted by cellular levels of cAMP on the immune reactivity of normal spleen cells was tested using a one-way mixed lymphocyte culture (MLC) assay. These studies[122] demonstrated that a short preincubation of the cells with imidazole (an agent known to stimulate cAMP phosphodiesterase activity and shown to markedly decrease cellular cAMP levels in lymphocytes), completely abolished the proliferative response of the cells. In contrast, preincubation of the responding cells with THF or with DBcAMP for 1 h, significantly enhanced the response of these cells in the MLC assay over and above the response observed with untreated normal spleen cells. Moreover, the exposure of cells pretreated with THF to imidazole 5 min prior to antigenic stimulation, abolished the inductive effect of THF. These studies thus demonstrate that a certain initial level of cellular cAMP is required to allow the cells to respond to antigenic challenge and that this level determines the magnitude of the immune response.

However, the addition of THF or of DBcAMP to the assay medium simultaneously with the exposure of the responding cells to the antigenic stimulation delayed the manifestation of the augmented proliferative response of the cells. This is in agreement with the reports that high levels of cellular cAMP delay the mitotic activity of lymphoid cells.[128] Therefore, these results led to the conclusion that although initial levels of cellular cAMP are required to enable spleen cells to respond to allogeneic stimulation, cAMP should decrease in the course of the reaction to allow the cells to enter the proliferative phase. Indeed, it was found that such a reduction in cellular levels of cAMP occurs in both untreated spleen cells and in spleen cells which were preincubated with THF 20 h following the start of the MLC test. The addition of theophylline or of DBcAMP to the medium at this time point prevents the proliferative response of the cells in the assay.

The experiments summarized here demonstrate that the reduction in cellular levels of cAMP, which occur during the course of the MLC reaction and which is essential for the cells to enter into the mitotic phase, takes place at the same time interval in both untreated spleen cells and in spleen cells preincubated with THF. Moreover, kinetic experiments showed that both spleen cell populations seem to reach the peak of their proliferative response at the same time following the start of the assay. In addition, the exposure of the cells to THF at the start of the assay delays rather than enhances the incorporation of tritiated thymidine into DNA of the responding cells. These observations suggest that larger numbers of spleen cells attain the state of immunocompetence under the influence of THF.

9. CLINICAL EFFECTS OF THYMIC FACTORS

In the light of the extensive studies on the effect of THF on the immunological system, and as a result of the information gained in these studies, preliminary trials of THF in humans have been started. At the present stage the cases selected are those in which T cell deficiency clearly established before initiation of treatment was considered as an integral part of the physiopathology of such diseases. Quantitative and functional evaluation of T cells was carried out in terms of T cell cytotoxiy, E-rosettes, macrophage inhibiting factor (MIF) index, GvH reactivity of lymphocytes, and response to T cell lectins. In addition, skin test reactions to various antigens such as candida albicans, purified protein derivative of tuberculin (PPD), trichophyton, and streptokinase–streptodornase were performed. The first cases treated included patients with subacute sclerosing panencephalitis, juvenile rheumatoid arthritis, Down's syndrome and acute lymphatic leukaemia in children. After various injection schedules of THF, reconstitution of T cell function was observed in most cases. These encouraging results justify widening the scope of treatment and initiating controlled clinical trials with THF. In parallel, clinical trials have been initiated with thymosin in immunodeficiency diseases and cancer patients[90] and the encouraging effects so far observed have given impetus to large-scale clinical trials.

As mentioned in previous sections, glucocorticoid hormones and thymic preparations seem to act to a certain extent on the same target: immature T cells either located in the thymus or distributed in the lymphoid system. However, these two families of hormones exert opposite effects on this target. While cortisone and allied hormones destroy these immature lymphocytes, thymic hormones seek these cells and bring them to a higher level of maturity thus permitting fulfilment of the role characteristic of mature T cells. This interpretation might have strong implications for the therapeutic use of thymic hormones.

10. CONCLUDING REMARKS

We have here described studies that lead us to the conclusion that thymus hormones play a role in the maturation of developing T cells. Thymic hormones seem to be involved in the very early events of stem cell proliferation and differentiation and in the later stages of acquisition of immunological competence by lymphoid cells,[112] while effects have not been detected in fully mature immunocompetent T cells.

A new chapter has been opened which represents a genuine crossroads between immunology and endocrinology. The questions already answered have led to further queries which require continuing investigation. We still do not know whether thymus hormones act on target cells inside and/or outside the boundaries of the thymus gland. It is not clear what the relationship is between hormones extracted from thymic tissue and those detected in the circulation, and whether one thymic hormone or a family of hormones exist. Although some progress has been achieved in the understanding of effects of hormones on the thymus and vice versa, we are still ignorant about feedback mechanisms between the thymus hormone and other components of the endocrine system. Finally, the potential therapeutic use of thymic hormones in a variety of diseases is awaiting confirmation from adequate clinical trials.

11. REFERENCES

1. Dougherty, T. F. (1952). *Physiol. Rev.*, **32**, 379–401.
2. Dougherty, T. F., Berliner, M. L., Schneebel, G. L. and Berliner, D. L. (1964). *Ann. N.Y. Acad. Sci.*, **113**, 825–843.
3. Comsa, J. (1973). In *Thymic Hormones* (Luckey, T. D., Ed.), University Park Press, Baltimore, pp. 39–96.
4. Baroni, C. (1967). *Experientia*, **23**, 282–283.
5. Fabris, N., Pierpaoli, W. and Sorkin, E. (1971). *Clin. exp. Immunol.*, **9**, 227–240.
6. Pierpaoli, W., Baroni, C., Fabris, N. and Sorkin, E. (1969). *Immunology*, **16**, 217–230.
7. Pandian, M. R. and Talwar, G. P. (1971). *J. exp. Med.*, **134**, 1095–1113.
8. Pierpaoli, W. and Sorkin, E. (1968). *J. Immunol.*, **101**, 1036–1043.
9. Gisler, R. H. and Schenkel-Hulliger, L. (1971). *Cell. Immunol.*, **2**, 646–657.
10. Chatterton, R. T., Jr., Murray, C. L. and Hellman, L. (1973). *Endocrinology*, **92**, 775–787.
11. Arrenbrecht, S. and Sorkin, E. (1973). *Eur. J. Immunol.*, **3**, 601–604.
12. Schlesinger, M. and Mark, R. (1964). *Science*, **143**, 965–966.
13. Ernström, U. and Larssön, B. (1967). *Acta Path. Microbiol. Scand.*, **70**, 371–384.
14. Ishidate, M. and Metcalf, D. (1963). *Aust. J. exp. Biol. Med. Sci.*, **41**, 637–649.
15. Blomgren, H. and Andersson, B. (1971). *Cell. Immunol.*, **1**, 545–560.
16. Cohen, J. J. and Claman, H. N. (1971). *J. exp. Med.*, **133**, 1026–1034.
17. Trowell, O. A. (1958). *Int. Rev. Cytol.*, **7**, 235–293.
18. Weissman, I. L. and Levy, R. (1975). *Israel J. Med. Sci.*, **11**, 884–888.
19. Harris, A. W. (1970). *Exp. Cell Res.*, **60**, 341–353.
20. Weissman, I. L. (1973). *J. exp. Med.*, **137**, 504–510.
21. Kemp, R. G. and Duquesnoy, R. J. (1973). *J. Immunol.*, **114**, 660–664.
22. Ambrose, C. T. (1970). In *Hormones and the Immune Response* (Wolstenholme, G. E. W. and Knight, J., Eds.), Churchill, London, pp. 110–116.
23. Gyllenstein, L. (1962). *Acta Pathol. Microbiol. Scand.*, **56**, 29–34.
24. Ernström, U. and Larsson, B. (1966). *Acta Physiol. Scand.*, **66**, 189–195.
25. Cherry, C. P., Eisenstein, R. and Glucksmann, A. (1967). *Brit. J. exp. Path.*, **48**, 90–106.
26. Selye, H. and Masson, G. (1939). *Endocrinology*, **25**, 211–215.
27. Pierpaoli, W., Fabris, N. and Sorkin, E. (1970). In *Cellular Interactions in the Immune Response* (Rose, N. R. and Milgrom, F., Eds.), Karger, Basel, pp. 25–30.
28. Whitfield, J. F., Perris, A. D. and Youdale, T. (1969). *J. Cell. Physiol.*, **73**, 203–211.
29. Whitfield, J. F., Rixon, R. H., MacManus, J. P. and Balk, S. D. (1973). *In vitro*, **8**, 257–278.
30. Melmon, K. L., Weinstein, Y., Shearer, G. M., Bourne, H. R. and Bauminger, S. (1974). *J. Clin. Invest.*, **53**, 22–30.
31. Rosenau, W., Baxter, J. D., Rousseau, G. G. and Tomkins, G. M. (1972). *Nature* (*New Biol.*), **237**, 20–24.
32. Munk, A. and Brink-Johnson, T. (1968). *J. Biol. Chem.*, **243**, 5556–5565.
33. Ishizuka, M., Gafni, M. and Braun, W. (1970). *Proc. Soc. exp. Biol. Med.*, **134**, 963–967.
34. Mendelsohn, J., Multer, M. M. and Boone, R. F. (1973). *J. Clin. Invest.*, **52**, 2129–2137.

35. Melmon, K. L., Bourne, H. R., Weinstein, Y., Shearer, G. M., Kram, J. and Bauminger, S. (1974). *J. Clin. Invest.*, **53**, 13–21.
36. Bianchi, E., Pierpaoli, W. and Sorkin, E. (1971). *J. Endocrinol.*, **51**, 1–6.
37. Pierpaoli, W. and Sorkin, E. (1967). *Brit. J. exp. Path.*, **48**, 627–631.
38. Pierpaoli, W. and Sorkin, E. (1972). *Experientia*, **28**, 851–852.
39. Pierpaoli, W. and Sorkin, E. (1972). *Nature (New Biol.)*, **238**, 282–285.
40. Nishizuka, Y. and Sakakura, T. (1969). *Science*, **166**, 753–755.
41. Nishizuka, Y. and Sakakura, T. (1971). *Endocrinology*, **89**, 886–893.
42. Nishizuka, Y. and Sakakura, T. (1971). *Endocrinology*, **89**, 902–903.
43. Basedovsky, H. E. and Sorkin, E. (1974). *Nature*, **249**, 356–358.
44. Osoba, D. (1973). *Contemporary Topics in Immunobiology*, **2**, 293–297.
45. Bach, J.-F., Dardenne, M., Pleau, J. M. and Bach, M.-A. (1975). *Ann. N.Y. Acad. Sci.*, **249**, 186–210.
46. Miller, J. F. A. P. and Osoba, D. (1967). *Physiol. Rev.*, **47**, 437–520.
47. Schlesinger, M. and Yron, I. (1970). *J. Immunol.*, **104**, 798–804.
48. Schlesinger, M. and Yron, I. (1969). *Science*, **164**, 1412–1413.
49. Raff, M. C. and Wortis, H. H. (1970). *Immunology*, **18**, 931–942.
50. Takiguchi, T., Adler, W. H. and Smith, R. T. (1971). *J. exp. Med.*, **133**, 63–80.
51. Meuwissen, H. V., van Alten, P. A. and Good, R. A. (1969). *Transplantation*, **7**, 1–11.
52. Davies, A. J. S. (1969). *Transplant. Rev.*, **1**, 43–91.
53. Raff, M. C. (1971). *Transplant. Rev.*, **6**, 52–80.
54. Doenhoff, M., Davies, A. J. S., Leuchars, E. and Wallis, V. J. (1970). *Proc. Roy. Soc. Lond. B*, **176**, 69–85.
55. Weston, B. J., Cheers, C., Carter, R. L., Leuchars, E., Wallis, V. J. and Davies, A. J. S. (1972). *Int. J. Cancer*, **9**, 66–75.
56. Wilson, D. B., Howard, J. C. and Nowell, P. C. (1972). *Transplant. Rev.*, **12**, 3–29.
57. Miller, J. F. A. P. (1965). *Brit. Med. Bull.*, **21**, 111–117.
58. Martinez, C., Dalmasso, A. P. and Good, R. A. (1964). In *The Thymus in Immunobiology* (Good, R. A. and Gabrielsen, A. E., Eds.), Hoeber-Harper, New York, pp. 465–477.
59. Linker-Israeli, M. and Trainin, N. (1968). *J. Nat. Cancer Inst.*, **41**, 411–420.
60. Metcalf, D. (1965). *Nature*, **208**, 1336.
61. Taylor, R. B. (1965). *Nature*, **208**, 1334–1335.
62. Miller, J. F. A. P., de Burgh, P. M. and Grant, G. A. (1965). *Nature*, **208**, 1332–1334.
63. Cantor, H. and Boyse, E. A. (1975). *J. exp. Med.*, **141**, 1376–1389.
64. Rotter, V. and Trainin, N. (1974). *Cell. Immunol.*, **13**, 76–86.
65. Kerbel, R. S. and Eidinger, D. (1972). *Eur. J. Immunol.*, **2**, 114–118.
66. Simpson, E. and Cantor, H. (1975). *Eur. J. Immunol.*, **5**, 337–343.
67. Mosier, D. and Cantor, H. (1971). *Eur. J. Immunol.*, **1**, 459–461.
68. Andersson, L. C., Häyry, P., Bach, M.-A. and Bach, J.-F. (1974). *Nature*, **252**, 252–254.
69. Umiel, T. and Trainin, N. (1974). *Transplantation*, **18**, 244–250.
70. Resnitzky, P., Zipori, D. and Trainin, N. (1971). *Blood*, **37**, 634–646.
71. Zipori, D. and Trainin, N. (1973). *Blood*, **42**, 671–678.
72. Zipori, D. and Trainin, N. (1975). *Exp. Hemat.*, **3**, 1–11.
73. Zipori, D. and Trainin, N. (1975). *Exp. Hemat.*, **3**, 389–398.
74. Trainin, N. and Small, M. (1973). *Contemporary Topics in Immunobiology*, **2**, 321–337.
75. Trainin, N. (1974). *Physiol. Rev.*, **54**, 272–315.
76. Luckey, T. D. (1973). *Thymic Hormones* (Luckey, T. D., Ed.), University Park Press, Baltimore.
77. Stutman, O. and Good, R. A. (1973). *Contemporary Topics in Immunobiology*, **2**, 299–337.
78. Goldstein, A. L. and White, A. (1973). *Contemporary Topics in Immunobiology*, **2**, 339–350.
79. Trainin, N., Small, M. and Globerson, A. (1969). *J. exp. Med.*, **130**, 765–776.
80. Auerbach, R. and Globerson, A. (1966). *Exp. Cell Res.*, **42**, 31–41.
81. Small, M. and Trainin, N. (1971). *J. exp. Med.*, **134**, 786–800.
82. Lonai, P., Mogilner, B., Rotter, V. and Trainin, N. (1973). *Eur. J. Immunol.*, **3**, 21–26.
83. Bach, J.-F., Bach, M.-A., Charreire, J., Dardenne, M., Fournier, C., Papiernik, M. and Pleau, J.-M. (1975). In *Biological Activity of Thymic Hormones* (van Bekkum, D. W., Ed.), Kooyker Scientific Publications, Rotterdam, pp. 145–158.
84. Scheid, M. P., Goldstein, G., Hammerling, U. and Boyse, E. A. (1975). *Ann. N.Y. Acad. Sci.*, **249**, 531–540.

85. Basch, R. S. and Goldstein, G. (1975). *Ann. N.Y. Acad. Sci.*, **249**, 290–298.
86. Scheid, M. P., Hoffmann, M. K., Komuro, K., Hammerling, U., Abbot, J., Boyse, E. A., Cohen, G. H., Hooper, J. A., Schuloff, R. S. and Goldstein, A. L. (1973). *J. exp. Med.*, **138**, 1027–1032.
87. Komuro, K. and Boyse, E. A. (1973). *J. exp. Med.*, **138**, 479–482.
88. Rotter, V. and Trainin, N. (1975). *Cell. Immunol.*, **16**, 413–421.
89. Stobo, J. D. and Paul, W. E. (1973). *J. Immunol.*, **110**, 362–375.
90. Goldstein, A. L., Thurman, G. B., Cohen, G. H. and Hooper, J. A. (1975). In *The Biological Activity of Thymic Hormones* (van Bekkum, D. W., Ed.), Kooyker Scientific Publications, Rotterdam, pp. 173–197.
91. Bach, F. H., Bock, H., Graupner, K., Day, E. and Klostermann, H. (1969). *Proc. Nat. Acad. Sci.*, **62**, 377–384.
92. Umiel, T. and Trainin, N. (1975). *Eur. J. Immunol.*, **5**, 85–88.
93. Umiel, T., Altman, A. and Trainin, N. (1976). In *Immune Reactivity of Lymphocytes* (Feldman, M., and Globerson, A., Eds.), Plenum, New York, 639–643.
94. Rotter, V., Globerson, A., Nakamura, I. and Trainin, N. (1973). *J. exp. Med.*, **138**, 130–142.
95. Small, M. (1974). *Israel J. Med. Sci.*, **10**, 1180.
96. Trainin, N., Levo, Y. and Rotter, V. (1974). *Eur. J. Immunol.*, **4**, 634–637.
97. Cantor, H. and Asofsky, R. J. (1970). *J. exp. Med.*, **131**, 235–246.
98. Stutman, O. (1975). *Ann. N.Y. Acad. Sci.*, **249**, 89–105.
99. Loor, F. and Roelants, G. E. (1974). *Nature*, **251**, 229–230.
100. Roelants, G. E., Mayor, K. S., Hägg, L.-B. and Loor, F. (1976). *Eur. J. Immunol.*, **6**, 75–81.
101. Trainin, N., Small, M. and Kimhi, Y. (1973). In *Thymic Hormones* (Luckey, T. D., Ed.), University Park Press, Baltimore, pp. 135–158.
102. Carnaud, C., Ilfeld, D., Brook, I. and Trainin, N. (1973). *J. exp. Med.*, **138**, 1521–1532.
103. Small, M. and Trainin, N. (1975). *Int. J. Cancer*, **15**, 962–972.
104. Carnaud, C., Markowicz, O. and Trainin, N. (1974). *Cell. Immunol.*, **14**, 87–97.
105. Trainin, N., Carnaud, C. and Ilfeld, D. (1973). *Nature* (*New Biol.*), **214**, 253–255.
106. Small, M. and Trainin, N. (1975). *Cell. Immunol.*, **20**, 1–11.
107. Small, M. and Trainin, N. (1976). In *Immune Reactivity of Lymphocytes* (Feldman, M., and Globerson, A., Eds.), Plenum, New York, 659–664.
108. Krantz, S. B. and Jacobson, L. O. (1970). *Erythropoietin and the Regulation of Erythropoiesis.* University of Chicago Press, Chicago.
109. Kruisbeek, A. M. (1975). In *The Biological Activity of Thymic Hormones* (van Bekkum, D. W., Ed.), Kooyker Scientific Publications, Rotterdam, pp. 209–211.
110. Hooper, J. A., McDaniel, M. C., Thurman, G. B., Cohen, G. H., Schulof, R. S. and Goldstein, A. L. (1975). *Ann. N.Y. Acad. Sci.*, **249**, 125–144.
111. Kook, A. I., Yakir, Y. and Trainin, N. (1975). *Cell. Immunol.*, **19**, 151–157.
112. Trainin, N., Small, M., Zipori, D., Umiel, T., Kook, A. I. and Rotter, V. (1975). In *The Biological Activity of Thymic Hormones* (van Bekkum, D. W., Ed.), Kooyker Scientific Publications, Rotterdam, pp. 117–144.
113. Goldstein, G. (1974). *Nature*, **247**, 11–14.
114. Trainin, N. and Small, M. (1970). *J. exp. Med.*, **132**, 885–897.
115. Kook, A. I., Yakir, Y. and Trainin, N. (1976). In *Immune Reactivity of Lymphocytes* (Feldman, M., and Globerson, A., Eds.), Plenum, New York, **66**, 215–220.
116. White, A. (1975). In *The Biological Activity of Thymic Hormones* (van Bekkum, D. W., Ed.), Kooyker Scientific Publications, Rotterdam, pp. 17–23.
117. Kook, A. I. and Trainin, N. (1974). *J. exp. Med.*, **139**, 193–207.
118. Kook, A. I. and Trainin, N. (1975). *J. Immunol.*, **114**, 151–157.
119. Sutherland, E. W. and Robinson, G. A. (1968). *Circulation*, **37**, 279–283.
120. Sutherland, E. W. (1972). *Science*, **177**, 401–408.
121. Trainin, N., Kook, A. I., Umiel, T. and Albala, M. (1975). *Ann. N.Y. Acad. Sci.*, **249**, 349–361.
122. Kook, A. I. and Trainin, N. (1975). *J. Immunol.*, **115**, 8–14.
123. Pastan, I. (1972). *Scientific American*, **227**, 97–105.
124. Schainberg, A., Yagil, G. and Yaffe, D. (1971). *Develop. Biol.*, **25**, 1–29.
125. Yaffe, D. and Dym, H. (1973). *Cold Spring Harbor Symposium on Quantitative Biology*, **37**, 543–547.
126. Sheppard, R. J. (1972). *Nature* (*New Biol.*), **236**, 14–16.

127. Kreider, J. W., Rosenthal, M. and Lengle, N. (1973). *J. Nat. Cancer Inst.*, **50**, 555–558.
128. Millis, A. J. T., Forrest, G. and Pious, D. A. (1972). *Biochem. Biophys. Res. Commun.*, **49**, 1645–1649.
129. Raff, M. C. and Cantor, H. (1971). In *Progress in Immunobiology* (Amos, B., Ed.), Academic, New York, pp. 83–93.

Chapter 6

The Regulatory Role of Macrophages in Immune Recognition

G. E. ROELANTS

Abbreviations

BGG: bovine γ globulin
BSA: bovine serum albumin
C: complement
CFA: complete Freund's adjuvant
DNP-: dinitrophenol

FGG: fowl γ globulin
GPA: guinea pig albumin
HGG: human γ globulin
HSA: human serum albumin
Ig: immunoglobulin
KLH: keyhole limpet haemocyanin
LPS: bacterial lipopolysaccharide (endotoxin)
M-Ag: antigen bearing macrophages
MSA: mouse serum albumin
MSH: *Maia squinado* haemocyanin
OA: ovalbumin
PPD: purified protein derivative of *M. tuberculosis*
SRBC: sheep erythrocytes
TMV: tobacco mosaic virus

1. INTRODUCTION

Lymphocytes are genetically programmed to bind to and be activated by a specific antigenic determinant. Macrophages also bind antigen but do not display any discriminatory specificity. It is perfectly clear that if specificity rests exclusively on the lymphocyte, macrophages play a crucial role in the regulation of the immune response. This regulation is mediated in multiple ways: removal and catabolism of foreign substances, a property of paramount importance for resistance to foreign organisms which also modulates the establishment of immune activation or unresponsiveness; efficient and fast transport of antigen through the body; presentation of antigen to lymphocytes in enhanced immunogenic form; release or binding of soluble factors stimulating or suppressing lymphocyte proliferation and differentiation, a mechanism which mediates the activity of some adjuvants; finally mediation of T–B lymphocyte cooperation, antigenic competition and some forms of cell-mediated cytotoxicity.

We will not treat in detail all these various aspects of macrophage function but focus on the way these cells regulate antigen recognition by T and B lymphocytes. Many other aspects and cross references may be found in a recent book entirely devoted to mononuclear phagocytes.[1] Reference to specialized reviews will also be given in the text.

2. HANDLING OF ANTIGEN BY MACROPHAGES

2.1 Uptake and catabolism of antigen

One of the most spectacular properties of macrophages is their ability to take up and degrade large amounts of foreign material. Three main forms of endocytosis have been described depending essentially on the physical state of the material handled. They are briefly described here (for reviews see References 2 and 3).

Simple solutes with no apparent interaction with the membrane are endocytosed proportionally to their extracellular concentration. On the other hand compounds that bind to the membrane will first be locally concentrated and then to a large extent interiorized. A good comparison between the efficiency of both mechanisms is given by the example of peroxidase–antiperoxidase immune complexes which bind widely to the plasma membrane of macrophages and have a 4000-fold higher rate of interiorization than soluble peroxidase.[3] Both mechanisms proceed by the uptake of fluid droplets which is called pinocytosis. Most vesicles are formed on the undulating membrane of macrophage extensions and are relatively large (over 50 nm). Tiny vesicles only detectable by electron microscopy also exist but form a minority. Interiorized vesicles flow centripetally while fusing with each other to form larger vesicles. The process seems to involve microtubules and requires aerobic respiration and oxidative phosphorylation.[3]

Particle uptake is called phagocytosis in which at least two discrete phases can be distinguished: one of attachment to the plasmamembrane and one of ingestion *per se*.[4] Contrarily to pinocytosis, phagocytosis is dependent on energy provided essentially by the glycolytic pathway and seems to be far less dependent on membrane motion.[3]

It is very important to realize that these mechanisms of endocytosis involve localized areas of the plasmamembrane. Endocytosis of one membrane-bound particle does not influence the uptake of others (Reference 5, see also F. Loor, herein). Likewise uptake and handling of one antigen by a given macrophage does not interfere with the processing of another.[6–8]

Most endocytosed vacuoles rapidly fuse with lysosomes to form phagolysosomes where ingested material is catabolized. Ingested material is very extensively degraded, unless lysosomes lack specific enzymes. Only molecules of the size of, at most, di- or tri-peptides are released from lysosomes after uptake of protein.[9–11] We will see, however, that some antigen escapes catabolism by remaining in association with the plasmamembrane or by storage in an internal pool not submitted to lysosomal degradation.

2.2 Persistence of macrophage associated antigen

It was realized that after uptake of antigen, macrophage associated immunogenicity persists for long periods of time (Section 3). Weeks after uptake of protein antigen immunogenic material can be extracted from macrophages.[12–17] At one time it was suggested that this persisting antigen was associated with RNA[12–14,18–23] but this turned out to be an artefact.[24–27] Antigenicity depends most often on the tertiary structure, i.e. the overall conformation of antigen (e.g. Reference 28). Thus a search was made in macrophages for antigen that might escape catabolism and persist in essentially undegraded form for some time. Exceptions to catabolism are known, e.g. *Mycobacterium tuberculosis*,[29] lactic dehydrogenase virus,[30] pneumococcal polysaccharide[31] and synthetic polypeptides of the D-enantiomorphic configuration,[28] but are rather connected with tolerogenicity than immunogenicity.[32]

Although handling by macrophages is variable from antigen to antigen (a good example being the compared fate of DNP-BSA and BSA),[33] a basic mechanism has emerged from extensive studies involving antigens as different as albumin from bovine (BSA),[34,35] human (HSA),[35,36] mouse (MSA)[36] or egg (OA)[35] sources, lysozyme,[35] ferritin,[8] immunoglobulin (Ig),[37] *Maia squinado* haemocyanin (MSH),[38–40] keyhole limpet haemocyanin (KLH),[41–44] fd bacteriophage,[45] tobacco mosaic virus (TMV),[8] erythrocytes[37] and Ig bound to erythrocytes.[37,46]

Using radioactive labels the handling of antigen by macrophages could be expressed in quantitative terms. Thus most macrophages take up antigen and about 90% of the antigen taken up is rapidly catabolized within 2 to 6 hours.[34,38,47] The remaining 10% is retained for at least 2 days in a macromolecular form as detected by TCA precipitation and migration in polyacrylamide gel electrophoresis in SDS-urea. The retained antigen appears to be in a 'segregated' compartment for it is not diluted out by addition of an excess cold antigen.[34] Isopycnic centrifugation shows that most retained TCA precipitable antigen is located in a ρ 1.26 storage compartment and that it is resistant to solubilization by detergents[34] while most of the TCA soluble material is associated with lysosomes. Electron microscopy combined with autoradiography 4–24 hours after uptake reveals the presence of persisting antigenic material associated with lysosomes (40%), small pinocytic vacuoles near lysosomes (40%) and with the cell periphery either on or just continuous to the cell membrane (20%).[42] That 15%–25% of the antigen persisting after 24 hours is located on the macrophage plasmamembrane is also demonstrated by the release of that material by treatment with EDTA[49] or trypsin.[42,43] Moreover specific antibody is able to bind retained antigen on the macrophage plasmamembrane.[42,43] We will see in Section 3 that membrane associated antigen is very important for immunogenicity although it does not represent the exclusive immunogenic moiety.

Most of the studies just described were done either at the cell population level, without information about individual macrophages or at the electron microscope level which does not permit the scanning of large cell numbers. The paradox of having persisting intact molecules on a cell displaying intense phagocytic, pinocytic and catabolic machinery[3,48–51] was reinforced by findings on the dynamic state of cell plasmamembrane (F. Loor, herein). Under usual laboratory conditions, cross-linking of surface membrane components induces the formation of microprecipitates which move in the plane of the membrane, are redistributed in polar cap and endocytosed. Thus F. Loor and I have reinvestigated the binding and redistribution of antigen, as well as Ig and lectins, on live macrophages at the single cell level using immunofluorescence techniques.[8]

We found that mouse peritoneal macrophages have easily detectable surface Ig. As on lymphocytes it is distributed homogeneously at the level of immunofluorescence detection (ring staining by monovalent reagents) and it can be clustered when divalent anti-Ig reagents are used. The amount of surface Ig is quite variable from cell to cell. Ig remains associated with the plasmamembrane of small-to-medium macrophages for at least four days in culture. On large macrophages it is progressively lost by spontaneous (non anti-Ig induced) capping. Most of this surface Ig is rapidly redistributed in polar caps by divalent but not monovalent anti-Ig reagents. Complete clearance of the membrane is, however, difficult to obtain. *In vivo*, one observes gradient formation rather than real capping.

Binding and redistribution of concanavalin A, phytohaemagglutinin and poly-L-lysine on macrophages are as described for lymphocytes. However, pokeweed mitogen, which binds in a detectable way only to thymocytes and not to peripheral B or T lymphocytes, binds also to macrophages and redistributes in polar caps by itself.

TMV and ferritin were used in antigen. Both bind to the majority of normal or proteose peptone induced peritoneal macrophages *in vivo* and *in vitro*. There is a great heterogeneity in the degree of binding and large macrophages, presumably more 'activated', appear quite brighter than smaller ones. Binding of antigen to macrophage plasmamembrane is far more effective at 37 °C than at 4 °C in accordance with findings by other,[42] thus it appears to require temperature dependent membrane rearrangement (F. Loor, herein). It is not impaired by sodium azide, a metabolic inhibitor with multiple sites of action. There is no competition between TMV and ferritin for binding. When both antigens are added simultaneously to a macrophage suspension all combinations are found: bright fluorescence for both antigen, bright for one, dull for the other etc.—indicating that the degree of antigen uptake is not a function of the stage of development or the cell cycle.

After binding, polar redistribution and endocytosis take place. The two phenomena are inhibited by cold and sodium azide but occur rapidly at 37 °C. It is difficult to decide if capping always precedes endocytosis or if direct endocytosis at the site of binding occurs as well. Thus to the classical two stages of antigen uptake, i.e. binding and pinocytosis we may add a third intermediate one: polar redistribution, at least for macrophages in suspension.

In accordance with the work with radioactive antigen, part of TMV and ferritin, as well as Ig and lectins, escape complete removal from the cell surface and remain associated with the plasmamembrane for at least three days in culture in a largely undegraded form still recognizable by antibody. (Preliminary experiments using deep etching and electron microscopy showed intact particles of TMV at the surface of macrophages.) After endocytosis and removal of antigen persisting at the cell surface by EDTA, it can be shown that some antigen later reappears at the cell surface. Reutilization of antigen released by dead macrophages could be excluded using mixed macrophage populations. Thus antigen may persist at the cell surface not only because it is attached to 'privileged sites' escaping endocytosis but also because some endocytosed antigen escapes catabolism and is later re-exposed at the surface.

Macrophage surface Ig does not appear to play a role in TMV and ferritin uptake which remained unchanged after removal of Ig by capping. This is consistent with the reports that radioactive antigen binding is not impaired by treatment of macrophages with anti-Ig reagents.[36,52] However, we will see in Section 2.4 that coating macrophages with specific cytophilic antibody has a dramatic influence on the rate of uptake of the corresponding antigen. After uptake, redistribution and endocytosis of a first dose of antigen, macrophages are perfectly able to take up a second dose of the same or another antigen. This indicates that they do not have a special receptor for antigen, comparable to the receptor for Fc or complement (C),[53–55] but that antigen is attached to an usual ubiquitous non-specific constituent of the macrophage membrane. Thus binding of antigen to macrophages is quite different from binding to lymphocytes. Some of the differences are listed in Table 1.

Table 1. Binding of Ag to lymphocytes (L) and macrophages (M) of normal mice

	L	M
Frequency of binding cell	rare	most
Antigen specific	yes	no
Receptor	Ig	? not Ig
Saturation	easy	difficult
Temperature dependent	no	yes

From Reference 8; reproduced by permission of Verlag Chemie Weinheim.

2.3 Release of antigen

In addition to persistence of antigen in and on macrophages small amounts of macromolecular antigen is also released from these cells for long periods of time. A number of pathways exist for the escape of macromolecules from cells[56] and exocytosis has been repeatedly postulated to explain the release of lysosomal enzyme from macrophages and other cells.[57–61]

Exocytosis of antigen after uptake by macrophage has been observed for many antigens including HSA, MSA and BSA,[34,36] hapten conjugated bovine Ig, synthetic polypeptides,[34] γ2a myeloma protein 5563,[40] MSH,[39,40] KLH,[42,43] sheep erythrocytes (SRBC) and various IgG bound to SRBC.[37,46] It has also been postulated for the fd bacteriophage[45] but was not detected for horseradish peroxidase.[62] As for uptake and catabolism the degree of exocytosis is related to the nature and physical state of the antigen, particulate antigens making macrophages more leaky.[40] It was calculated that 1%–5% soluble antigen[36,39,40,42–44] and 5%–15% particulate antigen[37,40,46] taken up could be released within a few days of culture. In the negative report with peroxidase less than 2% release would not have been detected.[62]

The released antigen appears to be independent of the one persisting on the plasmamembrane; indeed treatment of macrophages with EDTA or trypsin to remove surface antigen does not influence antigen release.[36,40] Most released antigen is in macromolecular form still recognizable by antibody, i.e. having intact antigenic sites, albeit it usually has a lower molecular weight than the native protein indicating that it has been submitted to some cell catabolism and reinforcing the idea that it is exocytosed from inside the cell.

2.4 Cytophilic antibody, receptors for Fc and C

'Cytophilic' antibody, i.e. antibody able to stick somehow to the cell surface was described by Boyden and Sorkin.[63] Macrophage cytophilic antibody has been described in many species and may be easily visualized by immunofluorescence.[8,64] It attaches to the macrophage plasmamembrane through the CH3 domain of the Fc portion[65] hence the designation of an 'Fc receptor' although the nature of the macrophage surface component fixing antibody and its function as a receptor comparable for instance to hormone receptors, are not established.

Macrophage Fc receptors show some discriminatory power for the class of Ig they bind; in man γ_1 and γ_3 subclasses,[66] in guinea pig γ_2,[67] in mouse γ_{2a} preferably to other IgG subclasses.[68] In mouse IgM may also be taken up but unlike IgG this requires divalent cations.[69] This receptor is not removed from the cell surface by trypsin treatment and binding of antibody is independent of C.

Fc receptors may play an important role in modifying antigen handling in the presence of antibody. Indeed presentation of antigen as immune complexes instead of solute has a dramatic effect on uptake and catabolism: as mentioned above, horseradish peroxidase as immune complexes, for instance, is taken up several thousandfold more efficiently than the solute form.[70] When macrophages are coated with anti-MSH antibody their uptake of MSH is increased about tenfold.[39] Similarly only 0·5% human blood monocytes bind human erythrocytes but more than 90% bind these erythrocytes coated with human IgG.[65] We described above that uptake of particulate antigen is quite considerably more efficient than soluble antigen; this effect in turn is greatly amplified by binding of the Fc parts of immune complexes. It should be emphasized that cytophilic antibody has a great amplifying effect but is in no way necessary for antigen binding on macrophage membrane (e.g. Reference 8).

Avidity of macrophages for the Fc portion of Ig is quite heterogeneous from one cell to another.[8,71] It is greater (about sixfold) for inflammation activated macrophages[71] and is modulated by insulin and cyclic nucleotides.[72] For minimally stimulated rabbit alveolar macrophages the average number of Fc receptors per cell was calculated to be of the order of 10^6 with an average association constant with IgG of 8×10^5. For cells heavily stimulated with complete Freund's adjuvant (CFA) the respective values were about 2×10^6 and 9×10^5.[73]

A receptor binding complement (C) is also found on macrophages.[54,55,65,74] It is specific for C3b and unlike the Fc receptor is destroyed by trypsin and requires the presence of divalent cations. While complexes of homologous erythrocytes and IgM do not bind to the macrophage surface (with the exception of the mouse system) addition of C to these complexes will permit binding. However, no interiorization will follow unless large amounts of C are used.

Both Fc and C receptors mediate amplifying pathways for the uptake and catabolism of antigen and both may act synergistically. It should be noted that such receptors have also been found on cells such as lymphocytes (reviewed in Reference 75) which do not show intensive endocytosis and that possible physiological functions other than the increased removal of foreign material is a matter of conjecture.

3. IMMUNOGENICITY OF MACROPHAGE ASSOCIATED ANTIGEN

3.1 Implication of the reticulo-endothelial system in immune regulation

Early work on the fate and distribution of *in vivo* injected antigen has shown that antigen is generally rapidly cleared from the circulation and is catabolized; the breakdown products are excreted or reutilized. However, part of the antigen is trapped, sometimes for very long periods of time, by reticulo-endothelial cells in lymphoid organs (e.g. Reference 31). Thus the reticulo-endothelial system was thought to play a role in the regulation of immunity or tolerance both by removing excess antigen and by 'presenting antigen in a proper microenvironment for interaction with immunocompetent cells.[32] Moreover, continuous release of large amounts of non-catabolizable antigen or small amounts of persisting antigen could also play a role in maintaining immunological tolerance or memory.[32]

Attempts to substantiate these ideas by reticulo-endothelial blockade or the use of anti-macrophage sera *in vivo* (reviewed in References 2, 76) gave inconclusive results. Blockade is never complete (e.g. Reference 77) and the products used may be generally toxic while anti-macrophage sera pose specificity problems.[78]

3.2 *In vivo* studies

Numerous workers have compared the immunogenicity of antigen in free form and associated with macrophages (M-Ag). Most studies bear on antibody formation and the bulk of information shows that association with macrophages enhances dramatically the immunogenicity of weak antigens. For strong immunogens the effect is less pronounced or even reversed. In no case was M-Ag found to induce tolerance. M-Ag appears to be especially efficient in priming for a secondary response, i.e. building memory, rather than in eliciting a primary or a secondary response. M-Ag also seems to enhance cellular immunity but this aspect has been less thoroughly investigated. Table 2 gives a summary of some major studies. Several factors are worth discussing in more detail.

3.2.1 *The crucial role of lymphocytes*

Interaction of M-Ag with lymphocytes is absolutely required for the elicitation of an immune response. M-Ag transfer to normal recipient is effective but if the lymphoid system of the recipient is destroyed by irradiation M-Ag transfer does not elicit a response

Table 2. Potentiation (+) or suppression (−) of antibody formation *in vivo* by M-Ag compared to free Ag

Antigen	Elicitation response: primary	Elicitation response: secondary	Priming for a secondary response	References
MSH	0	+	+++	40, 79–81
KLH	−		−	41
BSA	0	−	+++	35, 82, 83
HSA	+		+++	82, 84
OA			++	82
HGG	+		+	85, 86
Lysozyme			++	82
Tetanos toxoid	variable		+	87
DNP-BGG		+		88
DNP-OA	variable		+	87
fd phage	−		+	45
Shigella	+			89
SRBC	+			37, 90
	+	+[a]	+	91

[a] After priming with subimmunogenic doses.

unless lymphocytes are also transferred. The necessity for M-Ag–lymphocyte interactions was elegantly demonstrated as early as 1966 by W. L. Ford, J. L. Gowans and P. J. McCullagh.[92] They showed that rat thoracic duct lymphocytes first exposed *in vitro* to M-erythrocytes and later, freed of macrophages, reconstituted the haemolysin response of irradiated recipients while M-erythrocytes themselves were unable to do so nor did sonicated thoracic duct lymphocytes transferred to normal recipients.

In this respect it is also important to note that macrophages from tolerant animals handle antigen and mediate immunogenicity as well as macrophages from normal animals.[35] Conversely tolerant lymphocytes cannot be stimulated to antibody formation by M-Ag from normal donors.[35,93]

3.2.2 *Requirement for live macrophages*

Killed M-Ag are still somewhat immunogenic but good enhancement of immunogenicity requires live cells.[35,40] Fast rejection of cells transferred across histocompatibility barrier[35] or across species[40] also results in diminished effectiveness. Killing of transferred cells appears to be responsible rather than a basic block of macrophage lymphocyte interaction across a histocompatibility barrier as has been postulated[94] but appears to be disproven[95] for T–B lymphocyte cooperation (see also H. G. Bluestein, and M. Nabholz and V. C. Miggiano, herein).

3.2.3 *B memory*

The most striking effect of M-Ag is the building of B memory (Table 2). The immunogenicity of M-Ag depends mainly upon the amount of antigen transferred rather than the cell number.[40] With MSH as antigen it was shown that M-Ag can prime for both 19S and 7S responses. Low amounts of M-Ag prime for 19S only, larger amounts for 19S and 7S.[40] Coating macrophages with cytophilic antibody or using immune complexes enhances antigen uptake and immunogenicity.[39]

Transfer experiments done under conditions limiting the expression of hapten-specific B memory cells clearly show that M-Ag enhances the number of anti-hapten B memory cells that will effectively home in the irradiated host and produce antibody (revealed by the increased number of clones detected by isoelectric focusing of recipient sera) as well as the expansion of these clones (revealed by their greater strength and production of larger amounts of antibody).[88]

3.2.4 *T activation*

Although most studies *in vivo* were made on B activation enhancement of T responses have been reported. Macrophage bound HSA injected in rat foot-pads established 28 days later a better stage of delayed hypersensitivity, judged by induration and histology, than free HSA.[84] It was, however, less effective than HSA mixed with CFA. In the same series of experiments it was shown that M-Ag but not free antigen was also inducing antibody. Preferential induction of delayed hypersensitivity versus antibody formation by M-Ag compared to free antigen was reported in guinea pigs.[87] Finally, very elegant experiments by G. C. B. Klaus[97] using hapten-carrier transfer system demonstrated that M-Ag was highly potent in immunizing carrier reactive T cells, a conclusion also inferred from the reconstitution of T deprived mice with peritoneal exudate cells.[80]

3.2.5 *T dependency*

The help of M-Ag does not overcome T dependency. Thus administration of M-Ag does not alleviate the necessity for T–B cooperation in the elicitation of an antibody response to thymus dependent antigens.[80,98]

3.2.6 *M-Ag and soluble antigen competition*

The increased response induced by M-Ag is depressed by pre-injections of soluble antigens (e.g. Reference 83). This has been interpreted as indicating that soluble antigen induces a combination of unresponsiveness and immunity and that presentation of antigen on macrophages shifts the balance towards immunity.

3.3 *In vitro* studies

The most widely used antigen for the study of *in vitro* immune responses is SRBC. Most authors found that the *in vitro* antibody response to this antigen requires the presence of macrophages in the culture (References 99–103 and reviewed in Reference 2). It was reported that this requirement for macrophages was especially strong for a primary response but decreased after priming, thus memory cells, primed for a secondary response would be less macrophage dependent or even completely independent.[100] However, others have reported that the cellular requirements were the same for a primary or secondary response.[103]

Studies with other antigens have shown that macrophage dependency for an *in vitro* response was not always the case. In a series of experiments using SRBC, solubilized SRBC and polymerized flagellin it was found that the two last antigens would induce antibody *in vitro* in lymphocyte cultures largely depleted of macrophages.[102–105] From this it was proposed that the size of the antigen determines the macrophage dependency of *in vitro* responses (reviewed in References 106, 107). This distinction between macrophage 'dependent' and 'independent' responses has been recently challenged by E. Diener and coworkers (unpublished) who found that by further depletion of macrophages from cultures used for *in vitro* immune induction all antigens became macrophage dependent.

Responses of primed cells to hapten-carrier conjugates was also investigated. M. Feldmann[108] found that DNP-fowl gamma globulin (DNP-FGG) primed cells freed of macrophages did not respond any more to DNP-FGG but responded to DNP-flagella an antigen apparently macrophage and thymus independent. The response to DNP-FGG was restored by addition of small number of anti-θ treated peritoneal exudate cells. These results suggested that macrophages are essential in mediating T–B cooperation and they have been greatly expanded later (reviewed in Reference 109, see also Sections 4 and 5). At variance with the preceding observations D. H. Katz and E. R. Unanue[110,111] reported that macrophage depletion had no effect on the *in vitro* secondary response to DNP-KLH at the level of both direct and indirect plaque-forming cells (PFC). Moreover M-DNP-KLH did not have any consistent effect on direct PFC as compared to free DNP-KLH or DNP-KLH on fibroblasts and the enhancement of indirect PFC was only marginal (two to fivefold). This weak effect is in good correlation with that seen with *in vivo* transfer in the elicitation of a secondary response.

T cell activation *in vitro* was also studied. The ability of free antigen or M-Ag to induce a proliferative response by primed lymphocytes gave somewhat contradictory results. Using antigen as varied as BSA, HSA, bovine gamma-globulin (BGG), HGG, OA, DNP-BSA, KLH, SRBC and poliovirus it appeared that addition of mouse alveolar or peritoneal macrophages to spleen cell suspensions *in vitro* inhibited the DNA synthesis induced by antigen.[112] On the other hand, work in guinea pigs using DNP-guinea pig albumin (DNP-GPA) and purified protein derivative (PPD) seem to indicate that triggering of a proliferative response by primed lymphocytes requires the binding of antigen by specific cytophilic antibody at the macrophage surface and its subsequent interaction with T cells.[113–116] This last interpretation is, however, quite surprising for we saw that if cytophilic antibody does indeed enhance antigen capture it is certainly not required for binding of antigen to the macrophage plasmamembrane (Section 2). As was shown in the *in vivo* situation macrophages taken from rabbits tolerant to BGG took up BGG and stimulated DNA synthesis in BGG primed spleen cells just as macrophages from normal rabbits did.[117]

In vitro work is quite a bit more extensive than reported in this short section which only intended to give a few contradictory examples. Detailed information about *in vitro* requirements will be found in the chapter by M. H. Schreier and A. A. Nordin herein.

3.4 The immunogenic moiety

We have seen in Section 2 that some antigen remains associated with the macrophage plasmamembrane, some with an internal store, and that some is released by exocytosis. Each of those pathways appears to play a role in the enhanced immunogenicity of M-Ag although for various antigens one pathway may be more important than the other. For instance, treatment of M-MSH with anti-MSH serum does not decrease their immunogenicity[39,40] while M-KLH have their immunogenicity reduced, but not abolished, by anti-KLH or trypsin treatment.[43] It appears that for 'sticky' proteins like KLH, membrane associated antigen plays a bigger role in immunogenicity.

That surface associated antigen does not represent the unique immunogenic moiety is also shown by transfer to recipients of M-Ag contained in a millipore diffusion chamber.[79] The priming of such recipients was only reduced three- to fivefold indicating that though intimate contact with lymphocytes is more effective, immunogenic material is also released from macrophages. The same conclusions were reached from studies with sheep erythrocytes.[37] Immunogenicity was found associated with both M-SRBC and with culture supernatants. However, while M-SRBC themselves were able to induce a primary response as well as to prime for a secondary response, supernatants were only able to prime for a secondary response.

It should be stressed that the immunogenic moiety of M-Ag persists for long periods. Indeed, when lethally irradiated animals are transferred with mixtures of lymphocytes and M-Ag an equally good response is obtained when the lymphocytes are injected three days after M-Ag.[38] Moreover, transfer of M-Ag done in sub-lethally irradiated animals (which have to reconstitute their lymphocyte pool from their own precursors) showed that immunogenic material persists for at least two weeks.[38]

Using the *in vitro* proliferative assay of primed lymphocytes as a measure for M-Ag immunogenicity it was found that when antigen is allowed to bind to the plasmamembrane but not to be endocytosed, the immunogenicity is completely removed by trypsin.[115] However, when endocytosis was also allowed, trypsin treatment removed only part of the immunogenicity.[115] Here again both antigen moieties seem to be immunogenic. Using the same assay, macrophages exposed to low doses of DNP-GPA (100 ng/ml) lost most of their inductive power after trypsin treatment while that of macrophages exposed to higher doses of antigen (100 μg/ml) was largely trypsin resistant.[116] It was also observed that, although the association of DNP-GPA with the macrophage plasmamembrane remained apparently unchanged for several days, the inductive power of these cells was progressively diminishing. Thus it was claimed that as far as guinea pig T cell activation is concerned the immunogenic critical fraction was not the surface associated one.[116] However, DNA synthesis in control spleen cells exposed to macrophages which did not bear antigen also decreased with time in the same preparation and the apparent decrease of inductive power may simply be due to some side effect of macrophages on culture conditions, e.g. release of thymidine.[118]

4. INTERACTION OF MACROPHAGES WITH B AND T LYMPHOCYTES

4.1 Visualization of macrophage–lymphocyte interaction

The anatomical feasibility of macrophage–lymphocyte interaction upon macrophage transfer was directly demonstrated using radio-label tagged cells.[119–121] It was calculated that about 5% injected pulmonary alveolar macrophages and 10% peritoneal macrophages would locate in the spleen.[120] This direct proof was obviously of great importance since so many studies on M-Ag immunogenicity rely on cell transfer and postulate such an interaction. Morphological experiments at the electron microscope level have revealed that macrophages and lymphocytes may be found in close approximation with each other and that extensive areas of close flattened surfaces may exist between the two cell types (e.g. Reference 122, reviewed in Reference 2). The anatomic location where endogeneous lymphocytes and reticulo-endothelial cells come in close contact was also studied in detail.[123] An interesting aspect is that the kinetics of lymphocyte recirculation is altered upon administration of antigen. Depending on the route of administration, lymphocytes become sequestered in spleen or draining lymph nodes (J. Sprent, herein). The extent of this phenomenon seems to be related to whether or not the antigen can be easily phagocytosed and it has been suggested that macrophages are primarily responsible for lymphocyte ‘trapping’.[124,125]

The demonstration of a dichotomy between T and B lymphocytes and of activation of both cell types by M-Ag has brought up the question whether macrophages would interact with T and B lymphocytes, together or separately, and in what order. To treat this point we will focus on the very elegant work of O Werdelin *et al.*[126] and M. H. Nielsen *et al.*[127] which, while in accordance with other earlier work, offer a peculiarly careful and in-depth study of the parameters and mechanisms of 'macrophage–antigen–T, B lymphocyte' interactions.

These authors showed that guinea pig peritoneal macrophages mixed with syngeneic lymph node cells and antigen (PPD, horseradish peroxidase or diphtheria toxoid) formed clusters typically composed of one macrophage surrounded by several lymphocytes. Using PPD and cells from *M. tuberculosis* immunized animals, 167 out of 1000 macrophages had seven or more lymphocytes attached; with cells from non-immune animals the frequency was 20/1000. Macrophages having contact with only one or two lymphocytes were quite abundant as may be expected by chance after adding lymphocytes to macrophage monolayers. Also the number of clusters containing less than five lymphocytes was identical whether cells were taken from primed or unprimed animals. These clusters were considered of dubious significance and the study was done on clusters made of one macrophage surrounded by 7 to 25 lymphocytes. It should be noted that in the same type of culture but using tenfold higher concentrations of lymphocytes and sixfold higher concentrations of macrophages P. E. Lipsky and A. S. Rosenthal[128] reported that about 50% of macrophages were binding an average of 1·5 lymphocytes per cell (the range was not given) even in the absence of antigen. The claim for a significance of this observation for the mechanism of macrophage–lymphocyte interaction and immune induction is questionable.

Kinetic studies on cluster formation showed that none was formed two hours after addition of lymphocytes to the macrophage monolayer but that plateau cluster formation was reached at 8 hours. Clusters remained stable for at least 20 hours, the last point examined. Cluster numbers increased with increased concentrations of antigen in the culture from 1 ng to 10 μg/ml. No increment was obtained with higher concentrations (20 and 50 μg/ml).

Immune cells formed high numbers of clusters only when the antigen was used to which the animal had been sensitized showing that the phenomenon was specific. Using mixtures of sensitized macrophages and normal lymphocytes and vice versa it was shown that the specificity laid entirely on lymphocytes. Thus immune lymphocytes were forming as high a number of clusters with immune or non-immune M-Ag and normal lymphocytes were not forming more clusters with immune M-Ag than with normal ones. The number of clusters produced was directly proportional to the number of immune lymphocytes present in the culture. Thus cluster formation involves specific lymphocytes, the corresponding antigen and macrophages which play a non-specific role as far as antigen is concerned.

The sequence of antigen binding, first to lymphocytes or first to macrophages, and the more precise role of macrophages in antigen presentation was further investigated. Either lymphocytes or macrophages were incubated with antigen, washed and mixed with the other cell type which had not seen antigen. No further antigen was added to the culture. M-Ag mixed with antigen-free lymphocytes made as many clusters as when antigen was added to the medium as usual. Antigen-exposed lymphocytes mixed with antigen-free macrophages gave cluster numbers slightly better than controls but were quite ineffective. Further demonstration that antigen bearing macrophages are the ones involved in cluster formation was provided by labelling M-Ag with latex particles and mixing them with latex-free, antigen-free macrophages. Clusters were found exclusively with latex-labelled macrophages. The reciprocal experiment with antigen-free, latex-labelled macrophages mixed with antigen-bearing, latex-free cells gave the reciprocal result. Quite clearly macrophages bind antigen and serve as a focus for the attraction of lymphocytes specific for that antigen.

The nature of bound lymphocytes and more precise analysis of cell relationships within a cluster were studied using transmission and scanning electron microscopy[127] (see Figure 1 (Plate XIII)). The crucial finding was that the cluster consisted of an outer envelope of small lymphocytes (7 to 25) with the morphology of T cells attached through a uropod *not to a central macrophage* but to one central lymphocyte showing B cell characteristics and early stages of blast transformation. This B lymphocyte in turn, and this one only, had a broad surface contact with a macrophage. It was already reported that macrophages would preferentially interact with B cells because of the stabilization provided by binding of lymphocyte surface Ig by macrophage Fc receptor.[129]

These studies do not prove that the cluster characterized do indeed function as cooperative units; however, we will see in Section 4.2 that this model of 'macrophage–antigen–B, T cell' interaction explains best most functional studies. Another much publicized model suggested an initial antigen-independent lymphocyte–macrophage binding later stabilized by antigen (References 87, 96, 113–116, 128, 130–136 and reviewed in References 137, 138). It appears less generally applicable and attractive for, as discussed above, the binding in absence of antigen was quite marginal, DNA synthesis in primed T cells was almost the only test used with little detailed morphological characterization and some features are in opposition with results obtained in other species.

4.2 Functional studies

The dichotomy of the lymphoid system in a B component responsible for antibody synthesis and a T component responsible for cellular reactions, and various subpopulations are described in detail by P. C. L. Beverley, herein. We have seen before that antigen-bearing macrophages could potentiate or were required for both B and T cell activation. It is well known that activation of B lymphocytes also requires T cell cooperation (reviewed in Reference 139). Thus T deprived animals do not make antibody to most antigens with the exception of those antigens called 'thymus-independent' (G. G. B. Klaus, herein). Moreover, small haptenic groups which do not induce antibody formation by themselves are able to do so when attached to an immunogenic carrier ('helper effect'). The anti-hapten response shows a remarkable carrier specificity in that a secondary response to the hapten will be elicited only when it is administered on the same carrier molecule that served for priming ('carrier effect'). There is ample evidence to assign this hapten-carrier cooperative response to B and T cell cooperation.[140–142]

It was shown that association with macrophage does not overcome the T dependency of the antibody response to thymus dependent antigens[80,98] nor does it overcome the carrier specificity.[88,110] Thus while macrophages may also 'help' B cells, the nature of this help is different from that provided by T cells; classical helper effect and carrier effect are a unique property of T lymphocytes.

Close contact is required physiologically for T–B cooperation.[143] Considering the small chances for specific T and B lymphocytes to meet it was suggested that antigen-bearing macrophages could serve as a focus for T–B cooperation. We saw in the preceding section that there was excellent morphological evidence for this possibility. Functional studies using hapten-carrier system also point in that direction.

The effect of antigen bearing macrophages on the expression of B memory cells specific for the DNP group was studied using a transfer system.[88] Briefly, irradiated mice received a low number of DNP-BGG primed spleen cells (which would not mount a good response to challenge by soluble DNP-BGG) and M-Ag. The response was determined both as antibody binding capacity of serum using an indirect Farr assay and as the number and expression of anti-DNP clones using isoelectric focusing. It was first found that macrophages carrying DNP on an heterologous carrier (MSH) or macrophages carrying the carrier BGG alone had no influence on the anti-DNP response while DNP-BGG carrying macrophages enhanced both clonal expression with resulting high levels of antibody production and the number of anti-DNP clones colonizing the spleen effectively. It was argued that if B cells recognizing DNP and T cells recognizing BGG were triggered independently by macrophages in this system, an enhanced response would also be expected after transfer of a mixture of one macrophage population carrying DNP-MSH and another carrying BGG. Some enhancement of the anti-DNP response was seen after such transfer of mixed hapten and carrier bearing macrophages but the effect was much lower than after transfer of DNP-BGG carrying macrophages. It was concluded that the simplest explanation of the results was that the enhancing effect of antigen-carrying macrophages is due to the attraction of T and B cells in spleen foci where they have an increased chance to interact.

A result apparently contradicting this conclusion was reported by Unanue and Katz.[111] They found that hapten and carrier determinants had to be on the same *molecule* on macrophages and that merely having the two determinants on the same macrophage was not sufficient. Although several explanations could be put forward (e.g. lack of proximity of the two determinants at the macrophage surface, role of exocytosed material where antigenic determinants for T and B cells would be separated . . .) it remained puzzling that macrophage bearing determinants for both T and B cells would not be effective for they should be attracting both B and T cells in a focus. An explanation of this finding may be found in the morphological studies of M. H. Nielsen *et al.*[127] reported above, which shows that only B lymphocytes have antigen-mediated contact with macrophages in a cell cluster but that T lymphocytes are in contact with the central B lymphocyte only, not the macrophage (see Figure 1 (Plate XIII)).* Thus hapten specific B lymphocytes binding the hapten on a heterologous carrier would not be able to interact with autologous carrier specific T lymphocytes even though the carrier molecule is present at the macrophage surface (Figure 2).

4.3 Interaction across a histocompatibility barrier

We saw that the immunogenicity of antigen-carrying macrophages was greatly diminished when transfer was done in allogeneic or xenogenic animals.[35,40] This was interpreted to be due to the fast rejection of histoincompatible cells. However, *in vitro* experiments in mice appear to show that allogeneic mixtures of macrophages and lymphocytes were as efficient as syngeneic mixtures.[110] The results are not quite clear for in these studies M-Ag were not effective in enhancing the direct-PFC and only marginal in enhancing indirect-PFC when compared to soluble antigen. Moreover, experiments done with lymphocyte populations depleted of macrophages showed that the response to soluble antigen was not macrophage dependent in this system! Studies in the guinea pig appear to show that histocompatibility is required for T cell activation by antigen-carrying macrophages.[130,131,134,137,138] A first antigen-independent phase of binding of lymphocytes by macrophages would occur in syngeneic and allogeneic combinations, but subsequent DNA synthesis would occur only if macrophages and lymphocytes show identical membrane histocompatibility linked determinants. We already discussed why this model appears unattractive. The recent finding by Schirrmacher *et al.*[144] of specific lymphocyte-activating

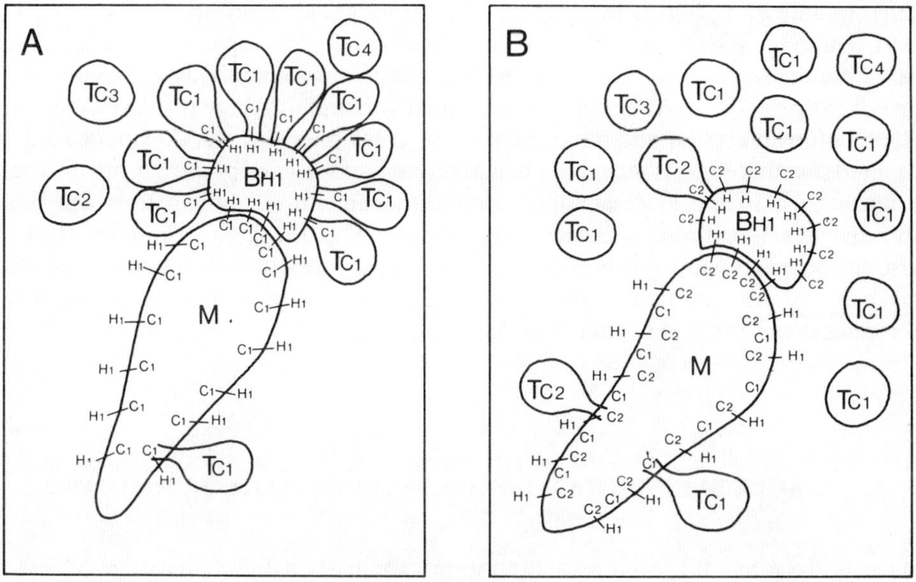

Figure 2. Schematic representation based on Nielsen *et al.*,[127] clusters (see Figure 1 (Plate XIII)) of interaction between B lymphocytes primed to hapten 1 (BH1), T lymphocytes primed to carrier 1 (Tc1) and macrophages bearing (A) hapten 1 on the homologous carrier (H1-C1) or (B) carrier molecule (C1). Only in (A) can primed T cells interact efficiently with primed B cells

determinants exposed on macrophage makes the interpretation of activation or not across histocompatibility barrier more intricated and difficult. It would appear to me that as for T–B lymphocytes cooperation across histocompatibility barrier, there is no intrinsic block to a cooperation between allogeneic cells[95] but that actual activation or suppression depends greatly on the experimental system used and is influenced by many other factors different from the histocompatibility difference itself (homing, allogeneic effect . . .). The role of histocompatibility region antigens in lymphocyte activation is treated in the chapters by H. G. Bluestein, and M. Nabholz and V. C. Miggiano, herein.

4.4 Macrophages and high or low responsiveness

High and low responder status are described for many antigens (reviewed in References 145, 146). Generally the defect is due to one gene, often but not always related to the main histocompatibility region. The cellular basis of the defect, i.e. whether it is expressed in T cells, B cells or macrophages or in the cooperative mechanism between those cells, is still a matter of controversy but the dogma has it that it would more often involve T cells. In the few cases where macrophages of high and low responder strains have been studied it was concluded that the genetic control of the immune response was not exerted at the level of the cells responsible for antigen handling.[26,147] It was, however, reported that genetically controlled low immune response to polyproline in DBA/1 mice could be phenotypically corrected by macrophages, probably due to increasing the chances of better T–B interaction.[148]

* Recent evidence was obtained that activated T cells may also be at the centre of a cluster (M. H. Nielsen, personal communication).

One notable exception is the high and low responder strains of mice selected by G. Biozzi. These mice were genetically selected for high or low response to sheep erythrocyte,[149] but the segregated genes were found to regulate the antibody response to a wide range of antigens.[150,151] Although it was originally thought that the primary defect was at the level of proliferation and differentiation of B lymphocytes, good evidence has been presented that this effect is secondary to a different handling and presentation of antigen in the mice.[152,153] In the low responder strains the uptake of antigen to which the low responsiveness is expressed is increased compared to the high responder strain, the lysosomal enzymes activity is higher, the intracellular digestion faster and the membrane antigen persistence greatly reduced. Thus high and low responder status may depend on macrophages but this is not usually the case.

5. SOLUBLE MEDIATORS OF MACROPHAGE FUNCTION

Besides antigen handling and presentation macrophages may influence lymphocytes by secreting a series of 'factors' or by binding 'factors' released by other cells. Proposed factors may be divided into two main types: the antigen specific ones believed to be T cell products binding to macrophages and those that are not antigen specific. These pharmacologically active products could be of great importance in the regulation of the immune response and mediate the activity of many adjuvants.

5.1 Mediation of T–B cooperation by T cell products binding to macrophages

It has been proposed that the sequence of T and B lymphocyte cooperation was (a) binding of antigen to T cell receptors, (b) release of soluble receptor–antigen complexes, (c) their fixation on the macrophage plasmamembrane and (d) presentation of antigen to B lymphocytes as a highly immunogenic lattice.

In one model the T cell factor was first described as 'carrier specific macrophage migration inhibition factor'[154] or 'cooperative antibody'[155,156] cytophilic for macrophages. Later, however, it was claimed that this factor was non-Ig but a product of the I histocompatibility region and that it reacted directly with B cells having a receptor site for it.[157–162] The early claim of macrophage requirement as an intermediary step in B cell activation was ignored in the later model-building games.

This model of lattice formation by T receptor–antigen complexes fixed on macrophages was also proposed by M. Feldmann and coworkers after an extensive series of experiments *in vitro*.[109,163–167] However, here the antigen specific T cell factor was presented as monomeric IgM.

J. W. Goodman, herein, reviews the controversy about the nature of T cell receptors and M. H. Schreier and A. A. Nordin, herein, critically evaluate immune responses *in vitro*. The elegant work of M. Feldmann undoubtedly shows that under the conditions used T–B cooperations may be mediated through soluble factors, however, the relevance of this mechanism *in vivo* for physiological T–B cooperation remains to be demonstrated. It should be recalled that allogeneic factors which are able to replace T cells for *in vitro* T–B cooperation are not systematically effective *in vivo*[143] and that antigen binding to macrophages does not require Ig.

5.2 Pharmacologically reactive factors

M. Hoffmann and R. W. Dutton reported that the defective *in vitro* response to SRBC of macrophage depleted spleen cell suspensions could be restored by the addition of macrophage culture supernatants.[171] Several explanations were offered including that macrophages would release an enzyme whose interaction with erythrocytes is required for immunogenicity. This would be in accordance with the abovementioned findings that the response to solubilized erythrocytes is macrophage independent.

Since then a whole battery of immunologically active factors has been reported to be released by macrophages and to either potentiate or suppress the immune response. P. Erb and M. Feldmann did a series of experiments on such factors.[172–178] In an *in vitro* system of generation of helper T cells they first showed that macrophage depleted T cell populations failed to yield helper cells and that restoration could be obtained with purified peritoneal macrophages showing similarities in the I–A part of the main histocompatibility region. They then found that T cell activation for helper activity did not require cell contact but was mediated by soluble factors. One of them, of importance with soluble antigens, is active exclusively when coming from I–A compatible macrophages; another, involved in the response to particulate antigens, does not require histocompatibility.

J. Calderon *et al*[179–181] reported that peritoneal exudate cell culture supernatants contained two biological activities. One had an inhibitory effect on DNA synthesis and proliferation of various cells in culture, whether EL-3 leukaemia cells, 3T3 cells or mitogen-stimulated spleen lymphocytes. The factor responsible had a molecular weight below 1400 daltons. The other factor induced proliferation of thymocytes and to a lesser extent spleen cells. It also increased the *in vitro* PFC response to a hapten-carrier protein. Its apparent molecular weight was between 15 000 and 21 000 daltons. Opitz *et al.*[182] also found a DNA synthesis inhibitory factor released by macrophages but on biochemical characterization found that it consisted of thymidine, a degradation product of cells dying in culture.[118]

This last finding raises a central problem about all these factors; that of their physiological significance. Indeed most of them have not been biochemically characterized and their activity is demonstrated under very peculiar conditions mostly *in vitro*. It is my feeling that so many factors modulate the immune response that the results are not sufficiently significant. Findings like that of Opitz *et al.*[118] and the evaluation of the requirements for *in vitro* responses analysed by M. H. Schreier and A. A. Nordin, herein, calls for considerable discretion in the interpretation of these results. Moreover, if pharmacologically active factors released by macrophages are indeed involved in the immune response their association with immunological phenomena may be accidental. Indeed macrophages are known to release a series of products which may act on neighbour lymphocytes, e.g. by modifying their plasmamembrane (F. Loor, herein) without being immunologically specific. They include lysozyme,[58,183] interferon,[84] elastase,[185] collagenase,[186] plasminogen activator,[187] cholesterol,[3] many undefined polypeptides,[3] aryl esterases[59] and other lysosomal enzymes.[3,60,188]

5.3 Macrophages and adjuvants

It is beyond the scope of this chapter to go into the vast literature on adjuvants and mechanism of adjuvanticity (for more information see Reference 189). Of interest here is that many adjuvants are thought to act via macrophages by labilizing lysosomal membranes and increasing hydrolytic activities: they include silica, vitamin A, beryllium,

Bordetella pertussis vaccine and bacterial endotoxins. Besides modification in antigen handling, activated macrophages may also increase synthesis and release of pharmacologically active products.

E. R. Unanue *et al.* have studied the mechanism of adjuvanticity of beryllium and *Bordetella pertussis* associated with macrophages on the immune response to MSH in a system of transfer of lymph node cells and macrophages to irradiated recipients.[190] They found increased immunogenicity with the sequence: soluble antigen, macrophage associated antigen and macrophage associated adjuvant plus antigen. Overall catabolism and retention of MSH was not altered by the presence of adjuvant; moreover a mixture of adjuvant-treated macrophages and antigen-treated macrophages had the same enhancing effect as when adjuvant and antigen were presented to the same macrophage population. Finally adjuvant treatment of lymph node cells, instead of macrophages before transfer had no effect. The results were interpreted to show that besides antigen presentation, activated macrophages may also secrete 'factors' stimulating neighbour lymphocytes.

Virtually the same kind of result was obtained with BSA as antigen and LPS as adjuvant.[83] Quite interestingly, here too treatment of lymphocytes with LPS before transfer had no effect on immunogenicity; only LPS treatment of macrophages was effective. It is interesting that LPS, which has been so widely used as a probe to study B cell activation, stimulates not only B cells but also macrophages[83,191] and through them T lymphocytes.[192,193] Since no lymphocyte population can be completely devoid of macrophages or macrophage precursors (see M. H. Schreier and A. A. Nordin, herein) or of T lymphocytes or T lymphocyte precursors, including in congenitally athymic nude mice,[194,195] it is a distinct possibility that LPS may act on B cells via macrophage–T cell activation as well as direct binding. The present considerations on multiple activation involving macrophages and macrophage products allow for an almost infinite variety of signals and make appear even shallower the recently much-advertised generalized one-signal model of lymphocyte activation (reviewed in References 196–199), challenged on other grounds by both G. G. B. Klaus and P. A. Bretscher, herein.

6. CONCLUSIONS

The presentation of the macrophage as a regulatory cell for lymphocyte activation is a matter of high bias. Indeed in phylogeny the reticulo-endothelial system appeared far earlier than the lymphoid system and, as pointed out by E. R. Unanue and J. Calderon,[200] clearly the immune system has arisen to help macrophages and not the reverse.

Nevertheless, within the optic chosen, macrophages have a crucial role in regulating the immune response. Lymphocytes are genetically determined to respond specifically to a single antigenic determinant and may be more intellectually stimulating to study, but it is largely antigen handling by macrophages which will determine whether an animal is going to mount a T or a B response or both, secrete antibody or build mainly memory, respond positively (immunity) or negatively (tolerance) or remain indifferent. This determination of the immune response mediated by macrophages takes place before any involvement of lymphocytes, although lymphocytes may in turn influence macrophages.[201] It modulates the immune response prior to other regulatory mechanisms, e.g. T–B cooperation (P. A. Bretscher, herein) or anti-idiotype cascade.[202]

The many facets of macrophage–lymphocyte interaction presented in this chapter provide for a very subtle modulation of lymphocyte responsiveness and should be remembered in any model of lymphocyte activation or inactivation.

7. REFERENCES

1. van Furth, R. (Ed.) (1975). *Mononuclear Phagocytes in Immunity, Infection and Pathology*. Blackwell, Oxford.
2. Unanue, E. R. (1972). *Advan. Immunol.*, **15**, 95–165.
3. Gordon, S. and Cohn, Z. A. (1973). *Int. Rev. Cytol.*, **36**, 171–214.
4. Rabinovitch, M. (1967). *Exp. Cell. Res.*, **46**, 19–28.
5. Griffin, F. M. and Silverstein, S. C. (1974). *J. exp. Med.*, **139**, 323–336.
6. Rhodes, J. M., Lind, I., Birch-Andersen, A. and Ravn, H. (1969). *Immunology*, **17**, 445–456.
7. Rhodes, J. M. and Lind, I. (1971). *Immunology*, **20**, 839–842.
8. Loor, F. and Roelants, G. E. (1974). *Eur. J. Immunol.*, **4**, 649–660.
9. Ehrenreich, B. A. and Cohn, Z. A. (1967). *J. exp. Med.*, **126**, 941–958.
10. Ehrenreich, B. A. and Cohn, Z. A. (1968). *J. Cell Biol.*, **38**, 244–248.
11. Ehrenreich, B. A. and Cohn, Z. A. (1969). *J. exp. Med.*, **129**, 227–243.
12. Haurowitz, F. (1960). *Ann. Rev. Biochem.*, **29**, 609–634.
13. Franzl, R. E. (1962). *Nature*, **195**, 457–458.
14. Campbell, D. H. and Garvey, J. S. (1963). *Advan. Immunol.*, **3**, 261–313.
15. Askonas, B. A. and Rhodes, J. M. (1964). In *Molecular and Cellular Basis of Antibody Formation* (Sterzl, J., Ed.), Publishing House of the Czechoslovak Academy of Sciences, Prague, pp. 503–512.
16. Askonas, B. A. and Rhodes, J. M. (1965). *Nature*, **205**, 470–474.
17. Uhr, J. W. and Weissman, G. (1965). *J. Immunol.*, **94**, 544–550.
18. Gottlieb, A. A., Glisin, V. R. and Doty, P. (1967). *Proc. Nat. Acad. Sci.*, **57**, 1849–1864.
19. Bishop, D. C., Pisciotta, A. V. and Abramoff, P. (1967). *J. Immunol.*, **99**, 751–759.
20. Fishman, M., Adler, F. L. and Holub, M. (1968). In *Nucleic Acids in Immunology* (Plescia, O. J. and Braun, W., Eds.), Springer-Verlag, New York, pp. 439–446.
21. Braun, W. and Cohen, E. P. (1968). In *Regulation of the Antibody Response* (Cinader, B., Ed.), Charles C. Thomas, Springfield, Mass., pp. 349–362.
22. Gottlieb, A. A. (1968). *J. Reticuloendoth. Soc.*, **5**, 270–281.
23. Raska, K. and Cohen, E. P. (1968). *Nature*, **217**, 720–723.
24. Roelants, G. E. and Goodman, J. W. (1968). *Biochemistry*, **7**, 1432–1440.
25. Roelants, G. E. and Goodman, J. W. (1969). *J. exp. Med.*, **130**, 557–574.
26. Roelants, G. E., Goodman, J. W. and McDevitt, H. O. (1971). *J. Immunol.*, **106**, 1222–1226.
27. Goodman, J. W., Roelants, G. E. and Byers, V. S. (1973). *Ann. N.Y. Acad. Sci.*, **207**, 288–300.
28. Sela, M. (1966). *Advan. Immunol.*, **5**, 30–129.
29. Lurie, M. B. (1942). *J. exp. Med.*, **75**, 247–268.
30. Porter, D. D., Porter, H. G. and Deerhake, B. B. (1969). *J. Immunol.*, **102**, 431–436.
31. Kaplan, M. H., Coons, A. H. and Deane, H. W. (1950). *J. exp. Med.*, **91**, 15–30.
32. Humphrey, J. H. (1969). *Antibiotica and Chemotherapia*, **15**, 7–23.
33. Rhodes, J. M. and Aasted, B. (1973). *Scand. J. Immunol.*, **2**, 405–415.
34. Kölsch, E. and Mitchison, N. A. (1968). *J. exp. Med.*, **128**, 1059–1079.
35. Mitchison, N. A. (1969). *Immunology*, **16**, 1–14.
36. Schmidtke, J. R. and Unanue, E. R. (1971). *J. Immunol.*, **107**, 331–338.
37. Cruchaud, A. and Unanue, E. R. (1971). *J. Immunol.*, **107**, 1329–1340.
38. Unanue, E. R. and Askonas, B. A. (1968). *J. exp. Med.*, **127**, 915–926.
39. Askonas, B. A. and Jaroskova, L. (1970). In *Developmental Aspects of Antibody Formation and Structure* (Sterzl, J. and Riha, I., Eds.), Academia, Prague, pp. 531–543.
40. Askonas, B. A. and Jaroskova, L. (1970). In *Mononuclear Phagocytes* (van Furth, R., Ed.), Blackwell, Oxford, pp. 595–610.
41. Unanue, E. R. (1969). *J. Immunol.*, **102**, 893–898.
42. Unanue, E. R., Cerottini, J.-C. and Bedford, M. (1969). *Nature*, **222**, 1193–1195.
43. Unanue, E. R. and Cerottini, J.-C. (1970). *J. exp. Med.*, **131**, 711–725.
44. Calderon, J. and Unanue, E. R. (1974). *J. Immunol.*, **112**, 1804–1814.
45. Kölsch, E. (1970). In *Mononuclear Phagocytes* (van Furth, R., Ed.), Blackwell, Oxford, pp. 548–560.
46. Cruchaud, A., Berney, M. and Balant, L. (1975). *J. Immunol.*, **114**, 102–109.
47. Wiener, E. and Curelaru, Z. (1973). *J. Reticuloendoth. Soc.*, **13**, 210–220.
48. Bowers, W. E. and de Duve, C. (1967). *J. Cell Biol.*, **32**, 339–348.
49. Bowers, W. E. and de Duve, C. (1967). *J. Cell Biol.*, **32**, 349–364.

50. Bowers, W. E. (1969). In *Lysosomes in Biology and Pathology*, Vol. 1 (Dingle, J. T. and Fall, H. B., Eds.), Elsevier, New York, pp. 167–191.
51. Jacques, P. J. (1969). In *Lysosomes in Biology and Pathology*, Vol. 2 (Dingle, J. T. and Fell, H. B., Eds.), Elsevier, New York, pp. 395–420.
52. Unanue, E. R., Schmidtke, J., Cruchaud, A. and Grey, H. (1970). In *Sixth International Symposium on Immunopathology* (Miescher, P. A., Ed.), Schwabe and Co., Basel and Stuttgart, pp. 35–51.
53. Huber, H., Polley, M. J., Linscott, W. D., Fudenberg, H. H. and Müller-Eberhard, H. G. (1968). *Science*, **162**, 1281–1283.
54. Lay, W. H. and Nussenzweig, V. (1968). *J. exp. Med.*, **128**, 991–1007.
55. Griffin, F. M., Bianco, C. and Silverstein, S. C. (1975). *J. exp. Med.*, **141**, 1269–1277.
56. Cohn, Z. A. (1975). *Fed. Proc.*, **34**, 1725–1729.
57. de Duve, C. and Wattiaux, R. (1966). *Annu. Rev. Physiol.*, **28**, 435–492.
58. Heise, E. R. and Myrvik, Q. N. (1967). *J. Reticuloendoth. Soc.*, **4**, 510–523.
59. Wiener, E. and Levanon, D. (1968). *Science*, **159**, 217.
60. Weissmann, G., Dukor, P. and Zurier, R. B. (1971). *Nature* (*New Biol.*), **231**, 131–135.
61. Henson, P. M. (1971). *J. exp. Med.*, **134**, 3, part 2, 114s–135s.
62. Steinman, R. M. and Cohn, Z. A. (1972). *J. Cell Biol.*, **55**, 186–204.
63. Boyden, S. V. and Sorkin, E. (1960). *Immunology*, **3**, 272–283.
64. Ferrarini, M., Munro, A. and Wilson, A. B. (1973). *Eur. J. Immunol.*, **3**, 364–370.
65. Huber, H. and Holm, G. (1975). In *Mononuclear Phagocytes* (van Furth, R., Ed.), Blackwell, Oxford, pp. 291–301.
66. Huber, H. and Fudenberg, H. H. (1968). *Int. Arch. Allergy Appl. Immunol.*, **34**, 18–31.
67. Berken, A. and Benacerraf, B. (1966). *J. exp. Med.*, **123**, 119–144.
68. Cline, M. J., Warner, N. L. and Metcalf, D. (1971). *Blood*, **39**, 326–330.
69. Lay, W. H. and Nussenzweig, V. (1969). *J. Immunol.*, **102**, 1172–1178.
70. Steinman, R. M. and Cohn, Z. A. (1975). In *Mononuclear Phagocytes* (van Furth, R., Ed.), Blackwell, Oxford, pp. 743–751.
71. Rhodes, J. (1975). *J. Immunol.*, **114**, 976–981.
72. Rhodes, J. (1975). *Nature*, **257**, 597–599.
73. Arend, W. P. and Mannik, M. (1975). In *Mononuclear Phagocytes* (van Furth, R., Ed.), Blackwell, Oxford, pp. 303–314.
74. Henson, P. M. (1969). *Immunology*, **16**, 107–121.
75. Nussenzweig, V. (1974). *Advan. Immunol.*, **19**, 217–258.
76. Unanue, E. R. and Cerottini, J.-C. (1970). *Semin. Hematol.*, **7**, 225–284.
77. Cruchaud, A. (1968). *Lab. Invest.*, **19**, 15–24.
78. Despont, J. P. and Cruchaud, A. (1969). *Nature*, **223**, 838–839.
79. Unanue, E. R. and Askonas, B. A. (1968). *Immunology*, **15**, 287–296.
80. Roelants, G. E. and Askonas, B. A. (1972). *Nature* (*New Biol.*), **239**, 63–64.
81. Askonas, B. A., Auzins, I. and Unanue, E. R. (1968). *Bull. Soc. Chem. Biol.*, **50**, 1113–1128.
82. Boak, J. L., Kölsch, E. and Mitchison, N. A. (1968). *Antibiotica et Chemoterapia*, **15**, 98–109.
83. Spitznagel, K. J. and Allison, A. C. (1970). *J. Immunol.*, **104**, 128–139.
84. Unanue, E. R. and Feldman, J. D. (1971). *Cell. Immunol.*, **2**, 269–274.
85. Cruchaud, A., Despont, J. P., Girard, J. P. and Mach, B. (1970). *J. Immunol.*, **104**, 1256–1261.
86. Schmidtke, J. R. and Dixon, F. J. J. (1972). *J. Immunol.*, **108**, 1624–1630.
87. Seeger, R. C. and Oppenheim, J. J. (1972). *J. Immunol.*, **109**, 244–254.
88. Askonas, B. A. and Roelants, G. E. (1974). *Eur. J. Immunol.*, **4**, 1–4.
89. Gallily, R. and Feldman, M. (1967). *Immunology*, **12**, 197–206.
90. Argyris, B. F. (1967). *J. Immunol.*, **99**, 744–750.
91. Ptak, W., Pryjma, J. and Moskalewski, S. (1974). *J. Reticuloendoth. Soc.*, **16**, 15–20.
92. Ford, W. L., Gowans, J. L. and McCullagh, P. H. (1966). In *The Thymus* (Wolstenholm, G., Ed.), Churchill, London, pp. 58–79.
93. Gowans, J. L. and McGregor, D. D. (1965). *Progr. Allergy*, **9**, 1–78.
94. Katz, D. H., Homaoka, T. and Benacerraf, B. (1973). *J. exp. Med.*, **137**, 1405–1418.
95. von Boehmer, H., Hudson, L. and Sprent, J. (1975). *J. exp. Med.*, **142**, 989–997.
96. Seeger, R. C. and Oppenheim, J. J. (1972). *J. Immunol.*, **109**, 255–261.
97. Klaus, G. G. B. (1974). *Cell. Immunol.*, **10**, 483–488.
98. Unanue, E. R. (1970). *J. Immunol.*, **105**, 1339–1343.
99. Mosier, D. E. (1967). *Science*, **158**, 1573–1575.

100. Pierce, C. W. (1969). *J. exp. Med.*, **130**, 345–364.
101. Dutton, R. W., McCarthy, M. M., Mishell, R. J. and Raidt, D. J. (1970). *Cell. Immunol.*, **1**, 196–206.
102. Shortman, K., Diener, E., Russel, P. and Armstrong, W. D. (1970). *J. exp. Med.*, **131**, 461–482.
103. Feldmann, M. and Palmer, J. (1971). *Immunology*, **21**, 685–699.
104. Shortman, K. and Palmer, J. (1971). *Cell. Immunol.*, **2**, 399–410.
105. Shortman, K., Williams, N., Jackson, H., Russel, P., Byrt, P. and Diener, E. (1971). *J. Cell Biol.*, **48**, 566–579.
106. Feldmann, M. (1974). *C.T. Mol. Immunol.*, **3**, 57–84.
107. Diener, E. and Feldmann, M. (1972). *Transplant. Rev.*, **8**, 76–103.
108. Feldmann, M. (1972). *J. exp. Med.*, **135**, 1049–1058.
109. Feldmann, M. and Nossal, G. J. V. (1972). *Transplant. Rev.*, **13**, 3–34.
110. Katz, D. H. and Unanue, E. R. (1973). *J. exp. Med.*, **137**, 967–990.
111. Unanue, E. R. and Katz, D. H. (1973). *Eur. J. Immunol.*, **3**, 559–563.
112. Parkhouse, R. M. E. and Dutton, R. W. (1966). *J. Immunol.*, **97**, 663–669.
113. Cohen, B. E., Rosenthal, A. S. and Paul, W. E. (1973). *J. Immunol.*, **111**, 811–819.
114. Cohen, B. E., Rosenthal, A. S. and Paul, W. E. (1973). *J. Immunol.*, **111**, 820–828.
115. Waldron, J. A., Horn, R. G. and Rosenthal, A. S. (1974). *J. Immunol.*, **112**, 746–755.
116. Ellner, J. J. and Rosenthal, A. S. (1975). *J. Immunol.*, **114**, 1563–1569.
117. Harris, G. (1967). *Immunology*, **12**, 159–163.
118. Opitz, H.-G., Niethammer, D., Jackson, R. C., Lemke, H., Huget, R. and Flad, H. D. (1975). *Cell. Immunol.*, **18**, 70–75.
119. Roser, B. (1965). *Aust. J. exp. Biol. Med. Sci.*, **43**, 553–562.
120. Russell, P. and Roser, B. (1966). *Aust. J. Biol. exp. Med. Sci.*, **44**, 629–638.
121. Gillette, R. W. and Lance, E. M. (1971). *J. Reticuloendoth. Soc.*, **10**, 223–237.
122. Gallily, R. and Ben-Ishay, Z. (1974). *Cell. Immunol.*, **11**, 314–324.
123. Nossal, G. J. V. and Ada, G. L. (1971). *Antigens, Lymphoid Cells and the Immune Response*, Academic, New York.
124. Frost, P. and Lance, E. M. (1974). *Immunology*, **26**, 175–186.
125. Frost, P. (1974). *Immunology*, **27**, 609–616.
126. Werdelin, O., Braendstrup, O. and Pedersen, E. (1974). *J. exp. Med.*, **140**, 1245–1259.
127. Nielsen, M. H., Jensen, H., Braendstrup, O. and Werdelin, O. (1974). *J. exp. Med.*, **140**, 1260–1272.
128. Lipsky, P. E. and Rosenthal, A. S. (1975). *J. Immunol.*, **115**, 440–445.
129. Schmidtke, J. and Unanue, E. R. (1971). *Nature* (*New Biol.*), **233**, 84–86.
130. Rosenthal, A. S. and Shevach, E. M. (1973). *J. exp. Med.*, **138**, 1194–1212.
131. Shevach, E. M. and Rosenthal, A. S. (1973). *J. exp. Med.*, **138**, 1213–1229.
132. Cohen, B. E. and Paul, W. E. (1974). *J. Immunol.*, **112**, 359–369.
133. Rosenstreich, D. L. and Rosenthal, A. S. (1974). *J. Immunol.*, **112**, 1085–1093.
134. Lipsky, P. E. and Rosenthal, A. S. (1975). *J. exp. Med.*, **141**, 138–154.
135. Rosenthal, A. S., Blake, J. T. and Lipsky, P. E. (1975). *J. Immunol.*, **115**, 1135–1139.
136. Oppenheim, J. J., Elfenbein, G. J. and Rosenstreich, D. L. (1975). In *Mononuclear Phagocytes* (van Furth, R., Ed.), Blackwell, Oxford, pp. 793–811.
137. Rosenthal, A. S., Lipsky, P. E. and Shevach, E. M. (1975). *Fed. Proc.*, **34**, 1743–1748.
138. Rosenthal, A. S., Lipsky, P. E. and Shevach, E. M. (1975). In *Mononuclear Phagocytes* (van Furth, R., Ed.), Blackwell, Oxford, pp. 813–823.
139. Davies, A. J. S. (1969). *Transplant. Rev.*, **1**, 43–91.
140. Rajewsky, K., Schirrmacher, V., Nase, S. and Jerne, N. K. (1969). *J. exp. Med.*, **129**, 1131–1143.
141. Mitchison, N. A. (1971). *Eur. J. Immunol.*, **1**, 10–17.
142. Mitchison, N. A. (1971). *Eur. J. Immunol.*, **1**, 18–27.
143. Rajewsky, K., Roelants, G. E. and Askonas, B. A. (1972). *Eur. J. Immunol.*, **2**, 592–598.
144. Schirrmacher, V., Pena-Martinez, J. and Festenstein, H. (1975). *Nature*, **255**, 155–156.
145. McDevitt, H. O. and Beneacerraf, B. (1969). *Advan. Immunol.*, **11**, 31–74.
146. Mozes, E. and Shearer, G. M. (1972). *C.T. Microbiol. Immunol.*, **59**, 167–200.
147. Cerottini, J.-C. and Unanue, E. R. (1971). *J. Immunol.*, **106**, 732–739.
148. Falkenberg, F. W., Sulica, A., Shearer, G. M., Mozes, E. and Sela, M. (1974). *Cell. Immunol.*, **12**, 271–279.

149. Biozzi, G., Stiffel, C., Mouton, D., Bouthillier, Y. and Decreusefond, C. (1968). *Ann. Inst. Pasteur*, **115**, 965–967.
150. Biozzi, G., Stiffel, C., Mouton, D., Bouthillier, Y. and Decreusefond, C. (1972). *J. exp. Med.*, **135**, 1071–1094.
151. Howard, J. G., Courtenay, B. M. and Desaymard, C. (1974). *Eur. J. Immunol.*, **4**, 453–457.
152. Weiner, E. and Bandieri, A. (1974). *Eur. J. Immunol.*, **4**, 457–463.
153. Mouton, D., Bouthillier, Y., Feingold, N., Feingold, J., Decreusefond, C., Stiffel, C. and Biozzi, G. (1975). *J. exp. Med.*, **141**, 306–321.
154. Lachmann, P. J. (1971). *Proc. Roy. Soc. Lond.*, **B176**, 425–426.
155. Taussig, M. J. and Lachmann, P. J. (1972). *Immunology*, **22**, 185–197.
156. Taussig, M. J. (1973). *C.T. Microbiol. Immunol.*, **60**, 125–174.
157. Taussig, M. J. (1974). *Nature*, **248**, 234–236.
158. Taussig, M. J., Mozes, E. and Isac, R. (1974). *J. exp. Med.*, **140**, 301–313.
159. Taussig, M. J. and Munro, A. J. (1974). *Nature*, **251**, 63–65.
160. Munro, A. J., Taussig, M. J., Campbell, R., Williams, H. and Lawson, Y. (1974). *J. exp. Med.*, **140**, 1579–1587.
161. Mozes, E., Isac, R. and Taussig, M. J. (1975). *J. exp. Med.*, **141**, 703–707.
162. Taussig, M. J. and Munro, A. J. (1975). In *Membrane Receptors of Lymphocytes* (Seligmann, M., Preud'homme, J. L. and Kourilsky, F. M., Eds.), North-Holland, Amsterdam, pp. 293–304.
163. Feldmann, M. and Basten, A. (1972). *J. exp. Med.*, **136**, 49–67.
164. Feldmann, M. and Basten, A. (1972). *J. exp. Med.*, **136**, 722–736.
165. Feldmann, M. (1972). *J. exp. Med.*, **136**, 737–760.
166. Feldmann, M. and Schrader, J. W. (1974). *Cell. Immunol.*, **14**, 255–269.
167. Feldmann, M., Schrader, J. W. and Boylston, A. (1975). In *Mononuclear Phagocytes* (van Furth, R., Ed.), Blackwell, Oxford, pp. 779–789.
168. Schimpl, A. and Wecker, E. (1972). *Nature* (*New Biol.*), **237**, 15–17.
169. Askonas, B. A., Schimpl, A. and Wecker, E. (1974). *Eur. J. Immunol.*, **4**, 164–169.
170. Kettman, J. and Skarvall, H. (1974). *Eur. J. Immunol.*, **4**, 641–645.
171. Hoffmann, M. and Dutton, R. W. (1971). *Science*, **172**, 1047–1048.
172. Kontiainen, S. and Feldmann, M. (1973). *Nature* (*New Biol.*), **245**, 285–286.
173. Erb, P. and Feldmann, M. (1975). *Nature*, **254**, 352–354.
174. Erb, P. and Feldmann, M. (1975). *Cell. Immunol.*, **19**, 356–367.
175. Erb, P. and Feldmann, M. (1975). *J. exp. Med.*, **142**, 460–472.
176. Erb, P. and Feldmann, M. (1975). *Eur. J. Immunol.*, **5**, 759–766.
177. Erb, P., Feldmann, M. and Hogg, N. (1976). *Eur. J. Immunol.*, **6**, 365–372.
178. Feldmann, M., Erb, P., Kontiainen, S. and Dunkley, M. (1975). In *Membrane Receptors of Lymphocytes* (Seligmann, M., Preud'homme, J. L. and Kourilsky, F. M., Eds.), North-Holland, Amsterdam, pp. 305–310.
179. Calderon, J., Williams, R. T. and Unanue, E. (1974). *Proc. Nat. Acad. Sci.*, **71**, 4273–4277.
180. Calderon, J., Kiely, J.-M., Lefko, J. L. and Unanue, E. R. (1975). *J. exp. Med.*, **142**, 151–164.
181. Calderon, J. and Unanue, E. R. (1975). *Nature*, **253**, 359–361.
182. Opitz, H. G., Niethammer, D., Lemke, H., Flad, H. D. and Huget, R. (1975). *Cell. Immunol.*, **16**, 379–388.
183. Gordon, S., Todd, J. and Cohn, Z. A. (1974). *J. exp. Med.*, **139**, 1228–1248.
184. Borecky, L., Lackovic, V., Fuchsberger, N. and Hajnicka, V. (1974). In *Activation of Macrophages* (Wagner, W.-H. and Hahn, H., Eds.), Excerpta Medica, Amsterdam, pp. 111–122.
185. Werb, Z. and Gordon, S. (1975). *J. exp. Med.*, **142**, 346–359.
186. Werb, Z. and Gordon, S. (1975). *J. exp. Med.*, **142**, 360–377.
187. Unkeless, J. C., Gordon, S. and Reich, E. (1974). *J. exp. Med.*, **139**, 834–850.
188. Welscher, H. D. and Cruchaud, A. (1976). *Advan. exp. Med. Biol.*, **66**, 705–710.
189. Wolstenholme, G. E. W. and Knight, J. (Eds.) (1973). *Immunopotentiation*, Ciba Foundation Symposium 18, new series. Associated Scientific Publishers, Amsterdam.
190. Unanue, E. R., Askonas, B. A. and Allison, A. C. (1969). *J. Immunol.*, **103**, 71–78.
191. Edelson, P. J., Zwiebel, R. and Cohn, Z. A. (1975). *J. exp. Med.*, **142**, 1150–1163.
192. Gery, I., Gershon, R. K. and Waksman, B. H. (1972). *J. exp. Med.*, **136**, 128–142.
193. Gery, I. and Waksman, B. H. (1972). *J. exp. Med.*, **136**, 143–155.
194. Raff, M. C. (1973). *Nature*, **246**, 350–351.

195. Loor, F. and Roelants, G. E. (1974). *Nature*, **251**, 229–230.
196. Coutinho, A. and Möller, G. (1974). *Scand. J. Immunol.*, **3**, 133–146.
197. Coutinho, A. (1975). *Transplant. Rev.*, **23**, 49–65.
198. Möller, G. (1975). *Transplant. Rev.*, **23**, 126–137.
199. Coutinho, A. and Möller, G. (1975). *Advan. Immunol.*, **21**, 113–236.
200. Unanue, E. R. and Calderon, J. (1975). *Fed. Proc.*, **34**, 1737–1742.
201. David, J. R. (1975). *Fed. Proc.*, **34**, 1730–1736.
202. Jerne, N. K. (1974). *Ann. Immunol. Inst. Pasteur*, **125c**, 373–389.

Chapter 7

An Evaluation of the Immune Response in vitro

M. H. SCHREIER
A. A. NORDIN

Abbreviations

AT cells: syngeneic activated T cells
CSF: colony stimulating factor
DAGG-: $(DNP\text{-}Ala\text{-}Gly\text{-}Gly)_n$-
DNP-: 2,4-dinitrophenyl
FBS: foetal bovine serum
Fe spleen cells: carbonyl Fe treated (macrophage depleted) spleen cells
HS: horse serum
LPS: bacterial lipopolysaccharide

ME: β-mercaptoethanol
MEM: minimum essential medium
NMS: normal mouse serum
PBS: phosphate buffered saline
PE(A) cells: peritoneal exudate (adherent) cells
PFC: plaque forming cells
SRBC: sheep red blood cells
UT cells: syngeneic control T cells (cf. AT cells)

1. INTRODUCTION

When Mishell and Dutton[1] reported the successful induction of the immune response totally *in vitro*, there were great expectations among immunologists that the cellular aspects of the immune response could now be completely understood. Few of us may remember that many scientists had made numerous attempts at *in vitro* stimulation of the immune response but failed. It is now difficult to imagine that these previous attempts met with such disastrous results since most investigators now can succeed after only limited trials. This high rate of success can only be attributed to the detailed description of the critical factors of the culture system as reported by Mishell and Dutton.[1,2] (We will not discuss the system described by Marbrook[3] since we have no experience with it.) These authors considered the following as critical features: (a) low O_2 tension, (b) gentle agitation of the cultures (rocking), (c) a selected appropriate foetal bovine serum (FBS), (d) adequate spleen cell density, (e) daily feeding of the cultures and (f) selected sheep erythrocytes that induced a high response. Under such defined conditions the *in vitro* system is comparable to the *in vivo* response since it is similar with respect to size, early kinetics, effect of antigen dose and the inhibitory effect of passively transferred antibody.[2]

Over the last decade the Mishell–Dutton *in vitro* culture system has been used by many investigators with only minor modifications. Even so, it is striking to note the marked quantitative difference not only from one laboratory to another but also from experiment to experiment within one laboratory over an extended period of time. Quantitative differences of at least three orders of magnitude, although troublesome, have not created the controversies encountered over qualitative differences. Most controversy has occurred in two broad areas, one being cellular requirements and the other being replacement of a cellular function by cell-products, cell-lines or defined chemical compounds.

During the past several years, we have attempted to analyse some parameters of the *in vitro* technique that we hoped would result in a more reproducible system. Because of the vast number of parameters that could be considered we have restricted the choice of antigens to sheep red blood cells (SRBC) and (DNP-Ala-Gly-Gly)$_n$-coupled Ficoll (DAGG-Ficoll),[4] the species of animal to the mouse, and have avoided any extensive cell separation procedures. For both antigens we used as assay for culture conditions the development of direct specific plaque-forming cells. The main emphasis in this paper is to consider the sources of variability within the system as opposed to mechanisms of cell functions. As such, this article is intended as a technical article rather than a review article as are the other contributions in this book. It is obvious then that it is impossible to quote all the reported studies of which there are more than 2000. In some instances the data we present will indeed be of a confirmatory nature and we will attempt to cite only the original observations.

2. CULTURE CONDITIONS

2.1 General culture conditions

For most of the experiments to be described here we used two minor modifications of the original system of Mishell and Dutton.[2] For reasons of simplicity these will be designated System A and System B, respectively.

System A differs very little from the original procedure: Eagle's minimum essential medium (MEM) is supplemented with Earle's salts, in which $CaCl_2$ has been reduced to 0·5 mM (see Section 2.2.1). The cultures are fed only twice instead of three times, i.e. after 24 and 48 hours. Since FBS was only slightly stimulatory, it was omitted from the 'nutritional mixture' in order not to interfere with serum-fractionation studies.

System B has the advantage of requiring no feeding. Instead of Eagle's MEM, RPMI 1640 is used and the final concentration of FBS is 20% instead of 5%.

In terms of size and early kinetics both systems give similar results and the number of PFC against SRBC is consistently higher as compared to the exact original method.

2.2 Medium

2.2.1 *Ionic conditions and pH*

The medium originally used by Mishell and Dutton,[2] namely Eagle's MEM, suspension type, is now widely replaced by MEM with Earle's salts or by RPMI 1640. The most remarkable difference between these three media is the content of Ca^{2+}. While the first contains no Ca^{2+} (and even a tenfold excess of Ca^{2+}-binding phosphates), Eagle's MEM with Earle's salts and RPMI 1640 are 1·8 mM and 0·5 mM in Ca^{2+}, respectively. Since several investigators including ourselves found RPMI 1640 superior to Eagle's MEM we titrated Ca^{2+} (see Figure 1) and found the Ca^{2+} optimum consistently at 0·5 mM. When we decreased the Ca^{2+} concentration from 1·8 to 0·5 mM in Eagle's MEM, both media supported the immune response *in vitro* to the same extent.

The system is remarkably resistant to changes in the ratio of the monovalent cations Na^+ and K^+ and even, to some extent, to their absolute concentration. We could increase K^+ from 5 to 25 mM keeping Na^+ at 105 mM or increase Na^+ from 105 to 145 mM at 5 mM

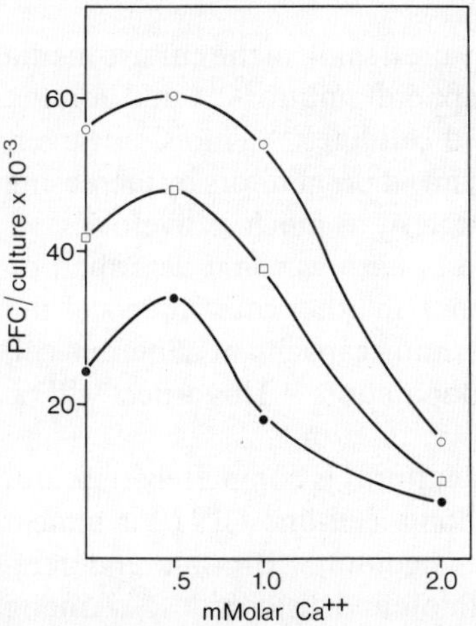

Figure 1. The effect of Ca^{2+} on the immune response *in vitro*. Eagle's MEM was prepared with Ca^{2+}-free Earle's salts and Ca^{2+} was added to the cultures at time zero as indicated on the abscissa. The endogenous Ca^{2+} (in FBS and cells) is ignored. PFC were determined on day 4 (●——●), day 5 (○——○) and day 6 (□——□). Culture system A

K^+ without affecting the number of PFC cultures (40 000 PFC/culture, data not shown).

The critical role of the pH on the immune response *in vitro* is widely established. The buffer system is bicarbonate—CO_2, which in many laboratories is supplemented with the organic buffer *N*-2-hydroxyethylpiperazine-*N'*-2-ethanesulphonic acid (HEPES) (0·015 to 0·02 M), which we routinely include in 'System A'. HEPES is not included in 'System B' which, owing to its higher protein content (20% FBS) has a higher buffering capacity. The concentration of bicarbonate is indeed very critical. We do not know to what extent this is due to consumption (carbon-donor) or to the buffer function.

In the original procedure the cultures were set up at 2·2 g of $NaHCO_3$ per litre and raised, because of the feeding schedule, by about 1 g per litre every day (consumption is ignored). We saw a slight advantage in starting the cultures at lower bicarbonate (1·0 g/l) and raising it only after 24 and 48 hours by 1 g/l, which corresponds to the old experience, that primary cultures should be started at low bicarbonate concentrations.

Bicarbonate and dextrose are in fact the most critical compounds in the feeding cocktail. The omission of these two compounds drastically reduces the number of PFC per culture, while the omission of non-essential and essential amino acids as well as glutamine has very little effect. Inversely, a feeding mixture containing only bicarbonate and dextrose is equivalent to the total 'nutritive mixture'.

2.2.2 *Purine and pyrimidine bases*

In 1972 Click and coworkers[5] showed a drastic stimulation of the immune response *in vitro* by the addition of a mixture of the four purine and pyrimidine bases (adenosine, guanosine, cytosine and uridine). This metabolic dependence should indeed be anticipated, since it has been shown in several species that bone marrow cells cannot synthesize a sufficient amount of purine-bases and are therefore dependent on an exogenous supply (from the liver).[6] Though the effect of this nucleotide mixture was not consistent in our hands, the influence of nucleotides on the anti-SRBC response *in vitro* was thoroughly tested in our Institute (Seegmiller and Schreier, in preparation). The addition of adenosine at high concentrations (1 mM) consistently increased the response by a factor of 2 to 10. The unphysiologically high doses required can be explained by considering that both FBS and mouse spleen cells contain adenosine-deaminase, which degrades this compound very rapidly.

2.2.3 *Thiols*

The inclusion of β-mercaptoethanol (ME), the most widely used thiol, in the culture media is stimulatory or even mandatory for both primary lymphoid cultures[7–10] and for the propagation of mouse leukaemic and neoplastic lymphoid cell lines.[11] Thiols, however, also have a significant effect on colony formation from erythroid progenitors by increasing the efficiency of erythropoietin-dependent colony formation by as much as fivefold.[12]

The stimulatory effect of ME on the immune response *in vitro* was first described by Click *et al.*[5] ME might never have received more attention than other constituents of the medium, had it not been shown to restore the antibody-forming capacity of adherent-cell depleted spleen cells to the size of an unfractionated cell population.[13] This aspect will be discussed in Section 4.2.4.

The mechanism of action of thiols on lymphoid cells in cultures is completely obscure. Based on the original observation that the mouse lymphoma cell-line L1210 is almost dependent on a high concentration of L-cysteine in the medium,[14] Broome and Jeng established precise structure activity relationships for 30 thiols and disulphides.[11] Among nine growth-promoting thiols the most active were α-thioglycerol and β-mercaptoethanol.[11] They are readily oxidized in tissue culture medium and the half-life of ME under these conditions is 5·9 hours.[11] The reduced state is not required for growth promotion. The stimulatory effect is only given in a very narrow concentration range, namely $2–5 \times 10^{-5}$ M. We will further discuss the effects of ME on the *in vitro* immune response as they occur throughout this chapter.

2.3 Serum

2.3.1 *Serum-free cultures*

A number of investigators have tried to bypass the problems involved with the use of growth-promoting homologous and heterologous sera by simply omitting them from the medium. Any results obtained in 'serum-free' cultures are confronted with two major objections. First, immune responses are notoriously low owing to poor cell survival and poor proliferation. Second, such cultures are free of only *exogenous* serum proteins, since it is well known that a considerable amount of serum protein attaches to the cell surface. The use of FBS during the preparation of the cell suspension or the removal of dead cells by centrifugation through a layer of FBS makes the term 'serum-free' an erroneous assumption.

2.3.2 *Foetal bovine serum*

Most investigators using the immune response *in vitro* agree on the pivotal role of the FBS, and from the very beginning much attention was given to the selection of the appropriate 'supportive', 'good' or 'positive' batch of FBS. This problem was judged important enough to result in a NIH-funded screening programme which was intended to establish commercially available lots of FBS which should guarantee scientists in this field successful *in vitro* immunization ('serum contract approved lots'). For any consideration of this problem, one has to be aware that marketed lots of FBS are pools of up to several hundred litres. Since a bovine foetus yields from very few millilitres to maximally 600 ml of serum,[15] a given lot represents a pool of 50 to 1000 individual sera.[16] Most randomly collected FBS (80% to 90%) support an *in vitro* immune response poorly or hardly at all and therefore have been termed 'deficient'.[16–18] The difference between supportive and deficient batches of FBS has been investigated in several laboratories including our own, but our present state of knowledge is hardly beyond the stage of speculation.

As widely different compounds as colony stimulating factor (CSF)[19] and bacterial lipopolysaccharide (LPS)[16] have been suggested as candidates to explain the difference between supportive and deficient sera. We could indeed confirm the observation that CSF, produced by the mouse cell-line JLSV5, significantly improves a deficient serum. However, we could never achieve a highly supportive serum. Moreover, when the CSF activity was enriched by several purification steps, there was no simultaneous enrichment of the supportive activity. This is in agreement with the observation of Metcalf (personal communication) that purified CSF from other sources had no effect on the *in vitro* immune response.

Supporters of any 'mitogen hypothesis' might feel encouraged by the recent findings of Shiigi and Mishell.[16] In a careful study based on more than two hundred samples of pooled FBS, the authors presented evidence that only those serum samples which were contaminated with bacteria during processing supported an immune response *in vitro*. These findings are in agreement with the observation that a deficient serum, after an intentional infection with Gram-negative psychrophilic micro-organisms, is converted into a moderately supportive serum via an unknown mechanism.

From several lines of evidence we tend to believe that the difference between a supportive and a deficient FBS cannot be attributed to the presence of any single factor but rather is of a much more complex nature.

(1) If a highly positive serum is fractionated, e.g. on a G-200 Sephadex column, and the fractions tested for their ability to support cultures containing a deficient FBS, some stimulatory effect can be found in every fraction. No single fraction, even at saturating input, can restore an optimal response.
(2) FBS can be fractionated by ammonium-sulphate precipitation into two interdependent fractions, F and A (F and A stand for the most abundant proteins of these fractions, fetuin and albumin, respectively (see Figure 2)). Neither fraction F (0–50% $(NH_4)_2SO_4$ precipitate), nor fraction A (60% to 80% $(NH_4)_2SO_4$ precipitate) can support an immune response *in vitro*. Upon recombination, the fractions become fully supportive. By titrating the optimal A:F ratio, we could achieve substantially better reconstituted sera as compared to the starting material. Criss-cross experiments, i.e. combination of fraction F from deficient sera with fraction A from a supportive batch of FBS and vice versa, combined with careful titrations (i.e. variation of the A:F ratio) revealed significant differences in the make-up of different deficient serum lots (Schreier and Vonzun, unpublished).

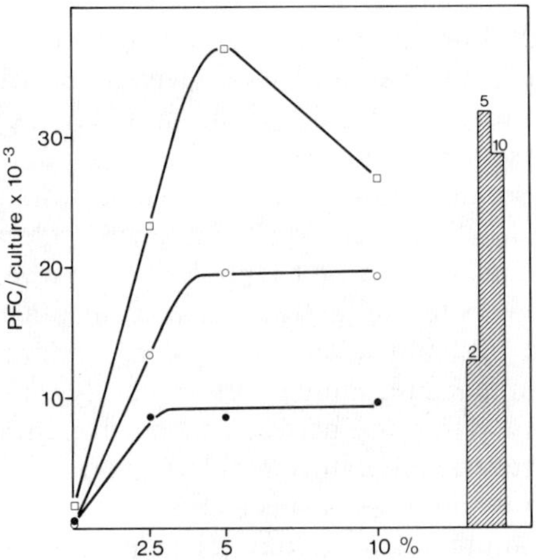

Figure 2. Reconstitution of a supportive FBS from the two interdependent Ammonium sulphate precipitates F (0–50%) and A (60–80%). The fractions were dialysed extensively and diluted to the original serum volume. Abscissa: Input of fraction F, corresponding to 2·5%, 5% and 10% FBS. Ordinate: Number of PFC formed upon addition of different amounts of fraction A, equivalent to (●——●) 2·5%, (○——○) 5% and (□——□) 10% FBS. The bars to the right, labelled 2, 5 and 10 indicate the number of PFC formed in the presence of 2·5%, 5% and 10% of the same unfractionated and undialysed lot of supportive FBS. Culture system A

(3) Addition of ME can improve the performance of deficient sera to a varying extent.
(4) The difference between a deficient and a supportive serum is often amplified by a difference in the kinetics of the response. Many investigators enumerate the number of PFC only on day 4 or day 5.
(5) An independent determination of endotoxin, using the limulus-assay,[20] shows no correlation between endotoxin activity and the supportive activity of the FBS (Table 1).

Table 1. The endotoxin content of 22 different lots of FBS, as measured in the limulus-assay,[20] and the ability of these sera to support an immune response *in vitro*[a]

Number of different serum lots	Limulus-assay[b]	Supportive	Deficient
5	−	3	2
9	+	5	4
6	++	2	4
2	+++	2	0

[a] Sera were tested in System B, using C57Bl/6 spleen cells and SRBC as antigen.
[b] The precipitation of an Endotoxin Standard by limulus extract gave the following results (in μg/ml)

< 0·1	−
1·0	±
10·0	+
50·0	++
100·0	+++

FBS has long been known to be extremely variable from lot to lot. Wide differences in concentration with respect to ions and especially hormones have been reported.[15,21] Moreover, the procedures presently used to collect FBS are far from being sterile and at least temporary contamination must be anticipated. The use of adult sera or, preferentially, the development of better defined media would therefore be highly desirable.

2.3.3 *Non-foetal serum source*

Normal mouse serum (NMS), as Mishell and Dutton[2] have shown, does not support a primary *in vitro* immune response to SRBC. If, however, mice were primed 4 days earlier with an immunogenic dose of SRBC *in vivo*, the *in vitro* response occurs nearly as well in NMS as in FBS.[2] Many investigators have reported an inhibitory effect by adding NMS to FBS-containing cultures. As shown in Table 2, a primary response to SRBC is much more affected than a secondary response, where the addition of small amounts of NMS gave repeatedly even a slight stimulation. This non-specific mouse serum inhibitor was shown to increase in concentration after immunization.[22] It did not bind to the antigen (SRBC), was soluble in 60% $(NH_4)_2SO_4$, and seems to migrate electrophoretically in the α-region. A similar inhibition was obtained with gerbil, rat and guinea pig sera, but not with human or rabbit serum.[22]

Table 2. The effect of NMS on the immune response *in vitro* PFC/culture

	Normal mice	Mice immunized 8 weeks earlier	Mice immunized 3 weeks earlier
No NMS	11 070	52 740	48 420
2% NMS	3 600	28 800	68 040
5% NMS	1 710	9 900	21 060
10% NMS	810	3 060	13 140

C57Bl/6J mouse spleen cells in culture system A (see Section 2.1).

Horse serum (HS), as a substitute for FBS, seldom elicits a significant *in vitro* immune response with unprimed spleen cells if measured only on day 4. We could, however, get excellent *in vitro* responses with delayed kinetics in the presence of 2% to 10% HS (Figure 3). The peak of the response can be even later in Falcon II microtitre plates (Figure 4). The

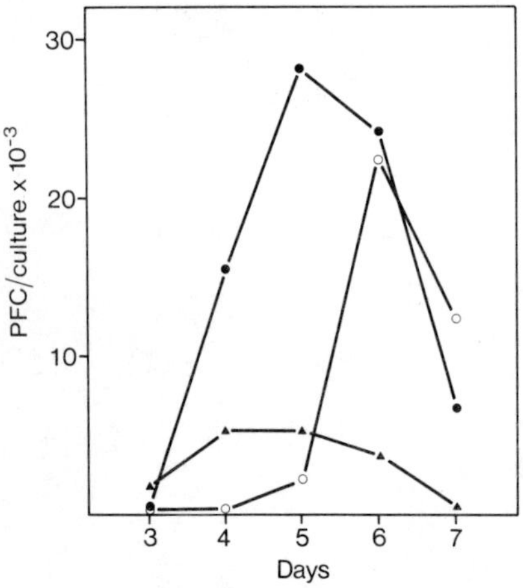

Figure 3. Kinetics of the immune response *in vitro* in the presence of different sera. ▲——▲ 5% FBS, ○——○ 2% HS, ●——● a combination of 2% FBS and 2% HS. Culture conditions: System A with 10^7 (C57Bl/6J × A/J)F_1 spleen cells/dish

inconvenient variability of the kinetics, for which we have no explanation yet, can be easily corrected by the addition of 2% FBS, which brings the peak of the response reliably to day 5 in the classical 1 ml cultures.

The mixture of HS and FBS has two remarkable advantages.

(1) The stimulatory effect of FBS and HS at 5% concentrations is more than additive, and at 2% each is almost interdependent (Table 3). This could possibly be exploited for the purification of the stimulatory activity in either HS or FBS.

Table 3. The immune response of mouse spleen cells to SRBC *in vitro*: Comparison of low dose of HS, FBS and a combination of both in culture system A

	PFC/culture Day 4	PFC/culture Day 5
2% HS	1 470	1 440
2% FBS	1 470	1 110
2% HS+2% FBS	8 970	23 220

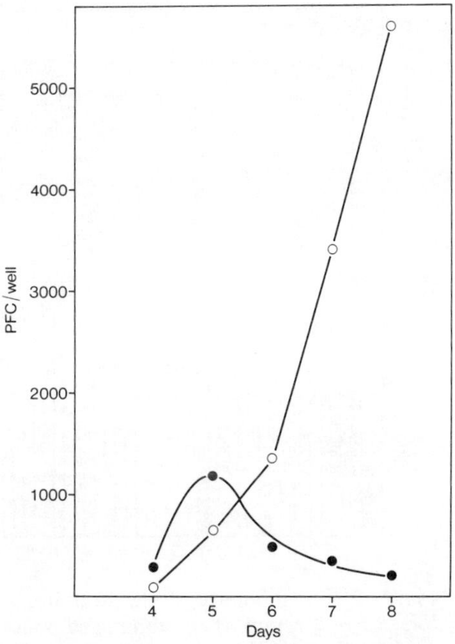

Figure 4. Kinetics of the immune response *in vitro* in Falcon II microtitre plates in the presence of FBS (●——●) and HS (○——○). The culture conditions are exactly as described for System A, i.e. each well contains, in a volume of 0·2 ml, 2×10^6 spleen cells and 2×10^6 SRBC. The cultures were fed with 20 μl 'feeding-mixture'[2] after 24 and 48 hours

(2) In combination with low doses of HS, all tested 'deficient' batches of FBS could be significantly improved; 30%–40% of the deficient sera in combination of 2% HS could well compete with an optimally supportive lot of FBS (Figure 5).

It is noteworthy that the majority of batches of HS show this potentiating effect. This is in agreement with the general experience that lot-to-lot variability is much less pronounced in HS than in FBS. The superiority of a serum mixture has been also observed for the growth of various continuous cell lines (G. Sato, personal communication).

2.4 Rocking and Feeding

These two parameters of the *in vitro* immune response have been widely varied or omitted by individual investigators. Based on valid observations different scientists have concluded that, for example, rocking is not necessary or feeding is not required. These convenient simplifications are often justified by other minor modifications, like the inclusion of ME or use of a different culture dish.

These trivial modifications can be controversial even within the same institute. In a large experiment, shown in Table 4, we varied only four out of an endless number of variables, namely rocking, feeding and the type of culture dish (Falcon 3001, Falcon 1008, Lux and Nunc) in the presence and absence of 5×10^{-5} ME. It has to be stressed that the two most

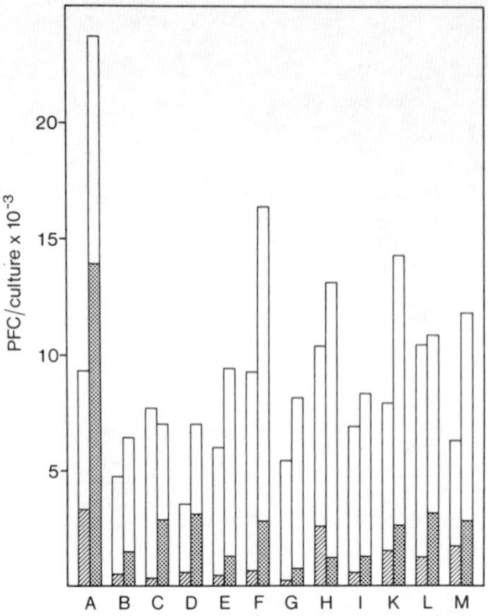

Figure 5. The influence of HS on different lots of FBS to support an immune response *in vitro* (System A). A supportive lot (A) and 11 deficient lots (B–M) were tested at 2% (hatched area) and 5% (stippled area) concentration. The superimposed white bars indicate the number of PFC upon addition of 2% HS. Number of PFC were measured on day 4. 2% HS alone gave rise to 750 PFC/culture

complex variables, namely the FBS and the cell source, were kept constant. One spleen cell pool from 20 mice was used for the whole experiment.

The results show clearly that much controversy can be already created with these four variables: investigator A assaying on day 5 finds no requirement for ME, while investigator B, who in contrast to A neither feeds nor agitates his cultures, finds them to be absolutely ME-dependent (lines 2 and 5, and 19 and 22, respectively).

Investigators C and D, who both rocked their cultures and included ME, will not agree on the necessity of feeding (lines 4 and 10, and 6 and 11 respectively), even though they exchanged spleen cell suspension and media, since C measured on day 4 and D on day 5. While investigator E, measuring on day 5, makes derogatory jokes about the value of rocking (lines 11 and 23), investigator F (lines 7 and 19) has a very short message, 'no rocking, no plaques'.

This simple experiment, in which few parameters within an otherwise controlled experiment are varied, shows impressively that *all* investigators A through F (and many others) are right.

3. ANTIGENS

As previously pointed out, Mishell and Dutton[2] listed the source of SRBC as one of the critical factors of the *in vitro* immune response. To emphasize again this important point, which has been forgotten or has gone unnoticed in some laboratories, we tested several

Table 4. The influence of rocking, feeding and ME on the immune response *in vitro* in different culture dishes (culture system A, anti-SRBC-response expressed as PFC/culture)

Line no.		Rocking	Feeding	ME	Falcon 3001	Falcon 1008	Lux	Nunc
1	Day 4				34 100	33 210	29 340	18 590
2	5	+	+	−	64 710	91 440	83 880	29 280
3	6				74 880	80 730	36 840	27 540
4	Day 4				53 060	39 510	40 050	22 050
5	5	+	+	+	86 760	76 770	74 340	31 680
6	6				59 490	46 620	16 980	16 980
7	Day 4				11 700	19 400	10 440	3 920
8	5	+	−	−	26 730	33 390	10 320	6 960
9	6				9 840	15 240	3 000	840
10	Day 4				45 410	40 010	33 300	20 970
11	5	+	−	+	16 560	6 930	14 040	6 540
12	6				11 340	5 700	1 380	1 920
13	Day 4				4 280	5 720	5 490	3 110
14	5	−	+	−	15 660	44 190	12 960	13 140
15	6				27 300	45 840	29 280	14 280
16	Day 4				23 360	12 920	17 550	15 440
17	5	−	+	+	52 110	47 640	40 800	42 300
18	6				69 570	45 600	40 680	33 960
19	Day 4				720	630	990	315
20	5	−	−	−	2 880	3 300	1 320	1 320
21	6				1 500	2 160	2 580	720
22	Day 4				18 270	14 090	15 800	13 860
23	5	−	−	+	45 450	48 240	31 620	37 260
24	6				13 560	8 160	2 880	10 920

individual sources of SRBC. Six sheep were bled aseptically into Elsever's solutions and each sample was tested for antigenicity on a constant source of spleen cells. The PFC generated in response to each sample of SRBC were assayed on all six sources of SRBC. The results are shown in Table 5. The influence that the SRBC exert on the *in vitro* immune response is easily appreciated. Sheep 103 was selected as the source of SRBC for all the experiments presented here, not only because of the high number of PFC generated but, equally important, because the number of PFC in non-stimulated cultures was low (<500

Table 5. Variation in the *in vitro* response of C57Bl/6 mouse spleen cells in culture system A to SRBC derived from different sheep (PFC/culture)

Sheep used for *in vitro* immunization	Sheep used for plaque assay					
	26	121	142	152	172	103
26	10 020	9 180	8 580	6 240	8 760	10 410
121	3 480	4 740	3 720	6 180	7 380	5 250
142	9 300	9 600	7 800	7 740	10 020	10 680
152	5 160	3 900	3 720	3 360	4 200	4 920
172	5 400	2 100	3 720	3 240	20 200	20 340
103	8 280	8 880	7 320	8 100	34 560	32 340

PFC/culture). The number of PFC generated in the *in vitro* immune response is directly related to the number of SRBC added (Table 6).

Table 6. Titration of SRBC from sheep 103 in culture system A

Number of SRBC added	Number of PFC/culture
None	480
10^5	4 890
10^6	18 750
10^7	34 800

The other antigen we used in these experiments was DAGG-Ficoll. The modification of Ficoll and the conjugation with *N*-(2,4-dinitrophenyl)-β-alanylglycylglycine Bor hydrazide, was done precisely as described by Inman.[4] The preparation that we used here contained 48 moles of DNP per mole of Ficoll. DAGG-Ficoll is highly immunogenic *in vivo*[23] and *in vitro*[24] and is a T-independent antigen.[23–25] Our studies confirm these findings. In addition, it is worth noting that DAGG-Ficoll did not behave as a mitogen (W. H. Adler, personal communication).

The *in vitro* immune response to DAGG-Ficoll was maximal 4 days after initiating the cultures. The PFC were detected by plating the cultured cells on SRBC conjugated with DAGG.[4] Properly prepared, DAGG-SRBC were stable for 3 weeks and were not subject to the spontaneous lysis frequently observed with SRBC conjugated with 2,4,6-trinitrobenzene sulphonic acid. Unstimulated cultures of spleen cells show a very low level of PFC when plated with either DAGG-SRBC or SRBC.

4. CELLS

4.1 The problem of cell density

An absolute requirement for successful *in vitro* immunization is a high cell density; the optimal concentration found by Mishell and Dutton[2] was $1\text{–}2\times10^7$ spleen cells per millilitre. This is in striking contrast to what is observed for the apparently T cell and adherent cell independent polyclonal activation of mouse B lymphocytes induced by bacterial lipopolysaccharide,[8] where the optimal initial cell density for both cell proliferation and differentiation is of the order of 10^5 cells/ml.[8,26] The requirement for high cell density has been a serious obstacle for numerous experimental approaches, such as limiting dilution analysis, and different approaches have been made to overcome this problem, e.g. the use of filler cells (irradiated spleen cells[27] or thymocytes).[28]

In Figure 6 we show a cell titration under 'optimal' and 'suboptimal' culture conditions, the only variable being the batch of FBS included in the medium. ME was added to all cultures to compensate for the dilution of adherent cells (see Section 4.2). As previously mentioned, the addition of ME to cultures containing deficient FBS results in moderate support of the *in vitro* response at standard cell density. Therefore, a considerable number of PFC is formed at 10^7 cells/ml under both conditions. However, there is a remarkable difference at lower spleen cell concentrations. With a highly supportive FBS we can work at lower cell density than with poorly supportive lots of FBS. Since all spleen cells come from the same pool and have the same number of B cell precursors per 10^6 cells, the

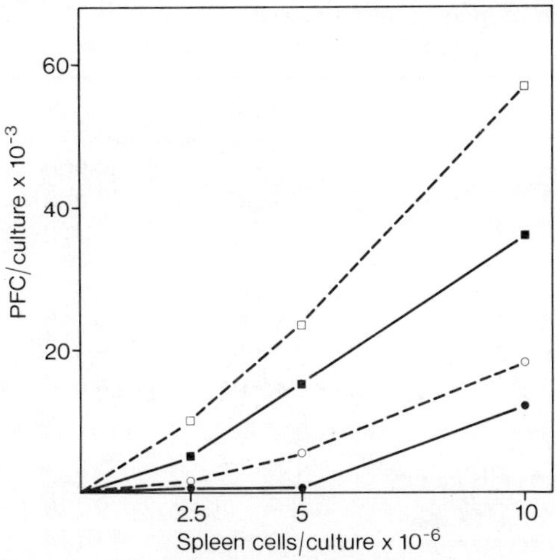

Figure 6. Titration of spleen cells in the presence of highly supportive (■, □) and poorly supportive FBS (●, ○) measured on day 4 (solid line) and day 5 (dashed line). All cultures (System A) contained 10^7 SRBC and ME (5×10^{-5} M)

difference in these titration curves suggests a significant difference in the induction of the helper T cell.

This notion is supported by several lines of evidence.

(1) Low dose priming of mice with SRBC (as few as 5×10^4 SRBC 3 days earlier) abolishes the difference between a highly supportive and a 'deficient' FBS (Table 7), suggesting that the increased responsiveness is a result of T cell priming.

Table 7. Number of PFC/culture with spleen cells from normal and preinjected* mice (culture system A tested at day 4)

	'Normal'	'Preinjected'*
5% 'positive' FBS	35 010	25 290
5% 'negative' FBS	1 650	27 900

* Preinjected with $1 + 10^6$ SRBC 3 days before.

(2) This situation can be mimicked *in vitro* by combining normal spleen cells with a source of syngeneic activated T cells (AT cells).[29] These are the spleen cells from mice that were irradiated (700 R) and injected with 5×10^7 thymocytes and 2×10^8 SRBC 7 days previously. If the thymocytes were injected without antigen (UT cells), no stimulatory effect was observed (Table 8).

(3) Since these findings suggest that the induction of the helper T cell is the limiting step in the immune response to SRBC *in vitro*, it is tempting to exploit this finding as a helper assay. To increase the sensitivity of the test system, we combined two unfavourable culture conditions, namely low cell density and 'deficient' FBS. As Table 8 and Figure 7 show, at a

Table 8. Titration of 'activated' (AT) and 'unactivated' (UT) T cells in culture system A (C57Bl/6J), (PFC/culture)

		Day 4	Day 5	Day 6
a	10^7 spleen cells alone	2 370	9 900	5 070
	$+2{\cdot}5\times10^5$ AT	5 100	21 060	24 210
	$+5\times10^5$ AT	11 370	26 370	34 800
	$+7{\cdot}5\times10^5$ AT	10 370	42 165	41 130
	$+2{\cdot}5\times10^5$ UT	3 810	9 150	7 260
	$+5\times10^5$ UT	2 400	6 360	6 150
	$+7{\cdot}5\times10^5$ UT	2 910	8 640	9 690
b	$2{\cdot}5\times10^6$ spleen cells alone*		240	
	$+2{\cdot}5\times10^5$ AT		9 720	
	$+5\times10^5$ AT		14 130	
	$+2{\cdot}5\times10^5$ UT		900	
	$+5\times10^5$ UT		930	

* All cultures contained 10^7 SRBC as antigen. Panels a and b are derived from two different experiments. For details see text.

density of 10^6 spleen cells/ml the induction of PFC became fully dependent on the addition of AT cells. If $2{\cdot}5\times10^5$ and 5×10^5 AT cells were co-cultured with $2{\cdot}5\times10^6$ spleen cells (Figure 7, panels a and b), the magnitude of the response did not increase significantly. Under these culture conditions, the AT cells determine the heights of the response and we are in fact working in B cell excess.

The situation is quite different if we perform the same experiment in the presence of a supportive lot of FBS; under these conditions, 1×10^6 spleen cells show a significant response with an earlier peak. The response increases non-linearly with increasing cell density and is maximal on day 5. The addition of AT cells, while essential at low cell density, has only little effect at higher cell input. The combined effect of added AT cells (which comes to bear at low cell density) and of increased cell density in the presence of supportive FBS, results in a nearly linear response. Here the number of PFC is proportional to the number of spleen cells, suggesting that all B cell precursors specific for SRBC are triggered to respond.

4.2 Adherent cells

In 1967 Mosier[30] reported that normal mouse spleen cells could be separated into two populations based strictly on the property of adhering to glass or plastic surfaces. Neither population upon *in vitro* stimulation with SRBC was capable of giving rise to PFC. However, after recombination of the two populations, the response to antigen was as good as with the unseparated spleen cell suspension. It is now widely accepted that the actual antibody-producing cells are derived from the non-adherent population and that the adherent population provides an essential accessory function for the *in vitro* induction of the immune response to most antigens. As a result, there have been numerous studies attempting to define this cell population. Although neither the precise cell type nor its

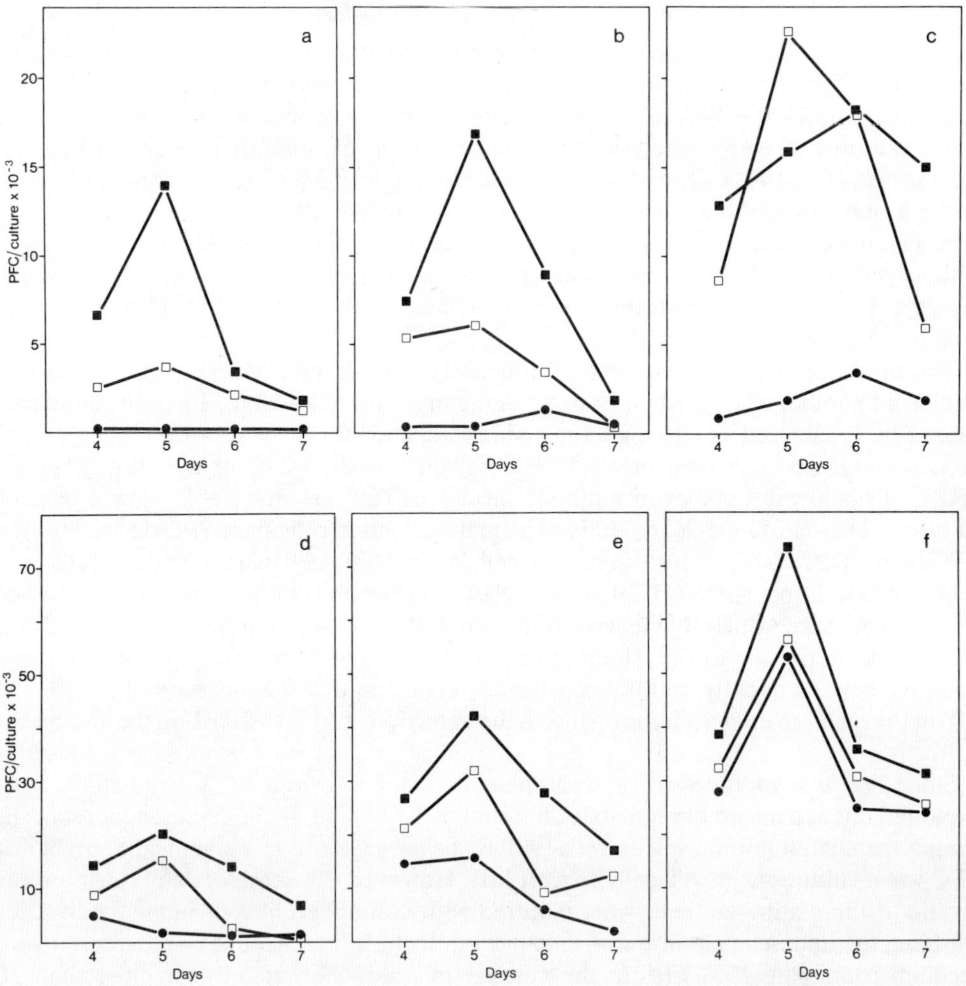

Figure 7. Kinetics of the stimulatory effect of activated T cells at different spleen cell concentrations in the presence of poorly supportive (panels a–c) and highly supportive (panels d–f) FBS. The spleen cell concentration was 1×10^6/ml (panels a and d), $2{\cdot}5\times10^6$/ml (panels b and e) or 5×10^6/ml (panels c and f). ●——● spleen cells only, □——□ spleen cells plus $2{\cdot}5\times10^5$ AT cells, ■——■ spleen cells $+5\times10^5$ AT cells. All cultures contain 5×10^{-5} M ME. Culture system A

function are known, there is some general agreement that the cell type involved is a macrophage but the function remains obscure. Excellent reviews[31–33] of these studies are available, and these findings will not be restated. We will instead adhere to our basic intent and discuss some technical problems relevant to accessory cells. These studies are not intended to define either the function or cell type(s) involved but to attempt to explain some reason(s) for controversies in this area.

Functionally active adherent cells for *in vitro* culture techniques are usually prepared from either spleen cells or peritoneal exudate cells. Since these represent two very different sources of accessory cells, each presenting different problems, we will deal with them separately.

4.2.1 *Adherent cells from spleen*

In experiments where we used adherent cells prepared from normal spleen, these cells were washed in medium without serum and were resuspended in Eagle's MEM, supplemented with 10% FBS. Tissue culture dishes (35 mm, Falcon 3001) containing 10×10^6 viable cells in 1 ml were placed in an atmosphere of the gas mixture used by Mishell and Dutton[2] (7% O_2, 10% CO_2 and 83% N_2) and incubated at 37 °C with rocking for 1 hour. After a standardized agitation of the dishes to detach the non–adherent cells, the supernatant fluid was discarded. The adherent layers were then washed twice with medium without serum with thorough agitation after each wash. One millilitre of medium containing 10% FBS was added and the adherent cells could then be kept at 37 °C in the gas mixture until needed.

This procedure results in a functional adherent cell layer which does not give rise to any significant number of PFC when cultured in the presence of antigen. However, this alone cannot be considered as evidence for a single functional cell type in the adherent cell population. If $0{\cdot}5 \times 10^6$ syngeneic AT cells are added to the adherent layer together with SRBC, a significant number of antibody producing cells are detectable after 4 days of culture.[34] These PFC are clearly derived from the adherent cells since PFC derived from a $(C57Bl/6 \times DBA/2)F_1$ mouse adherent cell layer, cultivated with C57Bl/6 cells, are sensitive to treatment with C57Bl/6 anti-DBA/2 serum and complement. Irrespective of the mechanism by which B cell activation occurs, there is no doubt that B cells or B cell precursors are present in the adherent cell layer prepared from spleen. Along the same lines, we have frequently, but not consistently observed that the inclusion of 5×10^{-5} M ME in the medium unexpectedly results in the detection of PFC to SRBC in the absence of AT cells.

Since the *in vitro* function of macrophages is not impaired by X-irradiation,[35] we exploited this as a means of reducing adherent B cell function. In our hands, adherent cells from X-irradiated normal spleen cells (900 R) never gave rise to a significant number of PFC when cultured with AT cells or with ME. However, the accessory cell function was erratic. If these adherent cells were cultured with non-adherent cells immediately after washing, the immune response was as expected. If the adherent cells were cultured with medium containing 10% FBS in the absence of non-adherent cells for more than 30 minutes, the accessory cell function deteriorated quite rapidly.

In summary, the B cell contamination of the adherent cell layer makes it difficult to interpret the results since the PFC can originate from both the adherent and non-adherent cell populations. A higher degree of complication occurs if ME is included in the culture medium. For this phenomenon we have no explanation.

4.2.2 *Adherent cells from the peritoneal cavity*

The other most commonly used source of adherent cells is peritoneal exudate (PE) cells. Such cells are harvested by flushing the peritoneal cavity of normal mice ('normal PE') or of mice previously injected with an irritant to increase the cell yield ('stimulated PE'). There is evidence that the cellular composition of the adherent cells prepared from normal and stimulated PE cells is different. Treatment of normal PE cells with anti-macrophage or anti-θ serum and complement kills more than 97% and less than 3%, respectively. The same treatment of adherent cells prepared from stimulated PE cells results in values of approximately 95% and 5%, respectively. Although PE cells contain T cells which increase when irritants are used, neither of the adherent cell populations contained cells binding anti-θ (K. Nakamura and H. Nariuchi, manuscript in preparation). Clearly both the source (normal or stimulated) and the type (adherent or unfractionated) of PE cells determines the ability of such preparations to supply various cell functions.

We use normal PE cells collected and washed under conditions selected to avoid cell clumping and adherence. They are obtained from 5–10 normal mice by flushing the peritoneal cavities with 3–4 ml of 0·34 M sucrose.[34] The washings are pooled into a siliconized conical tube kept in a melting ice bath. The cells are centrifuged and washed twice with 30 ml of Dulbecco's PBS without Ca^{2+} and Mg^{2+}. The PE cells are adjusted to the appropriate cell density with Eagle's MEM containing 10% FBS, and adherent cell layers (PEA) are prepared as described. The great majority, if not all, of the cells that remain attached after this procedure are of the macrophage type and do not give rise to any PFC when cultured under conditions that would have expressed PFC from normal spleen adherent cell layers.

4.2.3 *Adherent and phagocytic cell removal*

Treatment of normal spleen cells with carbonyl iron[36] provides a source of macrophage-depleted spleen cells by the following procedure: 25 mg of carbonyl Fe is carefully suspended in 20 ml of Eagle's MEM with 10% FBS and 10^8 spleen cells. This mixture is incubated at 37 °C for 40 minutes in a 5% CO_2 environment with occasional agitation. The carbonyl Fe is collected by magnetic attraction. The procedure is repeated once and a final removal of the carbonyl Fe is achieved by passing the cell suspension dropwise through a Pasteur pipette mounted next to a magnet. The resulting cell suspension is centrifuged and resuspended in RPMI 1640 supplemented with 20% FBS. It should be noted that during the incubation of the spleen cells with carbonyl Fe, adherent cells are also removed. Therefore, both the phagocytic and adherent properties of spleen cells are concerned by this removal procedure. Cell recovery is usually about 65%. The carbonyl Fe treated spleen cell suspension (referred to as 'Fe spleen cells') cannot be considered simply as an enriched lymphocyte suspension as compared to an unfractionated spleen cell suspension, since at least one population of cells, the adherent B cells, are also removed. At present the role of the adherent B cells is not known, but if, for example, this population regulates the *in vitro* immune response or has different requirements in culture then it would not be possible to equate the response of Fe spleen cells to normal spleen cells.

4.2.4 *Interdependence of adherent and non-adherent cells*

The controversies that have arisen over the requirement for macrophages for the primary or secondary *in vitro* immune response could be largely attributed to technical problems associated with the preparation of adherent cells and the 'macrophage'-depleted spleen cell population. For this reason, we gave the procedures for preparing both populations in some detail. In the following experiments, aimed at showing the necessity for adherent cells for the *in vitro* response to SRBC or DAGG-Ficoll, these procedures were rigidly followed.

When SRBC are used as antigen, 10^7 Fe spleen cells are cultivated either alone, with PEA cells prepared from 5×10^5 PE cells, or in the presence of 5×10^{-5} M ME. The cultures are assayed 5 days later (Table 9). As compared to unfractionated spleen cells, Fe spleen cells alone do not give rise to PFC. However, either PEA cells or ME restore the *in vitro* response of the Fe spleen cells above that seen with unfractionated spleen cells. This higher response may be due to enrichment of the Fe spleen cells with B and T cell precursors. It is noteworthy that PEA cells when added to unfractionated spleen cells resulted in a significant increase in PFC. Since PEA cells are saturating (see below), the increase in PFC from unfractionated spleen cells supplemented with PEA cells is most

Table 9. Reconstitution of Fe spleen cells with ME or PEA cells in culture system B

	PFC/culture		
	Dish 1	Dish 2	Average
10^7 Normal spleen cells	52 260	51 900	52 080
10^7 Fe spleen cells	60	480	270
10^7 Fe spleen cells + 5×10^5 PEA cells	73 680	83 880	78 780
10^7 Fe spleen cells + 5×10^{-5} M ME	69 660	76 440	73 050
10^7 Normal spleen cells + 5×10^5 PEA	228 240	186 300	207 730

likely due to a limiting number of macrophages in the unfractionated spleen cells and a loss of B cells during carbonyl Fe treatment. Microscopic inspection of the cultures reveals dramatic differences. While there is excessive cell death in Fe spleen cells cultured alone, there is very little cell death in cultures supplemented with PEA cells. Such cultures show extensive proliferation and many clusters of blasts and large cells. Fe spleen cells, when cultured in the presence of ME, show a similar picture; but there are significant numbers of dead cells. This difference is probably due to phagocytosis of cell debris by the PEA cells. One additional point is that very few adherent cells are observed in cultures of Fe spleen cells with or without ME.

Under the same culture conditions, we titrated the PEA cells using a constant source of Fe spleen cells (Figure 8). Three points should be stressed:

(1) Fe spleen cells cultured in the absence of PEA cells are unresponsive. In the *in vitro* immune response the generation of PFC from spleen cells is, as expected, completely dependent on adherent cell function. Treatment of spleen cells with carbonyl Fe sufficiently removes the adherent cells without adversely affecting the other non-adherent lymphoid cell types that are able to respond to SRBC *in vitro*.

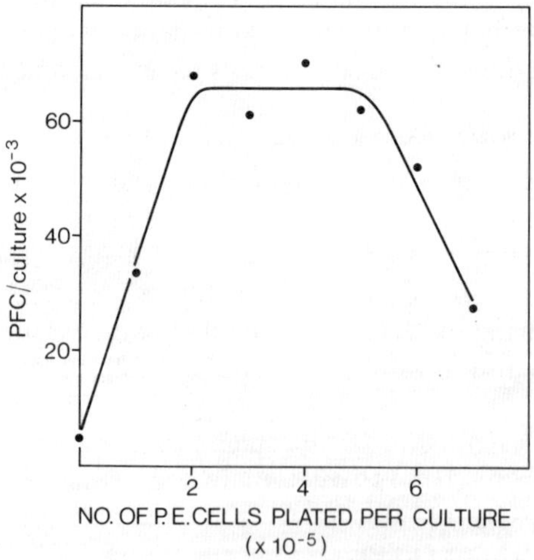

Figure 8. Titration of PEA cells derived from $1–7 \times 10^5$ PE cells, cultured with 10^7 Fe spleen cells and SRBC as antigen in system B. Both cell populations are from normal C57Bl/6J mice

(2) The generation of PFC from Fe spleen cells is controlled by the number of PEA cells. Although the other essential lymphoid cell types are obviously present, the interaction of these cells to yield PFC *in vitro* is limited by the number of PEA cells available. Therefore, any procedure that would result in the removal of a significant proportion of the adherent cell population will reduce the number of PFC even though the other potentially reactive lymphoid cell types are present.
(3) The maximum PFC response to SRBC occurs over a restricted range of PEA cell concentration. Excessive numbers of PEA cells can completely inhibit the development of PFC. The culture conditions after 5 days are highly acidic, and we attribute the failure of these cultures to metabolic acidosis. However, we do not follow the kinetics of the PFC development in cultures containing excessive numbers of PEA cells. Therefore, we are not able to determine if precursor cells are induced and then affected by adverse culture conditions or if induction fails.

The mechanism(s) by which the adherent cells influence the various events of the immune response is, of course, not known. It is most likely that the control exerted by the adherent cells is due to a complex variety of cellular functions rather than to any single function. These functions are nevertheless intrinsic to the adherent cells, since 3T3 fibroblasts in the same experiment cannot substitute for PEA cells. Also a continuously growing macrophage-like cell line, which was established (and unfortunately lost) in our laboratory and was shown to be phagocytically active, reconstituted only 10%–20% of the response of Fe spleen cells. It should be noted, however, that the presence of these phagocytic cells had a marked effect on the viability of the Fe spleen cells. Other investigators[37] report that established lines of macrophages are able to substitute completely for the adherent cell layer. Irrespective of the source of adherent cells, the main point is that these cells can drastically influence the generation of antibody-producing cells under conditions where T–B cell interactions should be unaffected unless such events are influenced by the adherent cells.

This pattern of the PFC response *in vitro* is as expected and fits well the suggestion that T-dependent antigens also require adherent cells. This strict requirement for adherent cells can, however, be overcome by including 5×10^{-5} M ME in the medium (Table 9). It should be recalled, however, that Fe spleen cells do not represent the full B cell population, since the adherent B cells are also removed. Moreover, it could also be argued that currently available methods do not completely remove adherent cells and that the few remaining cells then somehow benefit by the addition of ME so as to meet the requirements for the *in vitro* immune response. With respect to our data, this seems most unlikely. Another possibility is that the *in vitro* response to heterologous erythrocytes is unique and that macrophages assume a different role with respect to other antigens, i.e. soluble antigens.

Since it has been reported that T cell independent antigens are also macrophage independent,[38–42] we used the same experimental approach with DAGG-Ficoll as antigen. This antigen is soluble, a good immunogen *in vitro* and T cell independent.[43] Classical techniques were used to establish this as a T cell independent antigen, and our results using some of these criteria support these findings. For *in vitro* immunization, 0·001 μg of this preparation stimulated a maximal PFC response 4 days after culture when assayed on DAGG-SRBC. The conditions for *in vitro* culture involving DAGG-Ficoll were strictly identical to those described for SRBC. Various numbers of PE cells were plated as an adherent cell source and Fe spleen cells were used as a source of macrophage depleted spleen cells. The results of such an experiment can be seen in Figure 9. Although we do not

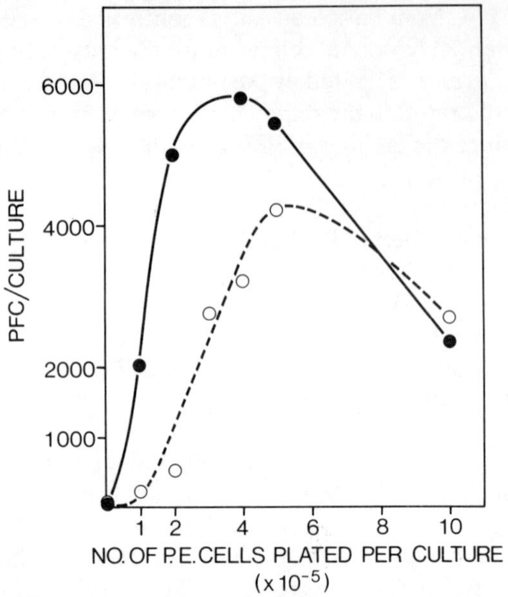

Figure 9. Reconstitution of the immune response to DAGG-Ficoll with various amounts of PEA cells in culture system B (see text). Two separate experiments (performed with C57Bl/6J mice) are shown

have as much experience with this antigen as with SRBC, results of five experiments consistently show that the response to DAGG-Ficoll is completely dependent on the presence of adherent cells. In addition, the number of adherent cells needed for the maximum response is very similar to that required for SRBC. These results support the findings of Oppenheim and Rosenstreich[33] that the T cell independent DNP derivative of the copolymer of D-glutamic acid and D-lysine requires adherent cells to stimulate an *in vitro* PFC response. Since there is an absolute requirement for adherent cells in the *in vitro* response to DAGG-Ficoll, we determined if ME could replace this function. As shown in Table 10, ME can in fact substitute for PEA cells. These data demonstrate that for both types of antigen the adherent cell can be replaced by ME.

We tend to interpret these findings in the following way. The adherent cells play a crucial role for cell survival, proliferation and/or differentiation via an unknown mechanism. Since this effect can somehow be mimicked or bypassed by ME, it is difficult to assign to the macrophage a central cell-associated function with respect to the *in vitro* induction of the immune response. If such a role for the induction could be convincingly shown for different

Table 10. The requirement of PEA cells or ME to elicit an anti-DAGG-Ficoll response *in vitro*. All cultures contain 10^7 Fe spleen cells from C57Bl/6J mice in system B

	PFC/culture		
Experiment no.	I	II	III
Fe spleen cells alone	15	0	60
Fe spleen cells + PEA	4215	5795	n.d.
Fe spleen cells + ME	5650	1893	3243

types of antigen, e.g. soluble protein antigens, we would regard this as a second, distinct function of the adherent cells.

4.3 The cell source: Strain differences and the health status of the mouse

It has been reported[5] that the magnitude of the immune response *in vitro* is dependent on the mouse strain used as spleen cell source. This suggests a genetic control of the immune response. Among three mouse strains compared by Click and coworkers,[5] C57Bl/6J was the highest responder *in vitro*. Balb/c was intermediate and CBA gave the lowest response. The *in vitro* response did not parallel the *in vivo* responses which even ranked in inverse order for these three strains. In respect to C57Bl and CBA mice, our own results support these findings. Over the last two years we have repeatedly compared different strains of mice under identical culture conditions. Each experiment revealed a certain order of responsiveness to SRBC *in vitro*, which can be remarkably constant for up to several months (Table 11). A comparison of those data over a period of several years,

Table 11. *In vitro* immunization of mouse spleen cells from four different strains (C57Bl/6J, Balb/c, C3H/HeJ and CWB/13) with four different lots of FBS in the presence and absence of 5×10^{-5} M ME in culture system A. Results are expressed as PFC/culture measured on day 5. Experiments 2 and 3 were performed 1 week and 6 weeks after experiment 1, respectively. Lot A83, R55 and C25 were known to be supportive from former testing; L45 was an unknown test sample. Sample A83 is only tested at 5% concentration. Note the difficulty in selecting the appropriate batch of FBS

		No ME			+ME		
		Exp. 1	Exp. 2	Exp. 3	Exp. 1	Exp. 2	Exp. 3
C57Bl/6J	5% A83	13 560	6 900	10 140	42 540	55 200	36 900
	5% R55	7 920	7 020	6 780	41 040	32 640	37 980
	10% R55	5 580	6 000	4 560	96 120	43 920	84 960
	5% L45	8 280	7 110	3 180	4 920	7 800	12 480
	10% L45	2 340	3 210	2 700	6 720	7 020	14 880
	5% C25	3 000	3 180	3 420	46 020	47 260	55 800
	10% C25	6 480	6 900	5 040	122 280	65 280	98 880
Balb/c	5% A83	2 280	1 800	780	1 020	4 230	4 320
	5% R55	720	1 230	1 140	240	1 710	1 860
	10% R55	120	2 190	660	660	3 030	6 120
	5% L45	120	2 070	540	180	3 030	2 340
	10% L45	480	630	240	780	1 140	3 540
	5% C25	420	420	600	120	1 980	1 920
	10% C25	780	1 290	2 940	2 460	3 390	4 320
C3H/HeJ	5% A83	180	960	300	2 160	2 220	2 400
	5% R55	480	330	420	1 680	2 220	1 620
	10% R55	660	390	420	4 260	4 230	2 400
	5% L45	60	90	540	780	780	2 760
	10% L45	180	60	600	1 380	1 020	1 680
	5% C25	60	60	360	1 620	720	1 860
	10% C25	120	270	1 200	2 820	8 400	3 960
CWB/13	5% A83	6 540	5 580	4 620	21 600	18 420	7 920
	5% R55	5 880	3 000	2 040	9 060	6 300	7 500
	10% R55	3 180	3 510	2 460	21 480	11 280	10 740
	5% L45	1 440	1 590	2 940	2 820	2 280	2 940
	10% L45	1 920	1 140	1 200	1 260	2 640	7 740
	5% C25	1 620	390	780	6 600	3 720	6 780
	10% C25	4 380	990	3 420	17 160	10 320	7 500

however, makes it difficult to accept the genetic control of the response as the sole or major determinant, and the low-responding Balb/c mice (Table 11) ranked equal to C57Bl/6 mice in similar experiments performed six months and one year earlier.

This leads to the most frustrating aspect of the immune response *in vitro*, namely the occasional failure to generate an adequate response for no apparent technical reason. The number of SRBC specific plaques, though remaining consistently between 30 000 and 50 000 PFC/culture over several months, slowly or suddenly declines while standard deviations and day-to-day variability increase. Although the culture system is carefully scrutinized for possible defects and several modifications are attempted, the *in vitro* system remains deficient for some time. Eventually, and for no obvious reason, the system reverts to normal and functions fully under the same culture conditions that were being used prior to the unexplainable failures. Although in no periods of such failures did the source of the mice change, in at least several instances we were able to show a possible cause related to the health of the mice. In one case, mice which by gross appearances were healthy were nonetheless infected by pinworms (*Enterobius vermicularis*). After a prescribed treatment with piperazine the mice were fully responsive while untreated mice remained deficient. During this period, the *in vivo* response was not affected and remained the same in both the treated and untreated mice. The spleen cells of infected mice could be restored to full reactivity by supplementing the cultures with PEA cells derived from either treated or untreated mice. Whether the pinworm infestation or the treatment by piperazine modified the activity of the macrophages is not certain, but the suggestion that the uncontrollable environment created in the host dramatically affects the *in vitro* response of spleen cells seems unavoidable.

Another similar incidence of the deleterious effects of infection was observed during an outbreak of Sendai virus in the animal colony. Different strains of mice show varying degrees of susceptibility to Sendai virus although all strains seem to be infected. C57Bl/6 mice showed a good serum titre to the virus but otherwise appeared generally healthy. However, the *in vitro* response to sheep erythrocytes declined drastically. Because of the severity of the problem created by a source of Sendai virus within a mouse colony, all mice had to be sacrificed and we were unable to gain any further information.

In both cases it is important to stress that these infections could have gone undetected. As stated, the mice appeared healthy and the spleens showed no gross abnormalities. The fluctuation that is observed in the *in vitro* immune response between various strains as well as within a single strain may in some cases be the consequence of unknown, subclinical infections.

A failure of the immune response *in vitro*, though troublesome, is far less worrying than inconsistent or even contradictory results obtained with the same mouse strain under identical culture conditions. Such a striking, though rare, example is given to demonstrate that somehow abnormal spleen cell suspensions can give the expected number of PFC if cultured under standard conditions, but that gross differences are revealed if the cells are subjected to fractionation. This is illustrated in Figure 10. Spleen cells, of a certain lot of 3–4-month-old C57Bl/6 mice, were treated with carbonyl Fe with the intention to titrate PEA cells as shown in Figure 8. Though the experimental conditions were identical for both experiments, the Fe spleen cells behaved completely differently in culture. The most striking point is that the Fe spleen cells yielded a maximal PFC response in the absence of any added PEA cells. As few as 5×10^4 PE cells caused a significant inhibition. The actual numbers of PFC/culture before and after carbonyl iron treatment were 60 480 and 55 640, respectively.

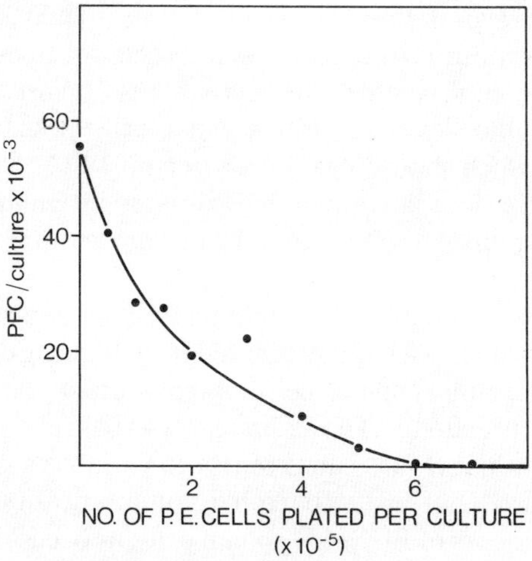

Figure 10. Titration of PEA cells performed under the same experimental conditions as for the experiment depicted in Figure 8, but with a different batch of mice (seemingly normal C57Bl/6 mice). The difference with results shown in Figure 8 is discussed in the text.

This result, which was highly consistent in five consecutive experiments with the same lot of mice, seems to exemplify the difficulties encountered in preparing a macrophage-depleted spleen cell population. The failure of the carbonyl Fe treatment to deplete the spleen cells of the adherent function can be attributed to the fact that macrophage precursor cells are not removed by this treatment. There is evidence to suggest that the precursor cells are not phagocytic and are less adherent than the mature cells.[43] Maturation of these cells *in vitro* could then supply even an excess of the essential adherent cell function. These problems, together with the observations that precursor cells do not express antigens present on PE cells, make it essentially impossible to prepare a macrophage-free spleen cell suspension even by treatment with anti-macrophage serum. However, the failure to remove adherent cells by carbonyl iron occurs infrequently and seems to correlate with particular lots of mice. It is likely that these mice have recently undergone some subclinical infectious process that would stimulate the activity of the adherent cell populations.

The continued decline in PFC with increasing numbers of PEA cells (Reference 44 and Figures 8, 9 and 10) is most likely due to adverse culture conditions resulting from excessive numbers of adherent cells. These cultures, as well as those that contained PEA prepared from more than 5×10^5 PE cells (Figure 1), were strongly acidic after 5 days in culture.

5. CONCLUDING REMARKS

The immune response of a mouse to an injection of an immunogenic dose of SRBC is predictable: 4 days later, its spleen will contain approximately 10^5 PFC, the actual numbers varying between 62 000 and 216 000.[45] The outcome of an *in vitro* immunization is far less certain. A basic question is how many PFC should be expected from an *in vitro* culture? The simple extrapolation from the *in vivo* data would give around 10^4 PFC/culture, since each culture contains 10^7 spleen cells, or one-tenth of the total spleen. However, various factors might increase or decrease the *in vitro* response compared to the *in vivo* response, such as the lack of recruitment of specific precursors from the recirculating lymphocyte pool[46] which should give lower *in vitro* PFC responses. On the contrary, since 'the *in vitro* response does not appear to be limited by whatever mechanisms [which] regulate the *in vivo* response',[2] spleen cells in *in vitro* culture should give higher PFC numbers.

A clue to the aforementioned problem might be found if approached in a different way. The frequency of precursors to SRBC is estimated from limiting dilution experiments to be $1–2 \times 10^{-5}$.[47] The typical primary spleen cell culture, with 10^7 initial cells, should therefore contain at least 100–200 precursors. Clone size analysis has revealed that, even under poor culture conditions, individual clones can give rise to variable numbers of PFC, from very few to several hundred (I. Lefkovits, personal communication). Therefore, it is quite possible that the low numbers of PFC/10^7 spleen cells, as occasionally reported, originate from less than 1% to only a few per cent of the precursors present in the culture. When such a minor fraction of the cells under investigation is actually responding, the serious question arises whether meaningful conclusions can be made from such quantitatively unsatisfactory results. Our own experience showed throughout that very low responses (<1000 to 5000 PFC/culture) are poorly reproducible.

Although it appears almost impossible to observe no anti-SRBC response whatever the culture conditions, the actual number of PFC can vary by three orders of magnitude. This variation can be due to different culture conditions and/or to the difference in animals from which the spleen cells are obtained. The culture conditions can be standardized to a high extent, at least within one laboratory, since the media can be prepared quite reproducibly and the sera keep their activity for at least a year. Minor variations of only one parameter, however, can seriously affect both the size and the kinetics of the response. A meaningful study of the basis for these differences will not be possible unless we can reliably dissociate induction, proliferation and maturation of the immunocompetent cells. The optimal culture conditions for these different phases of an immune response *in vitro* might be quite different.

As pointed out in Section 4.3, we regard the spleen cells, i.e. the mice themselves, as the most important and least controllable source of variability in the *in vitro* culture system. Periods of failures and poor or high responses are often parallel in different laboratories which derive their mice from the same breeder or keep them in the same animal colony. Clearly, the *in vitro* response seems to be much more sensitive to the changes in the health status of a mouse than is the *in vivo* response. Many scientists in this field have made this kind of observation, though these are hardly ever documented in writing. Shiigi and Mishell[16] have coined the term 'abnormally responsive spleen cell pool'. According to their long experience, about 15% of pooled spleen cell suspensions are 'abnormally reactive' in the sense that they give rise to an immune response after 4 days *in vitro* with batches of FBS which normally would not support an immune response *in vitro*. These authors protect themselves against erroneous conclusions by including controls with deficient sera in all experiments, but in other laboratories they will go undetected, including our own.

The existence of such 'abnormally responsive spleen cells', taken together with the other sources of variability discussed in this chapter, leaves ample room for controversial findings, far more than the literature reveals. This seems to be a formal proof that the results of *in vitro* experiments are interpreted cautiously and critically by those who apply this system.

6. REFERENCES

1. Mishell, R. I. and Dutton, R. W. (1966). *Science*, **153**, 1004–1006.
2. Mishell, R. I. and Dutton, R. W. (1967). *J. exp. Med.*, **126**, 423–442.
3. Marbrook, J. (1967). *Lancet*, **2**, 1279–1281.
4. Inman, J. K. (1975). *J. Immunol.*, **114**, 704–709.
5. Click, R. E., Benck, L. and Alter, B. J. (1972). *Cell Immunol.*, **3**, 264–276.
6. Pritchard, J. B., Chavez-Peon, F. and Berlin, R. D. (1970). *Amer. J. Physiol.*, **219**, 1263–1267.
7. Fanger, M. W., Hart, D. A., Wells, J. V. and Nisonoff, A. (1970). *J. Immunol.*, **105**, 1043–1045.
8. Kearney, J. F. and Lawton, A. R. (1975). *J. Immunol.*, **115**, 671–676.
9. Metcalf, D., Warner, N. L., Nossal, G. J. V., Miller, J. F. A. P., Shortman, K. and Rabelino, E. (1975). *Nature*, **255**, 630–632.
10. Cerottini, J.-C., Engers, H. D., MacDonald, H. and Brunner, K. (1974). *J. exp. Med.*, **140**, 703–717.
11. Broome, J. D. and Jeng, M. W. (1972). *J. exp. Med.*, **138**, 574–592.
12. Iscove, N. N. and Sieber, F. (1975). *Exp. Hematol.*, **3**, 32–43.
13. Chen, C. and Hirsch, J. G. (1972). *Science*, **176**, 60–61.
14. Broome, J. D. and Jeng, M. W. (1972). *J. Nat. Cancer Inst.*, **49**, 579–581.
15. Taylor, W. G. (1974). *J. Nat. Cancer Inst.*, **53**, 1449–1457.
16. Shiigi, S. M. and Mishell, R. I. (1975). *J. Immunol.*, **115**, 741–744.
17. Mishell, R. I., Chan, E. L., Crabbe, L., Ly, I., Lucas, A. and Mishell, B. B. (1973). In *Specific Receptors of Antibodies, Antigens and Cells* (Pressman, D., Tomasi Jr., T. B., Grossberg, A. L., and Rose, N. R., Eds.), S. Karger, Basel, pp. 340–353.
18. Watson, J. and Thoman, M. (1972). *Proc. Nat. Acad. Sci.*, **69**, 594–598.
19. Watson, J. and Pritchard, J. (1972). *J. Immunol.*, **108**, 1209–1217.
20. Leven, J., Jomasulo, P. A. and Oser, R. S. (1970). *J. Lab. Clin. Med.*, **75**, 903–911.
21. Honn, K. V., Singley, J. A. and Chavin, W. (1975). *Proc. Soc. exp. Biol. Med.*, **149**, 344–347.
22. Veit, B. and Michael, J. G. (1973). *J. Immunol.*, **111**, 341–351.
23. Paul, W. E., Sharon, R. A., Kask, A. M., Owens, J. D. and McMaster, P. R. B. (1974). *Fed. Proc.*, **33**, 756 (abstract).
24. Mosier, D. E., Johnson, B. M., Paul, W. E. and McMaster, P. R. B. (1974). *Fed. Proc.*, **33**, 756 (abstract).
25. Sharon, R., McMaster, P. R. B., Kask, A. M., Owens, J. D. and Paul, W. E. (1975). *J. Immunol.*, **114**, 1585–1589.
26. Melchers, F., Coutinho, A., Heinrich, G. and Andersson, J. (1975). *Scand. J. Immunol.*, **4**, 853–858.
27. Osoba, D. (1969). *J. exp. Med.*, **129**, 141–152.
28. Stocker, J. W. (1976). *Immunology*, **30**, 181–187.
29. Chan, E. L., Mishell, R. I. and Mitchell, G. F. (1970). *Science*, **170**, 1215–1217.
30. Mosier, D. E. (1967). *Science*, **158**, 1573–1575.
31. Dutton, R. W. (1967). *Advan. Immunol.*, **6**, 253–336.
32. Unanue, E. R. (1972). *Advan. Immunol.*, **15**, 95–165.
33. Oppenheim, J. J. and Rosenstreich, D. L. (1976). *Prog. Allergy*, **20**, 65–194.
34. Loughman, B. E., Farrar, J. J. and Nordin, A. A. (1974). *J. Immunol.*, **112**, 430–432.
35. Roseman, J. (1969). *Science*, **165**, 1125–1127.
36. Goldstein, O., Schirrmacher, V., Rubin, R. and Wigzell, H. (1973). *Cell. Immunol.*, **9**, 211–225.
37. Mocarelli, P., Palmer, J. and Defendi, V. (1973). *Immunol. Commun.*, **2**, 441–447.
38. Shortman, K., Diener, E., Russell, P. and Armstrong, W. D. (1970). *J. exp. Med.*, **131**, 461–482.
39. Shortman, K. and Palmer, J. (1971). *Cell. Immunol.*, **2**, 399–410.
40. Feldman, M. (1972). *J. exp. Med.*, **135**, 1049–1058.
41. Feldman, M. and Palmer, J. (1971). *Immunology*, **21**, 685–699.
42. Mosier, D. E., Johnson, B. M., Paul, W. E. and McMaster, P. R. B. (1974). *J. exp. Med.*, **139**, 1354–1360.
43. Cline, M. J. and Sumner, M. A. (1972). *Blood*, **40**, 62–69.
44. Hoffmann, M. (1970). *Immunology*, **18**, 791–797.

45. Jerne, N. K., Henry, C., Nordin, A. A., Fuji, H., Koros, A. M. C. and Lefkovits, I. (1974). *Transplant. Rev.*, **18**, 130–191.
46. Sprent, J. and Lefkovits, I. (1976). *J. exp. Med.*, **143**, 1289–1298.
47. Quintans, J. and Lefkovits, I. (1973). *Eur. J. Immunol.*, **3**, 392–397.

Chapter 8

Structure and Dynamics of the Lymphocyte Surface

F. Loor

Abbreviations

α-A: alpha-actinin
$\beta 2\mu$: beta-2-microglobulin

Con A: concanavalin A
Fab: monovalent antibody fragment
$F(ab')_2$: bivalent antibody fragment
H-2: mouse histocompatibility antigen
HLA: human histocompatibility antigen
(H)MM: (heavy) mero-myosin
IFM: immunofluorescence microscopy
(m)Ig: (membrane) immunoglobulin
MF: microfilaments
MT: microtubules
MV: microvilli
PM: plasma membrane
S(I)EM: scanning (immunor-) electron microscopy
T(I)EM: transmission (immuno-) electron microscopy
Tla: thymus leukaemia determined antigen
TMV: tobacco mosaic virus
θ: Thy-1 locus determined antigen

1. INTRODUCTION

The plasma membrane (PM) of any cell represents its proper limit of self and non-self. It is the site where there is an extensive exchange of ions and simple metabolites and of biologically structured molecules with the foreign milieu. Though it is not the only possible mechanism, it is a widely accepted belief that the surface of the cell is the site where recognition of, and by, the foreign milieu is taking place.

Those physiological activities of the cell membrane which only play a role in cell survival will not be considered here, and the only aspects which will be developed concern activities of the cell PM which, at the present time, seem to be important for the sociological behaviour of lymphocytes, especially for their immunological activity.

Thus, the cell surface is the site for a variety of biological recognition phenomena: recognition of, and by, other cells, and recognition of, and by, appropriate inducer or selector molecules during embryogenesis and individual life until death. As far as lymphocytes are concerned a series of these cell membrane functions can be listed: histocompatibility linked recognition, recirculation and histotypic localization, sensitivity to various hormones, to mitogenic molecules and, by extension, to any agents which can modify upon interaction with the lymphocyte membrane the cycle of the lymphocyte or its level of differentiation along its pathway, i.e. the expression of its genome.

2. GENERAL CHARACTERISTICS OF THE CELL MEMBRANE

The structure of the cell PM must be such that upon contact with a ligand on its outer face (i.e. the foreign milieu), the inner face (i.e. the cell itself) is specifically informed. However, the structure of the membrane must also be such that it constitutes an efficient barrier to an intermixing of self and non-self. Thus it must be a highly organized cell organelle, and the understanding of its structure and of its physiology is basic for cell biology. The membrane itself is extremely thin: at equal proportions, it would have only ~1 mm thickness for a lymphocyte of ~1 metre diameter! But as will be seen there are various structures in the cortical cytoplasm of the cell which are organized as a cytoskeleton conferring shape and rigidity to the cell (see the schema, in Section 6). General information about the composition, the structure and the function of cell membranes can be found in a series of recent books and reviews on the topic.[1–13] Specific references will be given only for more recently published information and for data which seems to us particularly relevant to this chapter.

2.1 Components

Biochemical analysis shows that in most functional membranes, proteins constitute the largest fraction (from one-half to three-quarters of total membrane weight); then come the lipids (from one-quarter to one-half) while the carbohydrate component is much less represented (usually less than 10%). The membrane lipids belong to three distinct categories: (a) neutral lipids which are mostly represented by cholesterol, (b) glycolipids which are mostly glycosphingolipids, and (c) phospholipids which are the most abundant of the membrane lipids. Most phospholipids are zwitterionic at physiological pH, except phosphatidyl serine which carries a negative charge.

The membrane carbohydrate moieties are essentially present in the form of oligosaccharide side chains (usually not more than 15 residues) on glycoproteins and to a lesser extent on glycolipids, and also in some cases in the form of mucopolysaccharides (proteoglycans). It is probably not pure chance that among the numerous different monosaccharides which exist in nature, only a few are found in membranes, especially sialic acid, glucose, glucosamine, galactose, galactosamine, mannose and fucose. The great variety of branching and linkages they constitute might provide complex patterns for recognition functions, if this variability is actually controlled by the cell, which is not known until now. The main part of the net negative surface charge of all mammalian cell PM is contributed by sialic acid residues, which are mostly terminal, as shown by their easy removal by neuraminidase treatment and the concomitant decrease of the cell zeta potential.[3]

The membrane proteins can operationally be classified into two broad overlapping classes: peripheral and integral. *Peripheral* proteins which represents less than 30% of total membrane protein are loosely bound to the membrane; they do not really form an 'integral' part of the membrane in that one does not need to disrupt the whole membrane to extract them: they can be solubilized by mild treatments such as a change in the ionic strength of the medium or the use of metal ion chelating agents; they come in solution free of lipids and when analysed further they show overall structural characteristics rather similar to the soluble proteins found in biological fluids. It is difficult to know whether by extension, one should also consider as peripheral proteins all proteins which secondarily associate with cell membranes, like serumalbumin, polypeptide hormones, immunoglobulins, antigens, etc. Interestingly enough, a series of the peripheral proteins seem to have integral proteins as receptors in the membrane.[6]

Integral proteins which represent 70%–80% of the total membrane proteins, are recognized as 'integral' in that they appear to be a true structural component of the membrane: they are strongly linked to it, their extraction requires the use of agents able to break hydrophobic bonds, i.e. ionic and non-ionic detergents, organic solvents or chaotropic agents; after extraction they still remain associated with membrane lipids and finally, when analysed further, they show an amphipathic structure (see below). Among integral proteins one also finds the majority of the membrane-associated enzymes, of the membrane antigenic proteins, of the hormone receptors, of the transport proteins and at least a fraction of the glycoproteins on which sugar binding reagents (phytolectins, for example) bind.

2.2 Structural organization

There will be no review of the variety of models which have emerged and disappeared as more experimental information was becoming available. The 'fluid mosaic model of membrane structure'[6–8] is the most recent and most widely accepted concept of how biological membranes are organized. This model fits the basic prerequisite that the genesis and the function of biological membranes should cost the cell as little free energy as possible though giving a highly thermodynamically stable structure. A review of experimental evidence which supports it is quite outside the scope of this chapter. To summarize its basic features, the primary matrix of the membrane would be made of a bilayer of phospholipids oriented to constitute an hydrophobic membrane interior (their aliphatic chains) and two hydrophilic membrane faces (their ionic groups), thus to be an hydrophobic barrier between two aqueous solutions. This basic bilayer structure is spontaneously formed when phospholipid molecules are put in water, forming an artificial membrane. Its thickness is 4·0–4·5 nm only, although the average total thickness of cell membranes is ~7·5 nm.[1] Roughly half (rather less) of the membrane surface area is occupied by the lipids, the remainder being occupied by proteins.

Membrane proteins would belong to the two broad classes mentioned above, either integral or peripheral. Like the phospholipids, integral proteins would have an *amphipathic* structure, i.e. they would be constituted of well differentiated domains, one hydrophobic and one (or more) hydrophilic. This amphipathic structure would thermodynamically fit their mode of insertion in the phospholipid membrane matrix. Thus, the hydrophobic domains, made of non-polar amino acid residues and of highly α-helical secondary structure, must be sequestered from contact with the water, while the highly hydrophilic region(s) of the protein, owing to the presence of many charged residues, need(s) to be in direct contact with the water. This hydrophilic interaction can be found either on the exterior faces of the membrane or by cooperation with hydrophilic domains of one or several other membrane components. Such aggregates made of integral protein 'subunits' might span the whole membrane. Furthermore each subunit might contribute one hydrophilic domain to generate through the membrane a continuous pore that could be closed or open either through conformational changes in protein subunits or through quaternary rearrangements of the subunits among themselves. Well documented studies on the red cell PM (e.g. References 14–16) strongly suggest that such oligomers of amphipathic integral proteins are the major constituents of the membrane intercalated particles. These intriguing intramembranous particles are components of heterogeneous sizes (mostly 8·5 nm) which are present in all membranes studied so far and are revealed by an electron microscopy method (freeze-fracture technique) which cleaves the membrane along its plane, into its two faces. The membrane particles have been shown to be associated with a variety of transmembrane functions (essentially permeation of small molecules or ions), thus their recent renaming of 'permeaphores'.[16] Though this is not yet convincingly demonstrated for other cell types, like lymphocytes, it is quite likely that the membrane particles of other cells are also oligomers of integral protein or permeaphore. 'Transmembrane glycoproteins' are single integral proteins which have an hydrophobic domain large enough to span the whole membrane and leave the hydrophilic extremities on both sides of the membrane.[12]

Peripheral proteins, in contrast to the amphipathic integral proteins, would show a more uniform distribution of ionic groups on their outer surface and their association with the membrane would thus not be through strong hydrophobic binding to membrane lipids or hydrophobic regions of integral proteins but rather through weaker non-covalent bonds (e.g. ionic, hydrogen . . .) to some of the integral proteins. A few well-known examples of such proteins are the cytochrome *C* present on the inner mitochondrial membrane and the tektin or spectrin present on the cytoplasmic face of the mammalian erythrocyte PM. In the latter case, the association of spectrin with a permeaphore has been convincingly demonstrated in that the perturbation of the organization of the spectrin on the inner face of the red cell membrane provokes a perturbation of other sites on the outer face of the membrane[17] and vice versa.[18] Thus peripheral proteins, though not spanning the membrane, might also transmit information through the membrane, presumably with the help of permeaphores or of transmembrane glycoproteins. Finally the carbohydrate moieties of membrane glycoproteins and glycolipids must be located essentially in the hydrophilic milieu.[1]

2.3 Asymmetry

The components of biological membranes are disposed across the membrane in an asymmetrical way (see References 6 and 10). This is indicated by a variety of methods which give concordant results: (a) the comparison of membrane components which become chemically or enzymatically labelled, either from intact cells or from disrupted cells or ghosts; (b) the comparison of the membrane components detectable on the surface of right-side-out and of inside-out vesicles prepared from isolated membranes; and (c) the use of electron microscopy, together with either histochemical methods able to localize *in situ* specific membrane enzymes, or electron dense labelled reagents (antibodies, lectins, hormones . . .) which are specific for a variety of cell membrane components. The asymmetrical disposition of phospholipids is supported by the analysis of the red cell membrane only and not yet of other cell membranes.[9,10] The carbohydrate moieties are exclusively localized to the non-cytoplasmic face of cell membranes, in the case of the plasma membrane (PM) on the outer cell surface.[19] Finally the distribution of peripheral proteins among the two membrane faces is markedly asymmetrical, as is also the orientation of the integral proteins.[1–12] The biological significance of asymmetry is unknown, but it is presumably essential for the asymmetry of exchanges at the level of the membrane.

2.4 Mobility of the components

The consequence of the model which is most frequently stressed is that it is a *fluid* mosaic of proteins and lipids. The 'fluidity' of the membrane—i.e. a liquid crystalline structure as opposed to a 'solid' gel structure—implies that some, if not all, membrane components can 'cis-rotate' (i.e. along an axis perpendicular to the plane of the membrane) and also move laterally, and that any condition which alters the fluidity of the lipid matrix will alter the rate of cis-rotation and lateral mobility of the membrane components. On the other hand, the asymmetry of the membranes shows that glycoproteins and glycolipids do not trans-rotate at any significant rate under physiological conditions (no flipping), i.e. they do not rotate from one face of the membrane to the other. This would be difficult to reconcile with the 'save energy' character of the thermodynamically built membrane.

The cis-mobility of membrane components does not imply, however, that the whole membrane would be in a fluid state. For instance, on sperm cells, the membrane seems to be 'rigid' in the acrosomal region, 'fluid' in the post-acrosomal region and 'rigid' again on the tail[20] and there are indeed many examples of membranes where an ordered structure or a limited mobility seems to exist (reviewed in Reference 6). This relative membrane rigidity may be due to interactions over long distances within the membrane among very specialized types of integral protein molecules. Such structures might constitute the barriers to segregate specialized cellular membrane areas that should show differential fluidity or localization of some membrane components. Another cause of restriction to the apparent fluidity of a whole cell membrane can also be due to the anchorage of integral membrane proteins to non-membranous sites having a restricted mobility, either directly or through peripheral membrane proteins, as is the case of the red cell membrane spectrin.[17,18] The lipid composition of the bilayer also determines its fluidity and thus the phospholipid heterogeneity may be a factor allowing the modulation of the overall membrane fluidity.[10] This can also be influenced by others factors, i.e. the presence of the proteins and also, in homeothermic animals, the presence of large amounts of cholesterol which can abolish the temperature-induced phase separation of the membrane phospholipids into liquid areas and gelified areas.[21] The mobility of cell membrane components can also be modulated upon extrinsic interactions with other cells or factors like those which seem to play the role of 'cement' for cells which have to be specifically organized as a tissue (e.g. Reference 22). Finally, the structural and dynamic organization of cell membranes is also under the control of cytoskeletal matrices made of various fibrillar structures, microtubules (MT) and microfilaments (MF), which are involved in a variety of functions of virtually all eukaryotic cells (see Section 2.6).

2.5 Membrane biogenesis

These properties of cell membranes put restrictions on how a membrane can be built, i.e. how new components are integrated into the membrane. Especially in the case of the plasma membrane (PM), the exclusive localization of carbohydrate moieties on the outer cell surface and the lack of trans-rotational motion makes it unlikely that glycoproteins and glycolipids might traverse a pre-built hydrophobic phospholipid bilayer. The membrane components have to be inserted in the growing membrane directly in the appropriate orientation.

The essential characteristic of the model suggested by Palade for the biogenesis of PM,[23] a concept which has experimental support,[24] is that the PM is not synthesized *in situ* (i.e. at the level of the cell surface) nor are its main components inserted there. On the contrary, there would be some kind of assembly-line process starting from intracellular precursor membranes. The assembly line would start at the level of the rough endoplasmic reticulum where the first membrane elements would be assembled. The integral protein moieties would be directly inserted in the phospholipid bilayer as they come out of the ribosomal machinery, although the possibility of their transient passage as soluble cytoplasmic proteins has also been suggested.[9,10] Further, one would find these first membrane elements in the smooth endoplasmic reticulum and Golgi apparatus. These Golgi elements would be pinched off into the precursor membrane vesicles which, referring to the PM, would be (cytoplasmic) inside-out vesicles. New PM would be generated by fusion of such vesicles from the precursor pool with the already existing PM. If there is no local restriction to membrane fluidity at the point of fusion, then the newly inserted PM components should be able to intermix with the other ones by translational diffusion. The covalent attachment of the carbohydrate moieties of the membrane components would proceed, in a stepwise fashion. The asymmetry would start in the rough endoplasmic reticulum, where the first saccharide residues would be transferred only to the cisternal side of the membrane (opposite to the side of ribosome attachment). These carbohydrate structures would then progressively increase in complexity all along the assembly line (and again it would be surprising that these complex structures simply arise at random).

An important aspect of this model is the concept of the existence of a pool of precursor membrane vesicles. In some cells, at least, this pool might be large enough to allow a continued incorporation of new proteins in the PM for several hours after the synthesis of new cytoplasmic proteins was blocked by cycloheximide treatment.[25]

In addition to this endogenous origin, it should be realized that, in the case of the PM, there may be also some material of exogenous origin which might contribute to its normal constitution. There might be glycoproteins, mucopolysaccharides or glycolipids absorbed from the environmental milieu. The best established case concerns the Lewis blood group antigen, which is a glycosphingolipid exclusively acquired from the plasma and integrated in the red cell membrane.[26] It would not be surprising if some cell–cell cooperation also implied some exchange of membrane components.

There is no definite model for how old membrane components are eliminated from the cell membrane, if they are. If old and new membrane components can continuously intermix, it is difficult to imagine how the cell could selectively eliminate the old ones only. This would be made easier for the cell if they segregated in some specialized areas of the cell; but what might make them segregate? The simplest and most likely alternative is that the rate of synthesis of new membrane is much higher than the rate of degradation of the membrane components themselves, so that large amounts of newly built membrane elements would continuously dilute out the older ones. There would be a continuous exocytosis and shedding or endocytosis of membrane vesicles. All membrane elements, whether of endogenous origin or of exogenous origin, would be split into elements to be re-utilized in the formation of new cell components, including membranes, a process which would not be wasteful for the cell since the catabolism of membrane (-adsorbed) components would also provide at least part of the energy required by the cells.

2.6 The cytoskeleton

Only what may be relevant to the cell surface is considered here (for reviews, see References 12, 27–31; for a schematical view, see Section 6).

2.6.1 *Microtubules (MT)*

MT are rather straight structures of various lengths and of 24 nm outer diameter.[28,30] They have a dense cortex and a hollow 15 nm core. The cortex is made up usually of 13 adjacent protofilaments, 4–5 nm wide, which are organized along the MT axis with a pitch of at least 10°. Furthermore, there are many lateral cross-bridges (2–5 nm thick by 10–40 nm long) between adjacent MT, but which can also occur between a MT and all kinds of membranous components, including membrane vesicles and PM. In some cases at least, such MT–membrane association has been found to be extremely stable.[31] Besides their structural roles, the cross-bridges are supposed to contain an ATPase which, by using ATP, would provide energy to generate motion relative to the tubule surface. MT protofilaments are built of subunits made of tubulin dimers (110 000–120 000 daltons), which give, upon denaturation, two slightly different monomers, α and β, whose organization is yet unknown. Each subunit has one GDP and GTP bound and can bind one molecule of colchicine and one molecule of vinblastine at separate sites. Some other plant alkaloids like podophyllotoxin seem to bind competitively to the colchicine site, but the action of some others, like griseofulvin, is less well defined (see also References 32–35). Colchicine reversibly disrupt MT, as does vinblastine, but the latter drug further induces a crystallization of tubulin. Low temperature, hydrostatic pressure and high calcium concentration (>10^{-5} M) can induce a reversible MT dissolution into subunits. Though tubulins seem to be highly conserved proteins throughout all eukaryotic cells, it should be emphasized that MT can exhibit quite different sensitivities to disruption by the aforementioned treatments, e.g. 10^{-4} M colchicine is needed to block division in a cultured plant cell, while 10^{-7} M is sufficient to block mitosis in the HeLa cells.[31] Similar differences exist for various MT from various cell compartments, which might depend on the degree of cross-bridging of the MT. The equilibrium of assembly–disassembly between MT and tubulin can be shifted to MT formation *in vitro* by D_2O, which supports the concept of an hydrophobicity-driven association of the subunits. Furthermore, the polymerization *in vitro* requires very low calcium concentrations (<$1 \cdot 10^{-6}$ M) and an initiator still not well characterized but thought to be a single disc of ~13 tubulin subunits. It is yet unclear how the tubulin-bound nucleotides can be involved in the polymerization process.

In the cell, MT formation also requires initiation; this would start in some specialized area, the MT organizing centres, usually characterized by the presence of amorphous or fibrous material. Such centres are essential since MT can no longer be formed upon destruction of these centres by special treatments. Such structures are not found in the cortical cytoplasm, although MT might be frequent there, and it thus remains unclear whether MT organizing centres are associated with the PM. Because of the aforementioned requirements for MT formation *in vitro*, it is usually speculated that MT organizing centres would function by locally reducing the calcium concentration.

Microtubules play a major role in the control of cell shape. In plant cells, MT in the cortical layer of the cytoplasm underneath the PM can strictly control the pattern of cellulose microfibrils of the cell wall (which is extracytoplasmic!). In animal cells, MT might act as structural cytoskeletal elements in the formation and maintenance of the shape of the cells. Indeed, high hydrostatic pressure or colchicine disruption of MT destroys the shape of the cells, a process which shows simple reversibility, in that no protein synthesis is needed to reacquire the normal cell shape. MT might mechanically influence the PM in two ways. The simplest is a polymerization–depolymerization process. The other would be to generate a motion of cellular structures (MT, membrane vesicle, membrane component) relative to the MT surface, as a result of forces generated by the ATPase cross-bridges.

The recent use of immunofluorescence (IFM) with anti-tubulin antibody to detect MT inside various cells gives very promising results. MT appear as separate fibres of uniform thickness which run throughout the cytoplasm. The fibres disappear after treatment by colchicine or low temperature,[36] they are converted to some kind of crystals after treatment with vinblastine.[37] This is accompanied by dramatic changes of the cell shape.

2.6.2 *Microfilaments*

Microfilaments are fine structures of various lengths and diameters.[27,29,38] The 5-8 nm thick MF are made of actin, while the 13–25 nm thick MF are made of myosin; in addition many animal cells contain intermediate size filaments (10 nm thick) (tonofilaments and neurofilaments) which cannot be related so far either functionally, structurally or biochemically to either MF or MT protofilaments.

Actin MF are made of fibrillar actin (F-actin) which is composed of two helically wound chains. F-actin is itself the polymer of globular actin (G-actin, 43 000–46 000 daltons). At low ionic strength in presence of ATP, G-actin polymerizes into F-actin with one bound ADP per monomer. Like tubulin, actin has very conserved composition and characteristics throughout all eukaryotic cells, of which they often constitute an important portion of the total cell protein (10%–20%).

Myosin MF are made of more variable molecules (180 000–460 000 daltons) which show a Ca^{2+} dependent ATPase activity and a Mg^{2+} dependent ATPase activity and can self-assemble into filamentous bipolar aggregates. However, the conditions for this polymerization can greatly differ for myosin proteins from various origins, as do their solubility properties. The Mg^{2+}-ATPase activity of myosin may be activated upon interaction with actin MF, but to variable degrees depending on the myosin. The specific binding to actin of heavy meromyosin (HMM) (which is the globular ATPase portion of skeletal muscle myosin obtained by trypsin digestion) is used in electron microscopy to 'decorate' actin MF into typical 'arrowhead' configurations, due to the double helical nature of actin MF, and to allow one to identify their polarity.

In situ the organization of most MF, visible in the cells, can be disrupted by cytochalasins ('A'–'E'), which are fungal antibiotics,[39] leaving patches of amorphous material. This is also a simple reversible process, since, after removal of the cytochalasin, the MF can reappear without any need for new protein synthesis.

This general figure is somewhat complicated by the fact that many animal non-muscle cells contain several species of MF. In the best studied cases, cultured fibroblasts and glial cells, which *in vitro* are mobile cells, at least two classes of MF systems were found (e.g. Reference 27). Some MF constitute a loose network of short interconnected MF forming a lattice localized at the anterior part of the cell, just beneath the PM, and may be associated with it. They are disrupted by cytochalasin but they do not seem to bind HMM. The other set of MF consists of parallel MF, organized as a sheaf or as bundles of fibres in the posterior part of the cell; they are not disrupted by cytochalasin, but they bind HMM.

It is likely that the two sets of MF have to assume separate, complementary functions in the cell, among which are gross cell locomotion (probably mediated by the posterior fibres), amoeboid movement, cytoplasmic streaming and cell shape changes (probably mediated by the anterior network), among which are surface ruffling and microvilli formation (MV) (there is an axial bundle of MF within each MV).[40] All these movements are reversibly stopped by cytochalasin. Several investigations seem to support the idea that MF may be anchored to membranes; especially in fibroblasts, both sets of MF are intimately linked to the PM. Frequently the sites of insertion are localized at some dense areas as is the case in MV. It would appear from the pattern of decoration with HMM that all the MF would have the same polarity which would be determined by the attachment to the membrane.

Unlike those of colchicine and vinblastine, the actual effects of cytochalasin are still being debated (e.g. References 31, 39); and it is yet unclear whether it affects only MF and how, in some cases, instead of the dissolution it induces a higher degree of aggregation. The drug may bind to actin, but to variable extents depending on the local environment; it may affect structures other than actin, bind to membrane or to the MF attachment sites in the membrane and therefore affect the MF structure or polarity indirectly, which is supposed to be important for mobility. Indeed in the skeletal muscle, the polarity of actin MF is essential for contraction and remains constant with regard to the Z bands of the sarcomere. Thus in non-muscle cells, membrane can play the role of controlling the MF polarity. Actin MF are by themselves not contractile. In the skeletal muscle motion is generated by the interaction of actin MF and myosin MF which slide past one another, the energy being provided by the myosin-ATPase cross-bridges. In non-muscle cells, there is no regular structural packing of actin and myosin. However, cytoplasmic preparations obtained from a variety of cells can stream in the presence of ATP. It is likely that the mechanism involves interactions of actin MF with 'adsorbed' or associated cytoplasmic myosin, as well as with tropomyosin; the latter protein would have by analogy to the muscle model the role of regulation of the Ca^{2+} dependent interactions of actin and myosin. (The level of bivalent cations, especially Ca^{2+}, is indeed important for MF activity[41,42] and in that respect the presence of calcium pumps in the fibroblast PM[43] is interesting.)

Recently, the localization of MF within fibroblasts was done by IFM by use of antibodies specific either for actin[44,45] or for myosin[46] or for tropomyosin.[45,47] The fluorescence was found primarily associated with long fibres[44–47] which span the whole cell length or often converge to 'focal points' and 'are intimately associated with the plasma membrane'.[45,47] These fibres probably correspond to the MF bundles seen in electron microscopy. The data obtained so far are consistent with the hypothesis that actin, myosin and tropomyosin would all be present in any given fibre; however, though the actin would be distributed without detectable interruptions along the fibres, the pattern of distribution of myosin would show striations, as is also the case for tropomyosin, which would further show some periodicity. The molecular arrangement of the three types of molecules is still unknown. The regular presence of actin-filaments in the cortical cytoplasm of rat mast cells which was also reported recently[48] is interesting in that many were found fixed to the inner surface of the PM. At the area of attachment there are often 'points of increased density on both sides of the membrane'.[48] Perhaps this should be related to the surprising finding that myosin-like protein were detected on the outer cell surface.[49–51] However, it should be pointed out that such outer surface localization was not found by others using IFM[52] or immunoelectron microscopy.[53] Furthermore, recent studies on the binding of tritiated cytochalasins,[54–56] especially by autoradiography,[57] are consistent with a reversible binding of the drug mainly to some membrane-proteins which 'are located either at the inner surface of the membrane or so embedded in the membrane as to be protected from proteolysis restricted to the outer cell surface'.

2.6.3 *MF–MT coordination*

The PM cytoskeleton would thus be constituted of two types of elements having somewhat antagonistic functions, MT having principally a structural role and MF having principally a contractile role. How these two types of functions get coordinated in the cell is a matter for speculation. Cyclic nucleotides have been proposed as modulators of MT and MF activity. However, since they have pleiotypic effects,[58,59] of which a basic one is to modulate the level of intracellular calcium, their effect on various expressions of MT and MF activity might be quite secondary. (Cyclic nucleotides are wonderful drugs for cell biologists: they *always* do something.) Indeed it has also been proposed that the activity of MF could be controlled by Ca^{2+} flow across the membrane.[60] The antagonistic effects of Ca^{2+} on MT (they need low Ca^{2+} to polymerize from tubulin) and on MF (Ca^{2+} is required for the myosin–actin interactions which are needed to generate motion) could even be exploited further, as suggested hereafter. If, in one domain of the cell, MT and MF actually act in an antagonistic way, e.g. MT constituting some kind of barrier to motion of a cell membrane or organelle by MF, this barrier could be crossed if the pressure exerted by the MF on the MT is high enough to locally dissolve the MT into tubulin. Of course, this would also require that the calcium concentration is high enough to activate the MF to contract and to make the MT more susceptible to depolymerization by pressure into tubulin, and to avoid an immediate repolymerization of the tubulin. The whole system (i.e. MT+MF) would rather function as a *thixotropic gel*: it gets liquid when and as long as a force is applied and solidifies immediately after.

3. LYMPHOCYTE PLASMA MEMBRANE (PM)

The overall molecular organization of lymphocyte plasma membrane is not different from that of other membranes from other cells.

3.1 Membrane 'antigens'. Definition

It results from the general organization of the PM that what is recognized by antibodies and sugar-binding phytolectins as 'membrane antigens' belongs essentially to the lymphocyte glycocalix area, i.e. the peripheral membrane proteins, the hydrophilic domains of integral proteins which protrude from the membrane, and the carbohydrate moieties of glycoproteins and glycolipids. This cell coat thus constitutes the panel of structural elements where recognition phenomena are taking place, and where subtle variability may also generate polymorphism.[61]

Most lymphocyte PM antigens can be split by proteolytic enzymes; then the antigenicity of the cell surface is markedly reduced, if not completely lost, and the antigenic sites can occasionally be found among the solubilized split products. Extraction from the membrane of whole components bearing the same antigenic sites requires the aforementioned procedures which are known to solubilize integral membrane proteins.[11,61,62]

Though having an amphipathic structure, their hydrophobic foot embedded in the membrane may be rather small, e.g. it is estimated that the hydrophobic foot which keeps the H-2 molecule in the membrane has a mass of ~3000–6000 daltons (the molecular weight of histocompatibility antigen monomer being ~45 000). Such a characteristic is especially important to remember with regard to the usual—and probably wrong—representation of the receptor immunoglobulins: the size of the Fc of an IgM or IgD can be estimated to be about 6–9 nm long,[63] which should be enough to span the 4·5 nm thick phospholipid bilayer; however, unless the actual receptors would belong to a novel immunoglobulin class,[64] their Fc is too hydrophilic to be deeply inserted in the membrane and indeed most of IgM or IgD Fc determinants are exposed on the surface of the B lymphocytes (except perhaps those of the fifth (carboxy-terminal) loop).[65] The size of this loop might be 2–3 nm long. Thus if receptor immunoglobulins have to transmit a signal through the membrane, the signal apparently cannot pass uniquely through their Fc, and cooperation with another integral protein (the 'Fc receptor'?) is probably needed (References 62, 64 and R. M. E. Parkhouse and E. R. Abney, herein). If the fifth domain were not folded, then it might span the phospholipid bilayer, but it is unlikely (though not formally excluded) that an unfolded polypeptide chain might transduce a signal.

3.2 Membrane antigens topography: General view

It has been recently—and it still is—a matter of controversy what the actual natural distribution of membrane antigens is over the lymphocyte surface; and it is difficult to escape an historical approach to the problem. Various membrane antigens which were at first described as highly organized in specific clusters over the cell surface[66] were later considered to be diffusely distributed at random, but now they are sometimes described as loosely clustered in a non-random fashion. This evolution, which corresponds to the 'discovery' by immunologists of the concept of membrane fluidity and of its restrictions, also corresponds to various refinements of the methods used to detect the membrane antigens. Let us consider for instance the distribution of B cell membrane immunoglobulins (mIg). They can be visualized by microscopy by using antibody to immunoglobulin (anti-Ig) tagged with a variety of optical or electron microscopy markers (for a review, see Reference 67): e.g. fluorochromes for immunofluorescence microscopy (IFM), enzymatic markers, radioactive labels, electron dense molecules or small particles (like ferritin, haemocyanin, plant viruses, latex beads) for immuno-electronmicroscopy (IEM) either by transmission (TIEM) or by scanning (SIEM). All these various methods give concordant results, in that the binding to the cell surface of antibody which is able to cross-link the mIg is followed by a redistribution phenomenon known as the 'ring → spots → cap → endocytosis and/or shedding' transition.[68–72]

The four steps can be summarized as follows:[69] immediately after coating lymphocytes with anti-Ig antibody and washing away unbound or loosely bound antibody, the mIg-anti-Ig complexes appear homogeneously distributed on the cell surface, then there is a clustering of the antibody into spots or patches (the spotting process), then the patches are collected in one cap at one cell pole (the capping process), and finally the complexes disappear from the cell surface by endocytosis and/or shedding. Though this will be essentially treated with reference to lymphocyte mIg, it should be insisted that the phenomenon is *not* an exclusive property of either mIg or normal lymphocytes. Spotting and capping have been also shown for transformed lymphocytes,[74] normal and transformed fibroblasts,[75–78] hepatoma cells,[79] macrophages,[80] chick embryo neural retina cell,[81] monkey kidney cells,[82] basophiles,[83,84] and amoeba.[85]

Such a redistribution of membrane antigens cannot be obtained on intact mature mammalian erythrocytes.[6,7,69,78] However, it is easy to induce spotting (but no capping) of B locus histocompatibility antigens on the membrane of intact chicken erythrocytes (which are nucleated cells) (F. Loor and J. R. L. Pink, unpublished). It seems possible to obtain some clustering of A blood group antigens on adult human erythrocytes only when they become extensively crenated.[86]

This complex redistribution phenomenon will be analysed further and it depends on many variables, i.e. the capacity of the antibody to cross-link membrane components, the fluidity of the membrane, the metabolic activity of the cell and probably the integrity of a PM associated cytoskeleton.

3.3 Natural distribution

The demonstration of the native distribution of cell surface antigens requires the use of monovalent ligands and possibly a prefixation of the cells prior to their labelling with the ligands (though fixation procedures are themselves a potential source of artefacts).

The various methods used (IFM, TIEM, SIEM)[67] show that the distribution of mIg on B cells from various species is essentially uniform when their detection is obtained either by use of monovalent antibody fragments (Fab) on (un)fixed cells or of bivalent antibody (fragments) ($F(ab')_2$ or IgG) on fixed cells.[68–74,87–91]

There are, however, two recent exceptions.[92,93] In one case, freeze-etched replicas of the labelled cells suggested that the mIg would be 'distributed in small clusters with interconnecting networks'.[93] In another report, an approach combining SEM and the immunolatex method describes the cells as showing 'a loosely clustered pattern of surface Ig', which 'appears to be specifically associated with surface microvilli'.[92] Thus in both of these cases the authors describe the existence of definite areas of unlabelled membrane, and the distribution of the mIg is definitely different from random. It is important—but it cannot be decided at present—to know what actually is the natural arrangement of lymphocyte surface antigens, especially the mIg, since a non-random organization might be associated with their function, i.e. uptake of antigen and transmission of information to the cell. However, even a non-random distribution of mIg over the cell surface might simply reflect either a cell surface remodelling during the preparation of the cell sample or a routine aspect of cell membrane growth. For instance, the fusion of new membrane vesicles with the PM might not be completely or only slowly followed by diffusion and intermixing of the new and old membrane components, and this might be due either to intrinsic differences in the fluidity of new and old membrane pieces or to restrictions to the diffusion of membrane components as aforementioned. A similar interpretation could be given of the preferential location (either exclusive or gradient-like) of mIg at one pole of the cell, which was observed on some bursa lymphoblasts in the developing chicken embryo,[94] and to gradient-like distribution of H-2 and Tla determinants on leukaemic lymphoblasts.[74] Another possible cause for non-random distribution of membrane antigens might be the formation of various invaginations (pinocytic vesicles) or evaginations (microvilli, MV) which are probably associated with localized variation of the membrane fluidity. The apparent association of the mIg with MV[92] is controversial since it was not observed by other authors who described a similar presence of Ig on both smooth and villous parts of the membrane.[87–89] It might be explained by the extreme ability of the MV to expand or retract rapidly, depending on environmental conditions (see Section 3.9).* A non-uniform distribution may be found if the rate of this process is higher than the rate of diffusion of mIg in the membrane plane, i.e. they will appear more concentrated on an MV which is retracting and less concentrated when it is expanding. This does not appear unlikely regarding the membrane lipid heterogeneity (see Reference 10) and the high rate of diffusion of lipids (at least when studied in artificial membrane), which is significantly higher than the rate of diffusion of membrane proteins (see References 3, 6, 12, 95). Differential rates of diffusion of the various membrane lipids and proteins might lead to phase separation of membrane components.[20,96–99] A deviation from random distribution will arise if protein antigens are sequestered from the gelifying lipid areas to the liquid phases. Combined spin-label and freeze-fracture studies on artificial lipid–protein bilayers,[97] and on membranes from *Escherichia coli*[97,98] and from yeasts,[99] show that when lateral phase separation of lipids into fluid and solid areas occurs, the membrane particles (which very probably represent protein components) tend to associate with the fluid lipid phase. As far as other cell surface antigens are concerned, the sites to which various lectins can bind have been found to be homogeneously and randomly distributed over the cell surface,[18,78,100–104] but since they are not as specific as antibodies, and since there is a large variety of glycoproteins on the cell surface, the chance of finding a random binding is much higher.

Membrane alloantigens were also found diffusely dispersed over the cell surface,[74,104] with the exception of the θ antigen on mouse thymocytes showing a uropod:[104] the θ antigen was 'scarce or absent from the constriction, posterior to the nucleus' and 'poorly represented on the uropod', both of which areas would appear to be less fluid parts of the membrane 'presumably immobilized by interaction with a cytoplasmic contractile system'.[104]

* The methods used to prepare lymphocyte suspensions are markedly unphysiological, e.g. tissue teasing which brings much cell debris into contact with the PM of other cells; sedimentation by centrifugation; low protein content of the medium used to wash the cells; (!) cells kept at or near 0 °C, etc.

Random distribution of membrane components is by no means a general property of cell PM and there are several examples where, on the contrary, a marked segregation of the components on distinct areas of the cell membrane has been undoubtedly demonstrated (for a review see References 1–6, 10 and 12). Whether natural or not, the non-random distribution of some antigens on the lymphocyte membrane is quite different from the spots and patches and the cap distribution, which are known to be the results of membrane rearrangements either passive or active (see hereafter).

3.4 Ligand induced clustering: A passive redistribution

The redistribution is passive in that the formation of spots and patches does not require the cell to be alive or metabolically active, e.g. it can be induced on cell ghosts. For the best-studied membrane antigen, the receptor immunoglobulin, one has shown that clustering requires cross-linking by the ligand: thus spots are formed upon binding of bivalent $F(ab')_2$-anti-Ig,[69] but not after binding of the monovalent Fab[68–74] or of the bivalent $F(ab')_2$ when binding of the latter is immediately followed by reduction of the disulphide bridge covalently linking the two antibody fragments.[69] The need for cross-linking is further proved by the formation of spots when cells coated with homogeneously distributed Fab-anti-Ig are further treated by a bivalent anti-Fab reagent.[69] Thus the formation of spots–patches can be visualized as a microprecipitin reaction between the antibody and the membrane antigen, which would occur at a two-dimension level by diffusion and collision of the complexes in the plane of the membrane. As any antigen–antibody precipitin reaction, the clustering of mIg by anti-Ig antibody is inhibited when the cells are kept in presence of an excess of bivalent antibody.[69]

Finally, the spotting process depends on all environmental factors which can influence the mobility of the membrane antigens. Thus, besides the influence of the cytoskeleton (see Section 3.8), the formation of the spots shows a marked temperature dependence, being much slower at or near 0 °C than at 37 °C or room temperature,[68–74] which probably reflects a change in the viscosity of the membrane lipids. There is, however, no real 'freezing point', and this is presumably due to the heterogeneity of the membrane lipids and the presence of cholesterol.[21]

The rate of spot formation is also modulated by insertion in the membrane of components which change its fluidity. Recently a whole series of 'membrane mobility agents' have been described that modulate the aggregation process.[105] Other agents, described as crenators or cup-formers on intact red cells,[106] have also been tried. Local anaesthetics and tranquillizers, which are cationic amphipathic molecules and cup-forming agents,[106] interfere with the spotting process in that the mIg-anti-Ig complexes get distributed into multiple small spots instead of a few large patches.[107] The reduction of the rate of the spotting process,[69,103] and the need for more 'membrane mobility agents'[105] when the lymphocytes are kept in a medium rich in serum or in serum albumin (non-defatted), might be due to the presence of large amounts of free fatty acids, which are anionic amphipathic molecules and crenating agents.[106] The exact way by which cup-forming and crenating agents modulate the rate of aggregation of membrane components is not clearly understood, nor is the reason why acidic pH can reversibly inhibit patching.[108] As mentioned above the PM is remarkably asymmetrical and the two halves of the bilayer could show a differential fluidity which might also be differentially affected by our manoeuvres.

The spotting process is also inhibited on cells coated with large amounts of phytolectins.[69,91,100,103,108–112] This might be due partly to cross-linking of the whole cell glycocalix by the lectins into a dense network entrapping cell membrane antigens. However, the main reason for their reduced mobility is very probably due to still undefined interactions with a cytoskeleton network associated with the inner face of the PM, which would either constitute a rigid mesh or contribute some immobile membrane sites to which mIg would be cross-linked via lectin molecules (see Section 3.8). Finally the rate of spot formation depends on the distribution, the concentration and the nature of the membrane antigen, especially on the total number of antigenic determinants which can combine simultaneously with the ligands (antibodies or lectins). With some membrane antigens like θ, H-2, $\beta 2\mu$ the formation of spots is not easily obtained by the anti-θ, anti-H-2 or anti-$\beta 2\mu$ alone, and requires a second layer of bivalent anti-Ig antibodies able to cross-link the first ones.[68–74,102–104,113–120] This lack of lattice formation by the first bivalent antibody is usually interpreted as showing that the corresponding antigen molecule behaves *in situ* in the membrane as an independent monomeric mobile unit.[74,104] For other antigens, like Tla, one can obtain variable results depending on the cell source and their processing for coating with anti-Tla antibodies.[74,104]

One should stress here that great caution must be taken in IFM to avoid confusing with real spots some structures like retracted microvilli or pinocytic vesicles containing some of the reagents.

Interestingly enough, whether they are few or multiple, the spots never fuse in a single mass on the membrane of lymphocytes whose metabolism is blocked. This indicates restriction of their mobility (see below).

3.5 Ligand induced capping: An active redistribution

When lymphocytes coated with anti-Ig antibody are kept in physiological conditions, the microprecipitates of mIg-anti-Ig do not remain distributed as spots all over the membrane but are concentrated at one pole of the cell forming a polar cap.[68–74] This is basically different from the spotting process, since it requires a metabolically active cell. However, all the aforementioned factors which influence the spotting process also influence the capping process in the same way, e.g. it requires cross-linking of the membrane antigen by the ligand and it is influenced by the membrane viscosity. Interestingly enough, capping (and resynthesis) of mIg on lymphocytes from the skate *Raja naevis* can readily be obtained even at +4 °C, a temperature at which that fish commonly lives.[121] It might be interesting to study the membrane composition of cells from poikilothermic animals, which have to live at variable temperatures.

The spotting process can be easily dissociated from the capping process by use of various inhibitors of glycolysis and oxidative phosphorylation;[68–70] the most commonly used is 10^{-2} M NaN_3, which reversibly inhibits capping.[68,69] Capping is not detectably affected either by inhibitors of protein synthesis[115] or by any depletion of extracellular calcium and magnesium[68] or by a depolarization of the membrane by high concentrations of potassium.[73] (It should be noted, however, that the methods used to deplete intracellular Ca^{2+} were insufficient to give any inhibition of fibroblast motility, which can be obtained only by more drastic means of Ca^{2+} depletion.)[41]

The capping process can be transiently associated with changes in the cell shape and stimulation of cell mobility,[122,123] and the cap is usually found over the tail of the mobile lymphocyte,[68–72] though exceptions have been reported.[74,124] The contact of the cell with a solid substrate is not obligatory for capping to occur.[68] Cells that have a restricted mobility, owing to their adherence to ligand-coated substrate, can also form caps.[103] Furthermore, capping and mobility can be dissociated by a drug that stops the mobility of the cells without affecting the capping process itself.[122] Capping appears to affect specifically the membrane components (e.g. Ig) that have been cross-linked by the ligands, since the distribution of other membrane components (e.g. H-2) is not affected[68–70,74,115,116,125] (see also Section 4.2).

There is a definite polarity of the capping process: the cap is almost always formed over the area of the cell which contains the Golgi complex, the centrosome and most cell organelles.[71] Furthermore, when various membrane components are induced to cap either simultaneously or successively, they migrate always to the same cell pole.[69–71] Finally, when lymphocytes adhere to a surface (plastic, glass, lectin-coated beads, or simply another cell), the cap shows a strong tendency to be localized close to the area of contact with the substrate,[69,126] though one exception has been reported.[127] Such cases, where there seems to be no polarity,[74,124,127] are probably more representative of a complete aggregation of all spots in one single mass than of an active, cell-controlled segregation of the spots.

The actual mechanism of capping is not clear yet, though hypotheses have been put forward following its discovery.[68–70,74,87,90,91,100–104,108–112,115,122–130] It seems to be closely linked to the activity of the cytoskeleton (see Section 3.8).

3.6 Membrane bound ligand endocytosis and degradation

The accumulation at one pole of the cell of the agglutinated ligand–membrane antigen complexes is rapidly followed by their endocytosis by the cell; this again is an energy-requiring process.[68–70] TEM studies of the cap area show the presence of frequent invaginations and of intracellular vacuoles of irregular shape and of rather large size (200–400 nm in diameter).[71] These internalized pieces of PM comprise both ligand-bound area and ligand-free area, suggesting that the lymphocyte is not able to phagocytose selectively its ligand-disturbed membrane areas only.

However, the internalization of membrane bound ligand does not require membrane component clustering or capping. This is especially clear when a monovalent ligand, like Fab-anti-Ig is used: it can become interiorized into small pinocytic vesicles distributed all over the lymphocyte periphery[72,90,91,131] and might sometimes be confused with true spots on the cell surface. This process is not blocked by NaN_3, although this metabolic inhibitor inhibits a cap-like migration of the pinocytic vesicles towards one cell pole.[91] Interestingly enough, vinblastine efficiently blocks pinocytosis,[91,129] while colchicine cannot.[91] If this shows the involvement of a microtubular system in pinocytosis, the different effects of the two plant alkaloids might be a result of their binding to different sites of the tubule protein subunit.[34]

After endocytosis, the ligand is rapidly degraded despite the fact that lymphocytes are apparently poor in lysosomal enzymes.[70,131–133] It is likely, as mentioned above, that both self-membrane components and membrane-adsorbed components are digested and re-utilized, but this remains to be demonstrated.

3.7 Re-expression of membrane components after capping

When appropriate amounts of anti-Ig antibody are used, all mIg can be capped. After endocytosis the cell is thus deprived of mIg. When put in appropriate culture conditions, it will re-express new Ig on its membrane. The mIg molecules start to be detectable within 3 hours of the beginning of the capping process and the re-expression can be completed by

6–8 hours from capping, though sometimes more time (up to 20 hours) seems to be required.[69,134–136]

The use of protein synthesis inhibitors does not completely block, but significantly reduces the expression of new membrane components;[115,118,137] it is difficult to know if this is due to a blockage of the synthesis of mIg itself or of a short-lived protein that would be needed either for the transport of preformed PM vesicles or for their fusion with the membrane. The re-expression of mIg is blocked by colchicine;[137–138] thus the transport of vesicles might depend on the integrity of a microtubular system as recently suggested.[60]

Besides mIg, re-expression of membrane antigen after capping has been shown only for H-2[74,139] and for Tla.[74,140]

In contrast to what happens in adult animals for mature B cells, the capping of mIg on early B cells (either from embryonic sources or from adult ones) is not followed by a regeneration of new mIg.[141,142] This might be related to a series of immunological phenomena such as tolerance (either self or induced in immature and mature animals) and suppression (either of a class of Ig, or of an allotypic specificity or of an idiotypic specificity). Curiously enough, even monovalent Fab-anti-Ig-mIg complexes are redistributed as caps on early B cells, and they can achieve the mIg modulation.[142] It seems to be a rather general property of early cells to show higher rates of capping and even spontaneous capping,[94,107] which might be due to a balance of MT and MF activities quite different from that of mature cells.

3.8 Involvement of the cytoskeleton in capping

This idea initially originated (a) from the surprising finding that in metabolism-inhibited cells the spots of microprecipitate do not fuse in a single mass,[69] (b) from the inhibitory effects of concanavalin A (Con A) on spotting and capping,[69,91,100,103,108–112,126–130] and (c) from a good dose of speculative analogies about the role of spectrin in the restricted mobility of red cell membrane components.

The classical use of the aforementioned 'anti-MT' and 'anti-MF' drugs gave further support to the involvement of the cytoskeleton in capping; indeed (a) drugs known to disrupt MT like vinblastine and colchicine,[28,32–34] do not detectably impair capping even when used at high dose ($\sim 10^{-4}$ M);[68,91,110,112,130] (b) cytochalasin B,[38,39] a drug affecting MF integrity, when used at high dose ($\sim 10^{-4}$ M) has a partial-to-total capacity to reversibly inhibit capping;[68,74,91,102,103,112,129,130] (c) the two classes of drugs show a definite synergistic effect for capping inhibition;[112,129,130] (d) cytochalasin B alone can very efficiently reverse the capping process, i.e. release the spots from being sequestered at one cell pole.[112,129] With respect to this latter point, the reversion of caps has also been obtained with metabolic inhibitors[143] or even spontaneously,[103] presumably after the cells had exhausted a reserve of energy.

Though generally accepted, the inhibitory effects of Con A on membrane antigen redistribution are reported by various authors in somewhat controversial terms. The reasons for this controversy are not understood yet, but might relate to slight differences of the cell processing or of the nature of the Con A reagent used. The essential data are reported hereafter.

(1) While high doses of Con A (e.g. 100 μg/ml) definitely inhibit membrane antigen redistribution, low doses, especially in the mitogenic range (e.g. 5 μg/ml), do not show such an effect.[103]

(2) The inhibition is obtained by use of native Con A which is tetravalent but not

by use of succinyl-Con A which is a bivalent derivative, unless the latter is further cross-linked by bivalent anti-Con A antibody (the monovalent Fab being ineffective).[144]
(3) The inhibition is observed when the cells are coated with Con A at 37 °C, or at 0 °C and warmed up at 37 °C without removing the excess of lectin. It is not observed when the cells are coated with Con A at or near 0 °C, washed to remove unbound or loosely bound lectin and only then warmed up to 37 °C.[91,109] This last point is controversial, at least in that variable results will be obtained depending on the Con A dose:[103] upon rewarming to 37 °C, capping might be observed but other antigens will be co-capped with and/or by Con A.[103,112]
(4) It is claimed that locally concentrated Con A, interacting with only a fraction of the total cell surface can, like soluble Con A, efficiently inhibit membrane antigen redistribution;[126,127] this is not always observed.[103]
(5) The lack of Con A redistribution cannot be reversed by cytochalasin;[110] however, 'anti-MT' drugs might favour the capping either with Con A itself or of other membrane antigens despite the presence of Con A bound to the cell membrane in inhibitory conditions;[109–111] this also remains controversial.[81,112,130] A definite modulation of Con A receptors mobility by colchicine requires a high concentration of the drug (10^{-4} M),[110] which is at least 100 times higher than what is needed to block mitotic events in most animal cells.[38] It is thus possible that the action of the drug would not be due to MT disruption, but rather to its incorporation in the PM with a consequent modulation of its fluidity, since plant alkaloids bind to membranes.[145] However, the modulation of Con A receptor mobility cannot be obtained[108] by use of an ultraviolet light-induced isomer of colchicine, i.e. lumicolchicine, which shows rather similar membrane binding properties to colchicine, but which does not bind to MT[145] and does not inhibit mitosis.[35]

The association of some sites on the outer cell surface with a MT-like system is a reasonable hypothesis. But there are no data to support the speculation that the restriction to the mobility of the membrane is actually *induced* by binding of Con A to some of its sites expressed only at 37 °C but not at 0 °C.[110] An alternative hypothesis is that such sites *always* have a restricted cytoskeleton-controlled mobility. Moreover, when cells are coated with Con A at various temperatures, the PM might not only show a modulated expression of its determinants for Con A binding, but the Con A itself might also undergo reversible quaternary structure transitions. Though the tetrameric form is predominant at 37 °C, it dissociates at lower temperatures and is almost completely in dimeric form at or near 0 °C.[146,147] Therefore, it should not be surprising that coating cells at 0 °C with native Con A is equivalent to coating them at any temperature with succinyl-Con A.

However, the lack of capping by Con A unless colchicine is added or the coating is done at 0 °C might imply that some sites available for Con A binding at 37 °C, but not 0 °C, are linked to a MT-network which would be organized on the cytoplasmic face of the membrane and have a restricted mobility. The Con A molecules might then cross-link a number of membrane glycoproteins (mIg among them) to these immobile or poorly mobile sites, leading to a colchicine-reversible, Con A-induced membrane 'freezing'; this cross-linking would become more efficient as the valency of Con A increases.

The inhibition of capping and its reversion by cytochalasin B suggest that MF structures are involved in moving all the spots to the cap area and in keeping them there, which requires energy. The reversion, cap→spots, further shows that there is no extensive cross-linking of the spots in the cap area. It is not known if the capping process is induced by the clustering of the membrane antigens. It might represent the normal cycling of some membrane components that would be glued in the micro-agglutinated membrane antigens. Though further speculations on the possible mechanisms and interactions between these two antagonistic systems (the one 'mobilizing', the other 'freezing') have already been presented (e.g. References 90, 103, 112, 126, 130), they seem to us rather too premature to be developed here. Some data will be further analysed in the next section but a complete review of available information will be presented elsewhere (F. Loor, *Progr. Allergy*, in preparation). The data fit well with our speculation (see Section 2.6.3) that a contractile MF system, anchored directly or via intermediate pieces to some membrane components and responsible for their polar migration, would be antagonized by an MT network also associated with some membrane components and conferring to the cortical cytoplasm of the cell thixotropic gel-like properties.

3.9 Lymphocyte microvilli (MV)

3.9.1 *Distribution and nature*

The lymphocyte surface is not smooth, and special attention was recently drawn to the expression of MV since various SEM reports seemed to indicate that B lymphocytes were covered with MV (hairy cells), while T lymphocytes had none or very few (bald cells).[87,148,149] This was not unlikely, considering the different panels of membrane antigens expressed on B and T cells (Reference 150 and P. C. L. Beverley, herein) and especially the different overall negative charge displayed by these cells.[151] The findings that B cells were more villous than T cells rapidly became controversial, since other authors reported the contrary.[152,153] The truth is probably that both B and T cells can be more or less villous.[88,89,91,154,155] The difficulty in showing this comes from the extreme lability and sensitivity of MV to various environmental conditions,[156,157] e.g. one culture medium allowing their expression, another one not.[91,158] There are, however, a series of reliable, interesting observations.

It was found that on motile lymphocytes 'the greater portion of the cell is devoid of MV whereas the terminal portion of the uropode is studded with MV'.[159] Similarly they were found as 'usually asymmetrically concentrated to one side of the cell'.[160] On the other hand, MV regularly distributed are found all around lymphocytes when their metabolism is blocked, e.g. by 10^{-2} M NaN_3,[91] and more generally when the level of ATP of the cell is decreased by inhibitors of respiration and glycolysis (S. de Petris, unpublished). Increased lengths of MV in the presence of NaN_3 have also been reported.[89] After removal of the inhibitor, the localization of MV at one cell pole is again observed, showing that the effect is reversible.[91] This was reminiscent of the aforementioned ring→spots→cap transition and this did not escape notice.[87,91]

When cells are fixed before labelling with anti-Ig, the mIg are found distributed homogeneously all over the surface whether smooth or villous,[87,88] although sometimes it is claimed that mIg show a preferential location on the MV.[92] When fixation follows anti-Ig labelling, the mIg-anti-Ig complexes are found at the base of the MV.[161] However, it is not clear if clusters of true two-dimension membrane antigen–ligand aggregates move to the base of the MV,[161] if individual MV shrivel upon membrane antigen clustering,[91] or if several MV are glued together as is the case for Con A which can find multiple sites for binding.[77]

The association, mIg spots–MV, is further supported by findings suggesting that they are mutually exclusive, the cells showing either spots or MV and that MV are progressively converted into spots under conditions favouring cross-linking.[91] By IFM on viable lymphocytes, the MV are best detected when the cells are labelled with monovalent reagents like Fab-anti-Ig.

MV are quite sensitive to chilling; they are less expressed on lymphocytes which have been kept at or near 0 °C than on lymphocytes which have been processed at 37 °C.[91,157] Actually, it is evident but not useless to emphasize that 0 °C is quite an unphysiological temperature for mammalian cells, and it is not surprising that various artefactual surface reorganizations take place. Processing the cells for anti-Ig labelling at 37 °C does not always lead to the detection of MV, however. In some cases, no MV at all are expressed and, on the contrary, one finds part of the membrane bound ligand in small pinocytic vesicles just underneath the membrane, even when monovalent ligands are used.[91] The expression of MV and the formation of pinocytic vesicles seem to be mutually exclusive.[87,91] For reasons still mostly unknown, it seems that ligands labelled with peroxidase[90,131] are regularly pinocytosed, and cells labelled with such reagents lose their MV.[87] It should be indicated that not all the ligand is pinocytosed, however, and with Fab-anti-Ig some is left homogeneously distributed as a ring on the outer cell surface.[91] Thus, quantitative studies on the amounts of reagent pinocytosed, remaining on the cell membrane, or shed in the medium after labelling[131] have to be considered with caution; their relative proportions might vary with the reagent used, one being more readily pinocytosed, the other one more readily shed.

The cap-like localization of MV on mobile lymphocytes, like the typical capping process and the polar concentration of pinocytic vesicles, is not only affected by NaN_3 but also by cytochalasin. The presence of this drug abolishes almost entirely the expression of MV on lymphocytes,[91,162] as well as on other cells;[162–164] when applied on cells expressing long MV, cytochalasin makes them collapse.[91] On the contrary, colchicine and vinblastine, which do not affect the formation of spots and caps, also do not detectably affect the expression and the distribution of villous structures on lymphocytes[91,162] and other cell types.[162]

Linear striations arranged radially in cross-sections can be detected by TEM within each MV present on the uropod of motile lymphocytes.[159] It remains unclear and controversial if these structures are of MT and/or of MF origin. The presence of MT within 'microspikes' has been reported in the past.[165] MT do not appear to be associated with the lymphocyte PM;[111] they were recently found under the PM but mostly located in the uropod.[112] MF are regularly observed under the lymphocyte PM and are more concentrated at the tail of lymphocytes capped with anti-Ig antibody.[166] Thick bundles of MF are frequently observed in MV and pseudopode-like structures.[111]

The existence of an MF backbone inside the lymphocyte MV is further supported by their sensitivity to cytochalasin, and by the binding of anti-smooth muscle antibodies to lymphocytes, either from normal blood or from lymphoid cell-lines.[167] Fixation makes the membrane permeable and the cell interior available for the antibody which can detect an actin-like component inside MV. No binding is detected on live lymphocytes[160,167] even when long hairy villous projections can be seen in phase contrast[164,167] and can be stained by anti-species antibodies.[164] The actin-like components present inside lymphocyte MV are thus not available for antibody binding on the actual cell surface.

If this MF complex in the cortical cytoplasm participates in the various capping-like phenomena, i.e. polar migration of true spots, of MV or of pinocytic vesicles, it must be anchored to some integral membrane components, either directly or through intermediate peripheral structures.*

* The bridges between MF and PM might be of a nature similar to the MT cross-bridges which can also link MT to various membranous components of the cell (see Section 2.6.1). Since this part of the chapter was written, papers have appeared which suggest that α-actinin (the main constituent of the Z-band of the sarcomere) is the molecule linking MF to the PM (References 230, 231, and see also the schema of the cell surface drawn after the most recent data (Section 6)).

3.9.2 *Microvilli and recognition*

It is difficult to clarify what the possible recognition and immunological functions of MV could be, because their fundamental function in cells other than lymphocytes is not clearly established either. In other systems, like differentiation and morphogenesis, important roles have for long been attributed to MV, although it is still unclear if the relations between inductor cell and target cell actually require MV mediated cell–cell contact or if they involve short-range diffusion of inductor molecules from the MV (for a review see Reference 168). The constant motility of MV, their short existence, and their general microscopic appearance have suggested that they might be cellular structures aimed to investigate, to 'taste' the environment.[169,170] But MV might simply be a reserve of PM for cell division,[40] for cell spreading,[171] or for modulating the surface of exchange between the cell and its milieu. From studies on Chinese hamster ovary cells in culture, it was postulated that MV might contain pumps for an active transport of sugars, rapidly growing cells expressing many MV to facilitate the diffusion of glucose across the membrane. This might also be the reason why MV are not equally expressed throughout the cell cycle, being profuse in G_1 and again in the late S phase, and much less numerous at other times.[172,173] Though less well-documented, similar observations were recently made on human lymphoid cell lines.[160] If the expression of many MV is a sign of high cell activity, it is quite surprising that lymphocytes, which spend most of their lifetime in G_0, are described as having so many MV. This characteristic is not due to membrane ligand binding but could be due to the culture media commonly used for their isolation, which might activate MV formation. This concept finds some support in recent SEM observations made on lymphocytes *in situ*.[174] They suggest that the quiescent state of B or T lymphocytes (G_0 phase) is characterized by a smooth surface. This seems to be the case when, coming from the blood, these cells have reached their specific B or T homing area in the lymphoid organs. To do this, they have to pass between the high endothelial cells forming the wall of the post-capillary venules, which are the main site of entry of lymphocytes in lymph nodes (see J. Sprent, herein). In these venules, they are found, however, quite villous, whether B or T. They seem to use these MV to adhere to the endothelial wall, they progressively lose the MV as they traverse the wall, and they arrive smooth in their specific homing area. The expression of many MV on recirculating lymphocytes might correspond to an 'activated state' making the cell capable of recognizing the endothelial cell wall and adhere to it.

The concept of MV as means of intercellular recognition receives some support from observations showing that the *initiation* of intercellular contact during cell aggregation is mediated by MV contact.[175,176] Similar conclusions were drawn from studies showing a positive correlation between the density of MV on cells and the extent of cell–cell aggregation, either spontaneous[172] or induced by low doses of lectins.[49,102] In the latter case, cytochalasin B could inhibit agglutination.[102] MV might, however, provide a *useful or necessary, but not sufficient* step for aggregation, since a metabolic inhibitor which increases MV formation actually inhibits cell aggregation.[102] This might be a reason why the involvement of MV in aggregation is a concept which is still being disputed.[12] Recent SEM studies on the cluster aggregation of dispersed embryonic neural retina cells indeed showed that very long MV (which can have several-fold the cell length) first 'bridged and connected distant cells and became shorter as the cells came together and formed aggregates'.[176] Later, only short MV were left on the cells. This is a process which requires active metabolism.

MV formation seems to be activated during formation of T cell rosettes, since they increase with time on these cells but not on the other, non-rosetting, cells in the suspension.[156] Various data also suggest the involvement of MV in cell adherence to glass and plastic surfaces,[177] in the control of intercellular adhesion,[178] and in their relation to immune adherence.[179] The interactions of lymphocytes with their target cells in physiological conditions also seem to be mediated by MV.[180–182] In this respect, it is interesting to remember that lymphocyte-mediated cytotoxicity is sensitive to cytochalasin, presumably by interfering at the level of the recognition phase (Reference 183, and J.-C. Cerottini and K. T. Brunner, herein).

Finally, the organization of the 'macrophage–B cell–T cells' clusters, which show antigen specificity *in vitro*, is remarkable (References 184, 185 and G. E. Roelants, herein); smooth T cells adhere to one central, villous B cell, the latter being attached to the macrophage; each T cell adheres through a uropod which contains most of the MF of the cell, suggesting again a preferential orientation of MV toward the B cell surface. It is possible that this would correspond to a specific concentration of some T cell membrane components having recognition or effector function.

3.9.3 *Microvilli and shedding*

Shedding is a term employed to describe the release by the cell of PM antigens. It is different from secretion in that secreted components were never integrated or bound to membrane, although a clear operational distinction between shedding and secretion is actually lacking.

How shedding happens is yet unknown; it may be either by release of membrane antigens upon enzymatic action of membrane-bound or cell-released proteases or glycosidases, or by release of whole PM fragments.

An example of the former form of shedding is given by the release by chick embryo fibroblasts in culture of biosynthetically labelled ^{3}H- or ^{14}C-glucosamine containing glycans, which show ion-exchange chromatographic properties similar to products that can be cleaved by trypsin from the surface of intact cells.[186] An example of the second form of shedding would be the shedding of mIg from lymphocytes, whose membrane components were labelled by the lactoperoxidase-catalysed iodination procedure:[187,188] the mIg is apparently still associated with a fragment of PM[188] and in a state similar to that obtained by use of chaotropic agents like urea.[187] Cellular respiration and protein synthesis would be required for shedding to occur,[187] although it occurs at a rapid rate,[187,188] two points which appear difficult to reconcile. Shedding of mIg happened within minutes at 37 °C after labelling lymphocytes with peroxidase-labelled Fab-anti-Ig reagents, but after a few minutes the release of mIg was markedly slowed down and this mIg was complexed with the anti-Ig reagent.[131] It thus appears that this is a ligand-induced shedding which differs from spontaneous shedding as far as kinetics are concerned.

It has been proposed that mIg shedding could happen by pinching off of MV[62] and one would then expect to find a variety of membrane antigens to be shed, besides mIg. Curiously enough, the rate of mIg shedding is reported to be different from that of other membrane components,[187] especially of H-2 antigen that would not be detectably shed from lymphocytes.[62,188] This is in contrast with other claims that human lymphocytes shed HLA antigens[189] and that mouse kidney cells and thymus cells shed a material able to inhibit the cytotoxic activity of an anti-H-2 antiserum.[186]

Caution has to be exercised when interpreting the results from some of the aforementioned experiments[131] which involve the use of peroxidase labelled Fab-anti-Ig reagents, for it has been indicated that MV are never observed with peroxidase labelled reagents[85] and that peroxidase has the property of being pinocytosed without actually binding to the membrane.[190] Thus, it appears that such properties might be responsible for a faster than normal shedding. Loss of MV tips as a usual cause of normal shedding appears possible, however. *In vitro*, cultured L cells have been shown to leave such bits of MV behind when they detach from the substrate,[191] which is presumably the origin of the 'antigenic zone' surrounding the cells.[192,193]

4. SOME APPLICATIONS OF CAPPING

4.1 Capping, turnover and cytophilia

Attempts to study the turnover of membrane components by following their resynthesis after capping have to be treated with caution—and the same is true when it is done after proteolytic enzyme digestion. Indeed the actual rate of turnover of membrane components might be quite dissimilar from the values obtained by such methods; and one should consider that the cell after capping a membrane component might either show a normal rate of re-expression, or show a compensatory and transient increase of its turnover to replace the removed component (e.g. if the cell can 'count' that antigen on its surface) or show a decreased capacity to replace it if its removal has made survival difficult.

Furthermore, and however good culture conditions may become, one has to face their usual pitfalls since it is evident that they cannot provide the same microenvironmental conditions which exist *in situ*.

A good example is the re-expression of the Tla antigens after removal by capping; while leukaemic cells can re-express them *in vitro*, thymus lymphocytes do not, although they can re-express H-2 antigens.[74] A possibility to consider is that expression of Tla antigens by lymphocytes requires an inducer from the thymus epithelium, not provided by the medium.

However, it is also tempting to consider that a lack of re-expression of a given membrane antigen after capping might show that it was not the actual product of the cells which had it, but a cytophilic component adsorbed on the cells, though being synthesized and secreted or shed by other cells. For example, one may speculate that the Tla antigens present on thymus lymphocytes are cytophilic and actually made by another cell type in the thymus.

Among others, the capping–resynthesis procedure was successfully used to show that (a) the mIg present on the membrane of thymus lymphocytes from adult fish and amphibian larvae was indeed made by these cells;[121,134] (b) the antigen receptors capped by anti-Ig on mouse T cells could be re-expressed by the cells;[136] (c) the IgD and IgM molecules expressed simultaneously on many B lymphocytes of the human newborn were both synthetic products of these cells;[194] (d) the lymphocytes from a patient with chronic lymphocytic leukaemia, which showed on their membrane μ, γ, κ, and λ chains, actually synthesized only a membrane IgM showing a rheumatoid activity and picking up the other chains from the serum of the patient;[195] and (e) the mIg present on the macrophage membrane is definitely cytophilic.[80]

4.2 Capping and membrane component interrelationships

Attempts have been made to study the possible relationships of various components in a cell membrane by a method of 'differential redistribution'. Ideally, if two membrane components A and B were associated *in situ* in the membrane, the capping of A should also bring B into the cap (co-capping); and, on the contrary, if they are not linked the capping of A should leave the distribution of B unaffected. There are, however, serious limitations to the interpretation of such experiments, which are due, for example, to the following pitfalls.

(1) A and B might be normally associated in the membrane, but the binding of anti-A might induce their dissociation, e.g. by a conformational change induced in A by anti-A binding.
(2) The association of A and B may be an equilibrium, which will be displaced if the affinity of anti-A for A is better than the affinity of B for A.
(3) A and B might be normally dissociated in the membrane, but the binding of anti-A on A makes it associatable with B, e.g. through quaternary structure rearrangement. This might be unspecific 'stickiness' or specific recognition between membrane components.
(4) There may be a stable association of a fraction of the A and B components in the membrane but the binding of anti-A interferes with possible binding of anti-B, e.g. if A and B are two determinants closely associated on the same membrane component.
(5) There may be a great excess of A over B. The agglutination of A by anti-A might entrap unspecifically all B components in spots, and leave no B on the surface after capping, even if A and B never actually became associated. On the contrary, the capping of B by anti-B might leave a large amount of A on the surface even if all B components were associated with A components.

Furthermore, since most studies using co-capping were made in IFM, other pitfalls are (a) the possible confusion of polar migration of either pinocytic vesicles or surface MV with real capping, (b) the danger of using several layers of antibodies to detect a membrane antigen, since it is known how sticky the Fc can be for lymphocyte membrane, and (c) the assumption (though frequently the appropriate controls are missing!) that the link between the membrane antigen and the first ligand is not affected by the further deposition of new ligands (see also Reference 67).

These restrictions show that great caution is needed in *interpreting* the data that were obtained by the aforementioned methods, i.e. the independent capping of HLA or H-2 antigens and mIg,[68,125] of homologous or non-homologous H-2 antigens either in *cis* or in *trans* positions,[119,128,196] of H-2 and θ,[74] of θ and Tla,[74] of $\beta 2\mu$ and mIg,[117] and on the contrary the co-patching or co-capping of HLA or H-2 with $\beta 2\mu$,[118,119,197] of aggregated mIg with 'Fc receptors',[198,199] the 'cooperative' effect of the capping of *Agaricus bisporus* lectin on the capping of *Lens culinaris* lectin B though the two lectins bind to different membrane molecules.[200] The present state of confusion, which concerns for instance a possible association of Ia determinants with other membrane components (e.g. References 199, 201, 202), is at least partly due to the use of methods *ejusdem farinae*.

This should stimulate us to revise our ways of tackling the problem of membrane antigen relationships. A better approach would be to use bifunctional reagents, like imidoesters[203] or bearing light-activatable groups like ketene and nitrene groups,[204,205] to cross-link neighbouring membrane components. Besides parameters affecting the membrane itself, one could vary the distance separating the two reactive groups (ranging from ~0·5 nm to ~1·5 nm usually) and the time allowed for reaction with the membrane components. These could then be isolated and analysed and some significant cross-bridging of some membrane components would imply their topographical relationships. Such methods have been used recently with good success for the study of erythrocyte membrane component relationships.[206–208] The most elegant studies were performed with cleavable bifunctional reagents allowing separation of the cross-bridged membrane components for analysis.[208] The components were isolated from the cell surface after reaction with an imidoester that could be cleaved by reduction of a disulphide bridge. Before cleavage, they were first separated by gel electrophoresis in one dimension, then reduced to cleave the bridging reagent and finally re-electrophorized in the other dimension (diagonal mapping electrophoresis method). Thus, membrane antigens that were not bridged map on a diagonal, while those that were bridged by the reagent map as spots outside the diagonal. They can be identified by their distance of migration during the second separation (after bridge cleavage). The existence of close relationships between some membrane components, and of various oligomers of some of them, can be demonstrated and the frequency of such associations can be evaluated.

The variety of proteins present in the erythrocyte PM is much more limited than in the lymphocyte PM, and in the latter case the mapping might become too complicated to interpret. In this respect, immunological analysis[209] might be helpful in the future. The results from such analysis would then be more valuable than those provided by our still 'superficial' surface antigen redistribution studies.

4.3 Capping and antigenic modulation

Antigenic modulation is the phenomenon by which the phenotypic expression of a membrane antigen can be (reversibly) suppressed by the related antibody. It is different from a selection of cells lacking the membrane antigen. It has long been known to occur in paramecia exposed to antibody directed against membrane constituents.[210] Most extensive studies concern the Tla determinants.[211,212] Briefly, Tla^+ thymocytes or leukaemia cells are lysed by treatment with anti-Tla antibody and complement; however, if they are first incubated *in vitro* or *in vivo* with anti-Tla antibodies at or near physiological temperatures in the absence of complement, the cells rapidly become resistant to anti-Tla antibody mediated cytolysis. For leukaemia cells, this modulation is reversible upon removal of antibody. Studies on H-2 antigens further show that requirements for modulation may vary among various cells.[213] This is reminiscent of the capping and resynthesis of mIg, and it was postulated that modulation simply represents the removal of the membrane antigen by capping. However, matters are not so simple. Indeed, cells can be modulated by treatment with a low concentration of anti-Tla antibody (however, in the presence of complement, that concentration would be high enough to be uniformly cytotoxic); they resist complement-mediated lysis, though they still show easily detectable Tla determinants on their surface.[74,140] Further, both modulated and non-modulated cells can be shown to synthesize Tla determinants at the same rate.[140] Similarly, in the case of the modulation of H-2 antigens, a large amount of H-2 determinants and even of anti-H-2 antibody can remain on the membrane of cells insensitive to the complement.[139]

Thus, a cell can become insensitive to antibody and complement-mediated killing even though no significant loss of antigen from the membrane has occurred. Some Tla^+ leukaemia cells are fully resistant to lysis by anti-Tla antibody and complement.[214] Some membrane active agents like anaesthetics can modulate the susceptibility of θ^+ cells to lysis by anti-θ antibody and complement.[215] On the other hand, one also finds some cells which cannot undergo modulation, i.e. Tla^+ somatic hybrids of Tla^+- and Tla^--leukaemic cells,[216] as well as cell-lines which cannot be induced to form caps of H-2.[217]

Thus it appears that a number of factors, e.g. membrane antigen density, membrane viscosity and efficiency of the cytoskeletal structures, can vary from cell to cell and influence the rate of killing of cells by antibody and complement, in ways which are not well understood. Since antigenic modulation is demonstrated by cytotoxicity methods, one could only *speculate* on its mechanism. But clearly it cannot simply be a clearance by capping of all the relevant antigens from the cell surface. Therefore, the killing of H-2 modulated target cells by T cells[218] cannot be taken as evidence that the T cell recognize something other than H-2 on the target, since T cell-mediated killing might show different requirements from antibody- and complement-mediated killing.

Antigenic modulation is one of the possible ways by which tumour cells could escape destruction (recently discussed in Reference 12). Furthermore it is also possible that the irreversible modulation of mIg on early B cells (discussed in Section 3.7) is related to the immunological phenomena of tolerance and suppression.

5. MEMBRANE CONTROLS OF LYMPHOCYTE FUNCTIONS

It is likely that the plasma membrane can exert some control on the activity of the cell, but this cannot yet be put in true molecular terms. Furthermore, the biological meanings of words such as 'receptors', 'triggering', 'activation', 'suppression', etc., have recently shown accelerated degeneracy; frequently one wonders to what the cell actually gets 'triggered'. Moreover, most speculations on triggering and the rest come either from studies on late immune reactivities *in vivo* (where little can be said about molecular processes) or from studies on immune-like responses (*x*, *y*, *z* as probes for . . .) by cells cultured *in vitro*, which are still black boxes as experimental systems, since they are definitely not under control (see M. H. Schreier and A. A. Nordin, herein). The response of the cell is simply studied too far away from the 'signal' to allow understanding of its nature in molecular terms.

I have the feeling that to speculate now about conditions for lymphocyte 'triggering' and to build up possible mechanisms would only be exciting intellectual *divertimenti*. This is due essentially to the present paucity of solid information about the biochemical events associated with lymphocyte 'triggering' and to the difficulty in deciding which of these events are essential steps for cell 'activation' and which are unrelated or accessory events. Too many logical models are still possible, which argues against their actual value as a frame for an oriented experimental approach.

> Chacun tourne en réalités,
> Autant qu'il peut, ses propres songes.
> (Jean de La Fontaine, dans *Le Statuaire et la Statue de Jupiter*)

A definition of cell triggering might be 'to bring about a modification in the cell cycle or the level of differentiation, i.e. the expression of the cell genome'. Thus, for lymphocytes this would include, for instance, signal(s) given by thymus epithelial cells to T cell precursors, signal(s) to recirculate or to stop in the homing area, mitogenic responses to lectins, blast transformation by anti-allotype antibodies, signal(s) to start synthesizing Ig for secretion, signal(s) to divide to memory cells.

As already stated in the introduction, it is a bias to believe that a signalling membrane ligand does not need to penetrate the cell; this is surely not the only possible mechanism. However, the fact that ligands like hormones, lectins or antibody can affect the cell behaviour, even when cross-linked to a non-phagocytable support, does indeed fit the concept of transmembrane information transfer. Our present knowledge of the structure of biological membranes only allows one to present reasonable speculations on how a ligand bound to the outer face of the plasma membrane can transmit a signal to the cytoplasmic face. However, actual mechanisms rapidly become too complicated to imagine. For instance, how could mIg, upon specific antigen binding, transmit the information inside the cell? As discussed before, the piece of the mIg inserted in the membrane is probably too short to traverse it, and cooperation with another membrane component is likely required. One could suggest that the antigen-induced aggregation of mIg (by virtue of its polymeric character, or with help of macrophages, T cell or any kind of 'factor') is followed by a recognition of clustered mIg by a Cl_q-like component present in the membrane. This would lead to a whole chain of reactions going on in the membrane which would be similar to complement activation pathways. Such reactions could lead to a number of possible modifications of the cell membrane, quite able to transmit signals. Many hypotheses have been proposed, but their review is outside the scope of this contribution (for reviews see References 7, 68–70, 103, 137, 138, 150, 219–225, and herein). The panel of mechanisms at the disposal of the cell is broad and varied. There can be lateral phase separation increasing the rate of some components encounters or decreasing the interaction of some others; quaternary rearrangements of permeaphores, e.g. to allow the rapid entry of Ca^{2+}; conformational changes in transmembrane glycoproteins, e.g. with a consequent activation of membrane enzymes located on the cytoplasmic face of the membrane.

The identity of some of these enzymes, i.e. adenylate and guanylate cyclase, ATPase, phosphodiesterase and adenylate kinase, gives some idea of the hypotheses which can be built up.

Thus, in general terms, a ligand might be a signal for a cell if it is able to modulate a series of membrane specific functions, like permeability to ions, transport of sugars, activation or inhibition of membrane bound enzymes (see References 7, 221, 225).

How all this actually works is unclear, but the discovery of the membrane fluidity (which allows cooperative interactions over long distances even for molecules that are rare) and of other mechanical properties of the cell surface, has been a major step towards an understanding of how the cell membrane can exert a control on cell behaviour.

6. THE LYMPHOCYTE SURFACE, SCHEMATICALLY

Figure 1 is a highly schematic representation of the cell surface, which was inspired from Nicolson's drawings,[12] as an attempt to draw *to scale* the various structures which determine the shape and activity of the cell surface. Our drawing includes a sectioned microvillus (MV) with actin microfilaments (MF), α-actinin molecules (α-A), and myosin molecules (MM) arranged as a filament; the plasma membrane (PM), one area of which shows some heterogeneity of distribution of membrane proteins and another which shows some receptor immunoglobulins (mIg); and the cortical cytoplasm with cross-bridged (c-b) microtubules (MT). The shape of the various elements has been kept as little arbitrary as possible, and their actual dimensions are given in the text.

Enlargment can be estimated from the size of the tobacco mosaic virus particle (TMV), the author's favourite antigen,[226,227] which is ~300 nm long and ~18 nm broad. It is drawn after Caspar's data,[228] and while the 2130 identical protein subunits (16 per turn) could not be shown, their helical arrangement is indicated. It is interesting to compare the size of the TMV with the size of the PM and that of mIg (which are drawn with their fifth domain only inserted in the PM and with their Fab extended). This shows quite well what multiple binding might be and how unspecific subsequent interactions of large polymeric antigens with the PM might become (cf. some models of lymphocyte activation). The density of mIg represented on that area of the PM is a few times higher than can be expected for a ratio of 10^5 mIg per lymphocyte. Peripheral and integral membrane proteins in other areas of the PM are drawn in the typical Singer–Nicolson mode.[6–8] Their density should not be too far from reality (more than 50% surface occupancy).

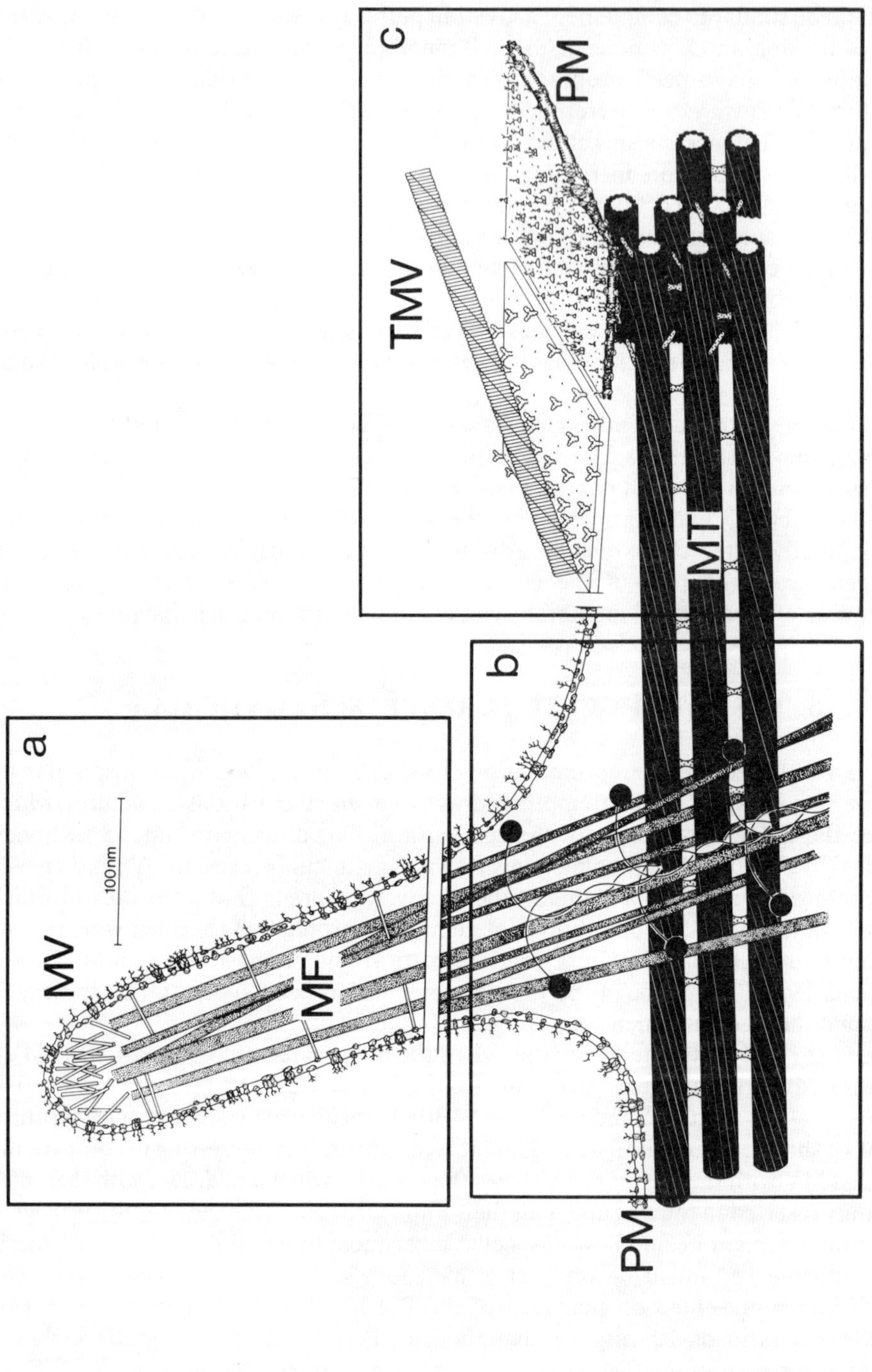

Figure 1. General view (From *Nature*, **264**, pp 272–273)

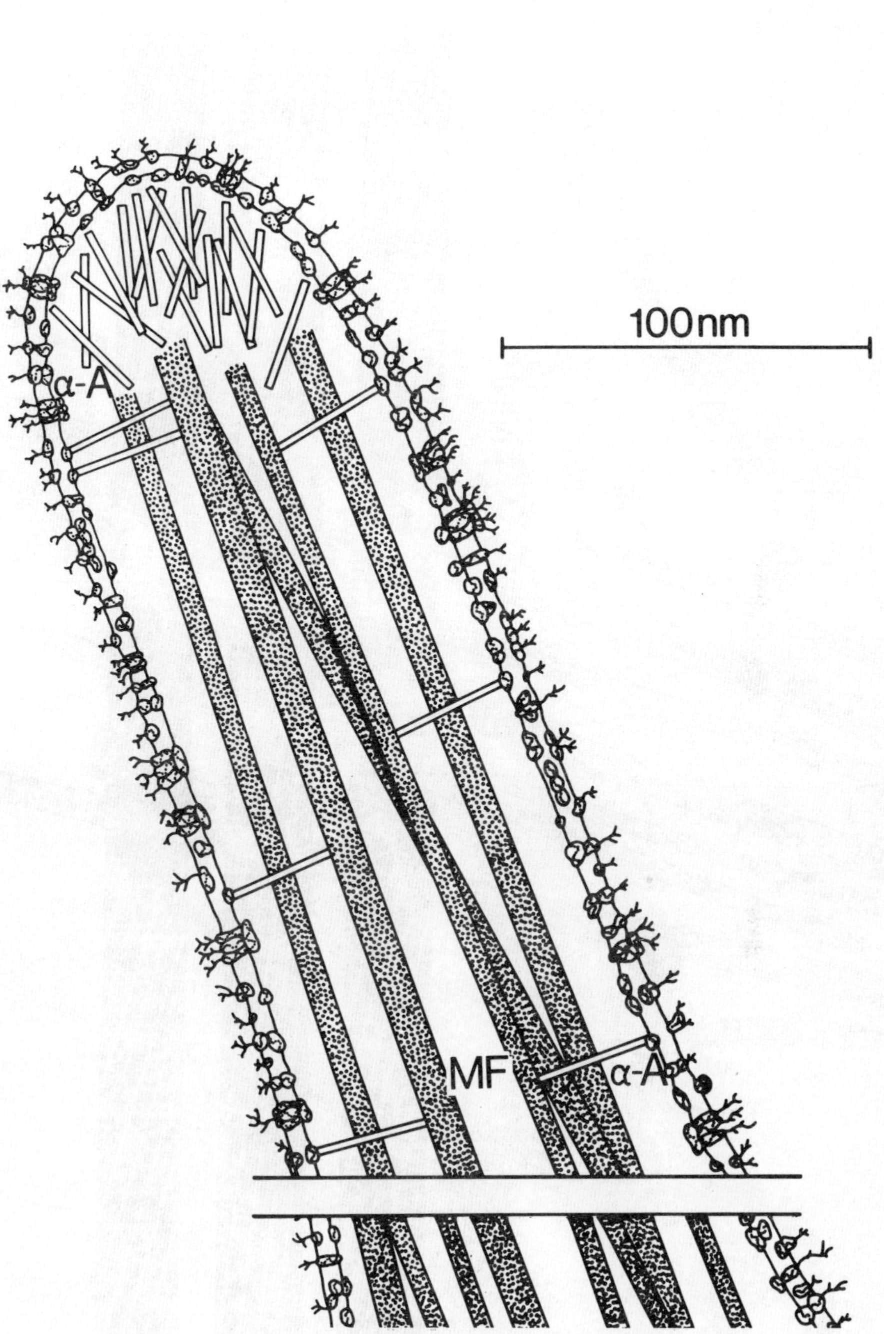

Figure 1(a)

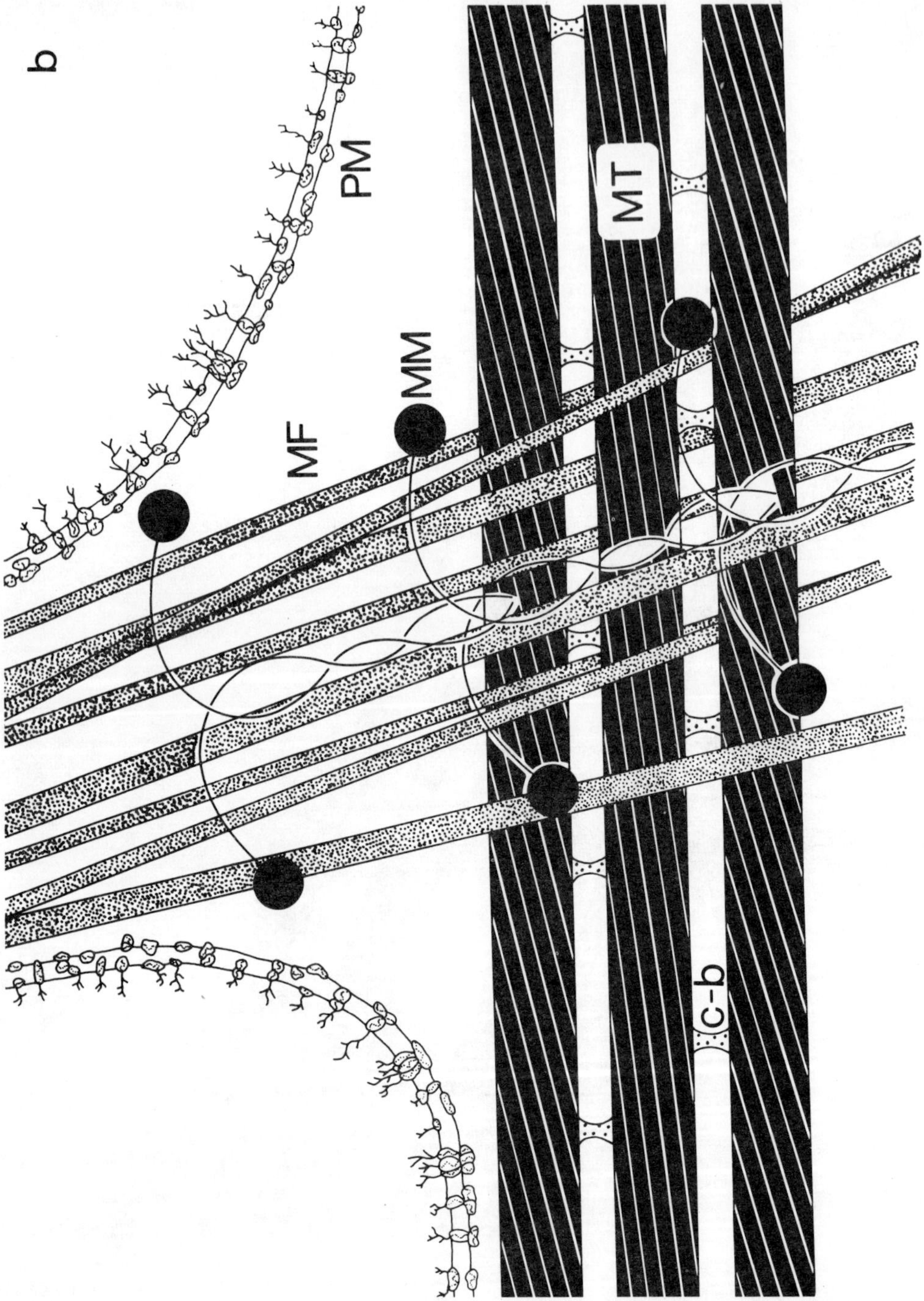

Figure 1(b)

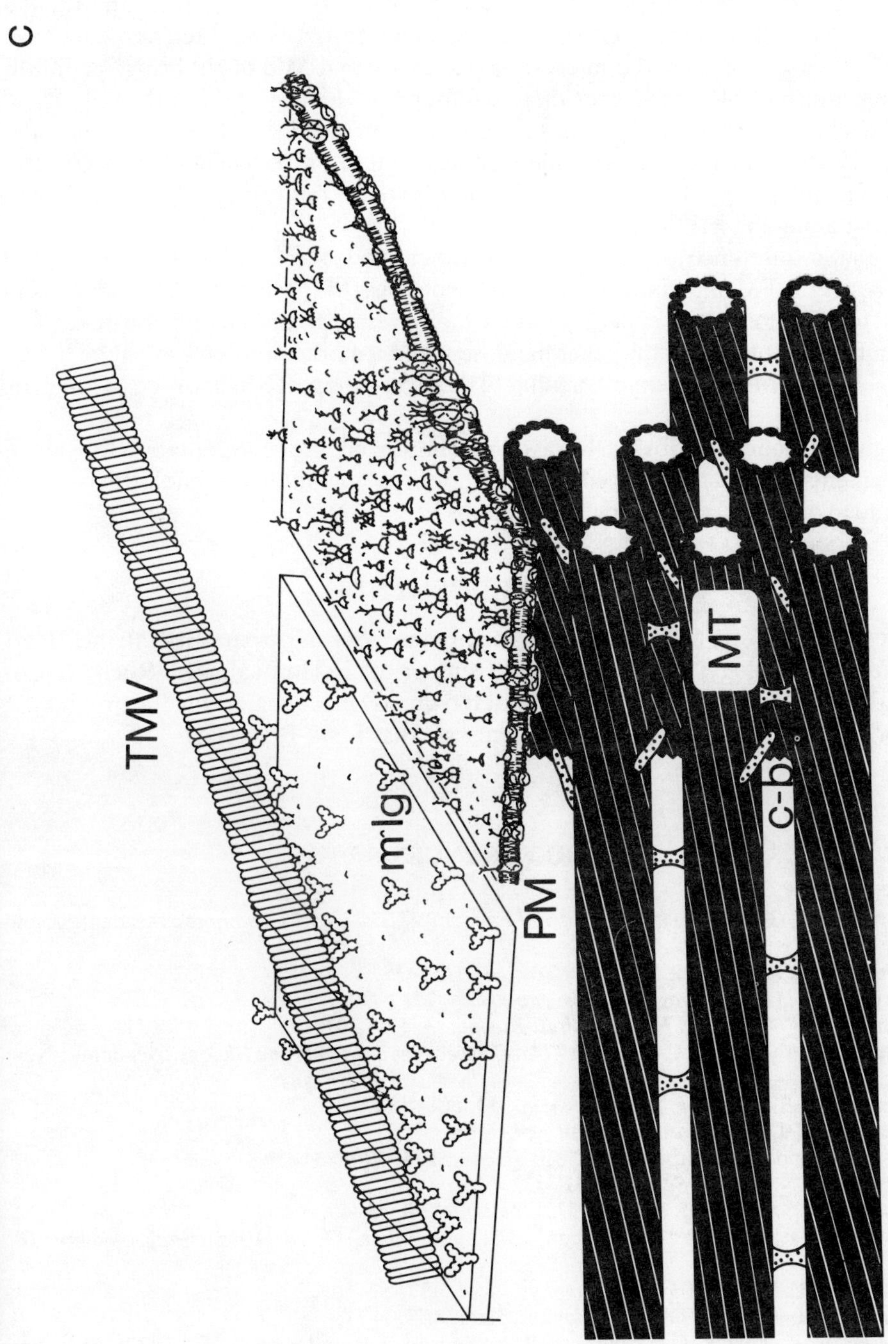

Figure 1(c)

At the time our model was drawn, the organization of MM (drawn after Reference 229) in non-muscle cells was not known; thus, MM have been put somewhat at chance on the top of actin MF. They have their light MM parts arranged so as to constitute an MF, and their heavy MM parts with the ATPase moiety extending apart. Since then new data have come out[230,231] which suggest the interesting possibility that MM might bridge actin MF from closely adjacent MV; moreover these data support the concept that the polarity of MF is determined by their attachment to the PM, in an area of high α-actinin density, inside the MV tips. The latter area would be equivalent to the Z-band of the sarcomere, and α-actinin molecules (which are rods of 2 nm × 30 nm) would also bridge the actin MF to the PM all along the MV length.

The dramatic difference between this and other models (cf. 12, 111) concerns the relative sizes of the PM and of the cytoskeletal structures (MT, MF and MM). Indeed, and in contrast to other models, it appears unlikely that several MF or MT could be linked to a single membrane protein. On the other hand, several membrane proteins might be linked to one single MT or MF, via elements similar to the cross-bridges (c-b) between MT and/or via α-actinin (α-A).

Although this model will obviously have to be improved, the representation to scale of various elements which are involved in the lymphocyte membrane physiology should at least help to avoid some misleading speculations.

ACKNOWLEDGEMENTS

The author's studies on the lymphocyte membrane were all performed at the Basel Institute for Immunology, with the generous support of F. Hoffmann–La Roche & Co. Limited. This review could be written without too greatly inconveniencing the progress of the author's researches, thanks to the efficiency of Miss Lena-Britt Hägg.

7. REFERENCES

1. Rothfield, L. I. (Ed.) (1971). *Structure and Function of Biological Membranes*. Academic, New York.
2. Guidotti, G. (1972). *Annu. Rev. Biochem.*, **41**, 731–752.
3. Mehrishi, J. N. (1972). *Prog. Biophys. Mol. Biol.*, **25**, 1–70.
4. Edidin, M. (1974). *Annu. Rev. Biophys. Bioeng.*, **3**, 179–201.
5. Estrada-O, S. and Gitler, C. (Eds.) (1974). *Perspectives in Membrane Biology*. Academic, New York.
6. Singer, S. J. (1974). *Annu. Rev. Biochem.*, **43**, 805–833.
7. Singer, S. J. (1974). *Advan. Immunol.*, **19**, 1–66.
8. Singer, S. J. and Nicolson, G. L. (1972). *Science*, **175**, 720–731.
9. Bretscher, M. S. (1973). *Science*, **181**, 622–629.
10. Bretscher, M. S. and Raff, M. C. (1975). *Nature*, **258**, 43–49.
11. Ladoulis, C. T., Gill, T. J. III., Chen, S.-H. and Misra, D. N. (1975). *Prog. Allergy*, **18**, 205–288.
12. Nicolson, G. L. (1976). *Biochim. Biophys. Acta*, **457**, 57–108.
13. Nicolson, G. L. (1976). *Biochim. Biophys. Acta*, **458**, 1–71.
14. Marchesi, V. T., Tillack, T. N., Jackson, R. L., Segrest, J. P. and Scott, R. E. (1972). *Proc. Nat. Acad. Sci.*, **69**, 1445–1449.
15. Tillack, T. N., Scott, R. E. and Marchesi, V. T. (1972). *J. exp. Med.*, **135**, 1209–1227.
16. Pinto Da Silva, P. and Nicolson, G. L. (1974). *Biochim. Biophys. Acta*, **363**, 311–319.
17. Nicolson, G. L. and Painter, R. G. (1973). *J. Cell Biol.*, **59**, 395–406.

18. Ji, T. H. and Nicolson, G. L. (1974). *Proc. Nat. Acad. Sci.*, **71**, 2212–2216.
19. Nicolson, G. L. and Singer, S. J. (1974). *J. Cell Biol.*, **60**, 236–248.
20. Nicolson, G. L. and Yanagimachi, R. (1974). *Science*, **184**, 1294–1296.
21. Oldfield, E. and Chapman, D. (1972). *FEBS Letters*, **23**, 285–297.
22. McDonough, J. and Lilien, J. (1975). *Nature*, **256**, 416–417.
23. Palade, G. (1959). In *Subcellular Particles* (Hayashi, T., Ed.), Ronal Press, New York, pp. 64–83.
24. Hirano, H., Parkhouse, B., Nicolson, G. L., Lennox, E. S. and Singer, S. J. (1972). *Proc. Nat. Acad. Sci.*, **69**, 2945–2949.
25. Ray, T. K., Lieberman, I. and Lasing, A. I. (1968). *Biochim. Biophys. Res. Comm.*, **31**, 54–58.
26. Marcus, D. M. and Cass, L. E. (1969). *Science*, **164**, 553–555.
27. Wessells, N. K., Spooner, B. S. and Luduena, M. A. (1973). *Locomotion of Tissue Cells*, Ciba Foundation Symposium 14 (new series). Elsevier Excerpta Medica, North-Holland, Amsterdam, pp. 53–82.
28. Olmsted, J. B. and Borisy, G. G. (1973). *Annu. Rev. Biochem.*, **42**, 507–540.
29. Pollard, T. D. and Weihing, R. R. (1974). *CRC Critical Rev. Biochem.*, **2**, 1–65.
30. Olmsted, J. B., Marcum, J. M., Johnson, K. A., Allen, C. and Borisy, G. G. (1974). *J. Supramol. Struct.*, **2**, 429–450.
31. Hepler, P. K. and Palevitz, B. A. (1974). *Annu. Rev. Plant Physiol.*, **25**, 309–362.
32. Borisy, G. G., Olmsted, J. B. and Klugman, R. A. (1972). *Proc. Nat. Acad. Sci.*, **69**, 2890–2894.
33. Marantz, R., Ventilla, M. and Shelanski, M. (1968). *Science*, **165**, 498–499.
34. Bryan, J. (1972). *Biochemistry*, **11**, 2611–2616.
35. Wilson, L., Bamburg, J. R., Mizel, S. B., Grisham, L. M. and Creswell, K. M. (1974). *Fed. Proc.*, **33**, 158–166.
36. Weber, K., Pollack, R. and Bibring, T. (1975). *Proc. Nat. Acad. Sci.*, **72**, 459–463.
37. Weber, K., Bibring, T. and Osborn, M. (1975). *Exp. Cell Res.*, **95**, 111–120.
38. Wessells, N. K., Spooner, B. S., Ash, J. F., Bradley, M. O., Luduena, M. A., Taylor, E. L., Wrenn, J. T. and Yamada, K. M. (1971). *Science*, **171**, 135–143.
39. Carter, S. B. (1972). *Endeavour*, **31**, 77–82.
40. Follett, E. A. C. and Goldman, R. D. (1970). *Exp. Cell Res.*, **59**, 124–136.
41. Gail, M. H., Boone, C. W. and Thompson, C. S. (1973). *Exp. Cell Res.*, **79**, 386–390.
42. Hawkes, R. B. and Holberton, D. V. (1973). *J. Cell. Physiol.*, **81**, 365–370.
43. Perdue, J. F. (1971). *J. Biol. Chem.*, **246**, 6750–6759.
44. Lazarides, E. and Weber, K. (1974). *Proc. Nat. Acad. Sci.*, **71**, 2268–2272.
45. Lazarides, E. (1975). *J. Histochem. Cytochem.*, **23**, 507–528.
46. Weber, K. and Groeschel-Stewart, V. (1974). *Proc. Nat. Acad. Sci.*, **71**, 4561–4564.
47. Lazarides, E. (1975). *J. Cell Biol.*, **65**, 549–562.
48. Röhlich, P. (1975). *Exp. Cell Res.*, **93**, 293–298.
49. Willingham, M. C., Ostlund, R. E. and Pastan, I. (1974). *Proc. Nat. Acad. Sci.*, **71**, 4144–4148.
50. Willingham, M. C. and Pastan, I. (1975). *Proc. Nat. Acad. Sci.*, **72**, 1263–1267.
51. Kemp, R. B., Jones, B. M. and Gröschel-Stewart, U. (1973). *J. Cell Sci.*, **12**, 631–639.
52. Allison, A. C. (1973). In *Locomotion of Tissue Cells.* Ciba Foundation Symposium 14 (new series). Elsevier Excerpta Media, North-Holland, Amsterdam, pp. 109–148.
53. Painter, R. G., Sheetz, M. and Singer, S. J. (1975). *Proc. Nat. Acad. Sci.*, **72**, 1359–1363.
54. Puszkin, E., Puszkin, S., Lo, L. W. and Tanenbaum, S. W. (1973). *J. Biol. Chem.*, **248**, 7754–7761.
55. Lin, S., Santi, D. V. and Spudich, J. A. (1974). *J. Biol. Chem.*, **249**, 2268–2274.
56. Tannenbaum, J., Tanenbaum, S. W. and Godman, G. C. (1975). *Biochim. Biophys. Acta*, **413**, 322–327.
57. Tannenbaum, J., Tanenbaum, S. W., Lo, L. W., Godman, G. C. and Miranda, A. F. (1975). *Exp. Cell Res.*, **91**, 47–56.
58. Kram, R., Mamont, P. and Tomkins, G. M. (1973). *Proc. Nat. Acad. Sci.*, **70**, 1432–1436.
59. Kram, R. and Tomkins, G. M. (1973). *Proc. Nat. Acad. Sci.*, **70**, 1659–1663.
60. Durham, A. C. H. (1974). *Cell*, **2**, 123–136.
61. Nathenson, S. G. and Cullen, S. E. (1974). *Biochim. Biophys. Acta*, **344**, 1–25.
62. Vitetta, E. S. and Uhr, J. W. (1975). *Biochim. Biophys. Acta*, **415**, 253–271.
63. Metzger, H. (1970). *Advan. Immunol.*, **12**, 57–116.

64. Marchalonis, J. J. (1975). *Science*, **190**, 20–29.
65. Fu, S. M. and Kunkel, H. G. (1974). *J. exp. Med.*, **140**, 895–903.
66. Aoki, T., Hämmerling, U., de Harven, E., Boyse, E. A. and Old, L. J. (1969). *J. exp. Med.*, **130**, 979–1001.
67. de Petris, S. (1977). In *Methods in Membrane Biology* (Korn, E., Ed.), Plenum, New York.
68. Taylor, R. B., Duffus, P. H., Raff, M. C. and de Petris, S. (1971). *Nature* (*New Biol.*), **233**, 225–229.
69. Loor, F., Forni, L. and Pernis, B. (1972). *Eur. J. Immunol.*, **2**, 203–212.
70. Unanue, E. R. and Karnovsky, M. J. (1973). *Transplant. Rev.*, **14**, 184–210.
71. de Petris, S. and Raff, M. C. (1972). *Eur. J. Immunol.*, **2**, 523–532.
72. de Petris, S. and Raff, M. C. (1973). *Nature* (*New Biol.*), **241**, 257–259.
73. Raff, M. C. and de Petris, S. (1973). *Fed. Proc.*, **32**, 48–54.
74. Loor, F., Block, N. and Little, J. R. (1975). *Cell. Immunol.*, **17**, 351–365.
75. Edidin, M. and Weiss, A. (1972). *Proc. Nat. Acad. Sci.*, **69**, 2456–2459.
76. Comoglio, P. M. and Guglielmone, R. (1972). *FEBS Letters*, **27**, 256–258.
77. de Petris, S., Raff, M. C. and Mallucci, L. (1973). *Nature* (*New Biol.*), **244**, 275–278.
78. Weller, N. K. (1974). *J. Cell Biol.*, **63**, 699–707.
79. Leonard, E. J. (1973). *J. Immunol.*, **110**, 1167–1169.
80. Loor, F. and Roelants, G. E. (1974). *Eur. J. Immunol.*, **4**, 649–660.
81. McDonough, J. and Lilien, J. (1975). *J. Cell Science*, **19**, 357–368.
82. Sundqvist, K. G. (1972). *Nature* (*New Biol.*), **239**, 147–149.
83. Sullivan, A. L., Grimley, P. M. and Metzger, H. (1971). *J. exp. Med.*, **134**, 1403–1416.
84. Becker, K. E., Ishizaka, T., Metzger, H., Ishizaka, K. and Grimley, P. M. (1973). *J. exp. Med.*, **138**, 394–409.
85. Pinto da Silva, P., Martinez-Palomo, A. and Gonzalez-Robles, A. (1975). *J. Cell Biol.*, **64**, 538–550.
86. Gordon, J. A. and Marquardt, M. D. (1975). *Nature*, **258**, 346–347.
87. Reyes, F., Lejonc, J. L., Gourdin, M. F., Mannoni, P. and Dreyfus, B. (1975). *J. exp. Med.*, **141**, 392–410.
88. Lipscomb, M. F., Holmes, K. V., Vitetta, E. S., Hämmerling, U. and Uhr, J. W. (1975). *Eur. J. Immunol.*, **5**, 255–259.
89. Molday, R. S., Dreyer, W. J., Rembaum, A. and Yen, S. P. S. (1975). *J. Cell Biol.*, **64**, 75–88.
90. Antoine, J. C., Avrameas, S., Gonatas, N. K., Stieber, A. and Gonatas, J. O. (1974). *J. Cell Biol.*, **63**, 12–23.
91. Loor, F. and Hägg, L.-B. (1975). *Eur. J. Immunol.*, **5**, 854–865.
92. Linthicum, D. S. and Sell, S. (1975). *J. Ultrastruct. Res.*, **51**, 55–68.
93. Abbas, A. K., Ault, K. A., Karnovsky, M. J. and Unanue, E. R. (1975). *J. Immunol.*, **114**, 1197–1204.
94. Tao-Wiedmann, T. W., Loor, F. and Hägg, L.-B. (1975). *Immunology*, **28**, 821–830.
95. Edidin, M. and Fambrough, D. (1973). *J. Cell Biol.*, **57**, 27–37.
96. Petit, V. A. and Edidin, M. (1974). *Science*, **184**, 1183–1185.
97. Kleemann, W., Grant, C. W. M. and McConnell, H. M. (1974). *J. Supramol. Struct.*, **2**, 609–616.
98. Kleemann, W. and McConnell, H. M. (1974). *Biochim. Biophys. Acta*, **345**, 220–230.
99. James, R., Branton, D., Wisnieski, B. and Keith, A. (1972). *J. Supramol. Struct.*, **1**, 38–49.
100. Yahara, I. and Edelman, G. M. (1972). *Proc. Nat. Acad. Sci.*, **69**, 608–612.
101. Karnovsky, M. J., Unanue, E. R. and Leventhal, M. (1972). *J. exp. Med.*, **136**, 907–930.
102. Loor, F. (1973). *Exp. Cell Res.*, **82**, 415–425.
103. Loor, F. (1974). *Eur. J. Immunol.*, **4**, 210–220.
104. de Petris, S. and Raff, M. C. (1974). *Eur. J. Immunol.*, **4**, 130–137.
105. Kosower, E. M., Kosower, N. S., Faltin, Z., Diver, A., Saltoun, G. and Frensdorff, A. (1974). *Biochim. Biophys. Acta*, **363**, 261–266.
106. Seeman, P. (1972). *Pharmacol. Rev.*, **24**, 583–655.
107. Ryan, G. B., Unanue, E. R. and Karnovsky, M. J. (1974). *Nature*, **250**, 56–57.
108. Yahara, I. and Edelman, G. M. (1973). *Nature*, **246**, 152–155.
109. Yahara, I. and Edelman, G. M. (1973). *Exp. Cell Res.*, **81**, 143–155.
110. Edelman, G. M., Yahara, I. and Wang, J. L. (1973). *Proc. Nat. Acad. Sci.*, **70**, 1442–1446.
111. Yahara, I. and Edelman, G. M. (1975). *Exp. Cell Res.*, **91**, 125–142.

112. de Petris, S. (1975). *J. Cell Biol.*, **65**, 123–146.
113. Davis, W. C. (1972). *Science*, **175**, 1006–1008.
114. Lengerová, A., Pokorná, Z., Viklický, V. and Zelený, V. (1972). *Tissue Antigens*, **2**, 332–340.
115. Kourilsky, F. M., Silvestre, D., Neauport-Sautes, C., Loosfelt, Y. and Dausset, J. (1972). *Eur. J. Immunol.*, **2**, 249–257.
116. Bernoco, D., Cullen, S., Scudeller, G., Trinchieri, G. and Ceppellini, R. (1973). *Histocompatibility testing* (*1972*) (Dausset, J., Ed.), Munksgaard, Copenhagen, pp. 527–537.
117. Bismuth, A., Neauport-Sautes, C., Kourilsky, F. M., Manuel, Y., Greenland, T. and Silvestre, D. (1974). *J. Immunol.*, **112**, 2036–2046.
118. Poulik, M. D., Bernoco, M., Bernoco, D. and Ceppellini, R. (1973). *Science*, **182**, 1352–1355.
119. Solheim, B. G. and Thorsby, E. (1974). *Nature*, **249**, 36–38.
120. Stackpole, C. W., Jacobson, J. B. and Lardis, M. P. (1974). *Nature*, **248**, 232–234.
121. Ellis, A. E. and Parkhouse, R. M. E. (1975). *Eur. J. Immunol.*, **5**, 726–728.
122. Unanue, E. R., Ault, K. A. and Karnovsky, M. J. (1974). *J. exp. Med.*, **139**, 295–312.
123. Schreiner, G. F. and Unanue, E. R. (1975). *J. Immunol.*, **114**, 809–814.
124. Stackpole, C. W., De Milio, L. T., Hämmerling, U., Jacobson, J. B. and Lardis, M. P. (1974). *Proc. Nat. Acad. Sci.*, **71**, 932–936.
125. Preud'homme, J. L., Neauport-Sautes, C., Piat, S., Silvestre, D. and Kourilsky, F. M. (1972). *Eur. J. Immunol.*, **2**, 297–300.
126. Yahara, I. and Edelman, G. M. (1975). *Proc. Nat. Acad. Sci.*, **72**, 1579–1583.
127. Rutishauser, U., Yahara, I. and Edelman, G. M. (1974). *Proc. Nat. Acad. Sci.*, **71**, 1149–1153.
128. Neauport-Sautes, C., Lilly, F., Silvestre, D. and Kourilsky, F. M. (1973). *J. exp. Med.*, **137**, 511–526.
129. de Petris, S. (1974). *Nature*, **250**, 54–56.
130. Unanue, E. R. and Karnovsky, M. J. (1974). *J. exp. Med.*, **140**, 1207–1220.
131. Antoine, J. C. and Avrameas, S. (1974). *Eur. J. Immunol.*, **4**, 468–474.
132. Knopf, P. M. (1973). *Transplant. Rev.*, **14**, 145–162.
133. Engers, H. D. and Unanue, E. R. (1973). *J. Immunol.*, **110**, 465–475.
134. Du Pasquier, L., Weiss, N. and Loor, F. (1972). *Eur. J. Immunol.*, **2**, 366–370.
135. Elson, C. J., Singh, J. and Taylor, R. B. (1973). *Scand. J. Immunol.*, **2**, 143–149.
136. Roelants, G. E., Rydén, A., Hägg, L.-B. and Loor, F. (1974). *Nature*, **247**, 106–108.
137. Unanue, E. R., Ault, K. A., Schreiner, G. F. and Sidman, C. L. (1975). In *Membrane Receptors of Lymphocytes* (Seligmann, M., Preud'homme, J. L. and Kourilsky, F. M., Eds.), North-Holland, Amsterdam, pp. 363–372.
138. Unanue, E. R. (1974). *Amer. J. Path.*, **77**, 2–20.
139. Lesley, J. and Hyman, R. (1974). *Eur. J. Immunol.*, **4**, 732–739.
140. Yu, A. and Cohen, E. P. (1974). *J. Immunol.*, **112**, 1296–1307.
141. Sidman, C. L. and Unanue, E. R. (1975). *Nature*, **257**, 149–151.
142. Raff, M. C., Owen, J. J. T., Cooper, M. D., Lawton, A. R. III, Megson, M. and Gathings, W. E. (1975). *J. exp. Med.*, **142**, 1052–1064.
143. Sällstrom, J. F. and Alm, G. V. (1972). *Exp. Cell Res.*, **75**, 63–72.
144. Gunther, G. R., Wang, J. L., Yahara, I., Cunningham, B. A. and Edelman, G. M. (1973). *Proc. Nat. Acad. Sci.*, **70**, 1012–1016.
145. Stadler, J. and Franke, W. W. (1974). *J. Cell Biol.*, **60**, 297–303.
146. Gordon, J. A. and Marquardt, M. D. (1974). *Biochim. Biophys. Acta*, **332**, 136–144.
147. Huet, C., Lonchampt, M., Huet, M. and Bernadac, A. (1974). *Biochim. Biophys. Acta*, **365**, 28–39.
148. Polliack, A., Lampen, N., Clarkson, B. D., De Harven, E., Bentwick, Z., Siegal, F. P. and Kunkel, H. G. (1973). *J. exp. Med.*, **138**, 607–624.
149. Polliack, A., Hämmerling, U., Lampen, N. and de Harven, E. (1975). *Eur. J. Immunol.*, **5**, 32–39.
150. Greaves, M. F., Owen, J. J. T. and Raff, M. C. (1973). *T and B Lymphocytes*. Excerpta Medica, Amsterdam.
151. Zeiller, K., Hannig, K. and Pascher, G. (1971). *Hoppe-Seyler's Z. Physiol. Chem.*, **352**, 1168–1170.
152. Linthicum, D. S., Sell, S., Wagner, R. M. and Trefts, P. (1974). *Nature*, **252**, 173–175.
153. Kay, M. M. B. (1975). *Nature*, **254**, 424–426.
154. Polliack, A. and de Harven, E. (1975). *Clin. Immunol. and Immunopathol.*, **3**, 412–430.

155. Alexander, E. L. and Wetzel, B. (1975). *Science*, **188**, 732–734.
156. Lin, P. S. and Wallach, D. F. H. (1974). *Science*, **184**, 1300–1301.
157. Lin, P. S., Wallach, D. F. H. and Tsai, S. (1973). *Proc. Nat. Acad. Sci.*, **70**, 2492–2496.
158. Jerrells, T. R. and Hinrichs, D. J. (1975). *J. of Reticuloendothel. Soc.*, **17**, 84–94.
159. McFarland, W. (1969). *Science*, **163**, 818–820.
160. Fagraeus, A., Lidman, K. and Biberfeld, G. (1974). *Nature*, **252**, 246–247.
161. Wofsy, L., Baker, P. C., Thompson, K., Goodman, J., Kimura, J. and Henry, C. (1974). *J. exp. Med.*, **140**, 523–537.
162. Norberg, R., Lidman, K. and Fagraeus, A. (1975). *Cell*, **6**, 507–512.
163. Norberg, R., Fagraeus, A. and Lidman, K. (1975). *Clin. exp. Immunol.*, **21**, 284–288.
164. Fagraeus, A., Lidman, K. and Norberg, R. (1975). *Clin. exp. Immunol.*, **20**, 469–477.
165. Taylor, A. C. (1966). *J. Cell Biol.*, **28**, 155–168.
166. Raff, M. C. and de Petris, S. (1974). In *The Immune System. Genes, Receptors, Signals* (Sercarz, E. E., Williamson, A. R. and Fox, C. F., Eds.), Academic, New York, pp. 247–257.
167. Fagraeus, A., Nilsson, K., Lidman, K. and Norberg, R. (1975). *J. Nat. Cancer Inst.*, **55**, 783–789.
168. Müller-Berat, N. (Ed.) (1976). *Proc. of the 2nd International Conference on Differentiation.* ASP Biological and Medical Press B.V., Amsterdam, Holland.
169. Taylor, A. C. and Robbins, E. (1963). *Develop. Biol.*, **7**, 660–673.
170. Fisher, H. W. and Cooper, T. W. (1967). *J. Cell Biol.*, **34**, 569–576.
171. Trinkaus, J. P. and Erickson, C. A. (1974). *J. Cell Biol.* (*Abstr.*), **63**, 351a.
172. Porter, K., Prescott, D. and Frye, J. (1973). *J. Cell Biol.*, **57**, 815–836.
173. Elander, D., Tobey, R. A. and Scott, T. (1975). *Exp. Cell Res.*, **95**, 396–404.
174. Van Ewyk, W., Brons, N. H. C. and Rozing, J. (1975). *Cell. Immunol.*, **19**, 245–261.
175. Pugh-Humphreys, R. G. P. and Sinclair, W. J. (1970). *J. Cell Sci.*, **6**, 477–484.
176. Ben-Shaul, Y. and Moscona, A. A. (1975). *Exp. Cell Res.*, **95**, 191–204.
177. Weiss, L. (1972). *Exp. Cell Res.*, **74**, 21–26.
178. Rubin, R. W. and Everhart, L. P. (1973). *J. Cell Biol.*, **57**, 837–844.
179. O'Neill, C. H. and Follett, E. A. C. (1970). *J. Cell Sci.*, **7**, 695–709.
180. Ax, W., Malchow, H., Zeiss, I. and Fischer, H. (1968). *Exp. Cell Res.*, **53**, 108–116.
181. Able, M. E., Lee, J. C. and Rosenau, W. (1970). *Amer. J. Pathol.*, **60**, 421–428.
182. Wekerle, H., Lonai, P. and Feldmann, M. (1972). *Proc. Nat. Acad. Sci.*, **69**, 1620–1624.
183. Cerottini, J. C. and Brunner, K. T. (1972). *Nature* (*New Biol.*), **237**, 272–273.
184. Werdelin, O., Braendstrup, O. and Pedersen, E. (1974). *J. exp. Med.*, **140**, 1245–1259.
185. Nielsen, M. H., Jensen, H., Braendstrup, O. and Werdelin, O. (1974). *J. exp. Med.*, **140**, 1260–1272.
186. Kapeller, M., Gal-Oz, R., Grover, N. B. and Dolyanski, F. (1973). *Exp. Cell Res.*, **79**, 152–158.
187. Cone, R. E., Marchalonis, J. J. and Rolley, R. T. (1971). *J. exp. Med.*, **134**, 1373–1384.
188. Vitetta, E. S. and Uhr, J. W. (1972). *J. exp. Med.*, **136**, 676–696.
189. Ceppellini, R., Bernoco, D. and Jacot-Guillarmot, H. (1976). *Contemporary topics in Molecular Immunology* (Inman, F. I., Ed.), Plenum Publishing, New York.
190. Steinman, R. M. and Cohn, Z. A. (1972). *J. Cell Biol.*, **55**, 186–204.
191. Price, P. G. (1970). *J. Membrane Biol.*, **2**, 300–316.
192. Weiss, L. and Coombs, R. R. A. (1963). *Exp. Cell Res.*, **30**, 331–338.
193. Weiss, L. and Lachman, P. J. (1964). *Exp. Cell Res.*, **36**, 86–91.
194. Rowe, D. S., Hug, K., Forni, L. and Pernis, B. (1973). *J. exp. Med.*, **138**, 965–972.
195. Preud'homme, J. L. and Seligmann, M. (1972). *Proc. Nat. Acad. Sci.*, **69**, 2132–2135.
196. Neauport-Sautes, C., Silvestre, D., Lilly, F. and Kourilsky, F. M. (1973). *Transplant. Proc.*, **5**, 443–446.
197. Neauport-Sautes, C., Bismuth, A., Kourilsky, F. M. and Manuel, Y. (1974). *J. exp. Med.*, **139**, 957–968.
198. Krammer, P. H. and Pernis, B. (1976). *Scand. J. Immunol.*, **5**, 199–204.
199. Unanue, E. R. and Abbas, A. K. (1975). *Membrane Receptors of Lymphocytes* (Seligmann, M., Preud'homme, J. L. and Kourilsky, F. M., Eds.), North-Holland, Amsterdam, American Elsevier, New York, pp. 281–285.
200. Ahmann, G. B. and Sage, H. J. (1974). *Cell. Immunol.*, **13**, 407–415.
201. Krammer, P. H. and Pernis, B. (1976). *Scand. J. Immunol.*, **5**, 205–212.
202. Dickler, H. B. and Sachs, D. H. (1974). *J. exp. Med.*, **140**, 779–796.
203. McElvain, S. M. and Schroeder, J. P. (1949). *Amer. Chem. Soc. J.*, **71**, 40–46.

204. Converse, C. A. and Richards, F. F. (1969). *Biochemistry*, **8**, 4431–4436.
205. Thorpe, N. O. and Singer, S. J. (1969). *Biochemistry*, **8**, 4523–4534.
206. Ji, T. H. (1974). *Proc. Nat. Acad. Sci.*, **71**, 93–95.
207. Ji, T. H. (1974). *J. Biol. Chem.*, **249**, 7841–7847.
208. Wang, K. and Richards, F. M. (1974). *J. Biol. Chem.*, **249**, 8005–8018.
209. Converse, C. A. and Papermaster, D. S. (1975). *Science*, **189**, 469–472.
210. Beale, G. H. (1954). *The Genetics of* Paramecium aurelia. Cambridge University Press, London.
211. Boyse, E. A., Stockert, E. and Old, L. (1967). *Proc. Nat. Acad. Sci.*, **58**, 954–957.
212. Old, L. J., Stocker, E., Boyse, E. A. and Kim, H. H. (1968). *J. exp. Med.*, **172**, 523–539.
213. Takahashi, T. (1971). *Transplant. Proc.*, **3**, 1217–1220.
214. Yu, A., Liang, W. and Cohen, E. P. (1975). *J. Nat. Cancer Inst.*, **55**, 299–307.
215. Lee, S. K., Singh, J. and Taylor, R. B. (1975). *Eur. J. Immunol.*, **5**, 259–262.
216. Liang, W. and Cohen, E. P. (1975). *Proc. Nat. Acad. Sci.*, **72**, 1873–1877.
217. Yefenof, E. and Klein, G. (1974). *Exp. Cell Res.*, **88**, 217–224.
218. Edidin, M. and Henney, C. S. (1973). *Nature* (*New Biol.*), **246**, 47–49.
219. Ling, N. R. and Kay, J. E. (Eds.) (1975). *Lymphocyte Stimulation* (2nd ed.), North-Holland, Amsterdam, American Elsevier, New York.
220. Raff, M. C., Freedman, M. and Gomperts, B. (1975). In *Membrane Receptors of Lymphocytes* (Seligmann, M., Preud'homme, J. L. and Kourilsky, F. M., Eds.), North-Holland, Amsterdam, American Elsevier, New York, pp. 393–398.
221. Cuatrecasas, P. (1974). *Annu. Rev. Biochem.*, **43**, 169–214.
222. Greaves, M., Janossy, G., Feldmann, M. and Doenhoff, M. (1974). In *The Immune System. Genes, Receptors, Signals* (Sercarz, E. E., Williamson, A. R. and Fox, C. F., Eds.), Academic, New York, pp. 271–297.
223. Watson, J., Epstein, R. and Cohn, M. (1973). *Nature*, **246**, 405–409.
224. Möller, G. (Ed.) (1975). *Transplant. Rev.*, Vol. 23, Munksgaard, Copenhagen, Denmark.
225. Wedner, H. J. and Parker, C. W. (1976). *Prog. Allergy*, **20**, 195–300.
226. Loor, F. (1967). *Virology*, **33**, 215–220.
227. Loor, F. (1971). *Immunology*, **21**, 557–564.
228. Caspar, D. L. D. (1963). *Advan. Protein Chem.*, **18**, 37–121.
229. Young, M., King, M. V., O'Hara, D. S. and Molberg, P. J. (1973). *Cold Spring Harbor Symp. Quant. Biol.*, **37**, 65–76.
230. Mooseker, M. S. and Tilney, L. G. (1975). *J. Cell Biol.*, **67**, 725–743.
231. Lazarides, E. and Burridge, K. (1975). *Cell*, **6**, 289–298.

Chapter 9

The Specificity Repertoire and Antigen Receptors of T and B Lymphocytes

J. W. GOODMAN

Abbreviations

Ars: azophenylarsonate
DNFB: dinitrofluorobenzene
DNP: 2,4-dinitrophenyl
Ir genes: immune response genes
Lac: lactoside
MHC: major histocompatibility complex
TGAL: (Tyr, Glu)-Ala–Lys

1. INTRODUCTION

Although the dualistic character of the immune process has been recognized for a number of years, many of the detailed interrelationships underlying its organization have been unravelled only recently (reviewed in References 1 and 2 and herein). Humoral antibody (Ab) is produced by bone marrow derived (B) lymphocytes, which express surface antigen-specific receptors in the form of immunoglobulin (Ig) molecules. Cellular immune reactions are mediated by thymus derived (T) lymphocytes, which, in addition, regulate the induction of Ab formation by functioning as helper or suppressor cells. The molecular nature of the antigen-recognizing apparatus of the T cell has been a controversial issue and its identity remains in doubt.

The differentiation of T and B lymphocytes can be considered to proceed in two stages. The first stage is antigen-independent and leads to the appearance of individual lymphocytes with antigen receptors which are presumably restricted to one specificity. A persuasive body of evidence indicates that an individual antibody-producing cell at a given point in time is committed to the production of a single species of antibody (reviewed in Reference 3), although sensitive immunohistochemical techniques suggest the possibility that immature B cells may express receptors of more than one specificity.[4] In this context, it was found that tadpoles with only about 10^6 lymphocytes produced anti-hapten Ab with binding heterogeneity ostensibly comparable to that of Ab from mature mammals.[5] This can be accounted for either by multi-specific B cell precursors or by the generation of a heterogeneous population of specific cells among the progeny of restricted precursors by somatic diversification. The precise point in the differentiation pathway at which a B lymphocyte becomes committed to a single specificity has not been identified, but it is conceivable that it may occur only after initial contact with antigen.

Be that as it may, most investigators believe that the specificity of membrane antigen receptors is related in a simple, direct way to the specificity of the antibody produced by the B cell. Contact with antigen signals the second stage of lymphocyte differentiation: antigen-driven clonal selection and expansion, yielding clones of specific antibody-producing cells as well as reservoirs of memory B and T lymphocytes. Thus, the vast spectrum of immune specificities which vertebrates are capable of generating is represented by the antigen receptors on their lymphocyte populations, a concept postulated with remarkable clairvoyance by Paul Ehrlich at the turn of the century.[6]

Two major questions concerning the receptors themselves remain unanswered as of this writing and will comprise the focus of our attention. These have to do with the nature of the T cell receptor for antigen and the comparative repertoire of specificities exhibited by B and T lymphocytes. The mechanism of the generation of diversity in T and B lymphocytes is intimately connected with these questions, for if the receptors differ in their molecular nature or specificity repertoire, then diversity may be generated independently in the two cell populations.

2. ANTIGEN SPECIFICITY OF T AND B LYMPHOCYTES

T and B lymphocytes manifest a high degree of antigen specificity, reflected in the specificities of cellular and humoral immunity, respectively. Their comparative specificities have been the subject of vigorous investigation, the results of which are somewhat controversial and difficult to interpret (Table 1). Some studies indicate a fundamental similarity between the two systems,[7–9] while others reveal differences.[10–23] A number of variables in the experimental systems make it difficult to draw simple comparisons or straightforward conclusions.

First, the antigens differ markedly from one study to another. Some have been small, relatively simple molecules which bear a limited number of structural determinants (epitopes).[8,23] Others were complex proteins which carry a large, undetermined number of epitopes,[7,9–11] or chemically modified or denatured proteins[12–18] which may aggregate, possibly forming structures with differentially altered reactivity toward B or T lymphocytes, independent of the specificity of their antigen receptors. The fourth class of antigens,

Table 1. Comparative antigen specificities of B and T lymphocytes

Antigen(s)	Animal	Specificity assay: B cell	Specificity assay: T cell	Reference
1. Serum albumins	Mouse	Ab binding	Helper activity	7
2. DNP[a]-oligolysine conjugates	Guinea pig	Ab binding	Skin sensitivity Lymphocyte proliferation	8
3. Hapten–G.P. albumin conjugates	Guinea pig	Ab binding	Lymphocyte proliferation	9
1. Hen egg white lysozyme and α-lactalbumin	Guinea pig	Ab binding	Skin sensitivity Lymphocyte proliferation	10
2. Ascaris and mammalian collagens	Guinea pig	Ab binding	Skin sensitivity MIF[a] test Lymphocyte proliferation	11
3. Native and heat denatured serum albumins	Guinea pig	Ab binding	Skin sensitivity	12
4. Lysozyme and S-carboxymethylated lysozyme	Guinea pig	Ab binding	Skin sensitivity Lymphocyte proliferation	13
5. Flagellin and aceto-acetylated flagellin	Mouse	Ab binding	Skin sensitivity	14–17
6. Native and methylated bovine serum albumin	Mouse	Ab binding	Helper activity	18
7. Erythrocytes	Mouse	Ab binding	Helper activity	19
8. Erythrocytes	Mouse	Ab binding (plaque formation)	Helper activity	20
9. Erythrocytes	Mouse	Ab binding (plaque formation)	Helper activity	21
10. Erythrocytes	Mouse	Ab binding (plaque formation)	Rosette formation	22
11. Glucagon	Guinea pig	Ab binding	Skin sensitivity MIF test Lymphocyte proliferation	23

[a] Abbreviations: DNP: dinitrophenyl; MIF: migration inhibitory factor.

erythrocytes,[19–22] are multi-molecular particles and the membrane antigens to which T and B cell specificities are directed have not been delineated.

A second complication resides in the uncertainty that the various T cell activities which are assayed, such as helper activity, delayed hypersensitivity, antigen-induced transformation and rosette formation, are manifestations of the same cell population. Indeed, there is recent persuasive evidence that T cells which provide helper activity are distinct from those which mediate cytotoxicity,[24] and additional subdivisions are likely (P. C. L. Beverley, herein).

Variables other than the antigens and T cell activities may also have contributed to the lack of coherence in this story. For example, Rajewsky's detailed analysis of the specificity of helper cells in the murine response to the hapten 4-hydroxy-5-iodo-3-nitrophenacetate (NIP), coupled to either bovine or sheep serum abumin, led to the conclusion that the cross-reactivity of helper cells for the two carriers paralleled the cross-reactivity of antibody.[7] On the other hand, Hoffmann and Kappler assessed helper activity for erythrocytes from a variety of species and found marked disparities between the patterns of cross-reactivity of this T cell function and of antibody.[19] In the former case, helper activity was not titrated and may not have been a limiting factor in the experimental system, thereby conceivably exaggerating relatively weak cross-reactivities. In the latter case, the experiments were performed under conditions in which it was established that helper activity was limiting. This does not necessarily explain the discordant findings, but it does underscore the difficulties inherent in making direct comparisons.

In most of the instances where differences in specificity have been noted, the cross-reactivity of cellular immunity has been broader than that of antibody,[10–22] an observation which has been interpreted in some quarters to signify that reactions mediated by T cells are less specific than those mediated by conventional antibody. There are two major difficulties with this interpretation.

First, T and B cells may not necessarily recognize and respond to the same epitopes on an antigen molecule in a given set of circumstances. This has been lucidly demonstrated in studies with glucagon,[23] a polypeptide of only 29 amino acids, which therefore bears a limited number of epitopes, yet induces both cellular and humoral immunity in guinea pigs. The major haptenic determinant, against which antibody specificity is directed, proved to be located in the amino-terminal region of the molecule, whereas the major immunogenic determinant, against which T cell specificity is directed, resided in the carboxy-terminal portion. This dissection of the molecule along functional lines should not be construed to mean that B cells are incapable of recognizing epitopes carried in the carboxy-terminal region. In fact, guinea pigs immunized with glucagon-protein conjugates made antibodies directed against epitopes in that part of the molecule, and in conjugated form glucagon did not induce cellular immunity but instead behaved like a conventional hapten.[25] Thus, the physical nature of the antigen and the manner of its presentation play critical roles in the responses of B and T cells to particular epitopes. Antigens which do not share common haptenic determinants will not cross-react at the humoral level but might do so at the level of cellular immunity, without necessarily implying a qualitative difference in the nature of specificity of the two systems. For example, if the carboxy-terminal sequence of glucagon was joined to two different amino-terminal peptides, the antigens might cross-react at the cellular level due to the common portion, but not at the antibody level in view of the different amino-terminal sequences.

The second major difficulty in comparing the specificities of antibodies and T cells is that the former can be measured simply and directly by immunochemical techniques, whereas T cells, unlike B cells, lack a readily identifiable secretory product simulating their antigen receptors. T cell receptor specificity is therefore assessed indirectly, in terms of one or another cellular activity. Cell activation clearly involves much more than simply the binding of receptor and ligand, which can be considered an initial step in a complex process. Indeed, the reactions of B and T lymphocytes with lectins demonstrate that membrane binding does not necessarily lead to activation.[26] Phytohaemagglutinin (PHA) and concanavalin A (Con A) apparently bind equally well to B and T cells but activate only the latter. However, B lymphocytes can be activated by the lectins adsorbed to a solid surface. It is also worth noting that haptens bind readily to antibodies and to B lymphocytes but do not stimulate the cells.[1] The properties of lectins and haptens illustrate that the functional activation of lymphocytes depends not merely on receptor–ligand interaction, but also on the physical properties of the ligand. Cell–cell interactions represent still another factor in the equation; the *in vitro* stimulation of T cells by antigen requires the participation of macrophages (References 27 and 28 and G. E. Roelants, and M. H. Schreier and A. A. Nordin, herein). Hence, comparing binding and activation may be like comparing methane and elephants; one is an essential antecedent of the other, but the two have an obviously different order of complexity.

In full cognizance of the above limitations, it is nonetheless fair to say that investigations using antigens of defined structure clearly indicate that the specificities of cellular and humoral immunity are comparably discriminatory.[29] Referring again to glucagon, a series of synthetic peptides initiated at the carboxy-terminal position—the locus of the immunogenic determinant—were assayed for transforming activity with lymphoid cells from animals immunized with the native hormone. The cells discriminated sharply between members of the series.[23,29]

In another system, guinea pigs responsive to poly-L-lysine develop cellular immunity when immunized with polymers larger than the hexapeptide.[8] A sequence of at least 7 L-lysine residues, uninterrupted by a residue of the D configuration, is required for immunogenicity. Substitution of a residue of D-lysine causes loss of activity in a nonapeptide with the sequence L_4DL_4, whereas peptides of sequence L_7DL and LDL_7 are fully active. A comparison of the responses to a series of DNP-nonalysine conjugates in which the positions of the DNP group and D-lysine residues varied showed that lymph node cells from responder animals immunized with members of the series of peptides could readily discriminate between them and made maximal proliferative responses to the homologous immunogen. These responses presumably originated from T cells, but since the animals make antibody responses to the conjugates, it is not possible to completely exclude the participation of B cells in the *in vitro* responses. This criticism would, of course, apply to all experiments in which cultures containing both types of lymphocytes are used.

Another low molecular weight compound, L-tyrosine-*p*-azobenzenearsonate (ABA-L-Tyr), induces cellular immunity in guinea pigs without appreciable antibody.[30–32] An investigation of the structural requirements for immunogenicity of this molecule showed that other charged moieties could substitute for the arsonate group without loss of immunogenicity.[33] However, each of the analogues had a distinctive specificity on the basis of the absence of cross-reactivity in delayed hypersensitivity reactions between compounds bearing arsonate, sulphonate, carboxylate, acetamide, sulphonamide, nitro, or trimethylammonium substituents. Moreover, lymph node cultures from guinea pigs immunized with ABA-L-Tyr or one of the *p*-azobenzenearsonate-substituted tetrapeptides ABA-L-Tyr-$(L\text{-}Tyr)_3$, ABA-L-Tyr-$(L\text{-}Ala)_3$, and $(L\text{-}Tyr)_3$-L-Tyr-ABA readily discriminated in proliferative responses between the homologous and heterologous compounds.[32]

Thus, in the rare instances in which T lymphocyte responses have been induced by immunogens small enough to permit the elucidation of structure–specificity relationships, these responses appear to have a specificity comparable in discriminatory power to that of antibody.

3. REPERTOIRE OF SPECIFICITIES OF T AND B LYMPHOCYTES

3.1 Antigen-binding cells

In light of the preceding discussion, it is apparent that the only meaningful way to compare the spectra of T and B lymphocyte epitope specificities is by receptor–ligand binding rather than cell activation. Superficially, it might appear that since haptens generate antibody but not cellular immunity, B cells but not T cells recognize such epitopes, indicating a marked restriction in the diversity of T cell specificities. However, when it is recalled that B cells are not directly activated by haptens, but require the intercession of T lymphocytes, the logic of the argument crumbles.

The binding of antigens by lymphocytes has been extensively investigated by a variety of techniques[2,34] notably rosette formation (immunocytoadherence), autoradiography with radiolabelled antigens, and cell-binding to solid surfaces (antigen-coated beads or fibres). The upshot of the findings is that antigen-binding B cells can be easily demonstrated by all available methods, while the demonstration of antigen-binding by T cells has been relatively difficult.[2] However, the binding of heterologous erythrocytes[35] as well as soluble proteins[36] by lymphocytes which bear the Thy-1 (theta) membrane antigen characteristic of T lymphocytes has been firmly established. In addition, T lymphocyte functions have been specifically 'suicided' with antigens labelled with ^{125}I to very high specific radioactivities.[2,17,36] T lymphocytes which bind haptens have recently been reported,[37–39] which is particularly germane to the question of T cell receptor diversity. However, functional activity of antigen-binding T lymphocytes has been demonstrated only in the antigen suicide experiments. A promising approach to this issue was the demonstration that Thy-1 positive cells from mice primed with DNP-haemocyanin could bind to nylon fibres coated with DNP-albumin.[38] Binding was specifically inhibited by DNP-lysine, but functional activity of recovered cells has not yet been established.

The relative difficulty in demonstrating antigen-binding T cells has been most frequently attributed to a sparser distribution of receptors on T than on B cells. While it is true that in most studies in which antigen-binding by T cells was observed, B cells bound more antigen molecules than did T cells;[2] under conditions of antigen saturation the maximum numbers of antigen molecules bound by the two cell types were comparable.[36] This suggests that T cell receptors may have a lower avidity for antigen than receptors on B cells, a conclusion supported by other studies showing an absence of maturation in binding affinity of T cell receptors, in contrast to those of B lymphocytes.[37,40]

The low frequency of T cells which bind the multichain, branched synthetic polypeptide antigen (Tyr, Glu)-Ala–Lys (TGAL)—about 2×10^{-4}—and the large numbers of antigen molecules bound by individual cells (ranging up to about 5×10^{4}) offers a forceful argument for the intrinsic origin of T cell receptors,[36] as opposed to their acquisition by adsorption of antibodies or receptors shed by B cells. Similarly, the absence of change in affinity of T receptors with time, in contrast to antibody and B cell receptors,[40] speaks against passive acquisition. Additional evidence of specificity or avidity differences between T cell and B cell antigen binding derives from the work of Hämmerling and McDevitt,[41] who showed that T cells which bound TGAL could be inhibited only by the homologous cold antigen, whereas B cell binding was also inhibited by structurally related polypeptides. In contrast to other findings, this points in the direction of greater T cell receptor avidity, but of course the results are also explicable in terms of specificity differences. The sum of the evidence leaves little reason to doubt the authenticity of antigen-binding T lymphocytes.

3.2 Hapten-specific helper effects

In addition to evidence supporting the existence of hapten-binding T lymphocytes, there is now a substantial body of literature concerning the functional recognition of haptenic determinants by helper cells cooperating with B lymphocytes.[42–47] In view of the numerous independent demonstrations of the effect, there is little reason to question its reality, but the underlying mechanism of most of the experiments is still uncertain. The design of a majority revolves around the generation of helper activity by skin painting mice with dinitrofluorobenzene (DNFB), a chemically reactive compound which induces an apparent hapten-specific cellular immunity, as well as carrier activity by the hapten in recipients of spleen cells from sensitized mice.[42–46]

The earliest demonstration of a hapten helper effect utilized the observation that mice usually cannot make anti-idiotype antibody when immunized with isogeneic myeloma proteins, unless the immunoglobulin allotypes of the host and the myeloma protein differ.[48] In such a case, the allotypic markers may be regarded as carrier determinants and the idiotypic markers in the variable region as haptenic determinants. Iverson overcame the requirement for allotypic differences by skin painting animals with DNFB prior to immunization with a DNP conjugate of a myeloma protein bearing identical allotypic specificities to those of the host.[42] Under these circumstances, anti-idiotype antibodies were produced and the results were interpreted as signifying the cooperation of DNP-specific helper T cells with idiotype-specific B cells.

A different protocol was employed by Mitchison to demonstrate a hapten-specific helper effect.[43] In his experiments, the donors of helper cells were lethally irradiated mice reconstituted with either syngeneic thymus or spleen cells and immunized with DNP-protein conjugates. The recipients were treated with anti-lymphocyte serum to deplete them of T cells and were immunized with DNP-heterologous carrier conjugates. Helper activity was based on the anti-carrier antibody response. A hapten-specific helper effect was discernible only when spleen cells, rather than thymic cells, were used as helpers, unless the donors of thymic cells were skin painted with DNFB instead of being immunized with DNP–protein conjugates.

A serious objection to skin painting with chemically reactive compounds is the formation of covalent conjugates with an indeterminate number of proteins in the tissues of the painted animal, which become immunogenic by virtue of chemical modification. Indeed, it has been established that haptenated autologous proteins are immunogenic and that the immunochemical specificity is directed at new determinants of the carrier arising as a consequence of modification, as well as at the hapten itself.[49] Consequently, it is manifestly impossible to exclude as the cause of the helper effect cross-reactivity at the T cell level between one or more of the myriad of modified autologous proteins, which may be the inducers of cellular immunity, and the conjugate used to assess helper activity, rather than T cell specificity directed at the hapten itself.

A second serious objection to these experiments derives from the presence in most instances of hapten-specific B lymphocytes, since it has been repeatedly shown that antibody to one determinant can augment the antibody response to other determinants on the same antigen molecule.[50–58] The effect of passive antibody is difficult to predict and, hence, adequately control, because under some circumstances antibody may suppress the immune response, while under others it apparently enhances it.[50–59] The net effect may hinge on quantity, affinity and immunoglobulin class of the antibody.[50,51] In some cases, small amounts of passively administered antibody were more effective than larger quantities in producing a carrier effect.[51] It is obvious, then, that simulation of conditions in control and experimental animals would prove difficult, if not impossible.

A way around this dilemma is to remove hapten-specific B cells from pools of helper cells before transfer into irradiated recipients, an objective accomplished by passing the cells through anti-immunoglobulin affinity columns.[46] The effluent cell population, which contained very few immunoglobulin-bearing cells, was just as effective as the unfractionated pool in promoting the antibody response of primed B cells. On the other hand, treating the cells with anti-theta antiserum and complement severely diminished augmentation, indicating that activity resided in T cells. Be that as it may, removal of B cells from the skin painting scenario does not circumvent the objection raised earlier: namely, that cross-reactivity at the T cell level between hapten-modified proteins might account for the effect. Furthermore, abolition of the helper effect by eliminating T cells does not necessarily absolve antibody of a role in the process, since it has been recently shown that T cells, though they need not be antigen specific, are required for antibody-mediated helper activity.[60]

A different protocol has been employed by Henry and Trefts to induce T cell helper activity *in vitro* to the azophenylarsonate determinant.[47] Mice injected with the hapten covalently coupled to mycobacteria by diazotization (Ars–Mb conjugates) exhibited delayed skin reactivity and *in vitro* proliferative responses elicited by Ars conjugates of unrelated proteins. Spleen cells from sensitized mice were assayed for Ars-specific helper activity by their ability to restore the *in vitro* secondary response to a lactoside (Lac) hapten of T depleted anti-Lac memory B cells. Helper activity was generated when a double conjugate containing both Lac and Ars determinants was added to the cultures. Using column-purified cells and anti-T cell antiserum, T cells rather than antibody were implicated as the agent responsible for the helper activity. The major limitation of these very carefully executed experiments is that L-tyrosine-*p*-azobenzenearsonate has been shown to be immunogenic and to induce helper activity in the guinea pig,[29,31,33] and there is recent as yet unpublished evidence from the author's laboratory that this is also true in mice. Therefore, the Ars determinant coupled to a tyrosine residue, as it would be in the diazotization procedure, cannot technically be considered a hapten and may represent a special, perhaps unique, case. If reciprocal helper activity could have been demonstrated, for example, by sensitizing mice with Lac–Mb conjugates and thereby inducing helper cells specific for the Lac hapten, then an impressive argument might have developed for the generalized existence of hapten-specific T cell functional activity. As it stands, however, the work is an elegant demonstration of helper activity specific for a defined determinant, but does not really address the question of the repertoire of T lymphocyte specificities.

How does antibody exert its helper activity? The mechanism of augmentation is not clearly understood, but the requirement for normal T cells in at least some instances leads to seemingly plausible postulates.[60] The demonstration of Fc receptors on activated T cells[61–64] suggests that such receptors might bind antigen–antibody complexes, serving to implement helper effects mediated by antibody. The hypothesis that antibody-mediated helper effects operate through T cells which function by binding immune complexes by means of Fc receptors is consistent with the available evidence, but remains unproven.[60] It should be emphasized that this non-specific mechanism does not represent the major pathway of T cell activation, since it differs in important respects from typical cell-mediated helper activity.[60] Rather, it should be considered an ancillary pathway to that of specific T cells for the recognition of antigen and the mediation of cell cooperation, and may play a major role only in particular situations in which the 'carrier epitope' is recognized poorly or not at all by specific T cells, as appears to be the case for conventional haptens such as DNP.[60]

3.3 Conclusion

Can a definitive statement be made about the comparative repertoire of specificities of B and T lymphocytes? The B lymphocyte repertoire is clearly enormous, encompassing epitopes of all kinds. It has been possible to determine this simply and directly by antibody binding assays. Unfortunately, no such antigen-specific T lymphocyte product has been unambiguously identified, and therein lies the dilemma. Assays for T cell specificity have frequently relied upon cell activation by antigen, which is not directly equatable with ligand binding. There are very few examples of T cell activation by small, structurally defined immunogens about the size of an antigenic determinant; noteworthy exceptions include heptalysine and several azobenzenoid derivatives of L-tyrosine. Studies of antigen binding by lymphocytes indicate that T as well as B cells may bind a diverse assortment of haptens, although the cells are not activated in any perceptible way by such ligands. On the basis of the frequency of such cells and the numbers of antigen molecules they bind, it is probable that the antigen-binding receptors are synthetic products of the cells themselves, rather than passively acquired products of B cells. Thus, in terms of receptor specificity the diversity of the T cell library may be comparable to that of the B cell. However, there is yet to appear an unassailable example of specific T cell functional activity induced by conventional haptens.

4. ANTIGEN RECEPTORS OF T AND B LYMPHOCYTES

The molecular nature of antigen receptors on T and B lymphocytes has been extensively investigated and the question of the B lymphocyte receptor appears to be firmly settled. The T cell receptor for antigen is another matter altogether, and on its identity pivots one of the liveliest immunological controversies extant. In a nutshell, the issue has been polarized into two schools, one which holds that the receptor is immunoglobulin or an immunoglobulin-like molecule, and another which maintains that it is unrelated to Ig (Table 2), but may instead be a molecule coded for by a locus within the major histocompatibility complex, probably the product of the Ir gene locus. A third point of view is somewhat conciliatory and combinatorial, postulating that the receptor is comprised of both Ig and histocompatibility components[2] (see also R. M. E. Parkhouse and E. R. Abney, herein).

4.1 The B lymphocyte receptor for antigen

The majority of B lymphocytes in spleen, lymph nodes and blood bear surface Ig, which is readily detectable by immunofluorescence and autoradiography, as well as other techniques (reviewed in References 1 and 2). The identification of Ig-bearing cells as B lymphocytes stems principally from the labelling of bursal cells from birds or B cell-enriched lymphoid populations from mammals (nude or neonatally thymectomized mice or anti-Thy-1 plus complement treated spleen cells), the inability to label T lymphocytes by these methods, and the reciprocal correlation between Thy-1-positive and Ig-positive cells in various tissues of mice.

Estimates of the number of Ig molecules on B cells have been made by various investigators using several different techniques.[2] Considering the assumptions which must be made, whether uptake of labelled anti-Ig antibodies or direct cell membrane labelling is being studied, a remarkably consonant picture has emerged. The results indicate a tenfold range of 20 000 to 200 000 molecules per cell, with a nominal average of 10^5. These numbers are in reasonably good agreement with those derived from antigen-binding assays.[36]

Table 2. The nature of the T cell receptor for antigen

Immunoglobulin Finding	Reference	Non-immunoglobulin Finding	Reference
1. Extraction of Ig	96, 97	1. Failure to extract Ig	98–101
2. Inhibition of antigen-binding with anti-Ig sera	78–80	2. Difficulty in detecting Ig by immunofluorescence	1, 2
3. Release of carrier-reactive IgM-like molecules by activated T cells	93–95	3. Presence of Fc receptors on T cells	61–64
4. Capping antigen-binding receptors with anti-Ig sera	91, 92	4. Bursa-dependence of T cell surface Ig in chickens	73–75
5. Long-exposure autoradiography with labelled anti-Ig antibody	70–72	5. Inhibition of T cell activation by alloantisera	118
6. Inhibition of 'antigen suicide' of helper activity by anti-L chain sera	81	6. Release of antigen-specific, non-Ig helper factors by activated T cells	109, 122
7. Inhibition of helper activity by anti-L chain sera	82–84	7. Inhibition of antigen binding by alloantisera	117
8. Inhibition with anti-L chain sera of transfer of graft-versus host and tuberculin responses	85–89		
9. Suppression of delayed hypersensitivity in bursectomized chickens with anti-L chain sera	90		
10. Antigenic (idiotypic) similarities between antibodies and T cell receptors	126–128		

The most convincing evidence equating Ig molecules with antigen receptors on B cells emanates from the inhibition of antigen-binding by anti-Ig reagents (reviewed in Reference 2). In general, pretreating B cells with polyvalent anti-Ig serum totally inhibits antigen uptake, whereas similar treatment with anti-lymphocyte serum or anti-H-2 serum has little effect, indicating that the mere binding of antibody to the cell surface does not non-specifically hinder antigen-receptor linkage. While the possibility that B cell surface Ig is positioned much closer than other membrane markers to a non-Ig antigen receptor, and may even be physically associated with it, has not been rigorously excluded, experiments in which virtually all the surface Ig of cells specific for a flagellar antigen was capped by the antigen speaks persuasively against such a possibility.[65]

While there is general agreement that B cell antigen receptors are classical immunoglobulins, an understanding of the Ig classes to which they belong is in a state of flux.[2] It is evident that class shifts occur during B cell maturation, but the precise order of progression has not been definitely settled. Substantial quantities of 'IgD-like' immunoglobulin have been found on B cells, and studies in humans led to the suggestion that IgD is the primordial B cell antigen receptor.[66] In the mouse, however, it has recently been found that only IgM is present on bone marrow cells and that IgM precedes the appearance of IgD on neonatal splenocytes.[67] Moreover, blast cells in the spleen have primarily IgM on their surface, whereas more differentiated small lymphocytes carry IgD as well as IgM.[68] These findings suggest an IgM to IgD switch during the differentiation of B lymphocytes, which, since it also occurs in nu/nu and germ-free mice, appears to be thymus- and antigen-independent.[67]

Following exposure to antigen, there is a further shift to IgG-type receptors, although quantitation of individual classes indicates that many cells simultaneously express multiple heavy chain classes.[2] This story has not been finalized in its detail, but for all practical purposes it may be concluded that B lymphocyte antigen receptors are classical immunoglobulins.

4.2 Surface immunoglobulin on T cells

In view of the antigen specificity of B and T lymphocytes, the antigen-recognizing properties of immunoglobulins and the readily demonstrable immunoglobulins on B cells, it seemed both logical and aesthetic that immunoglobulins should prove to be universal antigen receptors for both types of lymphocytes. Despite the *a priori* reasons for this hypothesis, it is clear from the above discussion of B cell receptors that B and T cells can be easily distinguished on the basis of surface Ig. This has been variously attributed to a relative paucity of antigen receptors on T cells, a different disposition of Ig molecules in the T cell membrane, rendering reaction with anti-Ig sera more difficult, or to a class of Ig on T cells which is not represented in serum. Against the first of these proposals stands the observation that under conditions of antigen saturation T and B cells bind comparable amounts.[36] A telling argument against the second proposal was the failure to detect kappa chain variable region determinants on T cells under conditions which easily revealed them on B cells.[69] If T cells bind antigen molecules with Ig receptors, their variable regions should surely be exposed to antibody. The third proposal is more difficult to rebut, depending on the magnitude of the assumed differences between the distinctive 'IgT' and other classes (i.e. distinctive light as well as heavy chains), and requires isolation of the putative material.

The traces of Ig (usually IgM) which have been detected on T cells by long exposure autoradiography[70–72] are consistent with a range of several hundred to several thousand molecules per cell, perhaps 1% of that found on B cells. The strict dependence of labelled T cells in the chicken on the bursa of Fabricius[72–74] and the detection of Fc receptors on activated T cells[61–64] renders suspect the biosynthetic origin of T cell surface Ig. The results with bursectomized chickens indicate that this immunoglobulin is either endogenously synthesized under direct bursal control or, more probably, is a product of B lymphocytes which adheres to T cells. Indeed, it has been shown that chicken T cells are able to adsorb *in vitro* a cytophilic antibody produced by B cells, which then enables them to bind antigen.[75]

Evidence of a similar nature in other species has been rapidly accumulating. After injection into F_1 hybrids, murine blast cells derived from T lymphocytes of a parent strain displayed host allotypic markers as well as molecules with anti-alloantigen specificity, both of which were absent if the injected cells had been previously depleted of B lymphocytes.[76] A subpopulation of thoracic duct lymphocytes in rats which bound 100 to 2000 molecules of anti-Ig antibody (presumptive T cells) did not manifest allelic exclusion and bore both host and donor allotypes after transfer into irradiated recipients of a different allotype, in contradistinction to a second population which bound 10 000 to 100 000 molecules of anti-Ig antibody (presumptive B cells).[77]

Anti-Ig sera have also been used to inhibit antigen-binding[78-80] and functional activities[81-90] of T cells (Table 2). In general, the reagents which have been successful in these indirect demonstrations of surface Ig had specificities directed against light or mu chains. None of the investigations established the endogenous nature of the T cell Ig which was detected.

Experiments in which anti-Ig sera have been used to redistribute antigen receptors on the T cell surface provide perhaps the most forceful evidence for indigenous T cell Ig.[91,92] Spleen cell suspensions from mice immunized against protein antigens were pretreated with concentrations of polyvalent anti-Ig, anti-kappa chain or anti-mu chain sera insufficient to inhibit antigen-binding, but sufficient to induce capping of B cell membrane Ig. Under these conditions, subsequently introduced radiolabelled antigen was also localized to one pole of the cells which bound it, as expected. However, a comparable proportion of Thy-1-positive cells also bound the antigen in a cap, although Ig was not directly demonstrable on these cells by immunofluorescence.[91] The anti-Ig reagents were unable to produce the effect after absorption with purified mouse Ig, demonstrating that it was probably not caused by antibody directed against a contaminant in the materials used to prepare the reagents.

In experiments designed to establish the biosynthetic origin of the antigen receptors on T cells cappable by anti-Ig sera, normal rather than immune lymph node cells were cultured under capping conditions for extended periods of time.[92] After an hour in culture, B and T cells alike had lost their antigen receptors by endocytosis, but antigen-binding cells of both types reappeared 6 to 18 hours later, which could again be capped by the anti-Ig reagents. Crucial elements of these experiments were the use of lymphocytes from normal mice and the absence of antigen during the culture period, which make it difficult to explain the results in terms of passively acquired antibody. Nevertheless, the case for indigenous T cell Ig would have been strengthened by using enriched T cell preparations rather than unfractionated cells.

Feldmann and his coworkers have described an antigen-specific factor released by sensitized T cells which resembles monomeric IgM and is highly cytophilic for macrophages. Macrophages armed with the factor were then able to trigger specific B cell responses in the presence of the antigen.[93–95] Although it was shown that the IgM from T cells differed from B cell membrane IgM in its cytophilicity for macrophages, comparisons with serum IgM, which might be more germane, were not conducted. While this intriguing molecule may appear to be a promising candidate for the elusive T cell receptor, particularly in light of the anti-Ig capping experiments of Roelants, it should be noted that the Feldmann experiments describe a product released by T cells, the biosynthetic origin of which remains in question.

Efforts to chemically extract immunoglobulins from T lymphocyte membranes have given equally conflicting results. On the one hand, Marchalonis and his colleagues claim that in the mouse these cells contain as much membrane IgM as B cells,[96,97] while, on the other, several independent groups find little or no T cell IgM in the face of readily demonstrable B cell protein.[98–101] Parties to both sides of the debate have radio-iodinated cell surface proteins by the lactoperoxidase technique,[96–100] but their methods of solubilizing cells have differed, and therein may lie the explanation for the discrepancy. Marchalonis has employed acid-urea for this purpose, whereas others have resorted to non-ionic detergents. A direct comparison of the two procedures indicates that the acid-urea technique is less efficient, giving poorer recoveries of Ig than detergent extraction,[99–101] and it has been suggested that the finding of equivalent amounts of IgM on T and B cells may be more a failure to recover B cell material than the discovery of T cell

IgM;[101] the latter could conceivably derive from contaminating B cells, adherence of Ig to T cells through Fc receptors, or other mechanisms to be described below.

The evidence favouring an Ig T cell antigen receptor is thus, at best, insecure, resting principally on the observations of Roelants and Feldmann. In particular, it is difficult to reconcile the binding, under appropriate conditions, of large numbers of antigen molecules by T cells with the difficulty of detecting even variable region Ig markers on such cells, or recovering Ig molecules in appreciable quantities from their surfaces.

4.3 Ir genes and Ia antigens

The immune response to various antigens has been shown to be controlled by genes in the major histocompatibility complex (MHC), which have been designated immune response (Ir) genes.[102,103] These genes are probably operable in all vertebrates and, although phenotypic differences in the responses to relatively few antigens have been directly observed, it seems reasonable to assume that they regulate immune responses to all thymic-dependent antigens. The total involvement of the MHC in immune responses appears to be very complex, encompassing, in addition to B cell – T cell interaction, those occurring between T cells and macrophages[104] and between cytotoxic T cells and their targets.[105] Our concern will be confined to aspects which relate directly to the T cell receptor for antigen.

Initially, it was believed that Ir genes exerted their effect exclusively on T lymphocytes, since the responses of low-responder animals often resembled those made by normal animals against thymic-independent antigens,[102] characterized by a primary IgM response which fails to convert to IgG or to manifest anamnesis. In addition, low responders made normal antibody responses if the antigen was presented on a carrier capable of engendering T cell recognition, again implicating the T cell as the defective component of the system. Hence, attention naturally focused on whether Ir genes code for the T cell antigen receptor. However, other studies incriminated B cells[106,107] or both B and T cells[108,109] as the cells expressing the Ir gene defect.

Recently, alloantigens believed to be the products of Ir genes have been detected on lymphocyte surfaces.[110–113] The expression of these Ia antigens is greatest on B lymphocytes,[112,113] but lesser amounts have been detected on T cells by sensitive techniques.[114] Very little is known as yet about Ia antigens, which appear to comprise a multigenic, polymorphic system. Preliminary immunochemical studies characterize them as proteins or glycoproteins of about 30 000 molecular weight.[115,116]

Are Ia molecules T cell receptors for antigen? If so, why are they expressed more heavily on B cells? Are B and T cell Ia antigens identical? These questions are impossible to answer in a definitive way in view of the as yet fragmentary characterization of Ia antigens, but evidence is rapidly accumulating which may eventually solve the riddle. It has been noted earlier that anti-Ig sera have in some instances blocked antigen-binding by T cells, although results from different laboratories have varied. It is significant that inhibition of specific binding of antigen to mouse T cells has also been achieved using anti-H-2 alloantisera,[117] and in guinea pigs alloantisera have inhibited antigen-specific T cell activation.[118] While the antisera used in these studies were unquestionably heterogeneous and activity against Ia structures was not proven, the possibility that they acted by virtue of anti-Ia activity is both appealing and plausible.

More dramatic developments centre on soluble helper factors released by T cells which can replace the cells themselves in cooperative interactions with B lymphocytes.[109,119,120] The factors have molecular weights in the neighbourhood of 40 000 to 50 000, bear no serological resemblance to immunoglobulins, and in at least one case the factor was specific for the antigen which induced it.[109] Furthermore, these factors, which have thus far been demonstrated only in mice, react with antisera specific for antigens coded by the I region of the MHC,[109,121] within which the Ir genes are located. These properties mark the T cell factor as an obvious candidate for an Ir gene product, and it might be anticipated that in at least those genetically deficient immune responses in which the defect appears to be expressed in T cells, low responders would lack the ability to make the factor. This was investigated by Taussig and his colleagues in high and low responder strains of mice to the synthetic polypeptide TGAL.[109,122] They found that some low responder strains made the antigen-specific T cell factor as efficiently as high responders, but the factor would act effectively only on B cells of high responder origin. Such B cells could completely absorb the factor, whereas low responder B cells were unable to do so. It was shown that the factor taken up by B cells was biologically active in that bone marrow cells which had adsorbed factor in the presence of antigen and then were transferred into lethally irradiated recipients made good responses without further addition of factor. The defect, then, in these low responder strains was expressed in B cells rather than T cells. The B cells apparently lack an acceptor site for the T cell factor, and it would, of course, be of immense interest to know if the acceptor is coded by I-region genes. While this has not yet been definitely established, it has been reported that functional binding of factor to the acceptor site can be blocked by antisera specific for high responder H-2 antigens, but not by antisera raised against low responder haplotypes.[109] More precise mapping of the responsible genes within the MHC is needed before the acceptors can be positively designated as I-region gene products, but this prospect seems highly likely as well as intellectually appealing.

In other low responder strains, T cells were unable to produce the factor, and a third group of low responders exhibited defects in both cell populations, the T cells being unable to produce factor and the B cells unable to bind it. Thus, Ir gene defects can be expressed in either or both types of cells, in each case the defect being due to the absence of a product which is probably coded by I-region genes, albeit at different loci within the I region of the chromosome. Supporting evidence that two distinct Ir genes are required for the immune response to a single antigen has accrued from analyses of F_1 hybrids.[123–125] In a particularly noteworthy study, two genes which control the response to a synthetic polypeptide in mice were shown to map in different I subregions.[125]

The existence of I-region-determined T cell antigen-specific factors and B cell acceptors accords with the presence of Ia markers on both types of cells. The data are consistent with a greater density of acceptor on the B cell than of factor on the T cell, but this is not the only plausible explanation for the failure of antisera to detect more Ia antigen on T cells. The disposition of Ia molecules may be different in the membranes of the two cell types; or the acceptor may be more immunopotent that the factor, the latter cross-reacting weakly with anti-Ia sera. At any rate, if non-Ig antigen-recognizing molecules do indeed exist and mediate the antibody responses of B cells, the case for T cell immunoglobulins is further weakened. *A priori*, the postulate that a given cell would evolve two distinct systems for recognizing antigen is conceptually unattractive.

4.4 Idiotypic markers on T lymphocytes

Idiotypic determinants are epitopes on Ig molecules dictated by and situated within or near the antigen-binding site. They are, therefore, characteristic of particular antibody specificities, although populations of antibodies reactive with a particular epitope may display a variety of idiotypes, which is one manifestation of antibody heterogeneity. If antigen-recognizing structures on T lymphocyte have active sites which are similar to those of antibody molecules directed against the same epitope, then it might be expected that specific anti-idiotype reagents would interact similarly with both. In fact, data have been generated by several laboratories which suggest that B and T lymphocytes specific for the same alloantigens may share idiotypic determinants.[126–128]

Antisera raised in hybrid rats against either alloantibody or T lymphocytes from one of the parental strains displayed activity with both reactants, suggesting that B and T lymphocytes draw from the same genetic pool to construct the binding sites of their antigen receptors. It is important to note that, although T cells purified by filtration through anti-Ig affinity columns were used, the biosynthetic origin of the cell surface molecule bearing idiotypic determinants was not established in these studies. Substantiation of the findings using antigens other than alloantigens would be of more than passing interest, in view of the uncertainty that allogeneic recognition, which engages a disproportionately large fraction of the T cell population, and conventional antigen recognition are identical. Using a different approach, Eichmann and Rajewsky[128] found that guinea pig IgG_1 anti-idiotype antibody raised against mouse strain A/J antibody specific for streptococcal group A carbohydrate activated T helper cells as well as B precursor cells. Of particular interest, the helper cells engendered by anti-idiotype antibody cooperated only in anti-carbohydrate B cell responses which expressed that particular idiotype. These findings strengthen the evidence that B and T cell antigen receptors possess similar or identical active sites, although here too the biosynthetic origin of the T cell idiotypic molecule remains in question.

It is, perhaps, also worth mentioning that even should antibodies and T cell antigen receptors prove to have similar or identical active sites, the hypervariable regions which form antigen-binding sites could conceivably be incorporated into different molecular frameworks by genetic translocation of episomal insertion mechanisms.[129] This would be consistent with recent studies which suggest that the variable region of rabbit heavy chains may be encoded by two or more genes, one specifying framework residues and another the idiotype.[130]

4.5 Model-building

Is the indigenous T cell receptor for antigen an Ig-like molecule or a fundamentally different structural entity? There is obviously a great deal of data on each side of this vital question, and it is manifestly impossible to reconcile all the conflicting and diverse findings. The essential elements which must be accounted for by any model include: (a) the detection of minute quantities of Ig on T cells, but, thus far, only in the presence of functional B cells; (b) the apparent antigen specificity of T cell Ig, as indicated by the capping experiments of Roelants, in which receptors specific for a given antigen were capped by anti-Ig sera only on very infrequent T cells; (c) the absence of quantitative parallelism between Ig on T cells and antigen bound by T cells, when the latter was estimated under optimal conditions; (d) the description of non-Ig, antigen-binding T cell products.

At this point in time, it appears to this reviewer that the available information best fits a model in which the indigenous T cell receptor is not an Ig molecule, but more likely a product of the I-region of the major histocompatibility complex. Antigen molecules would be bound specifically by the receptors of appropriate cells and could, in turn, bind humoral antibody by means of unoccupied epitopes, forming an 'antigen sandwich' between the T cell receptor and antibody (Figure 1(A)). Considerable latticing of antigen and antibody can be visualized, possibly facilitating functional activities mediated by T cells, such as cooperation (Figure 1(B)), by providing additional epitopes on multideterminant antigens with which B cell receptors may combine. The antigen sandwich would be expected to play an important amplifying role primarily in weak helper situations or in antibody-mediated helper effects, which require the participation of T cells. The model can account for the antigen specificity of T cells, the small quantities of Ig detected on T cell surfaces, the variable inhibition by anti-Ig reagents of antigen-binding and functional activities of T cells reported by various investigators (Table 2), as well as the accounts of relatively small, non-Ig, antigen-specific T cell factors.

In order to account for the apparent antigen specificity of T cell surface Ig, the model stipulates that binding of Ig occurs principally via antigen rather than Fc receptors. However, Fc receptors on activated T cells may bind additional Ig non-specifically and

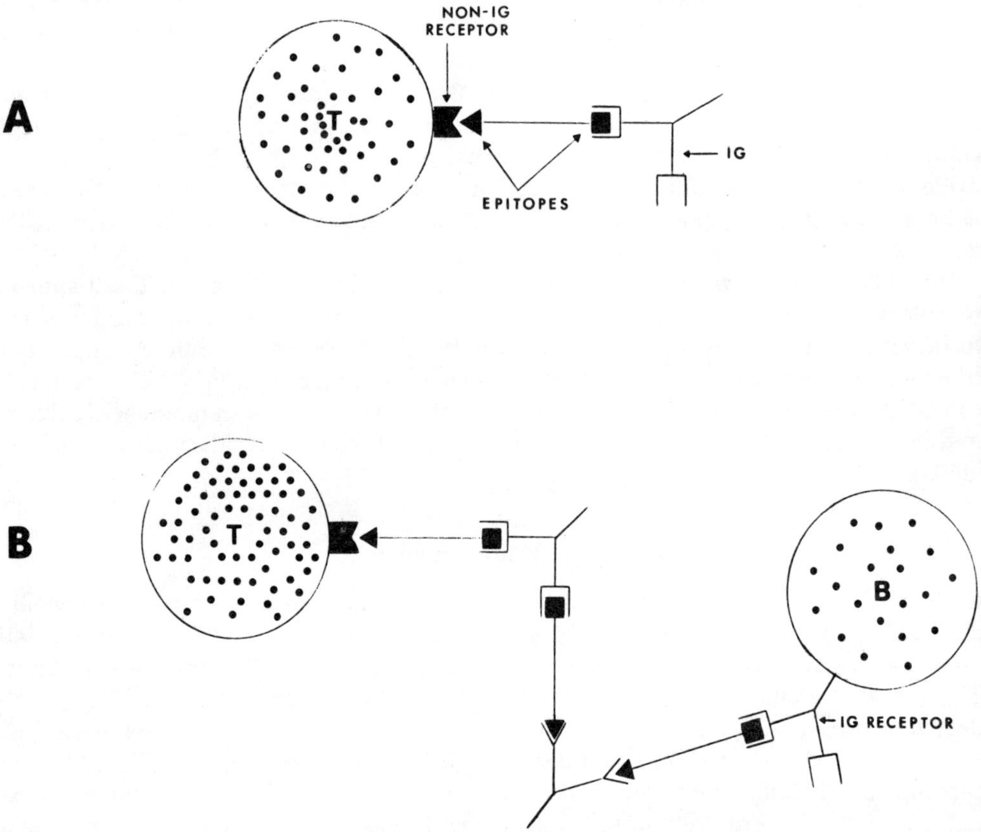

Figure 1. A hypothetical model to account for the detection of small quantities of Ig on T cells (A) and the facilitation by Ig of T-mediated functional activity (i.e. helper activity) (B). The model can account for the specificity of antigen-binding by T cells, as well as variable inhibition by anti-Ig reagents of antigen-binding and functional T cell activity. The 'antigen sandwich' may amplify weak helper activity and be a factor in antibody-mediated helper effects. Macrophages and Fc receptors on T cells have been omitted from the model for the sake of clarity, but may obviously contribute to antigen presentation and the stabilization of Ig on T cell surfaces. In addition, although the model stipulates that the indigenous T cell receptor is not an Ig molecule, it may share active sites with Ig molecules, which would be manifested as common idiotypic specificities

could play an even more important role in antibody-induced helper effects,[60] or serve to stabilize Ig molecules bound by the mechanism suggested here.

According to the model, anti-idiotype reagents could detect exogenous antibody bound to T cell surfaces through the antigen bridge, underscoring the necessity of establishing the biosynthetic origin of T cell idiotypic markers. Alternatively, the indigenous, non-Ig T cell receptor and antibody may have similar active sites, through the sharing of common, short hypervariable sequences, which would be manifested as idiotypic cross-reactivity.

The model is severely challenged by the experiments of Roelants,[92] in which TGAL-binding T cells in a normal lymph node cell population had their antigen receptors capped by anti-Ig sera of scrupulously proven specificity and subsequently regenerated them. The new receptors could again be capped with the same anti-Ig reagents. Although these experiments were performed with unfractionated cells, the prospect that specific B cells in the unprimed population, which on the basis of antigen binding occurred with a frequency of $54–108 \times 10^{5}$, might have released sufficient antibody during a 3- to 18-hour incubation period in the absence of antigen to account for the recapping and loss of receptors on infrequent ($23–43 \times 10^{-5}$) T cells strains credibility. However, it was stated at the outset that no hypothesis can satisfactorily explain all the observations which have appeared in the literature.

The issue of the T cell receptor for antigen will not be settled with finality until a biosynthetically proven product has been structurally characterized. Only then will the relationship of the receptor to immunoglobulin molecules, which can only be crudely approximated by serological means, be fully understood.

5. REFERENCES

1. Greaves, M. F., Owen, J. J. T. and Raff, M. C. (1974). *T and B Lymphocytes*, Excerpta Medica, Amsterdam.
2. Warner, N. L. (1974). *Advan. Immunol.*, **19**, 67–216.
3. Siskind, G. W. and Benacerraf, B. (1969). *Advan. Immunol.*, **10**, 12–20.
4. Decker, J. M., Clarke, J., Bradley, L. M., Miller, A. and Sercarz, E. E. (1974). *J. Immunol.*, **113**, 1823–1833.
5. Haimovich, J. and du Pasquier, L. (1973). *Proc. Nat. Acad. Sci.*, **70**, 1898–1902.
6. Ehrlich, P. (1900). *Proc. Roy. Soc., Ser. B.*, **66**, 424–448.
7. Rajewsky, K. and Mohr, R. (1974). *Eur. J. Immunol.*, **4**, 111–119.
8. Schlossman, S. F. (1972). *Transplant. Rev.*, **10**, 97–111.
9. Paul, W. E. and Siskind, G. W. (1970). *Immunology*, **18**, 921–930.
10. Maron, E., Webb, C., Teitelbaum, O. and Arnon, R. (1972). *Eur. J. Immunol.*, **2**, 294–297.
11. Michaeli, D., Senyk, G., Maoz, A. and Fuchs, S. (1972). *J. Immunol.*, **109**, 103–109.
12. Gell, P. G. H. and Benacerraf, B. (1959). *Immunology*, **2**, 64–70.
13. Thompson, K., Harris, M. and Benjamini, E. (1972). *Nature* (*New Biol.*), **238**, 20–21.
14. Parish, C. R. (1971). *J. exp. Med.*, **134**, 1–20.
15. Parish, C. R. (1971). *J. exp. Med.*, **134**, 21–47.
16. Cooper, M. G. (1972). *Scand. J. Immunol.*, **1**, 237–246.
17. Cooper, M. G. and Ada, G. L. (1972). *Scand. J. Immunol.*, **1**, 247–253.
18. Schirrmacher, V. and Wigzell, H. (1972). *J. exp. Med.*, **136**, 1616–1630.
19. Hoffmann, M. and Kappler, J. W. (1973). *J. exp. Med.*, **137**, 721–739.
20. Cunningham, A. J. and Sercarz, E. E. (1971). *Eur. J. Immunol.*, **1**, 413–421.
21. Playfair, J. H. L. (1972). *Nature* (*New Biol.*), **235**, 115–117.
22. Haritou, H. and Argyris, B. (1972). *Cell. Immunol.*, **4**, 179–181.
23. Senyk, G., Williams, E. B., Nitecki, D. E. and Goodman, J. W. (1971). *J. exp. Med.*, **133**, 1294–1308.
24. Cantor, H. and Boyse, E. A. (1975). *J. exp. Med.*, **141**, 1376–1389.

25. Senyk, G., Nitecki, D. E., Spitler, L. and Goodman, J. W. (1972). *Immunochemistry*, **9**, 97–110.
26. Loor, F. (1974). *Eur. J. Immunol.*, **4**, 210–220.
27. Hanna, N. and Leskowitz, S. (1973). *Cell. Immunol.*, **7**, 189–197.
28. Goodman, J. W., Bellone, C. J., Hanes, D. and Nitecki, D. E. (1974). In *Progress in Immunology II*, Vol. 2 (Brent, L. and Holborow, J., Eds.), North-Holland, Amsterdam, pp. 27–37.
29. Goodman, J. W. (1975). In *The Antigens*, Vol. 3 (Sela, M., Ed.), Academic, New York, pp. 127–187.
30. Leskowitz, S., Jones, V. and Zak, S. (1966). *J. exp. Med.*, **123**, 229–237.
31. Alkan, S., Nitecki, D. E. and Goodman, J. W. (1971). *J. Immunol.*, **107**, 353–358.
32. Becker, M. J., Levin, H. and Sela, M. (1973). *Eur. J. Immunol.*, **3**, 131–135.
33. Alkan, S. S., Williams, E. B., Nitecki, D. E. and Goodman, J. W. (1972). *J. exp. Med.*, **135**, 1228–1246.
34. Roelants, G. E. (1972). *Curr. Top. Microbiol. Immunol.*, **59**, 135–165.
35. Ashman, R. F. and Raff, M. C. (1973). *J. exp. Med.*, **137**, 69–84.
36. Roelants, G. E. (1975). In *Molecular Approaches to Immunology* (Smith, E. E. and Ribbons, D. W., Eds.), Academic, New York, pp. 55–75.
37. Möller, E., Bullock, W. W. and Mäkelä, O. (1973). *Eur. J. Immunol.*, **3**, 172–179.
38. Rutishauser, U. and Edelman, G. M. (1972). *Proc. Nat. Acad. Sci.*, **69**, 3774–3778.
39. Polak, L., Ryden, A. and Roelants, G. E. (1975). *Immunology*, **28**, 479–484.
40. Smith, E., Hammarström, L. and Möller, E. (1974). *Scand. J. Immunol.*, **3**, 61–70.
41. Hämmerling, G. T. and McDevitt, H. O. (1974). *J. Immunol.*, **112**, 1726–1733.
42. Iverson, G. M. (1970). *Nature*, **227**, 273–274.
43. Mitchison, N. A. (1971). *Eur. J. Immunol.*, **1**, 68–75.
44. Taylor, R. B. and Iverson, G. M. (1971). *Proc. Roy. Soc. (London), B*, **176**, 393–418.
45. Doughty, R. A. and Klinman, N. R. (1973). *J. Immunol.*, **111**, 1140–1146.
46. Rubin, B. and Wigzell, H. (1973). *Nature*, **242**, 467–468.
47. Henry, C. and Trefts, P. E. (1974). *Eur. J. Immunol.*, **4**, 824–830.
48. Potter, M. and Lieberman, R. (1967). *Advan. Immunol.*, **7**, 91–145.
49. Rubin, B. (1973). *Eur. J. Immunol.*, **3**, 26–32.
50. Henry, C. and Jerne, N. K. (1968). *J. exp. Med.*, **128**, 133–152.
51. Pincus, C., Miller, G. and Nussenzweig, V. (1973). *J. Immunol.*, **110**, 301–304.
52. Dennert, G. (1971). *J. Immunol.*, **106**, 951–955.
53. Rubin, B. (1972). *Scand. J. Immunol.*, **1**, 125–134.
54. Taussig, M. J. and Lachmann, P. J. (1972). *Immunology*, **22**, 185–197.
55. Walker, J. G. and Siskind, G. W. (1968). *Immunology*, **14**, 21–28.
56. McBride, R. A. and Schierman, L. W. (1973). *J. Immunol.*, **110**, 1710–1712.
57. Janeway, C. A., Jr. and Paul, W. E. (1973). *Eur. J. Immunol.*, **3**, 340–347.
58. Janeway, C. A., Jr. (1973). *J. Immunol.*, **111**, 1250–1256.
59. Uhr, J. W. and Möller, G. (1968). *Advan. Immunol.*, **8**, 81–127.
60. Janeway, C. A., Jr., Koren, H. S. and Paul, W. E. (1975). *Eur. J. Immunol.*, **5**, 17–22.
61. Yoshida, T. O. and Anderson, B. (1972). *Scand. J. Immunol.*, **1**, 401–408.
62. Grey, H. M., Kubo, R. T. and Cerottini, J. (1972). *J. exp. Med.*, **136**, 1323–1328.
63. Hunt, S. V. and Williams, A. F. (1974). *J. exp. Med.*, **139**, 479–496.
64. van Boxel, J. A. and Rosenstreich, D. L. (1974). *J. exp. Med.*, **139**, 1002–1012.
65. Raff, M. C., Feldmann, M. and de Petris, S. (1973). *J. exp. Med.*, **137**, 1024–1030.
66. Rowe, D. S., Hug, K., Forni, L. and Pernis, B. (1973). *J. exp. Med.*, **138**, 965–972.
67. Vitetta, E. S., Melcher, U., McWilliams, M., Lamm, M. E., Phillips-Quagliata, J. and Uhr, J. W. (1975). *J. exp. Med.*, **141**, 206–215.
68. Goodman, S. A., Vitetta, E. S., Melcher, U. and Uhr, J. W. (1975). *J. Immunol.*, **114**, 1646–1648.
69. Grey, H. M., Colon, S., Solomon, A. and McLaughlin, C. L. (1973). *J. Immunol.*, **111**, 1923–1925.
70. Bankhurst, A. D., Warner, N. L. and Sprent, J. (1971). *J. exp. Med.*, **134**, 1005–1015.
71. Nossal, G. J. V., Warner, N. L., Lewis, H. and Sprent, J. (1972). *J. exp. Med.*, **135**, 405–428.
72. Hudson, L., Thantrey, N. and Roitt, I. M. (1975). *Immunology*, **28**, 151–159.
73. Kincade, P. W., Lawton, A. R. and Cooper, M. D. (1971). *J. Immunol.*, **106**, 1421–1423.
74. Rabellino, E. and Grey, H. M. (1971). *J. Immunol.*, **106**, 1418–1420.
75. Webb, S. R. and Cooper, M. D. (1973). *J. Immunol.*, **111**, 275–277.

76. Hudson, L., Sprent, J., Miller, J. F. A. P. and Playfair, J. H. L. (1974). *Nature*, **251**, 60–62.
77. Hunt, S. V. and Williams, A. F. (1974). *J. exp. Med.*, **139**, 479–496.
78. Greaves, M. F. and Hogg, N. M. (1971). In *Progress in Immunology I* (Amos, B., Ed.), Academic, New York, pp. 111–126.
79. Hogg, N. M. and Greaves, M. F. (1972). *Immunology*, **22**, 967–980.
80. Dwyer, J. M., Warner, N. L. and Mackay, I. R. (1972). *J. Immunol.*, **108**, 1439–1446.
81. Basten, A., Miller, J. F. A. P., Warner, N. L. and Pye, J. (1971). *Nature* (*New Biol.*), **231**, 104–106.
82. Cheers, C., Breitner, J. C. S., Little, M. and Miller, J. F. A. P. (1971). *Nature* (*New Biol.*), **232**, 248–250.
83. Mitchison, N. A. (1971). *Eur. J. Immunol.*, **1**, 18–27.
84. Lesley, J. F., Kettman, J. R. and Dutton, R. W. (1971). *J. exp. Med.*, **134**, 618–629.
85. Mason, S. and Warner, N. L. (1970). *J. Immunol.*, **104**, 762–765.
86. Riethmüller, G., Rieber, E. P. and Seeger, I. (1971). *Nature* (*New Biol.*), **230**, 248–250.
87. Cole, L. J. and Maki, S. E. (1971). *Nature* (*New Biol.*), **230**, 244–246.
88. Tyan, M. L. (1971). *J. Immunol.*, **106**, 586–588.
89. Rouse, B. T. and Warner, N. L. (1972). *Cell. Immunol.*, **3**, 470–477.
90. Theis, G. A. and Thorbecke, G. J. (1973). *J. Immunol.*, **110**, 91–97.
91. Roelants, G. E., Forni, L. and Pernis, B. (1973). *J. exp. Med.*, **137**, 1060–1077.
92. Roelants, G. E., Rydén, A., Hägg, L.-B. and Loor, F. (1974). *Nature*, **247**, 106–108.
93. Feldmann, M. (1972). *J. exp. Med.*, **136**, 737–760.
94. Feldmann, M., Cone, R. E. and Marchalonis, J. J. (1973). *Cell. Immunol.*, **9**, 1–11.
95. Cone, R. E., Feldmann, M., Marchalonis, J. J. and Nossal, G. J. V. (1974). *Immunology*, **26**, 49–60.
96. Marchalonis, J. J., Cone, R. E. and Atwell, J. L. (1972). *J. exp. Med.*, **135**, 956–971.
97. Marchalonis, J. J. and Cone, R. E. (1973). *Transplant. Rev.*, **14**, 3–49.
98. Vitetta, E. S., Bianco, C., Nussenzweig, V. and Uhr, J. W. (1972). *J. exp. Med.*, **136**, 81–93.
99. Grey, H. M., Kubo, R. T. and Cerottini, J.-C. (1972). *J. exp. Med.*, **136**, 1323–1328.
100. Lisowska-Bernstein, B., Rinuy, A. and Vassalli, P. (1973). *Proc. Nat. Acad. Sci.*, **70**, 2879–2883.
101. Jensenius, J. C. and Williams, A. F. (1974). *Eur. J. Immunol.*, **4**, 98–105.
102. Benacerraf, B. and McDevitt, H. O. (1972). *Science*, **175**, 273–279.
103. Benacerraf, B. and Katz, D. H. (1974). *Advan. Cancer Res.*, **21**, 121–173.
104. Rosenthal, A. S., Lipsky, P. E. and Shevach, E. M. (1975). *Fed. Proc.*, **34**, 1743–1748.
105. Zinkernagel, R. M. and Doherty, P. C. (1975). *J. exp. Med.*, **141**, 1427–1436.
106. Mozes, E. and Shearer, G. M. (1972). *Cur. Top. Microbiol. Immunol.*, **59**, 167–200.
107. Katz, D. H., Hamaoka, T., Dorf, M. E. and Benacerraf, B. (1973). *Proc. Nat. Acad. Sci.*, **70**, 2624–2628.
108. Shearer, G. M., Mozes, E. and Sela, M. (1972). *J. exp. Med.*, **135**, 1009–1027.
109. Munro, A. J. and Taussig, M. J. (1975). *Nature*, **256**, 103–106.
110. Hauptfeld, V., Klein, D. and Klein, J. (1973). *Science*, **181**, 167–169.
111. David, C. S., Shreffler, D. C. and Frelinger, J. A. (1973). *Proc. Nat. Acad. Sci.*, **70**, 2509–2514.
112. Sachs, D. H. and Cone, J. L. (1973). *J. exp. Med.*, **138**, 1289–1304.
113. Hämmerling, G. J., Deak, B. D., Mauve, G., Hämmerling, U. and McDevitt, H. O. (1974). *Immunogenetics*, **1**, 68–81.
114. Fatham, C. G., Cone, J. L., Sharrow, S. O., Tyrer, H. and Sachs, D. H. (1975). *J. Immunol.*, **115**, 584–589.
115. Cullen, S. E., David, C. S., Shreffler, D. C. and Nathenson, S. G. (1974). *Proc. Nat. Acad. Sci.*, **71**, 648–652.
116. Vitetta, E. S., Klein, J. and Uhr, J. W. (1974). *Immunogenetics*, **1**, 82–90.
117. Hämmerling, G. J. and McDevitt, H. O. (1974). *J. Immunol.*, **112**, 1734–1740.
118. Shevach, E. M., Green, I. and Paul, W. E. (1974). *J. exp. Med.*, **139**, 679–695.
119. Schimpl, A. and Wecker, E. (1972). *Nature* (*New Biol.*), **237**, 15–17.
120. Armerding, D. and Katz, D. H. (1974). *J. exp. Med.*, **140**, 19–37.
121. Armerding, D., Sachs, D. H. and Katz, D. H. (1974). *J. exp. Med.*, **140**, 1717–1722.
122. Taussig, M. J., Mozes, E. and Isac, R. (1974). *J. exp. Med.*, **140**, 301–312.
123. Zaleski, M., Fuji, H. and Milgrom, F. (1973). *Transplant. Proc.*, **5**, 201–204.
124. Rüde, E. and Günther, E. (1974). In *Progress in Immunology II*, Vol. 2 (Brent, L. and Holborow, J., Eds.), North-Holland, Amsterdam, pp. 223–233.
125. Dorf, M. E., Stimpfling, J. H. and Benacerraf, B. (1975). *J. exp. Med.*, **141**, 1459–1463.

126. Ramseier, H. and Lindenmann, J. (1972). *Eur. J. Immunol.*, **2**, 109–114.
127. Binz, H. and Wigzell, H. (1975). *J. exp. Med.*, **142**, 197–211.
128. Eichmann, K. and Rajewsky, K. (1975). *Eur. J. Immunol.*, **5**, 661–666.
129. Wu, T. T. and Kabat, E. A. (1970). *J. exp. Med.*, **132**, 211–250.
130. Capra, J. D. and Kindt, T. J. (1975). *Immunogenetics*, **1**, 417–427.

Chapter 10

Biochemical Approaches to Receptors for Antigen on B and T Lymphocytes

R. M. E. PARKHOUSE
E. R. ABNEY

1. INTRODUCTION

The classical experiments of Gowans and his colleagues[1] established that the lymphocyte was the key cell that reacts upon contact with antigen and becomes, as a result of a complex and ill-understood series of differentiative events, the effector cell of immune responses, both humoral and cellular. A feature of the immune response is its specificity and this is explained by selectivity of antigen, as predicted by proponents of selective theories of immunity.[2–4] Thus receptors for antigen on an individual lymphocyte are of one specificity only, but the specificity varies from cell to cell within the total lymphocyte population of an organism. Consequently, when an immunogenic determinant is introduced into a vertebrate animal only the very few cells with appropriate receptor structures are able to interact with the antigen, and they are therefore selected. Following interaction with the antigen those cells divide and differentiate into effector cells. During this process some cells are set aside as memory cells, thereby forming an expanded population so that subsequent exposure to the same antigen provokes the well documented secondary immune response.

Lymphocytes, however, although they may all look quite similar when viewed under the light microscope, are a heterogeneous population of cells. The simplest and broadest division is into B lymphocytes and T lymphocytes, although it is true that a small proportion of cells with the morphological characteristics of lymphocytes cannot be assigned to either category. The latter may possibly be precursors to B lymphocytes or cells which function in antibody-mediated cytotoxicity reactions. The foundation for this division rests upon experiments with chickens.[5,6] Removal of the thymus was found to impair cellular immunity whereas removal of the bursa of Fabricius profoundly depressed development of humoral immunity. In mice similarly, removal of the thymus depressed cellular immunity.[7] To this day it has not been possible to pinpoint a mammalian equivalent of the chicken bursa, and recent evidence suggests that it is multifocal.[8] However, it was possible to show that in the mouse the precursors of antibody-secreting cells are the B lymphocytes and that their maturation frequently depends upon interaction with T lymphocytes.[9] This concept of an interaction between T and B lymphocytes originally stemmed from experiments in which mixtures of both cell types were considerably more efficient in transferring immune responses than either cell type alone.[10] Also clear was that the T lymphocyte, although it proliferates when presented with antigen, does not differentiate into an antibody-secreting cell.[11–13] The jigsaw was then completed with the advent of cell surface markers for B and T lymphocytes[14] and the demonstration that in many immune responses whilst the B lymphocyte may make the antibody, the T lymphocyte is necessary for induction ('help').[15] The most popular cell surface markers are immunoglobulin and Thy-1 for B and T lymphocytes respectively. Thus B lymphocytes carry surface immunoglobulin and differentiate into B memory and antibody-secreting cells when exposed to antigen, and usually with the collaborative influence of T lymphocytes. T lymphocytes alone are responsible for the various phenomena of cell-mediated immunity. Superimposed upon this already complex scheme are: possible roles for macrophages in immune induction and positive (help) and negative (suppression) effects of T lymphocytes on T and B lymphocyte responses (References 16 and 17, and G. E. Roelants, herein).

In this article, recent exciting and unexpected findings pertaining to receptors for antigen on lymphocytes will be considered in detail. For a thoroughly detailed review of the knowledge up to 1974 the treatise of Warner[18] is enthusiastically recommended.

2. METHODOLOGY

All procedures which attempt to define receptors for antigens on lymphoid cells ultimately depend on using an antiserum. For B lymphocytes antisera raised against the varied heavy and light chain isotypes are the reagents, whilst for T lymphocytes a recent development has been the use of sera with anti-idiotype specificity.

Reactivity between a given antibody and plasma membranes is usually taken as evidence for the presence of the relevant antigen on the cell surface. Whatever the system used to detect reactivity, it is obviously crucial that the antibodies used in the investigation be strictly characterized. In principle, sera should be monitored for unwanted specificities by techniques at least as sensitive as the methodology followed in the investigation. For example, absence of antibody activity to light chains as judged by Ouchterlony analysis cannot guarantee that the antibody sample will not detect light chains in more sensitive procedures, e.g. fluorescence or radio-immune precipitation.

The customary procedure for rendering an anti-immunoglobulin serum specific for one isotype is to pass it through solid phase immunoabsorbents bearing all other isotypes. After such treatment, however, it is still possible for the antiserum to contain unwanted specificities. These may be against components not normally routinely tested for, such as α2-macroglobulin, which is in fact found on the surface of most B lymphocytes.[19] Alternatively, an anti-immunoglobulin serum could contain antibodies directed against variable region determinants shared by some, but not all, representatives of most immunoglobulin classes. Such specificities would frequently escape detection in many testing systems. The possibility of antibodies to immunoglobulin cross reacting with β2-microglobulin is also raised by the finding of sequence homology between these two entities.[20] Since β2-microglobulin is found in association with histocompatibility antigens on the surfaces of most cells, antibodies of this type would create obvious, but perhaps unforeseen, problems. Removal of unwanted specificities must be done with solid phase immunoabsorbents to avoid artefacts arising from the presence of antibody–antigen complexes in sera. A major problem is heterophile antibodies to cell surfaces. Whilst many of these can be removed by absorption with membranes from appropriate organs (e.g. liver and kidney for anti-immunoglobulin), on occasions it is not feasible to carry out the appropriate absorption.

The ultimate test for specificity is actually to isolate the molecule on the cell membrane that the antiserum combines with, and then to carry out biochemical characterization. This is possible if the cell surface is first labelled with radioactive iodine using the enzyme lactoperoxidase.[21] Then the cell can be dissolved in dissociating solvents[22] or detergent.[23] The total mixture of radioactive cell surface molecules is reacted with the test antibody and the resulting precipitate is characterized, usually by electrophoresis in polyacrylamide gels containing dissociating (sodium dodecyl sulphate) reagents. By doing this one is not only recording the fact that an antiserum reacts with a cell but also defining what the antiserum reacts with.

In the simplest system the antibodies are tagged with fluorescent or radioactive molecules, mixed with living lymphocytes and then visualized on the surface by microscopy. Thus the presence of immunoglobulin on the surface of B lymphocytes is readily demonstrated using fluorochrome-coupled anti-immunoglobulin.[24,25] It is important to emphasize, however, that this type of approach can only identify molecules, and identification is not rigorous proof of a functional role. When whole IgG antibodies or divalent $(Fab)_2$ fragments are used, the surface components usually aggregate and then 'cap' at one end of the cell (References 26–29, and F. Loor, herein). The process of cap formation is an energy dependent phenomenon, probably involving microfilament activity, but does depend on cross-linking of the membrane located molecules since monovalent (Fab) antibody does not result in capping. Thus capping does not occur in the cold or in the presence of such metabolic inhibitors as sodium azide. In addition, capping is selective, only molecules recognized by the antiserum or strongly associating with the target of the antiserum are capped. This finding is in agreement with the fluid mosaic model of the cell membrane, where the membrane proteins are not fixed, but are free to move within the plane of the lipid bilayer.[30,31] Of great practical importance is the fact that cells can be screened for two surface antigens by using the two appropriate antisera sequentially; the first is tagged, for example, with a green fluorochrome and allowed to cap and the second, tagged with a red fluorochrome, is reacted under conditions where capping cannot occur. The observer can then record cells with green caps, red peripheral staining, or both. Also useful is that once a membrane component is capped by an antiserum it is lost from the cell surface, either by internalization through pinocytosis or by shedding of aggregates to the exterior milieu. This process, often termed modulation, is followed by reinsertion of newly synthesized units in the membrane but does give a finite period of time when the cell surface is denuded of a given membrane component. From the rate at which a modulated molecule reappears on the cell surface, estimates of synthetic and turnover times can be gathered.

Unfortunately, although the identification of cell surface components by reacting living cells with specific antisera is undoubtedly a powerful tool, there are major pitfalls that can be encountered, particularly with lymphocytes, even when the antisera are undoubtedly specific. The first, and most obvious, is whether identification by this type of approach necessarily means that the surface component visualized is actually an endogenous product of the cells in question. Both B and T lymphocytes can have receptors on their surfaces for the Fc portion of immunoglobulins[32–40] and it is therefore quite feasible for lymphocytes to passively absorb autologous immunoglobulin *in vivo* or the anti-immunoglobulin reagents used for studying the cells *in vitro*. These problems can be dealt with by first treating the cells in such a way that the cell surface component under study is lost. This may occur simply by incubating the cells *in vitro*, but can also be done by modulation with the antibody or by enzyme treatment. Having allowed sufficient time for resynthesis, the cells are then reacted with the chosen antibody, but in its $(Fab)_2$ or Fab form in order to avoid passive uptake mediated by Fc receptors. Ideally, the cell population used should be homogeneous. For example, in a mixture of cells it is possible that only one cell type resynthesizes a certain membrane component, but this is secreted or shed into the surrounding medium and subsequently taken up by other cell types. It is only very rarely that such stringent experimental conditions are followed and, as a result, the literature abounds with conflicting claims.

An alternative approach for identifying receptors for antigen has been to inhibit function of the lymphocytes with anti-immunoglobulin.[41] This approach has the advantage that it does have a functional correlate, but is clearly subject to all the problems discussed above. Basically, anti-immunoglobulin has been added to almost all systems possible; from injecting whole animals to attempts to inhibit the binding of antigen by lymphocytes.

The least ambiguous technique for identifying cell surface proteins with antibodies is by radiolabelling cells. In principle, the labelled cells are solubilized and cell surface components are coprecipitated by addition of appropriate antibody. The antibody will combine with whatever structures it has specificity for, and precipitation of these complexes can be effected by adding non-radioactive antigen (direct precipitation) or an antibody to the first antibody (indirect precipitation). For example, when rabbit anti-mouse immunoglobulin is used with labelled mouse B lymphocytes, precipitation can result from the addition of non-radioactive mouse immunoglobulin or goat anti-rabbit immunoglobulin. The precipitates are then washed and the amount of radioactivity they contain is expressed as a fraction of the total input radioactivity in macromolecular material (estimated by precipitation with trichloracetic acid). Since these specific immunological precipitates invariably contain non-specifically entrapped or adsorbed radioactivity, it is essential to do a non-specific control, with a parallel sample. This simply entails the formation of an immunological precipitate using an antibody lacking specificity for any of the radioactive components of the solubilized radioactive cells. For obvious reasons, the mass of precipitate formed in the specific and non-specific systems should be comparable. Although it is common practice to take the difference between the amounts of radioactivity found in the specific and non-specific precipitates, this does not necessarily give an accurate estimate of specific membrane molecules. Once again we must consider the specificity of the antisera employed. Two other important factors are: the total yield of radioactivity in the specific precipitate, and the choice of reactants for the non-specific control. Most membrane proteins individually account for a small percentage of the total membrane. Immunoglobulin, for example, constitutes about 2% of splenic lymphocyte membranes and this fraction is accordingly the maximum recovery possible in the specific precipitate. It is not uncommon to find 1% of the input radioactivity in non-specific precipitates, and so one is placed in the position of taking the difference between two numbers of fairly similar value. This is obviously a source of error. Far more serious, however, is the variable presence in normal and immune sera of heterophile antibodies to cell surface components. Suppose an antiserum to immunoglobulin has a high titre of such antibodies and suppose it is reacted with extracts of radioactive cells which do not have surface immunoglobulin, but which do have antigens recognized by the heterophile antibodies. Now suppose that the non-specific precipitation system is one with little or no heterophile antibody content. The result will be a higher recovery of radioactivity in the specific (anti-immunoglobulin) precipitate than in the non-specific precipitate. In this case, however, the higher recovery is not due to the presence of immunoglobulin on the surfaces of the cells examined. It is in fact an artefact of the coprecipitation system. Such artefacts are more likely when direct precipitation is done, and in particular in this case when the non-specific control system is chosen because of its low yield of radioactivity. There is in fact no truly satisfactory control for direct precipitation systems that yield relatively small amounts of radioactivity, e.g. 2% total radioactivity in macromolecular material. Clearly a precipitate formed between a rabbit antibody and mouse immunoglobulin will not necessarily collect the same quantity of non-specific radioactivity (i.e. not mouse immunoglobulin) as one formed between fowl immunoglobulin and a rabbit antibody. This will depend upon the presence or absence of contaminating or unsuspected heterophile type antibodies in the reagents. In this respect, the indirect precipitation system has a major advantage, the non-specific control can be similar to the specific system simply by using normal rabbit immunoglobulin with the same second reagent, e.g. goat anti-rabbit immunoglobulin. Even this, however, is not perfect since the unsuspected specificities in the rabbit antibody and normal immunoglobulin samples can be different. Indeed the yield of radioactivity from labelled spleen cells using normal rabbit immunoglobulin and goat anti-rabbit immunoglobulin does vary from one rabbit serum to another (Abney and Parkhouse, unpublished work). The best control is to use exactly the same reagents added in the specific system, but with the addition of sufficient purified antigen to block all the combining sites of the rabbit antibody. Finally, and as a result of all the above considerations, the radioactivity precipitated by the specific system should always be characterized by analysis in dissociating gels.

There are two ways to radiolabel cells, externally or internally. In the first procedure[21–23] the external proteins of the cell membrane are labelled with ^{125}I using lactoperoxidase as the catalyst. A major advantage of this procedure is that the label is confined to the cell membrane. Consequently the cells can simply be solubilized and the extract analysed by coprecipitation with appropriate antibodies. The drawback is that a positive identification does not necessarily guarantee endogenous synthesis by the cell. The method is in principle subject to the same problems as immunofluorescence, but does give the molecular characterization when gel analysis is performed.

For direct proof of endogenous synthesis cells must be labelled internally *in vitro* with radioactive amino acids (or sugars). The problem here is that most of the cell is labelled and that the plasma membrane, which carries receptor for antigen, comprises a small percentage of the total cell protein, about 5% of pig lymphocytes[42,43] or mouse lymphocytes (Abney and Parkhouse, unpublished work). If we assume the receptor for antigen accounts for 5% at most of the cell membrane, then the theoretical yield from internally labelled cells must be very low. In this situation it would appear futile to solubilize labelled cells and directly look for receptors by coprecipitation. Nevertheless, this has been attempted and with some success.[44,45] In order to distinguish between that fraction of total cellular labelled immunoglobulin on the membrane as opposed to that contained within the cell, the following strategy was adopted. Labelled spleen cells were incubated with or without anti-immunoglobulin, lysed with detergent, centrifuged and then immunoglobulin was estimated by coprecipitation. The basis for this is that surface immunoglobulin will be cross-linked on the cells treated with anti-immunoglobulin and in consequence will sediment when the lysate is centrifuged. This will not occur in the control cell sample and so in principle the difference between the two determinations yields the fraction of radioactive immunoglobulin which is surface associated. The remaining intracellular pool of immunoglobulin will be composed of material destined for the cell membrane and that which will be actively secreted—most lymphocyte suspensions contain some high-rate secreting cells. Apart from the fact that non-specific precipitation is particularly high in this system, rigorous attention to controls is essential. A particular source of error is that the anti-immunoglobulin bound to the cell surface is rarely saturated and, unless special precautions are followed, will therefore combine with intracellular immunoglobulin once the cell is lysed.[45] The failure to saturate both combining sites of anti-immunoglobulin absorbed to the cell suspension is not only possible as a result of only one of the two Fab portions interacting with surface immunoglobulin. It is also possible for added anti-immunoglobulin to interact with lymphocytes and macrophages via Fc receptors, in which case both combining sites of the antibody would be totally free to interact with intracellular immunoglobulin released by lysis. In view of these problems, the claim that mitogen-activated B lymphocytes display 10^2–10^3 times more surface immunoglobulin than the unstimulated cell,[44] should be considered as an exaggerated statement. As an alternative to using whole lysates of internally labelled cells, it is possible to prepare the plasma membrane fraction from the cells and then use this as the starting material for coprecipitation.[46] The obvious disadvantage of this procedure is that it takes a long time, although its merit is that a great deal of the 'noise' is thrown away prior to the coprecipitation step.

3. ISOTYPES, ALLOTYPES AND IDIOTYPES

The structural basis of antibody heterogeneity is at the broadest level, the consequence of different heavy and light chain isotypes (classes and subclasses), each of which is specified by a structural gene for the constant region of the polypeptide chain. In many cases there are allelic alternatives (allotypes) at these constant gene loci but in an allotypically heterozygous animal, although both allotypes are found in the serum, individual plasma cells synthesize only one of the allelic alternatives (allelic exclusion).

Isotype and allotype heterogeneity do not, however, explain the enormous repertoire of antibody combining site specificities that can be expressed by an animal. The antibody combining site is formed by the juxtaposition of the variable regions of heavy and light chains, and therefore the uniqueness of a homogeneous antibody is the consequence of its variable region amino acid sequences. It is these unique variable region sequences that define the idiotype. The clonal theory predicts that a precursor lymphocyte will bear receptors of one idiotype, and that its clonal progeny, derived by antigenic selection, will secrete antibodies of the same idiotype. Normally, when an animal responds to an immunogen an enormous number of idiotypes with specificity to the antigen appear in the serum, i.e. many precursor lymphocytes bearing a range of idiotypes are selected by antigen. If specific antibodies could be made to each idiotypic variant, then they could be used as tools to study the genetics of the immune response. For example, what is the pattern of idiotype inheritance. It is possible to raise antibodies specific for the idiotypic determinants of a homogeneous antibody like the phosphorylcholine binding mouse myeloma protein, and in this case the anti-idiotype has proved extremely useful since the normal immune response to phosphorylcholine in BALB/c mice is remarkably homogeneous. Similarly, when immune responses are pauciclonal and the serum antibody is restricted in its number of idiotypes, preparation of specific anti-idiotype antibodies is possible. The major example here is the anti-idiotype antibody with specificity for the antibody formed by mice injected with streptococcal Group A carbohydrate. Very exciting results have come from this system which, with others, will be discussed in detail in later sections.

4. B LYMPHOCYTES

The characteristic property of B lymphocytes in all species examined is their possession of easily detectable surface immunoglobulin. Each lymphocyte presents about 10^5 immunoglobulin molecules on its surface.[18] Questions of interest, therefore, are: what is the evidence that the surface immunoglobulin serves as a receptor for antigen and which classes of immunoglobulin are expressed on B lymphocytes?

That the immunoglobulin on the surface of B lymphocytes acts as receptor for antigen, there is absolutely no doubt. The most direct evidence comes from studies on antigen binding cells (reviewed in Reference 18). In most of these experiments an antigen is usually labelled with ^{125}I, mixed with lymphocytes and following a period of incubation to allow interaction between antigen and cell, the cells are washed and then examined for labelled cells by autoradiography. With certain exceptions, in unimmunized animals the number of antigen-binding cells range about a mean of about 0·05% of cells examined. Very few lymphocytes will bind two unrelated antigens, as would be expected from the clonal selection theory, and the frequency of antigen binding cells increases when animals are immunized. The frequency of 0·05% may seem rather high for one antigen, but this does not represent the number of cells specific for one antigenic determinant, nor the number of cells bearing exactly the same variable region pair for creation of a unique combining site (idiotype). Apart from the fact that most of the antigens used in antigen binding studies have multiple antigenic sites (e.g. bovine serum albumin), the repertoire of antibodies produced in response to a simple hapten is vast. By transferring limiting numbers of hapten (NIP)-primed cells from primed mice and then analysing the isoelectric heterogeneity of antibodies formed in recipients, it is possible to estimate that CBA strain mice can synthesize 3000–16 000 different antibodies to the one hapten.[47] The frequency of cells with one defined idiotypic specificity will therefore fall by 2–3 orders of magnitude, becoming somewhere between 50 and 500 cells per mouse spleen. In fact, the frequency of mouse lymphocytes binding phosphorylcholine, an antigen which elicits an immune response of restricted clonality, is in this range, being about 1 in 100 000.[48] The potential of the system is further expanded by the possibility of one antibody combining site having specificity for more than one antigen.[49] This consideration derives from the description of clonal antibodies which can bind two apparently unrelated determinants (e.g. dinitrophenol and menadione), and there is evidence for the occurrence of similar double specificities in the course of a normal immune response.

Having shown that there are small numbers of cells which will bind antigens in a specific way two questions arise. What structure on the cell membranes is binding the antigen, and what are the antigen-binding cells precursors of those cells which eventually synthesize and secrete the relevant antibody?

The suspicion that the receptor was immunoglobulin was clearly demonstrated by the failure to find antigen-binding cells when the lymphocytes were pretreated with anti-immunoglobulin.[50] The most elegant experiments of this type are with the phosphorylcholine system.[48] Here it was possible to examine individual antigen-binding cells and to show that they all possessed surface immunoglobulin, IgM in fact. In addition, the immunoglobulin receptors for phosphorylcholine on B lymphocytes of BALB/c mice possessed the same idiotype as a phosphorylcholine-binding myeloma protein, thereby demonstrating identical specificity of surface and secreted immunoglobulin as well as the restricted nature of this response in BALB/c mice. Further, when an anti-idiotype (anti-(anti-phosphorylcholine)) was used in conjunction with externally radiolabelled lymphocytes it was found to be reacting with IgM by the stringent criteria of isolation and gel analysis.[51] One note of caution relates to the last experiment, and that is the high yield of immunoglobulin precipitated by the anti-idiotype (about 10% of total).

Additional evidence for identification of the receptor as immunoglobulin came from experiments based on the antibody induced capping of surface immunoglobulin.[26–29] Using red blood cells it was shown that caps of red blood cells and surface immunoglobulin were contiguous.[52] Along similar lines, but with flagellin as the antigen, it was shown that at least 95% of the surface immunoglobulin of antigen-binding cells is drawn into a cap when antigen is bound.[53] To some extent the latter experiment shows that all of the immunoglobulin receptor molecules on the surface of a lymphocyte have the same specificity for antigen. A similar conclusion has been reached more recently using anti-idiotype antisera and chronic lymphatic leukaemia (CLL) cells.[54,55] Here, anti-idiotype raised against a monoclonal IgM from the patient's serum was found to cap all of the immunoglobulin on the CLL cell surface.

Having demonstrated that B lymphocytes can bind antigen via cell surface immunoglobulin, we must now ask whether the antigen-binding cells demonstrated *in vitro* are functional. In other words, are they the precursors of high-rate antibody secreting cells to the antigen they bind? One immediate problem is that the frequency of antigen-binding cells to a given antigen is higher than the frequency of precursor cells, as determined by measuring the number of antibody-secreting clones developed upon exposure of a mouse spleen to antigen.[56] Of course, it is possible that the difference is entirely the result of methodological limitations, but since there is a difference it is comforting to find that there is evidence to suggest that some, at least, of the antigen-binding cells are in fact precursors. The earliest experiments which bear on this issue have been termed 'suicide' experiments,[57,58] although Michael Sela is terminologically correct to say that they should more properly be described as 'murder' experiments. In these experiments the antigen-binding cells are formed using a ^{125}I-labelled antigen of very high specific activity. The cells are washed, transferred into an X-irradiated host, challenged with antigens and an antibody response to the relevant and to control, irrelevant, antigens is measured. If the specific activity of the original radiolabelled antigen is sufficiently high, then the antibody response is specifically abolished, the antigen-binding cell having received a lethal radiation dose from the bound radioactive antigen. Further evidence for the immunocompetence of antigen-binding cells comes from selective depletion or enrichment of precursor cells by passing lymphocytes through columns of antigen bound to inert supports,[59–61] or by separation in a fluorescence activated cell sorter following treatment of the cells with fluorescent labelled antigen.[62]

Some beautifully conceived work with anti-idiotype antibodies is apposite at this point. Since anti-idiotypes specifically react with the variable region of antibodies, frequently at or close to the combining site, could they not mimic the action of antigen? The answer is that they do. What is more, depending on the experimental system they can either inhibit induction of an immune response ((?)tolerance),[63–66] or stimulate production of the appropriate idiotypic antibody.[67,68] Using the streptococcal Group A carbohydrate system, anti-idiotypic antisera were raised (in guinea pigs) with specificity for the anti-Group

A carbohydrate antibody synthesized by A/J mice.[66,67] The guinea pig anti-idiotype was separated into the two major subclasses, IgG_1 and IgG_2, found in guinea pig serum. Whereas the IgG_2 fraction inhibited induction of the idiotype by the antigen,[66] the IgG_1 fraction was able to prime B lymphocytes so that they would respond vigorously to the streptococcal Group A carbohydrate when tested in a cell transfer system.[67] In A/J mice responding to the Group A carbohydrate antigen about 25% of the antibodies react with the anti-idiotype, i.e. the response is highly restricted. Nevertheless, upon priming with the anti-idiotype essentially all of the antibody subsequently synthesized is of the idiotype in question. There is therefore a selective expression of the one idiotype when stimulation is provoked by the relevant anti-idiotype. Even more impressive was when the same experiment was done with C57L mice, a strain in which this particular idiotype is undetectable. Upon priming with the anti-idiotype, however, there was production of a cross-reacting idiotype, now in easily detectable quantities. Eichmann and Rajewsky[67] draw three important conclusions from this work. That their and similar data[68] 'leave hardly any doubt that the precursors of idiotype-secreting B lymphocytes express the idiotype as functional receptor molecules. This is close to formally proving the basic assumption of selective theories of antibody formation'. The next point comes from the anti-idiotype provoked expansion of a previously undetectable clone in C57L mice. They point out that the idiotype produced here is cross-reactive with the A/J mouse product (i.e. it is not exactly the same) and that similar experiments with, for example, CBA mice failed to cause the expression of either a completely or incompletely cross-reacting idiotype. These observations 'support the view that the potential receptor *répertoire* of the immune system is strictly germ line controlled'. The inclusion of the word 'potential' in the sentence offers two alternative interpretations: that somatic diversification is preprogrammed or that antibody diversity is carried in the germ line. Many proponents of the germ line theory will almost certainly take satisfaction from the second interpretation and pay scant attention to the first. Meanwhile those favouring the somatic generation of antibody diversity will point out that there is no final conclusion possible. Nevertheless, what is obvious is that further experimentation of this nature will certainly contribute evidence, perhaps decisive evidence, to allow a final decision on whether antibody diversity is encoded in the germ line or somatically generated. The third conclusion is a practical one: 'in certain instances anti-idiotypic antibody may thus serve as a new type of vaccine'.

At this point it might seem churlish to ask the question whether immunoglobulin is actually synthesized by the B lymphocytes. It is, nevertheless, a question which has been posed and which has been answered in the affirmative. The simplest method has been to remove immunoglobulin by modulation with anti-immunoglobulin and then to record its reappearance after tissue culture (e.g. References 69 and 70). Additional evidence has also been presented which is based on internally labelling lymphocytes *in vitro* with radioactive protein and carbohydrate precursors. Surface immunoglobulin is then either defined by treatment of the cells with anti-immunoglobulin,[44,45] or by isolation from a purified plasma membrane fraction of the labelled cells.[46] The underlying methodology was described in an earlier section. The finding of radiolabelled immunoglobulin in these experiments constitutes unequivocal proof of synthesis by the cells under investigation. These and other studies[71–73] using cells externally labelled with ^{125}I also provide information on the turnover of membrane associated immunoglobulin. Most immunoglobulin released from the surface is found in the medium in association with fragments of plasma membrane, which is why the process has been termed shedding rather than secretion.[72] The analysis of surface immunoglobulin turnover is complicated by heterogeneity of surface immunoglobulin classes, for example the simultaneous presence of IgM and IgD on the same cell.[69,70] In addition there is a heterogeneity of B lymphocytes. Thus one subpopulation consisting of relatively large cells, releases cell surface immunoglobulin with a half-life of 1–3 hours. Another subpopulation, consisting of smaller cells, releases immunoglobulin at an appreciably slower rate, the half-life being 20–28 hours.[73]

The lymphocyte, unlike the plasma cell, does not contain the well developed membranous elements that characterize secreting cells. In plasma cells it is now clear that immunoglobulin destined for secretion is translated on membrane bound polyribosomes.[74] The nascent polypeptide chains are vectorially released from the polyribosomes into the cysternal spaces of the rough endoplasmic reticulum. From there, after their assembly into disulphide linked H_2L_2 structures, they pass to the exterior milieu via the Golgi apparatus. Some biosynthesis data obtained with B lymphocytes[45] are consistent with a similar transport route for membrane immunoglobulin, but the evidence is by no means conclusive. In the same series of experiments it was also possible to biosynthetically radiolabel cell surface immunoglobulin with labelled fucose and galactose, in contradiction to some earlier work which indicated that surface immunoglobulin was devoid of these sugars.[44]

There is therefore no doubt that the receptor for antigen on immunocompetent B lymphocytes is endogenous immunoglobulin, and that a given lymphocyte bears immunoglobulin of only one idiotype. Upon stimulation its clonal product then secretes antibodies of the same idiotype. Before passing to the vexed question of constant region expression on lymphocytes, it is worth mentioning the current argument over whether the receptor immunoglobulin acts as a signal to the cell when it binds antigen. At first sight this might appear to be a very silly question indeed. On the basis of experiments with the polyclonal B lymphocyte activator, lipopolysaccharide, Coutinho and Möller[75] have argued that the signalling site, is not immunoglobulin. The immunoglobulin merely serves to focus the antigen on the cell surface where it then interacts with another membrane protein, which is postulated and for which no concrete evidence exists. This membrane protein then provides the signal which directs the cell into a state of tolerance or immune induction, depending on the number of interacting sites: just right for induction and too much for tolerance. The major arguments against this 'one-signal' model are presented by Cohn and Blomberg in an article which also very fairly summarizes the Coutinho–Möller point of view.[76] The major point is that the one-signal model was designed to account for observations with lipopolysaccharide which is a very specialized system in that it deals with a polyclonal B cell activator, unlike most situations of B lymphocyte induction. There are therefore a number of observations which cannot be accounted for by the Coutinho–Möller model, e.g. maintenance of B cell paralysis to self; breaking of tolerance with material that cross-reacts with the tolerogen. Challenging and thought-provoking though the one-signal hypothesis may be, it more underlines our ignorance of basic mechanisms of immune induction than it explains them. The real truth is that we really do not understand the dual mystery of antigen recognition, the rendering of the target cell to a non-responsive or differentiative pathway.

Having said that, it is time to consider the expression of constant regions on immunocompetent cells. From a number of observations, and bearing in mind some apparent exceptions,[18] it is now clear that allelic exclusion occurs. The most conclusive experiments which demonstrate this are those of Jones, Cebra and Herzenberg.[77] These experiments are particularly instructive since they follow what is certain to be a model protocol which sidesteps problems due to cytophilic antibody. They took lymphocytes from allotypically heterozygous rabbits, stripped off endogenous and non-endogenous immunoglobulin by treatment with proteolytic enzymes *in vitro* and then placed the cells in tissue culture in order to allow for resynthesis of endogenous membrane proteins. Then, with the aid of fluorochrome-coupled specific anti-allotype antibodies they were able to sort the cells, on the basis of their surface immunoglobulin allotype, by using the fluorescence activated cell sorter. The results were quite compelling; lymphocytes of a given allotype went on to make that, and not the alternative allotype. This evidence taken together with that reviewed by Warner,[18] makes a persuasive case for allotypic exclusion at the level of constant genes on the lymphocyte surface.

In considering the constant regions expressed on B lymphocytes it would be as well to first define the questions that will be discussed. For example, does a given lymphocyte and its clonal progeny always synthesize one and the same constant region or can there be a switch from one constant region to another, either in ontogeny or upon an immunogenic challenge? In the first case, stem cells would give rise directly to lymphocytes bearing surface immunoglobulin of only one of the known classes, either IgM or IgG or IgA and so on. These, upon immunogenic induction, would eventually give rise to cells secreting exactly the same isotype represented on the original precursor cell surface. Thus, IgM- and IgG-secreting cells would derive from lymphocytes bearing IgM and IgG respectively. For this type of situation only one V–C gene integration event would be necessary. However, it is now almost certain that a model of this nature is untenable. Somewhere in the development of many individual B lymphocytes a change of C_H gene expression occurs. The major arguments therefore concern the sequence of heavy chain expression in lymphocytes and their progeny, and whether this 'switching' of C_H genes is antigen driven or not. Further, if exposure to antigen is responsible for switching of C_H genes, then at what level of differentiation does it occur? In an antigen independent model,[78] stem cells first develop into lymphocytes bearing surface IgM. Some of these cells are postulated to give rise to lymphocytes bearing IgA via a population with IgG on their membranes. This developmental progression, it is argued, occurs without the necessity for antigenic intervention, and involves two C_H gene switching steps, i.e. from $C\mu$ to $C\gamma$ and from $C\gamma$ to $C\alpha$. When antigen does interact with the cells, they differentiate into high-rate antibody secreting cells expressing the same isotype as the original precursor lymphocyte, e.g. cells with IgG receptors develop into IgG-secreting cells. The model was originally proposed by Cooper, Kincade and Lawton to explain a series of pioneering experiments with chickens.[78] By judicious timing of bursectomy and administration of anti-μ chain serum *in vivo* they were able to render the animals completely agammaglobulinaemic. A suitable delay in these manoeuvres could either cause appearance of IgM, but not IgG or IgA (bursectomy at 16–17 days), or selective suppression of IgA only (neonatal bursectomy). An important point is that once cells are seeded from the bursa to the periphery suppression of IgG and IgA requires a very prolonged regimen of anti-μ chain injections. It is therefore easier to inhibit the development of IgG expression with anti-μ when bursal B lymphocytes, rather than peripheral cells, are the target of the anti-μ chain reagent. Nonetheless, chronic administration of anti-μ chain *in vivo* can prevent the appearance of all antibody, both in mice and chickens,[78,79] and is strong evidence for the occurrence of C_H gene switching in lymphocyte differentiation. Other frequently quoted evidence for switching is the presence within one individual of myeloma protein, which is heterogeneous with respect to isotype, but homogeneous for the variable regions.[80,81] For example, in one case the myeloma protein was shown to consist of IgM, IgG and IgA, but all three isotypes shared the same idiotypic determinants.[81]

Taking the Cooper–Kincade–Lawton model as the starting point for discussion, we shall now pose some questions. In doing this, it should be pointed out that their model, first published in 1971,[82] has probably stimulated more alternative forms than any other attempted description of B cell development (cf. Reference 18). There are no serious objections to the idea of a switch in C_H expression during differentiation and development of lymphocytes. The major alternatives centre on whether the C_H switch is antigen independent or not, and whether the sequence of IgM→IgG→IgA occurs entirely on the lymphocyte surface so that the receptor and eventual secreted product are of the same isotype.

The original postulate of antigen independent switching was based on the bursa-dependence of the switch in chicken, and also the fact that deliberate administration of antigens to chicken embryos failed to influence the development of IgM- and IgG-containing cells in the bursa. On the other hand, unprimed mouse spleen cells, which respond to sheep red blood cells *in vivo* by synthesizing IgM, IgG and IgA antibody, fail to do so if treated with anti-μ.[83–85] Furthermore, exposure of normal spleen cells to anti-μ chain caused a marked suppression of IgG production by the cells when tested in an *in vivo* transfer system.[86] If the appropriate precursor cells were generated in an antigen independent way it is difficult to see how responses to IgG and IgA could be inhibited by anti-μ. An alternative explanation, and one that will be discussed below, is that the precursors of the IgG and IgA secreting cells rarely have IgG and IgA on their surfaces.

Having said that, this is as good as any time to say that the situation regarding C_H representation on the surface of B lymphocytes is thoroughly confused, as perusal of the review by Warner[18] will indicate. There are three very clear reasons for the confusion. First, much of the data comes from immunofluorescent staining and, as pointed out in an earlier section, there are many problems associated with material binding through Fc receptors, or perhaps even being non-specifically absorbed. What is clear, however, is that the number of B lymphocytes recorded positive for IgG and IgA steadily decrease to a very low number (1%–3%) as the experimenter takes more care to avoid these problems.[87–90] Second, and in particular in the mouse, when surface immunoglobulin is characterized by external labelling and immunochemical procedures there is little, if any, IgG or IgA detected.[22,23,91] Third, is the problem of distinguishing between a precursor lymphocyte that has not been in contact with antigen and one that has, and, as a result, has embarked upon an irreversible differentiation pathway leading to the high-rate antibody secreting cell. To pose an example: suppose a lymphocyte with surface IgM interacts with antigen and its clonal progeny secrete IgG, as could be inferred from data discussed above.[78,79,82–85] At some stage of the differentiation process the cells will begin to express IgG, possibly on their surfaces. If so, are these surface IgG molecules to be regarded as receptors for antigen? For the purposes of this article, the answer will be in the negative. Nonetheless, this type of cell would be positively recognized in an immunofluorescence procedure. It is also possible that a cell at this stage of differentiation could be stimulated into an abnormally extensive clonal expansion under certain experimental conditions, for example, transfer into an irradiated host together with antigen, and this would create obvious problems in the design and interpretation of experiments. In this context it is also important to point out that many IgG-secreting mouse plasmacytoma cells can be shown to have surface immunoglobulin,[92] but here the distinction is clear. The surface associated immunoglobulin molecules are not functional receptors for antigenic stimulation, but once again an immunofluorescence assay would record a positive result. Not all the B cells stimulated by antigen go on to become high-rate antibody-secreting cells. Some will be set aside as memory cells and the nature of their receptor immunoglobulin is more a matter for conjecture than informed discussion. There is one beautiful series of experiments which does suggest that the isotype of memory cell surface immunoglobulin corresponds to the eventual secreted product.[93] Memory cells and high-rate antibody-secreting cells from mice primed 1–6 months prior to sacrifice were retained on antigen-coated columns. Assay for memory was done by transfer of the cells to lethally irradiated recipient mice. Pretreatment of the cells with either anti-IgG_1 or anti-IgG_2 before passage down the insoluble antigen column caused a selective 'sneaking through' of presumed precursor cells for IgG_2 and IgG_1 synthesis respectively. Impressive as the experiments are, they were done well in advance of our current appreciation of the problems arising from cytophilic antibody and Fc receptors. They cannot therefore be regarded as in any way conclusive.

At present we are therefore faced with the embarrassment of a total lack of correlation between isotypes expressed systemically and on lymphocytes. Since the number of IgG and IgA bearing lymphocytes is low, it is reasonable to ask whether this minor population of cells has biological significance. It is also a matter of urgency to define the immunoglobulin on the vast majority of B lymphocytes.

In answer to the first question we can say that it is entirely possible that a small population of, for example, IgG bearing cells could be rapidly expanded by cell division in such a way as to account for all of the serum IgG. Indeed, and as mentioned above, there are experiments suggesting that IgG is present on the surface of some, but not necessarily all, precursors of IgG secreting cells.[93,94] With the prevailing paucity of information, however, it is fair to ask if this is the exception rather than the rule. The situation with IgA secreting cell precursors is more confused. These are reported to have or not to have surface IgA, depending on the antiserum used to make the identification.[95,96] A positive identification was made with an anti-IgA allotype antiserum (directed towards the Fd part of the heavy chain), whereas IgA could not be detected when a conventional anti-isotype antiserum was used. For reasons given below, we find the results with the anti-isotype antiserum more convincing, and therefore predict that IgA will not be on the surface of precursor lymphocytes destined to secrete IgA.

If most immunoglobulin on B lymphocytes is not absorbed, then what is it? For some time IgM has been recognized as a major component of B lymphocyte membranes. External radiolabelling procedures showed that surface IgM differs from the secreted variety by being the monomeric, H_2L_2, subunit rather than the fully assembled $(H_2L_2)5$ pentamer.[22,23] In these and other[91] similar studies, neither IgG nor IgA were detected (lower limit of detection about 5% of total cell surface immunoglobulin). A major breakthrough came with the discovery that IgD was, in addition, present on many human peripheral B lymphocytes.[97,98] This was quickly followed by the demonstration of its endogenous synthesis[69,70] and occurrence on many human chronic lymphatic leukaemia cells. Although some cells expressed either IgM or IgD, on many both isotypes were simultaneously present.[99,100] Using chronic lymphatic leukaemia cells and an anti-idiotype raised against their IgM product it was possible to show that where the two isotypes co-exist on the same cell, they share the same idiotype,[54,55] and hence presumably V regions. The simultaneous expression by one cell of two isotypes sharing the same V region has implications for the mechanism of V–C gene integration. Because of the very long life of chronic lymphatic leukaemia cells, it seems probable that in those cells expressing both IgM and IgD the genome contains integrated genes for both heavy chains. To explain this, reiteration of the V gene or a copy-choice[101] mechanism for V–C gene integration is required. These speculations are based on the long life of chronic lymphatic leukaemia cells, and conclusive evidence (e.g. simultaneous transcriptions of mRNA for δ and μ chains) is lacking. Having raised the possible existence of two integrated heavy chain genes in the chronic lymphatic leukaemia cell, however, it does suggest the following hypothetical scheme. At some time in the development of a B-lymphocyte there is integration of a unique V gene, with the constant region genes of all heavy chain isotypes. Provision of multiple copies of the same V gene could be germ line determined or somatically generated by making a series of copies (i.e. copy-choice) at the time of integration. The advantage of simultaneous integration of all C_H genes early in the ontogeny of B lymphocytes is that questions relating to isotype expression would then revolve entirely around differential gene activation and repression; the requirement to account for V–C gene integration events at later stages of differentiation would no longer exist.

The paucity of serum IgD contrasts with its high frequency on B lymphocytes and if a special role for this class of membrane immunoglobulin is to be entertained, then the importance of establishing the existence of a similar molecule in an animal species is obvious. To date, there have been claims for a similar immunoglobulin in the mouse[91,102,103] and suspicions of one in the rabbit.[104]

In the mouse, the candidate for IgD is a disulphide-linked H_2L_2 molecule which can be precipitated from lysates of externally labelled nude or normal mouse B lymphocytes with anti-light chains, but not antisera to any of the known mouse heavy chain isotypes. It is therefore clearly an isotype hitherto undescribed in the mouse. This candidate for IgD resembles the human counterpart in heavy chain size, marked susceptibility to proteolysis, and in its occurrence on lymphocyte surfaces. In the absence of sequence data this new immunoglobulin class in the mouse cannot be formally identified as IgD. However, for ease in presentation, and because of the similarities noted above, we shall henceforward refer to this molecule as IgD. In addition, the probability that sequence studies will confirm this assignment appears to be very high.

In both humans[105] and mice[91,103] IgM precedes IgD in ontogeny, IgD appearing about two weeks after birth in mice and at some time between 3 and 4 months of gestation in the human. Earlier assumptions that IgD precedes IgM in the human were based on comparisons between cord and adult blood[69,70,98] and appear to be incorrect. Interestingly, what little we know of the amino acid sequence of human IgD would suggest that it evolved some time after the gene for μ-chains.[106] Thus IgD appears subsequent to IgM in both evolution and ontogeny.

There is a marked difference between the relative amounts of IgM and IgD in murine spleen lymph nodes and Peyer's patches.[91,102,103,107] By external labelling techniques IgM and IgD were found in approximately equal amounts in the spleen; in lymph nodes and Peyer's patches, IgD accounted for 70% and 90%, respectively, of the total immunoglobulin. It is important to emphasize that in these studies IgM and IgD are the *only* immunoglobulins recovered.

The external labelling procedure can only give the total yield of IgM and IgD, and not their distribution on individual cells. Based on the rarity of immunoglobulins other than IgM and IgD on the surface of mouse B lymphocytes, fluorescent staining was used to indicate that in the mouse, as in the human, there are splenocytes bearing IgM or IgD, or both IgM and IgD.[108] In those experiments IgM was first capped with rhodamine-labelled anti-μ chain, and then subsequent ring staining with fluorescein-labelled anti-Fab in the presence of sodium azide was taken as evidence for the presence of IgD. Recently, an antiserum specific for mouse IgD has been developed,[109] and the above result has been confirmed. Using this antiserum it could be shown that the majority of B lymphocytes in lymph nodes and Peyer's patches were expressing only IgD, although some doubles (IgM and IgD) and an occasional IgM-bearing cell were seen. In the spleen, however, all three categories of cells were present in similar proportions. Inspection of the cells doubly stained for IgM and IgD suggested a considerable variation in the relative intensities of the two fluorescent markers, suggesting a variation in the IgM and IgD ratio from cell to cell.

This, together with the observation of the three cell types (i.e. IgM, IgD and IgM–IgD) and the fact that IgM precedes IgD in ontogeny, suggests a developmental sequence of immunoglobulin expression on B lymphocytes; from IgM to IgD via an intermediate cell type with both isotypes. Thus the B lymphocytes of lymph nodes and Peyer's patches would in the main constitute a more mature population of cells. Of interest in this context is the finding that large, immature splenocytes from mice are enriched for IgM, whereas the smaller splenic lymphocytes, thought to be more mature, are enriched for IgD.[110]

It is an extraordinary fact that IgD was not detected on B lymphocytes earlier. However, it is present in very low amounts in normal serum,[111] and thus its very extensive representation on B cells must imply a specialized role as receptor for antigen. This is easy to say but is not particularly illuminating. At present, we can only guess at possible roles for IgD. As discussed above, there may be a developmental sequence. According to this, immature lymphocytes bearing IgM will develop into cells with surface IgD via an intermediate cell type with both isotypes expressed. Do each of these three cell types exercise different immunological functions? Do all B lymphocytes pass through these three distinguishable stages, or are some arrested or diverted in other directions (e.g. IgG) at any stage of the differentiation pathway? Time, and experiments with the fluorescence activated cell sorter will almost certainly provide an answer. A most intriguing question is whether IgM and IgD give different signals when simultaneously present on one lymphocyte. In many ways this would seem unlikely since they both share the same variable region. However, a key question could be the exact mode of insertion of these two immunoglobulins in the membrane, e.g. whether they are attached to and/or associated with the same or different cell surface components. An important question is exactly when IgD arose in evolution. If this isotype is not found in, for example elasmobranch fishes, then some clues as to the function of IgD might be suggested. The marked susceptibility of IgD to proteolysis[111] could conceivably be a clue. In fact, a model for B lymphocyte stimulation which incorporates proteolytic cleavage of IgD as a crucial event has been suggested.[112]

An interesting observation has recently been made with a line of CBA mice (CBA/N) which have an X-linked defect and associated failure to respond to thymus-independent antigens. It appears that these mice have reduced but detectable levels of IgD,[113] and consequently it is possible that IgD plays a critical role as a receptor for thymus-independent responses. In the experiments reported the immune responses measured were of the IgM variety, and so there is some conceptual difficulty in interpretation. In addition, there are reduced numbers of B lymphocytes in CBA/N mice and so the failure to respond to thymus-independent antigens could be the result of defects other than at the IgD level.

Experiments in the rabbit provide circumstantial evidence for IgD bearing lymphocytes as the precursors of IgA secreting cells. Peyer's patch cells were stripped with a proteolytic enzyme and then cell surface immunoglobulin was allowed to regenerate *in vitro*. The cells were then sorted in the fluorescence cell sorter using fluorescent-labelled anti-μ chain to give μ-negative and μ-positive subpopulations.[95] Upon transfer or stimulation with pokeweed mitogen *in vitro*, the μ-positive cells developed primarily into an IgM secreting population. The μ-negative cells, which contained lymphocytes stained with anti-light chain but not heterologous anti-α chain, contained precursors of IgA secreting cells; it is reasonable to assume that the precursors were the light chain bearing cells. In another publication,[96] it was claimed that surface IgA regenerated upon culture of the cells *in vitro* after enzymic stripping. For identification of IgA an anti-allotype antiserum thought to be specific for the IgA class was used. No functional studies were done with the IgA-positive population. Nevertheless, the authors concluded that precursors of IgA secreting cells have IgA on their surfaces. However, when untreated rabbit Peyer's patch cells were externally labelled with ^{125}I, the immunoglobulin recovered resolved into three components after reduction and alkylation.[104] The major components corresponded to μ and δ chain, and in addition there was a minor fraction which appeared to be α chain. Unfortunately, the presumptive δ chain was not taken into consideration in these experiments. Furthermore, the radiolabelling was conducted on untreated Peyer's patch cells and these are known to bear cytophilic IgA. Two controls are suggested. First the specificity of the anti-allotype antiserum should be checked by the stringent criteria of external labelling and subsequent immunochemical analysis; the cells for labelling should of course be first stripped with enzyme and then allowed to regenerate cell surface components *in vitro*. Second, the same population of cells should be fractionated with appropriate anti-immunoglobulin reagents and the subpopulations submitted to functional testing. We predict that when this is done the majority of IgA precursor cells will be found to express IgD on their surfaces.

Finally, any model of B lymphocyte differentiation must be superimposed upon the major two surface immunoglobulins, IgM and IgD. Whether or not IgG secreting cells derive from a small population of IgG bearing lymphocytes, their origin is almost certainly a cell with IgM, IgD or both. Furthermore, expression of surface IgD is independent of thymic[91,103] or antigenic[103] influence.

Is there anything that distinguishes membrane-bound from secreted immunoglobulin? At first sight there are two clear differences. First, surface IgM is assembled as far as the H_2L_2 monomer, whereas secreted IgM is polymeric, consisting of five disulphide linked monomer subunits. Second, a large proportion of lymphocyte associated immunoglobulin is IgD, an isotype which is a very minor component of serum. These differences do not, however, immediately afford an understanding of the mode of membrane insertion. Thus monomeric IgM can be secreted in certain situations[114,115] and so this particular subunit structure does not necessarily guarantee insertion into the membrane. In addition, and although the comparison cannot be made for IgD, the known sequence of IgM[116] does not suggest a C-terminal region of marked hydrophobicity. As noted above, there is confusion over whether surface-associated and secreted IgM differ in their relative amounts of carbohydrate.[44,45] However, while small carbohydrate differences cannot be ruled out, it does seem that surface immunoglobulin does possess some, if not all, of the fucose and galactose incorporated into secreted IgM. Similarly, other small structural differences between surface-associated and secreted immunoglobulin cannot be dismissed. One possible line of enquiry would be to ask whether membrane bound immunoglobulin heavy chains do or do not have the extra N-terminal sequence of about twenty amino acids found in the intracellular precursors of secreted immunoglobulins.[117] It is not inconceivable that this sequence, normally cleaved from immunoglobulin polypeptide chains prior to secretion, could be a feature of surface immunoglobulins. The actual integration of immunoglobulins into the membrane structure is another intriguing unknown, the solution of which could well have implications for the mechanism of antigenic stimulation.

5. T LYMPHOCYTES

In one year's time the controversy over the molecular nature of the T cell receptor for antigen will almost certainly be solved by extension of recent studies using anti-idiotype antisera. There is therefore no useful purpose in a detailed review of past errors and successes in this extraordinarily murky area. Instead, this section will be brief and, as a result, incomplete. For a voluminous account of the many approaches by which attempts have been made—some apparently successful and others contradictory—to demonstrate immunoglobulin on T cells, the reader is referred to Warner's excellent review.[18] At the outset we should like to draw attention to the fact that for years Simonsen has maintained consistent scepticism towards the idea that conventional immunoglobulin is the T-cell antigen receptor.[118,119] There is now increasing evidence to justify his scepticism.

To start with the least contentious points: it can be asserted that T lymphocytes do not generally express high density immunoglobulin determinants on their membranes; they do not stain when treated with fluorochrome coupled anti-immunoglobulin. Furthermore, in most systems where immunoglobulin has been demonstrated on the surface of T cells, when appropriate experiments have been carried out the immunoglobulin can be shown to be exogenously derived. For example, chickens deprived of B lymphocytes by bursectomy and anti-μ chain treatment do not form rosettes with sheep red blood cells unless previously injected with an anti-sheep red blood cell serum obtained from a normal, immunized bird.[120] Similarly, the low density labelling of chicken T cells observed using radiolabelled anti-chicken light chain disappears when the source of serum antibody, the B cells, is removed by bursectomy.[121] Therefore in chickens there is excellent evidence for an absence of endogenous immunoglobulin on the surface of T lymphocytes. In mammals too there is good evidence for T cells with exogenously derived immunoglobulin. Thus using allotypic markers it has been convincingly demonstrated that T cells in the thoracic duct of mice[122] and rats[123] passively acquire immunoglobulin. From these experiments it appears that activated T cells acquire more surface immunoglobulin than resting T cells, although in both situations it seems likely that the uptake is due to Fc receptors.[35–40]

Of the many reports that anti-immunoglobulin can inhibit various manifestations of cell-mediated immunity, there are certainly an equivalent number of contradictory findings.[18,118] We have discussed at length in an earlier section the sorts of unwanted specificities that can occur in poorly characterized anti-immunoglobulin antisera and agree with an explanation offered by Simonsen:[118] namely that positive findings are the result of contaminating antibodies to cell surface structures other than immunoglobulin constant regions.

An enormous amount of work has been devoted to antigen-binding T lymphocytes,[18] and in many cases it has been inferred (through the use of anti-immunoglobulin) that antigen is bound to T lymphocytes via immunoglobulin. However, as Simonsen points out,[118] control absorptions of the test antiserum with T lymphocytes are never done. More recent studies using the synthetic copolymer (T, G)-A-L indicate major differences in the specificity and nature of antigen-binding by T and B lymphocytes.[124–126] The differences are in temperature requirements, metabolic state, susceptibility to anti-histocompatibility antisera and specificity. Antigen-binding to T cells occurs more efficiently at 37 °C than at 4 °C, so that more antigen-binding cells are observed at the higher temperature. Consequently more or less antigen-binding T cells can be obtained simply by shifting the temperature up or down. The reversible nature of this phenomenon could be inhibited by addition of a metabolic inhibitor (sodium azide). Thus if T cells were incubated at 37 °C and then shifted to 4 °C in the presence of azide, the frequency of antigen-binding cells was the same as that observed at 37 °C. With B cells, on the other hand, equal numbers of antigen-binding cells are found at either temperature. Another difference was that pretreatment with antisera against the K or D locus of H-2b abolished the formation of T, but not B, antigen-binding cells. Finally, the specificity of antigen-binding was quite different for T and B cells. Whereas binding of (T, G)-A–L to B cells was inhibited by an excess of (H, G)-A–L or (Phe, G)-A–L, this was not the case for T cells. The simplest conclusion from this series of experiments demonstrating marked differences in the characteristics of antigen binding by B and T lymphocytes is that the two cell lines bear different structures for antigen recognition. This is the same as saying that immunoglobulin is not the antigen receptor on T cells. Those who wish to quarrel with the conclusion can assert that there is immunoglobulin, and argue that the observed differences in antigen-binding characteristics are a consequence of differences in membrane organization between B and T cells.

The most elegant experiments designed to demonstrate the immunoglobulin nature of the receptor on antigen-binding T cells are those of Roelants and his colleagues.[127,128] Basically they showed that pretreatment of T lymphocytes with anti-immunoglobulin caused capping of the radiolabelled antigen subsequently bound to T cells. To be sure that they were not dealing with an artefact resulting from cytophilic antibody they used unimmunized mice, and incubated the cells *in vitro* after the treatment with anti-immunoglobulin. Under these conditions the receptor for antigen would be lost by internalization and, or, shedding as a result of the anti-immunoglobulin-induced capping. After a suitable time *in vitro* endogenous receptors would be synthesized and reinserted into the cell membrane. The prediction then is that at short times (1 hour) after treatment with anti-immunoglobulin there would be no antigen-binding T cells, and at long times (6–18 hours) there would be. In fact, the observations were in agreement with the prediction but the experiments are nonetheless subject to the criticisms raised above concerning the specificity and use of antisera. It is exactly this type of situation that demands biochemical characterization of the material on the cell surface reacting with the antiserum. Furthermore, the experiments were with unfractionated populations of lymphocytes (i.e. containing B cells) and so an artefact due to the contaminating B lymphocytes cannot be strictly ruled out.

By far and away the most compelling evidence for immunoglobulins on T cells appears to come from external labelling studies. Using the lactoperoxidase-catalysed procedure for radiolabelling cell surfaces, immunoglobulin has been recovered in similar[22] or even greater[129] yields from thymocytes as B lymphocytes. The former group[22] has extended their observation to other sources of T cells[130–132] with similar findings, but others have failed to confirm their observation with thymocyte or peripheral T lymphocytes.[133–141] We shall not discuss here the presence or absence of immunoglobulin on T cell lymphomas[132,142] since this type of cell is malignant and therefore not necessarily a precise model for an immunocompetent T lymphocyte.

This situation is disturbing and almost without precedent. Here we are dealing with precise biochemical techniques including gel analysis of the labelled cell surface material reacting with anti-immunoglobulin. On the one hand, it is claimed that thymocytes and T cells have immunoglobulin which is a disulphide-linked 7S 'IgM-like' molecule. What is more, in these studies the yield of 7S 'IgM-like' molecule is comparable when either normal thymocytes or splenocytes from nude mice are labelled.[22] On the other hand, however, using similar techniques this observation cannot be reproduced by a number of different investigators. The only possible conclusion that can be drawn is that one of the two groups of protagonists in the controversy is wrong. We take the view that there is not a disulphide-linked 7S 'IgM-like molecule' on thymocytes and T lymphocytes. This point of view has been arrived at for a number of reasons, other than the irreproducibility of the work.

In the first place, in none of the experiments where 7S 'IgM-like' immunoglobulin has been isolated from radiolabelled T lymphocytes have three crucial controls been done.

The first is to see if the anti-immunoglobulin does or does not precipitate the putative T cell receptor when cold immunoglobulin is added to block the combining sites of the first antibody used in an indirect coprecipitation test. As discussed in detail in an earlier section, this is the most exact control for non-specific precipitation. The second is to absorb the anti-immunoglobulin with thymocytes prior to the coprecipitation step. The third important control is to analyse the radioactive immune coprecipitates on dissociating gels without a prior chemical reduction. Any radioactive material then appearing in the 7S region of the gel should then be eluted, reduced and submitted for a further gel analysis to show that the radioactivity now migrates in the predicted positions of heavy and light chains. We predict that the 7S 'IgM-like molecule' on T lymphocytes will not stand up to examination by these controls. Of particular relevance here, is the finding that anti-mouse immunoglobulins antisera can contain antibodies reactive with a component of the thymocyte surface which co-electrophoreses with 7S IgM in gel electrophoresis.[143] This high molecular weight material, however, is certainly not immunoglobulin since it does not dissociate into heavy and light chains upon chemical reduction. Furthermore, it is not recognized by specific anti-immunoglobulin reagents.

Another reason for doubting that immunoglobulin is the receptor for antigen on T lymphocyte comes from recent work on T cell helper[144] and suppressor[145] factors. Both types of molecules have been shown to be biologically active and antigen specific. They react with anti-Ia antisera but not with anti-immunoglobulin reagents.

The final pertinent point is that the molecules on T lymphocytes recognized by anti-idiotypic reagents (see below) have been characterized and are clearly not 7S 'IgM-like'.[146] Instead, and without chemical reduction, the major material appears to have a molecular weight of about 35 000, although a 50 000 molecular weight component was also present.

Now we come to the recent excitement generated by the use of anti-idiotype antisera for the characterization of antigen receptors. The conclusion that will be drawn is that T and B lymphocytes with specificity towards the same antigen have recognition structures that share the same idiotypic determinants. However, such indications as there are, suggest that on T cells these idiotypic determinants are not associated with conventional immunoglobulin molecules. These conclusions come from two independent lines of study. One using classical anti-idiotype antisera in the mouse[67] and the other with an anti-idiotype system provided by the rat.[146]

The simplest system to describe is in the mouse, and this has been considered earlier in relation to stimulation of B lymphocytes with anti-idiotypic antibodies. In these experiments the crucial reagent was the IgG_1 fraction of a guinea pig anti-idiotype directed towards a monoclonal anti-streptococcal Group A carbohydrate antibody formed in the A/J strain of mice. The critical observation was that mice injected with small amounts of this reagent were subsequently found to have generated 'helper' T cells with specificity for the Group A carbohydrate. Since B cells were also sensitized to Group A carbohydrate with the anti-idiotype, the obvious conclusion is that in this restricted system both B and T lymphocytes recognize the streptococcal antigen via structures displaying the antigenic determinants defined by the anti-idiotype.

In the rat system the anti-idiotypic antibodies were raised against antibodies to the major histocompatibility Ag-B determinants. The principle was based on earlier observations in mice[147] and rats,[148] and this will be presented in a general, rather than a specific, way. Assume an F_1 hybrid is derived from the two parental strains A and B. Within the F_1 hybrid there will be mutual tolerance to histocompatibility antigens of both parental strains. Or, to say it the other way round, cells of either parental type with immunological reactivity towards the other half represented in the F_1 hybrid will be deleted. Thus, for example, A strain cells capable of recognizing B strain cells will be absent. If we now make the reasonable assumption that recognition takes place through surface located receptors, then receptor molecules of this specificity must also be absent. Therefore, should such receptors be introduced into the F_1 hybrid, they should be immediately recognized as foreign and, accordingly, an immune response towards them should result. However, the foreign antigenic determinants in this situation can only be related to that part of the molecule which combines with the antigen (i.e. the idiotypic determinants), and thus the humoral immune response must be the production of anti-idiotype antibody. To pose a simple example: the general structure of B lymphocyte antigen receptors is the same, it is immunoglobulin. Thus the major structural disparity between the various receptor immunoglobulins of different clones is in the idiotypic determinants. In principle, therefore, it should be possible to use F_1 hybrid animals to make anti-idiotype reagents with specificity for recognition structures expressed in either parental strain, but not the F_1 hybrid, i.e. anti-(receptor of those strain A lymphocytes which recognize B strain lymphocytes); anti-(receptor of those strain B lymphocytes which recognize strain A lymphocytes). It should also be possible to design the immunization schedules so that the anti-idiotypes are directed either towards the immunoglobulin receptor of the appropriate B lymphocytes or the recognition unit expressed by T lymphocytes of the chosen specificity. Having achieved this, then the anti-idiotypic reagents can be used as probes for antigen receptors of B and T lymphocytes.

Anti-idiotype directed towards the T cell receptor was made by injecting parental peripheral T lymphocytes (A) into F_1 hybrid animals (A.B). The resulting antiserum (which should be anti-(A anti-B)) bound to the surface of T lymphocytes of the appropriate parental strain (A), but not to cells from the other half of the F_1(B) or appropriately chosen third party controls. Moreover, the anti-idiotype would also specifically bind to immunoglobulin carrying the relevant idiotype: made, for example, by injecting B strain cells into A strain animals and purifying the alloantibody, i.e. A anti-B. Thus receptors towards the same antigenic determinants, whether expressed by T or B cells, share the same idiotypes. Using this type of anti-idiotype it was possible to inhibit graft versus host reactions or mixed lymphocyte responses, in an appropriately specific way, and thereby to deduce a functional receptor role for the idiotype-bearing T lymphocytes. It was also possible to directly visualize those T and B lymphocytes bearing the idiotypic determinants by immunofluorescence. In all of the studies appropriate specificity controls were included. When used in conjunction with radiolabelled T lymphocytes the anti-idiotype appeared to recognize a single chain molecule with a molecular weight of 35 000; some material with a molecular weight of 50 000 was also recovered. With radioactive B lymphocytes, the indications were that immunoglobulin was selected by the anti-idiotype.

There are therefore two different lines of evidence to suggest that both B and T lymphocytes share the same idiotypic determinants when they are directed towards the same antigen. In one case the anti-idiotype was prepared against serum antibody, and in the other it was raised against the T cell receptor for antigen. Preliminary characterization of the material on T lymphocytes suggests that it is not conventional immunoglobulin. A common interpretation of the data is that the T cell antigen receptor may consist of an immunoglobulin variable region gene integrated with a polypeptide coded for by a gene (or genes) in the major histocompatibility locus. The problem here is that there is no known linkage between genes specifying immunoglobulin and the histocompatibility locus. This could mean a subset of V genes exclusively for T cells. However, any speculation on this point is of very little value since it is quite clear that most of these problems will shortly be better understood by experimentation in the types of systems discussed above.

Acknowledgements

E. R. Abney thanks the Consejo Nacional de Ciencia y Tecnologia de Mexico for financial support.

REFERENCES

1. Gowans, J. L. and McGregor, D. D. (1965). *Progr. Allergy*, **9**, 1–78.
2. Jerne, N. K. (1955). *Proc. Nat. Acad. Sci.*, **41**, 849–857.
3. Talmage, D. W. (1957). *Annu. Rev. Med.*, **8**, 239–256.
4. Burnet, F. M. (1959). *The clonal selection theory of acquired immunity*. Vanderbilt University, Nashville, Tennessee.
5. Warner, N. L. and Szenberg, A. (1962). *Nature*, **196**, 784–785.
6. Cooper, M. D., Peterson, R. D. A., South, M. A. and Good, R. A. (1966). *J. exp. Med.*, **123**, 75–102.
7. Miller, J. F. A. P. (1961). *Lancet*, **2**, 748–749.
8. Owen, J. J. T., Raff, M. C. and Cooper, M. D. (1975). *Eur. J. Immunol.*, **5**, 468–473.
9. Miller, J. F. A. P. and Mitchell, G. F. (1969). *Transplant Rev.*, **1**, 3–42.
10. Claman, H. N., Chaperon, E. A. and Triplett, R. F. (1966). *Proc. Soc. exp. Biol. Med.*, **122**, 1167–1171.
11. Davies, A. J. S., Leuchars, E., Wallis, V. and Koller, P. C. (1966). *Transplantation*, **4**, 438–451.
12. Davies, A. J. S., Leuchars, E., Wallis, V., Marchant, R. and Elliot, E. V. (1967). *Transplantation*, **5**, 222–231.
13. Davies, A. J. S. (1969). *Transplant. Rev.*, **1**, 43–91.
14. Raff, M. C. (1971). *Transplant. Rev.*, **6**, 52–80.
15. Mitchison, N. A., Rajewsky, K. and Taylor, R. B. (1970). In *Developmental Aspects of Antibody Formation and Structure*, Vol. 2 (Sterzl, J. and Riha, I., Eds.), Academic, New York, pp. 547–562.
16. Gershon, R. K. (1975). *Transplant. Rev.*, **26**, 170–185.
17. Tada, T., Taniguchi, M. and Takemori, T. (1975). *Transplant. Rev.*, **26**, 106–129.
18. Warner, N. L. (1974). *Advan. Immunol.*, **19**, 67–216.
19. McCormick, J. N., Nelson, D., Tunstall, A. M. and James, K. (1973). *Nature* (*New Biol.*), **246**, 78–81.
20. Peterson, P. A., Cunningham, B. A., Berggard, I. and Edelman, G. M. (1972). *Proc. Nat. Acad. Sci.*, **69**, 1697–1701.
21. Philips, D. R. and Morrison, M. (1970). *Biochem. Biophys. Res. Commun.*, **40**, 284–289.
22. Marchalonis, J. J., Cone, R. E. and Atwell, J. L. (1972). *J. exp. Med.*, **135**, 956–971.
23. Vitetta, E. S., Baur, S. and Uhr, J. W. (1971). *J. exp. Med.*, **134**, 242–264.
24. Möller, G. (1961). *J. exp. Med.*, **114**, 415–434.
25. Raff, M. C. (1970). *Immunology*, **19**, 637–650.
26. Taylor, R. B., Duffus, W. P. H., Raff, M. C. and de Petris, S. (1971). *Nature* (*New Biol.*), **233**, 225–229.
27. de Petris, S. and Raff, M. C. (1972). *Eur. J. Immunol.*, **2**, 523–535.
28. Karnovsky, M. J., Unanue, E. R. and Leventhal, M. (1972). *J. exp. Med.*, **136**, 907–930.
29. Loor, F., Forni, L. and Pernis, B. (1972). *Eur. J. Immunol.*, **2**, 203–212.
30. Frye, L. D. and Edidin, M. (1970). *J. Cell Sci.*, **7**, 319–335.
31. Singer, S. J. and Nicolson, G. L. (1972). *Science*, **175**, 720–731.
32. Basten, A., Miller, J. F. A. P., Warner, N. L. and Pye, J. (1971). *Nature* (*New Biol.*), **231**, 104–106.
33. Dickler, H. B. and Kunkel, H. G. (1972). *J. exp. Med.*, **136**, 191–196.
34. Grey, H. M., Kubo, R. T. and Cerottini, J.-C. (1972). *J. exp. Med.*, **136**, 1323–1328.
35. Anderson, C. L. and Grey, H. M. (1974). *J. exp. Med.*, **139**, 1175–1188.
36. Basten, A., Miller, J. F. A. P. and Abraham, R. (1975). *J. exp. Med.*, **141**, 547–560.

37. Basten, A., Miller, J. F. A. P., Warner, N. L., Abraham, R., Chia, E. and Gamble, J. (1975). *J. Immunol.*, **115**, 1159–1165.
38. Gyöngyössy, M. I. C., Arnaiz-Villena, A., Soteriades-Vlachos, C. and Playfair, J. H. L. (1975). *Clin. Exp. Immunol.*, **19**, 485–497.
39. Santana, V. and Turk, J. L. (1975). *Immunology*, **28**, 1173–1178.
40. Stout, R. D. and Herzenberg, L. A. (1975). *J. exp. Med.*, **142**, 1041–1051.
41. Greaves, M. F. (1970). *Transplant. Rev.*, **5**, 45–75.
42. Allan, D. and Crumpton, M. J. (1971). *Biochem. J.*, **123**, 967–975.
43. Crumpton, M. J. and Snary, D. (1974). *Contemporary Topics in Molecular Immunology*, **3**, 27–56.
44. Melchers, F. and Andersson, J. (1973). *Transplant. Rev.*, **14**, 76–130.
45. Vitetta, E. S. and Uhr, J. W. (1974). *J. exp. Med.*, **139**, 1599–1620.
46. Parkhouse, R. M. E. and Abney, E. R. (1974). *Proc. 9th FEBS Meetings*, pp. 151–156.
47. Kreth, H. W. and Williamson, A. R. (1973). *Eur. J. Immunol.*, **3**, 141–146.
48. Claflin, J. L., Lieberman, R. and Davie, J. M. (1974). *J. exp. Med.*, **139**, 58–73.
49. Richards, F. F., Konigsberg, W. H., Rosenstein, R. W. and Varga, J. M. (1975). *Science*, **187**, 130–137.
50. Warner, N. L., Byrt, P. and Ada, G. L. (1970). *Nature*, **226**, 942–943.
51. Strayer, D. S., Vitetta, E. S. and Kohler, H. (1975). *J. Immunol.*, **114**, 722–727.
52. Ashman, R. F. (1973). *J. Immunol.*, **111**, 212–220.
53. Raff, M. C., Feldman, M. and de Petris, S. (1973). *J. exp. Med.*, **137**, 1024–1030.
54. Salsano, P., Froland, S. S., Natvig, J. B. and Michaelsen, T. E. (1974). *Scand. J. Immunol.*, **3**, 841–846.
55. Fu, S. M., Winchester, R. J. and Kunkel, H. G. (1975). *J. Immunol.*, **114**, 250–252.
56. Klinman, N. R. and Press, J. L. (1975). *Transplant. Rev.*, **24**, 47–50.
57. Ada, G. L. and Byrt, P. (1969). *Nature*, **222**, 1291–1292.
58. Unanue, E. R. (1971). *J. Immunol.*, **107**, 1663–1665.
59. Truffa-Bachi, P. and Wofsy, L. (1970). *Proc. Nat. Acad. Sci.*, **66**, 685–692.
60. Wigzell, H. and Mäkelä, O. (1970). *J. exp. Med.*, **132**, 110–126.
61. Davie, J. M. and Paul, W. E. (1971). *J. exp. Med.*, **134**, 495–516.
62. Julius, M. H., Masuda, T. and Herzenberg, L. A. (1972). *Proc. Nat. Acad. Sci.*, **69**, 1934–1938.
63. Cosenza, H. and Köhler, H. (1972). *Proc. Nat. Acad. Sci.*, **69**, 2701–2705.
64. Hang, D. A., Wang, A. L., Pawlak, A. L. and Nisonoff, A. (1972). *J. exp. Med.*, **135**, 1293–1300.
65. Strayer, D. S., Cosenza, H., Lee, W. H. E., Rowley, D. A. and Köhler, H. (1974). *Science*, **186**, 640–643.
66. Eichmann, K. (1974). *Eur. J. Immunol.*, **4**, 296–302.
67. Eichmann, K. and Rajewsky, K. (1975). *Eur. J. Immunol.*, **5**, 661–666.
68. Trenkner, E. and Riblet, R. (1975). *J. exp. Med.*, **142**, 1121–1132.
69. Knapp, W., Bolhuis, R. L. H., Radl, J. and Hijmans, W. (1975). *J. Immunol.*, **111**, 1295–1298.
70. Rowe, D. S., Hug, K., Forni, L. and Pernis, B. (1973). *J. exp. Med.*, **138**, 965–972.
71. Cone, R. E., Marchalonis, J. J. and Rolley, R. T. (1971). *J. exp. Med.*, **134**, 1373–1384.
72. Vitetta, E. S. and Uhr, J. W. (1972). *J. exp. Med.*, **136**, 676–696.
73. Melchers, F., von Boehmer, H. and Philips, R. A. (1975). *Transplant. Rev.*, **25**, 26–58.
74. Bevan, M. J., Parkhouse, R. M. E., Williamson, A. R. and Askonas, B. A. (1972). *Progr. Biophys. and Molec. Biol.*, **25**, 131–162.
75. Coutinho, A. and Möller, G. (1974). *Scand. J. Immunol.*, **3**, 133–146.
76. Cohn, M. and Blomberg, B. (1975). *Scand. J. Immunol.*, **4**, 1–24.
77. Jones, P. P., Cebra, J. J. and Herzenberg, L. A. (1974). *J. exp. Med.*, **139**, 581–599.
78. Lawton, A. R., Kincade, P. W. and Cooper, M. D. (1975). *Fed. Proc.*, **34**, 33–39.
79. Manning, D. D. and Jutila, J. W. (1972). *J. exp. Med.*, **135**, 1316–1333.
80. Wang, A. C., Wilson, S. K., Hupper, J. E., Fudenberg, H. H. and Nisonoff, A. (1970). *Proc. Nat. Acad. Sci.*, **66**, 337–343.
81. Grubb, A. O. and Zettervall, O. H. (1975). *Proc. Nat. Acad. Sci.*, **72**, 4115–4118.
82. Cooper, M. D., Kincade, P. W. and Lawton, A. R. (1971). In *Immunologic Incompetence* (Kagan, B. M. and Stiehm, E. R., Eds.), Yearbook Publ., Chicago, pp. 81–104.
83. Pierce, C. W., Solliday, S. M. and Asofsky, R. (1972). *J. exp. Med.*, **135**, 675–697.
84. Pierce, C. W., Solliday, S. M. and Asofsky, R. (1972). *J. exp. Med.*, **135**, 698–710.
85. Pierce, C. W., Asofsky, R. and Solliday, S. (1973). *Fed. Proc.*, **32**, 41–43.

86. Herrod, H. G. and Warner, N. L. (1972). *J. Immunol.*, **108**, 1712–1717.
87. Jones, P. P., Cebra, J. J. and Herzenberg, L. A. (1973). *J. Immunol.*, **111**, 1334–1348.
88. Kurnick, J. T. and Grey, H. M. (1975). *J. Immunol.*, **115**, 305–307.
89. Lobo, P. I., Westervelt, F. B. and Horwitz, D. A. (1975). *J. Immunol.*, **114**, 116–119.
90. Winchester, R. J., Hoffman, S. M., Fu, T. and Kunkel, H. G. (1975). *J. Immunol.*, **114**, 1210–1212.
91. Abney, E. R. and Parkhouse, R. M. E. (1974). *Nature*, **252**, 600–602.
92. Preud'homme, J.-L. and Scharff, D. (1974). *J. Immunol.*, **113**, 702–704.
93. Walters, C. S. and Wigzell, H. (1970). *J. exp. Med.*, **132**, 1233–1249.
94. Okumura, K., Julius, M. H., Tsu, T., Herzenberg, L. A. and Herzenberg, L. A. (1976). Eur. J. *Immunol.*, **6**, 467–472.
95. Jones, P. P., Craig, J. W., Cebra, J. J. and Herzenberg, L. A. (1974). *J. exp. Med.*, **140**, 452–469.
96. Jones, P. P. and Cebra, J. J. (1974). *J. exp. Med.*, **140**, 966–976.
97. Van Boxel, J. A., Paul, W. E., Terry, W. D. and Green, I. (1972). *J. Immunol.*, **109**, 648–651.
98. Rowe, D. S., Hug, K., Faulk, W. P., McCormick, J. N. and Gerber, M. (1973). *Nature* (*New Biol.*), **242**, 155–157.
99. Fu, S. M., Winchester, R. J. and Kunkel, H. G. (1974). *J. exp. Med.*, **139**, 451–456.
100. Kubo, R. T., Grey, H. M. and Pirofsky, B. (1974). *J. Immunol.*, **112**, 1952–1954.
101. Williamson, A. R. (1971). *Nature*, **231**, 359–362.
102. Melcher, U., Vitetta, E. S., McWilliams, M., Lamm, M. E., Philips-Quagliata, J. M. and Uhr, J. W. (1974). *J. exp. Med.*, **140**, 1427–1431.
103. Vitetta, E. S., Melcher, U., McWilliams, M., Lamm, M. E., Philips-Quagliata, J. M. and Uhr, J. W. (1975). *J. exp. Med.*, **141**, 206–215.
104. Craig, S. W. and Cebra, J. J. (1975). *J. Immunol.*, **114**, 492–502.
105. Vossen, J. M. and Hijmans, W. (1975). *Ann. N.Y. Acad. Sci.*, **254**, 262–279.
106. Spiegelberg, H. L. (1975). *Nature*, **254**, 723–725.
107. Vitetta, E. S., McWilliams, M., Philips-Quagliata, J. M., Lamm, M. E. and Uhr, J. W. (1975). *J. Immunol.*, **115**, 603–605.
108. Parkhouse, R. M. E., Hunter, I. R. and Abney, E. R. (1976). *Immunology*, **30**, 409–412.
109. Abney, E. R., Hunter, I. R. and Parkhouse, R. M. E. (1976). *Nature*, **259**, 404–406.
110. Goodman, S. A., Vitetta, E. S., Melcher, U. and Uhr, J. W. (1975). *J. Immunol.*, **114**, 1646–1648.
111. Spiegelberg, H. L. (1972). Contemporary Topics in Immunochemistry, Vol. 1 (Inman, F. P., Ed.), Plenum Press, New York, pp. 165–180.
112. Vitetta, E. S. and Uhr, J. W. (1975). *Science*, **189**, 964–969.
113. Finkelman, F. D., Smith, A. M., Scher, I. and Paul, W. E. (1975). *J. exp. Med.*, **142**, 1316–1321.
114. Solomon, A. and Kunkel, H. G. (1967). *Amer. J. Med.*, **42**, 958–967.
115. Marchalonis, J. and Edelman, G. M. (1965). *J. exp. Med.*, **122**, 601–618.
116. Putnam, F. W., Florent, G., Paul, C., Shinoda, U. and Schimizu, A. (1973). *Science*, **182**, 287–291.
117. Milstein, C., Brownlee, G. G., Harrison, T. M. and Mathews, M. B. (1972). *Nature* (*New Biol.*), **239**, 117–120.
118. Crone, M., Koch, C. and Simonsen, M. (1972). *Transplant. Rev.*, **10**, 36–56.
119. Simonsen, M. (1974). In *Allergology*, Excerpta Medica, Amsterdam, p. 146.
120. Webb, S. R. and Cooper, M. D. (1973). *J. Immunol.*, **111**, 275–277.
121. Hudson, L., Thantrey, N. and Roitt, I. M. (1975). *Immunology*, **28**, 151–159.
122. Hudson, L., Sprent, J., Miller, J. F. A. P. and Playfair, J. H. L. (1974). *Nature*, **251**, 60–62.
123. Hunt, S. V. and Williams, A. F. (1974). *J. exp. Med.*, **139**, 479–496.
124. Hämmerling, G. J. and McDevitt, H. O. (1974). *J. Immunol.*, **112**, 1726–1733.
125. Hämmerling, G. J. and McDevitt, H. O. (1974). *J. Immunol.*, **112**, 1734–1740.
126. Hämmerling, G. J., Lonai, P. and McDevitt, H. O. (1975). In *Membrane Receptors of Lymphocytes* (Seligmann, M., Preud'homme, J. L. and Kourilsky, F. M., Eds.), Elsevier, New York, pp. 121–126.
127. Roelants, G. E., Forni, L. and Pernis, B. (1973). *J. exp. Med.*, **137**, 1060–1077.
128. Roelants, G. E., Ryden, A., Hägg, L.-B. and Loor, F. (1974). *Nature*, **247**, 106–108.
129. Moroz, C. and Hahn, Y. (1973). *Proc. Nat. Acad. Sci.*, **70**, 3716–3720.
130. Marchalonis, J. J. and Cone, R. E. (1973). *Transplant. Rev.*, **14**, 3–49.

131. Marchalonis, J. J. (1974). *J. Med.*, **5**, 329–367.
132. Marchalonis, J. J. (1975). *Science*, **190**, 20–29.
133. Grey, H. M., Kubo, R. T. and Cerottini, J.-C. (1972). *J. exp. Med.*, **136**, 1323–1328.
134. Vitetta, E. S., Bianco, C., Nussenzweig, V. and Uhr, J. W. (1972). *J. exp. Med.*, **136**, 81–93.
135. Vitetta, E. S. and Uhr, J. W. (1973). *Transplant. Rev.*, **14**, 50–75.
136. Lisowska-Bernstein, B., Rinuy, A. and Vassalli, P. (1973). *Proc. Nat. Acad. Sci.*, **70**, 2879–2883.
137. Lisowska-Bernstein, B. and Vassalli, P. (1974). *Biochem. Biophys. Res. Comm.*, **61**, 142–147.
138. Jensenius, J. C. and Williams, A. F. (1974). *Eur. J. Immunol.*, **4**, 98–105.
139. Jensenius, J. C. (1976). *Immunology*, **30**, 145–155.
140. Abney, E. R. and Parkhouse, R. M. E. (1976). *Advan. exp. Med. Biol.*, **66**, 373–379.
141. Grey, H. M., Colon, S., Solomon, A. and McLaughlin, C. L. (1973). *J. Immunol.*, **111**, 1923–1925.
142. Haustein, D., Marchalonis, J. J. and Harris, A. W. (1975). *Biochemistry*, **14**, 1826–1834.
143. Santana, Y., Wedderburn, N., Abney, E. R. and Parkhouse, R. M. E. (1976). *Eur. J. Immunol.*, **6**, 217–222.
144. Munro, A. J., Taussig, M. J., Campbell, R., Williams, H. and Lawson, Y. (1974). *J. exp. Med.*, **140**, 1579–1587.
145. Takemori, T. and Tada, T. (1975). *J. exp. Med.*, **142**, 1241–1253.
146. Binz, H., Kimura, A. and Wigzell, H. (1975). *Scand. J. Immunol.*, **4**, 413–420.
147. Ramseier, H. and Lindenmann, J. (1972). *Transplant. Rev.*, **10**, 57–96.
148. McKearn, T. J. (1974). *Science*, **183**, 94–96.

Chapter 11

B Cell Maturation: Its Relationship to Immune Induction and Tolerance

G. G. B. KLAUS

Abbreviations

ABC: antigen-binding cells
AFC: antibody-forming cells
ALS: anti-lymphocyte serum
B mice: T cell deprived mice

DNP: 2,4-dinitrophenyl
Ig: immunoglobulin
Ig(?): Ig (class not identified)
KLH: keyhole limpet haemocyanin
LE: levan
LPS: *E. coli* lipopolysaccharide
PBA: polyclonal B cell activator
POL: polymerized *Salmonella* flagellin
PVP: polyvinylpyrrolidone
S3: type III pneumococcal polysaccharide
TD: T cell dependent
TI: T cell independent
(T, G)-A–L: (Tyr, Glu)-Ala–Lys
θ: Thy-1 locus determined antigen

1. INTRODUCTION

The rapid expansion of immunological research in recent years has been stimulated not only by the widespread biomedical implications of immune phenomena, but also by the evident complexity of the cellular and molecular control mechanisms of the immune system. In this context, the immunological apparatus provides a unique model of a complex differentiating system, unique in that the stimulus for the differentiation of precommitted clonal precursors to effector cells is provided by the 'invasion' of the individual by a foreign substance, the immunogen.

The discovery that two functionally distinct lymphocytes cooperate in the induction of antibody synthesis to many antigens (reviewed in Reference 1) represented a revolutionary conceptual advance, which has provoked an intensive effort to elucidate the mechanisms of cell cooperation (reviewed in References 2–4). It is thus established that lymphocytes which migrate through the thymus (T cells) provide an, as yet, undefined helper function for the induction of high-rate antibody synthesis in non-thymus-processed (bone marrow derived or bursal equivalent) B cells. Furthermore, both T and B lymphocytes carry antigen recognition structures (membrane receptors): the B cell receptor is immunoglobulin (reviewed in References 5 and 6) and the antigen-binding specificity of receptors on a particular B cell is identical to that of the antibodies secreted by the antibody-forming cell (AFC) progeny of that precursor. The nature of the T cell receptor is still hotly disputed (see J. W. Goodman, and R. M. E. Parkhouse and E. R. Abney, herein).

The binding of antigen to immunoglobulin receptors on B cells is widely, but not universally,[7] believed to constitute a *signal* to the cell and may (*per se* or in conjunction with other signals) instruct that cell to proliferate and differentiate into AFC (induction) or may render it unresponsive (tolerance). In conventional (T cell dependent (TD) or cooperative) induction T cells recognize *carrier* determinants on a multi-determinant immunogen such as a foreign protein, and generate an additional signal, thereby enabling B cells to respond to *haptenic* determinants on the antigen. B cells can also be stimulated by antigens which do not apparently stimulate T cells (T cell independent (TI) antigens).

The main objective of this chapter is to examine the possible *sequelae* of signal reception by B lymphocytes, in particular, to attempt to predict some of the properties of various B cell subsets on the basis of their response to antigen, and to formulate a (very rough) picture of the maturation pathways stimulated precursors traverse on their way to becoming AFC or memory cells.

Clearly, a detailed understanding of lymphocyte differentiation pathways is of paramount importance for the future success of selective enhancement or suppression of various effector mechanisms in immunity.

The concepts advanced in the following discussion represent a personal viewpoint of a rapidly evolving field. Much of the interpretation is highly speculative, reflecting the paucity of available data, and the literature review is by no means comprehensive.

2. VIRGIN B CELL INDUCTION AND TOLERANCE

2.1. Thymus independent (TI) and thymus dependent (TD) induction

2.1.1 *General aspects*

Primary antibody responses to many antigens are drastically reduced in T cell depleted rodents—usually adult-thymectomized, irradiated, stem cell-repopulated mice ('B mice'), or congenitally 'athymic' (nu/nu) mice. These TD antigens include heterologous erythrocytes, proteins and their hapten-derivatives (reviewed in Reference 2).

On the other hand, there are a number of antigens which elicit normal antibody responses in B or nu/nu mice. Such TI antigens are mostly bacterial polysaccharides, e.g. type III pneumococcal polysaccharide (S3), levan (LE), *E. coli* lipopolysaccharide (LPS), pneumococcal C-polysaccharide, and *L. mesenteroides* dextran[8–13] (reviewed in References 14 and 15). Certain synthetic polypeptides composed of D-amino acids,[16] and polymers such as polyvinylpyrrolidone (PVP)[9] also fall into this category. Large proteins such as polymerized *Salmonella* flagellin (POL) induce TI antibody formation *in vitro*,[17] but the extent of the T independence of the *in vivo* response to POL is controversial.[18]

It should be stressed that the definition of TI responses is purely operational. Conventional B mice are known to have some 10% of the normal complement of T cells,[19] and even nu/nu mice, which are reputedly devoid of T cells, have small numbers of cells with T cell surface markers,[20] and may even have some *functional* T cells reactive against potent antigens such as flagellin.[21]

TI antigens have been widely studied in recent years in the hope that they would provide clues to the *minimal* requirements for B cell triggering. Furthermore, just as the use of hapten-substituted proteins proved invaluable in the delineation of lymphocyte cooperation (e.g. Reference 22), so hapten (e.g. 2,4 dinitrophenyl (DNP)) derivatives of known TI antigens are being increasingly used to study TI induction. Such semisynthetic antigens permit the study of immune responses to a single defined determinant (epitope), the evaluation of the role of the structure of the 'carrier' molecule, and of the influence of factors such as epitope density on the immunological properties of the conjugate.

2.1.2 *Characteristics of TI antigens*

TI antigens have certain common characteristics, which may be important in their capacity to stimulate antibody production in the relative absence of T cells (reviewed in Reference 15).

(a) They are polymeric molecules: some are homopolymers (e.g. LE, PVP), others copolymers (e.g. S3). Some have a branched structure (LE, B1355 dextran) and others linear (e.g. S3). Many are of high molecular weight ($1–10 \times 10^6$ for LE),[23] and depolymerized preparations are often less immunogenic.[24,25] The overall conclusion is that these antigens have a *limited epitope heterogeneity*, unlike TD globular proteins which usually carry a very heterogeneous epitope array because of their complex structure.

(b) As a direct consequence of (a) TI antigens are capable of *multipoint high avidity* binding to Ig receptors on lymphocytes;[26,27] (see also Section 3.3.4).
(c) Many of these antigens are highly persistent.[10,16,28,29] Thus, S3 has a half-life of 35 days in the mouse,[29] presumably because the animal lacks enzymes to degrade it. In contrast, the bulk of an injected dose of heterologous protein or lightly substituted hapten-protein (e.g. Reference 30) is very rapidly catabolized and excreted, although of course immunologically significant amounts may remain sequestered in lymphoid tissues, or at the injection site if the antigen is given as a depot preparation.
(d) Some TI antigens have ancillary properties which may relate to their immunogenicity. Thus, LPS and LE cleave the third component of complement (C3), via the alternate pathway of C activation.[31] LPS is also a well-recognized 'polyclonal B cell activator' (PBA) (reviewed in References 32 and 33). In other words, both *in vivo* and *in vitro*, suitable doses of LPS induce not only anti-LPS antibody, but also a 'polyclonal' IgM response to non-cross-reacting epitopes. In spleen cell cultures LPS induces proliferation and antibody-forming cell (AFC) differentiation of a substantial proportion of murine B cells, i.e. it is a powerful B cell mitogen. It has been claimed[7,34] that all TI antigens are polyclonal B cell activators, but this is controversial.[35,36]

2.1.3 *TI nature of the IgM response*

It is well established that TI antigens stimulate principally, if not exclusively, IgM antibody formation, whereas the IgM response to TD antigens is usually rapidly superseded by an IgG response (reviewed in References 14 and 15). It is therefore not surprising that hapten-conjugates of S3, LE, LPS, and dextran also elicit essentially only IgM anti-hapten antibodies.[37–41] In addition, DNP-conjugates of putatively non-immunogenic polymers, such as hyaluronic acid, and poly-γD-glutamic acid,[35] and of immunologically 'inert' materials such as polyacrylamide and Sepharose[42–45] also elicit TI anti-DNP IgM antibodies.

There is in fact abundant evidence that IgM responses in general (whether elicited by TD or TI antigens) are far less T cell dependent than IgG responses.[46–50] As an example, nu/nu mice produce significant numbers (though lower than euthymic controls) of IgM AFC to sheep red blood cells (SRBC), but very few IgG AFC.[47–50]

Thus TI and TD antigens differ only in a *quantitative* fashion in their capacity to induce IgM antibodies in T deprived animals. The important question is why TI antigens *only* stimulate IgM antibodies in the absence of active T cells (see Section 2.1.4).

We know very little about the precursor cells (Bμ cells) that generate IgM AFC (Section 3.1.1). There is some evidence for the existence of TD and TI Bμ cells,[51,52] but this is not definitive. Both TI antigens and polyclonal B cell activators (PBA)[53,54] trigger Bμ rather than Bγ cells (precursors of IgG AFC). This does not necessarily mean that PBAs and TI antigens activate Bμ cells via identical initial triggering signals as has been claimed[1,34] but rather, probably reflects some (unknown) property of Bμ cells, which makes them unusually susceptible to direct, non-cooperative induction (or resistant to tolerization, Section 2.3). Since IgM is the earliest antibody to appear in phylogeny,[55] ontogeny[56] and during the normal primary response, this may be a useful feature, from a teleological point of view.

The present status of knowledge of the mechanisms of B cell triggering is extremely confused (reviewed in References 3 and 4), presumably reflecting our lack of understanding of a plethora or phenomenology. Our own working hypothesis of TI induction,[36] which is similar to one advanced by Feldmann[57,58] is that polymeric, high molecular weight antigens can establish very high avidity persistent binding to antigen-reactive Bμ cells (see also Section 3.3.4). This induces receptor rearrangement, and membrane perturbations which somehow generate a triggering signal. Tolerance can be regarded as a saturation of receptors producing a 'frozen membrane'.[59] The fact that we do not understand what epitope presentation means at the level of receptor–epitope interaction does not detract from the model. We do know that TI Bμ cell triggering requires special (undefined) features of the antigen, since neither high epitope density TD antigens,[60] nor non-specifically aggregated proteins[61] act as TI antigens. The mere statement that a cell can respond to an 'array' of epitopes bound to it may be naive, but available knowledge does not justify a more precise formulation, nor an attempt to integrate TI and TD induction under a common molecular mechanism.

As discussed elsewhere,[35,36] we do not accept the concepts that all TI antigens trigger B lymphocytes by virtue of being polyclonal B cell activators,[7,34] or by their capacity to activate C3.[62] These properties may well confer an inherent 'adjuvanticity' on the antigen, and thereby modulate its triggering properties, but at present it seems unlikely that either property is *obligatory* for TI B cell triggering.

2.1.4 *TD nature of the IgG response*

As stated above IgG responses (especially IgG_1 in the mouse) are highly T cell dependent.[46–50,63] It is therefore pertinent to ask whether TI antigens generally fail to induce IgG antibody synthesis because of their simple structure and/or their failure to stimulate T helper cells, or because of other factors (Section 2.3).

A partial answer to this question has emerged from studies using the 'allogeneic effect' (reviewed in Reference 64). Inoculation of an animal with allogeneic T cells (e.g. an F_1 hybrid mouse given parental spleen cells) at the time of immunization, often markedly enhances the primary response to that antigen and partially abrogates the requirement for carrier-specific T cells in secondary antibody responses.[65–74]* Allogeneic T cells probably react to histocompatibility antigens on host B cells, thereby providing a 'non-specific' helper function to the latter to a concurrently administered antigenic stimulus. The helper effect is probably medited by non-antigen-specific T cell factors (reviewed in Reference 75) elicited by the graft-versus-host reaction.

A striking result of the allogeneic effect is the induction of an IgG response to antigens which normally only elicit IgM antibodies,[67,74,76] or which are potent B cell tolerogens.[68,69,74,76] An example is the response to the multichain polypeptide (T, G)-A–L in the mouse which is determined by the H-2 linked Ir-1 locus (reviewed in Reference 77). High responder mice produce IgM and IgG antibodies to (T, G)-A–L, whereas in low responders (whose B cells probably fail to cooperate with (T, G)-A–L specific T cells),[78] the polypeptide acts as a TI antigen, and elicits only IgM antibodies. Administration of low responder parental spleen cells plus (T, G)-A–L to low responder F_1 mice induces the formation of IgG antibodies.[67]

Similarly, we have demonstrated the induction of IgG anti-DNP antibody synthesis in (CBA × C57Bl)F_1 mice given CBA spleen cells at the time of immunization with DNP-LE, DNP-S3 and DNP-hyaluronic acid.[35,76] Paradoxically, we could not detect IgG antibody in these same mice to the *polysaccharide* epitopes of either DNP-LE or DNP-S3 although we did find a (variable) enhancement of anti-S3 and anti-LE IgM responses.[76] This may, however, be a reflection of the mouse strain combination used.[74]

* Immunopotention via the allogeneic effect depends critically on various factors, such as the allogeneic cell dose and the time of cell transfer relative to immunization; under some conditions graft-versus-host reactions produce profound immunosuppression (e.g. Reference 73).

Other groups[65,68,69] have found that the allogeneic effect failed to affect *primary* responses to TD hapten-protein conjugates. However, in our experiments[76] primary responses to three different DNP-proteins were markedly enhanced by the graft-versus-host reaction. This casts doubt on the claim[69] that there are unique antigenic requirements for the expression of the allogeneic effect in unprimed B cells.

These data strongly suggest that TI antigens are not immunogenically 'deficient' in their capacity to stimulate Bγ cells, provided there is a source of T helper cells. This also argues that Bγ cells differ *qualitatively* from Bμ cells in their induction parameters; it remains possible that TI antigens effectively tolerize Bγ cells under conventional immunizing conditions (Section 2.3). We still do not know why these antigens fail to stimulate T helper cells (Section 2.2.1).

Appropriate allogeneic cell transfers also substantially enhance *IgM* responses to a variety of TI and TD antigens.[68,69,73,74,76] This may mean that T cell help increases the AFC 'burst size' per clone of stimulated Bμ cells,[79] or it may reflect the triggering (or the completion of later differentiation stages) of an additional T cell dependent[51,52] Bμ cell subset.

Recently, it has become clear that there are antigens which elicit both IgM and IgG antibodies in an operationally TI fashion. These include DNP-Ficoll[44,45] and DNP-*Salmonella* organisms.[80] Work in our laboratory (Klaus, G. G. B., Phillips, J. M., Humphrey, J. H., Dresser, D. W. and Cross, A. M. (1976), *Eur. J. Immunol.*, **6**, 429–433) has confirmed that DNP-Ficoll elicits a strong IgM response, and variable numbers of indirect plaque-forming cells (PFC), developed by polyvalent anti-Ig. It appears that some of these developed PFC are IgM AFC, whereas others are *bona fide* IgG AFC. Serum antibody titrations and isoelectric focusing analyses have demonstrated the presence of a polyclonal IgG response, which appears to be as TI as the IgM response. However, again administration of CBA spleen cells to (CBA$\times$C57Bl)F_1 mice given DNP-Ficoll enhances the peak IgM response, and also elicits a much stronger IgG response.

In conclusion, it appears likely that TI antigens elicit relatively 'simple' antibody responses, i.e. mostly IgM, perhaps of relatively homogeneous (low) affinity,[81,82] which reflects the 'baseline' response relevant clones of B cells can generate in absence of helper cells—although this baseline response depends on the nature of the antigen, e.g. Ficoll and LE are better 'carriers' for DNP than S3 or hyaluronic acid.[35,38] This agrees with estimates of low AFC burst sizes in clones stimulated by TI antigens or B cell mitogens, versus the much larger burst size induced by TD antigens.[79,81,83] Such baseline TI stimulation may extend to Bγ cells with antigens such as DNP-Ficoll, suggesting that the induction parameters of Bμ and Bγ cells differ only relatively, and not absolutely (Section 4.1).

2.2 The role of T cells in T independent induction

2.2.1 *Helper function*

The concept of TI immunity has engendered considerable criticism from Bretscher and Cohn,[83,84] who maintain that all lymphocyte induction requires two 'signals', signal ① being the binding of antigen to receptors, while signal ② is 'associative antibody' produced by the helper T cell. They do, however, accept that Bμ cell induction may require only *minimal* levels of signal ② (Reference 84, and P. A. Bretscher, herein).

As stated above, experiments performed with T cell deprived mice do not provide definitive evidence for totally TI B cell induction. However, there is substantial additional evidence to suggest that T cells are not involved to any significant extent in many IgM responses elicited by polymeric antigens. Firstly, T cells do not proliferate in response to antigens such as S3,[85] and there appear to be no θ-bearing antigen-binding cells (ABC) in mice immunized with LPS.[86] Admittedly, we do not know if there are T cells that can *bind* TI antigens, but perhaps fail to respond to them.

Secondly, IgM anti-DNP responses can be induced by conjugates of immunologically 'inert' materials such as hyaluronic acid, Sepharose and polyacrylamide.[35,42,43] It seems highly unlikely that T cells 'recognize' a self-component like hyaluronic acid, a universal constituent of connective tissue, and of identical structure in diverse species.[87]

Thirdly, it is well known that tolerance induction to the carrier moiety of a TD hapten-protein conjugate (i.e. T cell tolerance) specifically prevents a response to the hapten on that carrier.[88–91] Similar experiments with conjugates of TI antigens have shown that carrier-specific tolerance *has not effect on the antihapten response*.[38,39,92] For example, mice tolerized with LE made a normal anti-DNP response to DNP-LE; conversely, mice tolerized with DNP-S3 made a normal anti-LE response to DNP-LE.[38] These data indicate that DNP and LE immunize separate and *non-interacting* B cell populations, i.e. they effectively rule out concepts of B cell – B cell cooperation,[37] and of low levels of T cell – B cell cooperation (Reference 84, and P. A. Bretscher, herein) invoked to explain TI induction. Similar dissociative effects have been observed with a non-synthetic antigen. B1355 dextran carries two major epitopes, related to α 1–3 and α 1–6 saccharide linkages. Administration of a large dose of B1355 abolished the anti-α 1–6 response of mice to subsequent optimal challenge, but had minimal effects on the α 1–3 response.[58]

Finally, it is well recognized that recovery from tolerance to TD antigens is impaired in thymectomized animals,[93] presumably because of the lack of T cell regeneration. Similar studies with LE[58] and S3[94] failed to detect any effect of T cell deficiency on the maintenance of tolerance, as would be demanded by the Bretscher–Cohn model.

These considerations therefore provide compelling evidence for B cell activation in mice in the absence of T cell helper activity. The dissociation of immunity and tolerance to different epitopes on the same molecule is particularly cogent, and can only be refuted by rather obscure arguments.

2.2.2 *Suppressor function*

On the other hand, there is evidence that T cells suppress responses to certain TI antigens (and TD antigens; reviewed in Reference 95). Thus, both adult thymectomized (non-repopulated) and B mice give enhanced IgM responses to PVP,[96,97] and anti-S3 and anti-PVP responses are enhanced by treating mice with anti-lymphocyte serum (ALS, which is believed to kill T cells) (References 98–100, and P. C. L. Beverley, herein). These results, together with other data, have led to the proposal that two functionally distinct T lymphocyte populations (amplifier and suppressor cells) regulate the anti-S3 response.[99] However, it should be emphasized that many laboratories have failed to elicit enhanced responses in T cell depleted mice, using DNP-LE,[37,38] LE,[10] DNP-Ficoll (Reference 44; unpublished data of J. H. Humphrey), S3,[8,102,104] and LPS.[102,103]

The results of adult thymectomy experiments suggest the presence of a short-lived inhibitory T cell population modulating the response to PVP.[96,97] It may be relevant that Andersson and Blomgren[101] found that while the anti-PVP response of adult spleen cells was suppressed by added thymocytes, that of neonatal spleen cells was enhanced. They therefore suggest that the effect of T cells is related to the 'maturity' of the B cell population.

The potentiating effects of ALS on anti-S3 responses[98,99] are difficult to evaluate. It is assumed that ALS kills circulating T cells:[105] in fact its mechanism of action is unclear and the current position may be summarized as follows:

(a) Although various ALS preparations enhance the anti-S3 response, gammaglobulins prepared from such sera are often inactive, even though they still suppress TD responses (Reference 106; K. James, personal communication).
(b) The enhancing effect depends on the preparation of S3 employed (G. W. Warr, P. J. Baker and K. James, in preparation). It has become clear that S3 preparations from different laboratories differ quite substantially in their chemical purity, their molecular weight, and hence in their immunological properties (J. H. Humphrey and G. G. B. Klaus, unpublished data).
(c) The enhancing effect depends critically on the timing of the ALS relative to antigen: ALS given before S3 may be ineffective or even immunosuppressive (G. W. Warr, P. J. Baker and K. James, in preparation).
(d) The effect is T cell dependent, since ALS fails to enhance the anti-S3 response of nu/nu mice.[107]

A reasonable working hypothesis is that the effects of ALS are due to some form of (T cell dependent) B cell *stimulation*, rather than to removal of T cell suppression. It is known that ALS causes marked splenomegaly[104] and that it can act as a T cell mitogen.[108] Thus, perhaps ALS causes the release of non-specific T cell 'factors' (from T cells activated by the *in vivo* cytotoxic effect?) and thereby induces the equivalent of an 'allogeneic effect' (Section 2.1.4). This would explain the need to give ALS concurrently with antigen, and also the reported induction of *IgG* anti-S3 antibodies by ALS,[74,109] which has been observed in (some) studies with the allogeneic effect.[74] A further complication is that neither ALS[110,111] nor allogeneic lymphocytes[103] enhance the response to LPS.

In summary, T suppressor cells may well modulate responses to antigens such as PVP,[96,97] but they certainly do not appear to play a significant role in all TI responses.

2.3 Susceptibility of virgin Bμ and Bγ cells to tolerization

The outcome of a lymphocyte's encounter with antigen is believed to depend on the balance of induction and inactivation 'signals' generated by that encounter. In Section 2.1 we concluded that virgin Bμ and Bγ cells differ qualitatively in their requirements for induction signals. It is therefore appropriate to ask to what extent the induction of IgM and IgG responses are modulated by the susceptibility of precursor cells to tolerance signals.

A number of studies employing various models of B cell tolerance have shown a preferential suppression of IgG responses in partially tolerant animals.[112–116] This was most clearly illustrated by Hamilton and Miller[113] who found that hapten-coupled syngeneic erythrocytes markedly reduced the primary IgG response to a hapten-protein in mice, but had relatively little effect on the IgM response. We described a similar effect in mice tolerized with heavily coupled DNP-bovine serum albumin.[114] Moreover, both Fidler and Golub[117] and we[114] found that during spontaneous loss of tolerance the IgM response recovered more rapidly than the IgG response.

These results, although intriguing, require careful interpretation. It is now clear that there are multiple mechanisms of tolerance, which can effect T and/or B lymphocytes (reviewed in References 118 and 119). Partial T cell tolerance would be expected to preferentially suppress the IgG response (Section 2.1.4), and this has indeed been found.[120] Moreover, some forms of tolerance may result from T cell-mediated suppression (reviewed in Reference 95) and this may also inhibit IgG antibody formation more than IgM.[115,121,122]

Nonetheless, dissociation of IgM and IgG responses has been shown in 'true' B cell tolerance, i.e. tolerance not mediated by T cells.[113] This suggests that virgin Bγ cells are more susceptible to tolerance than Bμ cells. The complexity of the problem is, however, increased by the fact that the tolerance susceptibility of B cells (of all classes?) is modulated by helper T cells. This is well illustrated by the capacity of activated T cells (e.g. the allogeneic effect, Section 2.1.4) to convert a tolerogenic stimulus to an immunogenic one,[68,69,123] and by the susceptibility of athymic mice to B cell tolerance induction by TD antigens.[124]*

The role of T cells in determining the tolerance-induction decision of B cells has been discussed by Mitchell,[14,125] who emphasizes that in the absence of T cells *high avidity* B cells are especially susceptible to tolerance. This agrees with the well documented preferential suppression of high affinity antibodies in partially tolerant animals (reviewed in Reference 126) and also with the susceptibility of primed B cells to tolerance in some systems (Section 3.3.4).

The above resoning is probably relevant to an explanation of why most TI antigens elicit only IgM antibodies of low affinity (Section 2.1.3), and why activated T cells permit these antigens to induce an IgG response (Section 2.1.4), and increase antibody affinity in TD responses.[127,128] These arguments also suggest that virgin Bγ cells in general have a higher avidity for antigen than Bμ cells (see also Section 3.3.4).

3. ANTIGEN DEPENDENT B CELL MATURATION

Operationally, we can define two functional species of lymphoid cells that arise as a result of stimulation of a virgin B cell:

(a) High-rate AFC, which are fully differentiated to the secretion of antibody of a single class and specificity, and
(b) B memory cells, which can be regarded as resting cells, requiring further contact with antigen (and usually T cells) to induce them to proliferate into AFC. It will be obvious that our knowledge of the existence of memory cells is indirect, since these cells can only be assayed by their potential to give rise to AFC.

3.1. The genesis of antibody-forming cells (AFC)

3.1.1 *General aspects*

AFC comprise a morphologically heterogeneous population of lymphocytoid and plasmacytoid cells (e.g. References 129 and 130) which develop by clonal proliferation and differentiation from precursor B cells. It has been estimated that clonal precursors undergo 6–10 rounds of cell division,[131] and the fully differentiated AFC is generally regarded as an end cell.

A controversial issue is whether IgG and IgM AFC derive from separate, pre-existing precursors, or if the former derive from the latter via an intra-clonal, antigen-driven $\mu \rightarrow \gamma$ heavy chain 'switch'. For present purposes it can be stated that both classes of AFC derive from precursors which at some stage in their history carry IgM receptors (reviewed in

* It should be noted that such treatment can also result in *B cell priming*, depending on the experimental conditions used (Section 3.3.2).

Reference 56). This has been clearly shown by studies of B cell ontogeny in the avian bursa of Fabricius,[132] from experiments on the immunosuppressive effects of class specific anti-Ig antibodies in both chickens[133] and mice,[134,135] and also from other data.[136]

It has been proposed that in birds the $\mu \rightarrow \gamma$ switch occurs within the bursa prior to antigenic contact.[56] Thus far, definitive evidence for the existence of a mammalian bursa-equivalent is lacking, although the foetal liver is a likely site of B cell maturation in the mouse (Reference 137, and J. J. T. Owen, herein).

Most B cells in the murine spleen and lymph nodes carry surface IgM and/or IgD[138–142] (also Section 3.3.3). However, we cannot discount the possibility that some of these may be functionally Bγ cells, i.e. already programmed to secrete IgG.

3.1.2 *Immunoglobulin (Ig) receptors on AFC and their immediate ancestors*

A tacit assumption in current immunological theory is that the signals which predicate lymphocyte differentiation operate principally at the level of antigen-reactive precursor cells. Recently, it has become clear that under some circumstances cells of the B cell lineage remain susceptible to antigen-mediated control up to the level of the mature AFC. This has emerged from studies of 'antibody-forming cell blockade', a phenomenon first described by Baker *et al.*[143]

Administration of a tolerogenic dose of a polymeric hapten-conjugate, such as DNP-S3 (Section 2.1) to mice already making anti-DNP antibodies markedly reduces the rate of antibody secretion per AFC,[38,143–145] as shown by a decrease in the diameter of plaques produced by AFC in a haemolytic plaque assay. Brief incubation of normal AFC with multivalent antigen *in vitro* has a similar effect.[38,144,145] The inhibition (like ABC blockade, Section 3.3.4) requires multivalent (i.e. high avidity) binding of antigen to AFC,[145] and is almost certainly mediated by antigen–Ig receptor interaction.

It has been claimed that mature plasma cells carry little or no surface Ig.[36] However, there is good evidence that most AFC appearing early in the immune response carry receptors, but that as the response progresses there is a concomitant decrease in the proportion of Ig-bearing AFC.[146–148] In agreement with this anti-DNP AFC in both primary and secondary responses become less susceptible to blockade by DNP-S3 with increasing time after immunization.[145]

It has been suggested that the synthesis of surface Ig and secreted Ig are under separate genetic control,[149] so that once a B cell has been stimulated, receptor synthesis is shut off. The quantity of surface Ig on the surface of the progeny of a stimulated cell would thus be an indicator of their 'maturity', if residual receptors are diluted out by consecutive cell divisions.

In this context, we found that IgG AFC were far less susceptible to blockade than IgM AFC, both *in vivo* and *in vitro*.[145] This agrees with observations of some,[150–152] although not all,[147] investigators that IgM AFC in general carry more surface Ig than IgG AFC, and suggests that the latter are more 'mature' than the former, perhaps because they have undergone more cycles of cell division.

Blockade of IgM AFC may well complicate the analysis of B cell tolerance induced by multivalent, persisting antigens.[145] Thus, tolerizing doses of DNP-S3 given to mice prior to immunization with DNP-haemocyanin, produced similar degrees of suppression of IgM and IgG AFC.[145] However, as discussed in Section 2.3 it is likely that Bγ *precursors* are more susceptible to tolerance than Bμ cells. The simplest explanation of this apparent discrepancy is that after triggering, Bγ cells rapidly become insusceptible to tolerization, whereas the progeny of Bμ cells can be inhibited by persisting antigen such as DNP-S3 up to the level of IgM AFC.[145] The possible role of the dynamic loss of receptors on maturing pre-AFC in this process has already been mentioned.

3.2 Changes in antibody affinity during the immune response

3.2.1 *The cell selection hypothesis*

The strength of the reaction of an antibody combining site with a defined antigenic determinant (e.g. a hapten) is determined by the affinity of the antibody and can be expressed in thermodynamic terms. If the antibodies are directed against undefined determinants, e.g. on a protein, the strength of the antigen–antibody reaction is determined by the avidity of the antibody (which cannot be expressed in thermodynamic terms). In either case, a high affinity/avidity antibody population binds antigen more firmly than a low affinity/avidity population. It is commonly found that antibodies raised to conventional immunogens are markedly heterogeneous, both in combining site specificity, and in affinity. Thus, antibodies within a single serum can vary by 10 000-fold in their affinity for homologous hapten.[153]

Some years ago Eisen and his associates[153,154] performed a series of elegant experiments which showed that in rabbits immunized with a DNP-protein the early antibodies tended to be of low affinity, and with time the average affinity of antibodies progressively increased. This process, which has been widely studied (reviewed in Reference 155) is termed immunological maturation. The degree of maturation depends on several parameters, most notably the primary dose of antigen. Thus, rabbits and mice given a high dose of antigen show a slow increase in antibody affinity, while small priming doses produce much faster maturation.[154,156] Steiner and Eisen[154] showed that maturation was in fact due to a shift in the affinity of antibody synthesized, and not due to the neutralization of high affinity antibody by persisting antigen shortly after immunization.

These findings produced a logical extension of Burnet's[157] clonal selection hypothesis, namely the cell selection model,[155] which proposed that (a) precursor cells carry receptors of the same specificity and affinity as those of the antibody which progeny of these cells secrete, and (b) that, with time after immunization, there is selective proliferation of precursors best able to bind antigen, i.e. with high affinity receptors. In other words, decreasing antigen concentrations in the animal produce a selective gradient, in which cells with low affinity receptors cannot compete with high affinity cells for residual antigen: consequently, antibody affinity progressively increases with time.

3.2.2 *Evidence against the hypothesis*

It has become increasingly evident that the basic cell selection hypothesis is oversimplified and, in particular, that serum antibody affinity does not necessarily accurately reflect the clonal repertoire of the primed B cell population.

Firstly, it has been shown that there are several variants of this simple picture of affinity maturation (reviewed in Reference 158). The initial rise in antibody affinity may be followed by a later fall (e.g. Reference 159), or alternating waves of high and low affinity antibodies may occur (e.g. Reference 160). Again, with some antigens, antibodies remain functionally homogeneous from the inception of the response (e.g. References 161 and 162). These variants clearly do not obey the predictions of cell selection.

A major problem in interpreting maturation experiments is the difficulty in obtaining reliable antibody affinity measurements. It is assumed that antibody affinities in a heterogeneous population are normally distributed, although this is certainly not always the case.[163] In addition, it is technically difficult to accurately measure low affinity antibodies in the presence of excess high affinity antibody.[163]

The most relevant question for the present discussion is what happens to precursor (memory) cells during maturation, since the model implies that as high affinity antibody dominates the late primary response, low affinity precursors die off. In fact, there is considerable evidence that affinity restrictions at the level of serum antibody are *not* accompanied by similar restrictions at the level of memory cells, but that low affinity memory cells arise and survive in primed animals.[164–167] The degree to which they are stimulated, however, depends on a number of factors. Thus, Celada *et al.*[166] showed that large numbers of primed cells transferred to irradiated mice make a high avidity secondary response, while low numbers of cells make low avidity antibody. Furthermore, boosting adoptive recipients with supraoptimal doses of antigen (which would tolerize high affinity precursors) led to a low avidity response.

Macario *et al.*[164] cultured lymph node fragments taken from rabbits with homogeneous high affinity serum antibody to *E. coli* β-galactosidase. During the culture the high affinity antibody decayed, to be superseded by a heterogeneous, low affinity secondary response. Furthermore, cultures stimulated with high concentrations of antigen produced lower affinity antibody than those treated with low concentrations.[165]

These results strongly indicate that animals producing relatively homogeneous high affinity antibodies carry a heterogeneous population of memory cells, even though the demonstration of low affinity cells may require special conditions. The presence of the latter would explain the continuing synthesis of substantial amounts of low affinity antibodies throughout the immune response in both mice and rabbits.[156,163]*

Studies of affinity maturation at the cellular level have been interpreted to support the cell selection hypothesis. These have largely involved studying the 'avidity' of AFC, by measuring their inhibition by free antigen (monovalent hapten or multivalent antigen) incorporated into the plaque-revealing monolayer of the haemolytic plaque assay system. Such studies have shown that IgG AFC, in particular, and in some cases IgM AFC, appear to acquire higher avidity for antigen, with time after immunization.[167–169] The validity of this procedure as a measure of antibody affinity, has, however, been criticized[170] since it fails to consider differences in secretion rate of different AFC populations.†

Experiments by Davie and Paul[169] have demonstrated a correlation between the maturation of serum antibody affinity, AFC avidity (determined as above) and ABC avidity in guinea pigs immunized with DNP-protein. They showed that decreasing concentrations of hapten were required to inhibit DNP-binding cells with increasing time after immunization. Bell and DeLisi[172] in a theoretical discussion of these experiments, point out that they were probably performed under non-equilibrium conditions, since the presence of free hapten probably slows the approach to equilibrium between a multivalent hapten-protein and cell receptors. The significance of these and other similar data[167] is therefore uncertain, as is the extent to which selective proliferation of high affinity receptor-bearing precursors occurs during immune maturation. As discussed in Section 3.3.4 there is growing evidence that the avidity of precursors is determined not only by the affinity, but also by the number of receptors per cell.

* A conceptual difficulty in analysing maturation at the cellular level is to decide at what stage after priming memory cells become involved in the maintenance of the response. This is especially pertinent since the induction parameters of memory cells appear to be less stringent than those of virgin precursors (Section 3.3.4), and because in this type of experiment antigen is usually given as a depot preparation.

† There is in fact evidence for changes in secretion rate of IgM AFC at different times after immunization.[171]

3.3 B memory cells

3.3.1 *Generation of memory cells*

Immunological memory is carried by both T and B lymphocytes. Thus, in responses to TD hapten-protein conjugates both hapten-reactive ('B cells') and carrier-reactive ('T cells') must be from primed donors to give an optimal secondary response (e.g. Reference 22). Memory generation involves clonal expansions of virgin precursors following primary immunization, and we assume that a similar process replenishes the memory cell pool following re-exposure to antigen. Thus, memory generation involves a *quantal* increase in precursor pool size (discussed in Reference 131). However, as discussed below, there is much evidence that memory cells are also qualitatively different from virgin precursors.

Studies with monoclonal antibody responses strongly suggest that B memory cells and AFC arise from a common clonal precursor.[173] Early experiments indicated that while T memory cells arose within a week after priming, and could be elicited by very low doses of antigen, maximal development of B memory required several weeks, and higher antigen doses.[174,175] More recently, experiments employing B cell clones have shown that B memory cells arise synchronously with AFC in both primary and secondary antibody responses,[176,177] although the memory pool may continue to expand over a longer period, by further recruitment.[178]

3.3.2 *The role of T cells in B memory generation*

The function of T cells in primary antibody responses has been discussed in Section 2.1. Secondary antibody responses to TD antigens are also T cell dependent, as shown by the pioneering studies of Mitchison and others[22,179] (however, see Section 3.3.4).

Therefore, efficient *expression* of memory cells appears to require helper cells. On the other hand, there is evidence that B memory cells can be *generated* in the relative absence of T cells. This was first formally demonstrated by Roelants and Askonas,[180] who found evidence for primed B cells in T deprived mice (B mice) immunized with *Maia squinado* haemocyanin, a highly TD antigen. Similar priming effects in B mice have been observed by others,[181,182] and it has been shown that even nu/nu mice can develop memory to TD antigens.[183,184] It is important to emphasize that in many cases these memory cells were only expressed when the mice were repopulated with T cells. It is clear that the establishment of memory in B mice is accompanied by clonal proliferation,[180,182] although it is uncertain if the memory cells are qualitatively normal (Section 3.3.3).

These experiments do not permit us to say that B memory generation is *totally* T cell independent, since the potent antigens used may induce sufficient clonal expansion of the few residual T cells that may occur even in nu/nu mice.[21] They do, however, indicate that much less T helper function is required for the generation of memory cells, than for either their expression or for the generation of primary IgG responses.

We have observed an apparently related 'dissociation' of memory generation and antibody formation in mice tolerized with DNP-S3 (Reference 185; see also Section 3.3.4). Limiting doses of DNP-S3 suppressed anti-DNP antibody formation to DNP-haemocyanin much more dramatically than hapten-specific memory (re)generation, in both primary and secondary responses. A striking finding was that lightly substituted antigen ($DNP_{0\cdot6}$-S3, Section 3.3.4) which effectively inhibits secondary antibody responses,[186] had much less effect on memory generation than $DNP_{2\cdot7}$-S3.[185] This is probably a crucial observation for our understanding of how memory cells are generated.

Additional data suggested that the 'sparing' of memory was not due to preferential expansion of DNP-S3-resistant clones, but rather that many clones of memory cells proliferated *covertly* in tolerant mice, but failed to make antibodies until the cells were transferred to non-tolerant hosts.[185]

On the basis of the known T-independence of memory generation we[185] interpreted these data as indicating that DNP-S3 (a TI antigen,[38] and a potent B cell tolerogen)[27,186] suppresses TD antibody production by somehow impairing physiological T cell – B cell cooperation. It is known that DNP-S3 'blockades' DNP-binding lymphocytes,[27] and could thereby inhibit the binding of immunogen (DNP-haemocyanin). However, we postulate that this B cell blockade has an *even greater* effect on the reception of T cell helper 'factors' by that cell, which is due to an unknown amplifying mechanism operative in normal cooperation. The net result of this effective reduction of T cell helper activity would be to inhibit the differentiation of AFC, while permitting the generation of memory cells (Section 4.2).

One would therefore predict that TI antigens should induce B cell memory. On the contrary, studies with S3,[187] LE,[10] DNP-LE[38] and C polysaccharide[12] suggest that these antigens elicit little or no memory. LPS appears to be exceptional, in that even minute doses have been found to prime mice for a secondary response.[188] It could be argued that the apparent absence of memory to TI antigens is due to the fact that they elicit principally IgM antibodies (Section 2.1.3), while secondary responses to TD antigens are mostly IgG. However, there seems little doubt that IgM memory exists (e.g. References 175 and 189).

We have been intrigued by the fact that mice primed with antigens such as DNP-LE essentially fail to respond to secondary homologous challenge,[38] although spleen cells from them give a normal response when transferred to irradiated hosts, or an enhanced response if the mice are challenged with a TD antigen such as DNP-haemocyanin (G. G. B. Klaus, unpublished).

Perhaps TI antigens do elicit memory cells, but their expression may be inhibited by:

(a) circulating antibody,[190] due to the persistence of the antigen (Section 2.2.2),
(b) absence of T helper cells,
(c) tolerization of memory cells by re-exposure to a TI stimulus, in the absence of T cells (Section 3.3.4).

An additional complication is that some TI antigens (e.g. LE) appear to induce 'exhaustive clonal differentiation', viz. by stimulating all available precursors to IgM AFC, thereby rendering the animal functionally tolerant.[191]

3.3.3 *The nature of receptors on memory cells*

As discussed (Section 3.1) most virgin B lymphocytes carry surface IgM at some stage in their history. More recently, it has been shown that many normal B cells also express IgD, which appears later than IgM in ontogeny.[130,131] Recent data suggest that 30%–40% of splenic B cells in the mouse carry both Ig classes on their surface. 17%–30% have only IgM and 37%–45% only IgD (Reference 192, and P. C. L. Beverley, and R. M. E. Parkhouse and E. R. Abney, herein).

The effects of priming on the class distribution of B cell Ig receptors are still unclear. A number of studies have shown the appearance of non-IgM bearing antigen-binding cells (ABC) in immunized mice,[12,138,182] but there is evidence that some memory cells still carry IgM. Thus, Pierce *et al.*[193] found that while anti-IgM antibodies suppressed all classes of antibody formation in cultures of unprimed mouse spleen cells, this treatment had progressively less effect on IgG responses of (short-term) primed cells. Ten days after priming anti-IgM still suppressed the IgM and IgA secondary response, but had essentially no effect on the IgG response;[194] the latter, however, was still susceptible to suppression by anti-IgG.

The reported presence of IgG receptors on ABC from immunized animals[138,195] is difficult to evaluate, since it is uncertain if this is derived cytophilically or by *de novo* biosynthesis (see Reference 196). Hämmerling *et al.*[195] showed that mice which are genetic low responders to (T, G)-A–L (and produce only IgM antibodies to this antigen, Section 2.1.4) developed as many IgG-bearing ABC after primary immunization as high responders. This conflicts with recent data from Davie and Paul[182] who found that B mice produced normal numbers of ABC (but very little IgG antibody) to DNP-haemocyanin (DNP-KLH). However, while intact mice rapidly produced non-IgM bearing ABC (Ig class not identified), those in B mice were essentially all IgM-bearing cells. This suggests that T cells are required both for the generation of a primary IgG response (Section 2.1.4), and for a 'switch' in receptors from IgM to Ig(?) (but not for the generation of memory cells, Section 3.3.2).

Claflin *et al.*[12] also found that mice immunized with DNP-KLH developed non-IgM bearing ABC. In contrast, those given the TI antigen C-polysaccharide only produced IgM AFC and IgM-bearing ABC. However, since C-polysaccharide does not appear to generate memory,[12] and since ABC and AFC responses developed with the same tempo, it is unclear if the ABC in immune mice have precursor activity, because IgM AFC are known to bind antigen (Section 3.1.2).

At present we do not know if a switch in receptor class is important in the functional properties of memory cells (Section 3.3.4). The possibility that the Ig(?) on immune ABC ((?) memory cells) may be IgD[192] is consistent with the following:

(a) IgD appears later than IgM in neonatal life and during B cell development in the bone marrow;[141,142] earlier reports that IgD precedes IgM in ontogeny are probably incorrect.[192] This suggests that IgD is a marker of 'mature' B cells.
(b) IgD comprises most (80%–90%) of the surface Ig on lymph node and Peyer's patch B cells, and only 40%–50% of the total surface Ig in the spleen.[141,142] This agrees with the known tissue distributions of memory cells (Section 3.3.5), and also with the observations that unimmunized lymph node cells synthesize predominantly IgG, while spleen cells synthesize mostly IgM.[139]

3.3.4 *Avidity of memory cells for antigen*

It is well known that with many (e.g. protein) antigens the induction of a primary response requires concomitant administration of an adjuvant, while boosting can be achieved with aqueous antigen. Furthermore, it has been shown that primed cells respond to lower concentrations of antigen than unprimed cells,[197–199] and that this probably reflects the greater avidity of memory cells. It has also been claimed that primed B cells become increasingly avid with time after immunization, whereas T cells do not,[167] although the latter point is still controversial.[200,201]

The cell selection hypothesis (Section 3.2.1) predicts that the high avidity of memory cells for antigen is due to selective proliferation of high affinity receptor bearing cells. Arguments against this simple interpretation of immunological maturation were presented in Section 3.2.2. The crux of the problem lies in interpreting factors that contribute towards a cell's capacity to bind antigen. It has been estimated that the average B cell carries 5×10^5 Ig molecules[202,203] on its surface (although there may be cells with considerably lower numbers of receptors). It is clear that multivalent interaction of epitopes with receptors (paratopes) is essentially irreversible, whereas monovalent paratope–epitope interaction is rapidly reversible (for a theoretical discussion, see Reference 172). On *a priori* grounds the overall avidity of a cell for multivalent antigen is therefore determined by the interplay of at least three major factors:

(a) epitope density of the antigen;
(b) paratope affinity;
(c) paratope density, viz. the number of receptors per cell.

As an extreme example, a cell expressing only a few receptors, especially if these are of low affinity, will bind a high epitope density ligand poorly, and a low epitope density antigen not at all. Additional factors may be important such as the stereochemical arrangement (e.g. accessibility, spacing) of epitopes on the ligand, the class of Ig receptor, the degree of mobility of receptors within the plane of the membrane, the capacity of cells to resynthesize shed receptors, etc.

Proponents of the cell selection hypothesis have emphasized the importance of paratope affinity in maturation. Strong indications that other fundamental changes occur in primed precursor cells first emerged from Klinman's extensive investigations employing spleen fragment cultures from non-immune or (DNP-KLH) immune mice. He found striking differences in the parameters of virgin (primary), and primed (secondary) B cell stimulation, which can be summarized as follows (see also Reference 131):

(1) Primary B cells cannot be stimulated by soluble antigen in the absence of primed T cells, while secondary B cells are stimulated (albeit inefficiently) in the absence of carrier recognition.[204,205]
(2) A tenfold molar excess of free hapten inhibited DNP-KLH stimulation of 70% of primary cells, but had much less effect on stimulation of secondary cells.[204,206]
(3) Although the average affinity of antibody produced by secondary cells was fivefold higher than in the primary response, the affinity of the latter was more antigen dose-dependent than of the former.[131,204]
(4) Stimulation of primary precursors was highly specific, even with very high antigen concentrations: a mixture of TNP (2,4,6 trinitrophenyl) and DNP-KLH gave an additive response, i.e. TNP and DNP epitopes (which cross-react markedly) stimulated separate, non-overlapping primary precursors. In complete contrast, 30% of secondary anti-DNP precursors could be stimulated by TNP.[206,207]

Such cross-stimulation by closely related epitopes in immunized individuals was first described as the phenomenon of 'original antigenic sin',[208,209] when individuals exposed to a particular type of influenza virus give a higher antibody titre to the virus with which they were *first* infected, than to the challenge virus. Antigenic sin has recently been shown to be a property of B memory cells.[210]

Cross-stimulation by TNP and DNP haptens in immunized animals[211] was originally ascribed to the degeneracy of the immune response, whereby high affinity antibody (i.e. paratope) exhibits greater cross-reactivity than low affinity antibody. Cramer and Braun,[212] who reported similar cross-stimulation in the streptococcal A carbohydrate system, have challenged this explanation, and in accord with Klinman relate the phenomenon to other triggering properties of memory cells.

Klinman's interpretation of these findings is that primary and secondary B cells 'see' antigen in a basically different fashion.[204] He argues that primary B cells have a certain affinity threshold for triggering (and for tolerance, *vide infra*), and the ease with which the response can be inhibited by free hapten suggests that the paratope-epitope interaction is *functionally monovalent.* Primed B cells have a much higher avidity for antigen, and behave functionally as if they have multivalent receptors. The simplest interpretation is that secondary cells carry more receptors than primary cells.

Our interest in this area emerged from a study of the hapten-specific (B cell) tolerance induced by DNP-S3.[27,29,186] Experiments by Feldmann[213] showed that tolerance induction *in vitro* in primary B cells by DNP-POL was exquisitely dependent on the epitope density of the antigen. He found that conjugates with <2 DNP groups (per 40 000 daltons protein) failed to induce tolerance, i.e. were 'obligate immunogens', while those with >3 DNP groups were obligate tolerogens.

We extended these observations by assaying the effects of DNP-S3 on primary and secondary responses to DNP-proteins *in vivo*.[27,186] The results were as follows:

(1) In agreement with Feldmann[213] lightly substituted antigen ($DNP_{0\cdot6}$-S3: with 0·6 DNP groups per 50 000 daltons S3) failed to suppress primary anti-DNP ABC or AFC responses to DNP-haemocyanin in mice, whereas $DNP_{2\cdot5}$-S3 was highly effective.[27,186]
(2) Only high doses of $DNP_{2\cdot5}$-S3 suppressed the primary response, while approximately tenfold less tolerogen inhibited the secondary response,[27,29,186] and minute doses suppressed the *adoptive* secondary response.[29,186]
(3) In complete contrast to (1) small doses of $DNP_{0\cdot6}$-S3 suppressed 85%–90% of the *secondary* AFC response.[186]
(4) DNP-S3 'blockaded' DNP-binding cells from mice primarily immunized with DNP-KLH (primary ABC), and from boosted DNP-KLH primed mice (secondary ABC) *in vitro*. Thus, addition of $>10^{-9}$ M DNP as $DNP_{2\cdot7}$–S3 to immune spleen cells which were then washed before assay blocked 95% of DNP-specific ABC (Reference 27; see also Reference 26).
(5) $DNP_{2\cdot5}$-S3 was 500–1000-fold more effective (on the basis of molarity of DNP) in blocking primary ABC *in vitro* than $DNP_{0\cdot6}$-S3, while both conjugates blocked *secondary* ABC impartially.[27] It remains possible that blocking of primary ABC by high concentrations of $DNP_{0\cdot6}$-S3 was in fact due to the presence of contaminating higher epitope density molecules.
(6) Monovalent hapten (DNP-lysine), which of course cannot blockade ABC, because it can be washed free of the cells, was a much poorer inhibitor of ABC than DNP-S3, as would be expected. Most importantly, equal concentrations of hapten were needed to inhibit primary and secondary ABC,[27] suggesting that the affinity 'spectra' of receptors on the two cell populations were comparable.
(7) Similarly, both $DNP_{0\cdot6}$-S3 and $DNP_{2\cdot7}$-S3 protected primed functional precursor cells from DNP-specific antigen 'suicide' while only very high concentrations of $DNP_{0\cdot6}$-S3 protected virgin precursors.[214] These experiments were performed by incubating virgin or DNP-protein primed spleen cells with DNP-S3, washing and then exposing them to highly radioactive DNP-protein (to induce suicide), prior to transferring them to adoptive host mice.

These data underscore the marked differences in the avidity of primed and unprimed B cells for multivalent antigen. The high avidity of primed cells does not appear to be due principally to higher affinity receptors ((6) above), in agreeement with other evidence showing that there are memory cells with low affinity receptors (Section 3.2.2). Instead, these findings, in conjunction with those of Klinman's group, provide a forceful argument for a higher receptor density on primed other than virgin B cells.

The strict epitope density requirements for stable antigen binding (and tolerance induction) to primary cells is unexpected, since one would predict that even $DNP_{0\cdot6}$-S3 (molecular weight: $1–10 \times 10^6$)[29] should be capable of multipoint binding to lymphocytes with relatively few receptors, especially since the latter are freely mobile within the plane of the membrane.[215] Perhaps steric factors influence the maximal number of paratope–epitope bonds that can be established by such a molecule.

Direct proof of paratope density changes in B cells will doubtless be difficult to obtain, due to the probable heterogeneity within different B cell subsets. Nonetheless, Strober[216] has presented preliminary data obtained by sorting rat B cells labelled with fluorescent anti-Ig in a fluoresence-activated cell sorter. He found that depletion of brightly stained cells had little effect on the primary response, but essentially abolished the responsiveness of primed cells. This, admittedly fragmentary, evidence suggests that most memory cells have a relatively high density of surface Ig.

Finally, our results with DNP-S3[27,186] appear to contradict earlier work which suggested that it was difficult to tolerize primed animals.[217,218] It should, however, be emphasized that degradable antigens (e.g. gammaglobulins) cannot give meaningful data on tolerance thresholds in immunized animals, since the tolerogen becomes complexed with antibody and in consequence is rapidly eliminated and/or becomes immunogenic,[219] On the other hand, DNP-S3 is multivalent and non-degradable[29] and can therefore effectively bind to cells in the presence of circulating antibody (References 27 and 145; see also Reference 220). Katz *et al.*[123,221] have also shown that a non-degradable antigen (DNP-coupled D-polypeptide) readily suppresses antibody formation in primed animals, thereby reinforcing our conclusion[186] that secondary B cells are more easily tolerized by polymeric antigens than primary B cells. The potential applicability of this observation to the therapy of autoimmune or allergic diseases has already been explored in a preliminary fashion.[222–224]

3.3.5 *Other characteristics of memory cells*

Strober has conducted extensive studies of the changes in rat B lymphocytes after priming (reviewed in Reference 216). Some of these have already been discussed (Section 2.3.4), but in addition he has found that:

(1) Virgin B cells are relatively sessile, whereas primed cells recirculate from blood to lymph:[225] most memory cells can be mobilized into the thoracic duct lymph within 5 days.[226]

(2) Virgin B cells specific for ferritin turn over rather rapidly (within 48 hours), while primed cells have a much longer lifespan.[225] However, both 'virgin' and 'primed' cells responsive to sheep erythrocytes and DNP turn over slowly,[227] suggesting that adult rats may already have been naturally exposed to these antigens. There is additional evidence for the short-lived nature of virgin B cells in other species,[228] and it appears likely that the long-lived B cells in thoracic duct lymph in both mice[229] and rats[225] are largely B memory cells.

The tissue distribution of memory cells appears to reflect their migratory properties, since thoracic duct lymph contains 3–5 times as many memory cells per cell as spleen, while the spleen is a better source of virgin B cells.[216]

Finally, there is preliminary evidence for other changes in B cell surface properties (apart from antigen receptors) as a result of priming. Schlegel *et al.*[230] have separated primary and secondary precursor cells by cell electrophoresis. Memory B cells had predominantly low mobility, while virgin B cells showed faster mobility. They[231] also found that memory cells were less adherent to glass bead columns than virgin cells, a conclusion shared by Schrader.[232]

4. SYNTHESIS: A SPECULATIVE SCHEME OF B CELL MATURATION

In this section some of the evidence discussed above has been integrated into a speculative scheme, whose central tenet is that the properties of various B cell subsets is, in the main, governed by the quantity and/or quality of Ig receptors on these cells. This is based on the widely held belief that Ig receptors provide an *obligatory* initiating signal in both the establishment of tolerance, and in the expression of a lymphocyte's pre-programmed potential for differentiation.

4.1 Induction and tolerance

Committed B cells are assumed to arise independently of antigen in the hypothetical mammalian bursa equivalent (cf. Reference 56). During this stage in their maturation they progressively express increasing numbers of receptors,[223,234] and acquire functional heterogeneity, *which includes variations in avidity for antigen, resulting from differences in paratope affinity and/or density* (Section 3.3.4).

Antigen-reactive cells generated by this process are committed to eventual synthesis of a single species and class of antibody (Bμ, Bγ, etc. cells), even though they may at this stage still carry IgM receptors. Furthermore, virgin Bμ and Bγ cells are each assumed to be heterogeneous in their avidity for antigen, the avidity of Bμ cells in general being lower than that of Bγ cells (but the two populations almost certainly overlap).

In consequence, a certain proportion of (low avidity) Bμ cells can be induced to differentiate to IgM AFC by multivalent, persisting (TI) antigen *per se* (Section 2.1). As a corollary, these cells are resistant to tolerization. The major distinction between TI and TD antigens lies in their capacity to efficiently stimulate this B cell subset. It is assumed that polyclonal B cell activators (mitogens) such as LPS, also stimulate this Bμ cell population (for a discussion see Reference 235). The remaining (higher avidity) Bμ cells may be induced to *proliferate* by antigenic (and (?)mitogenic)[235] stimuli, but fail to develop into high-rate AFC unless additional (T cell derived) maturation signals are present (cf. Reference 236). These cells are therefore not expressed in TI responses (Section 2.1), but may be regarded as 'tolerant'.

In contrast, the differentiation of most Bγ cells is T cell dependent, because of their higher avidity (i.e. susceptibility to tolerance in the absence of T cells).[14,125] The fact that some TI antigens elicit a modest IgG response (Section 2.1.4) again suggests that a certain proportion of Bγ cells may proliferate to 'pre-AFC' stages of maturation in the absence of helper T cells.

It follows that primary responses to TD antigens therefore tend to be heterogeneous (in class and affinity of antibody), while responses to TI antigens are more homogeneous (IgM: probably of low affinity). Following stimulation the progeny of both Bμ and Bγ cells progressively lose Ig receptors during clonal expansion, and thereby become less susceptible to antigen-mediated control: AFC appearing late in the response may thus be essentially devoid of receptors (Section 3.1.2). The observation that IgM AFC carry more receptors than IgG AFC,[145,150–152] suggests that the latter are more 'mature', perhaps because they have undergone more cell divisions, than the former.

Finally, memory cells have higher avidity for antigen than virgin precursors (Section 3.3.4). They can therefore be stimulated, and tolerized by lower concentrations of antigen, and by lower epitope density antigen than virgin cells. It is uncertain if Bμ and Bγ memory cells differ significantly in avidity.

4.2 *Generation of memory cells*

If we adopt the terminology of Sercarz and Coons[237] in which X = virgin precursor, Y = memory cell, Z = AFC, then we can envisage the following maturation pathways in the generation of memory cells:

(a) Following primary immunization:

X --→ Y --→ Z ('linear' pathway)

or

$$X \begin{array}{l} \nearrow Y \\ \searrow Z \end{array} \quad \text{('branched' pathway)}$$

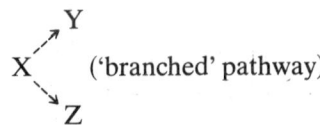

(The dashed arrows represent an unknown number of mitotic cycles, accompanied by maturational changes.)

In other words, Y cells are either an arrested stage in the development of Z, or both Y and Z derive from an 'asymmetrical division' occurring early in clonal expansion. A Z--→Y transition is highly unlikely.[175,238]

(b) Following secondary immunization (i.e. with pre-existing Y cells):

$$Y \dashrightarrow Y \dashrightarrow Z \quad \text{or} \quad Y \begin{array}{l} \nearrow Z \\ \searrow Y \end{array}$$

The 'branched' pathway is the more attractive alternative from existing evidence (Sections 3.3.1 and 3.3.2), and especially on the basis of differences, for example, in Ig receptor properties between Y and Z cells (see below).

What factors determine if a dividing cell becomes a Y or Z cell? Williamson *et al.*[177] have suggested that a major important factor is the ratio of antigen to T cell activity (i.e. signal ①: signal ② ratio, as defined by Bretscher and Cohn,[83] and P. A. Bretscher, herein). They argue that a high signal ①/signal ② ratio (e.g. in an immunized T deprived mouse, or in one tolerized with a TI antigen: Section 3.3.2) would preferentially channel dividing cells towards Y cells, whereas a low ratio favours the production of Z cells. We[185] prefer a simpler version of this model, whereby Y cells are generated normally in the presence of very low numbers of active T cells, insufficient to generate (T dependent) AFC. This is because (a) there is no evidence that T deprived mice develop *more* memory cells than euthymic mice and (b) because a putative signal ① alone (i.e. DNP-S3)[185] failed to induce memory. The latter issue is, however, complex, as discussed in Section 3.3.2.

A variant of this 'signal balance' model of memory generation states that once the presumptive asymmetrical division has occurred, cells destined to become Y cells require only one or a minimal number of 'pulses' of T cell signals to achieve maturity, while those destined to become Z cells require repeated pulses, perhaps at consecutive cell divisions.[185]

These arguments therefore predict that the increased cellular avidity characteristic of memory cells (*vide infra*) develops in a relatively T independent fashion. The 'switch' in receptor class distribution from IgM to Ig(?) ((?)IgD) may, however, be T cell dependent (Section 3.3.3).

4.3 Cellular basis of changes in antibody affinity

The stimulation of heterogeneous (high and low avidity) virgin precursors therefore generates a heterogeneous (high and low affinity) antibody response, and an *equally heterogeneous* population of memory cells (Section 3.2.2). In Section 3.3.4 we explored the likelihood that low affinity memory cells acquire high avidity for antigen because of an increased number of receptors per cell. A further prediction is that such an increase in receptor density also occurs in high affinity precursors. The *overall* avidity of B memory cells is consequently higher than that of the virgin cell population, but the *potential* capacity to synthesize antibodies of high and low affinity remains unchanged.

We can therefore envisage that as antigen levels fall in an immunized animal low avidity precursors (= low affinity receptors) are no longer stimulated as efficiently. This may result in a progressive restriction of the response to relatively homogeneous high affinity antibody (Section 3.2.1), although other patterns are possible (Section 3.2.2). The important point is that low affinity precursors persist, and if they acquire sufficient avidity (due to increasing receptor density) continue to synthesize low affinity antibody throughout the response (Section 3.2.2).

Re-exposure to antigen elicits a secondary response which may be restricted to high affinity antibody,[239] or which may be as heterogeneous as a primary response, if conditions permit the efficient triggering of low affinity memory cells, i.e. by removing them from endogenous control mechanisms (Section 3.2.2). In the latter case, comparable increases in antibody affinity may occur during the secondary response as in the primary response.[158]

The factors that govern the dynamic changes in antibody affinity throughout the immune response are undoubtedly extremely complex. Apart from the decay in antigen levels, other factors such as antigen-specific antibody feedback,[180] regulation by anti-idiotype antibody,[240] and the relative balance between T helper[116,117] and suppressor[241] functions may well determine the fraction of the total clonal repertoire that is expressed at any one time. In particular, the phenomenon of clonal dominance[242] may be extremely important. Although still poorly understood, this may be due to feedback by antibody produced by a rapidly proliferating clone, thereby suppressing the expression of less rapidly proliferating clones.[243]

Finally, it should be emphasized that the maintenance of a heterogeneous population of memory cells undoubtedly has important biological advantages.[158] Restriction of memory cells to a high affinity subpopulation would limit the responsiveness of a primed individual to low doses of antigen, because of the susceptibility of high affinity precursors in particular,[126] and of memory cells in general[27,29,123,186] to tolerization by multivalent antigen. It further follows that effective stimulation of low affinity memory cells (especially by paucivalent antigen) in the presence of circulating (maybe high affinity) antibody necessitates that these cells be of high avidity. This requirement is fulfilled in a teleological sense, by their high receptor density.

5. REFERENCES

1. Möller, G., Ed. (1969). *Transplant. Rev.*, **1**.
2. Katz, D. H. and Benacerraf, B. (1973). *Advan. Immunol.*, **15**, 1–46.
3. Möller, G., Ed. (1975). *Transplant. Rev.*, **23**.
4. Brent, L. and Holborow, J., Eds. (1974). *Progr. Immunol.*, Vol. 2, **3**, 63–120.
5. Wigzell, H. (1974). *Contemp. Topics Immunobiol.*, **3**, 77–96.
6. Warner, N. L. (1974). *Advan. Immunol.*, **19**, 67–92.

7. Coutinho, A. and Möller, G. (1974). *Scand. J. Immunol.*, **3**, 133–139.
8. Humphrey, J. H., Parrott, D. M. V. and East, J. (1964). *Immunology*, **7**, 419–439.
9. Andersson, B. and Blomgren, H. (1971). *Cell. Immunol.*, **2**, 411–420.
10. Miranda, J. J. (1972). *Immunology*, **23**, 829–842.
11. Möller, G. and Michael, G. (1971). *Cell. Immunol.*, **2**, 309–316.
12. Claflin, J. L., Lieberman, R. and Davie, J. M. (1974). *J. exp. Med.*, **139**, 58–73.
13. Howard, J. G., Courtenay, B. M. and Desaymard, C. (1974). *Eur. J. Immunol.*, **4**, 453–456.
14. Mitchell, G. F. (1974). *Contemp. Topics Immunobiol.*, **3**, 97–116.
15. Basten, A. and Howard, J. G. (1973). *Contemp. Topics Immunobiol.*, **2**, 265–291.
16. Sela, M., Mozes, E. and Shearer, G. M. (1972). *Proc. Nat. Acad. Sci.*, **69**, 2696–2700.
17. Feldmann, M. and Basten, A. (1971). *J. exp. Med.*, **134**, 103–119.
18. Mitchell, G. F., Mishell, R. I. and Herzenberg, L. A. (1971). *Prog. Immunol.*, Vol. 1 (Amos, B., Ed.), pp. 324–335.
19. Raff, M. C. and Wortis, H. H. (1970). *Immunology*, **18**, 931–942.
20. Loor, F. and Roelants, G. E. (1974). *Nature*, **251**, 229–230.
21. Kirov, S. M. (1974). *Eur. J. Immunol.*, **4**, 740–745.
22. Mitchison, N. A. (1971). *Eur. J. Immunol.*, **1**, 18–26.
23. Miranda, J. J., Zola, H. and Howard, J. G. (1972). *Immunology*, **23**, 843–855.
24. Howard, J. G., Zola, H., Christie, G. H. and Courtenay, B. M. (1971). *Immunology*, **21**, 535–546.
25. Howard, J. G., Vicari, G. and Courtenay, B. M. (1975). *Immunology*, **29**, 585–597.
26. Wilson, J. D. and Feldmann, M. (1972). *Nature*, **237**, 3–5.
27. Klaus, G. G. B. (1975). *Eur. J. Immunol.*, **5**, 366–372.
28. Britton, S., Wepsic, T. and Möller, G. (1968). *Immunology*, **14**, 491–501.
29. Mitchell, G. F., Humphrey, J. H. and Williamson, A. R. (1972). *Eur. J. Immunol.*, **2**, 460–467.
30. Klaus, G. G. B. and Mitchell, G. F. (1974). *Immunology.*, **27**, 699–710.
31. Pryjma, J., Humphrey, J. H. and Klaus, G. G. B. (1974). *Nature*, **252**, 505–506.
32. Andersson, J., Sjöberg, O. and Möller, G. (1972). *Transplant. Rev.*, **11**, 131–177.
33. Greaves, M. F. and Janossy, G. (1972). *Transplant. Rev.*, **11**, 87–130.
34. Coutinho, A. (1975). *Transplant. Rev.*, **23**, 49–65.
35. Klaus, G. G. B., Janossy, G. and Humphrey, J. H. (1975). *Eur. J. Immunol.*, **5**, 105–111.
36. Klaus, G. G. B. and Humphrey, J. H. (1975). *Transplant. Rev.*, **23**, 105–118.
37. del Guercio, P. and Leuchars, E. (1972). *J. Immunol.*, **109**, 951–957.
38. Klaus, G. G. B. and Humphrey, J. H. (1974). *Eur. J. Immunol.*, **4**, 370–377.
39. Desaymard, C. and Feldmann, M. (1975). *Cell. Immunol.*, **16**, 106–114.
40. Fidler, J. M. (1975). *Cell. Immunol.*, **16**, 223–236.
41. Jacobs, D. and Morrison, D. C. (1975). *J. Immunol.*, **114**, 360–364.
42. Feldmann, M., Greaves, M. F., Parker, D. C. and Rittenberg, M. B. (1974). *Eur. J. Immunol.*, **4**, 591–598.
43. Trump, G. N. (1975). *J. Immunol.*, **114**, 682–687.
44. Mosier, D. E., Johnson, B. M., Paul, W. E. and McMaster, P. R. B. (1974). *J. exp. Med.*, **139**, 1354–1360.
45. Sharon, R., McMaster, P. R. B., Kask, A. M., Owens, J. D. and Paul, W. E. (1975). *J. Immunol.*, **114**, 1585–1589.
46. Taylor, R. B. and Wortis, H. H. (1968). *Nature*, **220**, 927–929.
47. Kindred, B. (1971). *Eur. J. Immunol.*, **1**, 59–61.
48. Pantelouris, E. M. and Flisch, P. A. (1972). *Eur. J. Immunol.*, **2**, 236–239.
49. Reed, N. D. and Jutila, J. W. (1972). *Proc. Soc. exp. Biol. Med.*, **139**, 1234–1237.
50. Wortis, H. H. (1971). *Clin. exp. Immunol.*, **8**, 305–317.
51. Playfair, J. H. L. and Purves, E. C. (1971). *Immunology*, **21**, 113–121.
52. Gorczynski, R. and Feldmann, M. (1975). *Cell. Immunol.*, **18**, 88–97.
53. Andersson, J., Sjöberg, O. and Möller, G. (1972). *Eur. J. Immunol.*, **2**, 349–353.
54. Parkhouse, R. M. E., Janossy, G. and Greaves, M. F. (1972). *Nature* (*New Biol.*), **235**, 21–23.
55. Gray, H. M. (1969). *Advan. Immunol.*, **10**, 51–104.
56. Cooper, M. D., Lawton, A. R. and Kincade, P. W. (1972). *Contemp. Topics Immunobiol.*, **1**, 33–47.
57. Feldmann, M. (1972). *J. exp. Med.*, **135**, 735–753.
58. Feldmann, M., Howard, J. G. and Desaymard, C. (1975). *Transplwnt. Rev.*, **23**, 78–97.
59. Diener, E. and Paetkau, V. H. (1972). *Proc. Nat. Acad. Sci.*, **69**, 2364–2368.

60. Klaus, G. G. B. and Cross, A. M. (1974). *Cell. Immunol.*, **14**, 226–241.
61. Möller, G., Coutinho, A. and Persson, U. (1975). *Scand. J. Immunol.*, **4**, 37–45.
62. Dukor, P. and Hartmann, K. U. (1973). *Cell. Immunol.*, **7**, 349–356.
63. Torrigiani, G. (1972). *J. Immunol.*, **108**, 161–164.
64. Katz, D. H. (1972). *Transplant. Rev.*, **12**, 141–179.
65. Katz, D. H., Paul, W. E., Goidl, E. A. and Benacerraf, B. (1971). *J. exp. Med.*, **133**, 169–186.
66. Kreth, H. W. and Williamson, A. R. (1971). *Nature*, **234**, 454–456.
67. Ordal, J. C. and Grumet, F. C. (1972). *J. exp. Med.*, **136**, 1195–1206.
68. Hamilton, J. A. and Miller, J. F. A. P. (1973). *J. exp. Med.*, **138**, 1009–1014.
69. Osborne, D. P. and Katz, D. H. (1973). *J. exp. Med.*, **137**, 991–1008.
70. Osborne, D. P. and Katz, D. H. (1972). *J. exp. Med.*, **136**, 439–454.
71. Katz, D. H. and Osborne, D. P. (1972). *J. exp. Med.*, **136**, 455–465.
72. Rajewsky, K., Roelants, G. E. and Askonas, B. A. (1972). *Eur. J. Immunol.*, **2**, 592–598.
73. Byfield, P., Christie, G. H. and Howard, J. G. (1973). *J. Immunol.*, **111**, 72–81.
74. Braley-Mullen, H. (1974). *J. Immunol.*, **113**, 1909–1920.
75. Schimpl, A. and Wecker, E. (1975). *Transplant. Rev.*, **23**, 176–188.
76. Klaus, G. G. B. and McMichael, A. J. (1974). *Eur. J. Immunol.*, **4**, 505–511.
77. McDevitt, H. O. and Landy, M., Eds. (1972). *Genetic Control of Immune Responsiveness.* Academic, New York.
78. Taussig, M. J., Mozes, E. and Isac, R. (1974). *J. exp. Med.*, **140**, 301–312.
79. Quintans, J. and Lefkovits, I. (1974). *Eur. J. Immunol.*, **41**, 617–621.
80. Shinohara, N. and Tada, T. (1974). *Int. Arch. Allergy Appl. Immunol.*, **47**, 762–776.
81. Cosenza, H., Quintans, J. and Lefkovits, I. (1975). *Eur. J. Immunol.*, **5**, 343–349.
82. Baker, P. J., Prescott, B., Stashak, P. W. and Amsbaugh, D. F. (1971). *J. Immunol.*, **107**, 719–724.
83. Bretscher, P. A. and Cohn, M. (1970). *Science*, **169**, 1042–1049.
84. Bretscher, P. A. (1974). *Cell. Immunol.*, **13**, 171–195.
85. Davies, A. J. S., Carter, R. L., Leuchars, E., Wallis, V. and Dietrich, F. M. (1970). *Immunology*, **19**, 945–957.
86. Sjöberg, O. (1971). *J. exp. Med.*, **133**, 1015–1025.
87. Meyer, K., Davidson, E., Linker, A. and Hoffman, P. (1956). *Biochim. Biophys. Acta*, **21**, 506–518.
88. Green, I., Paul., W. E. and Benacerraf, B. (1968). *J. exp. Med.*, **127**, 43–53.
89. Henney, C. and Ishizaka, K. (1970). *J. Immunol.*, **104**, 1540–1549.
90. Paul, W. E., Thorbecke, G. J., Siskind, G. W. and Benacerraf, B. (1969). *Immunology*, **17**, 85–92.
91. Sanfilipo, F. and Scott, D. W. (1974). *J. Immunol.*, **113**, 1661–1667.
92. Feldmann, M. (1972). *Eur. J. Immunol.*, **2**, 130–137.
93. Taylor, R. B. (1964). *Immunology*, **7**, 595–602.
94. Howard, J. G., Christie, G. H., Courtenay, B. M., Leuchars, E. and Davies, A. J. S. (1971). *Cell. Immunol.*, **2**, 614–624.
95. Gershon, R. K. (1974). *Contemp. Topics. Immunobiol.*, **3**, 1–40.
96. Kerbel, R. S. and Eidinger, D. (1971). *Eur. J. Immunol.*, **2**, 114–118.
97. Rotter, V. and Trainin, N. (1974). *Cell. Immunol.*, **13**, 76–86.
98. Baker, P. J., Stashak, P. W. and Amsbaugh, D. F. (1970). *J. Immunol.*, **104**, 1313–1315.
99. Baker, P. J., Stashak, P. W., Amsbaugh, D. F., Prescott, B. and Barth, R. F. (1970). *J. Immunol.*, **105**, 1581–1583.
100. Kerbel, R. S. and Eidinger, D. (1971). *J. Immunol.*, **106**, 917–926.
101. Andersson, B. and Blomgren, H. (1975). *Nature*, **253**, 476–477.
102. Manning, J. K., Reed, N. D. and Jutila, J. W. (1972). *J. Immunol.*, **108**, 1470–1472.
103. Poe, W. J. and Michael, J. G. (1975). *Cell. Immunol.*, **15**, 255–262.
104. Warr, G. W., Ghaffar, A. and James, K. (1975). *Cell. Immunol.*, **17**, 366–373.
105. Lance, E. M. (1970). *Clin. exp. Immunol.*, **6**, 789–802.
106. Ghaffar, A. and James, K. (1973). *Immunology*, **24**, 1075–1085.
107. Baker, P. J., Reed, N. D., Stashak, P. W., Amsbaugh, D. F. and Prescott, B. (1973). *J. exp. Med.*, **137**, 1431–1441.
108. Janossy, G. and Greaves, M. F. (1972). *Clin. exp. Immunol.*, **10**, 525–536.
109. Barthold, D. R., Prescott, B., Stashak, P. W., Amsbaugh, D. F. and Baker, P. J. (1974). *J. Immunol.*, **112**, 1042–1050.

110. Barth, R. F., Singla, O. and Ahlers, P. (1973). *Cell. Immunol.*, **7**, 380–388.
111. Veit, B. R. and Michael, J. G. (1972). *J. Immunol.*, **109**, 547–553.
112. Hraba, T., Havas, H. F. and Pickard, A. R. (1970). *Int. Arch. Allergy Appl. Immunol.*, **38**, 635–647.
113. Hamilton, J. A. and Miller, J. F. A. P. (1974). *Eur. J. Immunol.*, **4**, 261–268.
114. Klaus, G. G. B. and Cross, A. M. (1974). *Scand. J. Immunol.*, **3**, 797–808.
115. Huchet, R. and Feldmann, M. (1974). *Eur. J. Immunol.*, **4**, 768–771.
116. Rittenberg, M. B. and Bullock, W. W. (1972). *Immunochemistry*, **9**, 491–504.
117. Fidler, J. M. and Golub, E. S. (1973). *J. Immunol.*, **11**, 317–323.
118. Weigle, W. O. (1973). *Advan. Immunol.*, **16**, 61–92.
119. Howard, J. G. and Mitchison, N. A. (1975). *Progr. Allergy*, **18**, 43–96.
120. Weber, G. and Kölsch, E. (1972). *Eur. J. Immunol.*, **2**, 191–193.
121. Tada, T. and Takemori, T. (1974). *J. exp. Med.*, **140**, 239–252.
122. Basten, A., Miller, J. F. A. P., Sprent, J. and Cheers, C. (1974). *J. exp. Med.*, **140**, 199–217.
123. Katz, D. H., Davie, J. M., Paul, W. E. and Benacerraf, B. (1971). *J. exp. Med.*, **134**, 201–221.
124. Mitchell, G. F., LaFleur, L. and Andersson, R. (1974). *Scand. J. Immunol.*, **3**, 39–49.
125. Mitchell, G. F. (1974). *Progr. Immunol.* (Brent, L. and Holborow, J., Eds.), Vol. 2, pp. 89–98.
126. Werblin, T. P. and Siskind, G. W. (1972). *Transplant. Rev.*, **8**, 104–120.
127. Gershon, R. K. and Paul, W. E. (1971). *J. Immunol.*, **106**, 872–874.
128. Elfenbein, G. J., Green, I. and Paul, W. E. (1973). *Eur. J. Immunol.*, **3**, 640–644.
129. Gudat, F. G., Harris, T. N., Harris, T. S. and Hummeler, K. (1970). *J. exp. Med.*, **132**, 448–474.
130. Thornthwaite, J. T. and Leif, R. C. (1974). *J. Immunol.*, **113**, 1897–1908.
131. Klinman, N. R. and Press, J. L. (1975). *Transplant. Rev.*, **24**, 41–83.
132. Kincade, P. W. and Cooper, M. D. (1971), *J. Immunol.*, **106**, 371–382.
133. Kincade, P. W., Lawton, A. R. and Cooper, M. D. (1970). *Proc. Nat. Acad. Sci.*, **67**, 1918–1924.
134. Lawton, A. R., Asofsky, R., Hylton, M. B. and Cooper, M. D. (1972). *J. exp. Med.*, **135**, 277–297.
135. Manning, D. D. and Jutila, J. W. (1972). *J. exp. Med.*, **135**, 1317–1333.
136. Pernis, B., Forni, L. and Amante, L. (1970). *J. exp. Med.*, **132**, 1001–1018.
137. Owen, J. J. T., Cooper, M. D. and Raff, M. C. (1974). *Nature*, **249**, 361–363.
138. Greaves, M. F. and Hogg, N. M. (1971). *Progr. Immunol.*, Vol. 1 (Amos, B., Ed.), 111–118.
139. Parkhouse, R. M. E. (1973). *Transplant. Rev.*, **14**, 131–144.
140. Vitetta, E. S., Baur, S. and Uhr, J. W. (1971). *J. exp. Med.*, **134**, 242–264.
141. Abney, E. R. and Parkhouse, R. M. E. (1974). *Nature* (*Lond.*), **262**, 600–602.
142. Vitetta, E. S., Melcher, U., McWilliams, M., Phillips-Quagliata, J., Lamm, M. E. and Uhr, J. W. (1975). *J. exp. Med.*, **141**, 206–215.
143. Baker, P. J., Stashak, P. W., Amsbaugh, D. F. and Prescott, B. (1971). *Immunology*, **20**, 481–492.
144. Schrader, J. W. and Nossal, G. J. V. (1974). *J. exp. Med.*, **139**, 1582–1598.
145. Klaus, G. G. B. (1976). *Eur. J. Immunol.*, **6**, 200–207.
146. McConnell, I. (1971). *Nature* (*New Biol.*), **233**, 177–178.
147. Nossal, G. J. V. and Lewis, H. (1972). *J. exp. Med.*, **135**, 1416–1421.
148. Bankert, R. B. and Wolf, B. (1973). *J. Immunol.*, **111**, 1790–1799.
149. Andersson, J., Buxbaum, J., Citronbaum, R., Douglas, S., Forni, L., Melchers, F., Pernis, B. and Stott, D. (1974), *J. exp. Med.*, **140**, 742–763.
150. Takahashi, T., Old, L. J., McIntire, K. R. and Boyse, E. A. (1971). *J. exp. Med.*, **134**, 815–832.
151. Pernis, B., Forni, L. and Amante, L. (1971). *Ann. N.Y. Acad. Sci.*, **190**, 420–429.
152. Bell, C. (1975). *Fed. Proc.*, **34**, 996 (Abstract).
153. Eisen, H. N. and Siskind, G. W. (1964). *Biochemistry*, **3**, 996–1008.
154. Steiner, L. A. and Eisen, H. N. (1967). *J. exp. Med.*, **126**, 1161–1183.
155. Siskind, G. W. and Benacerraf, B. (1969). *Advan. Immunol.*, **10**, 1–50.
156. Kim, Y. T. and Siskind, G. W. (1974). *Clin. exp. Immunol.*, **17**, 329–338.
157. Burnet, F. M. (1959). *The Clonal Selection Theory of Immunity*. Cambridge University Press, London.
158. Macario, A. J. L. and Conway de Macario, E. (1975). *Immunochemistry*, **12**, 249–262.
159. Doria, G., Schiaffini, G., Garavini, M. and Mancini, C. (1972). *J. Immunol.*, **109**, 1245–1253.
160. Kimball, J. W. (1972). *Immunochemistry*, **9**, 1169–1184.
161. Wu, W. H. and Rockey, J. H. (1969). *Biochemistry*, **8**, 2719–2723.

162. Haber, E. and Stone, M. (1969). *Israel J. Med. Sci.*, **5**, 332–337.
163. Kim, Y. T., Werblin, T. P. and Siskind, G. W. (1974). *Immunochemistry*, **11**, 685–690
164. Macario, J. L., Conway de Macario, E. and Celada, F. (1973). *Nature* (*New Biol.*), **241**, 22–23.
165. Macario, J. L. and Conway de Macario, E. (1974). *Immunochemistry*, **11**, 619–621.
166. Celada, F., Schmidt, D. and Strom, R. (1969). *Immunology*, **17**, 189–198.
167. Möller, E., Bullock, W. W. and Mäkela, O. (1973). *Eur. J. Immunol.*, **3**, 172–179.
168. Andersson, B. (1970). *J. exp. Med.*, **132**, 77–88.
169. Davie, J. M. and Paul, W. E. (1972). *J. exp. Med.*, **135**, 660–674.
170. North, J. R. and Askonas, B. A. (1974). *Eur. J. Immunol.*, **4**, 361–366.
171. Jaroszewski, J. (1973). *Z. Immun. Forsch.*, **144**, 472–481.
172. Bell, G. I. and DeLisi, C. P. (1974). *Cell. Immunol.*, **10**, 415–431.
173. Askonas, B. A., Williamson, A. R. and Wright, B. E. G. (1970). *Proc. Nat. Acad. Sci.*, **67**, 1398–1403.
174. Niederhuber, J. E. and Möller, E. (1973). *Cell. Immunol.*, **6**, 407–417.
175. Cunningham, A. J. and Sercarz, E. E. (1971). *Eur. J. Immunol.*, **1**, 413–421.
176. McMichael, A. J. and Williamson, A. R. (1974). *J. exp. Med.*, **139**, 1361–1367.
177. Williamson, A. R., McMichael, A. and Zitron, I. (1974). In *The Immune System: Genes, Receptors, Signals* (Sercarz, E. E., Williamson, A. R. and Fox, C. F., Eds.), Academic, New York, pp. 387–393.
178. Kreth, H. W. and Williamson, A. R. (1973). *Eur. J. Immunol.*, **3**, 141–147.
179. Rajewsky, K., Schirrmacher, V., Nase, S. and Jerne, N. K. (1969). *J. exp. Med.*, **129**, 1131–1143.
180. Roelants, G. E. and Askonas, B. A. (1972). *Nature* (*New Biol.*), **239**, 63–64.
181. Sinclair, N. R. St. C. (1967). *Immunology*, **12**, 559–566.
182. Davie, J. M. and Paul, W. E. (1974). *J. Immunol.*, **113**, 1438–1445.
183. Diamantstein, T. and Blitstein-Willinger, E. (1974). *Eur. J. Immunol.*, **4**, 830–832.
184. Schrader, J. W. (1975). *J. Immunol.*, **114**, 1665–1669.
185. Klaus, G. G. B. and Willcox, H. N. A. (1975). *Eur. J. Immunol.*, **5**, 699–704.
186. Klaus, G. G. B. and Humphrey, J. H. (1975). *Eur. J. Immunol.*, **5**, 361–365.
187. Baker, P. J., Stashak, P. W., Amsbaugh, D. F. and Prescott, B. (1971). *Immunology*, **20**, 469–480.
188. Reed, N. D., Manning, J. K. and Rudbach, J. A. (1973). *J. Inf. Dis.* (*Endotoxin Suppl.*), 570–574.
189. Nossal, G. J. V., Austin, G. M. and Ada, G. L. (1965). *Immunology*, **9**, 333–348.
190. Grantham, W. G. and Fitch, F. W. (1975). *J. Immunol.*, **114**, 394–398.
191. Howard, J. G. and Courtenay, B. K. (1974). *Eur. J. Immunol.*, **4**, 603–608.
192. Parkhouse, R. M. E. and Abney, E. R. (1975). In *Membrane Receptors of Lymphocytes* (Seligmann, M., Preud'homme, J.-L. and Kourilsky, F. M., Eds.), North-Holland, Amsterdam, pp. 51–56.
193. Pierce, C. W., Solliday, S. M. and Asofsky, R. (1972). *J. exp. Med.*, **135**, 675–697.
194. Pierce, C. W., Solliday, S. M. and Asofsky, R. (1972). *J. exp. Med.*, **135**, 698–710.
195. Hämmerling, G. J., Masuda, T. and McDevitt, H. O. (1973). *J. exp. Med.*, **137**, 1180–1200.
196. Vitetta, E. S. and Uhr, J. W. (1975). *Biochem. Biophys. Acta*, **415**, 253–271.
197. Tao, T.-W. (1968). *J. Immunol.*, **101**, 1253–1263.
198. Pierce, C. W. (1969). *J. exp. Med.*, **130**, 345–364.
199. Bullock, W. W. and Rittenberg, M. B. (1970). *J. exp. Med.*, **132**, 926–940.
200. Jokipii, A. M. M. and Jokipii, L. (1974). *Cell. Immunol.*, **13**, 241–250.
201. Wilson, J. D. and Feldmann, M. (1973). *Nature* (*New Biol.*), **245**, 177–180.
202. Rabellino, E., Colon, S., Grey, H. M. and Unanue, E. R. (1971). *J. exp. Med.*, **123**, 156–167.
203. Klein, E., Eskeland, T., Inoue, M., Strom, R. and Johansson, B. (1970). *Exp. Cell Res.*, **62**, 133–148.
204. Klinman, N. R. (1972). *J. exp. Med.*, **136**, 241–260.
205. Klinman, N. R. and Doughty, R. A. (1972). *J. exp. Med.*, **138**, 473–478.
206. Klinman, N. R. and Press, J. L. (1975). *J. exp. Med.*, **141**, 1133–1146.
207. Klinman, N. R., Press, J. L. and Segal, G. P. (1975). *J. exp. Med.*, **138**, 1276–1281.
208. Fazekas de St. Groth, S. and Webster, R. G. (1966). *J. exp. Med.*, **124**, 331–345.
209. Fazekas de St. Groth, S. and Webster, R. G. (1966). *J. exp. Med.*, **124**, 347–361.
210. Virelizier, J. L., Allison, A. C. and Schild, G. C. (1974). *J. exp. Med.*, **140**, 1571–1578.
211. Little, J. R. and Eisen, H. N. (1969). *J. exp. Med.*, **129**, 247–265.

212. Cramer, M. and Braun, D. (1973). *J. exp. Med.*, **138**, 1533–1544.
213. Feldmann, M. (1972). *J. exp. Med.*, **135**, 735–753.
214. Willcox, H. N. A. and Klaus, G. G. B. (1976). *Eur. J. Immunol.*, **6**, 379–382.
215. Taylor, R. B., Duffus, W. P., Raff, M. C. and de Petris, S. (1971). *Nature* (*New Biol.*), **233**, 225–229.
216. Strober, S. (1975). *Transplant. Rev.*, **24**, 84–112.
217. Dresser, D. W. (1962). *Immunology*, **5**, 161–168.
218. Dresser, D. W. (1965). *Immunology*, **9**, 261–268.
219. Von Felten, A. and Weigle, W. O. (1975). *Cell. Immunol.*, **18**, 31–40.
220. Bystryn, J. C., Siskind, G. W. and Uhr, J. W. (1975). *J. exp. Med.*, **141**, 1227–1237.
221. Katz, D. H., Hamaoka, T. and Benacerraf, B. (1972). *J. exp. Med.*, **136**, 1404–1429.
222. Eshhar, Z., Benacerraf, B. and Katz, D. H. (1975). *J. Immunol.*, **114**, 872–876.
223. Katz, D. H., Hamaoka, T. and Benacerraf, B. (1973). *Proc. Nat. Acad. Sci.*, **70**, 2776–2780.
224. Katz, D. H., Stechschulte, D. J. and Benacerraf, B. (1975). *J. Allergy Clin. Immunol.*, **55**, 403–410.
225. Strober, S. (1972). *J. exp. Med.*, **136**, 851–871.
226. Strober, S. and Dilley, J. (1973). *J. exp. Med.*, **137**, 1275–1292.
227. Strober, S. and Dilley, J. (1973). *J. exp. Med.*, **138**, 1331–1344.
228. Brahim, F. and Osmond, D. G. (1970). *Anat. Rec.*, **168**, 139–160.
229. Sprent, J. and Basten, A. (1973). *Cell. Immunol.*, **7**, 40–59.
230. Schlegel, R. A. and Shortman, K. (1975). *Cell. Immunol.*, **16**, 203–215.
231. Schlegel, R. A. and Shortman, K. (1975). *J. Immunol.*, **115**, 94–111.
232. Schrader, J. W. (1974). *Cell. Immunol.*, **10**, 380–393.
233. Osmond, D. G. and Nossal, G. J. V. (1974). *Cell. Immunol.*, **13**, 132–145.
234. Ryser, J. E. and Vassalli, P. (1974). *J. Immunol.*, **113**, 719–728.
235. Janossy, G. and Greaves, M. F. (1975). *Transplant. Rev.*, **25**, 177–236.
236. Askonas, B. A., Schimpl, A. and Wecker, E. (1974). *Eur. J. Immunol.*, **4**, 164–172.
237. Sercarz, E. E. and Coons, A. H. (1962). In *Mechanisms of Immunological Tolerance* (Hasek, M., Lengerova, A. and Vojtiskova, M., Eds.), Czechoslovak Academy of Science Press, Prague, pp. 73–82.
238. L'Age-Stehr, J. and Herzenberg, L. A. (1970). *J. exp. Med.*, **131**, 1093–1108.
239. Steiner, L. A. and Eisen, H. N. (1967). *J. exp. Med.*, **126**, 1185–1205.
240. Kluskens, L. and Köhler, H. (1974). *Proc. Nat. Acad. Sci.*, **71**, 5083–5088.
241. Takemori, T. and Tada, T. (1974). *J. exp. Med.*, **140**, 253–266.
242. Askonas, B. A. and Williamson, A. R. (1972). *Nature*, **238**, 339–341.
243. McMichael, A. J. (1974). *The Clonal Expression of Antibody Forming Cells*, Thesis, National Institute for Medical Research, London.

Chapter 12

The Biological Significance of the Mixed Leukocyte Reaction

M. NABHOLZ
V. C. MIGGIANO

Abbreviations

CML: cell-mediated lympholysis
GvH: graft-versus-host (reaction)
LCM: lymphochoriomeningitis (virus)
MHA: major histocompatibility antigens
MHC: major histocompatibility complex
MLC: mixed leukocyte culture
MLI: mixed leukocyte interaction
MLR: mixed leukocyte reaction
PHA: phytohaemagglutinin
TNP: trinitrophenyl

1. PREFACE

This chapter is divided into two almost independent parts. Rather than compiling an exhaustive review of the work related to the mixed leukocyte reaction (MLR), something which, we think, would be incredibly boring both to write and to read, we chose to give, in the first part, a fairly eclectic and inevitably biased interpretative view of the experimental data accumulated. In any case, we do not think we have said anything which is in stark contradiction with existing solid evidence. It is more in matters of emphasis in which our view differs from others. Our choice of references reflects our approach. Some apparent omissions, if they are not due to oversight, may have been made, for we think that the evidence, although solid and relevant when it was published, has been superseded by more direct demonstrations of the same point.

Quite often we have taken into account work which, although of obvious importance is, in our opinion, only preliminary and not conclusive. Mostly we have tried to indicate when we found ourselves confronted with this situation, which seems to become increasingly common in immunology.

The second part of the chapter presents the results of an attempt to integrate various experimental, conjectural and speculative features of T cell-mediated immune responsiveness and the major histocompatibility complex (MHC) into a unifying hypothesis. This involved forays into fields in which our knowledge is sometimes very limited. Thus we may have committed severe blunders in our interpretations and extrapolations. While we will be grateful if they will be pointed out to us we also hope that such errors 'in detail' will not deter those who recognize them from evaluating our hypothesis as a whole.

We make no claims to have been the first to have recognized the importance of the findings on which our hypothesis is based and we know, from discussions and preprints, that many of the mechanisms that we propose have also been thought of by others.* What seemed attractive to us was the possibility that emerged to incorporate them into a coherent structure, based principally on a modification of Jerne's hypothesis[2] on the generation of antibody diversity.

Building this framework we found that the one central prediction is that the genes determining the specificity of T cell recognition structures must be part of the major

* Shortly before this chapter was sent to the editors a paper[1] appeared which, in its reasoning and conclusions, follows many of the same lines as this text.

histocompatibility complex. This had, of course, been postulated previously, as a straightforward interpretation of the finding that MHC-linked genes control immune responsiveness. Here it is rather the common endpoint of different deductive pathways all starting from our modification of Jerne's hypothesis.

The ideas outlined here are, as is usual, to a large extent the fruit of continuous exchanges of views and interpretations with a large number of people. Of these we would like to mention Harald v. Boehmer, Antonio Coutinho, Bruce Elliott, Gary Fathman, Jon Sprent and Jorge Vives. Marie-Claire Herzog and Margaret Maraggiulo provided excellent and patient typing assistance.

2. MLR: THE PHENOMENON AND ITS PROPERTIES

2.1 The phenomenon and its measurement

When leukocytes from two unrelated individuals are mixed and cultured in a suitable medium (mixed leukocyte culture = MLC), a certain fraction of the cultured lymphocytes undergoes a series of changes, which include increases in the rate of synthesis of macromolecules, blast transformation and proliferation.[3,4] All these changes occur within a few days after the cultures are set up. Their sequence corresponds to that observed when lymphocytes are stimulated with unspecific mitogens such as phytohaemagglutinin or concanavalin A.[5] The ensemble of these phenomena is called the mixed leukocyte (or lymphocyte) interaction (MLI) or reaction (MLR). The occurrence of MLR has been demonstrated in all mammalian species investigated, as well as in birds and in certain amphibians.

Given the different aspects of the response, a number of techniques can be used to measure it more or less quantitatively. The most common ones are based on a direct estimation of the fraction of blast and/or dividing cells[3,4] in the culture or on measurements of the uptake, by the cultured cell population, of radioactive amino acids (^{3}H-leucine)[6] or DNA precursors (^{3}H-methyl-thymidine,[4,7] ^{125}iodo-deoxyuridine[8]) into acid precipitable material. DNA precursor uptake into single cells can be monitored through the use of autoradiography.[3]

Many different culture systems and media have been used. The source, preparation and numbers of cells used as well as cell harvesting techniques vary[3,4,9–13] and are usually chosen according to empirical determinations of optimal procedures for a given species and the purpose of the experiments. When untreated unrelated leukocytes are mixed with each other then both populations respond: one obtains a bidirectional reaction.[3,4] As this complicates the interpretation of results in many cases, reactions are made unidirectional in one of two ways, either by blocking DNA-synthesis of one cell population ('stimulators') irreversibly, usually by treatment with mitomycin C[7] or X-rays,[14] or by mixing parental ('responder') with F_1 ('stimulator') cells (see Reference 9 for review). The latter method is based on the findings that MLR is genetically controlled and follows the laws of transplantation. The genetic control of MLR is further discussed below, as well as the fact that the two types of unidirectional MLR may not always be equivalent.

The magnitude of the unidirectional response can be estimated in several ways (for review, see Reference 15). The value obtained from the reaction of the responder (R) to the stimulator Ⓢ cells (inactivated stimulator cells 'A' being denoted as Ⓐ), is denoted as v(RⓈ); in all cases, it has to be compared to the background result obtained in an autologous (e.g. man) or syngeneic (e.g. mouse) control MLC: v(R®). Activation can be measured as the difference between the two values usually called $\Delta = v(\text{R}Ⓢ) - v(\text{R}Ⓡ)$ or as their ratio, called the stimulation index,

$$\text{S.I.} = \frac{v(\text{R}Ⓢ)}{v(\text{R}Ⓡ)}.$$

The capacity of responder cells to react and of stimulator cells to activate should be checked in appropriate control mixtures. Significance of differences between experimental and control MLCs has to be ascertained by appropriate statistical tests. In genetic studies of human populations other measures (see Reference 16) have been introduced which are meant to account for the often considerable fluctuations of results obtained in different experiments even with the same responder–stimulator combination. The most common one is called 'relative response' and is derived by comparing the MLR of a given responder cell population to any particular stimulator with the activation of the same responder by a pool of stimulator cells from a number of unrelated donors.[17]

Compared to many other biological systems the problem of variable backgrounds is particularly severe, partly owing to the unspecific nature of the parameters used to measure responses.* Therefore, the sensitivity of different assay systems can vary widely and a particular responder–stimulator combination can easily give a negative result in one system and a positive one in another. As we will see later, such uncertainty is, in part, the source of arguments concerning the interpretations of the genetic analysis of MLR.

The time course of events depends on the system used. An increase in protein synthesis can usually be detected, using fresh cells from normal (unimmunized) donors, as early as 24 hours after initiation of cultures.[6] DNA synthesis and division begin to mount after two or three days and reach a peak after four to six days, after which time the values of these parameters tend to fall rapidly (for review, see Reference 9). The kinetics of the response are, however, quite variable depending on the system used, and even in a given system reproducibility requires rigorous controls of all parameters and may be very difficult to achieve. The reasons for the rapid decline of the response after its peak are largely unknown, but are probably, at least to some extent, artefacts dependent on the culture system.

2.2 Genetic control

2.2.1 *Association with the major histocompatibility complex*

The observation that leukocytes from one individual 'respond' to those of another, unrelated one, suggests that MLR may be controlled by genes at one or more polymorphic loci. Indeed, it was soon established that in man the reaction is associated with incompatibility at loci closely linked to the HLA system.[18] In the meantime studies on a number of other mammalian,[19–24] as well as avian,[25] and even amphibian,[26] species have demonstrated a striking homology: in all cases studied, with the exception of the mouse (see below), there seems to exist only one highly polymorphic genetic region controlling MLR,

* When, for example, DNA-precursor uptake is measured, the background incorporation is not necessarily, or even likely to be, due to cells belonging to the same subpopulation as MLR-responsive cells. In most experimental systems and laboratories it varies considerably, and therefore any presentation of MLR data not including the corresponding values is of highly questionable significance.

which is associated with the loci that control the major histocompatibility determinants as measured by graft rejection or alloantibody formation (major histocompatibility antigens = MHA). Today MLR has become one of the recognized attributes of the major histocompatibility complex (MHC), a segment of the genome which includes, at least in higher vertebrates, loci controlling specific immune responsiveness (chicken,[25,27,28] mammals),[29–35] levels and structural variation of complement components (in mouse and man, for review see Reference 36)[37–39] and the susceptibility to disease (mouse,[40–45] man (L. P. Ryder and A. Svejgaard, herein), chicken).[46]

The laboratory mouse is, to some extent, an exception: besides *H-2* associated loci controlling MLR, there is at least one other locus, not linked to *H-2*, the *Mls* locus (formerly M locus), differences at which lead to an often strong MLR.[47] The polymorphism at this latter locus seems to be lower than that of the *H-2* associated MLR loci and many *Mls* incompatibilities give rise to an MLR in only one direction.

2.2.2 *Genetic fine structure analysis of major histocompatibility complex associated MLR genes*

The finding that the loci controlling MLR were closely associated with the HLA system and that the reaction obeyed the laws of transplantation* is compatible with the hypothesis that a MLR represents the afferent (recognitive and proliferative) phase of an immune response to the major histocompatibility antigens on the stimulator cells: it appeared to be a partial *in vitro* analogue of a homograft rejection reaction. This notion seemed to be strengthened by the demonstration that, in the course of a MLR effector cells are generated which are able to specifically recognize and lyse target cells sharing major histocompatibility antigens with the stimulator cells.[50,51]

However, the straightforward interpretation seemed inconsistent with the observation that most combinations between cells from unrelated, HLA-identical individuals showed significant MLR incompatibility.[52–54] It became untenable when family segregation analysis provided evidence of recombination between the loci controlling HLA antigens and MLR determinants.[55] Further family studies in man (see, for example, References 56–58), the rhesus monkey[59] and dogs[60] and genetic analysis of inbred mouse strains carrying *H-2* haplotypes derived from recombination events within the *H-2* complex,[61,62] showed that MLR loci were definitely not identical with those controlling the classical ubiquitous major histocompatibility antigens (MHA). The relationship between MLR and MHC determinants leading to strong graft rejection is, at present, quite unclear.

As there are recent reviews on the genetics of the MHC and MLR[63,64] we have decided to attempt simply to summarize the current state of genetic analysis of MLR in man, the rhesus monkey and the mouse (Figure 1).

Like any such representation, ours is an interpretation rather than a summary of the available data. In this context it should be remembered, in particular, that the decision that a given responder–stimulator combination is MLR compatible (i.e. gives no response) is always to some extent an arbitrary simplification of data.

* This point of view is accepted by most investigators. There is, however, at least one group which has maintained for a long time that in the mouse cells from F_1 animals can respond to parental strain cells (for references, see Reference 48). An obvious explanation seemed to appear when the phenomenon of 'back stimulation' (see Section 2.4.1) was discovered. However, in a recent publication[48] proponents of the F_1 response of parental strains claim to show that this explanation is insufficient and that the F_1 response is specific. (For discussion, see Reference 49.) Our justification for ignoring this possibility is that in our own hands, and those of most workers in the field, it plays, if any, a very minor role and that, if it exists at all, it may be a quite unrelated phenomenon.

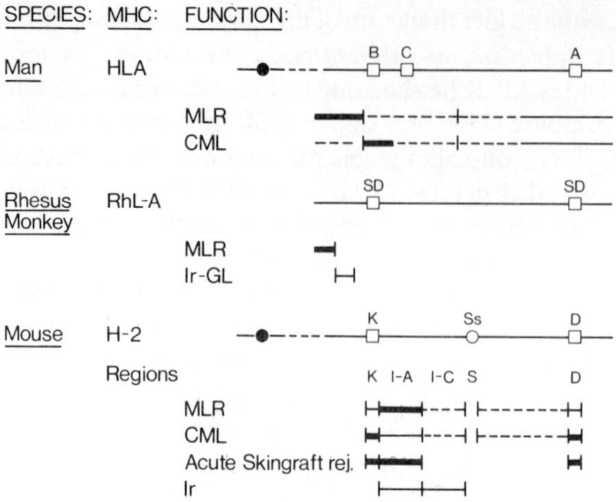

Figure 1. The position of loci controlling MLR-stimulating determinants and some other traits in the MHC of several mammalian species. (●: centromere.) Reference markers are the loci controlling the ubiquitous 'strong' cell surface alloantigens known as major histocompatibility antigens (□). For the mouse we have added the Ss-locus (○) which controls synthesis of the fourth component of complement. The location of genes controlling MLR and CML determinants is indicated as ━━ for 'strong', as ─── for 'weak' and as ---- for possible or probable loci. Vertical bars (—|—) indicate borders established by crossovers. Assignment of MLR genes to the *H-2K* and *H-2D* loci in the mouse is based on the analysis of *H-2* mutants (see text). The location of genes controlling specific immune responsiveness (Ir) is indicated only in as much as they have been mapped in relation to the other markers. (Ir-GL: gene controlling responsiveness to copolymer of L-glutamic acid and L-lysine)

2.2.2.1 *Man**

There is a single genetic region, HLA-D, which controls strong MLR. Whether this region contains one or more loci is as yet unknown. It lies outside the interval between two loci coding for the classical HLA antigens (HLA-B = SD2 = Four, HLA-A = SD1 = LA, HLA-C = SD3) (for review, see References 64 and 65) on the side of the HLA-B locus, about 1 centimorgan from the latter. In some populations very strong gametic associations (linkage disequilibria) between certain alleles of the HLA-B or the HLA-A locus and the HLA-C region have been found. Any estimate of the number of alleles in the HLA-C region has to make assumptions concerning the number of MLR loci it contains. If we assume that there is a single locus or several tightly clustered loci with alleles in very tight gametic association, then the low frequency of MLR compatible unrelated pairs indicates that the number of alleles at this locus or cluster of loci is comparable to that found at the MHA-loci. Evidence for a second weaker MLR-locus associated with the HLA-B locus is not yet conclusive.[58]

* The nomenclature adopted is that proposed by the WHO nomenclature committee after the Histocompatibility Testing Workshop, 1975.[66]

As has been mentioned above, in the course of an MLR cytotoxic T cells can be generated, which are able to specifically recognize and lyse suitable target cells, e.g. ^{51}Cr-labelled PHA-blasts. This latter system is called cell-mediated lympholysis (CML)[67,68] and has been used for a genetic analysis of the target determinants recognized by the cytotoxic cells. In man genes coding for CML-determinants are indistinguishable from the HLA-A and HLA-B loci, and it seems likely that the serologically recognized ubiquitous HLA antigens and the strong CML determinants are carried on the same molecules.[50,57,69–71]

2.2.2.2 *Rhesus*

The situation in the Rhesus monkey[24] is essentially the same as in man (see Figure 1) but in this species a crossover apparently separating an MHC-linked gene controlling immune responsiveness to one antigen from the region controlling strong MLR has been found.

2.2.2.3 *Mouse*

While genetic analysis of MLR and the other MHC associated traits in man and the rhesus monkey depends essentially on analysis of segregation patterns in families, in the mouse it has relied heavily on the existence of inbred congenic strains carrying *H-2* haplotypes which are the result of intra-*H-2* crossovers between the haplotypes of different inbred strains. The picture delineated in Figure 1 is the result of experiments*† in which the MLR is measured between strains which differ for defined segments of the *H-2* complex.[74,75] There is complete concordance between the incompatibilities giving rise to MLR and those giving rise to a graft-versus-host reaction (GvH), when the latter is measured as spleen enlargement in newborn F_1 mice after injection of parental (= responder) lymphocytes.[76] Therefore, in this chapter, MLR and GvH are considered equivalent. While it has been clear for some time that by far the strongest MLR incompatibilities are controlled by genes in the *I*-region (i.e. between the *K* and *Ss* loci) the number of MLR genes in this region is not known.[74,77] They are almost certainly at least two[75] and possibly many. *D* end incompatibility gives rise to a definite but much weaker MLR response. No conclusive information on the number and position of the *D* end associated loci is yet available.[78,79] Combinations differing only in the *K* region give rise to a very marginal response.[74,77] Whether this is due to the *K* locus difference itself or to closely linked loci is impossible to decide.

Most of the genetic analysis of CML determinants suggests that the 'strongest' determinants are carried by the same molecules as the *H-2K* and *H-2D* antigen specificities,[51,80,81] but *I* region incompatibility can be detected, to some extent, by cytotoxic T cells.[82,83] This analysis is complicated by the possibility that not all CML determinants may be expressed on a particular target cell type, e.g. PHA blasts.

* On the basis of Ia-typing and immune response differences the existence of a third, *I-B*, subregion of the *I*-region had been postulated (see Reference 63). The evidence for its existence was based entirely on the one *H-2* recombinant haplotype of B10.A (4R) and recent results cast some doubt on whether this recombinant is derived from a normal reciprocal crossover event (S. Cullen, personal communication).

† Frequency of recombination between *H-2K* and *H-2D* seems to vary considerably depending on the parental strains and/or *H-2* chromosomes involved (for review of data and discussion see Reference 72). An extreme genetic influence is exerted by certain alleles at the *T* locus which lies about 7 centimorgans proximal to *H-2* and may be phylogenetically related to it (for review and discussion see Reference 73).

The availability of inbred intra-*H-2* recombinant strains offers quite obviously enormous advantages; it introduces, however, certain biases into genetic analysis which should not be overlooked. To mention, as an illustration, two of these:

(a) From the experiments done it is difficult to rule out the existence of a major MLR locus to the left of the *K* locus. This is so because during production of congenic lines a 'contaminating' piece of adjacent genome of unknown length is transferred with the selected marker(s). This is, given the usual procedures for production of congenic strains,[84] an average of several centimorgans (which could contain several thousand genes) long.[85] Thus for any given strain we do not actually know whether any gene to the immediate left of *K* is derived from the 'background' strain or carried along with the transferred *H-2* haplotype.

(b) The number of possible combinations which can be used to test the importance of a certain type of incompatibility is limited by the available recombinants. There are, for instance, only two recombinants derived from a crossover between the *K* and *I-A* region. Both, in addition, involve one common haplotype, k, thus, all we can say with regard to *K* region incompatibility is that incompatibilities K^k-K^s and K^k-K^q lead to no or only marginal stimulation.*

One way to get around these and similar problems is the analysis of *H-2* mutants. This has, indeed, led to a number of interesting conclusions: all the mutants analysed so far have been detected as gain–loss mutations with regard to skin graft incompatibility, many of them in F_1 animals.[86,87]† The mutants fall into two classes, one carries mutations which alter the serologically defined specificities associated with *H-2D*, i.e. they are most likely mutations in the *H-2D* gene.[86] To the other class belong serologically undetectable mutations in the *K* end of the *H-2* complex.[87] Most of the *K* end mutations are difficult to map conclusively into the *K* or the *I* region but one at least definitely maps in the *K* region itself,[88] and for the others there is indirect evidence supporting the same conclusion.[87,89] Small peptide map differences have been detected in the H-2 molecules of one of the *D* region and one of the *K* region mutants (S. Nathenson, personal communication). Complementation tests between mutants of one class, derived from the same haplotype have so far always been negative.[87]

Many of these mutations have been analysed in MLR and they all result in incompatibility with the appropriate 'wild' *H-2* haplotype.[88,90–93] Thus, if we accept the assumption that each *H-2* mutant carries one point mutation then we have to conclude that small differences in the H-2K or the H-2D molecules of two strains can be sufficient to make them MLR incompatible.

While other studies strongly suggest the existence of a MHC in all mammalian species they do not yet provide any additional essential information on its fine structure.[19–21,23]

2.2.2.4 *Chicken*

Genetic analysis of inbred chicken strains has revealed that strong MLR incompatibility is associated with differences at the locus or loci controlling strong histocompatibility determinants, the B antigens, which, with regard to polymorphism and tissue distribution, resemble the major histocompatibility antigens of mammals, as well as genes controlling

* The relevant strain combinations are A.AL (K^k I-A^k I-C^k S^k D^d = kkkkd) – A.TL (skkkd)[74] and B10.AQR (qkkdd) – B10.A (kkddd).[77]

† The spontaneous rate of such mutations seems to vary considerably depending on the *H-2* haplotypes involved and can be very high: a recent estimate for K end associated mutations in the H-2^b haplotype of C57Bl/6 or (C57Bl/6 × Balb/c)F_1 mice is $5{\cdot}5 \times 10^{-4}$ mutations per gene per generation, while similar mutations in the H-2^d haplotype of the Balb/c strain seem to be very much rarer.[87]

immune responsiveness[27,94] (for review, see Reference 25). Recently one presumptive recombinant has been found which suggests that there are at least two loci controlling B-type antigens.[95] Results obtained with this bird and its offspring indicate that MLR incompatibility is controlled by loci associated with one of these (M. Vilhelmová, K. Hála, J. Hartmanová and V. C. Miggiano, in preparation).

2.2.2.5 *Amphibians*

Xenopus laevis is the only lower vertebrate species for which a MLR has unequivocally been demonstrated, and its genetic control been established. Analysis of the offspring of several pairs of outbred individuals has demonstrated the existence of a single genetic region responsible for the genetic control of MLR, acute skin graft rejection and strong alloantigens.[26] It is not known whether these traits are controlled by a single or several closely linked loci.

2.3 The nature of the stimulus

2.3.1 *Requirements on stimulator cells*

Although it is generally assumed that a MLR is the expression of a cellular immune response to cell surface determinants, this has not been conclusively demonstrated. The most compelling evidence comes from the studies on *H-2* mutants described above: if we accept the assumption that they are single-point mutations and take into account the evidence that the phenotypic effect of the mutations can be detected on the cell surface (by cytotoxic T cells and, in some cases, antibodies) then we have to conclude that differences in a single cell surface molecule are sufficient to trigger a MLR. Attempts at a physico-chemical characterization of the nature of MLR stimulating determinants have hitherto failed because it has remained impossible to induce a MLR with stimulating agents which are not living cells.

Dead lymphocytes or cellular fragments are not stimulatory (for references, see Reference 9). Stimulatory capacity can also be abrogated by irradiation with a germicidal UV-lamp,[96,97] and can be impaired with various inhibitors of RNA and protein synthesis.[98] The specificity of these effects has not been sufficiently investigated, but they suggest that stimulatory capacity depends on certain metabolic activities of the stimulator cells.*

That the apparently peculiar requirement for stimulating cells to be living may, in fact, apply to all T cell responses, including those to soluble antigens will be suggested below.

Not every type of cell is capable of stimulating a MLR response. While unfractionated leukocytes from various sources are the standard source of stimulating cells purified lymphocytes are equally efficient if the responder population itself is not depleted of macrophages.[103,104] Recent reports claim that in the guinea pig,[105] in the mouse[106] and in man,[104] macrophages also have stimulating capacity while in the last species neutrophils did not induce a response.[106] Other non-lymphoid cells are unable to stimulate (for references, see Reference 9) but there are reports from several species that epithelial cells from skin induce a MLR-like reaction.[107–109] It seems, however, that this response is not controlled by the same incompatibilities, and it may be of a nature different from a *bona fide* MLR.[109]

* This requirement, together with the apparent stimulation of F_1 lymphocytes by parental cells, has been the basis for a model postulating the existence of a novel type of recognition system operating in the induction of a MLR response in addition to specific recognition of allogeneic cell surface determinants by the responder cells.[99] There is no direct evidence for these postulated mechanisms and F_1 responses to parental cells are, most likely, due to non-specific 'back-stimulation' by factors released by immunocompetent irradiated parental cells, when they recognize the F_1 'responders'.[100–102]

That the capacity to stimulate a response to at least some MLR determinants is restricted to bone marrow derived blood elements is suggested by the pattern of responsiveness of adult F_1 mice resulting from a cross between *H-2* incompatible strains which have been lethally irradiated and reconstituted with parental bone marrow cells. In such mice several months after reconstitution no F_1 host lymphocytes can be found. If they have been reconstituted with bone marrow cells from both parents such tetraparental bone marrow chimeras will not respond in MLR to stimulator cells from either parent nor will they produce cytotoxic activity against them in a CML system.[110] Chimeras reconstituted with cells from one parent strain only behave differently: in MLR they do not respond to stimulator cells from the bone marrow donor strain, but give an albeit weaker than normal response to stimulators from intact F_1 animals or from the other parent. They are, on the other hand, unable to generate cytotoxic cells against either parental strain.[111] These findings suggest that in the latter animals, the reconstituting bone marrow population does not become tolerant to MLR determinants of the other parent because their expression is restricted to the host cell types eliminated by irradiation. But it is equally possible that tolerance induction to, as well as stimulation by, MLR determinants requires not only that MLR determinants are expressed, but also that they are present on bone marrow derived cell types.

Several investigators have dealt with the question which of the two classes of lymphocytes, bursa equivalent derived (B) or thymus derived (T) cells, are the most potent stimulators. From studies in the mouse it seems clear that T cells do not stimulate in a *Mls* incompatible situation.[112] B cells seem to be much more potent stimulators also in *H-2* incompatible combinations.[113,114] T cells may provide, in some situations, an adequate, if much weaker, stimulus but the supporting evidence is inconclusive. It has been reported that, when incompatibility is restricted to certain subdivisions of the *H-2-I* region, T cells are in fact better stimulators than B cells.[115] But as in these experiments, only one dose of stimulators was tested, the stimulator cells were not derived from F_1 animals and B cell and macrophage contamination was not completely eliminated, they remain unconvincing.

2.3.2 *The nature of MLR determinants*

While the studies on *H-2* mutants suggest that H-2K and H-2D molecules can carry MLR determinants, the recent discovery of antigens controlled by genes in the *I* region (Ia antigens) of the *H-2* complex (for review, see Reference 63) has led to the suggestion that the strong MLR elicited by *I* region incompatibility is due to recognition of Ia-antigen differences. The genetic correlation between Ia antigens and MLR incompatibility remains, by and large, complete even when detailed studies on intra-*I* region recombinants and the strain distribution of Ia antigen specificities are taken into account. In man (for references, see Reference 116), the rhesus monkey (H. Balner, personal communication) and the guinea pig,[117] MHC antigens with similar properties have been found and in these species, as well, there is some—albeit inconclusive—evidence for a genetic association between genes controlling Ia-like antigens and strong MLR determinants.

Ia antigens, which are controlled by at least two loci, in the *I* region of the *H-2* complex may have a similar tissue distribution as MLR determinants: they are most strongly expressed on B cells and possibly present on macrophages,[118,119] and only very weakly expressed on most T cells (for review, see Reference 63).

The most direct evidence that the two types of determinants may, at least in some cases, be carried on the same molecules comes from the demonstration that sera raised between *I* region incompatible strains (and containing anti-Ia antibodies) can specifically block the stimulatory capacity of the target cells when added to the responder–stimulator cell mixture.[120,121]

In conclusion, it should remain absolutely clear, that while some progress has been made towards identification of the determinants on whose recognition by the responder cells a MLR depends, the subsequent steps in triggering a responder cell are still completely unknown. We do not understand the significance of the requirement for living stimulator cells and we do not know whether a cell's capacity to stimulate depends only on the presence of the appropriate types of determinants on its surface.

2.4 The nature of the response

2.4.1 *MLR is a T cell dependent phenomenon*

From evidence obtained with mouse cells there is, today, no doubt that MLR in this species is a T cell dependent response, and most probably the specifically reactive cells are T cells themselves.* Some adherent accessory cells are required to obtain a response but these can be syngeneic with either the stimulating or the responding cells.[103] B cells are neither sufficient nor required among the responder cells but if they are present they can participate in an ongoing response.[49,122,123] While most proliferating or blast cells are thymus derived, some B cells are transformed, the percentage depending on the origin of the responder and stimulator cells,[49] and there is good evidence that participating B cells produce and secrete immunoglobulin, at least some of which seems to be antibody directed against MHC determinants of the stimulator cells.[124]

Recently, further characterization of mouse T cell populations has become possible through the use of antisera recognizing T cell specific alloantigens, in particular the Ly-1, Ly-2 and Ly-3 antigens (References 125, 126 and P. C. L. Beverley, herein). These are controlled by three loci, two of which, *Ly-2* and *Ly-3*, are very closely linked, but unlinked to *Ly-1*. None of them are linked to *H-2*. In adult mice of the strain C57Bl/6J all T cells carry the Thy-1 (θ) antigen, but they seem to belong to two largely non-overlapping subpopulations expressing either Ly-1, or Ly-2 and Ly-3 antigens respectively. Studies on mice in earlier stages of development suggest that these two subpopulations are derived from precursors which express all three Ly antigens.

When the responses of cell populations from which one or the other Ly antigen bearing class had been eliminated by complement-dependent antibody mediated lysis, were tested, results were obtained which support the following hypotheses:

(a) The cytotoxic cells recognizing H-2K and D determinants which can be generated in the course of a MLR, as well as their precursors carry Ly-2 and Ly-3 antigens but not Ly-1.
(b) The cells which respond by proliferation in a MLR between *I* region incompatible strains carry the Ly-1 antigen but are insensitive to anti-Ly-2 and anti-Ly-3 serum.
(c) In a MLR involving complete *H-2* incompatibility the proliferative response is due to both types of cells.

* For a long time, the evidence concerning the T cell dependence of MLR was conflicting. It was, for example, found that spleen cells from congenitally athymic (nu/nu) mice do 'respond' in MLR, to allogeneic lymphocytes from normal animals, but not from allogeneic nu/nu mice. Cells from nu/nu animals were perfectly good stimulators of normal allogeneic cells.[122,123] The answer to this puzzle is probably that, in an allogeneic cell mixture, made unidirectional by blocking DNA synthesis of the stimulator cells, the responder lymphocytes can be 'stimulated' non-specifically due to their recognition by the inactivated stimulator cells, the 'back-stimulation' phenomenon, which has already been referred to.

While there had been some previous indications that precursors of cytotoxic and MLR responsive cells might have different functional properties[127] the possibility to distinguish them on the basis of their surface antigens would, if confirmed, provide the first compelling evidence that they differ not only in their specificity but belong to distinct T cell subpopulations. In some of the earlier work on this subject these two aspects had been confused and the possibility that the same cell type could respond differently to different types of determinants was ignored.

2.4.2 *The specificity and the frequency of MLR-responsive cells*

If we assume that the clonal selection theory applies to MLR-responsive cells then the questions about the specificity and the frequency of MLR-responsive cells become connected through an apparent paradox: on the one hand the number of MLR-responsive cells responding in a particular allogeneic combination (complete haplotype incompatibility) seems very large: estimates obtained in different ways and in different species vary from 1% to 12%.* These figures are several orders of magnitude higher than estimates of the frequency of B or T cells specifically binding other antigens[135,136] or plaque-forming cell precursors (for references, see Reference 137) in unprimed animals. On the other hand, we must conclude from the available data that the number of distinct MLR alleles in most species must be very high. Consider, for instance, the extreme rarity of MLR compatible unrelated human individuals,[138] or the number of MLR alleles found in a limited sample of *Xenopus laevis*;[26] or combine the finding that all common *H-2* haplotypes carried in inbred strains are strongly MLR incompatible, with the results of serological analysis of wild mice populations which suggest that several hundred different *H-2* alleles may exist.[72,139] Are we then, in order to reconcile the high frequency of cells responding to one with the apparently large number of different haplotypes, forced to conclude that MLR-responsive cells are polyspecific?

Let us remember, at this point, that the experiments in support of one of the principal tenets of the clonal selection theory, namely that an antigen specific precursor cell is monospecific because it carries recognition structures of a single specificity, is, even for B cells, not entirely conclusive (for discussion, see Reference 140). What evidence is there to support the hypothesis, implicitly accepted in most discussions on MLR, that MLR-responsive or T cells reactive to MHC determinants in general are monospecific? Consider cytotoxic effector T cells and the blast cells generated in the course of a MLR: from cytotoxic effector cell populations specific for different *K*- and/or *D*-region incompatibilities a population reactive to one incompatibility can be absorbed by appropriate target cell monolayers without impairing reactivity to the other incompatibilities.[141,142]

* These estimates have been obtained by several different approaches: in the rat,[128] man[129] and the mouse[130] combination of isotope labelling of cells synthesizing DNA, autoradiography and colchicine induced block of mitosis was used to determine the accumulated number of cells responding in a MLR at various time points. Other estimates have been obtained in GvH systems: a very direct approach was the estimation of the minimal number of allogeneic lymphocytes required to produce splenomegaly in chicken embryos or proliferative lesions on the chorioallantoic membrane of eggs.[131] In the rat use was made of the demonstration that parental cells injected into a semi-allogeneic (F_1) host and recovered from its thoracic duct are specifically depleted of cells responding to the allogeneic F_1 haplotype.[132] Mixtures of ^{14}C-labelled parental cells and ^{3}H-labelled F_1 (control) cells were passed through a F_1 host and ratio of ^{3}H- and ^{14}C-counts in the input and output populations compared. From the relative reduction of label associated with parental cells the percentage of specifically retained parental cells was calculated.[133,134] The last, and most direct, approach is based on the use of so-called anti-idiotype sera and discussed below.

There are experiments which suggest that, in principle, the same can be done with the precursor cell population prior to exposure to stimulating cells.[143,144] These findings strongly suggest, but do not prove, that cytotoxic T cells and their precursors are monospecific.

The T blasts responding in MLR to a particular stimulator can, using immunofluorescence techniques, be stained with anti-H-2 antisera directed against this stimulator. It seems that when responder cells are stimulated with a mixture of two unrelated stimulators each of the resulting blast cells can be stained with only one of the two possible anti-stimulator sera: each responder T blast cell carries antigen derived from only one of the stimulators on its surface. These experiments were carried out using B depleted responder populations.[124] Furthermore, responder blasts treated with trypsin lose the stimulator antigen on their surface, but, after a few hours, regain their capacity to bind such antigen specifically (B. Elliott, Z. Nagy and M. Nabholz, in preparation). These results suggest that such T blast cells carry an endogenously synthesized receptor of restricted specificity.

When MLR cultures are kept beyond the peak of the primary response then the responder blasts revert to the size of small lymphocytes.[146] A small fraction of the responders survive for several weeks. This fraction seems to be very much enriched in cells responding specifically to the MLR determinants of the original stimulator: ten to twenty times less cells are needed to achieve the same amount of H^3-thymidine uptake in a 'secondary' response if and only if the stimulator shares the *I*-region with that used in the 'primary' MLR (D. Collavo, M. Nabholz, A. Rijnbeek and V. Miggiano, unpublished).

Experiments testing the specificity of MLR precursors have all suggested that cells reactive to one stimulator can be eliminated without impairing, to a significant extent, the response to cells from an unrelated second donor ('filtration' through F_1 host,[132] suicide of proliferating responders,[144,147] elimination of specific responders with anti-'idiotype' serum).[148] It should, however, be pointed out that in all of the systems employed a considerable overlap in the populations responding to two different stimuli, either because of precursors reactive to more than one specificity or because of unspecific recruitment of a small fraction of third party cells in each response, could remain undetected. A certain amount of overlap would, in any case, be expected owing to cross-reaction between different haplotypes. The major technical problems in such experiments have been (a) the unreliable quantitation and/or insensitivity of the measuring systems and (b) the absence of any system allowing to measure directly the activity of single precursor cells.

Most likely the final answer to this problem requires the isolation and characterization of the molecule(s) which form(s) the recognition structure on MHC reactive T cells. The approach which, at the moment, seems most likely to lead to this point involves the production of specific antisera against this structure. That it may be possible to raise such 'anti-T cell idiotype' antisera is suggested by the work of Binz and his collaborators: by injecting either parental lymphocytes or parental anti-F_1 alloantibody into F_1 animals, they have obtained sera which in the presence of complement are able to eliminate the precursors of the parental cells responding in MLR to the allogenic component on cells of the F_1 serum donor, without impairing the response to other unrelated stimulators.[148,149] One advantage of this approach is that, in contrast to all other experiments aiming at elimination of specific precursor cells, it does not involve exposure of the responsive cells to specific antigen and thus argues somewhat more strongly against models of semi-instructive T cell recognition.[133,134,140] Using indirect immunofluorescence, the number of parental T cells binding anti-(parental anti-F_1 alloantibody) serum was estimated to be

5%.* Tests using two or more antisera directed against different 'T cell idiotypes' have not yet been reported.

If these experiments are confirmed, they suggest that the 'active sites' or idiotypes of T cell recognition structures, for a particular set of MHC determinants, but not necessarily the biochemical properties of the receptor molecules, are similar to those of the corresponding B cell receptors. One complication in the interpretation of these experiments is, of course, the unknown number of MLR stimulating determinants in any particular responder–stimulator combination. In all cases where the frequency of responsive cells has been estimated, whole haplotype-incompatible or even completely unrelated strain combinations were used. Thus, for instance, the so-called anti-T cell idiotype sera should be directed not against a single type of recognition structure but against that set of parental T cell idiotypes specific for the ensemble of incompatible F_1 determinants.

In the context, the study of *H-2* mutants may again be of crucial importance. The MLR responses between two strains differing by what may be single-point mutations in the *K* region are usually somewhat, but not very much, lower than for complete *H-2* incompatibility.[88,92] Thus, assuming that the amount of ^{3}H-methyl-thymidine uptake is more or less related to the number of responding cells this suggests that the frequency of cells responding to changes on a single molecular species is still above 10^{-3}. But analysis using CML has shown that the T cell response of the 'wild type' strain to the mutant is still heterogeneous, i.e. T cells of more than one specificity recognize the changes caused by the mutational event.[89,92]

In summary: there is no direct evidence to conclusively establish or to conclusively repudiate the applicability of all aspects of the clonal selection theory to MHC reactive T cells. There is, however, a fair amount of indirect or circumstantial evidence which supports the view that precursors of MHC reactive cells are monospecific and upon stimulation transmit their specificity to their clonal offspring.

The 'paradox' of the high frequency of cells reacting specifically to a particular allogeneic stimulus does not, in our view, compel us to give up the model of clonally expressed specific receptors if we take into account (a) the possibility that the number of MLR determinants controlled by a MHC haplotype can be quite large (possibly in the hundreds) and (b) that MTR-responsive cells may be stimulated by determinants for which they have receptors of very low affinity. The conclusion that a very large fraction of T cells can specifically react to allogeneic MLR determinants is inevitable and consistent with arguments raised in the second part of this chapter.

2.5 Phenomena accompanying MLR

A MLR (or GvH) can be accompanied by a number of non-specific effects whose general consequence seems to be an enhancement or facilitation of the activity of other cells involved in an immune response. Most of these effects are mediated through elaboration of 'factors', the biochemical characterization of which is still very poor or non-existent. In other cases the active principle is unknown. While some of these activities have been discovered in a MLR system, others are known to be produced also in other systems involving proliferation or activation of lymphoid or T cells.

* It is interesting to note that the frequency of B cells stained in the same experiments was 1%, much higher than that binding 'conventional' antigens. Thus, advocates of polyspecific MLR responsive cells may be forced to take the same position for B cells recognizing allogeneic cell surface determinants. The interpretation of these experiments is, however, complicated by the fact that the strain combinations employed were not congenic for the MHC.

We want to mention these phenomena briefly because, as will become clear later, we believe that they offer important clues concerning the physiological function of MLR.

Operationally they can be grouped into three categories according to the processes which they influence: the best characterized of these activities is the so-called allogeneic factor which enhances the antigen-specific response of B cells. So far this effect has only been described for the mouse and genetic analysis using congenic and *H-2* recombinant strain combinations has shown that its production depends on the same incompatibilities as MLR (Reference 151; J. Kettmann and M. Nabholz, unpublished results). There is one report suggesting that this factor reacts with anti-Ia serum but the evidence is not yet convincing.[152]

Another effect is the MLR-dependent enhancement of the induction of cytotoxic T cells: in man and the mouse, the generation of cytolytic activity in MLC, although not directed against MLR determinants, is, but only in some systems, dependent on an ongoing MLR.[51,57,97] While this and related findings have given rise to a lot of rather far-fetched speculations concerning 'synergy' or 'cooperation' between MLR-responsive and precursors of cytotoxic cells, the most plausible possibility is that this activity may be mediated by the same or a similar factor as the allogeneic effect on B cell responses.[153]

The last group includes factors mediating or facilitating the non-specific phases of an inflammatory response: MLR is accompanied by production of macrophage migration inhibition factor and other factors chemotactic for, or altering the metabolic state of, macrophages.[154–156] The precise genetic requirements for the production of this type of activity have not yet been investigated.

If all these effects depend on non-specific activities of activated MLR-responsive cells, then one would expect to find them also in situations where the same cells are activated by other mechanisms, for instance by non-specific mitogens. In turn it seems not unlikely that all so-called lymphokines as well as interferon, can, given suitable culture conditions, be found in supernatants of a MLR.

2.6 Can 'modified self-determinants' induce a MLR response?

In the course of studies aimed at a dissection of the immune response to viral infection into its various components, ^{51}Cr-labelled target cells were exposed to lymphocytes from infected mice. The latter are specifically cytotoxic for targets infected with the same virus, and this cytolytic activity seems to be mediated by T cells. Surprisingly enough, cytotoxicity is, in addition, dependent on the genetic relationship between effector and target cells: specific ^{51}Cr-release is orders of magnitude greater when effector and target are *H-2* compatible, while compatibility at other genetic regions seems to be of no influence. Using the inbred *H-2* recombinant strains the requirement for compatibility could be narrowed down to the *H-2K* and *D*-regions.[157] When such cytolytic cell populations, generated in F_1 animals carrying two different *H-2* haplotypes are 'restimulated' by transfer into infected parental hosts, their cytolytic activity against targets from this parent strain increases while that against infected cells of the other parent decreases.[158]

While these findings were first made with lymphochoriomeningitis (LCM) virus[158] they have, in the mean time, been extended to ectromelia[157] and vaccinia virus infections.[159]

In addition, virtually identical genetic restrictions apply in a quite different system. They were discovered in the course of attempts to generate hapten-specific cytotoxic T cells by stimulating them in MLC with autologous lymphocytes modified by coupling trinitrophenyl (TNP) groups to their surface.[160,161]

At the moment the generally accepted interpretation of these findings is that the cytotoxic cells generated specifically recognize modifications of *H-2K* and *D* end determinants resulting from interactions between virus or the hapten and these 'self'-determinants. The only direct evidence to support this hypothesis comes from experiments suggesting that the cell-mediated cytotoxicity can be specifically inhibited with anti-*H-2* as well as with anti-viral antisera (Reference 159, but see also Reference 162).

One of the observations which have almost certainly influenced the interpretation of the results discussed is the close analogy with those genetic restrictions found when specific T cell dependent cytotoxic activity is induced by, and measured against allogeneic target cells: cytotoxic activity against *K* and *D* region determinants is much stronger than against *I* region or non-*H-2* determinants. Two non-exclusive interpretation of these differences are possible: (a) That the frequency of cytotoxic cell precursors specific for allogeneic or modified syngeneic *K* and *D* region determinants is much higher than that for other determinants. (b) But it is also possible, however, and certainly true in at least some situations, that cytotoxic T cell activity against, for example, allogeneic or modified syngeneic *I* region determinants is not detected because they are absent on the target cell type used.[82,83]

To complete the list of systems possibly involving T cell recognition of modified MHC determinants let us add that some, however weak, evidence suggests that recognition of tumour-specific determinants by syngeneic cytotoxic T cells can be specifically inhibited by anti-*H-2* antisera.[163] It seems, also, that the activity of cytotoxic cells raised in mice against (human × mouse) somatic cell hybrids carrying the SV40 genome may be, in part, directed against modified self-determinants (Trinchieri, Aden and Knowles, see Reference 199). Some biochemical evidence suggesting physical associations between tumour and *H-2* antigens had been reported.[164]

Furthermore, experiments measuring an *in vivo* response to *Listeria monocytogenes* infection can be interpreted to suggest that in this system, as well, cell-mediated immunity is directed against pathogen-induced modifications of H-2 determinants.[165]

Finally, it appears that cell-mediated cytotoxicity against non-*H-2* determinants is subject to similar constraints:[166] cytolytic T cells generated by immunization with *H-2* compatible allogeneic cells show activity only against targets which share, not only the non-*H-2* background but also either the *K* and/or the *D* end of at least one of their *H-2* haplotypes with the stimulator cells. Formally, these restrictions are the same as those discussed previously and can be interpreted in terms of modifications of MHC determinants, but it has been suggested, that it is unlikely that all of the many known minor histocompatibility loci represent genes whose products induce modifications of the MHC determinants. The alternative explanation proposed was that *H-2* linked genes can modify minor histocompatibility determinants, and this type of inversion of the modification hypothesis might indeed account for some of the other observations described above.

One very striking aspect of these constraints is their formal similarity with those found controlling interactions between T cells and macrophages,[167,168] as well as T and B cells[169,170] in responses to soluble antigens. Later on we will suggest that such genetic equivalence may reflect similarity of mechanisms.

Partly for *a priori* reasons discussed in the second part of this chapter we believe that at least viruses and certain other pathogens induce modification of MHC determinants on infected cells. If so we may ask whether infection of lymphocytes also induces alterations of MLR determinants. While this has not been measured in the systems involving viral infections, a definite, albeit rather weak, increase in ^{3}H-thymidine uptake was observed when lymphocytes are co-cultured with irradiated TNP-modified autologous stimulators.[171] But we do not know whether the apparent stimulation is a specific response to modified MLR determinants. The problem can most likely be solved by making use of the fact, previously referred to, that the cells obtained from allogeneic MLR cultures will, after several weeks, respond very strongly and more quickly to fresh stimulators but only if they share the *I* region with the original stimulator cells (D. Collavo, M. Nabholz, A. Rijnbeek and V. Miggiano, unpublished). Thus, the response of lymphocytes previously exposed to modified autologous stimulators to restimulation by similarly modified lymphocytes sharing defined segments of the *H-2* complex with the original stimulator could provide the answer to the question of the specificity of the proliferative response to TNP-modified syngeneic cells.

We suspect that the latter is most strongly stimulated by modification of *I* region determinants, although in view of the artificiality of the TNP system and considering the MLR data on *H-2* mutants (Section 2.2.2), we hesitate to make this prediction. It would obviously be more relevant and therefore crucial to answer the problem in one of the systems involving viral infection.

If the finding that the strong proliferative response to allogeneic *I* region determinants is mainly due to one subset of T cells, distinct from the precursors of cytotoxic cells specific for *K* and *D* region determinants, is confirmed as a general characteristic then it would become plausible to suggest that an analogous dichotomy holds for the T cell responses to modified autologous *H-2* determinants.

All this may seem to be a somewhat lengthy discussion of findings which have only a hypothetical relationship with the MLR phenomenon. The reasons for presenting it should become apparent below.

3. ON THE PHYSIOLOGICAL FUNCTION OF MLR

3.1 Is the major histocompatibility complex (MHC) a supergene?

3.1.1 *MLR, a polymorphism in search of a function*

When, about ten years ago, population studies in various species began to reveal that the average degree of polymorphism at loci coding for cytoplasmic enzymes as well as blood groups is much higher than previously suspected (simply taking into account electrophoretic variation of enzyme proteins, one arrives at the conclusion that in most species individuals are, on average, heterozygous at about 10% of these loci) an as yet unresolved debate on the meaning of these polymorphisms set in: do they represent adaptive variation (i.e. are they, or have they been, maintained by selection) or were most of them due to selectively neutral mutations which, through random genetic drift, had increased in frequency (for a review see Reference 172)?

A number of never entirely convincing attempts to provide an answer to this question have been made which were based on observations of a whole set of polymorphisms under various population genetic situations, without taking into account the known enzymatic function of the polymorphic proteins.[172] Today it seems that a definitive solution will probably depend on painstaking investigations of the functional differences between alleles and their possible impact on the fitness of the individual.

The only known 'function' of by far the most polymorphic systems known, those controlling major histocompatibility antigens (MHA) and MLR stimulating determinants are that their products are recognized as non-self by the immune system. In physiological situations an effective confrontation between allogeneic cells and the immune system occurs probably in the course of fertilization and pregnancy in viviparous animals and there exists some evidence that incompatibility at the MHC may indeed increase the survival of the offspring.[73] But as highly polymorphic systems controlling MLR and MHA exist in an amphibian species[26] where fertilization and ontogeny take place outside the parental animals such effects are unlikely to account for their evolutionary origin.

3.1.2 *Conservation of the MHC through evolution and the significance of linkage disequilibria*

Because the gene products of at least some of the loci in the MHC are definable only in terms of their polymorphisms we are forced to suppose that the allelic variations measured are related to the physiological function of the molecules expressing them, i.e. that the polymorphisms are the result of selective pressures. Alternatively we might dismiss the question by turning the argument upside down and suggesting that, indeed, the 'excessive' polymorphism observed at the MHA and MLR loci is an indication that these allelic variations can be tolerated exactly because they do not affect the physiological performance of the molecules affected, i.e. that they are selectively neutral.

Two related arguments against this contention have been raised:

(1) The observations that the existence of highly polymorphic linked systems controlling MHA and MLR determinants is not restricted to a few closely related species but extends to birds (where these systems are, as in mammals, known to be linked to Ir genes) and even amphibians, would, to most evolutionists, suggest that this association of genes has been maintained, and may have arisen several times, during evolution because of a functional interaction between the constituent loci.[174] But this argument supports essentially only a selective maintenance of the linkage between the MHC genes, it does not refer to the significance of the polymorphisms of the individual loci. In addition, it has been suggested that such associations of groups of genes might have arisen fortuitously and been maintained through evolution as 'frozen accidents' because it would be too costly, in terms of the genetic load, to break them up again.[175]

(2) There is, however, one further and fairly crucial type of observation which supports the contention that the MHC is a supergene, whose constituent loci functionally interact and that the respective polymorphisms must, at least to some degree, be selectively maintained: in human populations there exist strong linkage disequilibria between alleles at different loci of the HLA system.* These linkage disequilibria are the expression of preferential—negative or positive—association of two alleles with each other in the same HLA haplotype. They may be the reflection either of selective pressures favouring haplotypes made up of alleles the products of which interact in such a way as to increase the fitness of the carrier or of continued random effects of mutations, population structure and migration. Extensive analysis of different human populations suggests that at least some of the observed linkage disequilibria are a reflection of co-adaptation of alleles at various MHC loci. The most striking support for this interpretation comes from the growing number of cases of, sometimes very strong, associations between alleles controlling an HLA antigen and/or MLR determinants on the one hand and susceptibility to a particular

* If we know the gene frequencies, a_i, b_i, for the alleles of two linked loci A and B in a population we can calculate the haplotype frequencies $p_{ij} = a_i \times b_j$ that we expect if the population has reached an equilibrium with regard to the joint distribution of the alleles at these two loci. A linkage disequilibrium between the two alleles a_i, b_j is a significant deviation of the observed frequency of the haplotype a_ib_j from the expected one.

disease on the other (for references see References 116, 176 and 177, and L. P. Ryder and A. Svejgaard, herein; for discussion also Reference 178).

3.2 An hypothesis*

3.2.1 *The premises*

From what has been discussed so far, we conclude that it makes no sense to ask for the biological significance of MLR, if it is not in the context of a consideration of the function of the MHC as a whole. We now want to introduce interpretations of, and extrapolations from, available data, that are premises for an hypothesis on the biological role of MLR:

(1) T cell responses to allogeneic determinants conform to the tenets of the clonal selection theory (see Sections 2.4.1 and 2.4.2).
(2) In the MHC there are at least two types of highly polymorphic genes controlling determinants which elicit a reaction by allogeneic T cells (Section 2.2.2). The gene products carrying the determinants differ in their molecular properties and their expression in various tissues, but as it is likely that the genes coding for the two types of determinants are derived, by duplication, from common ancestors, it is not surprising that in some situations the distinction between the two types of determinants may become blurred. Allogeneic T cells respond to one type of determinant, called CML determinant and carried probably by the same molecules as the classical ubiquitous MHA determinants, mainly by the generation of cytotoxic effector cells. The precursors of these cells may belong to a subset of T cells ($Ly\text{-}1^- \; Ly\text{-}2^+ \; Ly\text{-}3^+$ in the mouse). MLR determinants, on the other hand (which may be carried by the same molecules as Ia determinants), elicit mainly a strong proliferative response possibly of another subset of T cells ($Ly\text{-}1^+ \; Ly\text{-}2^- \; Ly\text{-}3^-$ in the mouse) which may include specific helper cells[179] (Section 2.4.1).†

MLR determinants are expressed much more strongly, or presented much more effectively, on B lymphocytes and macrophages than on many other cell types (Section 2.3.2). Unprimed allogeneic T cells will respond to CML or MLR determinants only if they are present on living, metabolically active cells (Section 2.3.1).
(3) Probably both types of determinants can be modified at least through viral infection or coupling with haptens, possibly also as a consequence of some types of bacterial or protozoal infections or events involved in tumorogenesis. Such modifications give rise to new 'altered-self' MLR and CML determinants recognized by autologous T cells (Section 2.6).
(4) The MHC contains genes which affect immune responsiveness to a large variety of T cell dependent antigens. Responsiveness is dominant. These Ir loci seem to be more closely linked to those controlling MLR determinants than those coding for MHA (Section 2.2.2). For at least some antigens high responsiveness depends on the presence of responder alleles at more than one locus in the *I* region.[180–184]
(5) The MHC contains genes which affect susceptibility or resistance to a wide variety of diseases. It seems likely that most often susceptibility is dominant and in many cases an autoimmune aetiology of the disease is suspected.

*References in Section 3.2 are given only to evidence which has not been cited previously. To connect the first and the second part of the chapter we have tried to indicate, in the list of premises, the section of the former in which relevant evidence is cited.

†Assumption of the existence of two (antigenically distinct) T cell subsets is not crucial to the arguments developed below, but it simplifies their formulation and presentation. It may still turn out that the nature of T cell response is determined entirely by some as yet unknown features of the stimulating determinants or the stimulator cells.

There exist strong gametic associations between alleles at the loci controlling MHA, MLR determinants and disease susceptibility (Section 3.1.2).
(6) In the course of MLR mediators are produced which are involved in facilitating inflammatory and humoral antibody responses, as well as generation of cytotoxic T cells. Complement components or factors involved in complement activation are controlled by genes closely associated with the MHC (Sections 2.2.1 and 2.5).
(7) 'Antigen specific' suicide of T helper cells can only be achieved in the presence of antigen specific B cells.[185]

3.2.2 *Are MLR determinants the generators of diversity of T cell specificity?*

The argument that we want to develop can most easily be derived from Jerne's proposal[2] that the repertoire of antibody diversity is generated as a result of a physiological 'autoimmune' response to, and therefore to some extent determined by, the major histocompatibility antigens expressed in an individual: the germ line contains receptor genes specific for the major histocompatibility antigens of the species. A clone of cells expressing a receptor for a self-antigen will proliferate in the thymus and either be eliminated or, through somatic mutation of the expressed receptor gene, escape elimination and give rise to progeny reactive to non-self antigens.

A reconsideration of most of the phenomena on which this hypothesis is based—the so called allo-aggression phenomenon,* another name for MLR, the high rate of cell proliferation and death in the thymus, and the linkage of Ir genes to the MHC—in the light of our present knowledge of MLR and the structure of the MHC suggests that it applies, if at all, more likely to T cells and their precursors than to B cells. Thus we propose to adapt Jerne's hypothesis to our premises, by suggesting that the most important class of self-determinants driving somatic generation of T cell diversity are the MLR determinants. This would be consistent with the observations indicating that MLR may be one of the earliest reactions in the ontogeny of T cell dependent responses.[187]

Jerne did not deal with one important conceptual problem raised by his hypothesis: we think that no cell which synthesizes receptors recognizing its own determinants and displays them on its surface can be viable: self-tolerance to ubiquitous surface determinants is trivial. But Jerne's suggestion on the generation of diversity of recognition structures requires that 'self-recognizing' cells can survive and function for some time. This can, in our view, only be the case if the relevant self-determinants and the corresponding recognition structures are expressed on different cells, but cells which nevertheless must have a very high probability to encounter each other. Therefore, the evidence that expression of MLR determinants is restricted to blood cells other than T cells supports their candidacy as generators of diversity, and experiments determining the nature of the stimulus in the MLR-like response of newborn mouse thymocytes to syngeneic adult spleen cells[188] might strengthen our arguments further.

Jerne's hypothesis postulated that germ-line receptor genes coded for recognition structures corresponding to all the major histocompatibility antigens of the species. Bodmer[189] pointed out that no conceivably operating selection mechanism could ensure that, whenever a mutation generated a new histocompatibility antigen, a corresponding receptor gene would rapidly be acquired by all members of the species. This is not a serious

* Allo-aggression was a term used to describe the observation that MLR between allogeneic cells was much stronger than in xenogenic combinations. There has been considerable debate on this point but today there seems little reason to doubt that the strength of MLR and the phylogenetic distance between two species are inversely correlated.[186] This is, of course, very much what Jerne's hypothesis and our modification of it would predict.

impasse: we assume that the germ-line receptor genes of an individual correspond only to a relatively small subset of the species MLR determinants. This modification leads to the central prediction of our hypothesis.

3.2.3 *MHC-linked Ir genes code for MLR determinants or T cell receptors*

If our assumptions are correct then the possible set of functional receptors which can, in an individual, be generated by mutations and selection away from self-reactivity is limited by the combination of receptor genes and MLR determinants expressed. Therefore, immune responsiveness to a particular antigen may require, (a) the expression of a particular germ-line T cell receptor gene and (b) a corresponding MLR determinant. Thus, the immune response potential of an individual would be the result of the interaction between the set of genes coding for germ-line T cell receptors and the one controlling MLR determinants. Following the classical postulate of Fisher,[174] we therefore predict that, (a) the two sets are closely linked and (b) there is selection for haplotypes combining two co-adapted sets. Obviously MHC-linked *Ir* genes would be expected to fall into two classes.

Recently, several cases of immune responsiveness requiring the presence of responder alleles at at least two loci in the *I* region have indeed been found. If, as we would suggest, one of the loci involved controls MLR determinants, the other germ-line T cell receptors then, by combining the two in a tetraparental mouse derived from two complementing strains, this mouse should respond as well as the corresponding F_1 animals and all responsive T cells should be of the one genotype, which contributes the T cell receptor gene.*

One of the simplifications implied, until now, in our modification of Jerne's hypothesis is that we have ignored the problem raised by our assumption of a possible dichotomy in the T cell response to allogeneic cells: if there are two sets of T cells—MLR responders and cytotoxic cell precursors—differing in their functional properties as well as in their specificity spectrum, where is the ontogenic origin of this division? There are many possible answers to this question, but as our arguments do not crucially depend on them and the existing evidence is neither solid nor sufficient we do not want to pursue this aspect.

3.2.4 *We expect that most T cells respond to modified MHC determinants: Immune surveillance revisited*

If most antigen specific T cells are derived from precursors recognizing self-antigens controlled by the MHC, which, through a process of sequential mutations and selection, are driven away from self-reactivity, then one might expect that many or even most T cells recognize antigens only slightly different from the self-determinants recognized by their ancestors. This, besides the occurrence of germ-line gene coded receptors for allogeneic

* Jerne[2] pointed out that if his hypothesis in its original formulation was correct then in tetraparental mice resulting from fusion of embryos from two strains differing for a specific immune response gene the genetic low responder B cells should become phenotypically responsive. Subsequently McDevitt and his group[190] were using tetraparental mice to investigate the nature of *H-2*-linked specific immune response control. Their conclusion was that in mice derived from fusion of embryos from a high and a low responder strain the genotypic low responder B cells could, with the cooperation of T cells from the high responder strain, make antibodies against the particular antigen, and that this constituted proof that responsiveness was controlled at the level of T cells. But disregarding the possibility that if Jerne's concept might not be valid for the generation of the diversity of antibodies, it might still apply to T cell receptors they omitted to demonstrate that in the tetraparental mice the genetic low responder T cells had not become phenotypically responsive. According to our hypothesis this is not a necessary result but it should occur when strains carrying an appropriate germ-line T cell receptor gene and the corresponding MLR determinants are combined.

determinants, explains the strong response to allogeneic MHC determinants. The prediction is supported by the strong incompatibilities generated by—apparently single—mutational events and, most strongly of course, by the experimental systems which can be interpreted as strong T cell responses to modified self-MHC-determinants.

While such modifications have been observed after infection with different viruses, there are reports which may suggest similar effects as a consequence of bacterial infections and tumorogenic cell transformation. If these hints are corroborated by further evidence and, most crucially, if the modification affect MLR as well as CML determinants, then a unifying version of the immune surveillance hypothesis emerges: most T cells of an individual can be activated by such slight modifications of its MHC determinants as may be induced by viral and other pathogens and tumorogenic agents in their target cells.*

What then would be the respective role of MLR and CML determinants, or, in other words, MLR-responsive cells and precursors of cytotoxic cells? It seems to us most plausible that it is the role of the cells responding to modified MLR determinants to initiate and control the various aspects of an immune response to an invasion by pathogens through, (a) elaboration of chemotactic, mitogenic and arming factors involved in activating macrophages and the generation of an inflammatory response, (b) increased interferon production, (c) providing a source of specific helper cells able to trigger a B cell response and (d) the generation of cytotoxic T cells able to eliminate infected, i.e. modified, host cells. It seems likely that the tight linkage of several genes controlling levels of complement components or factors involved in its activation to the MHC is a reflection of the involvement of the complement system in inflammatory reactions, the induction of which is, as we now propose, controlled by other parts of the MHC.

Earlier we have given some reasons why their role in the ontogeny of immune responsiveness requires a restricted tissue distribution of MLR determinants (Section 3.2.2). Now we are led to ask whether their postulated function in the control of cellular immune responses to infection is compatible with their expression on particular cell types, notably macrophages and B lymphocytes. In discussing this problem we have to keep in mind that we do not know the mechanisms by which MHC determinants are modified, for instance, in the course of viral infection or oncogenic transformation, and that it is at least conceivable that in some such cases previously silent genes coding for MLR determinants are activated and modified.

But we think that the clue to the expression of MLR determinants on macrophages lies in their critical role in the control of infections: macrophages are the most likely cell type to interact with an infecting agent.[193–195] They may be attracted to the site of primary infection in the course of an initial unspecific inflammatory response. It may, in addition, be very important to prevent productively infected macrophages or monocytes from disseminating the infection to other parts of the body. Thus, a MLR-like response to the infected macrophages may, by inhibiting migration of the infected cells, by attracting further macrophages, inducing them to proliferation and arming them, be instrumental in restricting the infection to a controllable focus characterized by an inflammatory response. At the same time the generation of cytotoxic T cells directed against modified K and D determinants would provide a means of eliminating any type of infected cell throughout the body.

We will return later (Section 3.2.6) to a possible function of MLR determinants on B lymphocytes.

* That the 'self-modification' hypothesis has important implications for the immune surveillance concept is obvious and has been pointed out by others.[191,192] But we believe that it is an oversimplifying extrapolation from the available data[191] to interpret them as a verification of this connection.

3.2.5 *The origin of the MHC polymorphisms and their association with disease susceptibility*

According to our view the immune response to an infective agent will depend, (a) on the alterations of MHC self-antigens induced by the agents and (b) on the ability of T cells to recognize them. Thus, there should be a very high selective premium on mutants of the infective agent which induce no, or only unrecognizable, changes in MHC antigens. This tendency will be countered by the selection both for new antigen alleles susceptible to change by these mutants and for germ-line T cell receptor genes which give rise to T cells able to recognize these changes. This is, we propose, the evolutionary history of MHC haplotypes composed of co-adapted sets of alleles, coding for MHC determinants and for germ-line T cell receptors respectively.

Association of increased susceptibility to a disease with an MHC haplotype may reflect a better chance of the causal agent to escape the cell-mediated defence mechanism of the host through mutations of the type described above. In this case susceptibility should be recessive. However, in most recorded associations increased susceptibility is probably transmitted in a dominant fashion. An interpretation of this situation is suggested by the outcome of infection with lymphochoriomeningitis (LCM) virus in the mouse: (1) Viral infection and replication, as such, could be tolerated by the host without strong adverse effects. But, in immunologically competent animals, it leads to a T cell-mediated acute immune-like response to the infected cells, most likely to virus-induced modifications of self-determinants (see Section 2.6). Death is caused by progressive cellular infiltration of parts of the central nervous system. (b) Susceptibility to the lethal effects of LCM virus infection depends on *H-2* associated genes.[44] *H-2* heterozygote animals are as, or more, susceptible as the less resistant parent.[191] Thus, it may be that a particular combination of an MHC determinant and a T cell receptor gene leads, upon modification by a certain infective agent, to an 'excessive' or 'unbalanced' reaction against the modification, nevertheless unable to effectively eliminate the infection. The suspicion that in the aetiology of many of the diseases, susceptibility to which is associated with HLA haplotypes, autoimmune phenomena may play a role, fits well with this type of interpretation.

Although in most of the known examples of HLA-associated diseases susceptibility is probably dominant we would, of course, predict that cases of HLA-linked dominant resistance to a disease should be found. But as they would, probably, result in a decreased frequency of the HLA haplotypes conferring resistance among patients, such associations are even more difficult to detect. This is true in particular if, as may seem probable, the recessive alleles conferring susceptibility are rare.

In the mouse, of the two best characterized *H-2* linked loci controlling resistance to virus-induced leukaemia, one, the *Rgv-1* locus, which influences susceptibility to Gross virus, maps in the *K* end, relative resistance being dominant.[40] The other, which affects recovery from Friend virus disease, is associated with the *D* end and relative resistance is recessive.[45]

3.2.6 *A role for MLR determinants in T cell responses to 'soluble' antigens*

One of the findings which seemed to set MLR apart from other T cell responses to, apparently, soluble antigens was that only living stimulator cells were able to trigger a MLR response. But there is no experimental system which rules out the possibility that an effective 'T cell immunogen' is always associated with a living cell; on the contrary: (a) in the systems where this has been rigorously examined, the *in vitro* response of primed T cells to soluble antigens requires the presence of macrophages (see, for example, References 167, 196) and possibly the physical interaction between antigen, macrophage and T cell,[197] and (b) recent evidence suggests that antigen specific suicide of T helper cells may depend on presentation of the antigen on specific B cells.

From these findings we derive the speculative suggestion that, (a) a MLR is the prototype of a T helper cell response and that (b) helper cells do not react to a soluble antigen as such but to an 'immunogen' which is an antigen-mediated modification of MLR determinants on B cells or macrophages. Such modifications may be induced by the antigen directly, for instance through 'sticking' to the macrophages, or they may be a result of specific binding of antigen, for example by B cell receptors. Similar suggestions have been made by others (Reference 198, and H. G. Bluestein, herein).

Our speculation is, of course, in agreement with the finding that helper T cells may belong to the same subset as MLR-responsive cells. As in the case of modifications of MHC determinants produced by infection a response to a given antigen would require that, (a) it induces an appropriate change of MLR determinants and (b) that T cells can recognize this change. Thus, immune responsiveness to such an antigen would depend on three genetic conditions: presence of a particular germ-line T cell receptor gene and the corresponding MLR self determinant as well as a MLR determinant which is modified by the antigen in such a way as to be recognized by the descendants of the self-reactive T cells. The two MLR determinants could, of course, be identical.

If immunization with soluble antigen results in T cells primed not for the antigen itself but for the antigen induced modifications of MLR determinants, the experiments which were interpreted to suggest that successful interaction between primed T cells, on the one hand, and macrophages or B cells, on the other, required *I* region compatibility can be reinterpreted: in F_1 animals derived from a cross between high and low responder strains, T cells will, in many cases (namely in these situations, where the low responder does not share the crucial modifiable MLR determinant with the high responder), only recognize antigen-mediated modifications of the responder MLR determinants. They are therefore, only able to 'cooperate' with high responder B cells or macrophages and their response to these changes can, as has been demonstrated for T cell–macrophage interactions, be blocked by appropriate anti-*I* region alloantisera (H. G. Bluestein, herein). Such blocking still occurs when antisera are added up to many hours after addition of T cells to antigen-loaded macrophages; the same has been found when anti-*I* region alloantisera are used to block an allogeneic MLR (G. Fathman, personal communication; T. Meo, unpublished results).

4. SUMMARY: A UNIFYING HYPOTHESIS FOR T CELL-MEDIATED IMMUNE RESPONSIVENESS

T cells respond to antigen only if it is associated with a living 'stimulator' cell. Their specificity is genetically controlled by two sets of genes, the germ-line T cell receptor genes and those coding for MLR determinants. Antigen specific T cells are derived from self-reactive cells by somatic mutation and selection. Therefore a large fraction of T cells have receptors specific for modification of self-determinants controlled by the genes of the major histocompatibility complex.

Modifications of MLR determinants are induced by infection with viruses and some bacterial and protozoal pathogens on macrophages and by specific or non-specific binding of soluble antigens to macrophages and/or B cells. Activated by such modifications

MLR-responsive cells control, through non-specific mechanisms, other aspects of the host's response.

Non-modifying mutants of pathogens tend to escape detection by the immune system. This adaptation is countered by the host species through generation of new MLR determinants and/or germ-line T cell receptor alleles. As the immune response potential of the individual depends on two interacting sets of genes, they are linked in the major histocompatibility complex, and co-adapted T cell receptor and MLR alleles are associated in the same haplotype.

5. REFERENCES

1. Blanden, R. V., Hapel, A. J. and Jackson, D. C. (1976). *Immunochemistry*, **13**, 179–191.
2. Jerne, N. K. (1971). *Eur. J. Immunol.*, **1**, 1–9.
3. Bain, B., Vaz, M. R. and Lowenstein, L. (1964). *Blood*, **23**, 108–116.
4. Bach, F. H. and Hirschhorn, K. (1964). *Science*, **143**, 813–814.
5. Ling, N. R. and Kay, J. E. (Eds.) (1975). *Lymphocyte Stimulation*, North-Holland, Amsterdam.
6. Rosenberg, S. A., Levy, R., Schechter, B., Ficker, S. and Terry, W. D. (1972). *Transplantation*, **13**, 541–545.
7. Bach, F. H. and Voynow, N. K. (1966). *Science*, **153**, 545–547.
8. Elves, M. W. (1969). *Vox Sang.*, **16**, 81–96.
9. Sørensen, S. F. (1972). *Acta Path. Microbiol. Scand.*, Sect. B, Suppl. 230.
10. Hartzmann, R. J., Segall, M., Bach, M. L. and Bach, F. H. (1971). *Transplantation*, **11**, 268–273.
11. Sengar, D. P. S. and Terasaki, P. I. (1971). *Transplantation*, **11**, 260–267.
12. Miggiano, V. C., Meo, T., Birgén, I. and Nabholz, M. (1975). *Tissue Antigens*, **5**, 173–185.
13. Péna-Martinez, J. and Festenstein, J. (1975). *Transplantation*, **20**, 26–37.
14. Kasakura, S. and Lowenstein, L. (1965). In *Histocompatibility Testing* (Amost, B. and van Rood, J. J., Eds.), Munksgaard, Copenhagen, pp. 211–212.
15. Thorsby, E., Du Bois, R., Bondevik, H., Dupont, B., Eijsvoogel, V., Hansen, J. A., Jersild, C., Jørgensen, F., Kissmeyer-Nielsen, F., Lamm, L. U., Schellekens, P. Th. A., Svejgaard, A. and Thomson, M. (1974). *Tissue Antigens*, **4**, 507–525.
16. Thorsby, E. and Piazza, A. (1975). *Histocompatibility Testing* (Kissmeyer-Nielsen, F., Ed.), Munksgaard, Copenhagen, pp. 414–458.
17. Osoba, D. and Falk, J. (1974). *Cell. Immunol.*, **10**, 117–127.
18. Bach, F. H. and Amos, D. B. (1967). *Science*, **156**, 1506–1508.
19. Palm, J. and Wilson, D. B. (1973). *Transplant. Proc.*, **5**, 1573–1577.
20. Bradley, B. A., White, D. J. G. and Edwards, J. M. (1974). *Tissue Antigens*, **4**, 283–290.
21. van den Tweel, J. G., Vriesendorp, H. M., Termijtelen, A., Westbroek, D. L., Bach, M. L. and van Rood, J. J. (1974). *J. exp. Med.*, **140**, 825–836.
22. Seigler, H. F., Ward, F. E., Metzgar, R. S. and Stulting, S. M. (1974). *Transplantation*, **17**, 265–271.
23. Goldmann, S. F., Krumbacher, K., Schnappauf, H. P., Huget, R. P. and Flad, H.-D. (1975). *Tissue Antigens*, **5**, 155–164.
24. Dorf, M. E., Balner, H. and Benacerraf, B. (1975). *J. exp. Med.*, **142**, 673–693.
25. Pazderka, F., Longenecker, B. M., Law, G. R. J. and Ruth, R. F. (1975). *Immunogenetics*, **2**, 101–130.
26. Du Pasquier, L., Chardonnens, X. and Miggiano, V. C. (1975). *Immunogenetics*, **1**, 482–494.
27. Günther, E., Balcarová, J., Hála, K., Rüde, E. and Hraba, T. (1974). *J. Immunol.*, **4**, 548–553.
28. Karakoz, J., Krejči, J., Hála, K., Blaszczyk, B., Hraba, T. and Pekárek, J. (1974). *Eur. J. Immunol.*, **4**, 545–548.
29. Gasser, D. L. and Silvers, W. K. (1974). *Advan. Immunol.*, **18**, 1–66.
30. Günther, E., Rüde, E. and Štark, O. (1972). *Eur. J. Immunol.*, **2**, 151–155.
31. Levine, B. B., Stember, R. H. and Fotino, M. (1972). *Science*, **178**, 1201–1203.
32. Marsh, D. G., Bias, W. B. and Hsu, S. H. (1973). *Science*, **179**, 691–693.

33. Buckley, C. E., Dorsey, F. C., Corley, R. B., Ralph, W. B., Woodbury, M. A. and Amos, D. B. (1973). *Proc. Nat. Acad. Sci.*, **70**, 2157–2161.
34. Günther, E., Rüde, E., Meyer-Delius, M. and Stark, O. (1973). *Transplant. Proc.*, **5**, 1467–1469.
35. Greenberg, L. J., Gray, E. D. and Yunis, E. J. (1975). *J. exp. Med.*, **141**, 935–943.
36. Rosen, F. S. (1975). In *Immunogenetics and Immunodeficiency* (Benacerraf, B., Ed.), MTP, Lancaster, Chapter 6, pp. 229–257.
37. Meo, T., Krasteff, T. and Shreffler, D. C. (1975). *Proc. Nat. Acad. Sci.*, **72**, 4536–4540.
38. Curman, B., Österberg, L., Sandberg, L., Malmheden-Eriksson, I., Stålenheim, G., Raks, L. and Peterson, P. A. (1975). *Nature*, **258**, 243–245.
39. Lachmann, P. J., Grennan, D., Martin, A. and Démant, P. (1975). *Nature*, **258**, 242–243.
40. Lilly, F. (1971). In *Cellular Interactions in the Immune Response* (Cohen, S., Cudkowicz, G. and McCluskey, R. T., Eds.), Karger, Basel, pp. 103–108.
41. Tennant, J. R. and Snell, G. D. (1968). *J. Nat. Cancer Inst.*, **41**, 597–604.
42. Mühlbock, O. and Dux, A. (1971). In *Immunogenetics of the H-2 System* (Lengerová, A. and Vojtíšková, M., Eds.), Karger, Basel, pp. 123–128.
43. Vladutiu, A. O. and Rose, N. R. (1971). *Science*, **174**, 1137–1139.
44. McDevitt, H. O., Oldstone, M. B. A. and Pincus, T. (1974). *Transplant. Rev.*, **19**, 209–225.
45. Chesebro, B., Wehrly, K. and Stimpfling, J. (1974). *J. exp. Med.*, **140**, 1457–1467.
46. Pazderka, F., Longenecker, B. M., Law, G. R. J., Stone, H. A. and Ruth, R. F. (1975). *Immunogenetics*, **2**, 93–100.
47. Festenstein, H. (1974). *Transplantation*, **18**, 555–557.
48. Gebhart, B. M., Nakao, Y. and Smith, R. T. (1974). *J. exp. Med.*, **140**, 370–382.
49. Piguet, P.-F., Dewey, H. K. and Vassalli, P. (1975). *J. exp. Med.*, **141**, 775–787.
50. Trinchieri, G., Bernoco, D., Curtoni, S. E., Miggiano, V. C. and Ceppellini, R. (1973). In *Histocompatibility Testing 1972* (Dausset, J. and Colombani, J., Eds.), Munksgaard, Copenhagen, pp. 509–519.
51. Alter, B. J., Schendel, D. J., Bach, M. L., Bach, F. H., Klein, J. and Stimpfling, J. H. (1973). *J. exp. Med.*, **137**, 1303–1309.
52. van Rood, J. J. and Eijsvoogel, V. P. (1970). *Lancet*, **1**, 698–700.
53. Sørensen, F. S. and Staub-Nielsen, L. (1970). *Acta path. microbiol. scand.*, **78**, 719–725.
54. Eijsvoogel, V. P., Schellekens, P. Th. A., Breur-Vriesendorp, B., Koning, L., Koch, C., van Leeuwen, A. and van Rood, J. J. (1970). *Transplant. Proc.*, **3**, 85–88.
55. Yunis, E. J. and Amos, D. B. (1971). *Proc. Nat. Acad. Sci.*, **68**, 3031–3035.
56. Mempel, W., Albert, E. and Burger, A. (1972). *Tissue Antigens*, **2**, 250–254.
57. Eijsvoogel, V. P., du Bois, M. C. J. C., Melief, C. J. M., de Groot-Kooy, M. L., Koning, C., van Rood, J. J., van Leeuwen, A., du Toit, E. and Schellekens, P. T. A. (1973). In *Histocompatibility Testing 1972* (Dausset, J. and Colombani, J., Eds.), Munksgaard, Copenhagen, pp. 501–508.
58. Dupont, B., Good, R. A., Hansen, G. S., Jersild, C., Staub-Nielsen, L., Park, B. H., Svejgaard, A., Thomsen, M. and Yunis, E. J. (1974). *Proc. Nat. Acad. Sci.*, **71**, 52–56.
59. Balner, H. and Toth, E. K. (1973). *Tissue Antigens*, **3**, 273–290.
60. Grosse-Wilde, H., Vriesendorp, H. M., Wank, R., Mempel, W., Dechamps, B., Honauer, U., Baumann, P., Netzel, B., Kolb, H. J. and Albert, E. D. (1974). *Tissue Antigens*, **4**, 229–237.
61. Bach, F. H., Widmer, M. B., Segall, M., Bach, M. L. and Klein, J. (1972). *J. exp. Med.*, **136**, 1420–1444.
62. Meo, T., Vives, J., Miggiano, V. and Shreffler, D. (1973). *Transplant. Proc.*, **5**, 377–381.
63. Shreffler, D. C. and David, C. S. (1975). *Advan. Immunol.*, **20**, 125–196.
64. Thorsby, E. (1974). *Transplant. Rev.*, **18**, 51–129.
65. van Rood, J. J., van Leeuwen, A., Termiitelen, A. and Keuning, J. J. (1976). In *The Role of Products of the Histocompatibility Gene Complex in Immune Responses* (Katz, D. H. and Benacerraf, B., Eds.), Academic, New York, pp. 31–51.
66. WHO-IUIS Terminology Committee (1975). *Eur. J. Immunol.*, **5**, 889–891.
67. Lightbody, J., Bernoco, D., Miggiano, V. C. and Ceppellini, R. (1971). *J. bact. Virol. Immunol.*, **64**, 243–254.
68. Miggiano, V. C., Bernoco, D., Lightbody, J., Trinchieri, G. and Ceppellini, R. (1972). *Transplant. Proc.*, **4**, 231–237.
69. Grunnet, N., Kristensen, T. and Kissmeyer-Nielsen, F. (1975). *Tissue Antigens*, **6**, 205–220.
70. Kristensen, T., Grunnet, N. and Kissmeyer-Nielsen, F. (1975). *Tissue Antigens*, **6**, 221–228.

71. Kristensen, T., Grunnet, N. and Kissmeyer-Nielsen, F. (1975). *Tissue Antigens*, **6**, 229–236.
72. Klein, J. (1975). *Biology of the Mouse Histocompatibility-2 Complex*. Springer-Verlag, Berlin.
73. Bennett, D. (1975). *Cell*, **6**, 441–454.
74. Meo, T., Vives, G., Rijnbeek, A. M., Miggiano, V. C., Nabholz, M. and Shreffler, D. C. (1973). *Transplant. Proc.*, **5**, 1339–1350.
75. Meo, T., David, C. S., Rijnbeek, A. M., Nabholz, M., Miggiano, V. and Shreffler, D. C. (1973). *Transplant. Proc.*, **5**, 1507–1510.
76. Klein, J. and Park, J. M. (1973). *J. exp. Med.*, **137**, 1213–1255.
77. Widmer, M. B., Omodei-Zorini, C., Bach, M. L., Bach, F. H. and Klein, J. (1973). *Tissue Antigens*, **3**, 309–315.
78. Widmer, M. B., Peck, A. B. and Bach, F. H. (1973). *Tissue Antigens*, **5**, 1501–1505.
79. Widmer, M. B., Schendel, D. J., Bach, F. J. and Boyse, E. A. (1973). *Transplant. Proc.*, **5**, 1663–1666.
80. Nabholz, M., Vives, J., Young, H. M., Meo, T., Miggiano, V. C., Rijnbeek, A. and Shreffler, D. C. (1974). *Eur. J. Immunol.*, **4**, 378–387.
81. Abbasi, K. and Festenstein, H. (1973). *Eur. J. Immunol.*, **3**, 430–435.
82. Nabholz, M., Young, H., Rijnbeek, A., Boccardo, R., David, C. S., Meo, T., Miggiano, V. and Shreffler, D. C. (1975). *Eur. J. Immunol.*, **9**, 594–599.
83. Wagner, H., Götze, D., Ptschelinzew, L. and Röllinghoff, M. (1975). *J. exp. Med.*, **142**, 1477–1487.
84. Jackson Laboratory (1975). *Biology of the Laboratory Mouse* (Green, E. L., Ed.), 2nd ed., Dover Publications, New York.
85. Fisher, R. A. (1949). *The Theory of Inbreeding*. Oliver and Boyd, Edinburgh.
86. Egorov, I. K. (1974). *Immunogenetics*, **1**, 97–107.
87. Melvold, R. and Kohn, H. I. (1975). *Mutation Research*, **27**, 415–418.
88. Klein, J., Forman, J., Hauptfeld, V. and Egorov, I. K. (1975). *J. Immunol.*, **115**, 716–728.
89. Nabholz, M., Young, H., Meo, T., Miggiano, V., Rijnbeek, A., and Shreffler, D. C. (1975). *Immunogenetics*, **1**, 457–468.
90. Klein, J. and Egorov, I. K. (1973). *J. Immunol.*, **111**, 976–979.
91. Widmer, M. B., Alter, B. J., Bach, F. H., Bach, M. L. and Bailey, D. W. (1973). *Nature* (*New Biol.*), **242**, 239–241.
92. Forman, J. and Klein, J. (1975). *Immunogenetics*, **1**, 469–481.
93. Forman, J. and Klein, J. (1975). *J. Immunol.*, **115**, 711–715.
94. Benedict, A. A., Pollard, L. W., Morrow, P. R., Abplanalp, H. A., Maurer, P. H. and Briles, W. E. (1975). *Immunogenetics*, **2**, 313–324.
95. Hála, K., Vilhelmová, M., Hartmanová, J. (1976). *J. Immunogenet.* **13**, 97–103.
96. Lindahl-Kiesling, K. and Safwenberg, J. (1972). In *Proc. Sixth Leukocyte Culture Conf.* (Schwartz, M. R., Ed.), Academic, New York, pp. 623–638.
97. Lafferty, K. J., Misko, I. S. and Corley, M. A. (1974). *Nature*, **249**, 275–276.
98. Wagner, H. (1973). *Eur. J. Immunol.*, **3**, 84–89.
99. Lafferty, K. J., Walker, K. Z., Scollay, R. G. and Killby, V. A. A. (1972). *Transplant. Rev.*, **12**, 198–228.
100. Harrison, M. R. and Paul, W. E. (1973). *J. exp. Med.*, **138**, 1602–1607.
101. von Boehmer, H. (1974). *J. Immunol.*, **112**, 70–78.
102. Kennedy, J. C. and Ekpaha-Mensah, J. A. (1973). *J. Immunol.*, **111**, 1639–1652.
103. Rode, H. N. and Gordon, J. (1970). *J. Immunol.*, **104**, 1453–1457.
104. Rode, H. N. and Gordon, J. (1974). *Cell. Immunol.*, **13**, 87–94.
105. Greineder, D. K. and Rosenthal, A. S. (1975). *J. Immunol.*, **114**, 1541–1547.
106. Schirrmacher, V., Pena-Martinez, J. and Festenstein, H. (1975). *Nature*, **255**, 155–156.
107. Kountz, S. L., Cochrum, K. C. and Main, R. K. (1972). In *Proc. Sixth Leukocyte Culture Conference* (Schwartz, N. R., Ed.), Academic, New York, pp. 609–621.
108. Gillette, R. W., Cooper, S. and Lance, E. M. (1972). *Immunology*, **23**, 769–776.
109. Hirschberg, H. and Thorsby, E. (1975). *Tissue Antigens*, **6**, 183–194.
110. von Boehmer, H., Sprent, J. and Nabholz, M. (1975). *J. exp. Med.*, **141**, 322–334.
111. Sprent, J., von Boehmer, H. and Nabholz, M. (1975). *J. exp. Med.*, **142**, 321–331.
112. von Boehmer, H. and Sprent, J. (1974). *Nature*, **249**, 363–365.
113. Fathman, C. G., Handwerger, B. S. and Sachs, D. H. (1974). *J. exp. Med.*, **140**, 853–858.
114. Simpson, E. (1975). *Eur. J. Immunol.*, **5**, 456–461.
115. Lonai, P. and McDevitt, H. O. (1974). *J. exp. Med.*, **140**, 1317–1323.

116. Kissmeyer-Nielsen, F., Ed. (1975). *Histocompatibility Testing*, Munksgaard, Copenhagen.
117. Finkelman, F. D., Schevach, E. M., Vitetta, E. J., Green, I. and Paul, E. (1975). *J. exp. Med.*, **141**, 27–41.
118. Unanue, E. R., Dorf, M. E., David, C. S. and Benacerraf, B. (1974). *Proc. Nat. Acad. Sci.*, **71**, 5014–5016.
119. Hämmerling, G. J., Mauve, G., Goldberg, E. and McDevitt, H. O. (1975). *Immunogenetics*, **1**, 428–437.
120. Meo, T., David, C. S., Rijnbeek, A. M., Nabholz, M., Miggiano, V. C. and Shreffler, D. C. (1974). *Transplant. Proc.*, **7**, 127–129.
121. Meo, T., David, C. S. and Shreffler, D. C. (1976). In *The Role of Products of the Histocompatibility Gene Complex in Immune Responses* (Katz, D. H. and Benacerraf, B., Eds.), Academic, New York, pp. 167–178.
122. Wagner, H. (1972). *J. Immunol.*, **109**, 630–637.
123. Croy, A. B. and Osoba, D. (1973). *Transplant. Proc.*, **5**, 1721–1723.
124. Nagy, Z., Elliott, B., Nabholz, M., Krammer, P. and Pernis, B. (1976). *J. exp. Med.*, **143**, 648–659.
125. Cantor, H. and Boyse, E. A. (1975). *J. exp. Med.*, **141**, 1390–1399.
126. Cantor, H. and Boyse, E. A. (1975). *J. exp. Med.*, **141**, 1376–1389.
127. Stobo, J. D., Paul, W. E. and Henney, C. S. (1973). *J. Immunol.*, **110**, 652–660.
128. Wilson, D. B., Blyth, J. L. and Nowell, P. C. (1968). *J. exp. Med.*, **128**, 1157–1181.
129. Bach, F. H., Bock, H., Graupner, K., Day, E. and Klostermann, H. (1969). *Proc. Nat. Acad. Sci.*, **62**, 377–384.
130. Jones, G. (1973). *J. Immunol.*, **111**, 914–920.
131. Simonsen, M. (1967). *Cold Spring Harbor Symp. Quant. Biol.*, **32**, 517–523.
132. Ford, W. L. and Atkins, R. C. (1971). *Nature* (*New Biol.*), **234**, 178–180.
133. Atkins, R. C. and Ford, W. L. (1975). *J. exp. Med.*, **141**, 664–680.
134. Ford, W. L., Simmonds, S. J. and Atkins, R. C. (1975). *J. exp. Med.*, **141**, 681–696.
135. Roelants, G. E. (1972). *Current Topics Microbiology Immunology*, **59**, 135–165.
136. Roelants, G. E. (1975). In *Membrane Receptors of Lymphocytes* (Seligmann, M., Preud'homme, J. L. and Kourilsky, F. M., Eds.), North-Holland, Amsterdam, pp. 65–77.
137. Lefkovits, I. (1974). *Current Topics in Microbiol. and Immunol.*, **65**, 21–58.
138. Amos, D. B. and Bach, F. H. (1968). *J. exp. Med.*, **128**, 623–637.
139. Klein, J. (1973). In *International Symposium on HL-A Reagents* (Regaimy, R. H. and Spärck, J. V., Eds.), Karger, Basel, pp. 251–256.
140. Simonsen, N. (1973). *Proc. 8th Congress International Assoc. Allergology*, Excepta Media, Amsterdam, pp. 146–154.
141. Brondz, B. D. and Snegirova, A. E. (1971). *Immunology*, **20**, 457–468.
142. Golstein, P., Svedmyr, E. A. J. and Wigzell, H. (1971). *J. exp. Med.*, **134**, 1385–1402.
143. Wekerle, H., Lonai, P. and Feldman, M. (1972). *Proc. Nat. Acad. Sci.*, **69**, 1620–1624.
144. Clark, W. and Kimura, A. (1972). *Nature* (*New Biol.*), **235**, 236–237.
145. MacDonald, H. R., Cerottini, J.-C. and Brunner, K. T. (1974). *J. exp. Med.*, **140**, 1511–1521.
146. Zoschke, D. C. and Bach, F. H. (1971). *Science*, **172**, 1350–1352.
147. Hirschberg, H., Rankin, B. and Thorsby, E. (1974). *Cell. Immunol.*, **10**, 458–466.
148. Binz, H. and Askonas, B. A. (1975). *Eur. J. Immunol.*, **5**, 618–623.
149. Binz, H. and Wigzell, H. (1975). *J. exp. Med.*, **142**, 197–211.
150. Binz, H. and Wigzell, H. (1975). *J. exp. Med.*, **142**, 1218–1229.
151. Kennedy, J. C. and Ekpaha-Mensah, J. A. (1974). *J. Immunol.*, **110**, 1108–1117.
152. Armerding, D., Sachs, D. H. and Katz, D. H. (1974). *J. exp. Med.*, **140**, 1717–1722.
153. Altman, A. and Cohen, I. R. (1974). *Eur. J. Immunol.*, **5**, 437–444.
154. Godal, T., Rees, R. J. W. and Lamvik, J. O. (1971). *Clin. exp. Immunol.*, **8**, 625–637.
155. Dimitriu, A., Dy, M., Thomson, N. and Bona, C. (1974). *Clin. exp. Immunol.*, **18**, 141–148.
156. Ward, P. and Volkman, A. (1975). *J. Immunol.*, **115**, 1394–1399.
157. Blanden, R. V., Doherty, P. C., Dunlop, M. B. C., Gardner, I. D. and Zinkernagel, R. M. (1975). *Nature*, **254**, 269–270.
158. Zinkernagel, R. M. and Doherty, P. C. (1974). *Nature*, **251**, 547–548.
159. Koszinowski, V. and Thomssen, R. (1975). *Eur. J. Immunol.*, **5**, 245–251.
160. Shearer, G. M., Rehn, T. G. and Garbarino, C. A. (1975). *J. exp. Med.*, **141**, 1348–1364.
161. Forman, J. (1975). *J. exp. Med.*, **142**, 403–418.
162. Fischer Lindahl, K., Koszinowski, U. and Ertl, H. (1975). *Nature*, **258**, 550.

163. Germain, R. N., Dorf, M. E. and Benacerraf, B. (1975). *J. exp. Med.*, **142**, 1023–1028.
164. Fujimoto, S., Chen, C. H., Sabbadini, E. and Se Hon, A. H. (1973). *J. Immunol.*, **111**, 1093–1100.
165. Zinkernagel, R. M. (1974). *Nature*, **251**, 230–233.
166. Bevan, M. J. (1975). *J. exp. Med.*, **142**, 1349–1364.
167. Rosenthal, A. S. and Shevach, E. M. (1973). *J. exp. Med.*, **138**, 1194–1212.
168. Shevach, E. M. and Rosenthal, A. S. (1973). *J. exp. Med.*, **138**, 1213–1229.
169. Katz, D. H., Hamaoka, T., Dorf, M. E., Maurer, P. H. and Benacerraf, B. (1973). *J. exp. Med.*, **138**, 734–739.
170. Katz, D. H., Graves, M., Dorf, M. E., Dimuzio, H. and Benacerraf, B. (1975). *J. exp. Med.*, **141**, 263–268.
171. Shearer, G. M., Lozner, E. C., Rehn, T. G. and Schmitt-Verhulst, A.-M. (1975). *J. exp. Med.*, **141**, 930–934.
172. Lewontin, R. C. (1974). *The Genetic Basis of Evolutionary Change*, Columbia University Press, New York.
173. Palm, J. (1970). *Transplant. Proc.*, **2**, 162–173.
174. Fisher, R. A. (1930). *The Genetical Theory of Natural Selection*, Clarendon Press, Oxford.
175. Ohno, S. (1973). *Nature*, **244**, 259–262.
176. Möller, G., Ed. (1975). *Transplant. Rev.*, **22**.
177. Vladutiu, A. O. and Rose, N. R. (1974). *Immunogenetics*, **1**, 305–328.
178. Boettcher, B. (1975). *Immunogenetics*, **2**, 485–489.
179. Kisielow, P., Hirst, J. A., Shiku, H., Beverley, P. C. L., Hoffman, M. K., Boyse, E. A. and Oettgen, H. F. (1975). *Nature*, **253**, 219–220.
180. Günther, E. and Rüde, E. (1975). *J. Immunol.*, **115**, 1387–1393.
181. Dorf, M. E., Stimpfling, J. H. and Benacerraf, B. (1975). *J. exp. Med.*, **141**, 1459–1463.
182. Merryman, C. F., Maurer, P. H. and Stimpfling, J. H. (1975). *Immunogenetics*, **2**, 441–448.
183. Melchers, I. and Rajewsky, K. (1975). *Eur. J. Immunol.*, **5**, 753–759.
184. Munro, A. J. and Taussig, M. J. (1975). *Nature*, **256**, 103–106.
185. Basten, A., Miller, J. F. A. P. and Abraham, R. (1975). *J. exp. Med.*, **141**, 547–559.
186. Asantila, T., Vahala, J. and Toivanen, P. (1974). *Immunogenetics*, **1**, 272–290.
187. Miggiano, V. C., Meo, T., Nabholz, M., Trinchieri, G., Barrelet, V., Bossart, H. and Koch, R. (1973). In *Proc. First Internat. Congress of Immunology in Obstetrics and Gynaecology* (Centaro, A. and Carretti, N., Eds.), Excerpta Medica, Amsterdam, pp. 156–168.
188. von Boehmer, H. and Byrd, W. J. (1972). *Nature* (*New Biol.*), **235**, 50–52.
189. Bodmer, W. F. (1972). *Nature*, **237**, 139–145.
190. Bechtol, K. B., Freed, J. H., Herzenberg, L. A. and McDevitt, H. O. (1974). *J. exp. Med.*, **140**, 1660–1675.
191. Doherty, P. C. and Zinkernagel, R. M. (1975). *Nature*, **256**, 50–52.
192. Lennox, E. (1975). *Nature*, **256**, 7–8.
193. Mims, C. A. (1964). *Bacteriol. Rev.*, **28**, 30–71.
194. Allison, A. C. (1974). *Transplant. Rev.*, **19**, 3–55.
195. Nelson, D. S. (1974). *Transplant. Rev.*, **19**, 226–254.
196. Goodman, J. W., Bellone, C. J., Hanes, D. and Nitecki, D. E. (1974). In *Progress in Immunology II*, Vol. 2 (Brent, L. and Holborow, J., Eds.), North-Holland, Amsterdam, pp. 27–37.
197. Lipsky, P. E. and Rosenthal, A. S. (1975). *J. exp. Med.*, **141**, 138–154.
198. Klein, J. (1976). *Current Topics in Immunobiology*, **5**, 297–336.
199. Trinchieri, G., Aden, D. P. and Knowles, B. B. (1976). *Nature*, **261**, 312–314.

Chapter 13

The Role of Histocompatibility Region Antigens in Lymphocyte Activation

H. G. Bluestein

Abbreviations

AEF: allogeneic enhancing factor
BGG: bovine gammaglobulin
C.I.: control of immunocompetent cell interactions
DNP: 2,4-dinitrophenyl
Fab: monovalent fragment of antibody
$F(ab)'_2$: bivalent fragment of antibody
Fc: constant fragment of antibody
FGG: fowl gammaglobulin
GA: (L-Glu, L-Ala)
GAT: (L-Glu, L-Ala, L-Tyr)
GL: (L-Glu, L-Lys)
GLT: (L-Glu, L-Lys, L-Tyr)
GT: (L-Glu, L-Tyr)
GvH: graft-versus-host
(H, G)-A–L: (L-His, L-Glu)-L-Ala–L-Lys
I(r): immune (response)

KLH: keyhole limpet haemocyanin
LCM: lymphocytic choriomeningitis virus
MBSA: methylated bovine serum albumin
MHC: major histocompatibility complex
MIF: macrophage inhibition factor
MLC: mixed lymphocyte culture
PFC: plaque forming cell
(Phe, G)-A–L: (L-Phe, L-Glu)-L-Ala–L-Lys
PLL: poly-L-lysine
PPD: purified protein derivative of tuberculin
SRBC: sheep red blood cell
(T, G)-A–L: (L-Tyr, L-Glu)-L-Ala–L-Lys
TNP: 2,4,6-trinitrophenyl

1. INTRODUCTION

The major histocompatibility complex (MHC), as its name implies, is comprised of a group of linked genetic loci which determine the survival of tissue transplants exchanged between members of the same species (allografts). The structural organization of the histocompatibility region genes has been the subject of intensive investigation (see References 1 and 2 for recent reviews).

Studies in mice have been most revealing because the development of congenic-resistant strains having identical genetic make-up except at the MHC (named H-2 in mice) has permitted sophisticated genetic analyses. The entire H-2 complex spans a genetic map distance of approximately 0·5 centimorgans, i.e. there is a 0·5% crossover frequency between its boundaries. H-2 has been divided into four major regions, K, I, S and D reading from left to right, when the chromosome is oriented so that the centromere is to the left of the MHC. The K and D regions control the expression of the highly polymorphic classically defined transplantation antigens. The S region controls differences in a serum protein, Ss, which has not yet been shown to be related to cell membrane determinants. The I region contains genes governing the expression of cell membrane alloantigens which control a variety of traits that will be described in this chapter.

The biological role of the histocompatibility complex is yet to be defined with certainty. The grafting of foreign tissue is an 'unnatural' man-made phenomenon which could not have provided the selective pressures needed to maintain the integrity of the complex genetic structure of the MHC. It is only in the past decade that studies demonstrating the involvement of histocompatibility region gene products in immunological phenomena have led to some appreciation of their functional significance. Genes controlling specific immune responsiveness, genes controlling cooperative interactions between immunocompetent cells, and genes controlling the generation of immunoreactive cells to altered self-antigens are coded within the histocompatibility region. Thus the MHC is an important part of one of the major mechanisms for 'self-defence'. In the sections to follow some of the experimental evidence suggesting that products of MHC genes play a vital role in both the induction and expression of specific immunity will be discussed.

2. HISTOCOMPATIBILITY-LINKED IMMUNE RESPONSE GENES

In several mammalian species genes have been identified which control the ability of the animal to respond immunologically to specific antigens.[3,4] Two broad classes of immune response (Ir) genes have been identified. The major class, in terms of the numbers of genes thus far discovered, seems to exert its control on early events in the initiation of the immune response, since in its absence the animal is unable to make either cellular or humoral immune responses. Most of these genes have been located in the major histocompatibility region of the genome and have, therefore, been named histocompatibility or H-linked Ir genes.

Another type of Ir gene, thus far identified only in the mouse, seems to be exerting its action at the level of antibody production.[5–7] These genes, which are also antigen specific, do not control the ability of an animal to respond immunologically to that antigen but, rather, they control the idiotypic determinants or fine specificity of the antibody produced. These genes are linked to immunoglobulin structural genes which control allotypic determinants on heavy chains. Since these allotype-linked Ir genes are antigen specific it is likely that they are immunoglobulin variable region genes and thereby code for specific B cell receptors. Comparing these genes with the H-linked Ir genes leads to the conclusion that the latter are not coding for immunoglobulin type receptors. The heavy chain allotype-linked Ir genes are not linked to the major histocompatibility complex. Light chain structural genes, in man at least, also are not linked to the MHC. Since variable and constant region genes are thought to be linked in both heavy and light chains and since both heavy and light chain structural genes are not linked to the major histocompatibility complex, it is unlikely that histocompatibility-linked Ir genes code for structural determinants in immunoglobulin variable regions.

When challenged with foreign protein antigens, the mammalian immune system reacts with a complex series of cellular and humoral responses each of which may be directed at different parts of the immunizing molecule. One would not expect, therefore, that the ability to respond immunologically to such multideterminant antigens would be under simple genetic control. In fact, only by presenting limited antigenic challenge to the responding animal has it been possible to demonstrate the existence of unigenic loci controlling immune responsiveness to specific antigens. Three types of antigens have been used to provide the necessary restricted antigenic challenge: (1) Synthetic polypeptide antigens, whose structural heterogeneity can be limited by controlling the number or relative proportions of their constituent amino acids, have constituted the most useful tools for delineating Ir genes.[8] (2) Alloantigens which have limited structural differences from the immunized animal's own proteins have been used for the demonstration of genetic control of immune responses to their allotypic determinants.[9–11] (3) Complex protein antigens when used in limiting immunizing doses have also been used to identify specific Ir genes presumably because, when present in sufficiently low concentrations, only the most immunogenic determinant in the molecule is recognized by the immune system.[12,13]

A large number of H-linked Ir genes have now been identified in several animal species, including mice, guinea pigs, rats, and monkeys (reviewed in References 3 and 4). The characteristics of the functional control exercised by these genes is remarkably uniform from species to species. The PLL, GA and GT immune response genes of guinea pigs[14] and the Ir-1 locus in mice,[15] both of which have been defined with the use of synthetic polypeptide antigens, have been the most extensively investigated Ir gene systems. We will use them as models to discuss the characteristics of H-linked Ir gene control.

The guinea pig Ir genes have been named for the antigens used to identify them. The PLL gene was identified using the homopolymer poly-L-lysine,[16] but it also controls immune responses to the copolymer of L-glutamic acid and L-lysine (GL) and to the hapten conjugates of these polypeptides. The GA and GT genes control immune responses to linear random copolymers composed of approximately equal amounts of L-glutamic acid and L-alanine or L-glutamic acid and L-tyrosine, respectively.[17] When immunized with these antigens guinea pigs possessing the appropriate Ir gene respond with both cellular and humoral immune responses. The animals develop delayed hypersensitivity reaction to intradermal challenge with the antigen, and their lymphocytes respond to the specific antigen *in vitro* with increased synthesis of protein, RNA, and DNA as well as the production of migration inhibitory factor. They also produce high levels of specific serum antibody. Those animals that did not inherit the appropriate Ir gene do not develop any detectable cellular immune responses to the antigens either *in vivo* or *in vitro*. Furthermore, they develop either very low or absent levels of circulating antibody. Thus, the presence of the Ir gene is required for the development of T cell-mediated cellular immune responses.

In mice, where it is technically more difficult to directly assay for cell-mediated responses, Ir genes have been identified almost exclusively on the basis of the amount of specific antibody produced. The ability of mice to respond with either high or low amounts of antibody to each of three branched multichain synthetic polypeptides (T, G)-A–L, (H, G)-A–L, and (Phe, G)-A–L is under Ir gene control at a single locus, Ir-1.[15] Mice possessing the appropriate Ir-1 allele make high levels of circulating antibody, predominantly of the IgG class. Animals lacking the gene make much smaller antibody responses, and the antibodies they produce are predominantly of the IgM class. The non-responders also fail to develop memory cells as manifested by their inability to develop an anamnestic response upon secondary challenge with the antigen. The unique properties of the responder mice, like the unique cellular immune responses in responder guinea pigs, are T cell-mediated functions.

Ir genes are not randomly distributed with the genome. Rather, they form part of highly organized linkage groups which include multiple Ir gene loci, as well as other genes controlling the expression of lymphocyte membrane antigens. In mice, for example, as we have already mentioned, the ability to respond to each of the three branched multichain synthetic polypeptide antigens (T, G)-A–L, (H, G)-A–L, and (Phe, G)-A–L is controlled by alleles at the Ir-1 locus. The sophisticated genetic studies that can be conducted in mice because of the availability of large numbers of inbred strains, congenic strains, and inbred recombinant strains have permitted mapping of the Ir-1 locus within the major histocompatibility complex.[18] It is mapped in the I region of the MHC, and it is closely linked to many other mouse Ir genes which have also been mapped in the I region.

In guinea pigs, because of the relatively inefficient reproductive rate, precise genetic studies have not yet been possible. Two inbred strains are available, however, strain 2 and strain 13. The PLL and GA genes are found in strain 2 animals but not strain 13. The GT gene is present in strain 13 animals but not strain 2. Hybrid $(2 \times 13)F_1$ guinea pigs respond to all three antigens attesting to the dominant or codominant inheritance of the Ir genes. The progeny of back-cross matings of responder F_1 animals with the non-responder parental strain results in a 50/50 distribution of responder and non-responder animals. When F_1 animals were mated to strain 13 guinea pigs, for example, half the progeny responded to DNP-PLL and half did not. Thus the PLL gene is inherited as a classical mendelian autosomal dominant gene. All of the H-linked Ir genes identified so far, in all species, share that characteristic.

Among the back-cross progeny, the PLL responder animals all responded to GA while PLL non-responders were all GA non-responders as well. The PLL and GA responders also inherited the genes controlling the expression of strain 2 histocompatibility antigens on their cell surfaces.[19,20] This was demonstrated with a ^{51}chromium-release cytotoxicity assay. Lymphocytes from the PLL/GA responders were susceptible to lysis with an anti-strain 2 alloantiserum whil the non-responders were not. Thus, the ability to respond to PLL and GA was inherited together with the genes of the strain 2 histocompatibility type. Similar back-cross analysis of the inheritance of the GT gene revealed that the GT responders also possessed strain 13 histocompatibility antigens on their lymphoid cells while the non-responders did not. These studies provided the initial evidence for the linkage of the PLL, GA and GT Ir genes to the guinea pig MHC.

Breeding studies in inbred guinea pigs always showed the PLL and GA genes inherited together. The non-identity of those genes was demonstrated in random-bred guinea pigs, which also express Ir gene function. Although most GA responder, random-bred Hartley guinea pigs also respond to PLL, a small number of animals were identified which respond to GA and not to PLL and vice versa.[21] These animals probably result from a crossover between the PLL and the GA gene which has been maintained in the Hartley breeding colony. Thus, although they usually segregate together because they are closely linked, the GA and PLL Ir genes are separable. In the crossover animals, strain 2 histocompatibility genes are always inherited with the PLL gene.

The GT gene also does not segregate independently of the PLL or GA gene. Studies in random bred animals indicated that the guinea pigs were likely to respond to GA and PLL *or* to GT. The frequency of responder guinea pigs to all three antigens was much lower than would be predicted if the Ir genes segregated independently. This suggests that the GT gene functions as an allele or pseudoallele of the GA and PLL genes. This interpretation was confirmed in breeding experiments. When GT and PLL responder animals were mated to guinea pigs that were non-responders for both of those antigens, the progeny responded either to GT or to PLL but never to both and never to neither antigen.[20] Thus, in guinea pigs, as in mice, linkage groups composed of multiple Ir genes exist within the major histocompatibility complex.

3. ALLOANTISERUM-MEDIATED SUPPRESSION OF IR GENE CONTROLLED IMMUNE RESPONSES

The demonstration of linkage between histocompatibility and Ir loci is intriguing. The maintenance of this genetic relationship in many animal species[22] indicates that it provides some selective advantage and suggests a functional interaction between H antigens and Ir gene products. Involvement of cell surface antigens coded for in the major histocompatibility region in the development of specific immune responses has been suggested by the demonstration that Ir gene controlled responses can be suppressed *in vitro* by antibodies directed at the linked histocompatibility antigens. In guinea pigs, anti-strain 2 alloantisera, prepared by immunizing inbred strain 13 guinea pigs with the lymphoid cells from inbred strain 2 animals can specifically inhibit responses to the synthetic polypeptide antigens DNP-GL and GA, both of which are controlled by strain 2 histocompatibility-linked Ir genes. Conversely, anti-strain 13 alloantisera inhibit responses to GT which is a strain 13 associated immune response. The *in vitro* assay most commonly used in these studies has been antigen stimulated T cell proliferation measured as tritiated thymidine incorporation,[23,24] however, the alloantisera also suppress antigen-stimulated MIF production[25] as well as protein and RNA synthesis.[26]

The specificity of the suppression by anti-histocompatibility sera for immune responses controlled by linked Ir genes is most clearly demonstrated using cells from F_1 animals. Hybrid $(2\times 13)F_1$ guinea pigs possess the full complement of both strain 2- and strain 13-linked Ir genes. After immunization with both DNP-GL and GT, lymphoid cells from these animals respond to either of those antigens *in vitro* with increased tritiated thymidine incorporation (Table 1). The addition of an anti-strain 2 alloantisera to such cultures profoundly suppresses DNP-GL responses without affecting GT stimulation. GT stimulation, on the other hand, is suppressed by anti-strain 13 alloantisera at concentrations that do not appreciably affect the DNP-GL stimulation. Each of the lymphoid cells from the F_1 animals possess both strain 2 and train 13 lymphocyte membrane antigens. The exquisite specificity of the alloantiserum-mediated suppression for immune responses controlled by linked Ir genes, therefore, indicates a very close physical relationship between the site of action of the Ir gene product and the lymphocyte membrane antigen to which the alloantiserum was directed.

Table 1. Alloantiserum-mediated suppression of Ir gene controlled antigen stimulation *in vitro*

Antigen	Tritiated thymidine incorporation NGPS	Anti-2	Anti-13
None	876	661	787
DNP-GL	22 768	1 003	19 341
GT	12 515	12 719	1 322

4. IDENTIFICATION OF THE TARGET ANTIGEN

The mechanisms through which alloantisera suppress immune responses has been an area of considerable interest because it promises to provide some insight into the process of antigen recognition by T cells. Identification of the lymphocyte membrane antigen that serves as a target for the alloantisera has been an important first step. Antibodies directed at guinea pig histocompatibility antigens are known to be present in the alloantisera. The method of production of these antisera may also permit the induction of antibodies directed at Ir gene controlled lymphocyte membrane antigens. Since the alloantisera used to suppress a specific immune response were produced in animals that were genetic non-responders, their lymphoid cells might lack the specific antigen receptor or other Ir gene product. The immunizing lymphocytes, on the other hand, were obtained from genetic responders who must express those membrane markers on their cells. The anti-Ir gene product antibodies that might be present in such alloantisera could be responsible for the *in vitro* immunosuppressive effect.

Alloantiserum-mediated, complement dependent lymphocytotoxicity is due to antibodies directed at the major histocompatibility antigens. This has been demonstrated in analyses of back-cross progeny derived from matings between $(2\times 13)F_1$ guinea pigs and strain 13 animals which have shown the coinheritance of the susceptibility to lymphocytotoxicity by anti-strain 2 alloantisera and the ability to accept a transplant of leukaemia cells bearing strain 2 histocompatibility antigens.[27] Furthermore, absorption of the anti-strain 2 alloantisera by non-lymphoid tumour cells from strain 2 animals completely removes its cytotoxic capacity.[28] We, therefore, compared the cytotoxic capacity of an alloantiserum with its ability to suppress a histocompatibility associated immune response *in vitro*.[24] The curves obtained when plotting the cytotoxic capacity of an alloantiserum and its capacity to suppress antigen stimulation *in vitro* as a function of decreasing antiserum concentration are strikingly parallel (Figure 1). These results suggest that the same antibodies are responsible for both properties and thereby implicate the histocompatibility antigens as the target for the suppressive actions of the alloantisera. Further support for this interpretation is provided by the observation that absorption of the anti-strain 2 sera by strain 2 histocompatibility antigen bearing lymphoid cells depleted the cytotoxic capacity of the alloantisera in parallel with depleting its ability to suppress strain 2 histocompatibility-linked Ir gene controlled immune responses *in vitro*.[29]

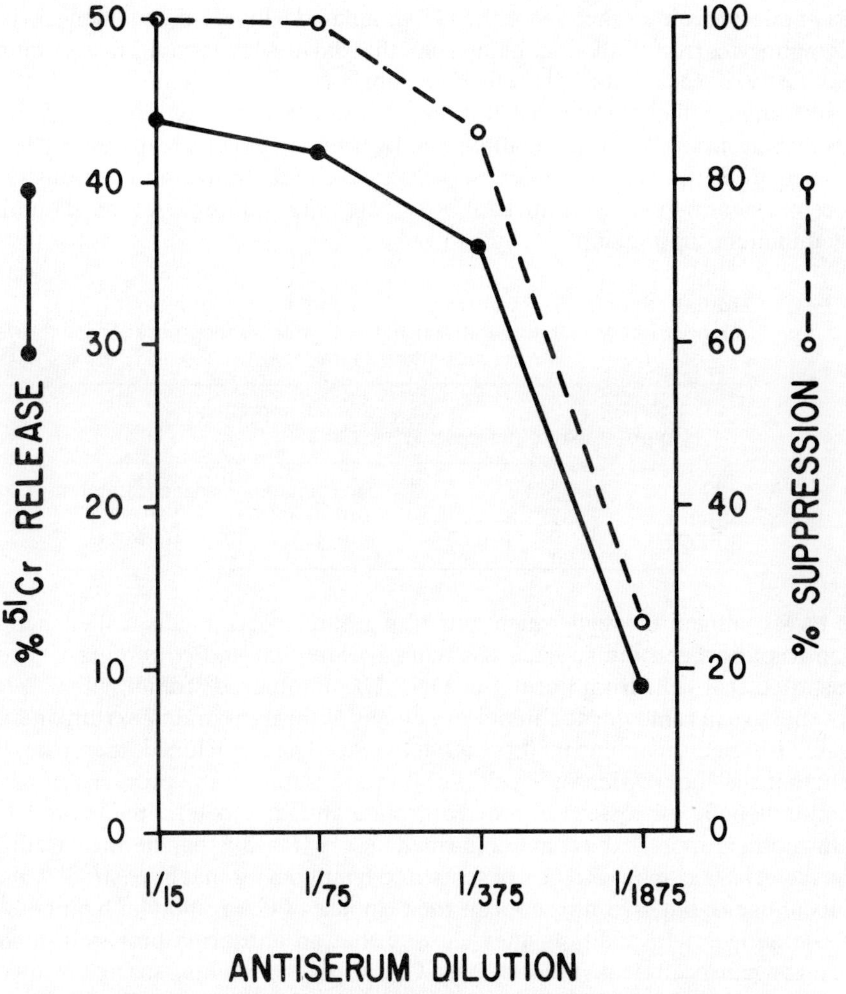

Figure 1. The effects of decreasing concentration of anti-strain 2 alloantiserum on its cytotoxicity to lymphocytes from $(2\times13)F_1$ guinea pigs and on its capacity to suppress the *in vitro* response of those cells to DNP-GL

The GA gene is linked to the strain 2 histocompatibility locus and to the PLL gene. In earlier studies of immune responses to GA in random-bred guinea pigs, we identified a small number of animals that responded to GA but not to DNP-PLL.[21] They do not have strain 2 histocompatibility antigens on their lymphocytes. We have made use of those animals, who presumably resulted from crossovers between the GA and PLL genes, to provide a more direct demonstration that anti-histocompatibility and not anti-Ir gene product antibodies are responsible for the suppression of specific immune responses *in vitro.*[24] Lymphocytes from random-bred Hartley guinea pigs that had responded to both GA and GT were characterized for the presence of strain 2 and strain 13 histocompatibility antigens. The effect of anti-strain 2 alloantisera on the proliferative response of those cells to both GA and GT was assessed *in vitro.* As shown in Table 2, suppression of GA stimulated responses by the anti-2 serum was observed in cultures of lymphoid cells bearing strain 2 histocompatibility antigens. The specificity of suppression for the strain 2 histocompatibility-associated immune response is verified in the random-bred animals by the lack of significant suppression of GT responses. The anti-strain 2 alloantisera had no significant effect on either the GA or the GT stimulated tritiated thymidine incorporation in the lymphocytes from responder guinea pigs that did not have strain 2 histocompatibility antigens on their cells. Thus, the effective suppression of Ir gene controlled immune responses requires the presence of the linked histocompatibility antigens. Since those animals lacking strain 2 histocompatibility antigens were genetic responders, the lack of suppression of their *in vitro* immune responses by the anti-strain 2 sera demonstrates that anti-receptor or anti-Ir gene product antibodies are not responsible for the alloantiserum-mediated immunosuppression.

Table 2. Requirement for strain 2 histocompatibility antigens for the suppression of GA stimulated tritiated thymidine incorporation by anti-strain 2 sera

H-type	No. of animals	% suppression GA	GT
		mean (range)	mean (range)
Strain 2+	8	60 (30–88)	14 (0–51)
Strain 2−	5	10 (0–21)	10 (0–25)

The target antigen through which guinea pig alloantisera mediate their suppressive function has also been the subject of studies by Shevach and coworkers.[29] They have demonstrated that cells from normal or DNP-GL immunized strain 2 guinea pigs were equally effective in removing the inhibitory effects of the anti-strain 2 serum on DNP-GL stimulated tritiated thymidine incorporation *in vitro.* They concluded, therefore, that it is unlikely that the suppression of T cell proliferation is due to the presence of antibodies specific for clonally expressed, Ir gene controlled antigenic determinants on T lymphocytes. In another report,[30] they have confirmed the observation that the anti-strain 2 serum inhibits the *in vitro* response to GA of sensitized lymphocytes that have strain 2 histocompatibility antigens, but does not suppress the response of GA responder lymphocytes that lack those antigens. In addition, they showed that an antiserum prepared in strain 13 guinea pigs against cells from random-bred GA responders lacking strain 2 histocompatibility antigens did not inhibit the GA response of cells from a $(2\times 13)F_1$ animal; while an antiserum prepared against GA responder cells possessing strain 2 histocompatibility antigens was capable of specifically inhibiting GA responses. These studies provide additional evidence that the alloantiserum-mediated suppression occurs via antibodies directed at histocompatibility antigens rather than via antibodies specific for Ir gene determined specificities.

5. MECHANISM OF SUPPRESSION

The mechanisms by which antibodies directed at histocompatibility antigens suppress Ir gene product function *in vitro* have not been adequately defined. Several models have been proposed, including (a) a steric hindrance model in which antigen access to its receptor is physically blocked by the antibody molecules attached to the very closely associated histocompatibility antigens; (b) the alloantisera may suppress stimulation by masking the histocompatibility antigens, thereby preventing optimum macrophage–T cell interaction and perhaps preventing presentation of the antigen to its receptor; or (c) rather than preventing antigenic stimuli from reaching the appropriate sites on the lymphocyte membrane, the antibodies attached to the histocompatibility antigens may interfere with the ability of the lymphocytes to respond to those stimuli. In an attempt to begin to distinguish among those alternatives, we have tested for the suppression of antigen stimulation *in vitro* by immunoglobulin fragments derived from the suppressive alloantisera.[31]

Fab, $F(ab)'_2$ and Fc fragments were prepared from the immunoglobulin fraction of a strain 13 anti-strain 2 alloantiserum. None of the fragments were cytotoxic for strain 2 histocompatibility antigen-bearing lymphocytes. The Fab and $F(ab)'_2$ fragments, however, retained their antigenic specificity as demonstrated by their ability to inhibit the cytotoxicity of the intact anti-strain 2 serum when lymphocytes from a $(2\times13)F_1$ guinea pig were used as target cells. The cytotoxicity of the anti-2 serum was unaffected by the addition of Fc fragments. None of the immunoglobulin fragments inhibited the cytotoxicity of an anti-strain 13 serum for the same target cells, thus demonstrating the specificity of the Fab and $F(ab)'_2$ fragments for strain 2 histocompatibility antigens.

The efficiency of binding of divalent antibody fragments to its antigen is considerably greater than the binding of monovalent fragments of the same antibody. In an attempt to define the difference in relative binding to strain 2 histocompatibility antigens by the Fab and $F(ab)'_2$ fragments prepared from anti-strain 2 sera, the effect of each of the fragments on cytotoxicity mediated by the intact antiserum was measured as a function of increasing competitor fragment concentration. As expected, there was considerable difference in their effectiveness. Essentially complete inhibition of the alloantiserum-mediated cytotoxicity was achieved with the $F(ab)'_2$ fragments at concentrations as low as 0·06 mg/ml. A concentration of 0·45 mg/ml of the Fab fragments was required to achieve the same degree of competitive inhibition. The divalent fragments, therefore, appear to bind to the histocompatibility antigens with approximately tenfold greater efficiency than do the monovalent Fab fragments.

We have tested the effectiveness of the immunoglobulin fragments as suppressors of antigen stimulation *in vitro*. The $F(ab)'_2$, Fab and Fc fragments were added to cultures of lymphocytes obtained from $(2\times13)F_1$ guinea pigs that had been immunized to DNP-GL and GT, and the effect on antigen stimulation ^{3}H-TdR incorporation was measured (Table 3). The $F(ab)'_2$ fragments completely suppress the proliferative response to DNP-GL but not to GT. The effect of the $F(ab)'_2$ fragments, therefore, like that of its parent intact antibody is specific for immune responses controlled by strain 2 histocompatibility-linked Ir genes. Unlike the divalent molecule, the monovalent Fab fragments did not suppress the

Table 3. The effect of immunoglobulin fragments derived from anti-strain 2 sera on *in vitro* antigen-stimulated tritiated thymidine incorporation

Ig fragment	None	Antigen DNP-GL	GT
	cpm	cpm	cpm
None	576	11 306	9833
$F(ab)'_2$	505	593	9612
Fab	619	10 527	8848
Fc	583	11 112	9866

response of the lymphocytes to DNP-GL. The Fc fragments also had no significant effect. Thus, although both the Fab and $F(ab)'_2$ fragments derived from the anti-strain 2 alloantisera have antigen combining sites specific for strain 2 histocompatibility antigens, only the $F(ab)'_2$ fragments retain the capacity to suppress the histocompatibility-linked Ir gene controlled immune response *in vitro*.

It was necessary to determine whether the differences observed between the Fab and $F(ab)'_2$ fragments are due solely to the more avid binding of the divalent fragment to its target antigen. Toward that end we tried to measure the magnitude of the difference between the highest non-suppressive concentration of Fab and the lowest effective concentration of $F(ab)'_2$ fragments. For practical reasons the highest concentration of Fab that could be achieved was 0·45 mg/ml. There was no significant suppression of DNP-GL stimulated responses at that concentration of monovalent fragment. Dilution studies with $F(ab)'_2$, however, have shown that the proliferative response of the lymphocytes to DNP-GL is exquisitely sensitive to the suppressive effects of the divalent fragments. Virtually complete suppression of the DNP-GL stimulated responses were observed at concentrations of $F(ab)'_2$ fragments as low as 60 μg/ml (Table 4). At 4·8 μg/ml the divalent fragments were still consistently effective, producing 23% mean suppression.

Table 4. Suppression of DNP-GL stimulated tritiated thymidine incorporation by anti-strain 2 serum and $F(ab)'_2$ fragments

Antiserum		$F(ab)'_2$ fragments	
Dilution	% suppression	% suppression	Concentration
			μg/ml
1/75	100	95	60
1/375	86	76	24
1/1875	25	23	4·8

Thus, there is a 100-fold difference between the lowest documented suppressive concentration of $F(ab)'_2$ and the highest Fab concentration that we tested. Obviously this is a minimum estimate since higher concentrations of Fab, had they been achieved, might also have been ineffective. In contrast, as we have shown, there is only a tenfold difference between the concentrations of Fab and $F(ab)_2'$ needed to compete equally with the intact antibody for binding to the histocompatibility antigens in the cytotoxicity assay. These results indicate that it is the divalent character of the $F(ab)_2'$ fragments, not their increased efficiency in combining with histocompatibility antigen, that is required for the effective suppression of antigen stimulation *in vitro*.

When we compared the intact anti-histocompatibility sera and their $F(ab)_2'$ fragments for their ability to suppress *in vitro* responses we found that mole for mole the divalent $F(ab)_2'$ fragments suppress as effectively as the intact antibody. The elimination of the Fc portion of the alloantibodies does not alter their effectiveness in suppressing antigen stimulation. This suggests that steric hindrance is an unlikely explanation for the mechanism of alloantiserum-mediated suppression. The Fc piece of the antibody is the portion that would be most likely to produce the steric inhibition and yet its removal has no effect on the efficiency of the molecule in suppressing antigen stimulation.

The inability of the Fab fragment to suppress antigen stimulation suggests that the mechanism of the suppression is not simply a masking of histocompatibility sites, and, thereby, preventing efficient macrophage–T cell interaction. The monovalent fragments, as shown by their inhibition of antibody-mediated cytotoxicity, effectively mask the histocompatibility antigens but they do not suppress antigen stimulation. These experiments, then, suggest that the alloantiserum does not prevent the antigenic stimulus from gaining access to the lymphocyte and its antigen receptor.

The possibility that the alloantisera suppress by preventing the interaction of sensitized lymphocytes with the antigenic stimulus has also been examined by studying the effects of delayed addition of the alloantiserum to the antigen stimulated cultures. The cultures were initiated by adding DNP-GL to lymphocytes from strain 2 guinea pigs that had been immunized to that antigen. Anti-strain 2 alloantiserum was added to the cultures at various times after their initiation. Effective suppression of DNP-GL stimulated tritiated thymidine incorporation does not require that the alloantiserum be present at the time of the initiation of the cultures. Essentially complete suppression (greater than 90%) was noted when addition of the anti-2 serum was delayed up to 3 hours. Adding the alloantiserum 6 hours after initiation of the cultures still resulted in a significant suppression although the degree of suppression observed is quite variable. After a 24-hour delay, however, there is only slight suppression, and by 48 hours after initiation of the cultures the alloantiserum had no effect on DNP-GL stimulated ^{3}H-TdR incorporation.

DNA synthesis is a relatively late event following lymphocyte stimulation. It is preceded by a number of membrane and intracellular biochemical events. We have been investigating the effects of anti-histocompatibility sera on some of those early events.[26] As shown in Figure 2, antigen stimulated ^{14}C-leucine incorporation into protein and ^{3}H-uridine into RNA was suppressed in a virtually identical fashion to ^{3}H-thymidine incorporation. Complete suppression of DNP-GL stimulated protein, RNA and DNA synthesis was observed when the anti-strain 2 histocompatibility serum was added at the initiation of the cultures. Addition of the antiserum 8 hours after addition of the antigen still suppressed all three responses more than 50%. After 18 hours in culture with antigen, the lymphocyte responses were only slightly altered by the anti-2 serum.

It seems likely that the interaction of sensitized lymphocytes with the antigenic stimulus is well under way within the first few hours in culture. It has been clearly demonstrated that there is significant uptake of antigen by macrophages after exposures as brief as five minutes.[32] In addition, *in vitro* lymphocyte proliferation has been observed when lymphoid populations have been exposed to antigen for only ten minutes.[33] The suppressive effects of the alloantisera when they were added several hours after initiation of the

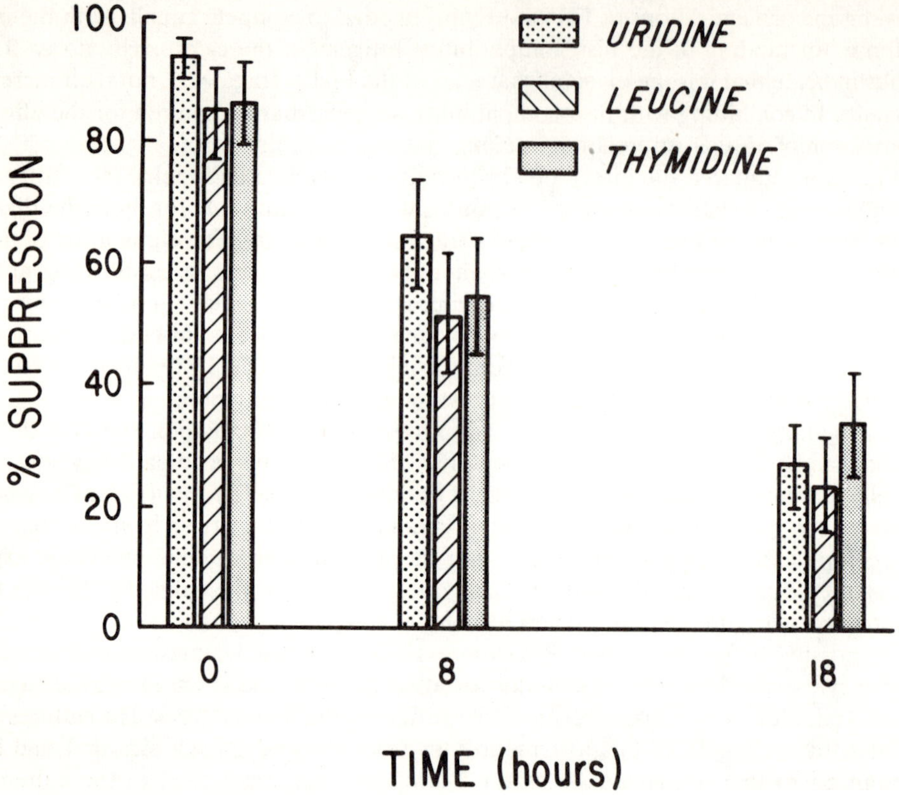

Figure 2. The effect of delayed addition of anti-strain 2 alloantiserum on the suppression of *in vitro* DNP-GL stimulation of ^{3}H-uridine, ^{14}C-leucine and ^{3}H-thymidine incorporation

cultures indicate, therefore, that the antisera can exert their effects after antigenic stimuli have already reached the lymphocyte membrane.

Any model attempting to explain the mechanism of alloantiserum-mediated suppression of Ir gene controlled responses must take into account both the requirement for divalent antihistocompatibility antibody and, also, the effective suppression with alloantiserum after the interaction of T cell with antigen has occurred. There are at least two models that may satisfy the requirements.

(1) The attachment of the alloantibody to its T cell membrane target antigen may induce a conformational change in the physically linked Ir gene product: i.e. an allosteric interaction, that leads to the release of the antigen from its receptor. In support of this model, studies of PHA stimulation have shown that the continued presence of the mitogen on the lymphocyte membrane is required for optimal stimulation.[34,35] By analogy it is likely that a similar requirement exists for the continued presence of antigen. Further support for this model can be derived from our observation that the apparent affinity of the GA receptor for its antigen differs depending on whether or not the responder lymphocytes have strain 2 histocompatibility antigen on their membrane.[36] One interpretation of that study is the suggestion that the receptor conformation may be altered by the presence of the histocompatibility antigen.

(2) The attachment of the antibody to the histocompatibility antigen may lead to a dynamic alteration of a localized region of the lymphocyte membrane, in effect 'depolarizing' it, thereby interrupting the transmisson of the signal being generated by the antigen in its receptor. This model fits especially well with the requirement for divalent alloantibody, since it has been shown that divalent antibody is required for the redistribution of lymphocyte membrane markers (References 37–39, and F. Loor, herein). At the present time there is no compelling reason to favour one of these models over the other. The end result of both mechanisms, however, is the same. The alloantisera suppress antigen stimulated T cell proliferation by a dynamic interaction with the lymphocyte membrane that interferes with the stimulus being generated by the antigen–receptor complex.

6. CELLULAR LOCALIZATION OF IR GENE FUNCTION

Histocompatibility-linked Ir genes are expressed in lymphoid cells. In guinea pigs the ability to respond to PLL can be passively transferred to irradiated non-responder recipients by the intravenous injection of lymphocytes from responder animals.[40] Transfer was most efficient with lymph node cells, a relatively T cell-enriched and macrophage-poor population. The cells actually making the immune responses in the chimeras carried the histocompatibility antigens of the responder strain providing the graft.[41] In mice, Ir gene controlled high responsiveness to (T, G)-A–L can be transferred to lethally irradiated low responder mice by the injection of cells from spleen or foetal liver, a rich source of immunocompetent cell precursors.[42] Another mouse H-linked Ir gene, Ir-GAT, which was first defined *in vivo*,[43] seems to exert its control in lymphoid cell cultures as well.[44] Dissociated spleen cell cultures derived only from those strains identified as GAT responders *in vivo* respond to that antigen *in vitro* with the development of GAT specific IgG plaque forming cells (PFC). Thus, immunocompetent cells are necessary and sufficient for the expression of histocompatibility-linked Ir genes.

The particular type of immunocompetent cell that expresses Ir gene function has been the subject of considerable investigation during the past five years. Because all of the antigens used to identify the H-linked Ir genes are 'thymus-dependent' and since, as we have already mentioned, development of T cell-mediated immune responses is absolutely dependent upon the presence of the specific Ir gene, T cells very early became a prime candidate for the site of Ir gene function. There is strong experimental support for that premise. Non-responder animals are perfectly capable of making specific antibodies when immunized with the antigen bound to an immunogenic carrier to which its T cells could respond. Strain 13 guinea pigs, for example, which do not possess the PLL gene and cannot respond to immunization with DNP-PLL, make excellent antibody responses to that antigen when immunized with DNP-PLL coupled to methylated bovine serum albumin (MBSA).[45] The strain 13 animals do not respond to skin testing with DNP-PLL nor do their lymphocytes respond to that antigen *in vitro*, but they do respond to the MBSA. Thus, if provided with T cell help non-responder guinea pigs have B cells capable of making specific antibody which does not differ in regard to immunoglobulin class or specificity from antibodies produced in responder animals.[46]

Functional studies in mice also suggest that Ir genes are selectively expressed in T cells. Low responder strains to (T, G)-A–L also produce high titres of specific antibody when immunized with that antigen complexed to MBSA.[47] In addition, more direct evidence of the importance of T cells for the expression of Ir gene function has been provided by studies of the response to (T, G)-A–L following thymectomy. When high and low responder mice are immunized with (T, G)-A–L in saline, both strains produce equal low titred IgM antibody responses. After secondary challenge, however, the high responders produce high titres of IgG antibodies while the low responders continue to demonstrate a low titred IgM response with no secondary peak.[48] After neonatal thymectomy both high and low responders continue to make the primary IgM response, but the high responders are no longer able to respond to secondary challenge with a high titred IgG response.[49] Furthermore, the antibody response of low responder mice can be altered by the non-specific stimulation of T cells produced by the induction of a graft-versus-host (GvH) reaction.[50] The induction of a GvH in low responder animals undergoing (T, G)-A–L immunization leads to the production of higher titred antibodies of the IgG class. These studies, like those in guinea pigs, emphasize that the H-linked Ir genes do not exert their influence directly on B cells. Instead, they seem to be expressed in T cells or perhaps at the level of T cell–B cell interaction.

Further support for the absence of Ir gene expression in B cells has been derived from the studies of responses to (T, G)-A–L in allophenic mice.[51] These tetraparental mice were produced by aggregating two 8 to 16 cell embryos, one of high responder genotype (CWB) and one of low responder genotype (C3H), to form chimeric blastocysts. After transplantation into the uterus of a pseudopregnant mouse the fused embryos completed their development. The resulting C3H–CWB chimeras are mosaic for cells of the two input strain genotypes. C3H and CWB are congenic strains differing at the major histocompatibility complex, including the Ir-1 locus, and also at the immunoglobulin heavy chain gene complex. Immunoglobulins of CWB mice have the Ig^b allotype, while C3H mice express the Ig^a allotype. Following immunization, the (T, G)-A–L specific antibodies generated by the tetraparental mice were tested for their allotype composition. Specific antibodies of both Ig^a and Ig^b allotypes were found in the serum of those mice that were high responders to TGAL. The fraction of the antibody response represented by each allotype showed a high degree of correlation with the percentage of Ig^b and Ig^a allotypes in the preimmunization sera. Thus, in tetraparental mice, B cells having the genetic constitution of either low or high responders respond equally well with specific antibodies to (T, G)-A–L.

Although there is considerable evidence that Ir gene function is expressed at the T cell level and not in B cells, this remains a controversial area. Shearer *et al.*[52] have concluded from limiting dilution studies that in some strains of mice, with certain antigens, the Ir-1 genes are expressed in both T cells and B cells; while with other strains and other antigens the defect is located only in bone marrow cells. Their technique involves the transfer of mixtures of bone marrow cells and thymocytes to irradiated recipients to reconstitute their response. Ir-1 control of (Phe, G)-A–L responses in SJL mice appears to be due to defects in both cell types; while the poor response to (T, G)-Pro–L in SJL and DBA/1 strains was localized solely to B cells.

More intriguing are the recent studies with a soluble factor derived from antigen stimulated presensitized thymocytes.[53] When incubated *in vitro* with (T, G)-A–L, thymocytes from high responder mice that had been pre-immunized to that antigen release a soluble factor into the medium which helps B cells respond to (T, G)-A–L. This help was demonstrated in cell transfer experiments into irradiated recipients. The soluble factor and bone marrow cells were injected together followed by the injection of the specific antigen. Twleve days later anti-(T, G)-A–L antibody producing cells were enumerated in the spleens of the recipients using a plaque-forming cell assay. In the key experiments pertinent to this discussion, the congenic resistant strains C3H.SW (high responders) and

C3H/HeJ (low responders) were used. Both strains were educated to (T, G)-A–L and the factor was produced from their thymocytes *in vitro*. Both high and low responder strain factors were tested for their ability to help both high and low responder bone marrow cells in their antibody response to (T, G)-A–L. T cell factors from both high and low responder strains were equally capable of cooperating with high responder B cells, suggesting that there is no Ir-1 determined T cell defect in the low responders. Furthermore, when the bone marrow cells of the two strains were compared for their ability to respond to (T, G)-A–L with the help of either high or low responder T cell factor, only the high responder B cells were able to make a response. These results suggest that the Ir-1 determined defect in (T, G)-A–L responsiveness, at least in these $H\text{-}2^{k}$ strains, resides in the bone marrow cells which are unable to respond to antigen and T cell signals. To complicate matters further, low responder SJL mice, which have the $H\text{-}2^{s}$ haplotype, seem to have defects in both T and B cells. No functional T cell factor could be recovered from cultures of (T, G)-A–L stimulated educated thymocytes of SJL origin, and their B cells did not respond to T cell factors of high responder animals.

The discrepancies between the experiments demonstrating normal B cell function in non-responders and those demonstrating defective B cell function cannot yet be resolved. It may be related to the characteristics of the experimental systems involved. In particular, the experiments suggesting a B cell defect involved transfer experiments into irradiated animals and the cells involved are not mature immunocompetent cells. Experiments showing normal B cell function in non-responders were mostly performed in non-irradiated animals with mature immunologic function. The significance of these differences in experimental detail remains to be elucidated.

A third hypothesis for the mechanism of Ir gene control of antibody production that does not ascribe non-responder status either to a defect in antigen recognition by T cells with the subsequent absence of helper T cells nor to a B cell defect has been proposed recently by Kapp and her associates.[54] They have been studying *in vitro* immune responses to the synthetic polypeptide antigen GAT, a random linear copolymer composed of L-glutamic acid, L-alanine and L-tyrosine. The ability of mice to develop antibody responses to that antigen is under the control of a histocompatibility-linked autosomal dominant Ir gene which maps in the I region of the major histocompatibility complex of mice. Dissociated spleen cell cultures from responder strains bearing the haplotypes $H\text{-}2^{a,b,d,k}$ respond to the addition of soluble GAT with the development of IgG plaque forming cell (PFC) responses specific for that antigen. No anti-GAT PFC responses can be detected after similar treatment of spleen cell cultures from non-responder strains ($H\text{-}2^{p,q,s}$). Spleen cells from both responder and non-responder mice develop PFC responses specific for GAT after incubation with GAT complexed to methylated bovine serum albumin (GAT-MBSA).

The Ir gene controlled response to GAT is thymus dependent. Pretreatment of spleen cells with anti-theta serum and complement destroys their ability to respond to that antigen with GAT specific PFC. Furthermore, *in vivo* immunization of responders with GAT leads to the development of GAT specific helper T cells which are X-ray resistant. These helper T cells can be recovered from the spleens of irradiated animals and are able to cooperate *in vitro* with normal B cells in the development of GAT specific PFC responses. No GAT specific helper T cells can be detected in the spleens of GAT primed X-irradiated non-responder mice.

In all of its characteristics thus far detailed genetic congrol of PFC responses to GAT is analogous to the genetic control of antibody responses to DNP-PLL in guinea pigs and (T, G)-A–L in mice. Subsequent studies of Ir gene control of GAT responsiveness, however, have provided some new insights. The injection of GAT into non-responder mice was followed not only by the absence of an antibody response to that antigen but, also, by the absence of subsequent anti-GAT PFC responses to a subsequent immunization with GAT-MBSA. This same phenomenon could be reproduced *in vitro*. Those studies clearly demonstrate that there is recognition of GAT at some level in the non-responder's immune system which results in the development of tolerance to GAT.

Studies of the immunocompetence of spleen cells from non-responder mice that had been primed with GAT reveal that their B cells are capable of making anti-GAT responses if they receive adequate T cell help. Column purified B cells from such mice respond well to GAT-MBSA when cultured with GAT-MBSA primed irradiated T cells. The unresponsiveness to GAT-MBSA induced by GAT priming, therefore, is not due to B cell tolerance.

The tolerance to GAT-MBSA induced by GAT-priming is a cell-mediated phenomena. The addition of tolerant spleen cells to GAT-MBSA stimulated cultures of spleen cells from syngeneic normal non-responder mice suppressed their anti-GAT PFC response. Treatment of the GAT-primed non-responder spleen cells with anti-theta serum and complement or with X-irradiation eliminates the suppressor effect. These studies demonstrate that GAT non-responder mice 'recognize' that antigen but they respond to it with the proliferation of a population of specific suppressor T cells.

Antigen-specific suppressor T cells is a relatively new concept, but one which has captured the imagination of immunologists. There is now compelling evidence that suppressor T cells may help to regulate both humoral and cellular immune responses.[55] It is not yet clear whether helper and suppressor T cells are induced from the same precursor under the influence of antigen, or whether they result from antigen stimulation of two different types of antigen-recognizing T cells. If the latter model is correct, one could place the function of the Ir-GAT gene at the level of helper T cells. That is, the non-responder strains lack the Ir gene necessary for the development of antigen specific helper T cells but are perfectly capable of developing specific suppressor T cells. Some recent data from Kapp and her coworkers, however, indicates that GAT-specific helper T cell function can develop in non-responder mice if GAT is presented to their lymphocytes in the appropriate manner.[56] Priming non-responder mice with GAT bound to syngeneic macrophages followed by X-irradiation permitted the demonstration of GAT specific helper T cells in their spleens when the spleen cells were combined with B cells from normal (responder × non-responder) F_1 mice. These studies suggest, then, that non-responder mice are capable of developing GAT specific helper as well as suppressor T cells.

There is a possible technical problem with these experiments, however, since the demonstration of effective T cell help required the cooperation of the non-responder T cells with semi-allogeneic F_1 B cells. Syngeneic B cells do not respond. Irradiation of the primed T cell population should prevent the development of an allogeneic response, and these observations, therefore, have been interpreted as indicating defects in both B and T cell function in non-responder animals. However, it is possible that a subtle allogeneic interaction is responsible for the GAT-specific plaque-forming cell responses observed with the F_1 B cells. With that caveat in mind, however, these studies suggest that Ir gene function may be responsible for determining to what degree antigen stimulation leads to the induction of helper or suppressor T lymphocytes. Studies of the Ir-GAT gene have thus led to the development of the concept that Ir genes function as regulatory genes rather than as genes coding for antigen recognition units. Whether this phenomenon can be generalized to other specific Ir genes has yet to be determined but this is an area that is currently under intensive investigation.

7. ROLE OF MHC ANTIGENS IN LYMPHOID CELL COOPERATION

In addition to genes controlling specific immune responsiveness, the I region of the MHC also contains genes coding for cell surface antigens that play a role in effective cooperation between immunocompetent lymphoid cells. Studies in guinea pigs and mice have demonstrated that optimum macrophage–T cell and T cell–B cell cooperation requires that both the interacting cell types share a common MHC antigen. These observations have important implications for the mechanisms of lymphocyte activation as well as Ir gene function.

The activation of presensitized T lymphocytes *in vitro* in response to antigen stimulation requires the presence of macrophages in the cultures.[57,58] In fact, pre-incubating macrophages with antigen, washing away the excess antigen, and presenting macrophage bound antigen to primed T cells is the most effective technique for inducing blast transformation and increased DNA synthesis. Using such a system in guinea pigs, Rosenthal and Shevach[59] investigated the importance of histocompatibility determinants, in the interaction of antigen pulsed macrophages with presensitized T lymphocytes. Inbred strain 2 and strain 13 guinea pigs were used for these experiments. Initial studies done with PPD (purified protein derivative of tuberculin) demonstrated that PPD bound to strain 2 macrophages effectively stimulated DNA synthesis in immunized strain 2 lymphocytes, but these same macrophages were relatively ineffective in stimulating immune strain 13 lymphocytes. Conversely, PPD bound to strain 13 macrophages provided a potent stimulus to the immune strain 13 lymphocytes but not to the strain 2 cells. PPD bound to macrophages from $(2\times13)F_1$ guinea pigs effectively stimulated sensitized lymphocytes from both strain 2 and strain 13 animals. In addition, both strain 2 and strain 13 macrophage-bound PPD stimulated immune F_1 lymphocytes. Thus, the optimum *in vitro* response to PPD, which is not under unigenic H-linked Ir gene control, requires that macrophages and lymphocytes share a common MHC antigen.

When antigen stimulated responses under the control of H-linked Ir genes was investigated, a more stringent MHC antigen requirement was observed. F_1 guinea pigs respond to DNP-GL (controlled by a strain 2 H-linked Ir gene) and to GT (linked to strain 13 MHC antigens). After pre-immunization with both of those antigens, $(2\times13)F_1$ lymphocytes were tested for their ability to respond to macrophage bound DNP-GL and macrophage bound GT. DNP-GL stimulation was nearly equivalent if that antigen were presented bound to either strain 2 or to F_1 macrophages, but if presented on strain 13 macrophages the response was only 10%–20% of that achieved with macrophages having strain 2 antigens on their surface.[60] Conversely, F_1 or strain 13 macrophage-bound GT were effective stimuli to the F_1 lymphocytes, while GT bound to strain 2 macrophages was a very poor stimulus. Thus, optimal macrophage–T cell interaction for H-linked Ir gene controlled immune responses requires that the cooperating cells share the *linked* histocompatibility antigen. Recently, Shevach *et al.*[61] have extended these observations to include antigen stimulated T cell responses other than DNA synthesis. Similar requirements for syngeneic macrophages sharing his histocompatibility region antigens linked to the Ir gene controlling the particular immune response was demonstrated for the stimulation of macrophage inhibitory factor and for the stimulation of protein synthesis.

The need for syngeneic or semi-allogeneic cells for successful macrophage–lymphocyte interaction has been more difficult to demonstrate in mice. Using a complex protein antigen, keyhole limpet haemocyanin, which is not under unigenic H-linked Ir gene control, Katz and Unanue[62] found that macrophages from syngeneic and allogeneic donors were equally successful in inducing both primary and secondary antibody responses in Mishell–Dutton culture system. Furthermore, Kapp *et al.*[63] studying the Ir-GAT H-linked Ir gene in Mishell–Dutton type cultures demonstrated a macrophage requirement for the primary response to GAT in responder animals, but non-responder strain macrophages were as effective as responder strain cells in satisfying that cellular requirement. More recently, however, Erb and Feldmann[64] have described an *in vitro* system for inducing unprimed T cells to differentiate into antigen specific helper cells. They incubated purified T cells from CBA mice with the soluble protein antigen KLH (keyhole limpet haemocyanin) for 4 days, together with peritoneal exudate macrophages. An aliquot of the KLH primed cells was subsequently cultured with normal spleen cells from CBA mice together with the antigen, trinitrophenylated KLH (TNP-KLH). After an additional 4 days in culture the anti-TNP plaque-forming cell response generated was enumerated. When syngeneic CBA macrophages were used during the initial KLH priming, considerable helper activity was induced, as manifested by the development of a TNP plaque-forming cell response in the second stage of culture. If allogeneic Balb/c macrophages were used during the priming stage instead of the syngeneic macrophages, no helper cell activity could be demonstrated. Semi-allogeneic (CBA × Balb/c)F_1 macrophages were also able to generate helper T cells, although not quite as efficiently as the syngeneic cells. Thus, antigen induced generation of helper T cells in mice, like antigen induced T cell activation in guinea pigs, requires macrophage–lymphocyte interaction between cells sharing a common histocompatibility region antigen.

A similar requirement of MHC antigens for the cooperative interaction of T and B cells was first suggested by the studies of Kindred and Shreffler.[65] They studied the ability of transplanted thymus cells to restore normal immune function to mice of the congenitally athymic 'nude' strain. Thymocytes syngeneic at the H-2 region, and semi-allogeneic F_1 thymus cells rendered nude mice capable of making immune responses to sheep red blood cells and to the bacteriophage T4. Nude mice that received allogeneic thymocytes remained unable to respond to those antigens. This study suggested, therefore, that a common MHC antigen is required for effective T–B cooperation. However, it does not rule out the possibility that lack of response in the allogeneic combination reflects activation of the transplanted T cells against the foreign MHC antigens of the host with a resultant non-specific suppressive effect on immune responses.

A more complex experimental protocol was devised by Katz *et al.*[66] in an attempt to come to grips with the problems of allogeneic T cell responses affecting physiologic T–B interactions. They used an adoptive cell transfer system, but the host animal for cell transfers was a semi-allogeneic e.g. (A × B)F_1 mouse, incapable of reacting against MHC antigens of either A or B parental strain. The normal F_1 mouse first received an intravenous injection of spleen cells from an A strain mouse that had been primed to bovine gamma globulin (BGG). After waiting 24 hours to allow the transferred cells to migrate to lymphoid organs, the F_1 mice were irradiated and then given a second intravenous injection of DNP-KLH primed spleen cells from either A or B strain mice that had been treated with anti-theta serum and complement to eliminate T cells. The animals were then challenged with DNP-BGG and their anti-DNP antibody responses measured 7 days later. If BGG primed spleen cells from strain A and DNP-KLH primed anti-theta treated spleen cells from the same strain were transferred in F_1 recipients, an excellent anti-DNP response was induced by subsequent challenge with DNP-BGG. If normal strain A spleen cells were used, rather than the BGG primed cells, no significant anti-DNP response resulted. Similar results were observed if both carrier and hapten primed cells were derived from strain B. If A strain BGG primed cells and B strain DNP-KLH primed cells were transferred or vice versa, there was no appreciable anti-DNP response following the DNP-BGG challenge. Studies of T–B cell cooperation *in vitro* produced similar results. The co-culture of semi-allogeneic cells, that is, F_1 carrier-primed helper T cells with B cells from either parental strain, or the combination of hapten-primed F_1 B cells with helper T cells from either parental strain, permits successful cooperative interaction in the generation of the anti-DNP response.

By using congenic resistant strains of mice in their *in vivo* adoptive cell transfer system, Katz and coworkers have clearly demonstrated that the genes coding for the factors required for cooperative interaction are located in the mouse MHC.[67] When cells from A strain mice were tested for their ability to cooperate with cells from the congenic resistant B10.A mice which have the A strain H-2 region superimposed on a different genetic background, successful cooperation between A strain T cells and B10.A B cells was obtained. B cells from the B10.A strain failed to cooperate with helper T cells from B10 donors who are genetically identical except at the H-2 region. Similar studies using inbred recombinant strains differing at known loci within the H-2 complex permitted more specific localization of the genes controlling T–B cell interaction to the I region of the histocompatibility complex.[68] Control of immunocompetent cell interaction and of the ability to develop specific immune responses, therefore, is a function of genes localized to the same subregion of the major histocompatibility complex. It is not yet clear whether the control of cellular interaction and immune response are both a function of the product(s) of the same or separate genes.

When cellular cooperation between T and B lymphocytes engaged in immune responses that are not under the control of an Ir gene were studied, F_1 T cells cooperated effectively with B cells from either parental strain. However, when the same types of studies were performed with a carrier molecule to which the immune response is under Ir gene control, a different observation was made. Responses to the synthetic polypeptide antigen GLT (a random linear terpolymer of L-glutamic acid, L-lysine and L-tyrosine) is under the control of the H-2 linked autosomal dominant Ir-GLT gene. Balb/c mice respond to GLT, but A/J strain mice do not. The F_1 hybrid mice are also responders. The *in vivo* adoptive cell transfer technique has been used to study the ability of GLT primed F_1 T cells to cooperate with DNP primed B cells from either Balb/c or A strain mice, both of which share common MHC genes with the F_1 heler T cells. Effective cell cooperation resulting in secondary antibody responses to DNP were observed only when the DNP specific B cells were derived from the GLT responder Balb/c mice. The GLT primed F_1 T cells did not provide effective help for DNP specific B cells from the non-responder A strain animals. Thus, in the development of immune responses under the control of H-linked Ir genes, optimum T–B cell interaction in mice, like macrophage–T cell cooperation in guinea pigs, requires that both cell types share the Ir gene linked histocompatibility antigens.

The studies described in this section provide compelling evidence for the control of immunocompetent cell interactions (C.I.) by genes located in the I region of the major histocompatibility complex. However, the conclusion that those observations are representative of a general physiological phenomenon is currently being challenged. The challenge is based on the results of studies from several laboratories which appear to show that T cells are able to cooperate with allogeneic B cells when the T cell population is depleted of immune reactivity to the alloantigens of the B cell donor. The usefulness of that experimental approach was first investigated by Bechtol and her associates with studies of Ir gene controlled anti-(T, G)-A–L responses in tetraparental mice.[51] The chimeras, derived from the fusion of an embryo of high responder genotype with one of a low responder genotype, respond to (T, G)-A–L immunization with specific IgG antibody of both high and low responder allotype. The (T, G)-A–L response is T cell dependent, and the experimental evidence reviewed earlier in this chapter suggests that low responder mice are incapable of providing T cell help for that antigen. The production of antibodies having both high and low responder allotypes, therefore, suggests that the high responder T cells were able to cooperate with the allogeneic low responder B cells in the generation of anti-(T, G)-A–L antibody. However, there are alternative interpretations that were not ruled out. A subtle allogeneic reaction between the chimeric cells may provide non-specific stimulation bypassing the need for T cell help. The absence of anti-(T, G)-A–L responses in chimeras produced by fusing embryos of two histoincompatible low responder strains provide inconclusive evidence against that possibility. Secondly, it is possible that low responder T cells acquire the functions of high responder cells in the chimeric environment.

A technique for producing chimeric mice with less equivocal T cell tolerance for the alloantigens of its component parts has been described by von Boehmer *et al.*[69] Lethally irradiated F_1 mice were repopulated with equal proportions of T cell-depleted bone marrow cells from both parental strains. Stem cells from the marrow are thus free to differentiate into T cells in the midst of a foreign environment. The resulting immunocompetent cells develop 'self-tolerance' to all of the histocompatibility antigens expressed on the cells of the F_1 host, although the lymphocytes themselves express only the antigens of the parental strain from which they were derived. Tolerance to H-2 antigens is not due to the generation of suppressor T cells, but rather appears to be the result of specific deletion of T cells reactive to those antigens. Carrier-primed T cells derived from these 'bone marrow chimeras' were studied for their ability to manifest T cell help in cooperation with allogeneic B cells.[70] T cells were obtained from the thoracic duct and lymph node lymphocytes of sheep red blood cell (SRBC) primed bone marrow chimeras produced in (CBA×DBA/2)F_1 hybrids. Treatment of the T cell population with anti-DBA/2 serum and complement resulted in a population of SRBC-primed CBA T cells tolerant to the histocompatibility antigens of the DBA/2 strain. Those cells were tested in an adoptive transfer system in X-irradiated F_1 hosts together with primed B cells. The B cells were derived from syngeneic CBA or allogeneic DBA/2 spleen cells treated with anti-theta serum and complement. The CBA T cells from the bone marrow chimeras effectively cooperated with both the syngeneic and allogeneic B cells, and the data suggested that T cell help across the histocompatibility barrier was as efficient as in the syngeneic combination.

Carrier-primed T cells derived from similarly produced bone marrow chimeras have also been tested *in vitro* for their ability to cooperate with allogeneic B cells. Waldmann *et al.*[71] mixed KLH-primed T cells with TNP-FGG (fowl gammaglobulin)-primed B cells in the presence of soluble TNP-KLH in tissue culture and assayed the resulting TNP plaque-forming cell response five days later. They found, as did Katz *et al.*,[66] that normal histoincompatible T and B cells will not cooperate *in vitro*. Balb/c (H-2^d) T cells cooperate with Balb/c B cells but not CBA (H-2^k) B cells. However, H-2^d T cells derived from (H-2^d+H-2^k) bone marrow chimeras cooperated efficiently with H-2^k B cells. Thus, in these *in vitro* experiments there appeared to be no difficulty in obtaining effective T–B collaboration across the histocompatibility barrier if the helper T cells were tolerant to the B cell alloantigens.

Cooperative T–B cell interaction across the histocompatibility barrier has also been suggested by Heber-Katz and Wilson.[72] They evaluated primary anti-SRBC plaque-forming cell responses obtained with combinations of syngeneic or allogeneic T and B cells derived from the thoracic duct lymph of inbred rat strains L and AUG. L strain T cells passaged through irradiated (AUG×L)F_1 rats, using a technique described by Ford and

Atkins,[73] were specifically depleted of detectable reactivity to AUG alloantigens in either mixed leukocyte culture (MLC) or graft-versus-host (GvH) assays. The AUG depleted T cells were able to cooperate with AUG B cells in the presence of SRBC to generate specific PFC. When studied as a function of the number of T cells added per culture, the allogeneic combination of depleted L strain T cells and AUG B cells gave quantitatively identical results to syngeneic T–B combinations. These results, like the results obtained using tolerant cells from bone marrow chimeras, demonstrate that effective T–B collaboration between allogeneic cells occurs when the T cells are prevented from reacting to alloantigens on the B cells. A corollary to this interpretation is the suggestion that the lack of collaboration in histoincompatible combinations results from interfering allogeneic interactions.

It is not yet clear how to reconcile the conflicts raised by the studies reviewed in this section. All of the experiments utilized complicated, contrived, experimental systems in an attempt to circumvent the problem of interfering allogeneic interactions. Differences in the results may reflect hidden allogeneic reactions operative in one of the systems. Allogeneic suppression, in those systems where no response is observed across the histocompatibility barrier or allogeneic stimulation, in those systems where cooperation does occur, could account for the differences observed. The studies of Katz and coworkers seem especially well controlled to rule out the likelihood that allogeneic suppression is responsible for non-cooperation. Most convincing are their experiments, already described, demonstrating that T cell help for a genetically controlled immune response requires that T cells and B cells share the Ir gene-linked histocompatibility antigen. In those experiments the F_1 T cells cooperated with only one of the parental strain B cells, although either parental strain is equally likely to have an allogeneic interaction with the F_1 cells. The T cells of the bone marrow chimeras appear tolerant of the histocompatibility antigens of their partner strain when tested in the classical assays MLC, GvH, and skin grafting. Nevertheless, they may produce a non-specific enhancing factor in the absence of other responses to the allogeneic cells.

8. HISTOCOMPATIBILITY ANTIGEN-CONTAINING SOLUBLE MEDIATORS

It has become apparent in the last several years that stimulated T cells produce biologically active soluble factors. The methods used to generate the active molecules have varied and so have the physico-chemical characteristics of the factors. Some of them contain histocompatibility region antigens as part of the structure. Both antigen-specific and non-specific factors are included in that group. Three histocompatibility antigen containing soluble mediators have been studied in considerable detail. Two of them are enhancing factors while the other is suppressive. Because of their relevance to our discussion of the role of histocompatibility region antigens in lymphocyte activation, they will be discussed here briefly. Detailed reviews of the characteristics of each have recently been published.[53,74,75]

A non-specific enhancing factor has been recovered from the supernatants of lymphocyte cultures in which primed T cells were mixed with histoincompatible target cells containing the alloantigen to which it had been presensitized. Armerding and Katz[76] showed that such supernatants replace helper cell function in cultures of primed T cell depleted spleen cells. They also enhance primary IgM anti-SRBC responses and reconstitute the primary anti-SRBC response of T cell depleted spleen cells *in vitro*. When the supernatant, whose active component is named 'allogeneic enhancing factor' (AEF), was fractionated by a combination of gel filtration, DEAE cellulose chromatography, and SDS-polyacrylamide gel electrophoresis, it was found to have a major AEF peak at a molecular weight of 47 000 daltons and a minor peak at 11 500 daltons. Since solubilized cell membrane antigens coded for in the mouse MHC also produce two peaks with similar molecular weights, the supernatants were analysed for the presence of histocompatibility antigens. The AEF activity of the supernatant was removed with anti-H-2 sera. More precise genetic analysis revealed that it was specifically antisera to the I region antigens that most effectively removed the activity. Antisera directed at immunoglobulin determinants had no effect on the supernatants' biological activity. The identification of a T cell product that contains histocompatibility antigen determinants and is capable of stimulating B cells prompted a study of the genetic requirements for the enhancing influence of the T cell factor. No clear-cut genetic restriction was found. Unlike the requirements for intact T cell–B cell interaction, AEF enhances the responses of allogeneic as well as syngeneic B lymphocytes.

An antigen-specific enhancing factor has been described by Taussig and Munro.[77] The factor is generated *in vitro* after a short-term incubation of (T, G)-A–L primed T cells with that antigen. Its biological activity is assayed as an adoptive cell transfer system *in vivo*. Bone marrow cells and the factor are injected together into lethally irradiated recipients followed by an injection of antigen. The factor is able to replace T cells for the generation of anti-(T, G)-A–L plaque-forming cells which are detected in the spleens of the recipient mice 12 days after cell transfer. It is relatively antigen-specific in that it does not stimulate the production of anti-SRBC PFC but it does exhibit some cross-reactivity with the related antigen (Phe, G)-A–L. The biological activity can be removed from the supernatant when it is passed over immunoabsorbent columns made by coupling the antigen (T, G)-A–L to sepharose. The active factor could not be removed by passage of the supernatant over anti-immunoglobulin immunoabsorbents but it could be removed with anti-histocompatibility reagents. The use of antisera specific for the various regions of the H-2 complex demonstrated that the factor is a product of the I region. Molecular weight of the factor, determined by Sephadex chromatography, is roughly 50 000 daltons. As described in detail earlier in this chapter (see Section 6) there are no restrictions on the ability of this factor to cooperate with allogeneic B cells.[78,79] Thus in all respects except its antigen-specificity this T cell factor is virtually identical to the enhancing factor described by Armerding and Katz.

Tada and colleagues[75] have studied a soluble extract obtained from thymocytes of KLH-primed mice. The extract was prepared by sonication of the thymocytes followed by ultracentrifugation. The supernatant was tested *in vivo* by studying the effect of its intravenous injection on DNP-KLH stimulated anti-DNP plaque-forming cell responses in normal mice. The extract significantly suppressed the development of IgG plaque-forming cells but not the development of IgM producers. The effect is carrier specific. The extract did not suppress anti-DNP responses stimulated by DNP-BGG. The active factor binds to antigen, since its ability to suppress was removed by a passage of the extract over a solid absorbent composed of antigen coupled to sepharose beads. The biological activity could not be removed with antibody to immunoglobulin but antibody directed at the histocompatibility haplotype of the cells producing the factor was able to remove the suppressor activity. Fractionation of the extract on Sephadex G-200 revealed its molecular weight to be less than 100 000 and more than 30 000 daltons.

The suppressor factor is like the enhancing factors described by Taussig and coworkers and by Armerding and Katz in its similar molecular weight and its histocompatibility antigen content. There are two major differences, however, between this factor and the two enhancing factors. The biological activity of the suppressing factor requires the presence of both T cells and B cells. It does not replace T cell function as do the two enhancing factors. In fact, this factor does not seem to act directly on B cells but, rather, may suppress the emergence of KLH-specific helper T cells. Furthermore, suppressor activity in thymocyte extract can be demonstrated only when tested in histocompatible animals. Suppressor factor from C57Bl ($H\text{-}2^b$) mice, for instance, did not suppress the response of Balb/c ($H\text{-}2^d$) animals, while under the same experimental conditions it dramatically suppressed responses of C57Bl recipients. Thus, the histocompatibility antigen containing, biologically active T cell factors share some striking similarities and some interesting discrepancies. The significance of the differences may become apparent when the various factors are compared in the same laboratory. Further study of these factors and their interaction with lymphocytes may provide some understanding of the molecular events that transpire during immunocompetent cell interaction.

9. THE ROLE OF HISTOCOMPATIBILITY ANTIGENS IN THE GENERATION OF CELL-MEDIATED IMMUNITY TO ALTERED CELL MEMBRANES

Cells whose membranes have been altered either by chemical modification or viral infection provide an effective antigenic stimulus to unmodified syngeneic lymphocytes which leads to the generation of cytotoxic T cells. Effective demonstration of these killer T cells requires that the target cells share two features in common with the immunizing cells. The targets must have been altered with the same agent and they must express a common cell surface antigen coded for in the major histocompatibility complex. Two experimental systems having those characteristics have been described in detail. In one, cytotoxic T cells are generated to virally infected autologous cells *in vivo*.[80] In the other, cells are chemically modified and used to stimulate unmodified autologous cells *in vitro*.[81]

Doherty and Zinkernagel demonstrated that the intracerebral injection of mice with lymphocytic choriomeningitis virus (LCM) lead to the generation of specifically sensitized splenic T cells that were cytotoxic to syngeneic LCM infected target cells.[80] Uninfected syngeneic cells and infected allogeneic cells were not killed by the immune spleen cells. Those findings suggested that the killer cells are sensitized to 'altered self'. Studies using congenic strains permitted the identification of the pertinent self-antigens as the determinants coded for in the K and D regions of the major histocompatibility complex. Identity between killer and target at either K or D was sufficient for specific lysis to occur. Differences elsewhere in the genotype were irrelevant.

Further support for the concept that the killer cells are sensitized to altered self has been provided by the demonstration that cytotoxic F_1 T cells have distinct receptors for each LCM infected parental cell type.[82] For example, the cytotoxicity of $(CBA\times Balb/c)F_1$ killer cells for LCM infected cells bearing the parental $H\text{-}2^k$ haplotype could be inhibited by other LCM infected cells bearing that haplotype but not by LCM infected cells bearing only the other parental haplotype ($H\text{-}2^d$). Conversely, the cytotoxicity of those same sensitized F_1 cells for infected cells with $H\text{-}2^d$ haplotype could not be inhibited by $H\text{-}2^k$ infected cells but were inhibited by cells with the $H\text{-}2^d$ haplotype. Similar experiments using cells from inbred recombinant mice have demonstrated that there are at least two specificities of LCM immune T cells in homozygotes associated with either K or D region antigens. It has been concluded from these studies that LCM infection induces changes in cell membrane molecules which produce unique immunogenic determinants. It is yet to be determined why the affected self-antigens are limited to those determined by the K or D histocompatibility regions. It may be that those are the only cell membrane antigens that are altered. Alternatively, perhaps the immune system is incapable of generating cytotoxic T cells to altered cell membrane antigens other than those at K and D.

Shearer *et al.*[81] have made similar observations in studies of the immunogenicity of chemically modified autologous cells. Cytotoxic T cells were generated in splenic lymphocytes by *in vitro* sensitization with trinitrophenylated (TNP)-autologous spleen cells. Target cells had to fulfil three criteria in order to be lysed by the sensitized T cells. They had to share histocompatibility antigens with the stimulating cells and with the killer cells, and they had to be TNP modified. Cells from congenic strains were used to identify the important histocompatibility regions. With some strains modified target cells sharing either K or D region antigens were effectively lysed. Those results are analogous to the findings in LCM modified cells. However, in other strain combinations TNP modified cells having homology only at the D region could not serve as effective target cells. In those cases modified target cells that shared K, or K plus I, region antigens were effectively lysed. This important observation, which has not yet been duplicated in the LCM system, has important implications for the interpretation of these observations.

B10.A cells sensitized to TNP-B10.A stimulating cells cannot lyse TNP-B10.D2 target cells which are homologous only at the D region. The TNP-B10.D2 cells, however, can sensitize unmodified B10.D2 cells which in turn can lyse TNP-B10.A target cells. Thus, the TNP-B10.D2 cells have been modified in the D region, demonstrating that the B10.A cells fail to respond to the modified antigenic determinant. The ability to react to that determinant is genetically controlled, and the genetic differences between the strains that recognize the determinant and those that do not, reside in the K plus I region. The ability to respond to that alteration of a D region antigen, then, is determined by a gene located in the region of the MHC that contains many of the histocompatibility-linked Ir genes.

Further evidence supporting the concept that histocompatibility-linked Ir genes control the ability to respond to modified self-antigens has been derived from comparisons between the responses of F_1 animals and the high or low responder parental strains from which they were derived.[83] Lymphocytes from an F_1 animal derived from the mating of B10.D2 mice who can respond to the modified D end products, and mice of the B10.A strain that cannot, generated a high level of cytotoxicity towards the modified D region of either parental strain. That was true whether the TNP modified stimulating cells were from either parental strain or autologous F_1 cells. Thus, responsiveness to the new antigenic determinant produced by D region modification is controlled by a dominant genetic trait, a characteristic of H-linked Ir genes. These studies suggest at least two requirements for the generation of cytotoxic cells to altered self-antigens on cell membranes. The offending agent must alter antigens coded for in the histocompatibility complex, and the animal must have the appropriate histocompatibility-linked Ir gene which controls immune responsiveness to the modified histocompatibility antigen.

On the basis of the studies into the role of histocompatibility region antigens in the generation of cytotoxic cells to altered self components, Doherty and Zinkernagel[84] have postulated that the major histocompatibility complex plays a major role in immune surveillance. According to their hypothesis the polymorphism of the histocompatibility region provides a substrate for the generation of a large number of new antigenic determinants when cells are affected by viruses or other foreign agents. This would lead to the rapid elimination of the altered cells thus providing an efficient mechanism for 'immune surveillance'.

10. THE MAJOR HISTOCOMPATIBILITY COMPLEX AND IMMUNE RESPONSIVENESS: AN HYPOTHESIS

The major histocompatibility complex is a relatively compact region comprised of genetic loci controlling several aspects of immune responsiveness. These include immune response genes governing the ability to manifest immune responsiveness to specific antigenic determinants, cell interaction genes regulating cooperation between immunocompetent cells in the development of immune responses, and genes controlling the expression of cell membrane antigens whose modification leads to the generation of cell-mediated immunity against the altered self-antigen. The products of these genes acting in concert provide a major defence mechanism essential to the survival of the species. It is likely that the necessity for the cooperative interaction of the products of these genes has provided the selective pressure responsible for the maintenance of the integrity of the MHC.

Although the importance of the MHC genes for the development of immune responses has been demonstrated, there is little known about the nature of the gene products or the mechanisms of their functional interaction. Studies of suppression of Ir gene function with anti-histocompatibility sera have suggested that the genetic linkage of histocompatibility and Ir loci has a counterpart in the phenotypic expression of the products of these genes on the lymphocyte membranes. The lymphocytes of hybrid F_1 animals bear histocompatibility antigens from both parental strains. Yet a given anti-histocompatibility serum suppresses only those immune responses controlled by Ir genes linked to that histocompatibility specificity. This suggests that the Ir gene product, as it is represented on the cell membrane, is intimately associated with or identical to the histocompatibility specificity.

The antigenic specificity apparent in Ir gene control, as well as the control of T cell function by Ir genes led to the suggestion that the Ir gene product is the antigen-specific T cell receptor. The nature of the T cell receptor remains controversial, although the evidence in support of the immunoglobulin nature of that molecule keeps growing stronger (reviewed in Reference 85). The demonstration of shared idiotypic determinants on T cells and B cells appears especially convincing.[86] As discussed earlier in this chapter, however, it is clear that Ir genes are not controlling classical immunoglobulin structure. The possibility remains that the T cell receptor is a more primitive type of immunoglobulin (the hypothetical IgT) which may share some variable region genes with the classical immunoglobulin system. The greatest difficulty with the notion of Ir gene products as T cell receptors comes from the studies demonstrating that similar numbers of responder and non-responder T cells have specific antigen receptors.[87,88]

If the Ir gene product is not the antigen-specific receptor, another mechanism must be invoked to account for the antigen specificity of Ir gene function. Postulating clonal expression of the Ir gene product in association with the antigen receptor provides the requisite specificity but does not account for the observation that one Ir gene may control responsiveness to several immunologically non-cross-reactive antigens.[45] In the absence of clonal expression, antigen specificity can be achieved through an obligatory interaction between the antigen and the Ir gene product. It is hypothesized, here, that the Ir gene product is a non-clonally expressed cell membrane structure coded for in the I region of the MHC. A steric interaction is envisioned between the antigen in its receptor and a specific I region molecule on the T cell surface. Complementarity between the interacting molecules results in the formation of a stable complex on the membrane which, as proposed by Edelman *et al.*,[89] is essential for T cell proliferation. This hypothesis, then, suggests that T cell activation is a two-step affair with the first step provided by the combination of antigen with its receptor and the second step by the interaction of the antigen–receptor complex with a specific I region molecule, which provides the proliferative signal. Non-response results if the complementary I region molecule needed to provide the second step of activation is absent.

By extrapolating liberally from the studies of the immunogenicity of modified histocompatibility antigens[80,81] the hypothesis can be extended to account from some of the observations regarding the role of the MHC in T cell–B cell collaboration. The interaction between the antigen and a specific I region molecule produces a new antigenic determinant on the cell membrane that is recognized as 'altered self' by other T cells. This clone of T cells with anti-modified I activity is stimulated to expand during the primary immune response. Antigen-receptor complexes on syngeneic B cells induce similar modifications in the complementary I antigens on their membranes. That altered self determinant is recognized by the clone of anti-modified I antigen T cells, whose response provides the stimulus for B cell activation.

The proposed model attributes T cell 'helper' effects to T cells reacting against altered-I and not to antigen-specific cells. The modifiable I region antigen must be present on both T and B cells accounting for the genetic restriction on physiologic cell interactions. It follows, then, that a cell membrane I antigen, the product of a structural gene located in the region of the MHC, performs the function of both Ir gene and C.I gene products, i.e. Ir gene = C.I. gene = I region structural gene.

The apparent lack of an I region homology requirement for successful T–B cooperation when cells from bone marrow chimeras are used[70,71] is compatible with the hypothesis that T cell help for B cells is provided by T cells responding to altered self antigens. In an (H-2^d + H-2^k) chimera, for example, the H-2^d T cells are tolerant of the H-2^k alloantigens. During antigen priming, however, interaction of antigen–receptor complexes with I region structures on H-2^k cells produces modified I^k antigens that are recognized by H-2^d T cells which then proliferate, expanding the clone. When cells from the primed chimeras are depleted of H-2^k cells, the remaining H-2^d T cells include those that recognize modified I^k. Effective T–B cooperation then follows from the interaction of these H-2^d T cells and antigen-bound H-2^k B cells.

The hypothesis presented in this section represents an attempt to reconcile some of the apparently contradictory experimental results reviewed in the rest of the chapter. Its main precepts, the formation of stable complexes between antigen and specific complementary I region molecules, the recognition of the complex as altered self, and the mediation of T helper activity by cells reacting to the modified I, are being put to the test in several laboratories. Our knowledge of the involvement of the major histocompatibility complex in the generation of immune responsiveness is the product of research done primarily in the last decade. During the past few years the tempo of research activity and the generation of new concepts has accelerated tremendously. It is likely that the major questions raised in this discussion will not need another decade to find their solution.

Acknowledgements

I am grateful to Dr Philip Greenberg for stimulating discussion and enthusiastic criticism, and to Ms Deborah Ann Frank for valuable editorial and secretarial assistance during the preparation of this manuscript.

Studies from the author's laboratory were supported by Grant AI 10931 from the National Institutes of Health.

11. REFERENCES

1. Klein, J. (1974). *Annu. Rev. Genet.*, **8**, 63–78.
2. Shreffler, D. C. and David, C. S. (1975). *Advan. Immunol.*, **20**, 125–195.
3. Gasser, D. L. and Silvers, W. K. (1974). *Advan. Immunol.*, **18**, 1–66.
4. Benacerraf, B. and Katz, D. H. (1975). *Advan. Cancer Res.*, **21**, 121–173.
5. Blomberg, B., Geckeler, W. R. and Weigert, M. (1972). *Science*, **177**, 178–180.
6. Eichmann, K. (1972). *Eur. J. Immunol.*, **2**, 301–307.
7. Pawlak, L. L. and Nisonoff, A. (1973). *J. exp. Med.*, **137**, 855–869.
8. McDevitt, H. O. and Benacerraf, B. (1969). *Advan. Immunol.*, **11**, 31–74.
9. Gasser, D. L. (1969). *J. Immunol.*, **103**, 66–70.
10. Lieberman, R. and Humphrey, W, Jr. (1972). *J. exp. Med.*, **136**, 1222–1230.
11. Lieberman, R., Paul, W. E., Humphrey, W., Jr. and Stimpfling, J. H. (1972). *J. exp. Med.*, **136**, 1231–1240.
12. Vaz, N. M. and Levine, B. B. (1970). *J. Immunol.*, **104**, 1572–1574.
13. Green, I. and Benacerraf, B. (1971). *J. Immunol.*, **107**, 374–381.
14. Benacerraf, B., Bluestein, H. G., Green, I. and Ellman, L. (1971). *Prog. Immunol.*, **1**, 485–494.
15. McDevitt, H. O., Bechtol, K. B., Grumet, F. C., Mitchell, G. F. and Wegmann, T. G. (1971). *Prog. Immunol.*, **1**, 495–508.
16. Levine, B. B., Ojeda A. and Benacerraf, B. (1963). *J. exp. Med.*, **118**, 953–957.
17. Bluestein, H. G., Green, I. and Benacerraf, B. (1971). *J. exp. Med.*, **134**, 458–470.
18. McDevitt, H. O., Deak, B. D., Shreffler, D. C., Klein, J., Stimpfling, J. H. H., and Snell, G. D. (1972). *J. exp. Med.*, **135**, 1259–1278.
19. Ellman, L., Green, I., Martin, W. J. and Benacerraf, B. (1970). *Proc. Nat. Acad. Sci.*, **66**, 322–328.
20. Bluestein, H. G., Ellman, L., Green, I. and Benacerraf, B. (1971). *J. exp. Med.*, **134**, 1529–1537.
21. Bluestein, H. G., Green, I. and Benacerraf, B. (1971). *J. exp. Med.*, **134**, 471–481.
22. Benacerraf, B. and McDevitt, H. O. (1972). *Science*, **175**, 273–279.
23. Shevach, E. M., Paul, W. E. and Green, I. (1972). *J. exp. Med.*, **136**, 1207–1221.
24. Bluestein, H. G. (1974). *J. Immunol.*, **113**, 1410–1416.
25. Ben-Sasson, S. Z., Shevach, E., Green, I. and Paul, W. E. (1974). *J. exp. Med.*, **140**, 383–395.
26. Greenberg, P. D. and Bluestein, H. G. (1975). *J. Immunol.*, **115**, 1206–1211.
27. Ellman, L., Green, I. and Benacerraf, B. (1971). *J. Immunol.*, **107**, 382–388.
28. Martin, W. J., Ellman, L., Green, I. and Benacerraf, B. (1970). *J. exp. Med.*, **132**, 1259–1266.
29. Shevach, E. M., Paul, W. E. and Green, I. (1974). *J. exp. Med.*, **139**, 661–678.
30. Shevach, E. M., Green, I. and Paul, W. E. (1974). *J. exp. Med.*, **139**, 679–695.
31. Bluestein, H. G. (1974). *J. exp. Med.*, **140**, 481–493.
32. Rosenstreich, D. L. and Rosenthal, A. S. (1973). *J. Immunol.*, **110**, 934–942.
33. Kasakura, S. (1969). *J. Immunol.*, **103**, 1078–1084.
34. Hausen, P. and Stein, H. (1968). *Eur. J. Biochem.*, **4**, 401–406.
35. Mendelsohn, J., Skinner, A. and Kornfeld, S. (1971). *J. Clin. Invest.*, **50**, 818–826.
36. Bluestein, H. G. (1974). *Fed. Proc.*, **32**, 985 (abstract).
37. Taylor, R. B., Duffus, P. H., Raff, M. C. and de Petris, S. (1971). *Nature* (*New Biol.*), **233**, 225–229.
38. Unanue, E. R. and Karnovsky, M. (1973). *Transplant Rev.*, **14**, 184–210.
39. Kourilsky, F. M., Silvestre, D., Neuport-Sautes,C., Loosfeld, Y. and Dausset, J. (1972). *Eur. J. Immunol.*, **2**, 249–257.
40. Foerster, J., Green, I., Lamelin, J.-P. and Benacerraf, B. (1969). *J. exp. Med.*, **130**, 1107–1122.
41. Ellman, L., Green, I. and Benacerraf, B. (1970). *Cell. Immunol.*, **1**, 445–454.
42. Tyan, M. L., McDevitt, H. O. and Herzenberg, L. A. (1969). *Transplant. Proc.*, **1**, 548–550.
43. Martin, W. J., Maurer, P. H. and Benacerraf, B. (1971). *J. Immunol.*, **107**, 715–718.
44. Kapp, J. A., Pierce, C. W. and Benacerraf, B. (1973). *J. exp. Med.*, **138**, 1107–1120.
45. Green, I., Paul, W. E. and Benacerraf, B. (1969). *Proc. Nat. Acad. Sci.* (*U.S.*), **64**, 1095–1102.
46. Levin, H. A., Levine, H. and Schlossman, S. (1971). *J. exp. Med.*, **133**, 1199–1218.
47. McDevitt, H. O. (1968). *J. Immunol.*, **100**, 485–492.
48. Grumet, F. C. (1972). *J. exp. Med.*, **135**, 110–125.
49. Mitchell, G. F., Grumet, F. C. and McDevitt, H. O. (1972). *J. exp. Med.*, **135**, 126–135.
50. Ordal, J. C. and Grumet, F. C. (1972). *J. exp. Med.*, **136**, 1195–1206.

51. Bechtol, K. B., Freed, J. H., Herzenberg, L. S. and McDevitt, H. O. (1974). *J. exp. Med.*, **140**, 1660–1675.
52. Shearer, G. M., Mozes, E. and Sela, M. (1972). *J. exp. Med.*, **135**, 1009–1027.
53. Taussig, M. and Munro, A. J. (1975). In *Immune Recognition* (Rosenthal, A. S., Ed.), Academic, New York, pp. 791–803.
54. Kapp, J. A., Pierce, C. W. and Benacerraf, B. (1975). In *Immune Recognition* (Rosenthal, A. S., Ed.), Academic, New York, pp. 667–682.
55. Katz, D. H. and Benacerraf, B. (1974). *Immunological Tolerance: Mechanisms and Potential Therapeutic Applications.* Academic, New York.
56. Kapp, J. A., Pierce, C. W. and Benacerraf, B. (1975). *J. exp. Med.*, **142**, 50–60.
57. Hersh, E. M. and Harris, J. E. (1968). *J. Immunol.*, **100**, 1184–1194.
58. Oppenheim, J. J., Leventhal, B. G. and Hersh, E. M. (1968). *J. Immunol.*, **101**, 262–270.
59. Rosenthal, A. S. and Shevach, E. M. (1973). *J. exp. Med.*, **138**, 1194–1212.
60. Shevach, E. M. and Rosenthal, A. S. (1973). *J. exp. Med.*, **138**, 1213–1229.
61. Shevach, E. M., Lee, L. and Ben-Sasson, S. Z. (1975). In *Immune Recognition* (Rosenthal, A. S., Ed.), Academic, New York, pp. 627–649.
62. Karz, D. H. and Unanue, E. R. (1973). *J. exp. Med.*, **137**, 967–990.
63. Kapp, J. A., Pierce, C. W. and Benacerraf, B. (1973). *J. exp. Med.*, **138**, 1121–1132.
64. Erb, P. and Feldmann, M. (1975). *Nature*, **254**, 352–354.
65. Kindred, B. and Shreffler, D. C. (1972). *J. Immunol.*, **109**, 940–943.
66. Katz, D. H., Hamaoka, T. and Benacerraf, B. (1973). *J. exp. Med.*, **137**, 1405–1418.
67. Katz, D. H., Hamaoka, T., Dorf, M. E. and Benacerraf, B. (1973). *Proc. Nat. Acad. Sci.*, **70**, 2624–2628.
68. Katz, D. H., Graves, M., Dorf, M. E., Dimuzio, H. and Benacerraf, B. (1975). *J. exp. Med.*, **141**, 263–268.
69. von Boehmer, H., Sprent, J. and Nabholz, M. (1975). *J. exp. Med.*, **141**, 322–334.
70. von Boehmer, H., Hudson, L. and Sprent, J. (1975). *J. exp. Med.*, **142**, 989–997.
71. Waldmann, H., Pope, H. and Munro, A. J. (1975). *Nature*, **258**, 728–730.
72. Heber-Katz, E. and Wilson, D. B. (1975). *J exp. Med.*, **142**, 928–935.
73. Ford, W. L. and Atkins, R. C. (1973). *Advan. exp. Med. Biol.*, **29**, 255–262.
74. Katz, D. H. and Armerding, D. (1975). In *Immune Recognition* (Rosenthal, A. S., Ed.), Academic, New York, pp. 727–753.
75. Tada, T. (1975). In *Immune Recognition* (Rosenthal, A. S., Ed.), Academic, New York, pp. 771–789.
76. Armerding, D. and Katz, D. H. (1974). *J. exp. Med.*, **140**, 19–37.
77. Taussig, M. J. and Munro, A. J. (1974). *Nature*, **251**, 63–65.
78. Taussig, M. J., Mozes, E. and Isac, R. (1974). *J. exp. Med.*, **140**, 301–312.
79. Mozes, E., Isac, R. and Taussig, M. J. (1975). *J. exp. Med.*, **141**, 703–707.
80. Doherty, P. C. and Zinkernagel, R. M. (1975). *J. exp. Med.*, **141**, 502–507.
81. Shearer, G. M., Rehn, T. G. and Garbarino, C. A. (1975). *J. exp. Med.*, **141**, 1348–1364.
82. Zinkernagel, R. M. and Doherty, P. C. (1975). *J. exp. Med.*, **141**, 1427–1436.
83. Schmitt-Verhulst, A.-M. and Shearer, G. M. (1975). *J. exp. Med.*, **142**, 914–927.
84. Doherty, P. C. and Zinkernagel, R. M. (1975). *Lancet*, **1**, 1406–1409.
85. Marchalonis, J. J. (1975). *Science*, **190**, 20–29.
86. Binz, H. and Wigzell, H. (1975). *J. exp. Med.*, **142**, 197–211.
87. Hämmerling, G. J. and McDevitt, H. O. (1974). *J. exp. Med.*, **140**, 1180–1188.
88. Kennedy, L. J., Dorf, M. E., Unanue, E. and Benacerraf, B. (1975). *J. Immunol.*, **114**, 1670–1675.
89. Edelman, G. M., Yahara, I. and Wang, J. L. (1973). *Proc. Nat. Acad. Sci.*, **70**, 1442–1446.

Chapter 14

Mechanism of T and K Cell-Mediated Cytolysis

J.-C. CEROTTINI
K. T. BRUNNER

Abbreviations

ADCMC: antibody-dependent cell-mediated cytolysis
CA: cytochalasin A
CB: cytochalasin B
cAMP: cyclic adenosine-3′,5′-monophosphate
cGMP: cyclic guanosine-3′,5′-monophosphate
CMC: cell-mediated cytolysis
CTL: cytolytic T lymphocytes
DFP: di-isopropylphosphofluoridate
DMSO: dimethyl-sulphoxide
EDTA: ethylene-diamine tetraacetic acid
E-RFC: spontaneous sheep erythrocyte rosette forming cells
K cells: killer cells
MLC: mixed leukocyte culture
PMN: polymorphonuclear cells
sIg: surface immunoglobulin

1. INTRODUCTION

In the last decade, cell-mediated cytotoxic reactions *in vitro* have been widely used in studies of the immune response to surface membrane structures such as histocompatibility, tumour and virus-associated antigens. Basically, antigen-bearing cells, which are referred to as target cells, are incubated with mononuclear cell populations from normal or immune donors, and the extent of destruction of the target cells is assessed at various time intervals thereafter. While various assay systems are used for the detection of cytotoxic reactions *in vitro*, it appears that one of the best defined systems available so far is based on the release of radioactive chromium (^{51}Cr) from labelled target cells.[1,2] This assay is based on the observation that ^{51}Cr, after diffusion through the cell membrane as chromate ion, is retained in the cytoplasm for a relatively prolonged period of time. Therefore ^{51}Cr release from a labelled target cell into the supernatant fluid does not occur unless the cell membrane is sufficiently damaged to allow the efflux of intracellular molecules. Moreover, the released ^{51}Cr is no longer the chromate anion and is not re-utilized by either lymphoid cells or target cells (reviewed in Reference 3). Since the assay is independent of target cell multiplication, the only parameter measured is direct lysis of the target cells, and the effector cells involved in this process may be referred to as cytolytic.

Evidence has recently been obtained that different effector cell types are involved in cell-mediated cytolysis (CMC) *in vitro*. This subject has been reviewed recently.[4] Within lymphoid cell populations, thymus-derived (T) cells from immune animals have been shown to specifically lyse appropriate target cells, in the absence of antibody. In contrast, IgG antibody-coated target cells may be lysed by certain non-T lymphoid cells present in both normal and immune donors. The two categories of lymphoid effector cells are usually referred to as cytolytic T lymphocytes (CTL) and killer (K) cells, respectively. In addition, there are lymphoid cells in normal donors which have been shown to effect the destruction of some cultured cell-lines in the apparent absence of antibody. According to their surface characteristics, these effector cells appear to belong to a subpopulation separate from CTL or K cells. Among non-lymphoid cells, monocytes and polymorphonuclear cells may lyse IgG antibody-coated target cells, particularly red cells, whereas macrophages activated by various specific and non-specific factors can be involved in lysis of nucleated cells in the absence of antibody. In view of the variety of effector cells involved in CMC *in vitro*, it is likely that the mechanisms of CMC also differ according to the models studied.

In the present review, we will discuss recent results regarding the mechanisms of CMC by two relatively well defined effector cell types: (a) CLT generated during the response to alloantigens in mice and (b) K cells present in human peripheral blood. Additional information on these two topics can be found in recent reviews.[3–8]

2. CTL-MEDIATED CYTOLYSIS

It is well established that the immune response either *in vivo* or *in vitro* to surface membrane antigens is characterized by the formation of CTL which specifically destroy target cells carrying the relevant antigens in the absence of antibody and/or complement. Most of the studies dealing with the mechanism of CTL-mediated lysis have used allogeneic model systems in mice, whereby the activity of CTL is assayed on ^{51}Cr-labelled tumour cells selected for their relatively high alloantigen content and/or susceptibility to lysis. From the data available, it is clear that (a) specific contact between a CTL and the relevant target cell has to take place, (b) this step is followed by the induction of a poorly defined target cells lesion and (c) the terminal stage is effector cell independent and involves membrane disintegration. However, the molecular pathway of CTL-mediated lysis remains to be defined.

2.1 General characteristics

Under appropriate conditions, target cell lysis as assessed by ^{51}Cr release may proceed at a rate as fast as that observed with antibody and complement, i.e. to become measurable within minutes.[4] Usually, however, one hour or more is required for the destruction of all target cells present in the reaction mixtures. As discussed below, there are at least two reasons for this apparent latent period: (a) despite irreversible membrane lesions occurring soon after contact, target cell lysis can require up to several hours for completion and (b) when the number of CTL becomes limiting, time is needed for one effector cell to damage several target cells successively. CTL-mediated lysis is temperature-dependent, as shown by the complete absence of target cell destruction below 15 °C.[3] Above 15 °C, the rate of lysis increases as a function of temperature up to 37 °C.[3,9] Recently, it has been demonstrated that lysis of relatively thermoresistant target cells by CTL is abolished at 44 °C due to effector cell inactivation.[10] As discussed below, these findings have been used to resolve the cytolytic pathway into discrete stages according to the temperature dependence.

It should be noted that the degree of lysis observed at room temperature may differ according to the source of effector cells. An example of such difference is illustrated in Figure 1. Increasing numbers of lymphoid cells containing CTL generated *in vitro* either in primary or in secondary mixed leukocyte cultures (MLC) were added to the appropriate ^{51}Cr-labelled target cells at 4 °C, the cell mixtures were then centrifuged in the cold and incubated thereafter for 3 hours at 15 °C, 20 °C or 37 °C, respectively. From the degree of ^{51}Cr release observed, it is evident that CTL formed in secondary MLC, in contrast to primary MLC effector cells, may be quite active at 20 °C.

It is established that CTL must be alive to be effective. Metabolic inhibitors, such as sodium cyanide,[11] sodium azide,[11,12] dinitrophenol,[11,12] iodoacetate,[11] antimycin A[13] and oligomycin,[13] all produce a dose-dependent inhibition of CTL-mediated lysis. In some instances, experiments using separate pre-incubation of either the effector lymphocyte population or the target cells with the metabolic inhibitor have ruled out the possibility that unimpaired energy at the target cell level is required for lysis. Therefore, it is very likely that the energy requirement for lysis is directly related to the metabolism of the CTL themselves. Indeed, recent studies indicate that inhibition of the energy metabolism prevents adequate CTL–target cell binding (see Section 2.3).

Sodium azide, an agent which blocks electron transfer, inhibits CTL-mediated lysis only when glucose is absent from the culture medium. It is thus likely that production of both aerobic and anaerobic energy has to be reduced to affect CTL activity. Along the same line is the observation that exogenous glucose is not required for lysis,[14] although 2-deoxyglucose, a non-metabolizable hexose analogue, appears to prevent target cell lysis by CTL when added at a high concentration to the culture medium.[13] These results thus suggest that the activity of the effector cells is not dependent on a specific energy pathway, but only requires a certain level of energy provided by aerobic oxidation and/or anaerobic glycolysis.

As expected from the rapid rate of lysis observed under optimal conditions, CTL function is unaffected by inhibitors of deoxyribonucleic acid (DNA) synthesis.[2] Actinomycin D, an antibiotic which inhibits DNA-dependent RNA synthesis, may cause a slight inhibition of lysis when present during the incubation period.

The role of protein synthesis in CTL-mediated lysis remains to be defined. As first reported by Brunner *et al.*,[2] the addition of cycloheximide, a reversible inhibitor of protein synthesis, to the cell mixtures causes a significant, but far from complete, reduction of

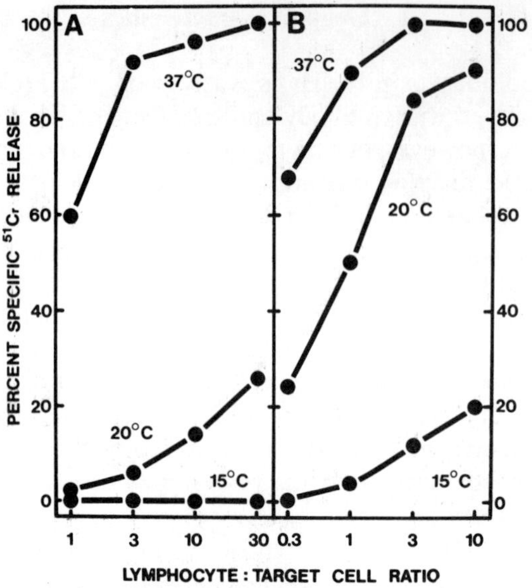

Figure 1. Temperature dependence of CTL-mediated cytolysis. CTL were generated in mixed leukocyte cultures (MLC) using C57Bl/6 spleen cells as responding cells and irradiated (1000 rads) DBA/2 spleen cells as stimulating cells. Responding cells were obtained from either normal mice (primary MLC) or from mice previously immunized against DBA/2 tumour cells (secondary MLC). On day 5 of culture, the cytolytic activities of the lymphoid cells recovered either from primary (panel A) or secondary (panel B) MLC were measured on ^{51}Cr-labelled DBA/2 target cells using a 3-hour assay system. (For further details, see Reference 94)

cytolysis. Similar results using puromycin have been reported.[13,15] It should be noted that the data obtained do not rule out the possibility that inhibition of protein synthesis at the target cell level may be responsible for the reduction of lysis. In support of this hypothesis is the observation that target cells cultured for a few hours in the presence of cycloheximide are subsequently less susceptible to lysis by CTL in the absence of the drug (K. Thomas, personal communication). Recently, complete abrogation of CTL-mediated lysis by pactamycin and emetine, two irreversible inhibitors of protein synthesis, has been reported.[16,17] Two observations, however, suggest that inhibition of cytolysis by these compounds is not necessarily related to inhibition of protein synthesis. First, complete inhibition of CTL-mediated lysis by pactamycin can be demonstrated without pre-incubation of CTL with the agent in a system in which lysis is detectable within 15 minutes in control uninhibited cultures.[17] The almost instantaneous inhibitory effect of pactamycin on cytolysis is difficult to reconcile with the demonstration that the compound inhibits the initiation rather than a late stage of polypeptide chain synthesis.[18] Second, the concentration of pactamycin needed for inhibition of cytolysis is 100-fold greater than that required for preventing incorporation of amino acids into cell-associated proteins.[16,19] Therefore, it is very likely that pactamycin and emetine inhibit CTL-mediated lysis by mechanisms unrelated to protein synthesis.

Organophosphorous agents, such as di-isopropylphosphofluoridate (DFP), which inactivate esterases with serine in their active site, inhibit target cell lysis by CTL.[20] The inhibition is time, temperature and concentration dependent, and can be demonstrated after pretreatment of the effector cell population, but not of the target cells. These findings thus suggest that fluoridates inhibit CTL-mediated lysis by irreversibly inhibiting an activated serine esterase, i.e. an esterase present in CTL in active form before contact with the target cells. While the role of this esterase in the lytic mechanism remains to be defined, it is worthy to note that a variety of cell functions including chemotaxis and phagocytosis by polymorphonuclear leukocytes, or histamine release from mast cells and platelets, appear also to be dependent on the presence of serine esterase.[21] In most instances, however, there is some evidence that the esterase becomes activated only after the appropriate stimulus has reacted with the cell. In contrast, the search for an activatable esterase in CTL has been inconclusive.[20]

The chelation of divalent cations with ethylene-diamine tetraacetic acid (EDTA) leads to complete inhibition of CTL-mediated lysis.[22,23] The inhibitory effect of EDTA can be reversed by adding excess calcium and magnesium ions to the medium. In more recent studies,[24,25] the divalent cation requirements have been investigated using media of defined cationic content. The results clearly indicate that, although calcium alone, but not magnesium, can support lysis, mixtures of both cations have synergistic effects owing to the existence of two distinct cation-requiring stages (see Section 2.3).

Inhibition of CTL-mediated lysis is observed in the presence of cytochalasin A (CA) and cytochalasin B (CB), two fungal metabolites.[14,20,26] With cytochalasin B (CB), the inhibitory effect is fully reversible, whereas CA-treated effector cells are no longer active after removal of the compound. As the cytochalasins interfere with numerous cell functions, the actual mechanism by which lysis is inhibited is not entirely clear. From the work of Bubbers and Henney,[14] however, it is evident that prevention of glucose transport through the cell membrane, a known effect of CB, is not involved in the inhibitory effect of the substance on cytolysis. Moreover, as discussed later, CB appears to inhibit intimate contact between CTL and target cells but has no effect on subsequent stages of the lytic pathway. It is unlikely that the compound exerts its inhibitory effect by preventing a secretory event such as has been demonstrated in antigen-induced lymphotoxin production.[16] In another lytic model system, there is indirect evidence that disruption of microfilaments by CB may play an important role in its inhibitory effect on cell-mediated lysis.[27] Further work is needed to determine whether impairment of microfilament function not only inhibits CTL motility but also prevents intimate contact between effector and target cells. Along this line, it is noteworthy that dimethyl-sulphoxyde (DMSO), a dipolar solvent, completely, but reversbly, inhibits CTL-mediated lysis when used at a concentration of 5%.[17,28] Wolberg *et al.*[28] have suggested that this effect is the result of a distortion of the cell membrane preventing intimate cell contact. As DMSO is used as a solubilizing agent for CB, its final concentration should be low enough to rule out any possible participation to the inhibitory effect observed of CB. The same consideration should apply when using other liposoluble substances.[29]

Colchicine and vinblastine are inhibitors of CTL-mediated lysis.[30,31] In contrast to the effect of CB, the inhibitory activity of these two compounds is irreversible and appears to affect a stage in the lytic cycle which takes place after the effector cell–target cell interaction. Direct evidence has been provided that the effect of colchicine and vinblastine is mediated through the effector cells, and not the target cells.[30] It is obvious that the antimitotic activity of these agents is not the basis of their inhibitory effect on cytolysis, but there remains the possibility that interference with the function of microtubules is germane to inhibition of CTL activity. In this respect, studies by Henney *et al.*[16] clearly indicate that antigen-induced lymphotoxin production is unaffected by colchicine or vinblastine under conditions in which CTL-mediated lysis is suppressed.

It is well-documented that the lytic process is inhibited by agents which increase the intracellular level of cyclic adenosine-3′,5′-monophosphate (cAMP) either by stimulation of adenyl cyclase activity or by inhibition of phosphodiesterase activity (reviewed in Reference 5). Studies using cholera toxin, which stimulates adenyl cyclase in a very protracted manner, have shown that pretreatment of the effector cell population, but not of the target cells, is associated with inhibition of lysis.[32,33] Under the same conditions, cholera toxin has no effect on the production of lymphotoxin.[16]

Adenosine, especially in the presence of an inhibitor of adenosine deaminase, is able to inhibit cytolysis.[34] As the concentrations of adenosine which inhibit lysis are also effective in causing an increase in cAMP levels of the effector cell population, Wolberg *et al.*[34] have suggested that the mechanism of inhibition involves an elevation of cAMP concentrations in CTL. Obviously, direct evidence for the inverse relationship between lytic activity and cAMP levels awaits the isolation of a pure CTL population, instead of the heterogeneous lymphocyte populations used in these studies. Also, little is known about the relative susceptibility to inhibition of effector cells taken at different stages of differentiation. In their studies of the inhibitory effect of histamine on CTL-mediated lysis, Plaut *et al.*[35] have observed a progressive increase in the degree of inhibition of effector spleen cells as a function of time after immunization. Moreover, CTL collected at the same time, but at two different sites, i.e. spleen versus peritoneal cavity, appear to differ as regards sensitivity to inhibition by histamine.[36] Since histamine appears to act through binding to specific receptors, the end result being an increase of cAMP levels following activation of adenyl cyclase, it is likely that the variable degree of susceptibility to inhibition by histamine reflects the existence of CTL containing different numbers of receptors.[36]

Contrasting with the inhibitory effects just reported, addition of cholinergic agonists, such as acetylcholine or carbamylcholine, has been found to enhance cytolysis.[37] Since cholinergic stimulation often results in increased intracellular levels of cyclic guanosine-3′,5′-monophosphate (cGMP), it has been suggested that the mode of action of the cholinergic compounds is mediated through augmentation of CTL cGMP levels. In support of this hypothesis is the observation that both exogenous 8-bromo cGMP and insulin, which are known to transiently elevate cGMP levels in tissue, also enhance lysis.[37,38] While these findings have been obtained in a rat allogeneic model system, similar attempts in the widely used murine systems have been unsuccessful so far. Whether the negative results reflect a species difference or inappropriate conditions is unclear. As pointed out by Strom *et al.*,[37,38] the augmentation of CTL-mediated lysis by cholinergic agonists or insulin depends upon a short period of pre-incubation of the effector cell population with these agents. Moreover, cholinergic enhancement is only observed with rat spleen cells collected relatively early after immunization (T. B. Strom, personal communication). As is the case for histamine receptors[39] the characterization of cholinergic[40] or insulin[38] receptors on effector lymphocytes is mainly based on pharmacologic manipulation of cytolytic activities. A better understanding of the biological significance of these various receptors must await the isolation of purified receptor-bearing cell populations.

2.2 CTL-independent stage

As already mentioned, target cells lysis, as assessed by release of intracellular ^{51}Cr into the medium fluid, is the result of permeability changes of the cell membrane allowing the efflux of cytoplasmic molecules. It is now evident that this event is not concomitant to the time when the CTL effects irreversible changes in the target cell. The latter effect is often referred to as 'lethal hit'. Several lines of evidence point out that the lethal hit may precede ^{51}Cr release, hence lysis, by several minutes up to a few hours.

Direct evidence for a CTL-independent stage during the lytic process has first been provided by the work of Martz and Benacerraf[41] using alloantibody and complement to destroy the effector cell population at different time intervals after the onset of the lytic assay. While no target cell lysis occurred in reaction mixtures containing pretreated lymphocytes, destruction of the effector cells after one hour of incubation with the target cells did not interfere with the subsequent lysis of a relatively large proportion of the target cell population. Experiments using inactivation of CTL by a short heat treatment gave similar results.[10,42]

As the CTL-independent stage does not require divalent cations,[41] these findings may explain the inability of EDTA to instantaneously inhibit target cell lysis when added to an ongoing cytolytic reaction.[22,23] It should be stressed, however, that while ^{51}Cr release in EDTA-treated reaction mixtures may proceed at nearly the same rate as in untreated control mixtures during the first 30 minutes after addition of the chelating agent, the lytic rate in the presence of EDTA decreases thereafter and eventually comes to a standstill in contrast to the progressive destruction of all target cells observed in mixtures without inhibitors.[23] As discussed below, addition of EDTA to a mixture of effector lymphocytes and target cells results in the immediate prevention of new CTL–target cell adhesions and leads, after a few minutes, to the dissociation of already established adhesions. Thus, it appears that ^{51}Cr release occurring after the addition of EDTA mainly reflects lysis of target cells which have been lethally hit prior to addition of the agent, although it has been suggested that a damaging effect of target cell-bound CTL may take place during the few minutes preceding the EDTA-induced dissociation of adhesions.[43]

When CB is added to a lytic reaction in progress, ^{51}Cr release also continues for a period of more than one hour before complete inhibition may be demonstrable.[26] The same pattern is seen with most, if not all, of the inhibitors mentioned above. It is thus likely that the CTL-independent stage does not require an active cell function to proceed. This stage, however, is still temperature-dependent (for references, see Reference 9), although to a lesser degree than the overall lytic reaction. In this respect, it is noteworthy that cooling to 0 °C of labelled target cells after induction of membrane damage by antibody and complement instantaneously prevents leakage of both electrolytes and macromolecules.[44] Whether or not CTL-induced membrane lesions are sealed at low temperatures is unknown, but it is clear that this arrest of lysis disappears as soon as the target cells, after inactivation of the effector cells, are placed again at 37 °C.[9,10]

It has been suggested that the CTL-independent stage is due to colloid osmotic lysis. As judged by time-lapse cinematography and light microscopy, the damaged target cells may enlarge before they disintegrate.[45–47] When the efflux of different markers from damaged target cells is followed as a function of time, molecules such as ^{86}Rb (a K^+ analogue), adenosine triphosphate or derivatives of nicotinamide, are released much more rapidly than ^{51}Cr.[44,48] These observations are reminiscent of the events described in haemolysis induced by hypotonic treatment or exposure to antibody and complement (for references, see Reference 49). In these instances, red cell damage is characterized by the early loss of cell K^+ with an accompanying entry of Na^+. As the intracellular concentration of colloid material is higher than the external concentration, there is an influx of water, the cell swells and eventually lyses as evidenced by haemoglobin loss. As the colloid osmotic phase of complement-mediated immune lysis may be prevented by high concentration of molecules of molecular weight above 40 000 in the external medium, it has been concluded that the reason for this effect is the relatively small size of the complement-induced membrane holes. If the macromolecules were unable to enter the membrane holes, the intracellular osmotic pressure should be counterbalanced by that of external medium, and target cell swelling and lysis should not occur.

Using a similar approach, Henney[50] and Ferluga and Allison[51] have found that, when high concentrations of dextran are added to ongoing cultures of effector lymphocytes and target cells, ^{51}Cr release is inhibited while ^{86}Rb efflux continues. Conflicting results, however, have been reported as regards the effect of dextrans of varying molecular size. In one case,[51] dextran of molecular weight 10 000 was found to be effective on CTL-mediated lysis, suggesting that the initial lesion was smaller than that produced by complement. On the other hand, Henney's studies[50] indicated that prevention of ^{51}Cr release from injured target cells required the presence of dextrans of molecular weight higher than 40 000. It should be stressed that the interpretation of these results is difficult. Kinetic studies of ^{86}Rb release used as an indicator of efflux of small molecules are limited by the rapid loss of the cation from the labelled target cells in the absence of lymphocytes.[48,51,52] Also, a comparative study of the release of different markers from CTL-injured target cells indicated that both small, with the exception of ^{86}Rb and nicotinamide, and large molecules were released with the same kinetics.[52] Finally, the use of external macromolecules to determine the size of membrane lesions has been seriously questioned.[49]

In view of these limitations, the possible role of osmotic lysis in the CTL-independent stage remains to be defined. It should be noted that the time required for the completion of this stage may vary from a few minutes in the presence of excess CTL to a few hours under less favourable conditions suggesting that the rate of lysis depends on the amount of damage sustained by the target cells. While no evidence to support the possibility of recovery or repair during this stage has been obtained, this point should be further investigated, particularly with target cells which are less susceptible to lysis than those used in the current model systems.

2.3 CTL-dependent stage

There is compelling evidence that the participation of CTL to the lytic pathway can be divided into two phases: (a) establishment of an intimate contact with the relevant target cell and (b) induction of a lethal hit to the target cell. Since multiple contact and hit events are likely to take place under the conditions used for optimal lysis, different means have been used to dissociate the two phases. The following procedures, employed alone or in combination, have been found to be most effective: incubation at different temperatures, prevention of cell interactions by high viscosity solutions, use of media of defined cationic content, addition of inhibitors.

Although it is well established that CTL-mediated lysis depends upon specific contact between the effector lymphocyte and the target cell, little is known about the exact nature of the CTL receptors and of the corresponding antigenic determinants on the target cell. Selective adsorption of CTL on target cell monolayers has been used to determine the factors influencing the formation of adequate adhesions (for references, see References 3 and 4). More recently, binding of CTL to target cells in suspension has been demonstrated by the formation of relatively stable conjugates under appropriate conditions.[53,54]

The formation of CTL-target cell adhesions is temperature-dependent, although to a lesser degree than the overall lytic process.[3] Since the phase of the lytic pathway which is the most temperature-dependent is the lethal hit step,[9,53] temperature conditions can be selected which allow binding, but no lethal hit, hence no lysis, to occur.

Several studies indicate that firm adhesions are not formed at 0–4 °C but are so at 15–20 °C.[53–55] The results of Wagner and Röllinghof,[42] however, suggest that some interactions may take place at low temperatures, although a valid interpretation of the data is complicated by the experimental design used. In this context, possible differences among CTL are suggested by recent observations made in our laboratory (Wyss, P., Cerottini, J.-C. and Brunner, K. T., manuscript in preparation). CTL populations were centrifuged together with ^{51}Cr-labelled target cells at 4 °C and incubated thereafter for 15 min at either 4 °C or 15 °C. After addition of CB, the temperature was raised to 37 °C and ^{51}Cr release determined after further incubation for 120 min. As discussed below, CB inhibits the formation of adhesions without affecting the induction of lethal hits, while the latter phase is very limited at 4 °C or 15 °C. Therefore, under the experimental conditions used, only those target cells which were bound to CTL at the time of CB addition could be lethally hit and lysed after raising the temperature. When either primary or secondary MLC culture populations were used as the source of CTL, a striking difference in the ability of establishing firm adhesions was observed. While CTL from primary MLC formed no, or very few, effective conjugates with target cells at 4 °C and 15 °C, respectively, firm adhesions between secondary MLC effector lymphocytes and target cells occurred at both temperatures. These observations, together with the results shown in Figure 1, suggest that studies of different sources of CTL, in addition to the commonly used peritoneal exudate lymphocytes, may be useful for the general understanding of the lytic process.

Under optimal conditions, effective adhesions between CTL and target cells can be demonstrated within 1 min.[53,56] This phase requires divalent cations as shown by the inhibitory effect of EDTA. Studies using media of defined cationic content indicate that either magnesium or calcium ions, when used alone, allow CTL–target cell interactions, although Ca^{2+} is less effective than Mg^{2+} [24,25,57] Once formed, adhesions dissociate within a few minutes upon addition of EDTA.[53,57] Metabolic inhibitors, such as sodium azide, sodium cyanide, dinitrophenol or iodoacetate, prevent the adsorption of CTL on target cell monolayers[12] and the formation of conjugates,[11] but have no dissociation effect on already established adhesions. DMSO[28] as well as benzyl or salicyl alcohol[58] prevent CTL-target interaction, as does CB.[59] The latter compound, however, has no apparent effect on formed conjugates (P. Wyss, personal communication).

Altogether, these results indicate that the establishment of firm adhesions is a rapid process which requires energy and presumably depends on CTL membrane integrity (for further discussion on the possible role of energy metabolism during this phase, see Reference 12). It should be stressed, however, that the demonstration of stable conjugates does not rule out the possibility that CTL, or a fraction thereof, establish only transient contacts with the target cells. In support of this contention is the demonstration that one effector lymphocyte can destroy several target cells successively, indicating that intimate contact is not irreversible.[4] Time lapse cinematography studies suggest that the CTL may dissociate from the target cell before the occurrence of morphologically detectable damage,[46] although this is not a general rule.[60] Since adsorption of CTL on killed, unfixed, target cell monolayers is negligible, it has been suggested that detachment of the effector lymphocyte from the target cell is a manifestation of its ability to recognize the initiation of irreversible target damage.[61] Further work is needed to establish whether or not induction of a lethal hit is a prerequisite for CTL detachment from the target cell.

A major problem related to the study of CTL is the lack of any means of identifying the effector lymphocytes at the single cell level. It is thus impossible to evaluate the efficiency of CTL binding, nor is it feasible to determine whether all target cell-bound lymphocytes are CTL. These limitations are particularly evident when the monolayer adsorption technique is used. Because of the relatively high number of non-specific adhesions, the binding of CTL can only be measured functionally, by assessing the lytic activity of non-adherent and adherent cell populations, while cell counting is meaningless. Along this line, recent studies regarding the formation of conjugates between immune peritoneal exudate lymphocytes and appropriate target cells in suspension suggest a more promising approach[53,54] Berke *et al.*[54] reported that up to 35% of the immune peritoneal lymphocytes were able to form conjugates with target cells under optimal conditions. While various types of conjugates were observed, the most common type consisted of one lymphocyte bound to one target cell. Direct evidence that the bound lymphocyte was indeed an effector cell was provided by micromanipulation experiments whereby individual conjugates were isolated and placed under culture conditions allowing target cell lysis.[47] For most conjugates, if not all, morphological evidence of target cell lysis could be obtained. It should be noted that non-specific adhesions may also occur as shown by the formation of conjugates between immune peritoneal exudate lymphocytes and irrelevant target cells, but the degree of non-specificity appears to be much lower than that observed with the monolayer adsorption technique.[53.54]

As already mentioned, the lethal hit phase is the most temperature-dependent step of the lytic process. In order to study this phase independently, it is necessary to prevent the formation of new adhesions. As shown by Martz,[53] this can be achieved by diluting the conjugates formed at 15 °C in a highly viscous dextran solution before raising the temperature to 37 °C. Under these conditions, a CTL which is firmly bound to a target cell at the time of addition of the dextran solution will be able to damage this target cell, but will be prevented from establishing any new contact with other target cells. As shown in Table 1, similar results are obtained when CB is substituted to the viscous dextran solution. In this experiment, mixtures of secondary MLC lymphocytes and ^{51}Cr-labelled target cells were brought into physical contact by centrifugation at 4 °C and then incubated for 15 minutes at 15 °C. Thereafter, a solution of 10% dextran of molecular weight 500 000, or CB (at a final concentration of 10 μg/ml), were added before a second incubation for 2 hours at 37 °C. In contrast to the very low lytic activity observed in cell mixtures containing the inhibitory agents during the whole incubation period, up to 70% of the target cells preincubated at 15 °C without dextran or CB lysed during the subsequent incubation at 37 °C. It is noteworthy that previous atempts using CB to demonstrate effective adhesions between CTL and target cells at room temperature have been unsuccessful:[17] In the light of more recent experiments, these negative results can be explained by the use of primary MLC effector lymphocytes which, although they have a high lytic activity under normal conditions, do not establish firm adhesions with target cells at 15 °C (P. Wyss, personal communication).

In contrast to the binding phase, the lethal hit step does not occur in the absence of Ca^{2+}.[24,25] It is thus evident that the two recognized phases of the CTL-dependent stages have distinct cation requirements. In the light of these results, both steps should be inhibited by EDTA. While it is clear that both formation and maintenance of firm

Table 1. Effects of temperature and dextran or cytochalasin B on CTL-mediated lysis[a]

First incubation[b]		Second incubation		Percent and specific ^{51}Cr release
Inhibitor	Temperature (°C)	Inhibitor	Temperature (°C)	
None	15	None	15	15
None	37	None	37	>95
Dextran	37	Dextran	37	7
None	15	Dextran	15	4
None	15	Dextran	37	60
CB	37	CB	37	11
None	15	CB	15	1
None	15	CB	37	70

[a] Source of CTL: day 5 secondary MLC C57Bl/6 spleen cells stimulated with irradiated DBA/2 spleen cells.
[b] Mixtures of MLC cells and ^{51}Cr-labelled L 1210 (DBA/2) target cells (lymphocyte to target cell ratio, 100 : 1) were prepared in the cold. After centrifugation at 4 °C, treatments indicated above were performed using either 10% dextran (mol. wt. 500 000) or 10 μg/ml cytochalasin B (CB) as inhibitors of the formation of CTL–target cell adhesions. Incubation time: 30 min (first) and 120 min (second).

adhesions can be affected by EDTA, it has been suggested that the chelating agent may not completely inhibit the activity of target cell-bound CTL.[43] It would appear that the concentration of chelating agent used may be of critical importance. In comparative studies of the role of inhibitors on CTL-mediated lysis, it has been claimed that EDTA only affects a relatively late step, occurring after the stage inhibited by CB (reviewed in Reference 5). In a similar study, we have been unable to confirm these findings.[17] As discussed in a recent paper by Plaut *et al.*,[25] these conflicting results may be explained at least in part by the different amounts of EDTA used in both studies, for the concentration of chelating agent required for inhibiting lymphocyte-target cell adhesions appears to be higher than that needed for preventing the lethal hit occurring.

There is evidence that the lethal hit phase can be completed within a few minutes.[42,53,56] Little is known about the nature of the lesions administered by the effector lymphocytes. In view of the accelerated ^{86}Rb release observed in mixtures of effector lymphocytes and labelled target cells,[48] it is tempting to speculate that the primary lesion is an alteration of the target cell membrane which becomes increasingly permeable to this electrolyte. Consistent with the hypothesis is the recent finding that the electrolyte permeability increase occurs concomitantly with the lethal hit step.[62] The possibility that lesions of another type are of more crucial importance cannot be ruled out, however.[52]

The mechanism by which CTL produce membrane lesions is not yet known. While secretion of a toxic factor which remains active at any distance from the effector lymphocyte is excluded,[4,63] there are a number of other possibilities which remain to be experimentally tested. Among others, the various hypotheses proposed include production of a local membrane shearing effect,[49] formation of persisting gap junctions,[51] secretion of a toxic factor into or onto the target cell,[5] local alteration of the transmembrane potential,[3] or insertion of channel-forming molecules into the membrane.[64] Obviously, these different possibilities are not mutually exclusive.

Whatever the mechanism might be, it is evident that CTL are unaffected by the lytic process. In agreement with several observations obtained at the population level, conclusive evidence that a single CTL can destroy several target cells sequentially in time has been provided by Zagury *et al.*[47] By using individual conjugates isolated by micromanipulation, it was possible to collect a single CTL which had lysed a target cell and to place it in contact with a new target cell. Most often, lysis following adhesion could be demonstrated. In a few cases, three sequential episodes of binding and killing were observed. In a more indirect approach, the results indicated that one CTL could lyse more than six target cells.[65]

3. K CELL-MEDIATED CYTOLYSIS

It is well established that non-immune effector cells can destroy appropriate target cells coated with antibody IgG molecules (see reviews in References 4, 6–8, 66). Several model systems using different sources of effector and target cells have been studied in animals and man. Attempts to determine the nature of effector cells have revealed considerable heterogeneity. Among other factors, the nature of the target cells may play a critical role in defining the effector cell type involved in antibody-dependent cell-mediated cytolysis (ADCMC). This heterogeneity is best illustrated by the results obtained in man using peripheral blood as the source of effector cells. When chicken red cells are used as target cells, polymorphonuclear cells (PMN), monocytes and lymphocytes are all found to behave operationally as effector cells.[67–70] IgG antibody-coated human red cells, however, are lysed in the presence of monocytes and PMN, but are unaffected by purified lymphoid cell populations.[68,71,72] In contrast, target cells such as lymphocytes or cell–lines of lymphoid or non-lymphoid origin appear not to be killed by monocytes and PMN in the presence of antibody (although there are exceptions, see Reference 73), while the same sensitized target cells are readily lysed by lymphoid effector cells.[68,70,73,74] In view of the effector cell heterogeneity, it is not too surprising that various conclusions have been reached as to the mechanism of ADCMC.

In this review, we will only consider the results obtained with the lymphoid-like effector cells present in human peripheral blood. As the nature of these effector cells is not completely established, they are referred to as K(= killer) cells.[8] It should be stressed that K cells, like CTL, are only defined functionally since specific markers of this effector cell type are not yet known. There is a general agreement that K cell activity is associated with a lymphocyte population devoid of surface Ig positive (sIg) lymphocytes[69,70,75–77] and of spontaneous sheep erythrocyte rosette forming cells (E-RFC),[66,69,70,77,78] but some controversy exists as to the presence of K cells also in sIg positive lymphocyte[77,78] and/or T cell populations.[79]

Studies of the structural requirements on antibody molecules necessary for ADCMC as well as inhibition experiments using antigen–antibody complexes clearly indicate that all K cells carry surface receptors for the Fc portion of IgG. However, Fc receptor bearing lymphocytes do not all behave as effector cells.

Kinetic studies of K cell-mediated lysis have usually documented that lysis, as assessed by ^{51}Cr release, is detectable within 1 hour and proceeds linearly as a function of time.[7] With chicken red cells, the lytic rate is relatively constant for several hours, whereas in other model systems using cell-lines as the source of target cells it has been reported that cytolysis decreases considerably after a few hours[73,80,81] (P. Wyss, J.-C. Cerottini and K. T. Brunner, manuscript in preparation). This is documented by the fact that, after incubation for about 4 hours, the rate of ^{51}Cr release ceases to exceed that observed in control cell mixtures containing no antibody. As this finding contrasts with the kinetics of lysis observed with CTL, we have tried to substantiate these observations by following the rate of ^{51}Cr release from the same population of antibody-coated target cells exposed either to human K cell or to mouse CTL (P. Wyss, J.-C. Cerottini and K. T. Brunner, manuscript in preparation). As shown in Figure 2, a clear difference in the cytolytic rate could be

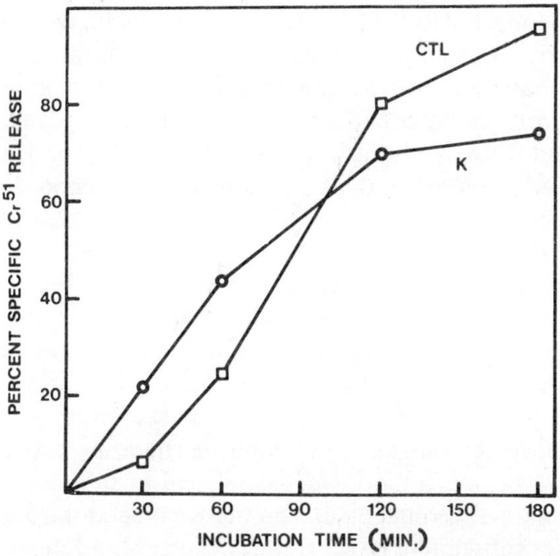

Figure 2. Comparison of the kinetics of CTL-mediated or K-cell-mediated cytolysis. Murine CTL were generated in secondary MLC using immune C57Bl/6 spleen cells as responding cells and irradiated DBA/2 spleen cells as stimulating cells (for details, see caption to Figure 1). Human peripheral blood lymphocytes were used as the source of K cells. The cytolytic activities of both effector cell types were measured on the same ^{51}Cr-labelled target cell population, namely L1210 (DBA/2) lymphoma cells, in the presence of a 1:5000 dilution of rabbit anti-L 1210 antiserum (while antibody was required for K cell activity, it produced no effect on the activity of CTL directed against H-2 associated antigens)

demonstrated, for K cell-mediated lysis progressively decreased with time, whereas the usual S-shaped curve was observed in CTL-mediated lysis. It should be stressed, however, that all target cells could be lysed in ADCMC when a higher lymphocyte–target cell ratio was used, indicating that the decrease in the lytic rate as a function of time (as shown in Figure 2) was not caused by inadequate antibody coating of a fraction of the target cells. Moreover, the lytic activity of the lymphocyte population cultured alone or in the presence of target cells without antibody was fully retained after incubation for several hours or even a few days. It therefore appears unlikely that cessation of lysis results from K cell inactivation under inadequate *in vitro* conditions.

K cell activity is temperature-dependent. In the model system described above, the temperature sensitivity of the lytic process mediated by K cells is very similar to that observed for secondary MLC CTL (as shown in Figure 1). While no lysis occurs at 4 °C, a significant activity can be demonstrated at 15 °C, although to a much lesser degree than at 37 °C. Several of the inhibitors effective in CTL-mediated lysis are also active in K cell-mediated lysis. These include metabolic inhibitors, such as antimycin A,[82] sodium azide,[83] or iodoacetate,[83] or agents such as CB,[7,74,84] DMSO[84] and EDTA.[84,85] A recent study of divalent cation requirement indicates that K cell activity, as assessed by lysis of antibody-coated Chang liver cells by human peripheral blood effector cells, can develop in the presence of Ca^{2+} alone, but not of Mg^{2+} alone.[86] Interestingly, lysis of antibody-coated sheep red blood cells by the same cell population is found to require preferentially Mg^{2+} or no cations, thus emphasizing again the heterogeneity of effector cells according to the nature of the target cells as described above. As is the case with CTL-mediated lysis, inhibition of DNA and RNA synthesis has no significant effect on K cell activity.[68,85] In a rat model system, in addition to the inhibitors already mentioned, agents which increase intracellular levels of cAMP tend to depress ADCMC,[31] whereas exogenous bromo cGMP may enhance it.[87] In this system, inhibitors of protein synthesis, such as cycloheximide, puromycin, emetine and pactamycin, appear to inhibit.[13] With the latter two agents, however, the inhibitory dose was higher than that required for preventing protein synthesis, as already observed in studies of CTL-mediated lysis (see Section 2.1). Pretreatment of effector lymphocytes, but not of target cells, with DFP resulted in ADCMC inhibition, suggesting the involvement of a serine esterase in rat K cell function.[13,88]

All available evidence thus suggests common mechanisms between CTL and K cell mode of action. However, much more work is needed to substantiate these similarities. While it is clear that intimate contact between the K cell and the antibody-coated target cell is a prerequisite for subsequent lysis,[82] little is known about the binding requirements. Studies in our laboratory have documented that effective contacts may occur at 4 °C, while they are prevented in the presence of CB (P. Wyss, personal communication). The latter agent, however, has no apparent effect on the post-adhesion stage leading eventually to lysis. Thus, a clear dissociation between formation of adhesions and the subsequent stages can be achieved by using a combination of initial incubation at low temperature followed by addition of CB and further incubation at 37 °C, as used in the system of CTL-mediated lysis (see Table 1). Other studies have documented the specific adhesion of human lymphocytes to monolayers of antibody-coated target cells.[89] However, such a method provides only indirect information regarding the adhesions actually involved in the lytic process, since lymphocytes other than K cells, but devoid of lytic potential, may bind to the antibody-coated target cells through their Fc receptor sites.

The recent work of Ziegler and Henney[80] has shown that increased efflux of ^{86}Rb precedes ^{51}Cr release, suggesting that actual lysis may follow the establishment of a primary lesion involving changes of the membrane permeability to electrolytes. While the data speak in favour of a relatively long time-interval between administration of the lethal hit and completion of the lytic process, other observations are in apparent conflict with this hypothesis. In studies dealing with the effect of EDTA on K cell-mediated lysis, it has been found (P. Wyss, J.-C. Cerottini and K. T. Brunner, manuscript in preparation) that addition of the chelating agent to ongoing cell interactions produced an immediate arrest of ^{51}Cr release (similar observations have been done by Trinchieri, personal communication). In parallel cultures containing the same target cell population incubated in the presence of CTL, complete inhibition of ^{51}Cr release was observed only one hour after addition of EDTA, in agreement with previous studies (see Section 2.2). The possibility thus exists that the stage following administration of lethal hits by K cells may be of a shorter duration than that observed in CTL-mediated lysis. In support of this hypothesis is the finding that lysis of up to 50% of the target cells can be demonstrated after 5 min of incubation under appropriate conditions (P. Wyss, J.-C. Cerottini and K. T. Brunner, manuscript in preparation).

Circumstantial evidence suggested that a single K cell may kill more than one target cell. Using a sensitive assay system, it has been found (P. Wyss, J.-C. Cerottini and K. T. Brunner, manuscript in preparation) that up to 50% of the target cells may be lysed after incubation for 4 hours at a lymphocyte to target cell ratio of 1 : 1. If one K cell killed only a single target cell, the percentage of K cells in the lymphocyte population would be equivalent to the percentage of target cells killed, i.e. 50%. Since in these experiments no attempt was made to select a lymphocyte subpopulation as the source of effector cells, these results would indicate that every other lymphocyte in peripheral blood was a K cell. As this is very unlikely, it appears that K cell can kill more than one target cell.

Recently, Perlmann *et al.*[79] have developed a plaque-forming cell assay which allows the detection of individual effector cells active against antibody-coated red cells. Direct evidence has been provided that about 5% of human peripheral blood lymphocytes act as effector cells against IgG antibody-coated chicken red cells and that one effector cell can kill several target cells.[79]

4. SUMMARY AND CONCLUSIONS

Based on the findings reviewed here, a tentative model for the pathway of CTL-mediated lysis can be outlined. Operationally, the lytic process can be resolved into three discrete steps. The first one is the establishment of an intimate contact between a CTL and the relevant target cell. Under optimal conditions, this may occur within one minute or so. This step is temperature and energy-dependent, requires the presence of magnesium ions (with Ca^{2+} as a less effective substitute), and depends on the membrane integrity of the effector lymphocyte. Little is known about the nature of CTL receptors and the corresponding target cell determinants (for discussion, see References 4 and 90). Once formed, the adhesions are relatively stable, even in the presence of metabolic inhibitors, but dissociate in the presence of EDTA.

The second phase in the lytic pathway involves a lethal hit administered by the CTL to the target cell resulting in an irreversible target cell damage. This is the most temperature-sensitive step and has an absolute requirement for Ca^{2+}. Under optimal conditions, only a few minutes are needed for completion of the lethal hit. The exact nature of the primary lesion is not yet known, although there is suggestive evidence that an increase in membrane permeability to electrolytes may be involved. There is no indication that the target cell actively participates in the reaction.

The third step is independent of the presence of the CTL and results in disruption of the target cell membrane. While this is the least temperature-sensitive stage in the lytic pathway, it may require a few hours for completion, but can occur within minutes under appropriate conditions. Most, if not all, of the substances inhibiting the lytic process have no effect on this late event. No evidence for recovery or repair has been found, but these negative findings may be related to the source of target cells chosen for these studies.

A major difficulty in the understanding of the lytic pathway in molecular terms has been the lack of available homogeneous cell populations, containing a majority of CTL. However, the demonstration that firm conjugates can be formed under conditions where cytolysis does not occur suggests the possibility of developing preparative separation techniques allowing the isolation of relatively large numbers of pure CTL. The effectiveness of this procedure may be dependent upon the proportion of effector lymphocytes present in the original population. In this context, recent studies on the *in vitro* generation of CTL indicate that a high degree of amplification of CTL may be obtained by repeated antigenic stimulation under appropriate conditions (for review, see References 91 and 92). Moreover, these studies demonstrate that the functional activity of CTL may vary depending on their differentiation stage. In particular, there is evidence that CTL may further differentiate into small inactive lymphocytes which, after appropriate antigenic stimulation, can rapidly regain activity in the absence of DNA synthesis.[93] Such CTL might represent a useful population to study whether reappearance of lytic activity is linked to the ability of the effector cell to establish intimate contacts and/or to inflict lethal hits.

While CTL are generated during the immune response to membrane-bound antigens, K cells are present in non-immune lymphocyte populations. On the basis of relative lytic potential, human peripheral blood is at present the best source of K cells. Although the exact nature of this effector cell type is not entirely defined, there is convincing evidence that some K cells are associated with sIg negative, E-RFC negative, Fc receptor positive lymphocyte subpopulation. It is also evident that Fc receptor positive lymphocytes do not all behave as K cells, and that PMN and monocytes are effective in some systems of ADCMC.

For lysis, K cells have to establish an intimate contact with IgG antibody-coated target cells. While little is known about the nature of the receptor sites, the corresponding structures on the IgG antibody molecules have been partially identified (for discussion, see References 6 and 8). From the little information available, it appears that the sequence of events in K cell-mediated lysis is similar to that described for target cell lysis by CTL. It would appear, however, that the time-interval required for lysis after administration of the lethal hit might be shorter with K cells than with CTL. Whether this reflects a qualitative rather than a quantitative difference in the lesion induced by one or the other effector cell type is unknown. While there is evidence that a single K cell may destroy several target cells, kinetic studies indicate that K cell activity, unlike that of CTL, becomes negligible after a few hours of incubation with antibody-coated target cells, suggesting a functional inactivation of the effector cells, perhaps due to physical or biological inhibition of the Fc receptor sites.

The difficulties mentioned above as far as the precise understanding of CTL-mediated lysis is concerned certainly do also apply to studies of K cell activity, since these effector cells appear to constitute a minority fraction among blood lymphocytes. Although no evidence is yet available that K cells can multiply *in vitro*, the possibility of increasing their number, either in absolute or in relative terms, cannot be excluded. Indeed, recent studies have shown that ADCMC activity increases in MLC.[84] By using cell separation techniques to obtain enriched effector cell populations, a much better understanding of the biological properties and the underlying lytic mechanism of the K cells should become available in the near future.

5. REFERENCES

1. Holm, G. and Perlmann, P. (1967). *Immunology*, **12**, 525–536.
2. Brunner, K. T., Mauel, J., Cerottini, J.-C. and Chapuis, B. (1968). *Immunology*, **14**, 181–196.
3. Berke, G. and Amos, D. B. (1973). *Transplant. Rev.*, **17**, 71–107.
4. Cerottini, J.-C. and Brunner, K. T. (1974). *Advan. Immunol.*, **18**, 67–132.
5. Henney, C. S. (1973). *Transplant. Rev.*, **17**, 37–70.
6. MacLennan, I. C. M. (1972). *Transplant Rev.*, **13**, 67–90.
7. Perlmann, P., Perlmann, H. and Wigzell, H. (1972). *Transplant. Rev.*, **13**, 91–117.
8. Perlmann, P. (1976). In *Clinical Immunobiology*, Vol. 3 (Bach, F. H. and Good, R. A., Eds.), Academic, New York.
9. Martz, E. and Benacerraf, B. (1975). *Cell. Immunol.*, **20**, 81–91.
10. Miller, R. G. and Dunkley, M. (1974). *Cell. Immunol.*, **14**, 284–302.
11. Berke, G. and Gabison, D. (1975). *Eur. J. Immunol.*, **5**, 671–675.
12. Todd, R. F., III (1975). *Transplantation*, **20**, 350–354.
13. Strom, T. B., Garovoy, M. R., Bear, R. A., Gribik, M. and Carpenter, C. B. (1975). *Cell. Immunol.*, **20**, 247–256.

14. Bubbers, J. E. and Henney, C. S. (1975). *J. Immunol.*, **115**, 145–149.
15. Calabresi, P., Brunner, K. T. and Mauel, J. (1967). *Fed. Proc.*, **26**, 478.
16. Henney, C. S., Gaffney, J. and Bloom, B. R. (1974). *J. exp. Med.*, **140**, 837–852.
17. Cerottini, J.-C., MacDonald, H. R., Engers, H. D., Thomas, K. and Brunner, K. T. (1974). *Advan. Biosciences*, **12**, 47–56.
18. MacDonald, J. S. and Goldberg, I. H. (1970). *Biochem. Biophys. Res. Commun.*, **41**, 1–8.
19. Thorn, R. M. and Henney, C. S. (1976). *J. Immunol.*, **116**, 146–149.
20. Ferluga, J., Asherson, G. L. and Becker, E. L. (1972). *Immunology*, **23**, 577–590.
21. Becker, E. L. and Henson, P. M. (1973). *Advan. Immunol.*, **17**, 93–193.
22. Mauel, J., Rudolf, H., Chapuis, B. and Brunner, K. T. (1970). *Immunology*, **18**, 517–535.
23. Henney, C. S. and Bubbers, J. E. (1973). *J. Immunol.*, **110**, 63–72.
24. Golstein, P. and Smith, E. T. (1976). *Eur. J. Immunol.*, **6**, 31–37.
25. Plaut, M., Bubbers, J. E. and Henney, C. S. (1976). *J. Immunol.*, **116**, 150–166.
26. Cerottini, J.-C. and Brunner, K. T. (1972). *Nature*, **237**, 272–273.
27. Kalina, M. and Hollander, N. (1975). *Immunology*, **29**, 709–717.
28. Wolberg, G., Hiemstra, K., Burge, J. J. and Singler, R. C. (1973). *J. Immunol.*, **111**, 1435–1443.
29. Kemp, A. and Berke, G. (1973). *Cell. Immunol.*, **7**, 512–515.
30. Plaut, M., Lichtenstein, L. M. and Henney, C. S. (1973). *J. Immunol.*, **110**, 771–780.
31. Strom, T. B., Garovoy, M. R., Carpenter, C. B. and Merrill, J. P. (1973). *Science*, **181**, 171–173.
32. Strom, T. B., Deisseroth, A., Morganroth, J., Carpenter, C. S. and Merrill, J. P. (1972). *Proc. Nat. Acad. Sci.*, **69**, 2995–2999.
33. Lichtenstein, L. M., Henney, C. S., Bourne, H. R. and Greenough, W. B. (1973). *J. Clin. Invest.*, **52**, 691–697.
34. Wolberg, G., Zimmerman, T. P., Hiemstra, K., Winston, M. and Chu, L.-C. (1975). *Science*, **187**, 957–959.
35. Plaut, M., Lichtenstein, L. M. and Henney, C. S. (1973). *Nature*, **244**, 284–287.
36. Plaut, M., Lichtenstein, L. M. and Henney, C. S. (1975). *J. Clin. Invest.*, **55**, 856–874.
37. Strom, T. B., Carpenter, C. B., Garovoy, M. R., Austen, K. F., Merrill, J. P. and Kaliner, N. A. (1973). *J. exp. Med.*, **138**, 381–393.
38. Strom, T. B., Bear, R. A. and Carpenter, C. B. (1975). *Science*, **187**, 1206–1208.
39. Plaut, M., Lichtenstein, L. M., Gillespie, E. and Henney, C. S. (1973). *J. Immunol.*, **111**, 389–394.
40. Strom, T. B., Sytkowsky, A. J., Carpenter, C. B. and Merrill, J. P. (1974). *Proc. Nat. Acad. Sci.*, **71**, 1330–1333.
41. Martz, E. and Benacerraf, B. (1973). *J. Immunol.*, **111**, 1538–1545.
42. Wagner, H. and Röllinghof, M. (1974). *J. Immunol.*, **4**, 745–750.
43. Martz, E. (1975). *Cell. Immunol.*, **20**, 304–314.
44. Martz, E., Burakoff, S. J. and Benacerraf, B. (1974). *Proc. Nat. Acad. Sci.*, **71**, 177–181.
45. Rosenau, W. (1968). *Fed. Proc.*, **27**, 34–38.
46. Koren, H. S., Ax, W. and Freund-Moelbert, E. (1973). *Eur. J. Immunol.*, **3**, 32–37.
47. Zagury, D., Bernard, J., Thiernesse, N., Feldman, M. and Berke, G. (1975). *Eur. J. Immunol.*, **5**, 818–822.
48. Henney, C. S. (1973). *J. Immunol.*, **110**, 73–84.
49. Seeman, P. (1974). *Fed. Proc.*, **33**, 2116–2124.
50. Henney, C. S. (1974). *Nature*, **249**, 456–458.
51. Ferluga, J. and Allison, A. C. (1974). *Nature*, **250**, 673–675.
52. Sanderson, C. J. (1976). *Proc. R. Soc. Lond. B*, **192**, 221–239.
53. Martz, E. (1975). *J. Immunol.*, **115**, 261–267.
54. Berke, G., Gabison, D. and Feldman, M. (1975). *Eur. J. Immunol.*, **5**, 813–818.
55. Golstein, P., Svedmyr, E. A. J. and Wigzell, H. (1971). *J. exp. Med.*, **134**, 1385–1402.
56. MacDonald, H. R. (1975). *Eur. J. Immunol.*, **5**, 251–254.
57. Stulting, R. D. and Berke, G. (1973). *J. exp. Med.*, **137**, 932–942.
58. Kemp. A. S. and Berke, G. (1973). *J. Immunol.*, **3**, 674–677.
59. Henney, C. S. and Bubbers, J. E. (1973). *J. Immunol.*, **111**, 85–90.
60. Sanderson, C. J. (1976). *Proc. R. Soc. Lond. B*, **192**, 241–255.
61. Stulting, R. D., Todd, R. E., III and Amos, D. B. (1975). *Cell. Immunol.*, **20**, 54–63.
62. Martz, E. (1976). *J. Immunol.*, **117**, in press.
63. Henney, C. S. (1975). *J. Reticuloendothel. Soc.*, **17**, 231–235.
64. Henkart, P. and Blumenthal, R. (1975). *Proc. Nat. Acad. Sci.*, **72**, 2789–2793.

65. Martz, E. (1976). *Transplantation*, **21**, 5–11.
66. Perlmann, P., Perlmann, H., Larsson, A. and Wahlin, B. (1975). *J. Reticuloendothel. Soc.*, **17**, 241–250.
67. Perlmann, P. and Perlmann, H. (1970). *Cell. Immunol.*, **1**, 300–315.
68. MacDonald, H. R., Bonnard, G. D., Sordat, B. and Zawodnik, S. A. (1975). *Scand. J. Immunol.*, **4**, 487–497.
69. Papamichail, M. and Temple, A. (1975). *Clin. exp. Immunol.*, **20**, 459–467.
70. Nelson, D. L., Bundy, B. M., Pitchon, H. E., Blaese, R. M. and Stober, W. (1976). *J. Immunol.*, **117**, in press.
71. Holm, G. and Hammarström, S. (1973). *Clin. exp. Immunol.*, **13**, 29–43.
72. Kovithavongs, T., Rice, G., Thong, K. L. and Dossetor, J. B. (1975). *Cell. Immunol.*, **18**, 167–175.
73. Gale, R. P. and Zighelboim, J. (1975). *J. Immunol.*, **114**, 1047–1051.
74. Dickmeiss, E. (1973). *Scand. J. Immunol.*, **2**, 251–260.
75. Wislöff, F. and Fröland, S. S. (1973). *Scand. J. Immunol.*, **2**, 151–157.
76. Perlmann, P., Wigzell, H., Golstein, P., Lamon, E. W., Larsson, A., O'Toole, C., Perlmann, H. and Svedmyr, E. A. J. (1974). *Advan. Biosciences*, **12**, 71–85.
77. Chess, L., MacDermott, R. P., Sondel, P. M. and Schlossmann, S. F. (1974). In *Progress in Immunology II*, Vol. 3 (Brent, L. and Holborow, H., Eds.), North-Holland, Amsterdam, pp. 125–132.
78. Zigelboim, J., Gale, R. P., Chiu, A., Bonavida, B., Ossorio, R. C. and Fahey, J. L. (1974). *Clin. Immunol. Immunopath.*, **3**, 193–200.
79. Perlmann, P., Biberfeld, P., Larsson, A., Perlmann, H. and Wahlin, B. (1975). In *Membrane Receptors of Lymphocytes* (Seligmann, M., Preud'homme, J. L. and Kourilsky, F. M., Eds.), North-Holland, Amsterdam, pp. 161–169.
80. Ziegler, H. K. and Henney, C. S. (1975). *J. Immunol.*, **115**, 1500–1508.
81. Trinchieri, G., de Marchi, M., Mayr, W., Savi, M. and Ceppellini, R. (1973). *Transplant. Proc.*, **5**, 1631–1646.
82. Perlmann, P. and Holm, G. (1969). *Advan. Immunol.*, **11**, 117–193.
83. Trinchieri, G. and de Marchi, M. (1975). *J. Immunol.*, **115**, 256–260.
84. MacDonald, H. R. and Bonnard, G. D. (1975). *Scand. J. Immunol.*, **4**, 129–138.
85. Dickmeiss, E. (1974). *Scand. J. Immunol.*, **3**, 817–821.
86. Golstein, P. and Fewtrell, C. (1975). *Nature*, **255**, 491–493.
87. Garovoy, M. R., Strom, T. B., Kaliner, M. and Carpenter, C. B. (1975). *Cell. Immunol.*, **20**, 197–204.
88. Trinchieri, G. and de Marchi, M. (1976). *J. Immunol.*, **116**, 885–891.
89. Yust, I., Smith, R. W., Dickler, H. B., Wunderlich, J. and Mann, D. J. (1975). *Cell. Immunol.*, **18**, 176–186.
90. Todd, R. F., III and Berke, G. (1974). *Immunochemistry*, **11**, 313–320.
91. Cerottini, J.-C., MacDonald, H. R., Engers, H. D. and Brunner, K. T. (1974). In *Progress in Immunology II*, Vol. 3 (Brent, L. and Holborow, H., Eds.), North-Holland, Amsterdam, pp. 153–160.
92. Engers, H. D. and MacDonald, H. R. (1976). In *Contemporary Topics in Immunobiology*, Vol. 5 (Weigle, W. O., Ed.), Plenum Press, New York, pp. 145–190.
93. MacDonald, H. R., Sordat, B., Cerottini, J.-C. and Brunner, K. T. (1975). *J. exp. Med.*, **142**, 622–636.
94. Cerottini, J.-C., Engers, H. D., MacDonald, H. R. and Brunner, K. T. (1974). *J. exp. Med.*, **140**, 703–717.

Chapter 15

T and B Cells in Cancer

N. A. MITCHISON

Abbreviations

CEA: carcinoembryonic antigen
EB: Epstein Barr
OFA: onco-foetal antigen
TSTA: tumour-specific transplantation antigens

1. INTRODUCTION

The central theme that has inspired tumour immunology through the years has been the hope that tumours can be controlled by immunological methods. The theme has its roots in the likelihood that immunological control, being the natural mechanism of control, would also be complete and permanent. This hope has not yet been realized. In the meanwhile, immunological methods are contributing usefully to the exploration of other topics in the biology of cancer, and cancer biology is in turn contributing to the exploration of the immune system. These contributions are reviewed in this chapter.

Section 2 surveys the achievements of immunotherapy. Specific immunotherapy, the raising of a protective immune response to antigens of the tumour cell surface, have been obtained with regularity only for a restricted range of tumours, namely those which have been induced in animals with strong carcinogens and oncogenic viruses. Non-specific immunotherapy, the use of adjuvants to enhance the response to these antigens, offers at present greater promise in the treatment of cancer in man. However, some at least of the forms of treatment which are practised as immunotherapy may not in fact operate via the immune system. New forms of treatment have been developed in animal systems and are expected to reach clinical trial before long.

Section 3 surveys the responses of T and B cells in cancer patients, and in animals bearing tumours. *In vitro* assays have been developed to the point where they can often detect T cell responses, but the relevance of these responses to tumour destruction *in vivo* is far from clear. On the way, useful information has been obtained about the nature and control of *in vitro* responses to cell surface antigens.

Section 4 surveys the present state of the theory of immunosurveillance. Surveillance, if it applies to tumours at all, seems to do so only for a restricted range of tumours. The bulk of the evidence comes from studies of immunosuppression, although the lack of association between immune response genes and cancer is also pertinent.

Section 5 surveys the use of antibodies to monitor tumour growth. The best examples of their use are in assays of carcinoembryonic antigen to monitor carcinoma of the large bowel, and of alpha-foetoprotein to monitor hepatoma.

Section 6 surveys the use of tumours to sample the immune system. This concerns mainly the definition of leukaemia and lymphoma cell phenotypes by means of secretory products and surface markers. Information acquired on the way tells us much about the workings of the immune system; it can also be used to classify disease, to guide therapy and to predict relapse.

In Section 7 the possibility is discussed that the products responsible for malignant transformation of cells can be recognized as antigens. Candidate tumour specific surface antigens of this type have been detected in avian virus systems.

2. IMMUNOTHERAPY

2.1 Specific immunotherapy

Modern immunology starts with the work of Foley[1] and Prehn and Main[2] which first established that chemically induced tumours could be prevented from growing in otherwise susceptible mice by immunization. If the immunization is performed in adequately inbred animals, the response is directed solely at tumour-specific transplantation antigens (TSTA). Although similar conclusions had been drawn from previous experiments, they were not generally accepted because of the confusion between TSTA and alloantigens.[3] The large number of animal experiments which stem from this work have been extensively reviewed and certain generalizations emerge.[4–6]

The antigens which operate in transplantation experiments do not generally cross-react between different chemically-induced tumours. The minimum number of unique antigens has not been established with any precision, and there are indications of limited cross-reactions.[6–9] Nevertheless, the number of different antigens is large enough to present a formidable problem in molecular biology. One possibility is that the unique antigens may be modified alloantigens, and somehow draw on the genetic information which is available for specifying alloantigens.[10,11] Spontaneous tumours may also display unique antigens, but these tend to be weaker than those borne by chemically-induced tumours. In general, those carcinogens which induce tumours most rapidly also generate tumours with the strongest TSTA; and within a range of tumours induced by the same carcinogens, those which appear earliest are the most immunogenic.

Tumours induced by viruses (e.g. simian virus 40,[12] murine leukaemia virus,[13] murine sarcoma virus,[14] feline leukaemia virus,[15] Epstein Barr virus[16]) share TSTA and/or surface antigens defined by antibodies. The extent of the cross-reactions between these antigens defines virus type-specific and virus group-specific antigens. These virus-induced antigens may or may not be virion proteins or glycoproteins, but all cross-reactive TSTA induced by virus are generally assumed to be virus-coded. Indeed the cross-reactions which are encountered between human tumours are often assumed to be viral in origin, in the absence of evidence to the contrary. The relationship between virus-coded antigens and malignant transformation is discussed separately below. Even though virus-induced tumours share antigens, they may also carry unique antigens which can be detected under special circumstances, e.g. when the host has been rendered tolerant of the viral antigens.[17]

Not all antigens on the surface of tumour cells operate as TSTA. An important category of shared cell surface antigen is the onco-foetal antigens (OFA) or embryonic antigens. These antigens can be detected by sera raised in multiparous females or by deliberate immunization with irradiated foetal tissue.[4,18] OFA either fail to operate as transplantation antigens, or need special techniques to demonstrate their activity in this respect. The reason why they normally fail to do so is puzzling, particularly since they can be effective targets for *in vitro* cytotoxicity tests, as discussed separately below. One possibility is that the reaction against them is blocked particularly effectively by free antigen present in serum.

The mechanisms which operate on TSTA to prevent survival of the transplant are essentially the same as those which operate on alloantigens.[19] Rejection normally takes place under the direction of T cells, although macrophages can be shown to play an important part in the final stages of rejection under certain circumstances.[20] Antibodies may also play a part, particularly in immunity to leukaemias and in the control of metastasis.

So clear is the evidence for T cell mechanisms, particularly cytotoxic T cells, that it has been suggested that B cell responses may impair the effectiveness of anti-tumour immunity via 'enhancement'.[21] This line of argument has become less convincing now that antigen seems to be more important than antibody in acting as serum blocking factor.[4,22] Indeed, production of antibody may actually favour the activity of cytotoxic T cells by helping to clear antigens from serum.[22]

Under circumstances where excellent immunity can be obtained against new transplants of a tumour it may still be difficult to prevent growth of an established tumour. Sometimes concomitant immunity can be demonstrated, i.e. an established tumour grows progressively but the growth of fresh transplants in other parts of the body is suppressed.

When one turns from inbred animals to man matters become much less satisfactory. Specific immunization against the antigens of tumours is, after all, the hub of cancer immunotherapy, and has been the goal of an enormous amount of painstaking clinical work. In spite of all this effort, there is no single instance where one can assert with any confidence that specific immunotherapy has been achieved. Yet one hesitates to draw the conclusion that human tumours lack rejection antigens. The difficulties of clinical investigation are immense. No study which has yet been undertaken satisfies all the requirements of an adequate test of specific immunotherapy.

2.2 Non-specific immunotherapy

Non-specific immunization is the equivalent in tumour immunology to the use of adjuvants in general immunology. In fact most of the adjuvants which are active for tumours were introduced first as general immunological adjuvants. In view of the importance of T cells in tumour immunity, the prospects are better for agents which favour differential activation of T cells. The principal agents which have been shown to enhance the response to TSTA in animal experiments are B.C.G.[23,24] and its methanol extract residue (MER),[25] *Corynebacterium parvum*,[26] synthetic polynucleotides[27] and levamisol.[28] Tubercle bacilli, the oldest and best-established member of this list, have for long been known to favour what are now termed T cell responses: their use goes back to the Dienes phenomenon, whereby the injection of a protein antigen into a tuberculous lesion produces delayed-type hypersensitivity.[29] In all systems which have been adequately tested the B.C.G. has to be located in the immediate vicinity of the tumour cells in order to obtain an optimum effect.[24,30]

The use of bacterial vaccines to obtain tumour immunotherapy in man goes back to Coley's vaccine.[31] This was an entirely empirical development in which bacterial cultures were inoculated into a wide range of human tumours. Much the same applies to Klein's uses of skin sensitizing against compounds such as fluorodinitrobenzene to treat widespread skin tumours in man.[32] The use of bacterial vaccines for treating leukaemia was introduced by Matthe *et al.*[33] and a series of careful and extensive trials of their use is now under way.[34,35] The results thus far are encouraging: in the trial organized by the Southeastern Cancer Study Group (USA), for example, the medium duration of remission was 45 weeks for the B.C.G./methotrexate group and 30 weeks for the methotrexate only group ($p < 0{\cdot}04$).[35]

From an immunological standpoint the difficulty in evaluating this kind of result is to know how much of the effect of the bacterial vaccine is immunological in nature. This applies with special force to the leukaemia trials, where it has been suggested that the B.C.G. acts rather as a 'bone marrow tonic' than upon the immune system.

2.3 New approaches to immunotherapy

2.3.1 *Heterogenization*

It might help the immune response to a weak antigen, such as a tumour-specific antigen, to have the antigen presented to the immune system in association with a strong antigen. Associative recognition could enhance the response through T–B or T–T cell collaboration. This hope has inspired a considerable amount of experimental work, particularly on associative responses between tumour and viral antigens. Tumour immunotherapy, according to these principles, with viruses has been the subject of several recent studies.[36–39] For a variety of viruses, including influenza, Newcastle disease, vesicular stomatitis, Friend leukaemia, and Semliki Forrest, membranes from infected tumour cells have proved more immunogenic than membranes from non-infected cells. They have not, however, proved more so than intact, irradiated tumour cells. It has been difficult to eliminate the alternative explanation that viral polypeptides may serve to stabilize the tumour specific antigens against degradation, although recent work has gone some way towards doing so.[36,37]

Similar experiments performed with hapten-conjugated cells have yielded less encouraging results,[40] as might have been expected from the inefficiency of hapten-specific helper T cells in tests with purified antigens.[41,42] Chemical derivatization of the tumour cell membrane has also been explored from a different point of view, namely that it may prove possible to alter antigenic determinants selectively so as to minimize the B cell response and maximize that of T cells.[43] This approach is based on studies of purified protein antigens where appropriate derivatization has had such an effect.[44,45]

The results which have been obtained so far are sufficiently encouraging to warrant further work of a more incisive character, possibly involving responses to cell surface alloantigens which are open to better characterization than is available for tumour antigens. The crucial question whether T–T cooperation depends to any appreciable extent on physically-linked antigenic determinants has not yet been entirely answered; as regards cell surface antigens the evidence on the whole suggests that a link is not necessary, a disappointing conclusion from the present viewpoint.

2.3.2 *Linkage of antibody to drugs*

Several reports have appeared of synergy between chemotherapeutic drugs and antibodies.[46–48] A more interesting matter is whether drugs can be covalently linked to antibodies and thereby have their efficiency increased beyond what could be expected of the mixture. Both phenylenediamine mustard[48] and daunomycin[145] have been linked to antibody with success in this way. These claims need to be treated with some scepticism until more is known about how these conjugated drugs succeed in reaching their site of action within the cell. Nevertheless, this kind of approach is potentially valuable, particularly because it can be applied with xeno-antibodies, e.g. antibodies to human leukaemia-specific antigens raised in rabbit.[49]

3. RESPONSES OF T AND B CELLS IN PATIENTS WITH CANCER

3.1 Responses of T cells

3.1.1 *The microcytotoxity test*

T cells from tumour-bearing individuals have been subjected to various tests in order to detect responses to the antigens of the tumour. The methods which have been used most extensively are based on the colony inhibition test of the Hellströms;[50,51] a version which has been much employed is the microcytotoxicity test, in which tumour cells are seeded as targets into the well of plastic dishes, where they are incubated for several days with lymphocytes and their viability then assessed.[52] The test thus measures the activity of the lymphocytes in killing target cells and in inhibiting their growth. By this and similar procedures it has been claimed that cytotoxic lymphocytes can be detected regularly in the blood of patients bearing a variety of tumours.[53–56] Cytotoxic lymphocytes also develop in rodents transplanted with chemically or virally induced tumours.[4] Difficulties have been encountered in confirming and extending this work in patients, mainly because lymphocytes from normal individuals sometimes prove active, so that the difference between patients and normals is reduced.[57,58]

The matter has not been entirely resolved, but certain points emerge from recent surveys:[59–61]

(1) Antigens detected by *in vitro* cytotoxicity tests are not all relevant to the growth of the tumour *in vivo*. This is illustrated by certain types of chemically-induced tumours in the rat which are weakly immunogenic *in vivo*. As judged by *in vitro* tests, these tumours generate cytotoxic lymphocytes as effectively as other tumours which are highly immunogenic *in vivo*.[62] These ineffective lymphocytes react against onco-foetal antigens, which make ineffective targets *in vivo* possibly because of the ease with which they are shed from the cell surface.

(2) The problem of high level activity in normal individuals is genuine. It can be overcome, in part, by using a large panel of normal controls, but the differences have become less clear-cut than had been hoped.

(3) There are technical problems in the preparation of both lymphocytes and tumour cell targets. Many investigators prepare lymphocytes from heparinized blood by flotation on Ficoll–Hypaque gradients, but this procedure has been questioned on the grounds that it generates non-specific activity. Tumour cell-lines have the great advantage over freshly prepared tumour cells that they are permanently available in a reproducible form. However, they are now being abandoned in most studies because of their poor performance as targets, although the human bladder carcinoma cell-line T24 may be an exception.

(4) More than one type of effector cytotoxic mechanism operates in the microcytotoxicity assay. When adequate fractionation procedures have been applied—passing lymphocytes through immunoglobulin-coated columns, or preparative enrichment by the sheep erythrocyte rosette method—both T cell and antibody-dependent non-T cell killing can be demonstrated.

(5) Microcytotoxicity can be blocked by serum blocking factors. These were at first thought to be antibody to antigens of the tumour cell surface, but more recent studies indicate that it is antigen or antigen–antibody complexes which can usually be demonstrated. The rise and fall of these blocking factors can be followed during the course of immunotherapy trials. In animal studies they have proved useful in analysing the specificity of T cell-mediated cytotoxicity, and it is reasonable to hope that they will find a use in standardizing microcytotoxicity measurements.

3.1.2 *The short-term isotope release test*

Assays of cytotoxic T cells in which the cells are incubated for a few hours with ^{51}Cr-labelled target cells have been much used for study of the allogeneic response. They are simpler and more reliable than long-term incubation assays, but less sensitive. They have been successfully applied in a few animal tumour systems, particularly in the response of rats and mice to tumours induced by murine leukaemia virus (MuLV).[63,64] These assays have enabled the kinetics of production of cytotoxic effector cells to be analysed in great detail. Both mice and rats have naturally-occurring cytotoxic T cells able to react specifically with the virus-controlled antigens, presumably as a consequence of harbouring endogenous viruses. Cytotoxicity attains very high levels during the regression of virus-induced tumours. The peak of activity is quite brief, and is followed by a longer period during which T cell activity is shut off (a) by blocking with antigen and (b) by suppressor cells which are discussed separately below. During this latter period non-T cell cytotoxic mechanisms operate.[65]

3.1.3 *Non-cytotoxicity tests*

Two other standard assays of T cell function have been applied to lymphocytes obtained from tumour bearers. The first of these is the proliferation assay. It has been developed in the form of an autologous tumour stimulation test, in which incorporation of ^{3}H-thymidine is measured during a 6-hour pulse at the end of a 6-day period during which human lymphocytes and tumour cells (or tumour extract) are incubated together.[66] The test is sensitive and reproducible, and comparable proliferation tests have been performed in animal virus tumour systems.[64]

The second standard assay of T cell function which has been applied to tumours is the migration inhibition assay.[67] Although the outcome of this test appears to correlate to some extent with tumour-bearer status, there is a good deal of variability. The test has not been claimed to be particularly sensitive.

Three other *in vitro* tests belong to a different category. All are claimed to detect cellular immunity in tumour bearers, and have indeed been designed for this purpose. Each has reached the stage where, after having been developed and tested by one group, it is ready for test by others. In each case there is some question whether the test in fact measures an immunological function rather than a non-immunological symptom of chronic disease. The tests in question are (a) the leukocyte adherence inhibition (LAI) test,[68] (b) the structured cytoplasm (SCM) test[69] and (c) the macrophage electrophoretic migration (MEM) test.[70] They should be used with caution until properly validated.

3.1.4 *Antigens recognized by T cells*

It has been proved difficult to determine precisely with which antigens of the tumour cell surface T cells react. Our ideas on this subject have been much changed by the discovery that cytotoxic T cells usually, and perhaps always, react with an interaction product between a foreign structure and the major transplantation antigens.[71] This principle, which was first encountered with cells infected with non-transforming viruses, has recently been verified in the murine sarcoma/leukaemia virus system.[146] The crucial experiment has been successfully performed of immunizing F_1 mice with viral leukaemias originating in the two parental strains, and demonstrating that the two types of tumour generate distinct (although slightly overlapping) populations of cytotoxic effector cell. These experiments with T cells are backed up by co-capping and other serological data which indicate that the antigenic determinants controlled by the virus and by the H-2 D and K end genes are carried on the same macromolecule.[72] Both transplantation tests and serological data[10,11] indicate that the same rules apply to the individual antigens of carcinogen-induced tumours in mice and rats, as has been discussed above.

Acceptance of these rules does not detract from the need to know what contribution other structures make to the antigenic determinant recognized by T cells. This has been analysed in some detail in the case of rats and mice reacting to syngeneic RNA virus tumours.[64] T cell specificities have been assigned to viral polypeptides, as judged by the following criteria: *in vitro* stimulation, inhibition of cytotoxicity, inhibition by antisera of serum blocking factors, and purification of T cells on gels.

3.1.5 *Enrichment of reactive cells*

An increase in the efficiency of reactive T cells has been obtained by a variety of procedures. Some of these aim at removing blocking factors (antigen or antigen–antibody complexes) from the effector cell surface. Others aim at removing non-specifically active cells. Still others aim at increasing the frequency of specifically activated effector cells. Blocking factors can be removed by pre-incubation in the absence of autologous serum, repeated washing of lymphocytes, or trypsinization.[55,73]

Non-specifically activated cells can be depleted by removing lymphocytes bearing complement receptors, by means of a rosette technique.[74,75] Another procedure which can dissect the reactivity of different effector cell populations is the use of media which have been depleted of Mg^{2+}.[76]

A solid tumour may act as a sponge, soaking up specific effector cells. Accordingly, efforts have been made to isolate effector cells from tumour tissue. Active lymphocytes have been successfully isolated from experimental[77] and human[78] tumours.

In vitro education of lymphocytes in the presence of tumour cells generates an extremely effective population of effector cells in certain experimental systems.[79–81] Application of the same procedure to human sarcomas has led to the production of non-specific effector cells,[82] possibly through mechanisms of T–T cooperation, briefly discussed above.

3.2 Responses of B cells

Antibodies to the antigens of certain tumours can be detected in tumour-bearing individuals, or normal individuals of the same species. The clearest examples are antibodies to viral antigens, such as those of EB virus in man, feline leukaemia virus, or the rodent leukaemia viruses. These are discussed separately below.

There are certain less well defined groups of human tumour antigens. One of these are the melanoma-associated antigens. Reactions of sera from melanoma patients have been reported with intracytoplasmic constituents of various melanoma cells, using mainly immunofluorescent methods.[83–85] Later reports question the specificity of these reactions, for sera from normal individuals appear to detect the same antigens.[86]

The second group of antigens are those associated with sarcoma. These antigens have been detected by various techniques including immunofluorescence and complement dependent lysis.[87–89] The antigens are apparently found only in fresh sarcoma cells and in cell-lines derived from sarcomas. Antibodies have been detected in a high percentage of sera from patients with various kinds of connective tissue neoplasm, and their relatives and associates. Transmission (for one passage) has been reported using cell-free extract *in vitro*.[89] This antigen may well be of viral origin.

It might be asked why these antibodies are not more effective in preventing tumour growth. The problem has been opened to experimental investigation by the recent discovery that the leukaemias of AKR mice can be successfully treated by infusion of serum, almost certainly because this replaces a deficiency of complement in the tumour bearers.[90] The treatment is hazardous, and has not yet been attempted in human patients.

3.3 Non-specific suppression of the immune response

There are numerous reports of suppression of T and B cell responses to mitogens in patients with cancer,[91,92] and of loss of delayed hypersensitivity reactions in patients with Hodgkin's disease.[93] T and B cells from laboratory animals with transplanted or virus-induced tumours are often hyporesponsive to mitogens.[94–96] Serum factors have been demonstrated, and recently a population of suppressor cells has also been shown to operate.[94,95,97] The suppressor cells have characteristics of B cells[94] and macrophages.[95] Suppressor cells have been demonstrated with similar properties for the allogeneic responses.[98]

4. IMMUNE SURVEILLANCE

4.1 The theory and its supports

The theory of immune surveillance postulates that a large number of tumours occur which are suppressed by the T cell system before they have grown to detectable size. This is an attractive idea because it offers a plausible account both for the presence of specific antigens on the surface of tumour cells, and for the evolution of the T cell system. It was originally introduced in order to account for the high incidence of cancer which is found among immunosuppressed individuals, particularly the recipients of kidney transplants.[99] It has given rise to a large amount of experimental work, which has been the subject of recent reviews.[100,101] Recently it has been suggested that the immune response may even have the contradictory effect of stimulating tumour growth.[102] The most cogent evidence comes from studies on immunosuppression.

The incidence of cancer is raised among individuals who are immunosuppressed, either congenitally or as a result of treatment. The nude mouse is particularly favourable material for this kind of inquiry, and it has now been established that the incidence of cancer does rise if the mice can be kept alive for a sufficiently long period in a germ-free colony (Reference 103, but see also References 104 and 105). However, both in man and mouse the type of tumour which arises is not what the surveillance theory would predict. Most of the tumours which occur belong to a narrow range of lymphoreticular tumours, and do not include the common types of carcinomas. It seems likely therefore that these tumours result either from the special conditions to which lymphoid cells are subject during immunosuppression, or possibly from viral infection.

Immune surveillance may apply only to viral oncogenesis. Nude mice and mice immunosuppressed by anti-lymphocyte serum are particularly susceptible to oncogenic viruses.[106,107] Marek's disease, a leukaemia of chickens induced by chicken herpes virus, can be controlled by immunization of susceptible chickens with turkey herpes virus.[108] Feline leukaemia can also be controlled by immunization against virus-controlled antigens under experimental conditions (W. Jarrett, unpublished data).

A qualified case can be made out for immune surveillance of Epstein Barr viral antigens.[16,109] EB virus transforms B cell-lines *in vitro*, and it is likely that the B cells which proliferate in infectious mononucleosis, a disease caused by the virus, have undergone a similar transformation. Normally this disease is effectively controlled by the immune response, just as the analogous proliferative disease induced by *Herpes saimirii* is controlled in the normal host of this virus, the squirrel monkey. According to the surveillance theory, Burkitt's lymphoma develops when this type of control does not operate effectively. Epidemiological studies have shown that the incidence of Burkitt's lymphoma correlates closely with malaria, and so it has been suggested that malaria may interfere with effective surveillance. In the same way, it gives rise to lymphoma when inoculated outside its normal host range into Owl monkeys, apparently because the immune response in this species is less well adapted to suppressing virus-transformed cells.

This attractively simple view of the cause of Burkitt's lymphoma has encountered two major difficulties. One is that the incidence of the disease ($\sim 10^{-5}$) even in the most susceptible areas of Africa is far lower than the incidence of either EB virus or malaria infection (both of which are practically ubiquitous). The other is that a high proportion of the lymphoma carry a chromosomal defect which has been identified as a translocation from chromosome 8 to 14.[110] This defect is not confined to lymphomas of this particular type, and does not occur in the B cell-lines transformed *in vitro* by EB virus. It follows therefore that some additional event, which results in the translocation, must occur in addition to infection with the virus and malaria. The nature of this crucial event is not understood.

4.2 Immune response genes and cancer

Immune response (Ir) genes probably play an important part in determining the transmission of oncogenic viruses. Susceptibility to mammary tumour virus and to murine leukaemia viruses is linked to H-2 in the mouse.[111,112] In both cases it is likely but not certain that the gene involved in the H-2 locus is Ir-1; the locus in question controlling susceptibility to leukaemia virus is called Rgv-1 pending resolution of this question. No doubt Ir genes are also important in determining species susceptibility to oncogenic viruses, e.g. the differences encountered among related monkeys species in the susceptibility of the newborn to *Herpes saimirii.*[113] Hence the normal host species is not susceptible to oncogenesis, perhaps because of past selection of appropriate Ir genes.

Despite claims which have been made from time to time, no instance of association between the occurrence of any form of human cancer and HLA has been established.[114] There may be an association between HLA type and the occurrence of nasopharyngeal carcinoma;[115] if so, this disease which is thought to be carried by EB virus would conform to the rule established in animal work that susceptibility to oncogenic viruses is under Ir control. On the other hand, an association has been found between HLA-type and overall survival in Acute Lymphoblastic Leukaemia[116,117] and Acute Myeloid Leukaemia.[118]

The idea that viruses to which the species is normally exposed select Ir genes receives support from studies made on naturally occurring cytotoxic cells. The occurrence of cells with specificities for murine leukaemia virus antigens appears to be under genetic control in the mouse.[119]

5. THE USE OF ANTIBODIES TO MONITOR TUMOUR GROWTH

Antibodies raised against tumour products which enter the bloodstream can be used to monitor tumour growth. One such product is carcinoembryonic antigen (CEA).[120,121] This high molecular weight glycoprotein is released mainly from carcinomas of the large bowel and pancreas, and to a lesser extent from inflammatory tissue of the gut and urinary tract. Sensitive radioimmuno-assays have been developed which enable the level of CEA in serum to be measured with precision. Unfortunately, the level of false positives is too high for the CEA test to be useful for screening purposes, but it is likely that it will find a useful place as a monitoring test.

Much the same applies to alpha-foetoprotein, which reaches high levels in the serum during hepatic carcinoma.[122] The most interesting recent development in this connection is a screen for alpha-foetoprotein in serum which is being run in China, which has already involved half a million people (reported at the XI International Cancer Congress, Florence, 1974).

Antigens associated with other types of tumour have been described, e.g. for carcinoma of the lung,[123] ovary,[124] and for astrocytoma.[125] These have not been developed as diagnostic tests nearly as far as CEA or alpha-foetoprotein, and it is not at present clear how far this type of approach can be pressed with other tumours.

6. TUMOURS AS MONOCLONAL SAMPLES OF THE IMMUNE SYSTEM

In a mixture of cells as complex as the immune system, tumours have an important contribution to make in helping to sort out cell types. Since tumours arise from single cells, they provide a homogeneous source of cells of a single type. Furthermore, tumours which occur in inbred strains can be passaged indefinitely and provide an unlimited supply of homogenous material. If a sufficiently large number of tumours are examined one can hope in this way to sample the entire immune system, and to a certain extent one can hope that the frequency of various types of tumour will reflect the frequency of their cells of origin.

The value of this approach has been most clearly shown with the plasmacytomas, particularly since the introduction of mineral oil as a reliable method of inducing this type of tumour in inbred mice.[126] An enormous amount of information has been gleaned from study of these tumours and their immunoglobulin products. Indeed, one can safely say that most of our information concerning the genetics and structure of immunoglobulins has been acquired through the use of plasmacytomas. In particular, the generalization based on these tumours that a single cell normally makes a single immunoglobulin product inspires the hope that T cell leukaemias will similarly contribute to our understanding of the T cell system.

Lymphomas are available in the mouse with both T and B surface markers.[127] An example of the use of T cell lymphomas is provided by recent work on an immunoglobulin-like product which some cell-lines secrete.[128] This material suppresses thymus-dependent responses *in vitro*, but has no effect on thymus-independent responses. It is believed to inhibit competitively the binding of complexes of T cell receptor and antigen to the macrophage surface. Since the T cell receptor has itself proved hard to characterize, this receptor-like material from tumour cell-lines may have a useful contribution to make to the chemistry.

6.1 Surface markers on human lymphoid tumours

Surface markers on human lymphoid tumours have important practical applications, in addition to their interest in defining samples of the human immune system. They can be used to refine the classification of leukaemias and lymphomas in terms of their probable origins. These refinements are likely in the future to guide the choice of therapy. Furthermore, markers able to distinguish malignant from normal cells may be used to predict relapse of leukaemias. For review and detailed references on human cell surface markers, see Reference 129.

Certain reservations about the use of tumour cells should be remembered. One is that tumours may display an abnormal phenotype, in which individual markers may be lost and unusual combinations of markers occur. Another is that a single patient may occasionally harbour more than one malignancy. A third, and probably the most important reservation is that tumours may display a condition of balance in which their cell population is in dynamic equilibrium between rapidly dividing stem cells and more slowly dividing daughter cell-lines. When this occurs the daughter cell-lines may display a different phenotype from the stem-line. This applies, for example, to B cell-lines transformed by EB virus, in respect of membrane antigen,[16] and to teratomas of the mouse in respect of the C, H-2 and Thy-1 antigens.[130] A somewhat similar situation exists in Chronic Lymphocytic Leukaemia (CLL) where the cells which appear in the blood during blast crisis have a surface phenotype different from those of the chronic disease. For all these reasons, the desirability of using panels of markers to determine the phenotype of leukaemias should be emphasized.

The most important markers for T cells in man are T lymphocyte antigen (HuTLA, probably the equivalent in man of the theta-equivalent recognized by rabbit antisera in the mouse), and E rosette receptor. For B cells they are B lymphocyte antigen (HuBLA), surface membrane immunoglobulin (SmIg), C receptors (probably a family of receptors for different complement components), and Fc receptor; the last two of these markers are present in small numbers on the surface of certain T cell subsets. The receptors for EB virus also provide a useful marker for human B cells, as discussed in the next section.

By the use of these surface markers, leukaemias in man have been divided into three main groups. One of these comprises a minor fraction (approximately one-quarter) of Acute Lymphoblastic Leukaemias (ALL): these show the T cell surface markers. A second group comprises CLL, hairy cell leukaemia, and plasma cell leukaemia, which carry B cell surface markers. A third group lack both T and B markers. This includes the major fraction of ALLs, which possess a unique leukaemia-specific surface antigen.[44] Lymphomas similarly include a class of those which carry T cell markers which includes lymphoblastic lymphoma, Sezary's syndrome, and thymoma. Those which carry B cell markers include Burkitt's lymphoma, some reticulum cell sarcomas and some lymphoblastic lymphomas, lymphocytic lymphoma, and multiple myeloma.

As regards the representation of subsets of T and B cells among tumours, the fullest information concerns the immunoglobulin products of plasmacytomas. Both in man and the mouse representative plasmacytomas are known for all the immunoglobulin classes and subclasses. Furthermore, there is a correlation between the frequency of the class in normal and malignant cells, e.g. IgE-secreting myelomas are known, but are extremely infrequent. It is clear also that certain secretors of classes of immunoglobulin are represented abnormally frequently among the malignancies, e.g. IgA-secretors among the tumours induced in mice by intraperitoneal injection of mineral oil. Much the same applies to the representation of antigen-combining sites and idiotypes: binding activity for the DNP group, for example, is probably present more often than would be expected of a strictly random malignant transformation process.

Representatives of the major T cell subsets have not yet been described among mouse or human T cell leukaemias. It is reasonable to expect that representatives of the T_H, T_{CS} and T_E subsets will be found with the appropriate Ly surface markers in the mouse (see P. C. L. Beverley, herein). They would prove valuable for absorption of antisera.

6.2 Selective tropisms of oncogenic viruses

The type-C RNA tumour viruses and the herpes viruses cause tumours of T and B cells in a large number of species. All the host–virus combinations which have been adequately studied are highly selective in that the tumours are predominantly of T or B cell type. In the mouse, for example, Moloney leukaemia virus and Gross virus (type-C viruses) induce tumours which originate in the thymus[131] and which carry the Thy-1 antigen.[132] Another type-C virus, Abelson virus, causes a lymphoma or leukaemia which does not involve the thymus;[133] some but not all of these tumours carry surface Ig.[134] The two groups of viruses therefore respectively cause tumours predominantly of T and B (or pre-B) cells.

Among the herpes viruses, EB virus transforms B cells, and Burkitt's lymphoma cells carry B cell markers.[135,136] On the other hand, *Herpes saimirii* virus in Owl monkeys causes tumours which carry T cell markers,[137] as does Marek's disease herpes virus in the chicken.[138]

The basis for this selectivity is not fully understood. Both C-type and herpes viruses do not grow solely in the tissues which they transform: indeed, the tissues which are the sites of transformation seem to represent an abnormal habitat for the viruses. In the case of EB virus an interesting relationship has been discovered with B cells. These cells have a receptor for the virus which T cells lack, and which may be identical with their C receptor. G. Klein suggests that this receptor may be involved in normal triggering of B cells in the immune response, and that transformation by the virus may therefore utilize a pathway shared in common with normal triggering. Whatever the role of the B cell receptor for EB virus may be, it certainly provides a useful marker for identifying B cells.[129]

Foetal liver cells can be transformed *in vitro* by Abelson virus.[139] This is the first instance of transformation in a completely *in vitro* system by a leukaemic virus, and it opens the way to further analysis of the selective basis of transformation.

7. CELL SURFACE ANTIGENS AND THE TRANSFORMED CELL PHENOTYPE

A simple idea about the origin of cancer runs as follows. Malignant transformation depends on an alteration of the cell surface. As a consequence of this alteration the cancer cell cannot perceive the signals from its environment which would otherwise control growth, division and invasiveness.[140] The alteration can itself be detected as a change in the reactions of the cell with lectins: receptors for lectins on the cell surface are either gained, lost or altered in their architecture.[141] Other ways in which the alteration can be detected are by the appearance of new antigens, and by the appearance of new membrane proteins which can be detected by electrophoresis. One of the attractions of this idea is that it suggests straightforward strategies for cure: cancer cells could be detected and eliminated by antibodies against the new antigens, or they could be persuaded to revert to normal by coating the altered cell surface receptors with the appropriate lectin.

Although this idea was popular a few years ago, it is no longer tenable in its simplest form. Differences between the membrane proteins of normal and transformed fibroblasts grown *in vitro* have been analysed in detail, with special attention being directed at the temperature-sensitive transformants which can be made to revert to a normal pattern of growth by raising the temperature. This analysis has so far failed to disclose new protein bands characteristic of the transformed phenotype.[142] The suggestion has therefore been made that malignant transformation may depend rather on an alteration—perhaps an uncoupling—of the connection between sub-membrane assemblies and the cell surface (G. Edelmann, personal communication).

So far as surface antigens characteristic of transformation are concerned, only certain kinds of information are relevant. The unique antigens of the chemically induced tumours for example are probably not, simply because their very variety makes it unlikely that they all affect the same surface receptor functions. Another argument which is too inconclusive to carry much weight concerns the lack of antigens on some chemically induced and many spontaneous tumours. If a complete lack of tumour specific antigens could be definitely established even in a few instances of malignant transformation, this would argue strongly against the concept of an obligatory alteration of the cell surface. The problem is that it can always be argued that a change has occurred which falls below the level of detection by the immune system.

Accordingly, more attention has been paid to virus-controlled antigens as candidates for transformation-specific products. For the SV40/polyoma viruses both the nuclear T antigen and the surface S antigen are possibilities. The problems are that the T antigen is present not only in transformed cells but also briefly during productive infection,[143] while the S antigen may be present in masked form on non-transformed cells.[12] For EB virus the strongest candidate is the nuclear antigen (EBNA), which fulfils most of the qualifications for a transformation-controlling product.[16] This antigen however has no obvious cell surface counterpart, and it is entirely unclear how the product defined by this antigen is linked to events at the cell surface.

For the RNA virus two kinds of non-virion, virus-induced antigens have been detected on the cell membrane. One of these is present on both transformed cells and infected, non-transformed cells; this category is exemplified by the feline oncornavirus-associated cell membrane antigen (FOCMA).[15] The second category is present on transformed but not on infected, non-transformed cells, and has therefore been defined as tumour specific surface antigens. These antigens have been well characterized as a 100 000 m.w. surface polypeptide with group-specific antigenic specificities in the chicken;[144] similar antigens have been detected in mice, where they display a spectrum of related specificities.[14] Thus far, these antigens appear to be specifically associated with malignant transformation. One should be cautious in drawing the further conclusion that they are dually responsible for maintenance of the transformed state, even though they may be the best available candidate for this role. Before that conclusion could be accepted, one would need to know a good deal more about the part they play in cell physiology.

8. REFERENCES

1. Foley, E. J. (1953). *Cancer Res.*, **13**, 835–837.
2. Prehn, R. T. and Main, J. M. (1957). *J. Nat. Cancer Inst.*, **18**, 769–778.
3. Woglom, W. H. (1929). *Cancer Rev.*, **4**, 129–214.
4. Baldwin, R. W. (1973). *Advan. Cancer Res.*, **18**, 1–75.
5. Mitchison, N. A. (1975). In *Clinical Aspects of Immunology* (Gell, P. G. H., Coombs, R. R. A. and Lachmann, P. J., Eds.), Blackwell Scientific, Oxford, pp. 599–621.
6. Old, L. J., Boyse, E. A., Clark, D. A. and Carswell, E. A. (1962). *Ann. N.Y. Acad. Sci.*, **101**, 80–106.
7. Baldwin, R. W., Glaves, D. and Pimm, M. V. (1971). In *Progress in Immunology* (Amos, B., Ed.), Academic, London, pp. 907–920.
8. Basombrio, M. A. (1970). *Cancer Res.*, **30**, 2458–2462.
9. Reiner, J. and Southam, C. M. (1969). *Cancer Res.*, **29**, 1814–1820.
10. Invernizzi, G. and Parmiani, G. (1975). *Nature*, **254**, 713–714.
11. Bowen, J. G. and Baldwin, R. W. (1975). *Nature*, **258**, 75–76.
12. Häyry, P. and Defendi, V. (1970). *Virology*, **41**, 22–29.
13. Aoki, T. (1974). *J. Nat Cancer Inst.*, **52**, 1029–1034.
14. Aoki, T., Stephenson, J. R., Aaronson, S. A. and Hsu, K. C. (1974). *Proc. Nat. Acad. Sci.*, **71**, 3445–3449.
15. Essex, M., Cotter, S. M., Hardy, W. D., Hess, P., Jarrett, W., Jarrett, O., Mackey, L., Laird, H., Perryman, L., Olsen, R. G. and Yohn, D. S. (1975). *J. Nat. Cancer Inst.*, **55**, 463–467.
16. Klein, G. (1975). In *Oncogenesis and Herpes Viruses*, Vol. 11 (de-The, G., Epstein, M. A. and zur Hausen, H., Eds.), Lyons International Agency for Reseach on Cancer, Scientific Publications, Lyons, pp. 293–309.
17. Vaage, J., Kalinovsky, T. and Olson, R. (1969). *Cancer Res.*, **29**, 1452–1456.
18. Alexander, P. (1972). *Nature*, **235**, 137–140.
19. Cerottini, J.-C. and Brunner, K. T. (1974). *Advan. Immunol.*, **18**, 67–132.
20. Evans, R. and Alexander, P. (1972). *Immunology*, **23**, 615–626.
21. Hellström, K. E. and Hellström, I. (1970). *Annu. Rev. Microbiol.*, **24**, 373–398.
22. Currie, G. A. (1973). *Brit. J. Cancer*, **28**, 25–35.
23. Old, L. J., Clarke, D. A. and Benacerraf, B. (1959). *Nature*, **184**, 291–292.
24. Bartlett, G. L., Zbar, B. and Rapp, H. J. (1972). *J. Nat. Cancer Inst.*, **48**, 245–257.
25. Yron, I., Weiss, D. W., Robinson, E. *et al.* (1973). *Nat. Cancer Inst. Monogr.*, **39**, 33–35.

26. Ghaffar, A., Cullen, R. T., Dunbar, N. and Woodruff, M. F. A. (1974). *Brit. J. Cancer*, **29**, 199–205.
27. De Clerq, E. and Stewart, W. E. (1974). *J. Nat. Cancer Inst.*, **52**, 591–594.
28. Chirigos, M. A., Pearson, J. W. and Pryor, J. (1973). *Cancer Res.*, **33**, 2615–2618.
29. Dienes, L. (1928). *J. Immunol.*, **15**, 141–152.
30. Baldwin, R. W. and Pimm, M. V. (1973). *Brit. J. Cancer*, **27**, 48–54.
31. Coley, W. B. (1911). *Surg. Gynacol. Obstet.*, **13**, 174–190.
32. Klein, E., Holtermann, O. A., Papermaster, B. W., Milgrom, H., Rosner, D., Klein, L., Walker, M. J. and Zbar, B. (1974). *J. Nat. Cancer Inst. Monogr.*, **39**, 229–240.
33. Mathe, G., Amiel, J. L., Schwarzenberg, L., Schneider, M., Cattan, A., Schlumberger, J. R., Hayat, M. and de Vassal, F. (1969). *Lancet*, **1**, 697–699.
34. Powles, R. L., Crowther, D., Bateman, C. J. T., Beard, M. E. J., McElwain, T. J., Russell, J., Lister, T. A., Whitehouse, J. M. A., Wrigley, P. F. M., Pike, M., Alexander, P. and Hamilton-Fairley, G. (1975). *Brit. J. Cancer*, **28**, 365–376.
35. Vogler, W. R. and Chan, Y.-K. (1974). In *Proc. XI International Cancer Congress* (Bucalossi, P. *et al.*, Eds.), Excerpta Medica, Amsterdam, pp. 351–354.
36. Boone, C. W., Paranjpe, M., Orme, T. and Gillette, R. (1974). *Int. J. Cancer*, **13**, 543–551.
37. Griffith, I. P., Crook, N. E. and White, D. O. (1975). *Brit. J. Cancer*, **31**, 603–613.
38. Lindenman, J. (1974). *Biochim. Biophys. Acta*, **355**, 49–75.
39. Mitchison, N. A. (1970). *Transplant. Proc.*, **2**, 92–103.
40. Bauminger, S. and Yachnin, S. (1972). *Brit. J. Cancer*, **26**, 77–83.
41. Henry, C. and Trefts, P. E. (1974). *Eur. J. Immunol.*, **4**, 824–830.
42. Leech, S. and Mitchison, N. A. (1976). *Eur. J. Immunol.* in press.
43. Benjamini, E. and Scibienski, R. J. (1974). In *Proc. XI International Cancer Congress* (Bucalossi, P. *et al.*, Eds.), Excerpta Medica, Amsterdam, pp. 327–332.
44. Parish, C. R. (1971). *J. exp. Med.*, **134**, 21–47.
45. Dailey, M. and Hunter, R. (1974). *J. Immunol.*, **112**, 1526–1534.
46. Ghose, T. and Nigam, S. P. (1972). *Cancer*, **29**, 1398–1400.
47. Rubens, R. D. and Dulbecco, R. (1974). *Nature*, **248**, 81–82.
48. Davies, D. A. L. and O'Neill, G. J. (1974). In *Proc. XI International Cancer Congress* (Bucalossi, P. *et al.*, Eds.), Excerpta Medica, Amsterdam, pp. 218–221.
49. Greaves, M. F., Brown, G., Rapson, N. T. and Lister, T. A. (1975). *Clin. Immunol. Immunopathol.*, **4**, 67–84.
50. Hellström, I. and Sjögren, H. O. (1965). *Exp. Cell Res.*, **40**, 212–215.
51. Hellström, K. E. and Hellström, I. (1969). *Advan. Cancer Res.*, **12**, 167–223.
52. Takasugi, M. and Klein, E. (1970). *Transplantation*, **9**, 219–227.
53. Wybran, J., Hellström, I. and Hellström, K. E. (1974). *Int. J. Cancer*, **13**, 515–521.
54. Hellström, K. E. and Hellström, I. (1974). *Advan. Immunol.*, **18**, 209–277.
55. Currie, G. A. and Basham, C. (1972). *Brit. J. Cancer*, **26**, 427–438.
56. O'Toole, C., Stejskal, V., Perlmann, P. *et al.* (1974). *J. exp. Med.*, **139**, 457–466.
57. Takasugi, M., Mickey, M. R. and Terasaki, P. I. (1974). *J. Nat. Cancer Inst.*, **55**, 1527–1538.
58. de Vries, J. E., Cornain, S., Rümke, P. (1974). *Int. J. Cancer*, **14**, 427–434.
59. Baldwin, R. W. (1975). *J. Nat. Cancer Inst.*, **55**, 745–748.
60. Herberman, R. B. and Oldham, R. K. (1975). *J. Nat. Cancer Inst.*, **55**, 749–753.
61. Stevenson, G. T. and Lawrence, D. J. R. (1975). *Int. J. Cancer*, **16**, 887–896.
62. Baldwin, R. W. and Embleton, M. J. (1974). *Int. J. Cancer*, **13**, 433–443.
63. Herberman, R. B., Ting, C. C., Holden, H. T., Glaser, M. and Lavrin, D. (1974). In *Proc. XI International Cancer Congress* (Bucalossi, P. *et al.*, Eds.), Excerpta Medica, Amsterdam, pp. 258–263.
64. Shellam, G. R., Knight, R. A., Mitchison, N. A., Gorczynski, R. and Maoz, A. (1975). *Transplant. Rev.*, **29**, 249–276.
65. Owen, J. J. T. and Seeger, R. C. (1973). *Brit. J. Cancer*, **28**, suppl. 1, 26–34.
66. Vanky, F., Klein, E., Stjernswärd, J. and Nilsonne, U. (1974). *Int. J. Cancer*, **14**, 277–288.
67. Cochran, A. J., Grant, R. M., Spilg, W., Mackie, R. M., Ross, C. E., Hoyle, D. E. and Russell, J. M. (1974). *Int. J. Cancer*, **14**, 19–25.
68. Maluish, A. and Halliday, W. J. (1974). *J. Nat. Cancer Inst.*, **52**, 1415–1420.
69. Cercek, L. and Cercek, B. (1975). *Brit. J. Cancer*, **31**, 252–253.
70. Field, E. J., Caspari, E. A. and Smith, K. S. (1973). *Brit. J. Cancer*, **28**, suppl. 1, 208–214.
71. Doherty, P. C. and Zinkernagel, R. F. (1974). *Transplant. Rev.*, **19**, 89–120.

72. Schrader, J. W., Cunningham, B. A. and Edelman, G. M. (1975). *Proc. Nat. Acad. Sci.*, **72**, 5066–5070.
73. Kontiainen, S. and Mitchison, N. A. (1975). *Immunology*, **28**, 523–533.
74. Svedmyr, E. and Jondal, M. (1975). *Proc. Nat. Acad. Sci.*, **72**, 1622–1626.
75. Cornain, S., Carnard, C., Silverman, D., Klein, E. and Rajewski, M. F. (1975). *Int. J. Cancer*, **16**, 301–311.
76. Golstein, P. and Fewtrell, C. (1975). *Nature*, **255**, 491–493.
77. Haskill, J. S., Yamamura, Y. and Rador, L. (1975). *Int. J. Cancer*, **16**, 798–809.
78. Jondal, M., Svedmyr, E., Klein, E. and Singh, S. (1975). *Nature*, **255**, 405–407.
79. Röllinghoff, M. (1974). *J. Immunol.*, **112**, 1718–1725.
80. Platta, F., Cerottini, J.-C. and Brunner, K. T. (1975). *Eur. J. Immunol.*, **5**, 227–233.
81. Bruce, J., Mitchison, N. A. and Shellam, G. (1976). *Int. J. Cancer* **17**, 342–350.
82. Martin-Chandon, M.-R., Vanky, F., Carnard, C. and Klein, E. (1975). *Int. J. Cancer*, **15**, 342–350.
83. Oettgen, H. F., Aoki, T., Old, L. J., Boyse, E. A., Deharven, E. and Mills, G. M. (1968). *J. Nat. Cancer Inst.*, **41**, 827–843.
84. Morton, D. L., Malmgren, R. A., Holmes, E. C. and Ketcham, A. S. (1968). *Surgery*, **64**, 233–240.
85. Lewis, M. G., Ikonopisov, R. L., Nairn, R. C., Phillips, T. M., Hamilton Fairley, G., Bodenham, D. C. and Alexander, P. (1969). *Brit. Med. J.*, **3**, 547–552.
86. Whitehead, R. H. (1973). *Brit. J. Cancer*, **28**, 525–529.
87. Eilber, F. R. and Morton, D. C. (1970). *J. Nat. Cancer Inst.*, **44**, 651–656.
88. Moore, M., Witherow, P. J., Price, C. H. G. and Clough, S. A. (1973). *Int. J. Cancer*, **12**, 428–437.
89. Giraldo, G., Beth, E., Hirshaut, Y., Aoki, T., Old, L. J., Boyse, E. A. and Chopra, H. C. (1971). *J. exp. Med.*, **133**, 454–478.
90. Kassel, R. L., Old, L. J., Carswell, E. A., Fiore, N. C. and Hardy, W. D. (1973). *J. exp. Med.*, **138**, 925–938.
91. Gatti, R. A. (1971). *Lancet*, **1**, 1351–1352.
92. Suciu-Foca, N., Buda, J., McManus, J., Thiem, T. and Reemtsma, K. (1972). *Cancer Res.*, **33**, 439–450.
93. Young, R. C., Corder, M. P., Bernard, C. W. and de Vita, V. T. (1973). *Arch. Intern. Med.*, **131**, 446–454.
94. Gorczynski, R. M. (1974). *J. Immunol.*, **112**, 1826–1838.
95. Kirchner, H., Chused, T. M., Herberman, R. B., Holden, H. T. and Lavrin, D. H. (1974). *J. exp. Med.*, **139**, 1473–1487.
96. Kilburn, D. G., Smith, J. B. and Gorczynski, R. M. (1975). *Eur. J. Immunol.*, **4**, 784–788.
97. Gorczynski, R. M., Kilburn, D. G., Knight, R. A., Norbury, C., Parker, D. C. and Smith, J. B. (1975). *Nature*, **254**, 141–143.
98. Eggers, A. E. and Wunderlich, J. R. (1975). *J. Immunol.*, **114**, 1554–1556.
99. Penn, I. and Starzl, T. E. (1972). *Transplantation*, **14**, 407–417.
100. Möller, G. and Möller, A. (1975). *J. Nat. Cancer Inst.*, **55**, 755–759.
101. Möller, G., Ed. (1975). *Transplant. Rev.*, **28**, 1–97.
102. Prehn, R. T. (1972). *Science*, **176**, 170–171.
103. Outzen, H. C., Custer, R. P., Eaton, G. J. and Prehn, R. T. (1975). *J. Reticuloendothel. Soc.*, **17**, 1–9.
104. Rygaard, J. and Povlsen, C. O. (1974). *Acta Pathol. Microbiol. Scand.*, **83**, 99–106.
105. Stutman, O. (1974). *Science*, **183**, 534–536.
106. Simpson, E. and Nehlsen, S. L. (1971). Clin. exp. Immunol., **9**, 79–98.
107 Allison, A. C., Monga, J. N. and Hammond, V. (1974). *Nature*, **252**, 746–747.
108. Okazaki, W., Purchase, H. G. and Burmeister, B. R. (1970). *Avian Dis.*, **14**, 413–429.
109. Henle, W. and Henle, G. (1975). In *Oncogenesis and Herpes Viruses*, Vol. 11 (de-The, G., Epstein, M. A. and zur Hausen, H., Eds.), Lyons International Agency for Research on Cancer, Scientific Publications, Lyons, pp. 216–224.
110. Stanley, N. F. (1974). In *Proc. XI International Cancer Congress* (Bucalossi, P. *et al.*, Eds.), Excerpta Medica, Amsterdam, pp. 270–274.
111. Muhlbock, O. and Dux, A. (1971). *Transplant. Proc.*, **3**, 1247–1250.
112. Lilly, F. (1972). *J. Nat. Cancer Inst.*, **49**, 927–934.
113. Stevens, D. A. (1973). *Nat. Cancer Inst. Monogr.*, **36**, 55–63.

114. Lawler, S. D. (1973). *Brit. J. Cancer*, suppl. 1, 243–249.
115. Simons, M. J., Wee, G. B., Chan, S. H., Shanmugaratham, K., Day, N. E. and de-The, G. (1975). In *Oncogenesis and Herpes Viruses*, Vol. 11 (de-The, G., Epstein, M. A. and sur Hausen, H., Eds.), Lyons International Agency for Research on Cancer, Scientific Publications, Lyons, pp. 249–258.
116. Lawler, S. D., Klouda, P. T., Smith, P. G., Till, M. M. and Hardisty, R. M. (1974). *Brit. Med. J.*, **1**, 547–548.
117. Rogentine, G. N. Jnr., Trapani, R. A. and Henderson, E. S. (1973). *Tissue Antigens*, **3**, 470–476.
118. Dellon, A. L., Rogentine, G. N., Jr. and Chretien, P. B. (1975). *J. Nat. Cancer Inst.*, **54**, 1283–1286.
119. Kiessling, R., Klein, E. and Wigzell, H. (1975). *Eur. J. Immunol.*, **5**, 112–117.
120. Thomson, D. M. P., Krupey, J., Freedman, S. O. and Gold, P. (1969). *Proc. Nat. Acad. Sci.*, **64**, 161–167.
121. Neville, A. M., Nery, R., Hall, R. R., Turberville, C. and Laurence, D. J. R. (1973). *Brit. J. Cancer*, **28**, suppl. 1, 198–207.
122. Ruoslahti, E., Pihko, H. and Seppala, E. (1974). *Transplant. Rev.*, **20**, 38–60.
123. McIntyre, K. B. and Sizaret, P. P. (1974). In *Proc. XI International Cancer Congress* (Bucalossi, P. *et al.*, Eds.), Excerpta Medica, Amsterdam, pp. 295–299.
124. Bhattacharya, M., Barlow, J.-T., Chu, T. M. and Piver, M. S. (1974). *Cancer Res.*, **34**, 818–822.
125. Coakham, H. (1974). *Nature*, **250**, 328–330.
126. Potter, M. and Boyce, C. R. (1962). *Nature*, **193**, 1086–1087.
127. Harris, A. W., Bankhurst, A., Mason, S. and Warner, N. L. (1973). *J. Immunol.*, **110**, 431–438.
128. Feldmann, M., Boylston, A. and Hogg, N. M. (1975). *Eur. J. Immunol.*, **5**, 429–431.
129. Greaves, M. F. (1975). *Progr. Haematol.*, **9**, 255–304.
130. Stern, P. L., Martin, G. R. and Evans, M. J. (1975). *Cell*, **6**, 455–465.
131. Siegler, R. and Rich, M. A. (1966). *Nat. Cancer Inst. Monogr.*, **22**, 525–547.
132. Shevach, E. M., Stobo, J. D. and Green, I. (1972). *J. Immunol.*, **108**, 1146–1151.
133. Abelson, H. T. and Rabstein, L. S. (1970). *Cancer Res.*, **30**, 2213–2222.
134. Sklar, M. D., Shevach, E. M., Green, I. and Potter, M. (1975). *Nature*, **253**, 550–552.
135. Jondal, M. and Klein, G. (1973). *J. exp. Med.*, **138**, 1365–1378.
136. Greaves, M. F., Brown, G. and Rickinson, A. B. (1975). *Clin. Immunol. Immunopathol.*, **3**, 514–524.
137. Waller, W. C., Neubauer, R. H., Rabin, C. and Cichanek, J. L. (1973). *J. Nat. Cancer Inst.*, **51**, 967–975.
138. Hudson, L. and Payne, L. N. (1973). *Nature* (*New Biol.*), **214**, 52–53.
139. Rosenberg, N., Baltimore, D. and Scher, C. D. (1975). *Proc. Nat. Acad. Sci.*, **72**, 1932–1936.
140. Holley, R. W. (1975). *Nature*, **258**, 487–490.
141. Rapin, A. M. C. and Burger, M. M. (1974). *Advan. Cancer Res.*, **20**, 1–91.
142. Kolata, G. B. (1975). *Science*, **190**, 39–40.
143. Butel, J. S., Brugge, J. S. and Noonan, C. A. (1974). *Cold Spring Harbor Symposia on Quantitative Biology*, **39**, 25–36.
144. Rohrschneider, L. R., Kurth, R. and Bauer, H. (1975). *Virology*, **66**, 481–491.
145. Hurwitz, E., Levy, R., Maron, R., Wilchek, M., Arnon, R. and Sela, M. (1975). *Cancer Res.*, **35**, 1175–1178.
146. Blank, K. J., Freedman, H. A. and Lilly, F. (1976). *Nature*, **260**, 250–252.

Chapter 16

B and T Lymphocytes in Autoimmunity

K. H. Fye
H. Moutsopoulos
N. Talal

Abbreviations

ADCC: antibody-dependent cellular cytotoxicity
CD: Crohn's disease
DNCB: dinitrochlorobenzene
DM: diabetes mellitus
EA: sheep erythrocyte-antibody
EAC: sheep erythrocyte-antibody-complement
HB(A): hepatitis-B (antigen)

HT: Hashimoto's thyroiditis
LATS: serum long-acting thyroid stimulator
LD: lymphocyte-defined (locus)
MG: myasthenia gravis
MIF: macrophage inhibiting factor
MLR: mixed lymphocyte reaction
MS: multiple sclerosis
PHA: phytohaemagglutinin
PPD: purified protein derivative of tuberculin
RA: rheumatoid arthritis
RF: rheumatic fever
SD: serologically-defined (locus)
SK-SD: streptokinase-streptodornase
(S)LE: (systemic) lupus erythematosus
SS: Sjögren's syndrome
UC: ulcerative colitis

1. INTRODUCTION

Autoimmunity is both a major immunologic phenomenon in clinical medicine and an important immunobiologic clue to the workings of the immune system. It appears in a wide range of clinical circumstances including ageing, response to viral and other infections, unusual reactions to a variety of drugs, organ-specific immunologic diseases (e.g. thyroiditis) and generalized systemic diseases (e.g. systemic lupus erythematosus). It can be transient and reversible, or persistent and life-threatening. An entire area of clinical medicine, the 'autoimmune diseases' (of which rheumatoid arthritis is an important example), is concerned with illnesses in which autoimmunity plays a dominant pathogenetic role.

As a phenomenon in immunobiology, autoimmunity provides an opportunity to study the normal regulation of the immune response through an examination of one of its major derangements. Immune reactions are mediated by effector cells such as B lymphocytes or their plasma cell progeny which secrete antibody, effector T lymphocytes which produce lymphokines and mediate cellular immunity, or monocytes and tissue macrophages. These effector cells are regulated by an intricate network of controlling mechanisms which include specific immunoglobulin receptors on lymphocyte surface membranes, secreted antibody itself which can compete for antigen with cell receptors, and specialized subpopulations of T lymphocytes which send 'on and off' signals to these effector cells. Autoimmunity is manifest as an abnormal or excessive activity on the part of effector cells, including the production of auto-antibodies by B lymphocytes and tissue infiltration and destruction by T lymphocytes and macrophages. These effector cells may be deranged primarily, or may be responding to an aberrant set of regulatory signals received from a disordered control system.

Animal models such as the New Zealand black (NZB) mouse offer experimental tools for the study of autoimmunity which complement the clinical experience. Two major lessons learned from the study of animal models are the importance of genetic factors and the role of chronic virus infection in the pathogenesis of autoimmunity.

The production of auto-antibodies by B lymphocytes may be closely related to the premalignant and malignant proliferation of lymphocytes seen in a variety of clinical and experimental circumstances. These circumstances include unusual responses to drugs and other environmental agents, chronic graft-versus-host disease, renal allografts and immunosuppression, autoimmune disorders and immunodeficiency diseases. These lymphocytic and plasma cell malignancies may also arise as a consequence of an imbalance in the immune regulatory system similar to that responsible for autoimmunity.

The discrimination of self from non-self is an essential and primary function of the immune system. Nevertheless, this important discriminatory recognition system is overcome quite frequently, suggesting the presence of an easily perturbable regulatory system normally functioning to prevent autoimmunity. Although several elements may contribute to this system, the major controlling factor appears to be the proper function of the T regulatory cells.

2. AUTOIMMUNITY AND T CELL REGULATION

Auto-antibodies are immunoglobulins produced by B lymphocytes through apparently normal mechanisms of protein synthesis and secretion. Auto-antibody producing B lymphocytes, like their more physiological counterparts, contain surface membrane immunoglobulins with antigen combining sites identical to their secreted antibody product. Since lymphocytes with the ability to bind auto-antigens (e.g. thyroglobulin or DNA) at their surface have been described in some normal individuals, one may question why the secretion of such antibodies does not attain clinical significance more often. Apparently, the proliferation of these cells and the expansion of their auto-antibody population is prevented by immunologic control mechanisms which act normally to regulate the immune response and prevent autoimmunity. The major control mechanism involves the action of T regulatory (helper and suppressor) cells.

T lymphocytes have two major functions in the immune system. T effector cells, in analogy with B lymphocytes, mediate cellular immunity through the production of certain soluble factors (macrophage inhibitory factor (MIF), blastogenic factor, etc.). In addition, T regulatory cells exert a second vital function, i.e. immunologic control. Although various factors (including antigenic load, antibody itself and accessory cells such as macrophages) exert effects on antibody formation, the major responsibility for regulating this complex phenomenon falls to the T regulatory cells.

There are two major populations of T regulatory cells, called helper and suppressor. Helper T cells promote antibody formation by sending stimulatory signals to the B cells. Suppressor T cells diminish antibody responses by sending inhibitory signals to the B cells. Some antigens require helper cells to produce an effective antibody response, whereas others do not. The former are often called 'thymic-dependent' antigens, the latter 'thymic-independent'. 'Thymic-independent' is not entirely correct, since many such antigens are under the control of suppressor T cells. The cellular level at which regulation occurs (B cell itself, macrophage or through T–T cell interaction), the number of cells involved and the nature of the regulatory signals are unknown. Whatever the exact mechanisms of regulation, the net effect at the level of the B cell results either in expansion or repression of the clone and either in increased or decreased antibody concentration.

It follows, therefore, that there are at least two ways to produce autoimmunity. Either a decrease in the number and/or function of suppressor T cells, or an increase in the number and/or function of helper T cells, leads to an unbalanced state of immunologic regulation that could cause the proliferation of B cell clones capable of producing auto-antibodies. A balance of T cell derived regulatory signals normally prevents the emergence of auto-antibodies and clinical autoimmune disease. Factors associated with autoimmunity (e.g. certain virus infections, a variety of different drugs, the ageing process) may act by upsetting this delicate balance and interfering with normal regulatory mechanisms.

Auto-antibody production can be seen as a problem of B cell escape from T cell regulation. In circumstances where autoimmunity may be followed by lymphoid malignancy (e.g. NZB mice and Sjögren's syndrome), this escape can lead to neoplastic proliferation.

3. AUTOIMMUNITY IN NEW ZEALAND BLACK MICE

The introduction of the New Zealand black (NZB) mouse and the related hybrid NZB/NZW F_1 (B/W) mouse into experimental medicine 20 years ago has contributed greatly to our understanding of the role of T and B cells in autoimmunity. Studies in these NZ mice support the concept that a deficiency in immunologic regulation is important in the pathogenesis of autoimmunity and lymphoid malignancy.[1] NZB mice spontaneously develop Coombs's positive haemolytic anaemia as their predominant clinical expression of disease. B/W mice develop lupus erythematosus cells, anti-nuclear factor and immune complex glomerulonephritis as their major clinical features. Antibodies to thymocytes are common in NZB mice, whereas antibodies to nucleic acids are more frequent in B/W mice. Female B/W mice have a more accelerated disease and die prematurely in renal failure, generally before one year of age. They are the closest model for acute lupus nephritis in young women. Male B/W mice survive longer and may develop lymphoma or macroglobulinaemia as they age. The incidence of lymphoid malignancy is increased if the mice are given immunosuppressive drugs.

Genetic, immunologic and virologic factors are all intimately involved in the pathogenesis of NZ mouse disease and, by analogy, in human SLE as well.[2] Several autosomal genes appear to influence the mouse disease, although exact genetic mechanisms are uncertain. Various immunologic abnormalities are seen in NZ mice, suggesting profound disturbances in several different components of the immune system (Table 1). NZ mice harbour C-type leukaemia viruses and produce antibodies to viral antigens. Some of these anti-viral immune complexes, in addition to anti-DNA immune complexes, deposit in the kidney and contribute to the glomerulonephritis.

Abnormalities of both suppressor and helper T cells are present in NZ mice and influence the time of onset and severity of the autoimmune disease. Suppressor cell function declines between 1 and 2 months of age, associated with a deficiency of circulating thymic hormone and a resistance to the establishment and maintenance of T cell tolerance.

Table 1

1. Premature development of competence in B and T cells
2. B cells make excessive antibody responses, particularly to nucleic acids and viral antigens
3. Loss of T suppressor cell function
4. T cells are unable to develop and maintain tolerance
5. Decreased serum 'thymic hormone' activity
6. Spontaneous production of thymocytotoxic antibody
7. Immune complex glomerulonephritis
8. Deficient T cell functions later in life
9. Malignant B cell lymphomas and monoclonal IgM

Significant titres of auto-antibodies, lymphocytic tissue infiltrates, haemolytic anaemia and glomerulonephritis develop progressively over the next several months. NZ mice manifest a progressive deficiency of other T cell functions over time. Antibodies cytotoxic to T lymphocytes may contribute to their T cell abnormalities. Ultimately, gross defects of T effector function, malignant lymphomas and monoclonal macroglobulinaemia may appear. These pathologic events can be viewed as progressive stages of lymphoid cell escape from T cell regulation.

We have explored the possibility that a deficiency of thymic hormone contributes to the progressive T cell abnormalities of NZ mice. Our experimental results suggest that a thymic extract (thymosin) can restore suppressor cell function in NZ mice. Nevertheless, treatment of NZB/NZW mice with thymosin has thus far produced only a modest delay in the onset of autoimmunity. The simple hypothesis of a deficiency in thymic hormone and defective suppressor T cells may be inadequate to explain the disease fully. An imbalance between suppressor and helper T cells may more accurately describe this condition.

We have observed an accelerated switch from IgM to IgG anti-DNA antibodies in female NZB/NZW mice who develop severe SLE glomerulonephritis and die earlier than males.[3] Male NZB/NZW mice have predominantly 19S anti-DNA when female mice have 7S IgG anti-DNA antibodies. Since the switch from IgM to IgG involves the action of T regulatory cells, this finding implies excessive helper activity for the DNA response in female NZB/NZW mice. It suggests that sex factors may act on T regulatory cells to influence the balance of controlling influences which regulate antibody responses by B cells.

We are accumulating additional evidence that T cell function may be hyperactive at some periods in the life of NZ mice. We have reported on a possible abnormal thymic microenvironment which resulted in increased T cell differentiation in X-irradiated bone marrow repopulated NZB mice.[4] The deficiency of circulating thymic hormone may be only one component of a more generalized abnormality of thymic epithelial cells. There may be other thymic epithelial cell products synthesized which can act locally in the thymus to cause accelerated T cell maturation and contribute further to an unbalanced state of T cell regulation.

4. IDENTIFICATION AND FUNCTION OF LYMPHOCYTES IN HUMAN AUTOIMMUNE DISEASES

4.1 Systemic lupus erythematosus (SLE)

Systemic lupus erythematosus is an acute and chronic inflammatory disease predominantly affecting young women and characterized by glomerulonephritis, central nervous system involvement and widespread vasculitic lesions. A variety of different auto-antibodies are produced which include antibodies to nucleoproteins and nucleic acids, antibodies to lymphocyte surface antigens, antibodies to platelets and to a variety of ubiquitous cellular constituents.

The exact composition and function of various lymphocyte populations in this disease is a subject of some controversy. In large measure, this controversy can be attributed to differences in experimental techniques employed in various laboratories and to the presence of anti-lymphocytic antibodies[5–8] which make it difficult to enumerate B lymphocytes by standard immunofluorescent methods. These antibodies, which have predominant specificity for T lymphocytes,[5] can coat T lymphocytes *in vivo* and make them appear as immunoglobulin-positive staining cells in fluorescent microscopy. These cells with coating immunoglobulin on their surface membranes may then be falsely enumerated as B lymphocytes. This technical problem is probably responsible for the elevation in B lymphocytes reported in patients with systemic lupus erythematosus.

An additional problem relates to the frequent presence of circulating immune complexes in the serum of lupus patients. These immune complexes may adhere to cells with Fc receptors which will then also appear as immunoglobulin-positive lymphocytes. Furthermore, some investigators have found that anti-lymphocytic antibodies can interfere with the binding of sheep erythrocytes to T lymphocytes and thereby inhibit T cell rosette formation.[9] Another explanation for differences observed from one laboratory to another relates to the variable nature of lupus, an illness characterized by fluctuations in disease activity. Experimental results may differ depending upon whether patients are studied in periods of disease remission or exacerbation, or in relation to the drug regimens employed.

In general, most workers have described a decrease in the number of T lymphocytes in the peripheral blood of patients with SLE, measured either by rosette formation or with specific anti-T cell antiserum.[10–13] The decrease in T rosette forming cells generally observed was approximately 20% compared to normals. By contrast, Winchester *et al.*[14] found T cells to be slightly increased in their patients with systemic lupus erythematosus (83% in SLE versus 78% in normals).

Messner *et al.*[11] prepared an anti-human T cell antiserum by immunizing rabbits with pooled foetal thymocytes and absorbing with lymphocytes from patients with chronic lymphocytic leukaemia. Employing an indirect immunofluorescent method, they found a decrease in peripheral blood T cells in most patients with active lupus (67% in SLE compared to 79% in normals).

The number of peripheral blood immunoglobulin-positive cells in patients with lupus varies greatly depending upon whether the study is performed on fresh lymphocytes or after overnight incubation. The number of such cells decreases greatly following incubation, presumably due to the presence of coating anti-lymphocyte antibodies or immune complexes which are shed from the cell surface into the incubation medium.[13–15]

A number of investigators[11,12] used sensitized sheep erythrocytes coated with complement to detect lymphocytes with C3 receptors on the cell surface. They found a decrease in the percentage and total number of such cells in patients with SLE. This finding could be attributed to circulating complement-containing immune complexes that block specific cell receptors, or to a decreased subpopulation of B lymphocytes containing surface complement receptors.

Lymphocyte function has been studied in lupus by skin testing for delayed hypersensitivity, by studying lymphocyte response to phytomitogens and to specific antigens, and by measuring immunoglobulin synthesis. Several investigators have found a diminished skin reactivity to the purified protein derivative of tuberculin (PPD) in patients with SLE. Block *et al.* found none of 20 patients reactive,[16] and Hahn *et al.*[17] found that only one of 39 patients had greater than 10 mm of induration. Abe and Homma also found decreased PPD responses in SLE.[18] Horwitz[19] studied 14 patients and found a markedly decreased response to Candida, streptokinase-streptodornase (SK-SD) and tricophytin, when compared to normal individuals or patients with tuberculosis. Decreased skin reactivity was found only in patients with active disease. By contrast, Goldman *et al.*[20] found no decrease of skin reactivity to PPD, Candida, tricophytin or histoplasmin. Some evidence for increased skin reactivity to DNA and nucleoproteins in SLE was presented in early studies.[21–23]

Although a decreased response to phytohaemagglutinin (PHA) has been reported in lupus,[24,25] most investigators have found normal reactivity when optimal concentrations of the phytomitogen were employed.[20,26–28] Response to pokeweed mitogen and concanavalin A also appears to be normal.[27] Lymphocyte stimulation by suboptimal concentrations of phytohaemagglutinin appears to be a particularly sensitive method for detecting abnormalities of lymphocyte function, and has revealed an apparent defect in some patients with lupus.[24] Lymphocyte response to mitogens and delayed hypersensitivity to test antigens may return towards normal as patients progress from active disease into clinical remission.[25]

Some laboratories have found increased lymphocyte proliferative responses to specific nuclear antigens,[20,27,28] whereas others have reported normal responses.[27–29] Since humoral antibodies to DNA and RNA are a hallmark of SLE, the status of cellular immunity to such antigens is an important question. There are some reports that DNA or synthetic polynucleotides can stimulate lymphocyte proliferation in occasional patients with SLE.[20,27,28] However, when compared to the very high titres of antibodies to nucleic acids found in this disease, the degree of T cell immunity to such antigens is relatively insignificant. This dichotomy supports the concept of an immune imbalance in SLE characterized by excessive B cell activity and decreased T cell immunity.[1,2]

The one-way mixed lymphocyte reaction is decreased in lupus patients,[15,30] but can be restored to normal after overnight incubation. Under these conditions, blocking factors are shed into the incubation medium.[15] Response to measles antigen as measured by MIF production is decreased in SLE.[31]

Synthesis of immunoglobulins by blood lymphocytes is markedly elevated in patients with active lupus, whereas patients with inactive lupus or rheumatoid arthritis show normal synthetic values.[32] The increased immunoglobulin synthesis declined in three patients with active SLE as they responded to treatment with a decrease in disease activity.

The ability of peripheral blood lymphocytes in SLE to mediate antibody-dependent cellular cytotoxicity (ADCC) has been studied recently using ^{51}Cr-labelled chicken erythrocytes as target cells.[33] This activity, dependent upon effector cells with free Fc receptors, was decreased in patients with active SLE. The decrease was attributed to both blocking immune complexes and to anti-lymphocyte antibodies. Serum from active SLE patients could block ADCC mediated by normal lymphocytes. This assay can be used to detect serum immune complexes.

In summary, B cell activity in lupus is increased, as evidenced by a wide variety of auto-antibodies produced and the increased immunoglobulin synthesis in peripheral blood. The exact status of T cell function in this disease remains unresolved, but the presence of circulating immune complexes and antibodies to lymphocytes can interfere both with accurate cell enumeration and function. The nature of these blocking factors will be discussed in greater detail in the next section of this chapter.

4.2 Rheumatoid arthritis (RA)

Rheumatoid arthritis is a systemic multi-system inflammatory disease involving primarily (but not exclusively) the synovial lining tissues of articular surfaces. The production of rheumatoid factor, an anti-IgG antibody generally belonging to the IgM immunoglobulin class, is a hallmark of the disease. Circulating immune complexes and systemic vasculitis may occur in some patients. Synovial fluid immune complexes comprised of both IgM and IgG, frequently including IgG rheumatoid factor, are a characteristic finding in RA. These complexes are found within synovial phagocytic cells and are implicated in pathogenetic mechanisms leading to joint destruction.

Williams and coworkers, using a rabbit anti-T cell antiserum, found that RA patients segregated into two groups according to whether they had normal or decreased numbers of T lymphocytes.[10] Patients with decreased T cells had more active disease. Winchester *et al.*[14] found T lymphocytes, as measured by rosette formation, increased in rheumatoid arthritis (85% in RA versus 78% in normals).

RA patients also have a decrease in percent of total immunoglobulin-positive cells as well as a decrease in cells containing surface C3 receptors.[10,34] Two laboratories have found an increase in immunoglobulin-positive lymphocytes in RA by direct immunofluorescent methods.[14,35] Winchester and coworkers also found an increase using fluorescein labelled aggregated IgG.[14]

The intense lymphocytic infiltration of synovium and synovial fluid offers another tissue source for studies of T and B cell numbers and function. The majority of mononuclear cells in RA synovial fluid are T lymphocytes.[36] Two laboratories[14,36] have found small numbers of immunoglobulin-positive cells in synovial fluid (6% and less than 1% respectively), whereas Mellbye and coworkers found that 29% of synovial fluid lymphocytes had detectable surface immunoglobulin as compared to 15% in the blood of these same patients.[34] These patients studied by Mellbye and coworkers had no increase in synovial fluid lymphocytes with complement receptors, presumably due to blocking by immune complexes present in the synovial fluid. In support of this concept, these authors found that RA synovial fluid was able to block C3 receptors on normal peripheral blood lymphocytes. Thus, many of the problems encountered in attempting to enumerate T and B lymphocytes in lupus serum are also present in studies of RA synovial fluid. An additional difficulty is the lack of suitable synovial fluid controls to permit meaningful comparisons between RA patients and normal individuals. A third population of lymphocytes lacking surface immunoglobulin or receptors for sheep erythrocytes may also be present in RA synovial fluid. Winchester *et al.*[14] found that the sum of B and T lymphocytes did not account for all of the mononuclear cells present in synovial fluid, suggesting that receptors are either turning over rapidly or blocked in such a way that they cannot be demonstrated. Beta-2 microglobulin concentrations are increased in RA synovial fluid compared to non-rheumatoid and non-inflammatory synovial fluids.[37]

As with synovial fluid, the predominant cell infiltrating rheumatoid synovium is a T lymphocyte. Tannenbaum and coworkers studied synovial tissue sections and found only a small number of lymphocytes that stained for immunoglobulin or that formed EAC rosettes.[38] They deduced that the major lymphoid cell present in the rheumatoid synovium was a cell lacking immunoglobulin or complement receptors (presumably a T cell or null cell). Van Boxel and Paget[39] used collagenase and deoxyribonuclease to digest the synovium of 5 patients with rheumatoid arthritis and showed that 70% to 85% of lymphocytes isolated from such digested tissue formed rosettes with sheep erythrocytes. Only 9% to 35% had surface immunoglobulin detectable by immunofluorescence. Morphologically, a mixture of lymphocytes, blast cells and plasma cells are seen, compatible with both cellular and humoral immune responses.[40] Large amounts of immunoglobulins[41] and rheumatoid factor[42] are produced locally in the inflamed synovial tissues, indicating the presence of active B cells.

Patients with RA showed decreased sensitization to both dinitrochlorobenzene (DNCB) and paranitrosodimethyl aniline.[43] Waxman *et al.*[44] studied 84 patients with rheumatoid arthritis and found that 42% showed skin anergy to 5 common test antigens (PPD, SK-SD, Candida, mumps, tricophytin). Patients with the most longstanding disease were anergic, although anergy was not related to disease severity or treatment. In this study, anergic patients also failed to become sensitized to DNCB.

The lymphocyte proliferative response to PHA is essentially normal in rheumatoid arthritis.[44] Because anti-immunoglobulin (rheumatoid factor) is a classic manifestation of rheumatoid arthritis, attempts have been made to demonstrate sensitization to IgG by studies of lymphocyte stimulation with native or aggregated, autologous or non-autologous, IgG. The incorporation of tritiated thymidine into DNA was not stimulated in this manner.[45,46] However, native and aggregated IgG can stimulate the production of MIF and migratory enhancement factor,[47] thereby suggesting that T cell sensitization to IgG may be present in rheumatoid arthritis.

Astorga and Williams[48] found that lymphocytes from RA patients failed to stimulate lymphocytes from other RA patients in a one-way mixed lymphocyte response, even though the RA lymphocytes responded to normal lymphocytes and could stimulate normal lymphocytes. This defect is probably related to a common genetic determinant present on RA lymphocytes and will be discussed in a subsequent section of this chapter. The sera of 8% of rheumatoid arthritis patients contained an IgG immunoglobulin that could suppress the mixed lymphocyte response (MLR) of normal lymphocytes. This suppression correlated with lymphocytotoxic antibody in individual sera.[49]

In summary, T cell function in peripheral blood may be impaired in some patients with rheumatoid arthritis, particularly when disease is severe and longstanding. Synovial fluid and synovial tissues contain large numbers of T lymphocytes, a finding which appears to be characteristic of rheumatoid arthritis. B lymphocytes synthesizing large amounts of immunoglobulin and producing rheumatoid factor are also present in the inflamed synovial tissues. Immune complexes are present in synovial fluid and may block Fc receptors present on synovial fluid lymphocytes.

4.3 Sjögren's syndrome (SS)

Sjögren's syndrome is an autoimmune disease defined by the triad of dry eyes (keratoconjunctivitis sicca), dry mouth and an associated connective tissue disease, most often rheumatoid arthritis. The deficiency of tears and saliva is associated with an extensive lymphoid and plasma cell infiltrate of the salivary and lacrimal glands. Histologic confirmation of the diagnosis can often be made by biopsy of the minor salivary glands of the lower lip. In some patients, a more aggressive and extensive lymphocyte proliferative process occurs with involvement of other exocrine glands and infiltration of major parenchymal organs such as the lungs, kidneys and liver. These lymphoid infiltrates may suggest malignancy; Waldenstrom's macroglobulinaemia, pseudolymphoma and reticulum cell sarcoma occur in some patients.[50] Hypergammaglobulinaemia and production of many different auto-antibodies is characteristic of SS. Hypogammaglobulinaemia and loss of serum auto-antibodies may be seen in patients with highly undifferentiated reticulum cell sarcoma. SS represents a human counterpart of the spectrum of pathology seen in the NZB mouse.[1]

The percentage of peripheral T lymphocytes in Sjögren's syndrome may be normal or decreased. Van Boxel and coworkers, using E-rosette formation with neuraminidase-treated sheep erythrocytes, found a normal percentage of T cells.[51] However, Talal and coworkers, using both rosette formation and fluorescent anti-T cell antiserum, found decreased T cells in 7 of 24 patients with SS.[52] A recent study[13] reported that 10 of 25 patients had a significant decrease in T rosette-forming cells (52% in SS versus 67% in normals) which was restored towards normal following *in vitro* incubation with thymosin fraction 5. Peripheral blood T cells may be decreased in SS because of preferential sequestration of T cells in infiltrated target organs and/or because of a defect in the differentiation of immature into mature T cells.

There is a modest increase in immunoglobulin-positive lymphocytes in peripheral blood.[51,52] This increase involved several immunoglobulin classes and was not directly correlated with the elevated concentrations of serum immunoglobulins. Van Boxel[51] stained lymphocytes with antiserum to heavy chain and found that the sum of cells staining with antiserum for all classes of heavy chain was greater than the number of cells staining with anti-Fab or with aggregated Ig. They suggested that many B cells in SS contained Ig of more than one heavy chain class. Their data may be explained by adherence of one class of immunoglobulin to the surface of cells producing a different class. For example, IgG might adhere to B cells producing IgM rheumatoid factor[53] or antibodies to lymphocytes may coat cells in occasional patients with SS.[54]

The nature of the cells infiltrating the minor labial salivary glands have been studied by immunofluorescence using anti-human T cell antisera.[52] The number of T cells was generally related to the extent of the lymphocytic infiltration. T cells were found in clusters close to the salivary ducts. Chused and coworkers, using EAC-rosette formation, found large numbers of B cells.[55] Tannenbaum and coworkers found EAC-positive cells in large clusters located between salivary ducts in five lip biopsy specimens.[38] A cell suspension prepared from the parotid glands of one patient contained 28%–30% B lymphocytes, 40%–44% T lymphocytes and 3%–13% histiocytes. They also found that 90% of the lymphocytes in a pseudolymphoma mass formed EAC-rosettes but not T or EA-rosettes, identifying them as B lymphocytes with complement receptors.

Great quantities of immunoglobulin, including rheumatoid factor, are synthesized by the lymphocytes infiltrating the salivary glands.[56,57] In some patients with co-existing Waldenstrom's macroglobulinaemia and SS, the monoclonal IgM is also produced by the infiltrating lymphocytes.[56] Thus, the mixed T and B lymphocytic infiltrate and the local production of rheumatoid factor in the salivary glands of patients with SS are analogous to immunologic abnormalities in inflamed RA synovial tissues.

Skin test responsiveness to DNCB is decreased in SS; the response to PHA may be either normal or impaired.[58] This defect is even more apparent when suboptimal concentrations of PHA are employed (Michalski *et al.*, in preparation). A parotid gland extract stimulated MIF production by peripheral lymphocytes in SS.[59]

Beta-2 microglobulin, a low molecular weight cell surface constituent non-covalently linked to the HLA antigen, is increased in the saliva and serum in SS.[37] The salivary increase correlated directly with the degree of lymphocytic infiltration present on labial gland biopsy.[60] Serum concentrations were particularly elevated in patients with associated renal or lympho-proliferative disease. Since foetal or neoplastic cells undergoing rapid proliferation secrete increased quantities of Beta-2 microglobulin, this increase in SS (and in RA synovial fluid) may reflect rapid cell proliferation and turnover in these inflamed tissues. However, since Beta-2 microglobulin may play a role in immune responses, it is possible that the increased concentrations found in autoimmune inflammatory fluids may reflect the disordered immunologic regulation thought to underlie these diseases.

In summary, T cell number and function in peripheral blood is impaired in some patients with SS. B cell number is modestly increased and function is excessive, as manifest by hypergammaglobulinaemia and auto-antibody formation. The salivary gland infiltrates are composed of mixtures of both B and T lymphocytes, with T lymphocytes predominating in the most extensive lesions. Immunoglobulins, rheumatoid factor and Beta-2 microglobulin are produced locally by these infiltrating lymphocytes. Lymphocyte aggressive disease can lead to pseudolymphoma or frank lymphoma. This spectrum of disease conversion from autoimmunity to lymphocyte malignancy is consistent with the hypothesis of B cell escape from T cell regulation.

4.4 Rheumatic fever (RF)

Rheumatic fever is an acute inflammatory disease whose major manifestations include polyarthritis, carditis, Sydenham's chorea, characteristic skin lesions and subcutaneous nodules. It develops after prior Group A streptococcal infection and is considered an autoimmune response to cross-reactive streptococcal antigens.

Lueker and coworkers studied T and B cells in 53 patients with RF.[61] The proportion and number of T cells was decreased during active disease, while the proportion and number of cells showing surface immunoglobulin was increased. Cytophilic antibodies were not detected.

Infections with Group A streptococci result in the production of a variety of different anti-streptococcal antibodies. These antibodies can be directed against structural antigens[62] or against streptococcal toxins. Sera from patients with RF contain antibodies against streptococcal cell wall antigens that cross-react with human heart tissue. Autoantibodies to human myocardial tissue have been demonstrated using a variety of serologic techniques.[63] The incidence of such auto-antibodies, as determined by immunofluorescence on perinatal human heart, appears related to disease activity. Such antibodies were present in 63% of patients with carditis, 27% without carditis, 16% of inactive patients and 3% of controls.[64]

Extracts containing streptolysin-S induced blastic transformation in normal lymphocytes.[65] The response to such extracts is decreased in patients suffering their first attack of RF, but is normal during subsequent attacks.[66] There is an increased T cell response to Group A streptococcal cell membranes and cell wall extracts, as determined by MIF production, in patients with RF.[67] The greatest response was seen in patients recently recovered from an acute attack.

Lymphocytes from RF patients showed no proliferative response to cardiac homogenates or myocardial fractions, even though the latter contained antigens cross-reactive with anti-cardiac and anti-streptococcal antibodies.[68] The response to PHA and streptococcal antigens was also normal. There is a decreased MLR response not related to steroid therapy in patients with acute RF.[69]

In summary, rheumatic carditis appears to be mediated primarily by antibodies produced in response to as yet undefined streptococcal antigens. These antibodies cross-react with myocardial auto-antigens and lead to the inflammatory changes characteristic of this disease.

4.5 Polymyositis

Polymyositis is an acute and chronic inflammatory disease in which the primary clinical manifestation is muscle weakness. It is called dermatomyositis when accompanied by a characteristic rash. The muscles contain focal lymphocytic and plasma cell infiltrates which may be sensitized to muscle antigens. No antibodies directed against muscle antigens have been found, suggesting that humoral mechanisms may not be important in polymyositis.

Lymphocytes isolated from patients with active polymyositis demonstrated a cytotoxic effect on muscle cell cultures.[70] This cytotoxic effect was mediated by a lymphotoxin produced when the lymphocytes were incubated with autologous muscle.[71] Cytotoxicity disappeared when patients were placed on immunosuppressive drugs or when they entered remission.[72] There was a poor correlation between the degree of lymphocyte-mediated cytotoxicity and the level of creatinine kinase, a muscle enzyme that correlates with disease activity.[72]

In summary, T cell-mediated cytotoxicity may be an important disease mechanism in polymyositis. T cells may recognize antigens in autologous muscle tissue and, in the absence of immunologic regulation, produce mediators (such as lymphotoxin) that lead to the clinical manifestations of disease.

4.6 Autoimmune liver disease

Immunological and viral mechanisms are important in a variety of diseases that effect the liver, including acute and chronic active hepatitis. Attention has focused in particular on the immune response to hepatitis-B virus. Viral hepatitis is usually an acute and self-limited illness. The intensity of the cellular immune response to viral infection in the liver appears to be the primary determinant of the extent and duration of liver damage.[73–77]

Although controversy exists, most investigators have found a depressed PHA response during acute viral hepatitis.[75,77] Patients with hepatitis B-antigen (HBA)-positive acute hepatitis who had recovered and cleared the HBA antigen from the blood had a higher PHA response than patients who developed a chronic carrier state.[75] The subsequent development of cirrhosis was also associated with a low PHA response.

HBA-positive patients produced MIF in response to serum and liver extracts containing HBA antigen but not in response to normal liver extracts, suggesting that the stimulating antigen was an HBA component not present in normal liver.[78]

Lymphocytes in acute and chronic hepatitis are cytotoxic to human Chang liver target cells,[79] as well as to autologous human liver cells.[80] Sera containing hepatitis-B antigen were able to depress this T lymphocyte-mediated spontaneous cytotoxicity, suggesting the presence of inhibitory serum factors.[79]

In summary, the autoimmune liver diseases provide an excellent example of the interplay between immunologic and virologic factors in inflammatory disorders. Adequate T cell function directed against hepatitis-B antigen may be associated with self-limited disease, whereas defective cell-mediated immunity is associated with more persistent or varying disease. Direct cytotoxicity against autologous liver cells has been demonstrated in chronic active hepatitis.

4.7 Inflammatory bowel disease

Crohn's disease (CD) and ulcerative colitis (UC) are chronic recurrent inflammatory diseases of the bowel which can be distinguished clinically, radiologically and histologically but may share an immunologic pathogenesis.

The number of T lymphocytes is decreased in CD and UC.[81] B cells with surface IgA or IgM are increased in both diseases. These changes in T and B cells are unrelated to disease activity, duration or mode of treatment. Decrease in T cells might be related to sequestration or increased loss in the gastrointestinal tract.

Although anti-colonic antibodies have been described, major attention has been given to evaluation of T lymphocyte immunity. There is probably no decrease in delayed skin hypersensitivity to most antigens, including PPD,[82] although sensitization to DNCB may be impaired in CD.[83] Although there is some controversy, most investigators have found decreased responses to optimal or suboptimal doses of PHA and Con A, particularly in patients with active disease.[84–86] Lymphocytes taken from lymph nodes directly draining diseased bowel showed lower responses to PHA than cells from nodes at the root of the mesentery, which in turn were lower than responses given by circulating lymphocytes.[85]

Peripheral blood mononuclear cells are cytotoxic to suspensions of allogeneic colonic epithelial cells.[87,88] This cytotoxicity can be eliminated by horse anti-human thymocyte serum,[88] and disappears following colectomy.[89] This lymphocyte-mediated cytotoxicity appears specific for colonic epithelium, and is not seen with suspensions of gastric or ileal epithelial cells.[89] Lymphotoxins have been demonstrated in supernatants when UC or CD lymphocytes are exposed to colonic epithelial cells.[90] Normal lymphocytes can be induced to develop such cytotoxicity for colonic epithelium if they are exposed to lipopolysaccharide from *E. coli.*[91] Colonic ulceration with bacterial contamination of sub-mucosal tissues may activate lymphocytes and induce a response to shared bacterial and colonic epithelial cell antigens.[92] Lymphocytotoxic and lymphocytophilic serum antibodies also occur.[93]

In summary, an increased permeability of colonic epithelium in susceptible individuals may predispose to inflammatory bowel disease.[92] Factors increasing permeability might include the nature of the diet in infancy, enterobacterial toxins, proteolytic enzymes of the faecal stream, and impaired function of secretory IgA. A cell-mediated immune response to shared bacterial and colonic antigens then takes place. Sensitized lymphocytes and lymphotoxin production are responsible for continued tissue destruction.

4.8 Autoimmune endocrine disorders

4.8.1 *Thyroiditis*

Chronic thyroiditis of the Hashimoto type is characterized clinically by thyroid inflammation, goitre formation and eventual hypothyroidism. During early stages of Hashimoto's thyroiditis (HT), some patients demonstrate features of Grave's disease including hyperthyroidism, ophthalmopathy and the presence of serum long-acting thyroid stimulator (LATS). Histologically, the thyroid shows epithelial cell abnormalities and lymphocytic infiltration. High titres of antibodies to thyroglobulin are characteristic. Autoimmune thyroiditis can be induced in experimental animals by the injection of thyroglobulin or thyroid extract in Freund's adjuvant. Moreover, spontaneous autoimmune thyroiditis has been described in OS chickens and in certain strains of rats. Thymectomy potentiates the induction of experimental thyroiditis, consistent with the removal of T cell regulatory mechanisms.[94]

Peripheral blood T cells may be increased in HT and in untreated Grave's disease.[95] The percentage of T cells was normal in patients in remission on therapy. There was a significant correlation between the elevation of T cells and the production of MIF in response to human thyroid extract, suggesting a role for cell-mediated immunity in the pathogenesis of both HT and Grave's disease.[95] MIF was produced in approximately 75% of patients with HT in response to thyroid extract. There was a modest decrease in T cells (57% versus 67% in normals) in HT patients who were euthyroid at the time of study.[96] B cells were not different from normal individuals. ^{125}I-thyroglobulin binding lymphocytes were found to be increased in the peripheral blood[94] and in a thyroid infiltrate[96] in patients with HT. Binding was not due to cytophilic antibody and could be inhibited by anti-light chain antiserum. There was no correlation between such antigen-binding lymphocytes and the titre of anti-thyroglobulin antibody. These results suggest an increase in specific antigen-binding B lymphocytes in peripheral blood and thyroid of HT.

Lymphocyte transformation in response to thyroglobulin,[97] direct cytotoxicity to human thyroid cells grown in monolayer culture,[98] and to thyroglobulin coated ^{51}Cr-labelled chicken erythrocytes,[99] have been described in HT. Antibody-dependent cellular cytotoxicity (ADCC) has also been shown when labelled target cells were incubated with HT serum and normal peripheral blood effector cells. There was a significant correlation between ^{51}Cr release and the titre of anti-thyroglobulin antibody.[99]

In summary, the autoimmune thyroid diseases are mediated both by classical B and T cell mechanisms and by an additional mechanism (ADCC) which may involve yet another lymphocyte population. The autoimmune thyroid diseases, therefore, involve various peripheral lymphoid populations which operate simultaneously to mediate inflammatory tissue destruction. This concept may prove to be pathogenetically significant in other autoimmune diseases as well.

4.8.2 *Addison's disease*

Idiopathic Addison's disease, or primary chronic adrenocortical insufficiency, is characterized pathologically by a mononuclear cell inflammatory infiltrate of the adrenal cortex. Eventually the cortex degenerates into an atrophic fibrotic shell with islands of necrotic cortical cells. Although the aetiology of idiopathic Addison's disease is unknown, an autoimmune pathogenesis is suggested by the association of other diseases of possible autoimmune origin, including hypothyroidism, pernicious anaemia, idiopathic hypoparathyroidism and gonadal insufficiency.

Anti-adrenal antibodies are found in up to 70% of patients.[100] The methods used to detect such antibodies include indirect immunofluorescence, complement fixation, tanned red cell agglutination and gel diffusion. Recent evidence suggests that these auto-antibodies recognize cell surface antigens shared by steroid producing cells in adrenal, ovarian and testicular tissues in several experimental animals.[101]

4.8.3 *Diabetes mellitus (DM)*

There are two clinically distinct forms of diabetes mellitus: 'juvenile onset', which is characterized by a dysfunction of pancreatic Beta-islet cells and low levels of serum insulin; and 'adult onset', in which serum levels of insulin are normal or elevated. Both forms are probably manifestations of autoimmunity, but precise pathogenetic mechanisms may be different.

Auto-antibodies to pancreatic islet cells[102] and to insulinoma cells[103] have been described in juvenile DM. Furthermore, peripheral lymphocytes from such patients produce MIF when stimulated with human pancreatic extract.[104] A mononuclear cell pancreatic infiltrate is present in up to 68% of patients with disease of recent onset.[105] Therefore, in 'juvenile onset' DM, an autoimmune attack on the pancreatic Beta-islet cells may result in decreased insulin production and hyperglycaemia.

In 'adult onset' DM, antibodies to islet cells are not generally seen and there appears to be peripheral resistance to the effects of insulin. Antibodies to insulin receptors on human peripheral blood monocytes have been detected in a few patients with severe insulin resistance, suggesting auto-antibody blockade of hormone action.[106]

4.9 Autoimmune neuromuscular diseases

4.9.1 *Myasthenia gravis (MG)*

Myasthenia gravis is characterized by weakness and easy fatiguability, particularly of muscles innervated by the cranial nerves. The basic pathogenetic defect appears to involve pharmacological blockade at the neuromuscular junction. Suggestive evidence for autoimmunity includes the frequent association with thymic hyperplasia or thymoma and the occasional association with other autoimmune disorders, particularly RA and SLE. Moreover, specific auto-antibodies against muscle tissue and thymic cells have been identified frequently in MG patients.

The percentage of T cells in the peripheral blood and thymus was normal, whereas the percentage of B cells was increased in thymic cell suspensions but not in peripheral blood.[107] Therefore, B cells may be part of an infiltrating lymphoid population present in the thymus.

The response of peripheral blood and thymocytes to PHA was normal in MG.[107] By contrast, the response to pokeweed mitogen was increased in thymocyte preparations, further supporting the hypothesis that B cells are present in the thymus. Thymocytes from patients with MG and thymic hyperplasia can stimulate autologous lymphocytes. Such stimulation was not seen in normals or in MG patients with thymoma.[107]

Lymphocytes from MG patients appear sensitized to muscle antigens. Thymocytes stimulated by PHA become cytotoxic to foetal muscle cells, and will produce a leukocyte inhibitory factor when stimulated with muscle or thymic extracts. These extracts did not induce lymphocyte transformation, however.[108]

Goldstein has suggested, based on the experimental induction of an MG-like syndrome in guinea pigs, that an autoimmune thymitis is a primary mechanism in MG.[109] The inflamed thymus releases a substance called thymin or thymopoietin which Goldstein believes responsible for the neuromuscular blockade. He has recently presented evidence that thymopoietin is also a thymic differentiation hormone.[109]

Recent studies using a radio-immunoassay have found antibodies to the acetylcholine receptor in over 70% of patients with MG.[110] Such antibodies might block neuromuscular transmission. Thus, MG may be another example of a more general immunopathologic phenomenon in which auto-antibodies to specific hormone receptors or to receptors for pharmacological mediators can induce specific physiological blockade and result in disease.

4.9.2 *Multiple sclerosis (MS)*

Multiple sclerosis is a chronic recurrent degenerative and demyelinating disease of the central nervous system. Evidence for an autoimmune response to a latent virus includes its similarity to subacute sclerosing panencephalitis in which measles virus has been implicated. Furthermore, an autoimmune allergic encephalitis can be induced by immunization of animals with brain antigens.[111]

T and B lymphocytes are present in normal percentages in MS.[112] Serum and spinal fluid immunoglobulin concentrations are increased, with some evidence of local oligoclonal production in the central nervous system.[113] Cellular immunity, as measured by skin testing and response to PHA, is essentially normal.[114] Peripheral blood lymphocytes do not produce MIF in response to whole human brain homogenates,[113] but do respond with MIF production to basic myelin A_1 protein during periods of active disease.[115] This finding suggests that cellular immunity to brain antigens may play a role in disease causation.

4.10 Other clinical considerations

The diseases chosen for discussion were selected because they illustrate specific aspects of immunopathogenesis as related to autoimmunity. Obviously, a large number of diseases of probable autoimmune pathogenesis have not been discussed because of space limitations.

The ageing process itself is associated with autoimmunity and decreased T cell responses to plant mitogens and allogeneic cells.[116] The percentage of T and B cells in aged individuals is normal. Immunodeficiency, sometimes limited to decreased IgA as the sole abnormality, is also associated with autoimmunity. Recent studies suggest abnormalities of T cell regulation and suppressor T cells in immunodeficiency patients with common variable hypogammaglobulinaemia.[117]

5. LYMPHOCYTE ABNORMALITIES AS A CONSEQUENCE OF AUTOIMMUNITY

A unique aspect of lymphocytes and autoimmunity arises as a consequence of autoimune processes that themselves interfere with lymphocyte function. One example is the presence of antibodies directed against cell surface antigens which coat lymphocyte membranes and interfere with cell function. The exact nature of these surface antigens has not yet been defined. Such antibodies are seen in over 80% of SLE patients[118,119] but also occur occasionally in RA and SS. Antigen-antibody complexes formed at the cell surface may be shed *in vivo* or under *in vitro* experimental conditions. Anti-lymphocyte antibodies are present in cryoprecipitates in SLE, perhaps complexed to antigen.[120]

As a result of such antibody coating of the lymphocyte surface, it may be difficult to accurately perform HLA typing in SLE.[15] Wernet and coworkers showed that MLR reactivity was impaired in SLE, but could be restored to normal by overnight incubation that allowed complexes to be shed from the cell membranes.[15] Exposure of either SLE or normal lymphocytes to the culture medium containing these shed 'blocking factors' interfered with MLR reactivity and HLA typing in these fresh cell populations. Decreased PHA responses in SLE, particularly seen with suboptimal concentrations of PHA[30] may also be related to anti-lymphocyte antibodies.

Some anti-lymphocyte antibodies can interfere with cell function without actually causing cell death, while others fix complement and result in lysis of the lymphocyte.[121] Cytophilic antibodies that fix complement are probably responsible for complement-coating of SLE lymphocytes.[122] Such lymphocytotoxins are associated with active lupus and with lymphopenia.[8,123]

Most anti-lymphocyte antibodies in SLE are cold-reactive IgM.[7] These antibodies are highly reactive with foetal thymocytes and much less reactive with bone marrow cells, implying T cell specificity.[5] However, Winfield and coworkers reported that they were equally reactive with B and T lymphocytes of both normal and SLE patients, and did not show HLA specificity.[7] Williams and coworkers found that the IgG fraction in 16% of SLE patients and in 17% of RA patients inhibited MLR and correlated strongly with lymphocytotoxic antibodies, suggesting that some anti-lymphocytic antibodies belong to the IgG class.[124] Winchester, using an immunofluorescent method and studying pregnancy sera, has found a high incidence of anti-lymphocyte antibodies reacting with a B cell determinant that could be the equivalent of the Ia antigen in the mouse.[125] Thus, cold-reactive IgM may be the prototype for an entire family of anti-cell antibodies arising in diverse situations (autoimmunity, pregnancy, chronic viral and inflammatory diseases, ageing). The aetiology, nature and significance of these antibodies is yet to be determined, but they may be a family as diverse and potentially important as the anti-nuclear antibodies in SLE.

Genetic and viral factors may contribute to the formation of anti-lymphocyte antibodies. There is an unconfirmed report of an antibody in SLE that reacts with antigenic determinants of an oncornavirus present on a mouse lymphoma line.[126] Lymphocytotoxic antibodies were found in 57% of asymptomatic relatives of SLE patients compared to 3% in control individuals.[6] These antibodies correlated with anti-RNA antibodies present in 25% of these asymptomatic relatives.[127] Both consanguinity and close household contact may be factors contributing to the development of anti-lymphocytic antibodies in these family members of SLE patients.[127]

Our laboratory has found that 'blocking factors' in active SLE can inhibit T rosette formation by normal lymphocytes.[13] Moreover, the ability of calf thymosin fraction 5 to stimulate T rosette formation is also blocked by SLE serum, suggesting possible interference with specific hormone receptors present on cell membranes.[13]

6. GENETICS AND AUTOIMMUNE DISORDERS

Several chapters in this book deal extensively with the association between histocompatibility antigens, immune response genes and disease. A brief discussion of this association should be made in relation to lymphocyte function and autoimmunity, because (a) the frequent association of specific HLA antigens and autoimmune disorders implies a genetic predisposition to autoimmunity and (b) the predisposition may be due to effects that surface determinants have on mechanisms for lymphocyte recognition of histocompatibility and/or viral antigens. Such effects on recognition mechanisms, as already discussed, may lead to the derangement of T cell regulatory processes that are thought to occur in the development of autoreactivity.

The strongest association between a specific HLA gene and human diseases is that between HLA B27 and ankylosing spondylitis and Reiter's syndrome. More than 90% of these patients carry the B27 allele, compared with less than 8% of controls. This association is curious because neither autoimmunity nor lymphocyte abnormalities are a feature of these diseases.

Not all individuals with HLA B27 develop ankylosing spondylitis. Clearly, the histocompatibility antigens must interact with other factors (e.g. bacteria, viruses, tissue antigens) to produce disease. An interesting example is the association of HLA B8 with a diverse group of organ-specific autoimmune disorders that includes coeliac disease, juvenile onset diabetes mellitus, chronic active hepatitis, dermatitis herpetiformis, myasthenia gravis, idiopathic Addison's disease and Sjögren's syndrome. Perhaps HLA B8 is linked to an autoimmune susceptibility gene, with the actual target organ being determined by other factors (such as latent virus infection).

An increase in genetic susceptibility to hypothetical infectious agents may be present in patients with SLE who have hereditary deficiencies of various complement components, most commonly C2.[128] C2 deficient individuals have an increased frequency of certain histocompatibility antigens such as A10 and B18.[128]

In some conditions, lymphocyte-defined loci (LD) detected by MLR show a stronger association with disease than do serologic loci (SD) defined by HLA typing sera. The association of multiple sclerosis with the LD locus 7a is an example. Some years ago, Astorga and Williams reported that lymphocytes from patients with RA failed to respond to each other in MLR although they reacted well with lymphocytes from non-RA individuals.[48] Stastny has recently presented evidence that a common LD locus may be present in a majority of patients with RA.[129] This very interesting report suggests that genetic susceptibility based on lymphocyte surface antigens may be present in this most common autoimmune disorder.

7. IMMUNOSUPPRESSION AND IMMUNOSTIMULATION

Autoimmunity is properly viewed as a two-stage process in which a breakdown in immunologic regulation unleashes immunologic effector mechanisms which result in auto-antibody formation and tissue destruction. The ideal immune therapy should be directed at achieving an immunologic balance that would activate regulatory mechanisms and suppress effector mechanisms, thereby restoring control and reducing tissue damage.

Corticosteroids and immunosuppressive drugs are commonly used to treat autoimmune diseases. The effects of these agents on the immune system are non-specific and potentially harmful. Whereas patients do benefit and autoimmune processes may subside, the drugs bring with them a host of side effects and complications, including increased susceptibility to unusual infections and malignancy.

Corticosteroids decrease lymphocyte response to phytomitogens, decrease cutaneous hypersensitivity reactions, and decrease immunoglobulin production. Both T and B lymphocytes decreased as an acute response to 60 mg of prednisone, but T cells were more severely effected so that the percentage of B cells actually increased.[130] In six patients with active SLE treated with prednisone for five days, response to mitogens and skin antigens improved in the four individuals who also showed a clinical response.[25] Although the exact way in which corticosteroids act is uncertain, they may influence immune reactivity by causing a redistribution of cells out of the circulation and into other body compartments.[131]

Cyclophosphamide is a potent immunosuppressive agent that inhibits both humoral and cellular immune responses. Patients with SLE, RA and scleroderma treated with cyclophosphamide showed a decrease in absolute numbers of both T and B lymphocytes.[132–134] RA patients treated with cyclophosphamide showed decreased rheumatoid factor titres and immunoglobulin concentrations, lymphopenia and decreased mitogen responses.[135,136] In SS, treatment with either corticosteroids or cyclophosphamide can decrease salivary gland lymphoid infiltrates and result in less immunoglobulin and rheumatoid factor production locally in the involved tissues.[57]

In contrast to corticosteroids, azathioprine may have a more specific effect on B lymphocytes rather than T lymphocytes. In six patients with MS treated for one year, there was no effect on T cell function but a depressed response to pokeweed mitogen and to heterologous anti-immunoglobulin antiserum.[137]

Immune depletion brought about by thoracic duct drainage has been studied in RA. Reduction in T lymphocytes was associated with clinical improvement and decrease in the size of rheumatoid nodules. However, disease recurred when drainage was stopped.

Studies currently under way are evaluating the possible benefits of immune potentiation in autoimmune diseases. Levamisole, an anti-helminthic drug that stimulates lymphocytes, macrophages and leukocytes, has been used to treat patients with RA. Transfer factor therapy has been employed in some patients with Bechet's syndrome and juvenile RA. In some instances, clinical improvement was associated with increased immune function. These random studies suggest that careful double-blind control investigations should be performed to evaluate adequately the role of these new modalities on T and B cell function and clinical status in autoimmune disorders.

8. CONCLUSIONS

Studies in NZB mice and in human autoimmune diseases suggest that abnormalities of both B and T lymphocytes occur frequently. Genetic, immunologic and viral factors combine to produce these disorders. Abnormalities of immunologic regulation (helper and suppressor function) as well as 'blocking' effects due to immune complexes and anti-lymphocyte antibodies are commonly encountered in autoimmunity, making it difficult to distinguish primary from secondary manifestations.

Techniques to study suppressor T cells and immunologic regulation in humans are currently being developed in several laboratories. As we learn more about these aspects of immune function in man, exact pathogenetic mechanisms involved in these diseases may become clearer.

Acknowledgements

This research was supported by grants USPHS AM16140 and CA 15684, and research support from the Veterans Administration and the Arthritis Foundation.

9. REFERENCES

1. Talal, N. (1974). *Progr. Clin. Imm.*, **2**, 101–120.
2. Talal, N. (1970). *Arthritis Rheum.*, **13**, 887–894.
3. Papoian, R., Pillarisetty, R. and Talal, N. (1976). (In press.)
4. Dauphinee, M. J., Palmer, D. and Talal, N. (1975). *J. Immunol.*, **115**, 1054–1059.
5. Lies, R. B., Messner, R. P. and Williams, R. C., Jr. (1973). *Arthritis Rheum.*, **16**, 369–375.
6. DeHoratius, R. J. and Messner, R. P. (1975). *J. of Clin. Invest.*, **55**, 1254–1258.
7. Winfield, J. B., Winchester, R. J., Wernet, P., Fu, S. M. and Kunkel, H. G. (1975). *Arthritis Rheum.*, **18**, 1–8.
8. Butler, W. T., Sharp, J. T., Rossen, R. D., Lidsky, M. D., Mittal, K. K. and Gard, D. A. (1972). *Arthritis Rheum.*, **15**, 231–238.
9. Aiuti, F., *et al.* (1975). *Clin. Immunol. Immunopathol.*, **3**, 584–597.
10. Williams, R. C., Jr., DeBoard, J. R., Mellbye, O. J., Messner, R. P. and Lindstrom, F. D. (1973). *J. Clin. Invest.*, **52**, 283–295.
11. Messner, R. P., Lindstrom, F. D. and Williams, R. C., Jr. (193). *J. Clin. Invest.*, **52**, 3046–3056.
12. Scheinberg, M. A. and Cathcart, E. S. (1974). *Cell. Immunol.*, **12**, 309–314.
13. Moutsopoulos, H., Fye, K. H., Sawada, S., Goldstein, A. and Talal, N. (1976). *Clin. exp. Immunol.* (in press).
14. Winchester, R. J., Winfield, J. B., Siegal, F., Wernet, P., Bentwich, Z. and Kunkel, H. G. (1974). *J. Clin. Invest.*, **54**, 1082–1092.
15. Wernet, P., Fotino, M., Thoburn, R., Moore, A. and Kunkel, H. G. (1973). *Arthritis Rheum.*, **16**, 137 (abstract).
16. Block, S. R., Gibbs, C. B., Stevens, M. B. and Shulman, L. W. (1968). *Ann. Rheum. Dis.*, **27**, 311–318.
17. Hahn, B. H., Bagby, M. K. and Osterland, C. K. (1973). *Amer. J. Med.*, **55**, 25–31.
18. Abe, T. and Homma, M. (1971). *Acta Rheum. Scand.*, **17**, 35–46.
19. Horwitz, D. A. and Cousar, J. B. (1975). *Amer. J. Med.*, **58**, 829–835.
20. Goldman, J. A., Litwin, A., Adams, L. E., Krueger, R. C. and Hess, E. V. (1972). *J. Clin. Invest.*, **51**, 2669–2677.
21. Azoury, F. J. and Hess, J. W. (1967). *J. Amer. Med. Ass.*, **201**, 97–101.
22. Azoury, F. J., Jones, H. E., Derbes, V. J. and Gum, O. B. (1966). *Ann. Intern. Med.*, **65**, 1221–1228.
23. Fardal, R. W. and Winkelmann, R. K. (1965). *Arch. Dermatol.*, **91**, 503–511.
24. Malave, I., Layrisse, Z. and Layrisse, M. (1975). *Cell. Immunol.*, **15**, 231–236.
25. Rosenthal, C. J. and Franklin, E. C. (1975). *Arthritis Rheum.*, **18**, 207–217.
26. Horwitz, D. A. (1972). *Arthritis Rheum.*, **15**, 353–359.
27. Senyk, G., Hadley, W. K., Attias, M. R. and Talal, N. (1974). *Arthritis Rheum.*, **17**, 553–562.
28. Patrucco, A., Rothfield, N. F. and Hirschhorn, K. (1967). *Arthritis Rheum.*, **10**, 32–37.
29. Hahn, B. H., Bagby, M. K. and Osterland, C. K. (1973). *Amer. J. Med. Sci.*, **266**, 193–201.

30. Suciu-Foca, N., Buda, J. A., Thiem, T. and Reemtsma, K. (1974). *Clin. exp. Immunol.*, **18**, 295–301.
31. Utermohlen, V., Winfield, J. B., Zabriskie, J. B. and Kunkel, H. G. (1974). *J. exp. Med.*, **139**, 1019–1024.
32. Jasin, H. E. and Ziff, M. (1975). *Arthritis Rheum.*, **18**, 219–228.
33. Feldmann, J. L., Becker, M. J., Moutsopoulos, H., Fye, K., Blackman, M., Epstein, W. V. and Talal, N. (1976). *J. Clin. Invest.* (in press).
34. Mellbye, O. J., Messner, R. P., DeBord, J. R. and Williams, R. C. Jr. (1972). *Arthritis Rheum.*, **15**, 371–380.
35. Papamichail, M., Brown, J. C. and Holborow, E. J. (1971). *Lancet*, **2**, 850–852.
36. Froland, S. S., Natvig, J. B. and Husby, G. (1973). *Scand. J. Immunol.*, **2**, 67–73.
37. Talal, N., Grey, H. M., Zvaifler, N., Michalski, J. P. and Daniels, T. E. (1975). *Science*, **187**, 1196–1198.
38. Tannenbaum, H., Pinkus, G. S., Anderson, L. G. and Schur, P. H. (1975). *Arthritis Rheum.*, **18**, 305–314.
39. Van Boxel, J. A. and Paget, S. A. (1975). *New Engl. J. Med.*, **293**, 517–520.
40. Ziff, M. (1974). *Arthritis Rheum.*, **17**, 313–319.
41. Smiley, J., Sachs, C. and Ziff, M. (1968). *J. Clin. Invest.*, **47**, 624–632.
42. Mellors, R., Keimar, R., Corcos, J. *et al.* (1959). *J. exp. Med.*, **110**, 875–886.
43. Epstein, W. L. and Jessar, R. A. (1959). *Arthritis Rheum.*, **2**, 178–181.
44. Waxman, J., Lockshin, M. D., Schnapp, J. J. and Doneson, I. N. (1973). *Arthritis Rheum.*, **16**, 499–506.
45. Kacaki, J. N., Bullock, W. E. and Vaughan, J. H. (1969). *Lancet*, **1**, 1289–1290.
46. Runge, L. A. and Mills, J. A. (1971). *Arthritis Rheum.*, **14**, 631–638.
47. Weisbart, R. H., Bluestone, R. and Goldberg, L. S. (1975). *Clin. exp. Immunol.*, **20**, 409–417.
48. Astorga, G. P. and Williams, R. C. Jr. (1969). *Arthritis Rheum.*, **12**, 547–554.
49. Williams, R. C. Jr., Lies, R. B. and Messner, R. P. (1973). *Arthritis Rheum.*, **16**, 597–605.
50. Anderson, L. G. and Talal, N. (1972). *Clin. exp. Immunol.*, **10**, 199–221.
51. Van Boxel, J. A., Hardin, J. A., Green, I. and Paul, W. E. (1973). *New Engl. J. Med.*, **289**, 823–827.
52. Talal, N., Sylvester, R. A., Daniels, T. E., Greenspan, J. S. and Williams, R. C. Jr. (1974). *J. Clin. Invest.*, **53**, 180–189.
53. Preud'homme, J. L., Seligmann, M. (1972). *Proc. Nat. Acad. Sci.*, **69**, 2132–2135.
54. Utsinger, P. D., Yount, W. J., Zvaifler, N. J., Bluestein, H. G. and Fallon, J. G. (1976). *Arthritis Rheum.* (in press).
55. Chused, T. M., Hardin, J. A., Frank, M. M. and Green, I. (1974). *J. Immunol.*, **112**, 641–648.
56. Talal, N., Asofsky, R. and Lightbody, P. (1970). *J. Clin. Invest.*, **49**, 49–54.
57. Anderson, L. G., Cummings, N. A., Asofsky, R., Hylton, M. B., Tomasi, T. B. Jr., Wolf, R. O., Schall, G. L. and Talal, N. (1972). *Amer. J. Med.*, **53**, 456–463.
58. Leventhal, B. G., Waldorf, D. S. and Talal, N. (1967). *J. Clin. Invest.*, **46**, 1338–1345.
59. Berry, H., Bacon, P. A. and Davis, J. D. (1972). *Ann. Rheum. Dis.*, **31**, 298–302.
60. Michalski, J. P., Daniels, T. E., Talal, N. and Grey, H. M. (1975). *New Engl. J. Med.*, **293**, 1228–1231.
61. Lueker, R. D., Abdin, Z. H. and Williams, R. C. Jr. (1975). *J. Clin. Invest.*, **55**, 975–985.
62. Pachman, L. M. and Fox, E. N. (1970). *J. Immunol.*, **105**, 898–907.
63. Kaplan, M. H. and Frengley, J. D. (1969). *Amer. J. Cardiol.*, **24**, 459–473.
64. Hess, E. V., Fink, C. W., Taranta, A. and Ziff, M. (1964). *J. Clin. Invest.*, **43**, 886–893.
65. Keiser, H., Kushner, I. and Kaplan, M. H. (1971). *J. Immunol.*, **106**, 1593–1601.
66. Cuppari, G., Quagliata, F., Ieri, A. and Taranta, A. (1972). *J. Lab. Clin. Med.*, **80**, 165–178.
67. Read, S. E., Fischetti, V. A., Utermohlen, V., Falk, R. E. and Zabriskie, J. B. (1974). *J. Clin. Invest.*, **54**, 439–450.
68. McLaughlin, J. F., Paterson, P. Y., Hartz, R. S. and Embury, S. H. (1972). *Arthritis Rheum.*, **15**, 600–608.
69. Lueker, R. D. and Williams, R. C. Jr. (1972). *Circulation*, **66**, 655–660.
70. Currie, S., Saunders, M., Knowles, M. *et al.* (1971). *Quart. J. Med.*, **40**, 63–84.
71. Johnson, R. L., Fink, C. W. and Ziff, M. (1972). *J. Clin. Invest.*, **51**, 2435–2449.
72. Dawkins, R. L. and Mastaglia, F. L. (1973). *New Engl. J. Med.*, **288**, 434–438.
73. Dudley, F. J., Fox, R. A. and Sherlock, S. (1972). *Lancet*, **1**, 723–726.
74. DeHoratius, R. J., Strickland, R. G. and Williams, R. C. Jr. (1974). *Clin. Immunol. Immunopathol.*, **2**, 353–360.

75. Giustino, V., Dudley, F. J. and Sherlock, S. (1972). *Lancet*, **2**, 850–853.
76. Wicks, R. C., Kohler, P. F. and Singleton, J. W. (1975). *Amer. J. Dig. Dis.*, **20**, 518–522.
77. Martini, G. A., Rossler, R., Havemann, K. and Dolle, W. (1970). *Scand. J. Gastroenterol. Suppl.*, **7**, 39–42.
78. Dudley, F. J., Giustino, V. and Sherlock, S. (1972). *Brit. Med. J.*, **4**, 754–756.
79. Wands, J. R., Perrotto, J. L., Alpert, E. and Isselbacher, K. J. (1975). *J. Clin. Invest.*, **55**, 921–929.
80. Wands, J. R. and Isselbacher, K. J. (1975). *Proc. Nat. Acad. Sci.*, **72**, 1301–1303.
81. Strickland, R. G., Korsmeyer, S., Soltis, R. D., Wilson, I. D. and Williams, R. C. Jr. (1974). *Gastroenterol.*, **67**, 569–577.
82. Fletcher, J. and Hinton, J. M. (1967). *Lancet*, **2**, 753–754.
83. Jones, J. V., Housley, J., Ashurst, P. M. and Hawkins, C. F. (1969). *Gut*, **10**, 52–56.
84. Aas, J., Huizenga, K. A., Newcomer, A. D. and Shorter, R. G. (1972). *Scand. J. Gastroenterol.*, **7**, 299–303.
85. Guillou, P. J., Brennan, T. G. and Giles, G. R. (1973). *Gut*, **14**, 20–24.
86. Asquith, P., Kraft, S. C. and Rothberg, R. M. (1973). *Gastroenterol.*, **65**, 1–6.
87. Watson, D. W., Quigley, A. and Bolt, R. J. (1966). *Gastroenterol.*, **51**, 985–993.
88. Shorter, R. G., Spencer, R. J., Huizenga, K. A. and Hallenbeck, G. A. (1968). *Gastroenterol.*, **54**, 227–231.
89. Shorter, R. G., Cardoza, M., Spencer, R. J. and Huizenga, K. A. (1969). *Gastroenterol.*, **56**, 304–309.
90. Shorter, R. G., Huizenga, K. A., Spencer, R. J. and Guy, S. K. (1972). *Amer. J. Dig. Dis.*, **17**, 689–696.
91. Shorter, R. G., Cardoza, M., Huizenga, K. A., ReMine, S. G. and Spencer, R. J. (1969). *Gastroenterol.*, **57**, 30–35.
92. Shorter, R. G., Huizenga, K. A. and Spencer, R. J. (1972). *Amer. J. Dig. Dis.*, **17**, 1024–1031.
93. Strickland, R. G., Friedler, E. M., Henderson, C. A., Wilson, I. D. and Williams, R. C. Jr. (1975). *Clin. exp. Immunol.*, **21**, 384–393.
94. Fagraeus, A. and Rose, N. R. (1974). *Progr. Immunol. II*, **5**, 325–328.
95. Farid, N. R., Munro, R. E., Row, V. V. and Volpe, R. (1973). *New Engl. J. Med.*, **288**, 1313–1317.
96. Urbaniak, S. J., Penhale, W. J. and Irvine, W. J. (1973). *Clin. exp. Immunol.*, **15**, 345–354.
97. Ehrenfeld, E. M., Klein, E. and Benezra, D. (1971). *J. Clin. Endocrinol. Metab.*, **32**, 115–116.
98. Laryea, E., Row, V. V. and Volpe, R. (1973). *Clin. Endocrinol.*, **2**, 23–35.
99. Calder, E. A. (1974). *Proc. Roy. Soc. Med.*, **67**, 502–506.
100. Nerup, J. (1974). *Acta Endocrinol.*, **76**, 142–158.
101. Kamp, P., Platz, P. and Nerup, J. (1974). *Acta Endocrinol.*, **76**, 729–740.
102. Bottazo, G. F., Florin-Christensen, A. and Doniach, D. (1974). *Lancet*, **2**, 1279–1282.
103. MacLaren, N. K., Huang, S. W. and Fogh, J. (1975). *Lancet*, **1**, 997–1000.
104. MacCuish, A. C., Jordan, J., Campbell, C. J., Duncan, L. J. P. and Irvine, W. J. (1974). *Diabetes*, **23**, 693–697.
105. Gepts, W. (1965). *Diabetes*, **14**, 619–633.
106. Flier, J. S., Kahn, C. R., Roth, J. and Bar, R. S. (1975). *Science*, **190**, 63–65.
107. Abdou, N. I., Lisak, R. P., Zweiman, B., Abrahamsohn, I. and Penn, A. S. (1974). *New Engl. J. Med.*, **291**, 1271–1275.
108. Armstrong, R. M., Nowak, R. M. and Falk, R. E. (1973). *Neurology*, **23**, 1078–1083.
109. Goldstein, G. (1975). *Ann. N.Y. Acad. Sci.*, **249**, 177–185.
110. Appel, S. H., Almon, R. R. and Levy, N. (1975). *New Engl. J. Med.*, **293**, 760–761.
111. Paterson, P. Y. (1973). *J. Chronic. Dis.*, **26**, 119–126.
112. Nowak, J. and Wajgt, A. (1975). *Clin. exp. Immunol.*, **21**, 278–283.
113. Kolar, O. J., Ross, A. T. and Herman, J. T. (1970). *Neurology*, **20**, 1052–1061.
114. Davis, L. E., Hersh, E. M., Curtis, J. E. *et al.* (1972). *Neurology*, **22**, 989–997.
115. Sheremata, W., Cosgrove, J. B. R. and Eylar, E. H. (1974). *New Engl. J. Med.*, **291**, 14–17.
116. Weksler, M. E. and Hutteroth, T. H. (1974). *J. Clin. Invest.*, **53**, 99–104.
117. Waldmann, T. A., Broder, S., Blaese, R. M., Durm, M., Blackman, M. and Strober, W. (1974). *Lancet*, **2**, 609–613.
118. Mittal, K. K., Rosen, R. D., Sharp, J. T., Lidsky, M. D. and Butler, W. T. (1970). *Nature*, **225**, 1255–1256.
119. Terasaki, P. I., Mottironi, V. D. and Barnett, E. V. (1970). *New Engl. J. Med.*, **283**, 724–728.

120. Winfield, J. B., Winchester, R. J., Wernet, P. and Kunkel, H. G. (1975). *Clin. exp. Immunol.*, **19**, 399–406.
121. Messner, R. P., Kennedy, M. S. and Jelinek, J. G. (1975). *Arthritis Rheum.*, **18**, 201–206.
122. Stastny, P. and Ziff, M. (1971). *Arthritis Rheum.*, **14**, 733–736.
123. Winfield, J. B., Winchester, R. J. and Kunkel, H. G. (1975). *Arthritis Rheum.*, **18**, 587–594.
124. Williams, R. C. Jr., Lies, R. B., Messner, R. P. (1973). *Arthritis Rheum.*, **16**, 597–605.
125. Winchester, R. J., Fu, S. M., Wernet, P., Kunkel, H. G., Dupont, B. and Jersild, C. (1975). *J. exp. Med.*, **141**, 924–929.
126. Lewis, R. M., Tannenberg, W., Smith, C. and Schwartz, R. S. (1974). *Nature*, **252**, 78–79.
127. DeHoratius, R. J., Pillarisetty, R., Messner, R. P. and Talal, N. (1975). *J. Clin. Invest.*, **56**, 1149–1154.
128. Schur, P. H. (1975). *Clin. Rheum. Dis.*, **1**, 519–544.
129. Stastny, P. (1974). *Tissue Antigen.*, **4**, 571–579.
130. Yu, D. T. Y., Clements, P. J., Paulus, H. E., Peter, J. B., Levy, J. and Barnett, E. V. (1974). *J. Clin. Invest.*, **53**, 565–571.
131. Fauci, A. S. and Dale, D. C. (1974). *J. Clin. Invest.*, **53**, 240–246.
132. Clements, P. J., Yu, D. T. Y., Levy, J., Paulus, H. E. and Barnett, E. V. (1974). *Arthritis Rheum.*, **17**, 347–354.
133. Ziff, M., Hurd, E. R. and Stastny, P. (1974). *Proc. Roy. Soc. Med.*, **67**, 536–540.
134. Hurd, E. R. and Giuliano, V. J. (1975). *Arthritis Rheum.*, **18**, 67–75.
135. Alepa, F. P., Zvaifler, N. J. and Siliwinski, A. J. (1970). *Arthritis Rheum.*, **13**, 754–760.
136. Hurd, E. R. and Ziff, M. (1974). *Arthritis Rheum.*, **17**, 72–78.
137. Abdou, N. I., Zweiman, B. and Casella, S. R. (1973). *Clin. exp. Immunol.*, **13**, 55–63.

Chapter 17

B and T Lymphocytes in Immunodeficiency and Lymphoproliferative Diseases

M. D. Cooper
M. Seligmann

Abbreviations

ADA: adenosine deaminase activity
ALL: acute lymphoblastic leukaemia
CLL: chronic lymphocytic leukaemia
MLR: mixed lymphocyte reaction
PHA: phytohaemagglutinin
PW: pokeweed mitogen
SCID: severe combined immunodeficiency
SmIg: surface membrane-bound immunoglobulin
WM: Waldenstrom's macroglobulinaemia
XLA: X-linked agammaglobulinaemia

1. INTRODUCTION

At present, the pathogenesis of most immunodeficiency and lymphoproliferative diseases can be best considered as defects at certain points in stem cell differentiation along the T and B cellular pathways.[1] The use of surface markers for identification and enumeration of human T and B lymphocytes has proved especially valuable for diagnosis and for pathophysiological considerations of these diseases. Relevant information on normal lymphoid development and differentiation in large part must be derived by extrapolation from animal studies. On the other hand, special insight is also provided by the developmental arrests seen among the immunodeficiency syndromes and exaggerated clonal proliferation at fixed stages in differentiation in lymphoproliferative diseases.[1–3] Investigation of such patients may prove to be particularly helpful in the definition of subclasses of human T and B cells. Concepts derived from this blend of information can sometimes be directly validated by repair of the immunodeficiency in patients by the appropriate transplantation of stem cells, embryonic thymus or even T and B cells.

Before beginning a consideration of human immunodeficiencies and lymphoid malignancies from the viewpoint of faulty T and B cell development, we will review briefly some of the features of B cell development in humans. This information will be used to outline an operational scheme of B cell differentiation. Although seen most easily as it unfolds during ontogeny, the pathway is probably used throughout life for clonal renewal. We therefore assume that the same scheme can be used as a guide for analysing both congenital and acquired defects in differentiation. The same reasoning can be applied to the analysis of T cell abnormalities but here less is known about the specifics of T cell ontogeny (and functional subpopulations) in humans so that we must rely more heavily on information from animal studies. This body of knowledge is reviewed in other chapters.

B cell differentiation can be detected first in human foetal liver by IgM expression, initially in the cytoplasm then later at the cell surface. After a lag of 2–3 weeks, some IgM bearing cells also begin to express surface membrane-bound IgD (SmIgD). As B cells seed into the circulation to settle in spleen and lymph nodes, the proportion of $SmIgM^{+}$ cells that also bear SmIgD reaches approximately 80%. By contrast, in foetal liver and later in bone marrow, tissues which are thought to be sites of B cell generation, the proportion of SmIgM+SmIgD doubles usually remains less than 50%. The expression of SmIgG and SmIgA on B lymphocytes also follows IgM expression during ontogeny (References 4 and 5, and Cooper, Gathings and Lawton, unpublished observations). As has been shown in birds and rodents, the sequential expression of genes for the heavy chain constant regions probably occurs in humans by a switch mechanism and allows each of the C region genes to be expressed with the same sets of V region genes coding for the variable regions which determine antibody specificity (reviewed in References 6 and 7).

Our synthesis of the available information on normal ontogeny, experimental manipulations and human diseases of B cell differentiation is represented in the differentiation scheme shown in Figure 1, which will be discussed in detail elsewhere.

It is generally accepted that lymphoid differentiation can be divided conveniently into two stages, an idea that is implicit in the clonal selection theory of Burnet. The initial stage in B cell differentiation, and probably in T cell differentiation as well, involves the antigen-independent generation of diversity. The current debate about whether the initial expression of Ig class diversity is or is not regulated by antigens with T cell help is beyond the scope of this discussion. In either case the second stage, which encompasses the differentiative events of the antibody response, is regulated by antigen triggering, T cell interactions, macrophages and circulating antibody, as is discussed elsewhere in this book. At a cellular level this stage begins with activation of resting virgin lymphocytes, continues with memory cell production and ends with plasma cell differentiation.

2. METHODS USED FOR THE IDENTIFICATION OF B AND T CELLS IN MAN

Currently available markers of human B and T cells are listed in Table 1. Although most of these techniques are now reasonably well standardized,[19] they are exposed to a number of pitfalls. The diagnostic value of these markers and hence the implication of such studies for the pathogenesis of immunodeficiency and lymphoproliferative diseases rely heavily on a critical evaluation of the methods.

Surface membrane bound immunoglobulins (SmIg) detectable by immunofluorescence constitute the most reliable marker of B cells. They give evidence for the genetic commitment of the cell when truly monospecific antisera to various Ig chains, subclasses and allotypes are used (reviewed in Reference 20). They may also represent a clonal marker of B cell proliferations. It should be emphasized however that the mere presence of Ig at the surface of a cell does not necessarily mean that they are produced by that cell. Erroneous interpretations may result from the attachment of circulating immune complexes or IgG aggregates to any cell carrying receptors for C3 or Fc, from an anti-IgG (rheumatoid factor) activity of membrane-bound IgM or from the presence of antibodies to any lymphocyte surface determinants. In these circumstances, *in vitro* experiments (such as trypsinization followed by short-term culture) may be required to ensure that surface Ig are synthesized by the cells under study.[21] The problem of extrinsic versus native surface Ig is especially important for IgG bearing cells because of the presence of receptors

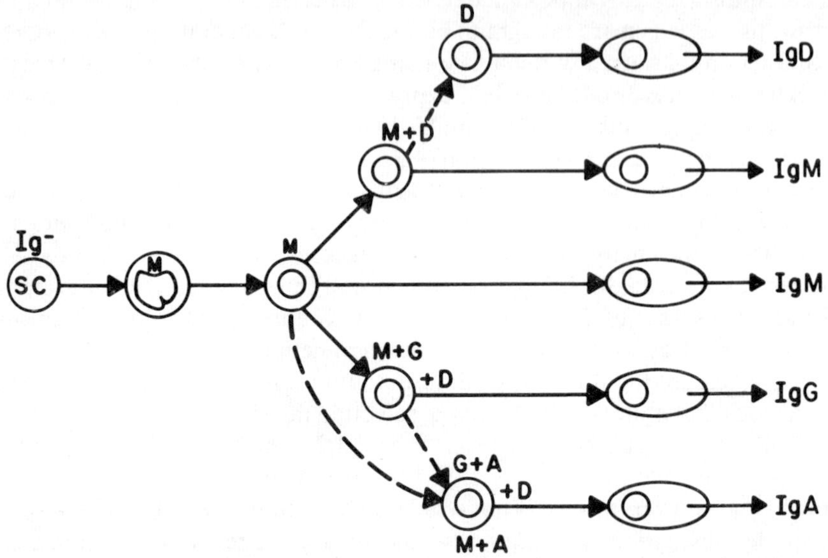

Figure 1. This diagrammatic scheme attempts to illustrate some of our views on possible sequences of generation of B cells expressing different classes of immunoglobulin (Ig). SC denotes the haemopoietic stem cells. The letters M, G, A and D represent classes of Ig inside or on the surface of B cells.

This model is based on convincing evidence which indicates that (a) B cells in man and other mammals initially express cytoplasmic IgM without stable surface IgM receptors (SmIgM), (b) the next stage in B cell differentiation is represented by cells expressing SmIgM only and (c) B cells producing other Ig classes are the progeny of cells that formerly expressed SmIgM only. A more complete version of this model will be presented elsewhere (Abney, E. A., *et al.*), but a brief discussion of experimental evidence supporting it has been published.[187]

SmIgD is expressed early in the development of all of the B cell sublines which are genetically committed to terminal plasma cell differentiation and production of a single class of Ig. As maturation proceeds, SmIgD is lost. The model also reflects the idea that B cells which switch from IgM to the expression of other Ig classes may continue to express multiple Ig classes on their surface; not shown in this diagram is the fact that their progeny may bear only the final Ig class selected for expression. The dotted line indicates the possibility that SmIgM positive cells can switch directly to IgA expression.

Evidence from immunodeficiency and lymphoproliferative diseases that is relevant to the possible pathways of differentiation outlined in this scheme is discussed in the text, but further evidence is needed to elucidate all of the correct sequences and mechanisms for switches in expression of constant region genes during differentiation of B cell clones.

for the Fc of IgG on monocytes and some T lymphocytes in addition to B lymphocytes. In fact, if one removes labile extrinsic IgG by preincubation at 37 °C[22] and performs the direct immunofluorescence test with the Fab$'_2$ fragment of rabbit antibodies to IgG,[23] very low figures for IgG bearing lymphocytes are found in normal peripheral blood. It should be stressed, however, that even after excluding extrinsic IgG, a small subpopulation of IgG producing B lymphocytes is definitely found in normal blood and tissues. This statement is in good agreement with the finding of monoclonal proliferations of IgG-synthesizing lymphocytes in some patients and with the inhibition of IgG precursors by antisera to γ chains. The presence of monocytic cells which may resemble lymphocytes in the cell suspensions accounts partly for the high figures reported by some investigators for IgG bearing lymphocytes in normal human blood. The important problem raised by these monocytic cells is partly solved by their identification by the endogenous peroxidase activity, the demonstration of which can be easily coupled to immunofluorescence tests,[24] and by the use of specific anti-macrophage antisera.[25]

Monocytes also express C and Fc receptors. Although commonly used as markers of B lymphocytes, these receptors cannot be considered as really specific for these cells since they are also found on a minor subset of T cells and on the so-called third population, a term coined by Fröland and Natvig[26] for some 'non-B non-T;, small to medium-sized, non-phagocytic mononuclear cells that may play an important role in antibody dependent cytotoxicity (K cells). Recent data suggests that these cells may be related either to the monocytic series[25] or to T cell lineage.[23]

Table 1. Currently available markers of human B and T cells

Marker	B lymphocytes	T lymphocytes
Easily detectable SmIg (if actual cell product)	+	–
C receptors[a] (EAC rosettes)	+ (or –)[b]	– (or +)[c]
Receptor for Fc or IgG		
EA rosettes	+ (or –)[d]	– (or +)[c]
Aggregated IgG binding	+	– (or +)[c]
EBV receptor[11,12]	+	–
Specific antigens recognized by anti-B heteroantisera	+	–
Mouse E rosettes[177]	+[b]	–
Sheep F rosettes	–	+[e]
Helix pomatia haemagglutinin[14]	–	+
Measles virus receptor[15]	–	+
Rosettes with lymphoblastoid[16] B cell-lines	–	+
Receptor for Fc of IgM[17]	–	+[b]
Receptor for C-reactive protein[18]	–	+[b]
Specific antigens recognized by anti-T heteroantisera	–	+

[a] Distinct receptors for C3b and C3d[8,9] and C4.[10]
[b] Only a subpopulation.
[c] Some T cells (mostly activated(?)) are positive for these markers.
[d] Results depending upon the nature of erythrocytes and the nature and dose of the sensitizing IgG antibodies.
[e] A modified technique (so called 'active' rosettes)[13] demonstrates binding of sheep erythrocytes by aproximately one-third of the T cells.

Many heteroantisera revealing specific B or T antigens can be used only in cytotoxicity tests because they are only relatively specific. This procedure is less satisfactory than immunofluorescence since it does not allow simultaneous checking of other markers or direct examination of the positive cells. Further problems arise when antisera to B or T cells are used for the study of neoplastic cells which may experience surface changes and lose some normal membrane antigens. Neoplastic cells may also express membrane antigens only at certain stages of the cell cycle or express foetal or tumour-associated antigens, and these occurrences may invalidate results obtained with antisera to foetal or leukaemic cells.

In view of these limitations and pitfalls, the need for using a panel of several B and T membrane markers in the study of immunodeficiency and lymphoproliferative diseases is obvious. Very little is known about the relationships between the expression sequence of currently available markers and the stages of differentiation of human B and T cells, and so far no well-defined markers exist for functional subpopulations of human T cells (see also 'Notes added in proofs').

3. IMMUNODEFICIENCY DISEASES

3.1 Deficiencies of T and B cells

3.1.1 *General features of severe combined immunodeficiency (SCID)*

Severe combined immunodeficiency (SCID) is a syndrome defined by gross functional impairment of both humoral and cell-mediated immunity.[27] Consequently, affected individuals have rampant infections with the whole range of microorganisms, including viruses, bacteria, fungi and poorly classified organisms such as *Pneumocystis carinii.* Lacking the ability to reject cells of any kind, some of these patients have had acute, fatal graft-versus-host disease on receiving viable histo-incompatible lymphocytes in blood transfusions. Unless the immunodeficiency can be repaired, affected individuals usually die within a few months. Mainly for this reason SCID is almost never seen in adults, unless both T and B cells are abolished by a catastrophic event such as whole body X-irradiation.

The syndrome can result from an extraordinary diversity of developmental flaws of the immune system, the classification of which rests on a pragmatic approach considering the cellular level of the defect in differentiation, the inheritance pattern and, in one instance, an associated enzyme deficiency.[1,27]

3.1.2 *SCID with general haemopoietic hypoplasia (De Vaal's syndrome)*

Apparently the most severe stem cell defect compatible with live birth is the syndrome of reticular dysgenesia in which affected newborns are deficient in granulocytic development as well as in all lymphoid elements.[28] These infants have died within a few hours of birth. It is likely that more severe stem cell defects are incompatible even with *in utero* survival.

3.1.3 *SCID with T and B cell deficiencies ('Swiss-type' agammaglobulinaemia)*

The classic example of SCID is characterized by a virtual absence of both T and B cell-lines, is usually inherited in an autosomal recessive pattern and may also occur sporadically.[29] This defect can be repaired promptly by the T and B cells in histocompatible bone marrow, and lasting correction may be achieved by donor stem cell differentiation along these lines.[30] This argues well for the stem cell nature of the inherited defect.

3.1.4 *SCID and adenosine deaminase deficiency*

A virtually identical pattern of SCID may occur in association with a deficiency of adenosine deaminase activity (ADA) which is inherited as an autosomal recessive defect.[31] This enzyme, an aminohydrolase that catalyses the conversion of adenosine to inosine, appears to be necessary for normal lymphoid differentiation especially along T cell-lines, for some ADA deficient infants have had substantial B cell development. The enzyme normally is widely distributed throughout the body, and its deficiency in other tissues may be responsible for other clinical manifestations, e.g. central nervous system (CNS) and bony growth abnormalities, that can be seen in ADA deficient infants. Enzyme supplied by the mother may support some lymphoid differentiation *in utero*, since even ADA deficient infants with the most severe form of SCID may have feeble thymic lymphopoiesis, transient circulation of lymphocytes and transient production of IgM. One ADA deficient infant having some poorly-responsive lymphocytes in his blood temporarily developed normal lymphocyte responsiveness and a visible thymic shadow on receiving enzyme replacement by transfusion with normal red cells.[32] The immunologic deficits in other SCID patients with ADA deficiency have been corrected, apparently in a complete and long-lasting fashion, by both foetal liver and histocompatible bone marrow transplants as sources for healthy stem cells capable of ADA production.[1,33,34]

3.1.5 *SCID with 'arrested' development of B lymphocytes*

Another phenotypic form of SCID is characterized by virtual absence of T lymphocytes but normal or near-normal numbers of circulating B lymphocytes and usually presence of some serum IgM.[1] These infants lack the stigmata of the third and fourth pharyngeal pouch syndrome and, when studied, have had usually normal ADA enzyme activity. This form can be inherited in either an X-linked or autosomal recessive fashion. The absence of T cells may be responsible for the failure of normal plasma cell differentiation which is reflected by very low serum levels of IgG and usually absent IgA. The class distribution of surface immunoglobulins can be distinctly abnormal (i.e. virtual absence of $B\gamma$ and $B\alpha$) in such patients.[35] This observation is not so easily explained unless one accepts an important regulatory role for T cells in the development of these B cell sublines. An interesting but at present unexplained finding in one such patient was that all of the circulating lymphocytes had either SmIgM and SmIgD or SmIgD only; lumphocytes with SmIgM only were not detectable although such $B\mu$ cells normally constitute ~20% of the circulating B lymphocytes.[35]

Theoretically, two sorts of cellular defects could account for this form of SCID with a primary deficiency of T cells. It could result either from faulty thymic epithelium or from a stem cell defect that precludes normal T cell differentiation. In support of the first possibility, cells from the bone marrow of an affected boy acquired T cell characteristics when cultured on normal thymic epithelium.[36] On the other hand, a girl with this syndrome had a lasting correction following transplantation with histocompatible bone marrow,[37] but the nature of the cells responsible for the repair is unknown.

3.1.6 *X-linked SCID*

Before surface markers for T and B cells were available, many patients with an X-linked inheritance of SCID were studied.[38,39] These boys had variable degrees of lymphopenia, ranging from severe to moderate, feeble thymic lymphoid development, small foci of lymphocytes scattered about in their lymph nodes and spleen and often significant IgM production. Whether X-linked SCID belongs to the previous category of primary T cell defects or represents a different sort of stem cell defect is presently unknown, but it is noteworthy that three boys affected with this form of SCID had near normal numbers of circulating B cells (Griscelli, Preud'homme and Seligmann, unpublished results). The immunological deficits in some boys with X-linked SCID have been corrected by bone marrow transplants, but here also insufficient information is available to decide if the primary repair is by mature T cells, T and B cells and/or stem cell development along one or both cell-lines.[30]

3.1.7 *SCID with restricted clonal diversity*

Two other phenotypic patterns have been described in infants with the syndrome of SCID. In one of these, serum Ig may be normal but severe lymphopenia, selective antibody responsiveness to a panel of antigens, selective rejection of skin allographs and limited electrophoretic heterogeneity of serum Ig all suggest development of a limited number of clones.[40] Such a defect could result from an early arrest in the generation of clonal diversity.

3.1.8 *SCID with 'arrested' development of T and B lymphocytes*

A few infants with SCID have been seen in whom both T and B cells were present in apparently normal numbers. Yet, in addition to functional unresponsiveness *in vivo*, lymphocytes from such patients have not responded to stimulation *in vitro* with the usual lymphocyte mitogens. Lymphocytes from one such patient were responsive to a calcium ionophore, and it has been suggested that the lymphocyte abnormality and associated blood cell abnormalities could have been due to a common membrane defect.[41]

3.1.9 *SCID induced by auto-antibodies* (?) (*immunologic amnesia*)

Lymphocytotoxins have been detected in a few immunodeficient patients having recurrent bacterial and fungal infections.[42] Most have had fluctuating lymphopenia. Both cellular immunity and specific antibody responses were impaired, although serum Ig levels were usually normal. This syndrome may be due to formation of auto-antibodies to lymphocyte surface antigens.

3.2 Primary T cell deficiencies

3.2.1 *General remarks*

It is somewhat artificial not to include here some of the syndromes described earlier which present as SCID. Similarly some of the antibody deficiencies to be discussed later, such as IgA deficiency, may be due to more subtle abnormalities of T cell function. Part of the difficulty in defining the full spectrum of T cell defects in humans has been the lack of information about T cell subpopulations, antigen receptors and differentiation antigens. Until very recently, laboratory assay of human T lymphocyte function were limited to correlates of cell-mediated immunity such as skin testing for contact and delayed hypersensitivity, skin allograft rejection, mixed lymphocyte reactions, proliferative responses to antigens, responses to T cell mitogens and MIF assays. With the development of appropriate methodology, abnormalities of T cells will almost certainly be revealed to be far more common and heterogenous than is appreciated at present. Already the means to identify and enumerate T cells, to examine the responsiveness of 'null cells' to thymic extracts and to begin to examine T cell interactions with B lymphocytes (to be discussed under B cell deficiencies) have led to recognition of T cell deficiencies that previously were unrecognized.

The kinds of infections that typically plague the patient deficient in T cells and cell-mediated immunity are progressive fungal, mycobacterial and viral invasions. *Pneumocystis carinii* is another common marauder, and candidiasis is so common as to be acclaimed the best diagnostician of T cell deficiency.

3.2.2 *Thymic hypoplasia* (*Di George's syndrome*)

The classic example of isolated T cell deficiency results from maldevelopment of structures derived embryologically from the third and fourth pharyngeal pouches.[43] Affected infants usually present with symptoms and signs referrable to abnormalities of the great vessels or to severe deficiencies in function of parathyroid and thymus glands. Severe lymphopenia is not a prominent early feature, but practically all of the lymphocytes are B cells.[2,44] Usually the class and anatomical distribution of B lymphocytes, plasma cells and all serum immunoglobulins appear normal. Antibody responsiveness is only partially impaired although this has not been evaluated fully.[45] These findings are somewhat puzzling in view of the need for T cell help in the antigen induced maturation of B cells. The explanation may rest with the fact that most if not all such patients may have a tiny thymus, histologically normal and usually in a ectopic location.[46] Affected infants who have been followed over several months eventually may spontaneously develop detectable numbers of functional T cells. Dramatic repair of the T cell deficiency has quickly followed transplantation of a foetal thymus in a few infants with the syndrome,[45,47] and in one instance when the thymus implant was enclosed within a cell impermeable diffusion chamber.[48] Thymus extracts may induce *in vitro* the appearance of cells with T surface markers.[188]

3.2.3 *Nezelof syndrome*

Another deficiency of T cells, which may be inherited in an autosomal fashion, is called the Nezelof syndrome.[49] Affected infants have had a small thymus, embryonic in appearance and lacking in Hassall's corpuscles and lymphocytes, lymphopenia and virtually no lymphoid population of thymus dependent areas in spleen and lymph nodes. On the other hand, plasma cells have been abundant, and serum Ig levels are usually normal although IgG and IgA may be slightly reduced. Although incompletely evaluated, antibody responses appear to be significantly impaired.

3.2.4 *Thymic hormone deficiency: Possible cause of T cell deficiency*

Recently, several children, usually evaluated immunologically because of gastrointestinal and pulmonary infections, have been found to have few circulating T cells but many 'null cells' lacking surface markers characteristic for either B or T cells (References 50 and 51, and Aiuti, personal communication). In many instances, *in vivo* and *in vitro* assays for T cell integrity showed impaired responsiveness while in others the deficiency in T cells was not so evident. Infants in this general category who have died usually had small embryonic-appearing thymuses. The most interesting finding has been that an active thymus extract may induce *in vitro* a significant increase in the numbers of cells displaying surface markers characteristic of T cells. This suggests that the primary deficiency is in production of thymic humoral factors. This idea receives additional support from improvement in T cell numbers and functions in such patients which have followed within hours the implantation of a normal foetal thymus or injections of thymic extracts.[50,51]

3.2.5 *T cell deficiency and related enzyme defects*

It was mentioned earlier that adenosine deaminase deficiency may sometimes be associated with a primary deficiency of T cells. A deficiency in the number of T cells has also been associated, in 3 patients, with an inherited deficiency of nucleotide phosphorylase activity.[52] This enzyme catalyses inosine conversion to hypoxanthine, only one metabolic step beyond adenosine conversion to inosine catalysed by ADA. It will be important to know if this T cell deficiency is correctable by enzyme replacement.

3.2.6 *Aberrant T cell development in ataxia-telangiectasia*

Ataxia-telangiectasia, inherited in an autosomal recessive pattern, is characterized by cerebellar ataxia, oculocutaneous telangiectasia and immunodeficiency (reviewed in References 53 and 54). The latter is frequently manifested by recurrent and chronic sinopulmonary infection. The two most frequent causes of death are chronic pulmonary disease and malignancy, usually lymphomas, although carcinomas may occur also. Maldevelopment of the thymus consistently occurs and may well be the primary basis for the immunological abnormalities in this disease. If found at all after death, the thymus has been very hypoplastic and always embryonic in appearance. Patients' T cells may be deficient in numbers, and reduced lymphocyte responsiveness to T cell mitogens, cutaneous anergy and delayed rejection of skin allografts are common. Although the numbers and class distribution of B lymphocytes are usually normal,[44] most affected individuals are deficient in serum IgA and IgE,[55–57] and a few are deficient in IgG as well. Serum IgM and IgD levels are usually normal or even elevated.[57] This constellation of features suggests a maturational impairment of B lymphocytes that is secondary to a deficiency in T cell function. The finding that lymphocytes from one ataxia-telangiectasia patient with no serum IgA were induced by pokeweed mitogen to become IgA secreting plasma cells is consistent with this hypothesis.[58]

Another interesting clue to the nature of ataxia-telangiectasia is the occurrence of elevated α foetoprotein levels.[59] This, like the embryonic appearance of the thymus, could reflect an abnormality of epithelial differentiation. On the other hand, evidence has recently been brought forth which indicates that α foetoprotein may have a powerful immunosuppressive activity, mediated perhaps through its attachment to membrane receptors present on a subpopulation of T cells.[60]

3.2.7 *Chronic mucocutaneous candidiasis*

The syndrome of chronic mucocutaneous candidiasis is different from the severe T cell deficiencies in that superficial candidiasis is usually the only major manifestation of immunodeficiency. These patients rarely develop systemic infection with Candida or other fungal agents and are not unusually susceptible to virus or bacterial disease. The syndrome is often familial and may be associated with various endocrinopathies, such as hypothyroidism, as well as iron deficiency. Treatment of the associated conditions may lead to improvement or even cure of the Candida infection.[61]

3.2.8 *Acquired deficiency of T cells*

Not much is known yet about late onset forms of T cell deficiency. Using the older parameters for appraising the T cell system, defects may not be found even months after abrupt cessation of T cell renewal and support by thymectomy. While this may be due in part to the long lifespan of some T cells, there is little doubt that many subtle defects will be elucidated by improved knowledge about generation and maintenance of the various subpopulations of T cells and the availability of assays for their functional heterogeneity.

3.3 B cell deficiencies

3.3.1 *Infantile X-linked agammaglobulinaemia (XLA) without B lymphocytes (Bruton's XLA)*

X-linked agammaglobulinaemia (XLA) characterized by susceptibility to recurrent bacterial infections, appears to represent a central failure of B cell development. Almost all affected males have very few immunoglobulin bearing lymphocytes in their circulation or lymphoid tissues (reviewed in References 62 and 63). They often have substantial numbers of small mononuclear cells with receptors for C3 and Fc of IgG.[64,65] Although touted by some as young B lymphocytes, these cells have recently been shown to have markers characteristic of the monocyte line and to lack receptors for Epstein–Barr virus and antigens that are specific for B lymphocytes.[24,66,67] Within the same family some affected males have had substantial serum IgM, IgG and IgA, while other family members have been nearly agammaglobulinaemia.[68] All of these patients have had very few circulating B lymphocytes. This suggests that the few B lymphocytes generated in such patients are fully capable of proliferation and plasma cell differentiation (see also 'Notes added in proofs').

3.3.2 *XLA with B lymphocytes*

A few patients with well documented, X-linked inheritance of panhypogammaglobulinaemia appeared to have normal numbers of Ig bearing lymphocytes.[69–71] If it can be shown that these are native Ig products of the cells that bear them, this would suggest the existence of two distinct forms of XLA, the rarest of which may represent an arrest at a B lymphocyte stage in differentiation.

3.3.3 *Agammaglobulinaemia correctable by vitamin B_{12}*

Arrested B lymphocyte differentiation occurred in one agammaglobulinaemic infant who had an inherited deficiency of transcobalamin II, a serum protein involved in transport of vitamin B_{12} (cobalamin).[72] Treatment with large doses of B_{12} was followed by normal Ig production and correction of maturational defects in other cell-lines (haemopoietic and intestinal epithelium). Normal B lymphocyte numbers were also observed in another severely hypogammaglobulinaemic infant whose maturational block later spontaneously disappeared rather suddenly as evidenced by massive increases in circulating Igs and the numbers of plasma cells in lymphoid tissues.[73] It is not yet clear, however, if this is a common phenomenon among infants with transient hypogammaglobulinaemia.

3.3.4 *IgA deficiency*

At the other end of the spectrum of B cell disorders is the isolated deficiency of IgA (reviewed in References 57, 63 and 74). This is the most frequently encountered immunodeficiency, occurring in approximately 1 in 600 individuals of European ancestry but perhaps less often in other groups of people such as American negroes. IgA deficiency occurs in association with autosomal dominant or recessive patterns of inheritance, sporadically, with deletions of chromosome 18, with congenital infections with rubella, cytomegalovirus and *Toxoplasma gondii* organisms and after hydantoin therapy. It can be a transient phenomenon, but more often is a permanent defect. Adults with isolated IgA deficiency may be healthy. As a group, however, individuals with IgA deficiency have increased numbers of respiratory infections of varying severity, systemic infections, chronic diarrhoeal diseases, autoimmune syndromes such as rheumatoid arthritis and systemic lupus erythematosus and atopic diseases such as asthma, hay fever and eczema. The latter conditions may be reflected by elevated levels of IgE, although a deficiency in

IgE sometimes accompanies the IgA deficiency.[56,57] IgA deficient patients often have a variety of auto-antibodies, and not infrequently form specific antibodies to human IgA.[75]

The pathogenesis of IgA deficiency, whether genetic or caused by environmental insult, apparently involves a defect in terminal differentiation of B lymphocytes. With one apparent exception[76] among more than 60 patients studied (reviewed in Reference 63), normal numbers of, IgA bearing lymphocytes have been found in the circulation. Moreover, in most instances patients' lymphocytes cultured with pokeweed mitogen (PW) could be induced to become mature plasma cells which secrete IgA.[58] The serum from one IgA deficient patient was capable of blocking PW induction of terminal differentiation of normal lymphocytes into plasma cells producing IgA. This patient had circulating antibodies to IgA and no detectable IgA bearing lymphocytes.[76] Defects of IgA secretion and IgA specific suppressor cells have been recently reported in patients with IgA deficiency.[189]

The frequent occurrence of IgA deficiency in ataxia-telangiectasia patients has already been mentioned as an example of associated, and possibly primary, T cell dysfunction. Although gross abnormalities of the T cell system are not usually seen in other IgA deficient patients, diminished numbers of circulating T cells and abnormal production of lymphokines by stimulated T cells have been described.[65,77] In one patient with deficiency of both T cells and IgA, the addition of Sezary cells (a malignancy of T cells to be discussed later) to the patient's cells allowed the latter to be induced by PW to become IgA secreting cells.[76] Thus, it seems likely that the failure of IgA bearing lymphocytes to mature into secretory plasma cells in some patients may reflect a subtle defect in T cell function. This primary defect, rather than the secondary IgA deficiency, may in part explain the frequent association with other immunologically related diseases.

3.3.5 *IgM deficiency*

Isolated IgM deficiency could be regarded as the obverse of IgA deficiency in the sense that IgM antibody responses, particularly to antigens with repeating determinants such as pneumococcal polysaccharides, and lipopolysaccharide, are relatively thymus independent. In this regard, it is interesting that serious infections with pneumococcal and Gram-negative organisms are unusually frequent in such patients.[78] If, as we think is possible, cells leading to IgM and IgD secretion belong to a separate subline of B cells, this defect could be limited in expression to that subline sparing the precursors of plasma cells producing the other Ig classes.

3.3.6 *Wiskott–Aldrich syndrome*

The immunological abnormalities in X-linked immunodeficiency with eczema and thrombocytopenia may also primarily reflect defective development of a B cell subline. This curious immunodeficiency is featured by deficient antibody responsiveness to polysaccharide antigens and low IgM levels, whereas antibody responses to protein antigens are usually intact.[79–81]

3.3.7 *X-linked defect of B cells in mice*

In a mutant strain of mice, CBA/HN, a similar immunodeficiency pattern has been defined as an X-linked defect involving B cells.[82] These mice produce antibodies to thymus dependent antigens quite well, but fail to respond to thymus-independent antigens and have low serum IgM. Very recently a reduced sIgD/sIgM ratio has been demonstrated in mice of this strain.[83] Careful appraisal of serum and membrane-bound IgD is clearly warranted in the human conditions featuring IgM deficiency.

3.3.8 *X-linked immunodeficiency with increased IgM*

In X-linked immunodeficiency with increased levels of IgM, IgD levels may also be high while IgG and IgA levels in serum are very low or even absent (reviewed in Reference 84). The numbers of B lymphocytes bearing SmIgM, SmIgG and SmIgA appear to be normal in affected boys.[2] The defect here could involve the mechanism for T cell help which is required for terminal differentiation of these B lymphocytes. Solid data in favour of or against this hypothesis is unavailable at present.

3.3.9 *Variable immunodeficiencies*

In the present WHO classification a variety of syndromes are included under the heading of variable immunodeficiencies because of insufficient information about patterns and causes.[27] Among this heterogeneous group of syndromes, which may be congenital or acquired, sporadic or familial, and occur in both males and females, two general phenotypic patterns commonly occur: those with antibody deficiency associated with panhypogammaglobulinaemia and others mainly showing deficiencies in IgG and IgA (reviewed in References 20 and 63).

Approximately one-third of both groups of patients have few or no detectable B lymphocytes in their blood, suggesting a central failure of B cell development as in XLA. Repeated evaluations are needed to establish this, however, because a substantial increase in circulating B lymphocytes may be seen in patients with variable immunodeficiencies during infections.[85] Most such patients have normal numbers and distribution of B lymphocytes. Consistent with the evidence that B lymphocytes in at least some of these patients can bind antigens and respond to antigens with proliferation but not plasma cell differentiation, these patients may have generalized germinal centre hyperplasia.[85–87] This may occur because of loss of feedback inhibition of B lymphocyte proliferation by antibodies.

By use of *in vitro* assays capable of measuring the capacity for terminal B cell differentiation under conditions requiring T cell help, three major types of defects leading to arrested B cell differentiation have been tentatively identified. First there is evidence that in some patients T cells may be unable to help B lymphocytes differentiate normally, while in others T cells, or their products, may actively suppress terminal differentiation of B lymphocytes (References 89 and 90, and Keightley, Lawton and Cooper, unpublished observations). However, while animal studies suggest that hyperactivity of suppressor T cells can cause agammaglobulinaemia,[91] it may still be premature to conclude that abnormal suppressor activity of T cells is the primary cause of immunodeficiency in these patients. Suppressor T cell activity can also be demonstrated by lymphocytes from some but not all boys with X-linked agammaglobulinaemia where B cell deficiency appears to be the primary lesion. Nevertheless, *in vitro* demonstration of suppressor T cell activity[89] stands as a landmark in emphasizing an important role of T cells as regulators of B lymphocyte differentiation in man. Furthermore, excessive activity of suppressor T cells could hinder attempts to correct B cell deficiencies.

A second general category of defects are those intrinsic to B lymphocytes. Even in the presence of normal T cells or their mitogenic products, B lymphocytes from some hypogammaglobulinaemic patients cannot be induced to achieve plasma cell maturation (Reference 92, and see Table 2). Indeed, under the best *in vitro* conditions available, their B lymphocytes cannot be triggered even to proliferate. B lymphocytes from other panhypogammaglobulinaemic patients can be induced *in vitro* to produce large amounts of

Table 2. Assessment of T and B cell functions in a boy with normal numbers of B lymphocytes, but low serum IgM and absent IgG and IgA[a]

T cells added ($\times 10^{-3}$)	Plasma cells/culture ($\times 10^{-3}$)		
	Normal B cells		Patient B cells
	Donor A	Donor B	
Donor A			
10	7	16	0
50	41	62	<1
100	159	28	<1
Donor B			
10	2	5	0
50	4	11	<1
100	15	13	<1
Patient			
10	0	1	0
50	3	5	0
100	2	9	0

[a] In this assay the plasma cell response of peripheral blood B lymphocytes (5×10^4) to pokeweed mitogen stimulation depends upon the help of added autologous or allogeneic T cells. It can be seen from the results in this assay that this boy's B lymphocytes were incapable of plasma cell maturation even when cultured with normal T cells in adequate numbers. On the other hand, his T lymphocytes were capable of helping normal B lymphocytes mature into plasma cells, although they were slightly less effective than T cells from the two normal donors (data from studies of Keightley, Lawton and Cooper).

cytoplasmic Ig as judged by immunofluorescence but fail to secrete their immunoglobulin.[92] The reason for this apparent secretory abnormality, important *in vivo* correlations and the fate of the cytoplasmic Ig have not yet been fully elucidated.

A third category is suggested by the failure of T cells from some patients to help or suppress normal B cell differentiation and the failure of differentiation of B lymphocytes from the same patients even with the help of normal T cells. An abnormality of both T and B cells interfering with the collaborative mechanism should be sought in such patients.

3.3.10 *Familial agammaglobulinaemia following infectious mononucleosis*

Following a severe illness thought to be infectious mononucleosis in three related males (two siblings and one maternal cousin), the two survivors became agammaglobulinaemic while retaining normal numbers in circulating B lymphocytes.[88] This suggests that EB virus may somehow cause arrested B lymphocyte differentiation in genetically susceptible individuals.

3.3.11 *Immunodeficiency with thymoma*

The recognition of immunodeficiency with thymoma provided one of the early clues to the important role of the thymus in immunobiology (reviewed in Reference 93). Remarkably, many patients with this syndrome appear to have normal numbers of T cells and cell-mediated immunity, but are very deficient in circulating B lymphocytes[63] and hypogammaglobulinaemic. When cultured for several days in the presence of normal serum, lymphocytes from one such patient acquired B cell markers[94] but this observation has not been confirmed in other such patients who appear instead to have a central failure in continued B cell production. Suppressor T cells have been reported in several patients with thymoma and hypogammaglobulinemia.[190] Since patients with thymoma also may have deficiencies of other haemopoietic cells, a stem cell defect has been suspected. How this may relate to the thymoma, usually spindle cell epitheliomas, has not been explained, although this could be a compensatory hyperplasia (see also 'Notes added in proofs').

3.3.12 *Deficiencies of kappa chains and IgG subclasses*

Other antibody deficiency syndromes have been reported which at the present cannot be fruitfully considered as cellular defects in differentiation because of insufficient information. These include deficiencies in kappa chain[95] or in IgG subclass production,[96,97] and undoubtedly more selective gaps in antibody responsiveness will also be found in the future.

It is evident, however, even from this brief consideration of selected examples of human immunodeficiencies that this is a clinically important group of extremely heterogeneous diseases the further elucidation of which will contribute significantly to our ultimate understanding of the immune system. In turn, these patients should be the prime beneficiaries from this knowledge.

4. LYMPHOPROLIFERATIVE DISEASES

4.1 B Cell proliferations

The B cell origin of several lymphoid proliferations such as Waldenstrom's macroglobulinaemia (WM), most chronic lymphocytic leukaemias (CLL) and most lymphoblastic lymphomas is now well documented.[1] Two main concepts have emerged from the study of lymphocyte membrane markers in B cell malignancies: the monoclonal nature of the proliferation and the fact that these monoclonal B cell proliferations vary considerably with respect to the level of the proliferating cells within the normal pathway of differentiation of a B cell clone.[3]

4.1.1 *Monoclonality of B cell proliferations*

The fact that B cell proliferations affect in most cases a single clone is supported by the 'monoclonal' character of the SmIg synthesized by the proliferating lymphocytes. Indeed these SmIg are homogeneous since in a given patient they are restricted with respect to the light chain type, the heavy chain class and subclass[98,99] and, for SmIgG, the allotype.[100] Moreover they have been shown to share the same idiotypic specificity[101–103] and the same antigen binding specificity.[99] Various antibody activities of the monoclonal lymphocyte SmIg have been recognized in some patients: rheumatoid factor activity,[21] anti-blood group I in cold agglutinin disease[99,104] and anti-Forssmann activity.[105] These findings, reflecting identical variable regions of the SmIg, further substantiate the monoclonal nature of B cell proliferations.

An exception to the rule of a single Ig product is the simultaneous presence of δ and μ chains on the proliferating B cells from many patients.[106–108] However, IgD and IgM molecules on the cells of a single proliferating clone share the same light chains, idiotypic specificity[102,103] and antibody activity.[109] Both molecules, therefore, apparently have the same variable regions and differ only in the constant part of the heavy chains. In view of these findings, the simultaneous presence of IgM and IgD does not argue against the monoclonal nature of most B cell proliferations. In fact most μ chain bearing lymphocytes from normal blood and peripheral lymphoid tissues also synthesize and express δ chains.[110] In contrast IgD molecules have rarely been found on the surface of IgG or IgA producing neoplastic cells[108]; the combination of SmIgD and SmIgA has been recorded once (Kanner, S. and Bull, D., personal communication).

The concept of SmIg monoclonality holds true even in those CLL cases where freshly drawn lymphocytes carry simultaneously μ, γ, κ and λ chains. This 'mixed staining pattern' occurred in 14% of 153 CLL patients (Table 3), and represents a false polyclonal appearance. Several reasons may account for this mixed staining:[99] rheumatoid factor activity of a SmIgM leading to the binding of serum IgG, coexistence of SmIg and immune complexes or antibodies bound to the cell surface. In all the cases with mixed staining where *in vitro* biosynthetic studies were performed, the actual cell product was a 'monoclonal' IgM[99,111] This finding strengthens the predominance of IgM (and IgD) as the major SmIg in CLL. In addition, in some cases where the freshly drawn cells apparently bore IgG, even with a single light chain type, the lymphocytes could be shown after stripping to synthesize a 'monoclonal' IgM.[99] However, CLL lymphocytic clones truly producing IgG or IgA do exist (Table 3). The actual synthesis of a monoclonal IgG or IgA has been proved in several patients.[99] Thus the distribution of heavy and light chains among CLL patients roughly reflects the distribution of SmIg classes on normal blood lymphocytes. The IgM predominance of SmIg in B cell proliferations is not restricted to CLL but is found also in other B cell malignancies such as Burkitt's tumour[112] and some non-Hodgkin malignant lymphomas.[113–115]

Table 3. Lymphocyte SmIg in 153 CLL patients (SmIgD not included)

Serum monoclonal Ig	Total number	μ		γ		α		Biclonal	Mixed staining	Not detectable
		κ	λ	κ	λ	κ	λ			
None	119	45	18	12[c]	7	0	1	1	17[a]	18[d]
IgM	13	7	3						3[a]	
IgG	19	1[b]	1[b]	8[c]	4			4	1	
IgA	2							2		
Total	153	75		31		1		7	21	18

[a] SmIgM synthesized *in vitro* in the seven cases studied.
[b] SmIgM synthesized *in vitro*, IgG being found on freshly drawn cells.
[c] IgG proven to be an actual cell product in six cases.
[d] T cell origin demonstrated in eleven cases (including two patients with Ig attached to but not synthesized by the cells).
From Preud'homme, Brouet and Seligmann, Lymphocyte membrane markers in human lymphoproliferative diseases, in *Membrane Receptors of Lymphocytes* (M. Seligmann, J. L. Preud'homme and M. F. Kourilsky, Eds.), North-Holland, Amsterdam, 1975, p. 420. Reproduced by permission of North-Holland.)

The monoclonal Ig in CLL usually cannot be detected inside the proliferating lymphocytes by direct intracytoplasmic immunofluorescence staining. Among exceptions to this rule were a few patients with intracytoplasmic rod-shaped crystalline inclusion bodies which were shown to be monoclonal Ig.[99,116–118] These Ig crystals are surrounded by ergastoplasmic membranes usually lined with ribosomes. Crystals of the three main classes of Ig have been found but a strange finding was that the light chains belonged most often to the λ type. The reasons why SmIg may accumulate in the cytoplasm as crystalline bodies are unknown.

4.1.2 *Differentiation capabilities of malignant B cells*

The second main concept drawn from studies of lymphocyte markers in B cell malignancies is that the proliferating monoclonal B cells may either be apparently 'frozen' at a given stage along the pathway of differentiation of the B line or conversely correspond to a clone of B lymphocytes that pursues uninterrupted maturation up to the Ig secreting plasma cell.[3]

The best example of the latter situation is WM, a disease featured by a pleomorphic lymphoid proliferation in the bone marrow and lymph nodes. All the pleomorphic lymphoid cells, ranging from the small lymphocyte to the plasma cell, belong to the same B cell clone and synthesize a monoclonal SmIgM identical in its light chain type, idiotypic specificity and antibody specificity to the serum monoclonal component which is secreted by the most mature cells.[119,120] In untreated patients, a high percentage of the blood lymphocytes also produce the monoclonal SmIgM. This disorder may therefore be considered as a leukaemic process despite the usual absence of an increase in blood lymphocyte counts.[119]

Monoclonal B cell proliferations with persistent differentiation are not restricted to WM since the same concept applies to rare cases of pleomorphic lympoid proliferation in the bone marrow, lymph nodes and spleen with serum and cell bound monoclonal IgG or IgA.[99] Apart from the difference in the class of secreted Ig, these patients have many features in common with WM.

In agreement with the concept of persistent maturation, the amount of SmIg greatly varies from cell to cell, leading to a strikingly heterogeneous immunofluorescence pattern in patients with WM and related conditions. Conversely in other B cell malignancies, a very uniform staining pattern of the SmIg on all the proliferating cells of a given patient, together with the absence of plasma cells and of serum monoclonal component, suggests a precise block in the maturation process of the proliferating clone. For instance in most cases of CLL without any serum monoclonal protein, the uniform and faint fluorescence pattern of all the leukaemic lymphocytes, with a SmIg density much lower than on normal lymphocytes, accords with the hypothesis that their development was consistently arrested at a relatively early stage.[99] In occasional patients, the arrest of maturation appears to take place at a level where SmIg are not yet detectable by immunofluorescence. The study of several other B and T cell markers is critical in order to distinguish such cases from T derived CLL. Conversely, in a few cases leukaemic lymphocytes may be heavily loaded with SmIg. This situation is encountered in most patients with the 'prolymphocytic'[121] type of CLL. In many cases of B type lymphomas (such as Burkitt's lymphoma, 'poorly differentiated lymphocytic' lymphoma, 'immunoblastic' sarcoma), the malignant cells show also a uniform SmIg pattern and the density of SmIg in these disorders is usually strikingly higher than on common CLL lymphocytes or even on normal lymphocytes. These sarcomatous cells may represent triggered B lymphocytes unable to differentiate further.

CLL with a serum monoclonal Ig usually correspond to a situation intermediate between common CLL and WM, i.e. an incomplete block in the maturation process. In many such cases (Table 3) the concept of a B cell proliferation with some degree of persistent maturation of the neoplastic clone into plasma cells is supported by the finding of the very same Ig chains on the leukaemic lymphocytes and in the serum monoclonal Ig secreted by the rare plasma cells.[99] This hypothesis was recently confirmed by experiments performed with anti-idiotypic reagents.[101,102] However, in a few instances the serum monoclonal Ig appears to be unrelated to the lymphocytic proliferation. In such cases one possibility is the involvement of a second malignant plasmacytic clone; an alternative possibility is that the monoclonal serum Ig may result from the conjunction of antigenic stimulation and immunodeficiency.[99]

The study of surface IgD in lymphoid malignancies provided additional information with respect to the maturation process of the proliferating B cell clone. WM offered a unique opportunity to study a B cell clone with persistent differentiation of IgD and IgM bearing cells into IgM secreting plasma cells. The density of surface IgD appeared clearly to decrease during this maturation process and the amount of IgD molecules on the membrane of IgM secreting cells was very low.[109]

Multiple myeloma obviously corresponds to the proliferation of the most mature cells of the B cell series. However, 'monoclonal' populations of B lymphocytes carrying the same Ig chains as the plasma cells have been reported.[122–124] In some of these cases, the actual synthesis of this monoclonal SmIg by the lymphocytes was proved, and shared idiotypic specificity between the serum and cell membrane bound monoclonal Ig was also established. Such monoclonal lymphocyte populations may be demonstrable in the bone marrow and/or blood of more than 30% of patients with IgG, IgA or Bence–Jones myeloma, or non-secreting myeloma.[125] These findings suggest that at least in these cases B lymphocytes or earlier cells, and not plasma cells, could represent the target for the neoplastic transformation in myeloma. Of special interest is the detection in one patient of μ and λ chains at the surface of myeloma plasma cells containing and secreting IgA λ.[122] This observation and the occurrence in the serum of another patient of a monoclonal IgM and a monoclonal IgA sharing individual antigenic determinants[126] and identical light chains,[127] but produced in separate cells[128] are consistent with the possibility of a direct switch from IgM to IgA synthesis.

4.1.3 *Double B cell proliferations*

The two concepts of monoclonality and of variability in the proliferation patterns with respect to the differentiation pathway of the B cell-line are especially useful in interpreting some unusual events in lymphoproliferative diseases.

Double lymphoid proliferations with separate cell populations or clones featured by distinct SmIg were found in several patients.[122,129] In only one of these patients all the features of multiple myeloma were associated with those of CLL. The clinical presentation in the other cases was similar either to WM or most often to CLL. Various patterns were encountered with regard to intracytoplasmic Ig staining and serum monoclonal Ig reflecting persistent maturation of the corresponding clone: both cell populations, only one (usually producing IgG or IgA) or none may pursue some maturation into plasma cells secreting their monoclonal Ig in the serum. In some patients the SmIg markers of the two clones differed in both heavy and light chains. In other instances, both proliferating populations shared the same type of light chains, suggesting, especially when λ light chains were found, that they may have originated from a common clone. In one such CLL patient[99] who had a monoclonal IgGλ in his serum, three populations of proliferating cells were demonstrated by sequential and double labelling immunofluorescence experiments: (a) lymphocytes with small crystalline $\mu\lambda$ inclusions and bearing SmIgMλ; (b) lymphocytes without detectable intracytoplasmic Ig and with surface positivity for μ, γ and λ chains together; (c) lymphocytes with surface positivity restricted to γ and λ chains, a few of those and a few plasma cells containing intracytoplasmic IgGλ. These findings are compatible with the hypothesis of an original $\mu\lambda$ clone shifting to IgGλ producing and secreting cells with a transitional population of cells producing the two classes of heavy chains. This hypothesis was not documented by a study of the idiotypic specificity of the SmIg. Such a switch mechanism involving a transition from IgG to IgA synthesis occurring at the B lymphocyte level was recently documented in another CLL patient[129] with a dual population of lymphocytes bearing either IgGκ or IgAκ associated with a single serum monoclonal IgGκ protein against which idiotypic antibodies were prepared. These idiotypic determinants were detected on the surface and in the cytoplasm of both the IgG and IgA bearing cell populations which thus presumably had a common clonal origin. The distribution and patterns of surface and cytoplasmic IgG and IgA staining in individual cells suggested that the direction of switching was from IgG to IgA synthesis; i.e. surface IgG was found on cells with either cytoplasmic IgG or IgA, whereas surface IgA was present on cells with cytoplasmic IgA but not on those with cytoplasmic IgG. Such transitions in individual clones from IgM to IgG and from IgG to IgA synthesis, characterized by a change in the C_H gene without an apparent change in the C_L, V_H or V_L genes, have been previously described at the plasma cell level in myeloma patients.[130–132]

A new and more severe type of proliferation may occur in some patients previously affected with chronic B cell malignancies. Examples of this situation are the rare acute blastic transformation of CLL and the so called 'reticulum cell' sarcoma supervening on CLL, WM or heavy chain diseases. The study of lymphocyte membrane markers in such cases has provided an answer to the question of whether or not the supervening malignancy with a totally different cell morphology corresponded to the emergence of a new clone. Indeed the superimposed proliferating cells were shown to be B cells synthesizing the same Ig chains (with the same antibody specificity) as the lymphocytes of the previous chronic phase, indicating that the two cell populations were derived from the same clone.[113,133]

4.2 T Cell proliferations

Neoplastic diseases featured by proliferating T cells are less common than B cell malignancies.

The origin of the large abnormal circulating cells described by Sezary in some patients with a malignant disease featured by erythrodermia had been disputed. These so called Sezary cells, characterized by their cerebriform and serpentine nucleus, were shown to lack B cell markers, to form E rosettes and to be killed by heteroantisera to T cells, indicating that the Sezary syndrome is a proliferative disorder of T cells.[134–136] Although doubtful results were recorded in a few patients,[137] the neoplastic cells were shown to be of T cell origin in 15 consecutive cases of a recent series.[125] As outlined in the section on immunodeficiency diseases, Sezary cells, at least from some patients, can function as helper T cells in a non-specific way permitting *in vitro* maturation of B cells from some patients with T cell and IgA deficiencies into plasma cells secreting all of the Ig classes (Reference 76, and personal communication from T. Waldmann). It is noteworthy that the immunological study of the neoplastic cells of a related dermatologic malignancy, mycosis fungoides, also point to their likely T cell origin,[138] thus leading to a unified concept of some cutaneous lymphomas and suggesting a preferential homing of T lymphocytes in the skin.

Until recently only scattered cases of T derived CLL had been published.[122,139–142] However, T derived CLL should no longer be considered as a very rare disorder since 11 consecutive cases were recently diagnosed in a single department.[143] Two of these patients had no unusual features except for a high white blood cell count and the prolymphocytic type of the leukaemic cells. The other patients exhibited a quite distinct clinical and haematological pattern with a massive splenic enlargement and skin involvement in several cases, frequently severe neutropenia and moderate but consistent invasion of blood and bone marrow by the leukaemic lymphocytes. The leukaemic T lymphocytes had a high content of lysosomal enzymes, and peculiar cytoplasmic azurophil granules were often readily apparent. Immunological study of the leukaemic cells in these patients pointed to their T cell nature. Much interest emerged from the results of the study of these leukaemic T lymphocytes with different antisera to T cells. As is the case for normal blood T lymphocytes, the cells from four patients reacted equally well with the antisera to peripheral T cells and to foetal thymocytes. By contrast the leukaemic cells from the other patients reacted strongly with only one or the other of these two antisera. The antiserum to human brain (which reacts only with a subset of peripheral normal T lymphocytes)[191] stained the majority of leukaemic lymphocytes in two cases and gave negative results in the three other cases studied. These findings provide evidence for the homogeneity of these populations of leukaemic T lymphocytes for which no monoclonal marker is presently available. Further studies of the antigenic and functional properties of the homogeneous subsets of human T lymphocytes provided by these patients should be of great interest. Preliminary results showed that the reactivity of the leukaemic T cells with allogeneic cells was preserved in some patients but not in others (Table 4).

4.3 Lymphoid proliferations with unusual surface markers

The investigation of membrane markers in lymphoproliferative diseases may lead to the individualization of minor subsets of normal lymphoid cells. Thus the 'hairy cells' of tricholeukocytic leukaemia (leukaemic reticuloendotheliosis) whose study has led to controversial interpretations regarding their lymphocytic[144] or monocytic[145] origin appear to represent peculiar Ig producing (IgM and IgD) phagocytic cells[146] for which a normal couterpart has not been defined so far.

In some patients the leukaemic cells may apparently exhibit markers of both the T and B series. In a CLL patient whose lymphocytes carried SmIg and formed rosette with sheep erythrocytes, the B nature of the leukaemic cells was demonstrated and the rosette formation was shown to be due to an anti-Forssmann antibody activity of the monoclonal SmIgM produced by the cells.[105] In a few other such cases, the leukaemic cells were T lymphocytes bearing cytophilic Ig.[143] It should be stressed that in humans only a single malignancy involving cells featuring both SmIg synthesis and T cell properties has yet been described.[178] However, in a few patients the proliferating cells appeared to truly express simultaneously markers thought at the time to be specific for the T and B lines[147–149] and similar surface membrane properties were found for a few normal human blood lymphocytes.[150–152] In fact all these reports dealt with cells displaying T cell properties and expressing C or Fc receptors which are now considered to be present on a subset of T cells.

4.4 Non-Hodgkin's lymphomas

Non-Hodgkin malignant lymphomas are now classified primarily according to morphology. The old, and numerous new, terminologies rely mainly upon the appreciation of morphological similarities between normal and neoplastic cells, although the normal couterpart of some lymphoma cells is not easy to define. The identification of B and T cell membrane markers on the malignant cells by means of cell suspensions or frozen tissue sections[153] provides a new tool for the study of these diseases but does not yet allow scientifically accurate or satisfactory operational classification.

Most non-Hodgkin lymphomas are monoclonal B cell malignancies.[113–115,122,153–155] This applies to 'well differentiated' lymphocytic lymphoma which is closely related to CLL and to most if not all cases of follicular (nodular) lymphomas with predominantly small or large cells. In the majority of cases of diffuse 'poorly differentiated' lymphocytic lymphomas in adults, the malignant cells carry B cell markers. Some cases of T cell origin have

Table 4. Immunological characterization of peripheral blood lymphocytes (% reactive cells)

Patient	Surface Ig[a]	Ig aggregates	Antiserum to B cells (CLL)	E rosettes	Antiserum to peripheral T cells	Antiserum to foetal thymocytes	Antiserum to brain	PHA (% control)	PW (% control)	MLR Response	MLR Stimulation ability
1	<1	<1	ND	78	95	ND	ND	100	ND	ND	ND
2	1	0	3	83	80	100	<1	0	0	0	0
3	5	4	10	40	60	76	ND	10	15	0	N
4	<1	<1	5	60	60	76	ND	15	15	N	N
5	5	5	10	70	85	90	ND	80	95	N	N
6	4	3	3	86	80	35	75	15	10	N	N
7	6	5	6	54	60	30	ND	ND	ND	ND	ND
8	2	1	8	66	35	93	3	25	35	N	N
9	1	1	3	65	30	90	<1	6	7	0	low
10	$\gamma\kappa$[b]	4	8	70	30	90	85	0	0	ND	ND
11	$\gamma\lambda$[b]	7	ND	66	75	ND	ND	ND	ND	ND	ND

[a] Expressed as the sum of lymphocytes positive for γ, μ and α chains or for κ and λ chains.
[b] No synthesis by the cells after *in vitro* culture following trypsinization.
PHA: phytohaemagglutinin. PW: pokeweed mitogen. MLR: mixed lymphocyte reaction. N: normal. ND: not done.
From J. C. Brouet *et al.*, Chronic lymphocytic leukaemia of T cell origin. An immunological and clinical evaluation in eleven patients. *Lancet*, 1975, **2**, 890–893. (Reproduced by permission of *The Lancet*.)

been encountered, mainly in children with the 'convoluted' cell type and mediastinal tumours. These 'lymphoblastic' childhood T lymphomas[156,157] appear in many respects to be related to acute lymphoblastic leukaemias of a T cell nature.

It is of interest that in those cases of lymphomas featured by a mixture of small and large malignant cells (so called 'mixed lymphocytic–histiocytic' lymphomas), both cell types share the same B surface markers or lack the same surface markers.[158]

The study of membrane markers provides evidence that the so-called diffuse 'histiocytic' lymphomas represent a heterogeneous group of neoplasias.[158] Only a very small number of such lymphomas featured by large cells may be truly related to the monocytic series. In three cases[113,159,160] the cells bore a receptor with an apparently high affinity for the Fc fragment of IgG and were devoid of other B or T cell markers. Although suggestive, such findings are not sufficient to definitely identify these malignant cells as monocytic. T cell derived 'histiocytic' lymphomas appear to be infrequent since they account for only 10% of the reported cases. The percentages of cases which were shown to be of B cell origin varies in the different series (reviewed in Reference 158). Two such patients were previously affected with CLL or WM.[158] In four other instances,[159] B cell type of 'diffuse histiocytic' lymphomas occurred in patients in whom B cell nodular lymphomas were diagnosed on biopsies performed previously or at another site of the body. One may therefore assume that many 'histiocytic' lymphomas of B cell origin are in fact closely related to another well documented B cell neoplasia. Nearly half of the cases of 'histiocytic' lymphomas remain unidentified with respect to the surface markers. The use of specific heteroantisera to B or T cells so far has not given further evidence for the cellular origin of these unclassified cases. These negative data do not exclude the lymphoid origin of these large malignant cells. However, in view of these negative findings, a new classification of 'histiocytic' lymphomas appears premature and further studies are warranted in order to elucidate the nature of the large malignant cells.

4.5 Acute lymphoblastic leukaemias (ALL)

A heterogeneous pattern has emerged from the study of the surface properties of blast cells from patients with ALL.

Acute leukaemias with B cell markers constitute a minor subgroup. No instances of B cell ALL were recorded in relatively large series[161–163] and only scattered cases were reported elsewhere.[122,164–166] Those patients with a 'monoclonal' B cell proliferation are usually not affected with common ALL; these B cell acute leukaemias belong mostly to two specific entities characterized by unusual clinical or cytological features. In three cases[133] the blastic proliferation supervened in patients previously affected with common CLL. In eight other cases, the patients showed the histological and cytological features of Burkitt's tumour.[167] Since the leukaemic presentation of this disease is distinctly unusual, the finding in all such patients of a B cell proliferation (seven with SmIgM and one with SmIgG) supports the view that these peculiar ALL are closely linked to Burkitt's lymphoma. Only three patients with a monoclonal B cell process belonged to a group of 100 unselected patients with common ALL.[168] In two of these three patients, the blast cells had unusual cytological features suggestive of poorly differentiated lymphocytic lymphoma. These findings suggest that most if not all cases of B cell ALL may in fact correspond to lymphomas with a leukaemic presentation.

The blast cells from 25% to 30% of unselected and untreated patients with common ALL (adults and children) show T cell surface features.[161–163,168–172] In the E rosette assay the percentage of rosette forming blast cells may be relatively low in some of these patients. E rosette formation occurs at 37 °C as well as at 4 °C, this property being similar to that of normal thymocytes and not of peripheral T cells.[192] The use of antisera to T cells may therefore be rather critical. In fact in a few patients, the blasts reacted with such antisera although they did not form obvious E rosettes. The results obtained with antisera to thymic cells must be carefully interpreted when evaluating the possible T origin of blast cells since such antisera may show a wide pattern of reactivity with leukaemic cells[168,173] and, until T cell antigens are more precisely defined and purified, the absolute specificity of T cell antisera will be difficult to ensure. However, some antisera to thymic cells previously absorbed with peripheral T cells were shown to react with T derived ALL cells.[193]

The largest group of patients with common ALL (close to 70%) is characterized by the absence of B or T markers on the blast cells. However, the blast cells from this subgroup of patients are featured by the presence of leukaemia associated antigens which are not found on T derived ALL cells. Two distinct antigenic systems have allowed this delineation of the non-T non-B subgroup of ALL. Some rabbit antisera to CLL B cells were shown to contain, in addition to anti-B cell antibodies, antibodies that reacted with the blast cells from most non-T non-B ALL patients but not with T ALL cells.[168] Interestingly enough, the corresponding antigen was absent on peripheral T cells but present on thymocytes. The other set of leukaemia associated antigenic determinants was revealed by a rabbit antiserum to non-T non-B ALL cells and appears to be specific for these blast cells.[174]

The question of the true cellular origin of non-T non-B ALL presently remains unsettled. The involvement of lymphoid progenitor cells devoid of markers of mature lymphocytes is a likely possibility which is supported by the presence on these blast cells of a leukaemia associated antigen shared by thymocytes[168] and of Ia-like antigens,[175] and by the finding of a thymic enzyme, terminal desoxynucleotidyl transferase, in the blast cells of most patients with ALL.[176] Whatever the true cellular origin of these unidentified blast cells, the question is raised whether T and non-T non-B ALL represent two distinct kinds of acute leukaemia. Several statistically significant differences between these two groups of patients have been noted. T cell ALL is more often featured by high leukocyte counts, tumoral presentation and thymic masses.[163,168] The blast cells with T properties usually exhibit a strong acid phosphatase positivity which rarely occurs in the other group of patients.[168,172] Although meningeal relapses were more frequent in patients with T cell ALL, no correlations were found in our series between the immunologic type of the blast cells and survival[168] but the follow up in this series was only 3 to 36 months and differences in the actuarial survival curves may well become apparent in the future. In fact, relapses appear to occur earlier in children affected with T derived ALL[163,193] which, as stated above, are associated with well-known poor prognosis features.

There is reason for hope that the immunologic subclassification of ALL will not only help to improve the accuracy and reproducibility of the classification of acute leukaemias but will also help to identify the target cell for the leukaemic process and to provide new tools for treating and monitoring patients with acute leukaemias.

Acknowledgements

We are especially indebted to our colleagues, Drs Jean-Louis Preud'homme, Jean-Claude Brouet, Alexander Lawton and Richard Keightley for sharing their ideas, criticisms and observations.

Support for the work cited from Laboratoire d'Immunochimie et Immunopathologie (INSERM U. 108) included grants from the délégation Générale à la Recherche Scientifique et Technique n°. 74.7.0607 and 75.7.0786, from Institut National de la Santé et de la Recherche Médicale n°. 10.74.31.3 and from Centre National de la Recherche Scientifique E.R.A. 239, and from the Spain Immunology Laboratory included National Institutes of Health Grant RR32, National Cancer Institute Grant CA 16673 and National Foundation Grant 1-354.

5. REFERENCES

1. Möller, G., Ed. (1973). *Transplant. Rev.*, **16**.
2. Cooper, M. D. and Lawton, A. R. (1972). *Amer. J. Pathol.*, **69**, 513–528.
3. Salmon, S. E. and Seligmann, M. (1974). *Lancet*, **2**, 1230–1233.
4. Lawton, A. R., Self, K. S., Royal, S. A. and Cooper, M. D. (1972). *Clin. Immunol. Immunopathol.*, **1**, 84–93.
5. Vossen, J. M. and Hijmans, W. (1975). *Ann. N.Y. Acad. Sci.*, **254**, 262–279.
6. Lawton, A. R., Kincade, P. W. and Cooper, M. D. (1975). *Fed. Proc.*, **34**, 33–39.
7. Hood, L., Campbell, J. H. and Elgin, S. C. R. (1975). *Annu. Rev. Genet.*, **9**, 305–354.
8. Ross, G. D, Polley, M. J., Rabellino, E. M. and Grey, H. M. (1973). *J. exp. Med.*, **138**, 798–811.
9. Eden, A., Miller, G. W. and Nussenzweig, V. (1973). *J. Clin Invest.*, **52**, 3239–3242.
10. Bokisch, V. A. and Sobel, A. T. (1974). *J. exp. Med.*, **140**, 1336–1347.
11. Jondal, M. and Klein, G. (1973). *J. exp. Med.*, **138**, 1365–1378.
12. Greaves, M. F., Brown, G. and Rickinson, A. V. (1975). *Clin. Immunol. Immunopathol.*, **3**, 514–524.
13. Wybran, J., Levin, A. S., Spitler, L. E. and Fudenberg, H. H. (1973). *New Engl. J. Med.*, **288**, 710–713.
14. Hammarström, S., Hellström, U., Perlmann, P. and Dillner, M. L. (1973). *J. exp. Med.*, **138**, 1270–1275.
15. Valdimarsson, H., Agnarsdottier, G. and Lachmann, P. J. (1975). *Nature*, **255**, 554–556.
16. Jondal, M., Klein, E. and Yefenof, E. (1975). *Scand. J. Immunol.*, **4**, 259–266.
17. Moretta, L., Ferrarini, M., Durante, M. L. and Mingari, M. C. (1975). *Eur. J. Immunol.*, **5**, 565–569.
18. Mortensen, R. F., Osmand, A. P. and Gewurz, H. (1975). *J. exp. Med.*, **141**, 821–839.
19. Aiuti, F., Cerottini, J. C., Commbs, R. R. A., Cooper, M., Dickler, H. B., Fröland, S., Fudenberg, H. H., Greaves, M. F., Grey, H. M., Kunkel, H. G., Natvig, J., Preud'homme, J.-L., Rabellino, E., Ritts, R E., Rowe, D. S., Seligmann, M., Siegal, F. P., Stjernswärd, J., Terry, W. D. and Wybran, J. (1974). *Scand. J. Immunol.*, **3**, 521–532.
20. Preud'homme, J.-L. and Seligmann, M. (1974). *Progr. Clin. Immunol.*, **2**, 121–174.
21. Preud'homme, J.-L. and Seligmann, M. (1972). *Proc. Nat. Acad. Sci.*, **69**, 2132–2135.
22. Lobo, P., Westervelt, F. B. and Horwitz, D. A. (1975). *J. Immunol.*, **114**, 116–119.
23. Winchester, R. J., Fu., S. M., Hoffman, T. and Kunkel, H. G. (1975). *J. Immunol.*, **114**, 1210–1212.
24. Preud'homme, J.-L. and Flandrin, G. (1974). *J. Immunol.*, **113**, 1650–1653.
25. Greaves, M. F., Falk, J. A. and Falk, R. E. (1975). *Scand. J. Immunol.*, **4**, 555–562.
26. Fröland, S. S. and Natvig, J. B. (1973). *Transplant. Rev.*, **16**, 114–162.
27. Cooper, M. D., Faulk, W. P., Fudenberg, H. H., Good, R. A., Hitzig, W., Kunkel, H. G., Roitt, I. M., Rosen, F. S, Seligmann, M. and Soothill, J. F. and Wedgwood, R. J. (1974). *Clin. Immunol. Immunopathol.*, **2**, 416–445.
28. DeVaal, O. and Seynhaeve, V. (1959). *Lancet*, **2**, 1123–1125.
29. Hitzig, W. H., Barandun, S. and Cottier, H. (1968). *Ergebn. Inn. Med. Kinderheilkd*, **27**, 79–154.
30. Good, R. A. *et al.* (1975). In *Immunodeficiency in Man and Animals*, Vol. XI (Bergsma, D., Ed.), Sinauer Associates, Sunderland, Mass., pp. 377–430.
31. Meuwissen, H. J., Pickering, R. J., Pollara, B. and Porter, I. H. (1975). *Combined Immunodeficiency Disease and Adenosine Deaminase Deficiency: A Molecular Defect*, Academic Press, New York.
32. Polmar, S. H., Wetzler, E., Stern, R. C. and Hirschhorn, R. (1975). *Lancet*, **2**, 743–746.
33. Keightley, R. G., Lawton, A. R., Cooper, M. D. and Yunis, E. J. *Lancet*, **2**, 850–853.
34. Parkman, R., Gelfand, E. W., Rosen, F. S., Sanderson, A. and Hirschhorn, R. (1975). *New Engl. J. Med.*, **292**, 714–719.
35. Preud'homme, J.-L., Clauvel, J. P. and Seligmann, M. (1975). *J. Immunol.*, **114**, 481–485.
36. Pyke, K. W., Dosch, H.-M., Ipp, M. M. and Gelfand, E. W. (1975). *New Engl. J. Med.*, **293**, 424–428.
37. Seligmann, M., Griscelli, C., Preud'homme, J. L., Sasportes, M., Herzog, C. and Brouet, J.-C. (1974). *Clin. exp. Immunol.*, **17**, 245–252.
38. Gitlin, D. and Craig, J. M. (1963). *Pediatrics*, **32**, 517–530.

39. Hoyer, J. R., Cooper, M. D., Gabrielsen, A. E. and Good, R. A. (1968). *Medicine*, **47**, 201–226.
40. Lawton, A. R., Wu, L. Y. F. and Cooper, M. D. (1975). In *Immunodeficiency in Man and Animals*, Vol. XI (Bergsma, D., Ed.), Sinauer Associates, Sunderland, Mass., pp. 28–32.
41. Kersey, J. H., Sabad, A., Vance, J. C., White, J. G. and Neely, A. N. (1975). *Clin. Res.*, **23**, 411A.
42. Gelfand, E. W., Parkman, R. and Rosen, F. S. (1975). In *Immunodeficiency in Man and Animals*, Vol. XI (Bergsma, D., Ed.), Sinauer Associates, Sunderland, Mass., pp. 158–162.
43. Di George, A. M. (1968). In *Immunologic Deficiency Diseases in Man*, Vol. IV (Bergsma, D. and Good, R. A., Eds.), The National Foundation, New York, pp. 116–121.
44. Gajl-Peczalska, K. J., Park, B. H., Biggar, W. D. and Good, R. A. (1973). *J. Clin. Invest.*, **52**, 919–928.
45. August, C. S., Rosen, F. S., Filler, R. M., Janeway, C. A., Markowski, B. and Kay, H. E. M. (1968). *Lancet*, **2**, 1210–1211.
46. Lischner, H. W. and Huff, D. S. (1975). In *Immunodeficiency in Man and Animals*, Vol. XI (Bergsma, D., Ed.), Sinauer Associates, Sunderland, Mass., pp. 16–21.
47. Cleveland, W. W. (1975). In *Immunodeficiency in Man and Animals*, Vol. XI (Bergsma, D., Ed.), Sinauer Associates, Sunderland, Mass., pp. 352–356.
48. Steele, R. W., Limas, C., Thurman, G. B., Schuelein, M., Bauer, H. and Bellanti, J. A. (1972). *New Engl. J. Med.*, **287**, 787–791.
49. Nezelof, C., Jammet, M. L., Lortholary, P., Labrune, B. and Lamy, M. (1964). *Arch. fr. Pediat.*, **21**, 897–920.
50. Wara, D. W., Goldstein, A. L., Doyle, N. E. and Amman, A. J. (1975). *New Engl. J. Med.*, **292**, 70–74.
51. Aiuti, F., Businco, L. and Gatti, R. A. (1975). In *Immunodeficiency in Man and Animals*, Vol. XI (Bergsma, D., Ed.), Sinauer Associates, Sunderland, Mass., pp. 370–374.
52. Giblett, E. R, Amman, A. J., Sandman, R., Wara, D. W. and Diamond, L. K. (1975). *Lancet*, **1**, 1010–1013.
53. Peterson, R. D. A., Cooper, M. D. and Good, R. A. (1966). *Amer. J. Med.*, **41**, 342–359.
54. Boder, E. (1975). In *Immunodeficiency in Man and Animals*, Vol. XI (Bergsma, D., Ed.), Sinauer Associates, Sunderland, Mass., pp. 255–270.
55. Amman, A. J., Cain, W. A., Ishizaka, K., Hong, R. and Good, R. A. (1969). *New Engl. J. Med.*, **281**, 469–472.
56. Polmar, S. H., Waldmann, T. A., Balestra, S. T., Jost, M. C. and Terry, W. D. (1972). *J. Clin. Invest.*, **51**, 326–330.
57. Buckley, R. H. (1975). In *Immunodeficiency in Man and Animals*, Vol. XI (Bergsma, D., Ed.), Sinauer Associates, Sunderland, Mass., pp. 134–141.
58. Wu, L. Y. F., Lawton, A. R. and Cooper, M. D. (1973). *J. Clin. Invest.*, **52**, 3180–3189.
59. Waldmann, T. A. and McIntire, K. B. (1972). *Lancet*, **2**, 1112–1115.
60. Murgita, R. A. and Tomasi, T. B., Jr. (1975). *J. exp. Med.*, **141**, 440–452.
61. Quie, P. G. and Chilgren, R. A. (1971). *Semin. Hemat.*, **8**, 227–242.
62. Good, R. A. (1973). *Harvey Lecture Series*, **67**, 1–107.
63. Cooper, M. D., Keightley, R. G. and Lawton, A. R., III (1975). In *Membrane Receptors of Lymphocytes* (Seligmann, M., Preud'homme, J.-L. and Kourilsky, F. M., Eds.), North-Holland Publishing Company, Amsterdam, pp. 431–442.
64. Yata, J. and Tsukimoto, I. (1972). *Lancet*, **2**, 1425.
65. Schiff, R. I., Buckley, R. B., Gilbertsen, R. B. and Metzgar, R. S. (1974). *J. Immunol.*, **112**, 376–386.
66. Hayward, A. R. and Greaves, M. F. (1975). *Clin. Immunol. Immunopathol.*, **3**, 461–470.
67. Hayward, A. R. and Greaves, M. F. (1975). *Scand. J. Immunol.*, **4**, 563–570.
68. Goldblum, R. M., Lord, R. A., Cooper, M. D., Gathings, W. E., Goldman, A. S. (1974). *J. Pediat.*, **85**, 188–191.
69. Siegal, F. P., Pernis, B. and Kunkel, H. G. (1971). *Eur. J. Immunol.*, **1**, 482–486.
70. Geha, R. S., Rosen, F. S. and Merler, E. (1973). *J. Clin. Invest.*, **52**, 1726–1734.
71. Litwin, S. D., Ochs, H. and Pollara, B. (1973). *Immunology*, **25**, 573–581.
72. Hitzig, W. H., Dohmann, U., Plüss, H. J. and Vischer, D. (1974). *J. Pediat.*, **85**, 622–628.
73. Fröland, S. S. and Natvig, J. B. (1973). *Transplant. Rev.*, **16**, 114–162.
74. Amman, A. J. and Hong, R. (1971). *Medicine*, **50**, 223–236.
75. Vyas, G. N., Holmdahl, L., Perkins, H. A. and Fudenberg, H. H. (1969). *Blood*, **34**, 573–581.

76. Waldmann, T. (1973). *Advan. exp. Med. Biol.*, **45**, 415–417.
77. Epstein, L. B. and Amman, A. J. (1974). *J. Immunol.*, **112**, 617–626.
78. Hobbs, J. R. (1975). In *Immunodeficiency in Man and Animals*, Vol. XI (Bergsma, D., Ed.), Sinauer Associates, Sunderland, Mass., pp. 112–115.
79. Cooper, M. D., Chase, H. P., Lowman, J. T., Krivit, W. and Good, R. A. (1968). *Amer. J. Med.*, **44**, 499–513.
80. Blaese, R. M., Strober, W., Brown, R. S. and Waldmann, T. (1968). *Lancet*, **1**, 1056–1060.
81. Ayoub, E. M., Dudding, B. A. and Cooper, M. D. (1968). *J. Lab. Clin. Med.*, **72**, 971–979.
82. Scher, I., Steinberg, A. D., Berning, A. K. and Paul, W. E. (1975). *J. exp. Med.*, **142**, 637–650.
83. Finkelman, F. D., Smith, A. H., Scher, I. and Paul, W. E. (1975). *J. exp. Med.*, **142**, 1316–1321.
84. Rosen, F. S., Craig, J. M., Vawter, G. and Janeway, C. A. (1968). In *Immunologic Deficiency Disease in Man*, Vol. IV (Bergsma, D. and Good, R. A., Eds.), The National Foundation, New York, pp. 67–70.
85. Preud'homme, J.-L., Griscelli, C. and Seligmann, M. (1973). *Clin. Immunol. Immunopathol.*, **1**, 241–256.
86. Cooper, M. D., Lawton, A. R. and Bockman, D. E. (1971). *Lancet*, **2**, 791–795.
87. Dwyer, J. M. and Hosking, C. S. (1972). *Clin. exp. Immunol.*, **12**, 161–169.
88. Provisor, A. J., Iacuone, J. J., Chilcote, R. R., Neiburger, R. G., Crussi, F. G. and Baehner, R. L. (1975). *New Engl. J. Med.*, **293**, 62–65.
89. Waldmann, T. A., Broder, S., Blaese, R. M., Durm, M., Blackman, M. and Srober, W. (1974). *Lancet*, **2**, 609–613.
90. Siegal, F. P. and Siegal, M. (1975). *Clin Res.*, **23**, 297A.
91. Blaese, R. M., Weiden, P. L., Koski, I. and Dooley, N. (1974). *J. exp. Med.*, **140**, 1097–1101.
92. Geha, R. S., Schneeberger, E., Merler, E. and Rosen, F. S. (1974). *New Engl. J. Med.*, **291**, 1–6.
93. Jeunet, F. S. and Good, R. A. (1968). In *Immunologic Deficiency Disease in Man* (Bergsma, D. and Good, R. A., Eds.), The National Foundation, New York, pp. 192–203.
94. Wernet, P., Siegal, F. P., Dickler, H., Fu, S. and Kunkel, H. G. (1974). *Proc. Nat. Acad. Sci.*, **71**, 531–535.
95. Bernier, G. M., Gunderman, J. R. and Ruymann, F. B. (1972). *Blood*, **40**, 795–805.
96. Terry, W. D. (1968). In *Immunologic Deficiency Disease in Man*, Vol. IV (Bergsma, D. and Good, R. A., Eds.), The National Foundation, New York, pp. 357–361.
97. Schur, P. H., Borel, H., Gelfand, E. W., Alper, C. A. and Rosen, F. S. (1970). *New Engl. J. Med.*, **283**, 631–634.
98. Grey, H. M., Rabellino, E. and Pirofsky, B. (1971). *J. Clin. Invest.*, **50**, 2368–2375.
99. Preud'homme, J.-L. and Seligmann, M. (1972). *Blood*, **40**, 777–794.
100. Fröland, S. S. and Natvig, J. B. (1972). *J. exp. Med.*, **136**, 409–414.
101. Schroer, K. R., Briles, D. E., Van Boxel, J. A. and Davie, J. M. (1974). *J. exp. Med.*, **140**, 1416–1420.
102. Fu, S. M., Winchester, R. J., Feizi, T., Walzer, P. D. and Kunkel, H. G. (1974). *Proc. Nat. Acad. Sci.*, **71**, 4487–4490.
103. Salsano, F., Fröland, S. S., Natvig, J. B. and Michaelsen, T. E. (1974). *Scand. J. Immunol.*, **3**, 841–846.
104. Feizi, T., Wernet, P., Kunkel, H. G. and Douglas, S. D. (1973). *Blood*, **42**, 753–762.
105. Brouet, J. C. and Prieur, A. M. (1974). *Clin. Immunol. Immunopathol.*, **2**, 481–487.
106. Kubo, R. T., Grey, H. M. and Pirofsky, B. (1974). *J. Immunol.*, **112**, 1952–1954.
107. Fu, S. M., Winchester, R. J. and Kunkel, H. G. (1974). *J. exp. Med.*, **139**, 451–456.
108. Preud'homme, J.-L., Brouet, J.-C., Clauvel, J. P. and Seligmann, M. (1974). *Scand. J. Immunol.*, **3**, 853–858.
109. Pernis, B., Brouet, J.-C. and Seligmann, M. (1974). *Eur. J. Immunol.*, **4**, 776–778.
110. Rowe, D. S., Hug, K., Forni, L. and Pernis, B. (1973). *J. exp. Med.*, **128**, 956–972.
111. Nies, K. M., Oberlin, M. A., Brown, J. C. and Halpern, M. S. (1973). *J. Immunol.*, **111**, 1236–1242.
112. Fialkow, P. J., Klein, E., Klein, G., Clifford, P. and Singh, S. (1973). *J. exp. Med.*, **138**, 89–102.
113. Brouet, J.-C., Labaume, S. and Seligmann, M. (1975). *Brit. J. Cancer*, **31**, suppl. 11, 121–127.
114. Leech, J. H., Glick, A. D., Waldron, J. A., Flexner, J. M., Horn, R. G. and Collins, R. D. (1975). *J. Nat. Cancer Inst.*, **50**, 11–20.
115. Aisenberg, A. C. and Long, J. C. (1975). *Amer. J. Med.*, **58**, 300–306.
116. Hurez, D., Flandrin, G., Preud'homme, J.-L. and Seligmann, M. (1972). *Clin. exp Immunol.*, **10**, 223–234.

117. Cawley, J. C., Barker, C. R., Britchford, R. D. and Smith, J. L. (1973). *Clin. exp. Immunol.*, **13**, 407–416.
118. Clark, C., Rydell, R. E. and Kaplan, M. E. (1973). *New Engl. J. Med.*, **289**, 113–117.
119. Preud'homme, J.-L. and Seligmann, M. (1972). *J. Clin. Invest.*, **51**, 701–705.
120. Wernet, P., Feizi, T. and Kunkel, H. G. (1972). *J. exp. Med.*, **136**, 650–655.
121. Galton, D. A. G., Goldmann, J. M., Wiltshaw, E., Catovsky, D., Henry, K. and Goldenberg, G. J. (1974). *Brit. J. Haematol.*, **27**, 7–23.
122. Seligmann, M., Preud'homme, J.-L. and Brouet, J.-C. (1973). *Transplant. Rev.*, **16**, 85–113.
123. Mellstedt, H., Hammarström, S. and Holm, G. (1974). *Clin. exp. Immunol.*, **17**, 371–384.
124. Abdou, N. I. and Abdou, N. L. (1975). *Ann. Intern. Med.*, **83**, 42–45.
125. Preud'homme, J.-L., Brouet, J.-C. and Seligmann, M. (1975). In *Membrane Receptors of Lymphocytes* (Seligmann, M., Preud'homme, J.-L. and Kourilsky, F. M., Eds.), North-Holland, Amsterdam, pp. 417–430.
126. Yagi, Y. and Pressman, D. (1973). *J. Immunol.*, **110**, 335–344.
127. Seon, B. K., Yagi, Y. and Pressman, D. (1973) *J. Immunol.*, **110**, 345–349.
128. Silverman, A. Y., Yagi, Y., Pressman, D., Ellison, R. R. and Tormey, D. C. (1973). *J. Immunol.*, **110**, 350–353.
129. Rudders, R. A. and Ross, R. (1975). *J. exp. Med.*, **142**, 549–559.
130. Wang, A. C., Wilson, S. K., Hopper, J. E., Fudenberg, H. H. and Nisonoff, A. (1970). *Proc. Nat. Acad. Sci.*, **66**, 337–343.
131. Levin, A. S., Fudenberg, H. H., Hopper, J. E., Wilson, S. K. and Nisonoff, A. (1971). *Proc. Nat. Acad. Sci.*, **68**, 169–171.
132. Todel, C. W., Franklin, E. C. and Rudders, R. A. (1974). *J. Immunol.*, **112**, 871–876.
133. Brouet, J.-C., Preud'homme, J.-L., Seligmann, M. and Bernard, J. (1973). *Brit. Med. J.*, **4**, 23–24.
134. Brouet, J.-C., Flandrin, G. and Seligmann, M. (1973). *New Engl. J. Med.*, **289**, 341–344.
135. Broome, J. D., Zucker-Franklin, D., Weiner, M. S., Bianco, S. and Nussenzweig, V. (1973). *Clin. Immunol. Immunopathol.*, **1**, 319–329.
136. Zucker-Franklin, D., Melton, J. W. and Quagliata, F. (1974). *Proc. Nat. Acad. Sci.*, **71**, 1877–1881.
137. Braylan, R., Variakojis, D. and Yachnin, S. (1975). *Brit. J. Haematol.*, **31**, 553–564.
138. Edelson, R. L., Kirkpatrick, C. H., Shevach, R. M., Schein, P. S., Smith, R. W., Green, I. and Lutzner, M. (1974). *Ann. Intern. Med.*, **80**, 685–692.
139. Dickler, H. B., Siegal, F. P., Bentwich, Z. H. and Kunkel, H. G. (1973). *Clin. exp. Immunol.*, **14**, 97–106.
140. Catovsky, D., Galetto, J., Okos, A., Galton, D. A. G., Wiltshaw, E. and Stathopoulos, G. (1973). *Lancet*, **2**, 232–234.
141. Sumiya, M., Mizoguchi, H., Kosaka, K., Miura, Y., Takaku, F. and Yata, J. I. (1973). *Lancet*, **2**, 910.
142. Yodoi, J., Takatsuki, K. and Masuda, T. (1974). *New Engl. J. Med.*, **290**, 572–573.
143. Brouet, J.-C., Flandrin, G., Sasportes, M., Preud'homme, J.-L. and Seligmann, M. (1975). *Lancet*, **2**, 890–893.
144. Catovsky, D., Pettit, J. F., Galetto, J., Okos, A. and Galton, D. A. G. (1974). *Brit. J. Haematol.*, **26**, 29–37.
145. Jaffe, E. S., Shevach, E. M., Frank, M. M. and Green, I. (1974). *Amer. J. Med.*, **57**, 108–114.
146. Fu, S. M., Winchester, R. J., Rai, K. R. and Kunkel, H. G. (1974). *Scand. J. Immunol.*, **3**, 847–851.
147. Shevach, E., Edelson, R., Frank, M., Lutzner, M. and Green, I. (1974). *Proc. Nat. Acad. Sci.*, **71**, 863–866.
148. Sandilands, G. P., Cooney, A., Grant, R. M., Gray, K., Browning, J. D., Andersson, J. R., Dagg, J. H. and Lucie, N. (1974). *Lancet*, **1**, 903–904.
149. Chiao, J. W., Pantic, V. S. and Good, R. A. (1974). *Clin. exp. Immunol.*, **18**, 483–490.
150. Dickler, H. B., Adkinson, N. F. and Terry, W. D. (1974). *Nature*, **247**, 213–215.
151. Haegert, D. G., Hallberg, T. and Coombs, R. R. A. (1974). *Int. Arch. Allergy Appl. Immunol.*, **46**, 525–538.
152. Mendes, N. F., Miki, S. S. and Peixinho, Z. F. (1974). *J. Immunol.*, **113**, 531–536.
153. Shevach, E. M., Jaffe, E. S. and Green, I. (1973). *Transplant. Rev.*, **16**, 3–28.
154. Peter, C. R., MacKenzie, M. R. and Glassy, P. J. (1974). *Lancet*, **2**, 686–688.
155. Jaffe, E. S., Shevach, E. M., Sussman, E. H., Frank, M., Green, I. and Berard, C. W. (1975). *Brit. J. Cancer*, **31**, suppl. 11, 107–120.

156. Smith, J. L., Barker, C. R., Clein, G. P. and Collins, R. D. (1973). *Lancet*, **1**, 74–77.
157. Kaplan, J., Mastrangelo, R. and Peterson, W. D. (1974). *Cancer Res.*, **34**, 531–525.
158. Brouet, J.-C., Preud'homme, J.-L., Flandrin, G., Chelloul, N. and Seligmann, M. (1976). *J. Nat. Cancer Inst.* **56**, 631–633.
159. Braylan, R. C., Jaffe, E. S. and Berard, C. W. (1975). *Pathology Annual*, Appleton-Century Crofts, New York, pp. 213–217.
160. Habeshaw, J. A. and Stuart, A. E. (1975). *J. Clin. Pathol.*, **28**, 289–297.
161. Brown, G., Greaves, M. F., Lister, T. A., Rapson, N. and Papamichael, M. (1974). *Lancet*, **2**, 753–755.
162. Belpomme, D., Dantchev, D., Du Rusquec, E., Grandjon, D., Huchet, R., Pouillart, P., Schwarzenberg, L., Amiel, J. L. and Mathe, G. (1974). *Biomedicine*, **20**, 109–118.
163. Sen, L. and Borella, L. (1975). *New Engl. J. Med.*, **292**, 828–832.
164. Gajl-Peczalska, K. J., Bloomfield, C. D., Nesbit, M. E. and Kersey, J. H. (1974). *Clin. exp. Immunol.*, **17**, 561–569.
165. Davey, F. R. and Gottlieb, A. J. (1974). *Amer. J. Clin. Pathol.*, **62**, 818–822.
166. Haegert, D. G., Stuart, J. and Smith, J. L. (1975). *Brit. Med. J.*, **1**, 312–314.
167. Flandrin, G., Brouet, J.-C., Daniel, M. T. and Preud'homme, J.-L. (1975). *Blood*, **45**, 183–188.
168. Brouet, J. C., Valensi, F., Daniel, M. T., Flandrin, G., Preud'homme, J. L. and Seligmann, M. (1976). *Brit. J. Haematol.*, **33**, 319–328.
169. Kersey, J. H., Sabad, A., Gajl-Peczalska, K. J., Hallgren, H. M., Yunis, E. J. and Nesbit, M. E. (1973). *Science*, **182**, 1355–1356.
170. Chin, A. H., Saiki, J. H., Trujillo, J. M. and Williams, R. C. (1973). *Clin. Immunol. Immunopathol.*, **1**, 499–510.
171. Brouet, J. C., Toben, H. R., Chevalier, A. and Seligmann, M. (1974). *Ann. Immunol. Inst. Pasteur*, **125C**, 691–696.
172. Catovsky, D., Goldman, J. M., Okos, A., Frisch, B. and Galton, D. A. G. (1974). *Brit. Med. J.*, **2**, 643–646.
173. Mohanakumar, T. and Metzgar, R. S. (1974). *Cell. Immunol.*, **12**, 30–36.
174. Greaves, M. F., Brown, G., Rapson, R. and Lister, A. (1975). *Clin. Immunol. Immunopathol.*, **4**, 67–84.
175. Fu, S. M., Winchester, R. J. and Kunkel, H. G. (1975). *J. exp. Med.*, **142**, 1334–1338.
176. McCaffrey, R., Harrison, T. A., Parkman, R. D. S., Baltimore, D. (1975). *New Engl. J. Med.*, **292**, 775–780.
177. Stathopoulos, G. and Elliot, E. V. (1974). *Lancet*, **1**, 600–601.
178. Hsu, C. C. S., Marti, G. E., Schrek, R. and Williams, R. C. (1975). *Clin. Immunol. Immunopathol.*, **3**, 385–395.
179. Ferrarini, M., Moretta, L., Abrile, R. and Durante, M. C. (1975). *Eur. J. Immunol.*, **5**, 70–72.
180. Ferrarini, M., Moretta, L., Mingari, M. C., Tonda, P. and Pernis, B. (1976). *Eur. J. Immunol.*, **6**, 520–521.
181. McConnel, I. and Hurd, C. M. (1976). *Immunology*, **30**, 835–839.
182. Gmelig-Meyling, F., Van der Ham, M. and Ballieux, R. E. (1976). *Scand. J. Immunol.*, **5**, 487–496.
183. Moretta, L., Ferrarini, M., Mingari, M. C., Moretta, A. and Webb, S. R. (1976). *J. Immunol.* (in press).
184. Moretta, L., Webb, S. R., Grossi, C. E., Lydyard, P. M. and Cooper, M. D. (1976). *Clin. Res.*, 448A.
185. Moretta, L., Mingari, M. C., Moretta, A. and Lydyard, P. M. (1977). *Clin. Immunol. Immunopathol.* (in press).
186. Vogler, L. B., Pearl, E. R., Gathings, W. E., Lawton, A. R. and Cooper, M. D. (1976). *Lancet*, **2**, 376.
187. Cooper, M. D., Kearney, J. F., Lawton, A. R., Abney, E. R., Parkhouse, R. M. E., Preud'homme, J. L. and Seligmann, M. (1976). *Ann. Immunol.* (*Inst. Pasteur*), **127C**, 573–581.
188. Touraine, J. L., Touraine, F., Dutruge, J., Gilly, J., Colon, S. and Gilly, R. (1975). *Clin. Exp. Immunol.*, **21**, 39–46.
189. Waldmann, T. A., Broder, S., Krakauer, R., Durm, M., Meade, B. and Goldman, C. (1976). *Trans. Assoc. Amer. Phys.* (in press).

190. Waldmann, T. A., Broder, S., Durm, M., Blackman, M., Krakauer, R. and Meade, B. (1975). *Trans. Assoc. Amer. Phys.*, **88**, 120–134.
191. Brouet, J. C. and Toben, H. (1976). *J. Immunol.*, **116**, 1041–1044.
192. Borella, L. and Sen, L. (1975). *J. Immunol.*, **114**, 187–190.
193. Gelfand, E. W. and Chechik, B. E. (1976). *New Engl. J. Med.*, **294**, 275–276.

Notes added in proof

(Section 2) Two distinct subpopulations of human T cells have now been defined: T cells bearing surface receptors for Fc of IgG (Tγ subpopulation) and ones expressing receptors with affinity for the pentameric Fc of IgM (Tμ) (References 17, 179–183, and unpublished observations of Moretta, L., Grossi, C. *et al.*). The Tμ cells have less cytoplasm and smoother surfaces than Tγ cells, are the predominant T cell type in blood, tonsils and lymph nodes, respond better to PHA, and can help B cells respond to PW mitogen. In contrast, Tγ cells have extensive cytoplasm containing acidophilic granules and numerous mitochondria, have a villous surface, are the prevalent T cell type in spleen, do not home to lymph nodes normally, respond poorly to PHA, and do not help B cells respond to PW mitogen; they can be activated via their Fc receptors to suppress the Tμ cell help of B cell responses. Very few thymic cells have detectable receptors either for IgM or IgC, and a variable proportion of blood T cells lack these receptors (T 'null').

Imbalances in the distribution of circulating Tμ and Tγ cells occur in patients with both congenital and acquired immunodeficiencies; deficiency of Tμ cells, with or without an increase in Tγ cells, is the most frequent aberration.[184] Leukaemic T cells (T ALL) consistently express IgM receptors and occasionally express receptors for IgG as well.[185]

(Sections 3.3.1 and 3.3.11) Recently, X-LA patients were found to have an arrest at a very early stage in cellular differentiation along the B cell pathway.[185] Affected boys have normal numbers of bone marrow lymphoid cells that contain intracytoplasmic IgM in small amounts but no SmIgM detectable by immunofluorescence (see Figure 1). Such cells, termed pre-B cells, may be missing entirely in the bone marrow of patients having immunodeficiency with thymoma.[186] This suggests that their stem cells have ceased to differentiate along the B cell axis.

It seems likely that the oncogenic process for many B cell malignancies will be evident as early as the pre-B cell stage in differentiation.

Chapter 18

Histocompatibility Antigens, Mixed Lymphocyte Reaction Genes and Transplantation

J. DAUSSET
D. FRADELIZI

Abbreviations

CML: cell-mediated lympholysis
LDA: lymphocyte dependent antibodies
MHC: major histocompatilibity complex
MLR: mixed lymphocyte reaction
PHA: phytohaemagglutinin
PW: pokeweed mitogen

1. INTRODUCTION

There is now a partial understanding of the role of lymphocytes and their subpopulations in the human allogeneic response. The subtle processes which govern the recognition of the non-self and the consequent production of weapons which leads to the rejection of grafts has begun to be explained. This has been made possible by two cellular immunological techniques:

(a) mixed lymphocyte reaction (MLR) which measures the *in vitro proliferative response* of the lymphocytes of one individual brought into contact with the lymphocytes of another individual;

(b) cell-mediated lympholysis (CML) which shows the *in vitro lytic action of effector lymphocytes* generated either *in vitro* during a MLR or *in vivo* by planned human alloimmunization.

These two tools have enabled an initial approach to the cellular events involved in the human allogeneic response to be made. However, we know that we are dealing with a rough model which as yet only barely indicates the extreme complexity of the mechanisms.

We shall consider the kinetics of the MLR and the CML using either normal unprimed lymphocytes or alloimmunized lymphocytes, obtained through *in vitro* sensitization or by planned *in vivo* immunization. We shall then see to what extent these data can be applied to human transplantation.

2. THE HUMAN ALLOGENEIC RESPONSE

The human allogeneic response can be divided arbitrarily into two processes, possibly taking place simultaneously, both governed by the 'major histocompatibility complex' (MHC) in man, the HLA complex.

The first process appears to correspond to a non-self recognition phenomenon which, at least *in vitro*, depends on the lymphocyte stimulatory product governed in man by genes of the HLA-D locus. During the second process, the specialized lymphocytes (T and B) are immunized against foreign structures they do not possess (namely, transplantation antigens) whether or not they are coded by the HLA complex. The first process would seem, at least *in vitro*, to be a necessary and sufficient prerequisite to the second.

2.1 The human mixed lymphocyte reaction (MLR)

2.1.1 *The primary MLR*

The proliferative reaction observed *in vitro* when one co-cultures the peripheral lymphocytes of two individuals[1–3] was originally considered to be triggered off by serologically detectable histocompatibility antigen differences.[4]

The interpretation of the MLR has subsequently been corrected, due to the study of informative families. In the vast majority of cases the MLR remains negative between identical HLA siblings. However, some exceptions have been noted by Bach and Amos[5] and by Yunis and Amos.[6] These informative families showed the existence of a special gene governing stimulatory products. This gene is situated at a certain distance from the genes governing the HLA-A, HLA-B and HLA-C markers, and as a result of recombination it may occasionally differ in two serologically identical HLA siblings. The validity of this concept has rapidly been confirmed by several teams.[7–9]

Table 1. The HLA complex in Caucasoids on chromosome VI. Nomenclature adopted during the VIth Histocompatibility Workshop Conference (Aarhus, 1975) is presented. The three loci coding for serologically defined antigens are called A, B, and C. The main locus coding for lymphocyte stimulatory product is called D. The most clearly defined alleles of these four loci are given

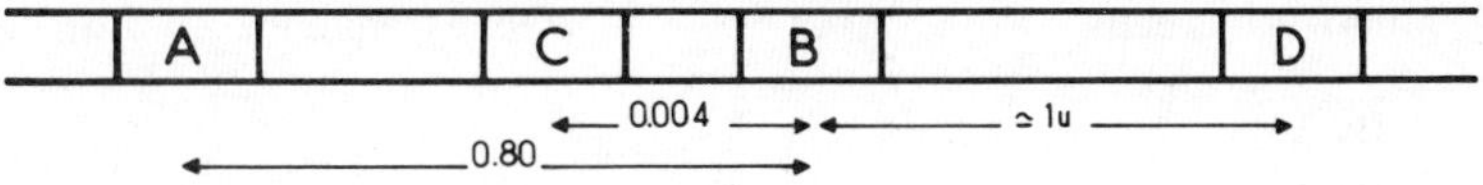

A	C	B	D
A1	CW1	B5	DW1
A2	CW2	B7	DW2
A3	CW3	B8	DW3
A9 {AW23, AW24}	CW4	B12	DW4
	CW5	B13	DW5
A10 {AW25, AW26}		B14	DW6
A11		BW16 {BW38, BW39}	
A28		BW17	
A29		BW21	
AW30		BW22	
AW31		BW35	
AW32		BW37	
AW33		BW40	
		BW41	

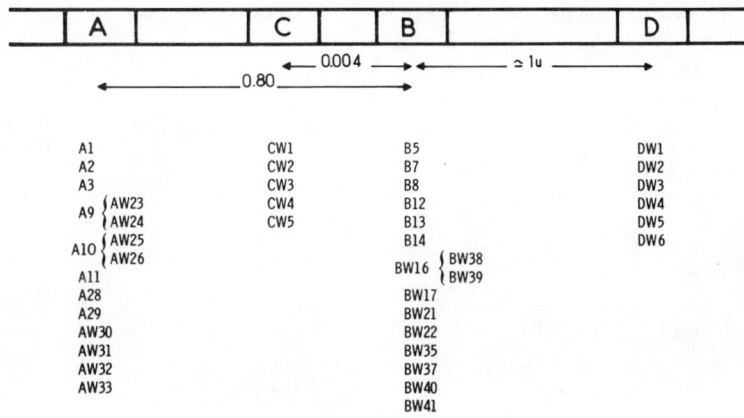

We now know that there exists in man a gene (or complex of genes) which codes for a lymphocyte stimulatory product mainly responsible for the development of a MLR. It constitutes what is called the D locus of the HLA system. It is polymorphic and produces an as yet unknown number of alleles. Six of them have already been given a provisional name (from Dw1 to Dw6).[10] The collection of genes still unknown has a frequency of 0·524 (Tables 1 and 2).

The situation in the mouse is apparently slightly different from that in man. Murine genes capable of causing the MLR are distributed along the H-2 complex. Nevertheless, the principal MLR genes are centred in the I-A region near the K region. In the mouse, these genes are within the interval between the K and D locus, while in man they are outside the interval between the A and B locus at about one recombination unit from the B locus. In man, accessory MLR genes present in the HLA complex have been postulated, one being near the A locus.[11–13]

Outside the MHC, the mouse possesses another locus capable of triggering off the MLR, the M locus,[14] to which there is so far no known equivalent in man.

The techniques for determining the alleles of the D locus (MLR typing) are still laborious. Cellular immunology techniques were used to start with, in the absence of lymphocytoxic antibodies capable of detecting the products of these genes. In certain families (especially in those where the two parents are related) one can occasionally affirm that certain children are homozygous for the MHC, including the D locus. Such families can be detected because they have a negative MLR between the parents and certain children, and unilaterally negative reactions occur between the parents and the homozygous child (Figure 1). The homozygous cells are valuable reagents. Used as stimulant cells they are incapable of triggering off a clear-cut positive response by a responding heterozygous cell (and, of course, by a homozygous cell) for the D allele they possess[15–19] (Table 3). This principle has been widely applied and, owing to an enormous international collaborative effort, has enabled the first alleles of the D locus to be defined (Table 2).

Table 2. Alleles of the HLA-D locus in Caucasoids. The six most clearly defined alleles of the HLA-D locus coding for lymphocyte stimulatory products are presented (from the *Joint Report of the VIth Histocompatibility Workshop Conference in Histocompatibility Testing 1975*, Munksgaard, Copenhagen, 1976, reproduced with permission)

	HLA-B typing of homozygous cells	Antigen frequency (%)	Gene frequency	Most significant Δ value with B and C on haplotypes
Dw1	Bw35 Bw35 B27 Bw35 Bw35 B27	19·3	0·102	Bw35 Cw4
Dw2	B7 B7 B7 B12	15·2	0·078	B7
Dw3	B8 B8 B8 B8 B8 B8	16·4	0·085	B8
Dw4	B7 B12 Bw15 Bw15 Bw15 Bw15	15·6	0·082	Bw15 Bw27
Dw5	Bw35 Bw16 — Bw16 B18 —	14·6	0·075	Bw16
Dw6	B12 Bw16 Bw15 B8 Bw15 Bw40	10·5	0·054	(Bw16) (Cw2)
Blank			0·524	

The other suggested methods of HLA-D typing involve antibodies. Some might be specifically able to block the MLR and others are thought to recognize the product of the D gene or of one very close by.

In the mouse, it is thought that certain anti-Ia antibodies, directed against the products of genes in the I region, may react with the lymphocyte stimulatory products.[20]

Study of the antibodies which could be the anti-Ia equivalents in man has already begun.[21–24] However, we do not yet know the exact relationship which may exist between

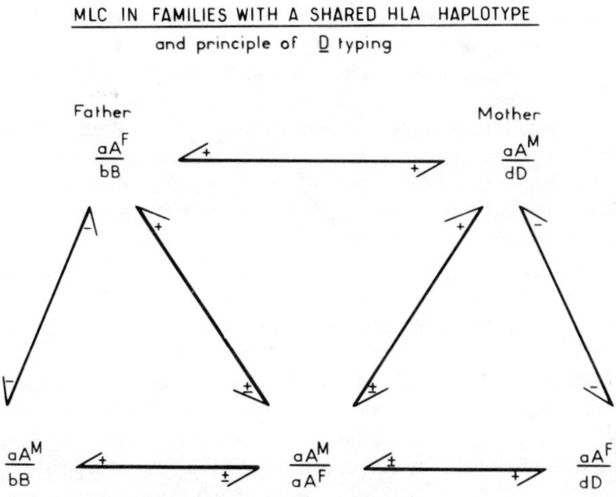

Figure 1. The MLR reactivity pattern observed in families in which parents share one haplotype is given here. The arrows indicate the direction of the reaction: Responding cell → 2000 R irradiated stimulating cell. a, b and d stand for HLA serologically detectable specificities, A, B and D are the HLA-D product specificities. The identy of the HLA-D specificity on the father's haplotype aA^F with the HLA-D specificity on the mother's haplotype aA^M is shown by the mutually negative MLR observed between the parents and the heterozygous sibs.

The child $\frac{aA^M}{aA^F}$ is HLA-D homozygous and therefore gives MLR partially negative results (typing response) when used as a stimulating cell with the parents or the heterozygous sibs as responding cells

the antigens they detect and the products of the D locus defined by the MLR. In some instances[22] they have a close statistical correlate, but this correlation may not be absolute. These antibodies could, therefore, be directed against the products of one or several genes near the HLA-D locus having a high linkage disequilibrium with the alleles of the D locus. Indeed, an 'Ag' locus very similar to HLA-D with already three serologically detectable alleles has been suggested. However, one serum seems to detect the Dw3 product.[25] The problem of HLA-D typing, therefore, has not yet been solved satisfactorily. Another approach has been suggested using the secondary MLR.[26,27] This will be discussed later.

The kinetics of the primary MLR *in vitro* is relatively constant, when standard technical conditions are used. One of the lymphocyte populations treated with mitomycine or X-ray irradiation, although alive, is unable to proliferate. The proliferation of the other population is therefore measured by radioactive thymidine incorporation, which indicates DNA synthesis. The proliferation starts around the third or fourth day, reaches a peak by the sixth or seventh day and ceases towards the tenth to thirteenth day (Figure 2).

The intensity of the proliferation depends on the number of HLA-D allele differences between the two cellular populations brought into contact. There is no proliferation between two identical HLA siblings. The proliferation observed between two semi-identical HLA siblings will usually be half as strong as between two unrelated individuals whose D alleles are different.[28] The type of reaction observed between a heterozygous

Table 3. HLA-D typing using homozygous cells. The cells to be typed (R1 to R5) are confronted with different varieties of irradiated homozygous typing cells, presenting different HLA-D alleles (HLA-Dw1, HLA-Dw2 and HLA-Dw3). The mean c.p.m. response is given in the table. Also included is the autologous control, and the response against a pool of randomly selected, unrelated, irradiated, stimulating control lymphocytes. The response of the cell to be typed against the typing cells will be compared with the response against the pool of unrelated control lymphocytes.

A clear typing response is obtained when the cell to be typed gives a negative or almost negative response against the homozygous typing cell as compared with the positive response it gives against the pool of unrelated lymphocytes: in the table, cell R1 against the HLA-DwI homozygous cell, cell R2 against HLA-Dw2, and cell R3 against HLA-Dw3. However, it is sometimes difficult to decide whether a response is positive or negative (in the table the response of cell R4 against the HLA-Dw3 homozygous cell). Rather than simply considering the stimulatory index and drawing an arbitrary line between the positive and negative response, one must take into account not only the proliferative ability of the cell to be typed as judged from its response against the pool of unrelated lymphocytes, but also the stimulating capability of one homozygous typing cell. An approach of this kind was used for the analysis of the Sixth Histocompatibility Workshop results: a standardized relative response value was calculated by considering both the responding ability of this cell and the stimulating capacity of the typing cell (double standardization) (Piazza, A. and Galfré, G., in *Histocompatibility Testing 1975*, (Kissmeyer-Nielsen, F., Ed.), Munksgaard, Copenhagen, 1975, pp. 552–556)

		Irradiated stimulating cells				
			Typing cells homozygous for:			
		Autologous control	HLA-Dw1	HLA-Dw2	HLA-Dw3	Pool of unrelated cells
Responding	R_1	1 700	4 200	27 500	23 600	41 900
cells	R_2	1 600	23 700	2 300	19 700	25 500
to be	R_3	1 600	25 300	41 900	4 200	27 500
HLA-D	R_4	1 900	8 300	7 200	4 300	9 400
typed	R_5	1 700	25 400	23 900	18 500	24 200

responding cell and a homozygous stimulating cell comes between the negative reaction and the response against a single D allele (one haplotype difference). This reaction is not completely negative but is nevertheless sufficiently weak to indicate the presence of an allele common to both populations (typing response) (Table 3). This weak reaction is perhaps due to the liberation of mitogenic product by the X-irradiated or mitomycin treated homozygous cell, in response to the incompatibility with one of the D alleles of the heterozygous cell.[12,29,30] It is not impossible, however, that this may indicate a more subtle type of recognition reaction.[31]

It has been possible to show that the proliferation observed in MLR was initially clonal, by using specific elimination techniques: BudR light inactivation[32] or hot radioactive suicide.[33,34] Each individual would possess a series of lymphocytic clones capable of specifically recognizing the different HLA-D alleles of its species.

It is interesting to note in this respect that the intensity of the mixed xenogeneic response decreases when the phylogenic distance between the two species increases indicating possible structural analogies in the products of MLR genes of related species.

2.1.2 *The secondary MLR*

Early studies where the MLR was evaluated on morphological criteria failed to demonstrate a clear MLR anamnestic response. Nevertheless, ten days after *in vivo* immunization a slightly increased reaction was observed when looking at day 3 or day 5 of the MLR. Even if immunization was continued, the MLR returned to its initial level.[35]

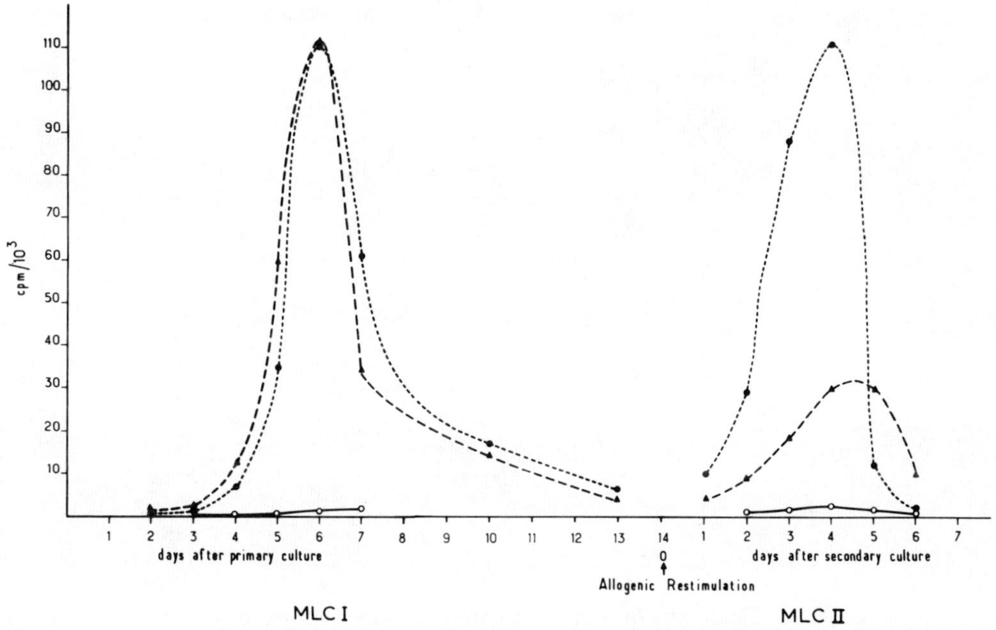

Figure 2. The kinetics of the primary MLR and secondary MLR are shown. On the left panel (MLC I) the kinetics of ^3H thymidine incorporation by A responding cells confronted in a primary *in vitro* culture with A (outline circles), B (solid circles) or C (triangles) irradiated stimulating cells are shown. On the right panel (MLC II) the proliferative response of AB_{14} cells (A cell primed with B irradiated cell by 14 days *in vitro* co-culture) confronted in a secondary culture with A (outline circles), B (solid circles) and C (triangles) irradiated cells is shown. The response of AB_{14} cells is earlier and higher against the specific cell B than against cell C. (From Reference 39; reproduced by permission of Verlag Chemie GmbH)

However, kinetic studies of the MLR response of *in vivo* immunized human lymphocytes[36] have shown that from day 7 to day 30 following the immunizing injection, an early proliferative response is seen with a peak on days 3–4. This has also been shown by Charmot *et al.*,[37] after one leukocyte injection (Figure 3). Interestingly enough, a father hyperimmunized by 16 weekly injections of his daughter's leukocytes showed MLR memory cells persisting 6 months after the last injection[38] (Figure 4).

More recent studies using *in vitro* alloimmunized human lymphocytes have shown the characteristics of the secondary MLR; the secondary proliferative response starts very early (detectable as soon as day 1 or 2 after restimulation); the peak of thymidine incorporation occurs on day 3–4 (1 or 2 days earlier than a primary MLR); the amplitude of this peak is about the same size as the sixth day peak response of fresh unprimed lymphocytes[39] (Figure 2).

The accelerated secondary response is specific and observed when the primed lymphocytes are confronted with the priming cell. Contact with a third party cell produces a proliferative response with different kinetics, comparable to the primary MLR of non-sensitized 14 days' culture lymphocytes, which have a response on day 4 or 5.

The serologically detected HLA antigens play no role in restimulation.[27,38,40] This has been shown by experiments made using cells with chromosomal recombination between HLA-B and D (Table 4).

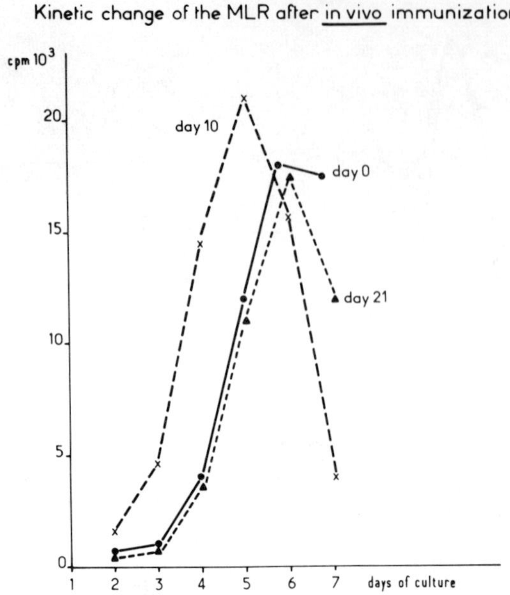

Figure 3. *In vivo* immunization was performed between a volunteer recipient A (A2, A3; B7, Bw22) who received 150×10^6 leukocytes subcutaneously once from a donor B (A3, A11; B5, B12). Blood was drawn from the recipient before and on various days after immunization. Each sample of lymphocytes was frozen on the day of collection (same experiment as in Figure 8). One month later the MLR reactivity of the different samples of A cells was tested in one single experiment against irradiated B cells. Circles: A cells drawn before immunization tested in MLR against B (day 0), Crosses: A cells drawn 10 days after immunization and tested in MLR against B (day 10). Triangles: A cells drawn 21 days after immunization and tested in MLR against B (day 21). The proliferative response of peripheral lymphocytes collected 10 days after *in vivo* immunization and confronted *in vitro* with the donor's cells peaks on day 5 while the response of lymphocytes collected on day 0 and day 21 peaks on day 6

It may therefore be possible to use these *in vitro* primed lymphocytes for HLA-D typing.[26,27] First attempts indicate that this method applies correctly to intra-familial situations with a difference of one HLA haplotype (Table 5), but, with some difficulties, between unrelated individuals where the two HLA-D alleles are usually different.[27,38]

The proliferative reaction obtained by mitogens such as PHA or pokeweed (PW) is non-specific and involves different mechanisms from the MLR. *In vitro* primed lymphocytes are nonetheless capable of proliferation with mitogenic stimulation.

2.2 The human cell-mediated lympholysis (CML)

2.2.1 *The primary CML*

The second process in the cellular response can be tackled experimentally through the study of cytotoxic effector cells which appear from the third day of the MLR. The

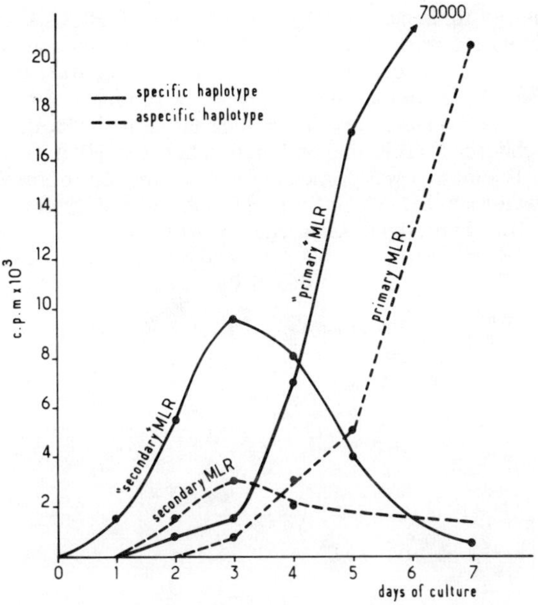

Figure 4. Kinetics of the 'primary' and 'secondary' MLR six months after *in vivo* immunization of a father against one maternal haplotype. A father (A3, B7, Dw2/Aw23, Bw21) was hyperimmunized *in vivo* by his daughter possessing the maternal haplotype Aw24, Bw40. Six months after the last leukocyte injection, the peripheral lymphocytes were collected and tested as for a primary and a secondary MLR against specific cells and non-specific cells bearing the other maternal haplotype. Both primary and secondary MLR using the priming haplotype were still modified: the proliferative response was much higher than against the non-specific haplotype, demonstrating the persistence of memory cells, six months after a strong *in vivo* immunization

technique was first described in the mouse by Haÿry and Defendi,[41] and modified for human systems by Bach and Solliday[42] and Lightbody *et al.*[43] The latter workers used PHA stimulated lymphoblasts as target cells.

Studies of human families in which there were chromosomal recombinations between the different HLA-A and B or B and D locus have clarified the mechanism of the CML reaction. Such studies have shown that *lymphocytic proliferation in MLR was a necessary preliminary to the appearance of cytotoxic effectors.* This was especially clear from one family studied by Eijsvoogel, in which there were eleven children, one of whom had a chromosomal recombination between HLA-B and D.[7,44] The cytotoxic effectors only appear in combinations where there are both positive MLR and HLA antigen differences. No killer cells appear in the combinations where the MLR was negative whether or not there were HLA antigen differences.

A non-specific lymphocytic proliferation, such as that obtained using the mitogen PHA, does not generate killer cells.[45] Not every proliferation, therefore, is in itself capable of inducing what has been described somewhat vaguely by the term 'lymphocyte activation'. This 'activation' is triggered off, at least *in vitro*, by differences at the HLA-D locus.

Table 4. Influence of HLA-D (but not of HLA-A or B) difference in secondary MLR. a, b and c are the antigens of the HLA-A and B loci: a = A28, B14; b = Aw30, B13; c = A3, Bw35. A, B and C are the HLA-D products. Only one haplotype is given, the other being shared by the mother and her children. a/B is a recombination between HLA-B and D loci. The mother was primed *in vitro* by either one of the three following sibs bB, aA and a/B. Sibs aA and a/B differ only by the HLA-D product. Restimulation is obtained by presenting the a/B cells, and not by aA cells, and is therefore due to the B specificity

Priming Combination		Restimulation		MLR
		bB	aA	a/B
cC 3, W35	- bB W30, 13	Specific 4. 781	919	5. 452
	- aA 28, 14	1. 171	Specific 2. 933	1. 419
	- a/B 28, 14	6. 966	1. 283	Specific 5. 479

a, b, c are the HLA-A and B antigens } only one haplotype is given, the other being
A, B, C are the HLA-D products } shared by the mother and her children.

a/B is a recombination between HLA-B and D.

Table 5. HLA-D typing using secondary MLR. a, b, c, d and x are the antigens of the HLA-A and B loci; A, B, C and D are the HLA-D products. Only one haplotype is given for the children, the other being shared with the father. The father aA was primed *in vitro* by the child cC. Restimulation was performed with every member of the family and an unrelated control cell which is mutually MLR negative with the priming cell. Only figures of early (second day) typing responses and peaks are given

Priming combination	Restimulation	Day 1	2	3	4	5	6
	aA/bB (father)	—	—	—	2 300	—	—
	cC/dD (mother)	—	6 200	10 188	—	—	—
	cC (priming cell)	—	5 344	9 124	—	—	—
aA + cC 3, 7, Dw2 W23, W21	cC	—	5 000	9 600	—	—	—
	dD	—	—	—	2 700	—	—
	dD	—	—	—	3 400	—	—
	dD	—	—	—	4 013	—	—
	control xC 3, 7, Dw2/W24, W40	—	4 950	8 291	12 800	—	—

The necessary HLA-D difference can be carried by a third party cell different from that carrying the incompatibilities for the transplantation antigens. This was shown using once again HLA-B/D recombinant cells: the responding cells were activated by a population differing only by one HLA-D gene but were immunized against the third population differing only by HLA-B (Table 6).[45,46]

Table 6. Relationship between lymphocyte activation and cytotoxic effector production in an HLA-B/D recombinant family. a, b, c and d are the HLA-A and B serologically defined antigens. A, B, C and D are the HLA-D gene stimulatory products. Sib C_3 (a/B/cC) being a recombinant between HLA-B and D is MLR+CML− with sib C_1 and MLR−CML− with sib C_2. However, co-culture of sib C_3 with a mixture of cells from C_1 and C_2 leads to activation, due to the HLA-D stimulating product of sib C_1 *and* production of effector killer cells directed against the HLA-A and B specificities presented by sib C_2. Same family as Table 4

Responding cells	Mitomycin treated stimulating cells	MLR (index)	CML on PHA targets (specific chromium release) C_1 aA/cC	C_2 bB/cC	Father aA/bB
	C_1 = aA/cC	34	8	7	1
C_3 = a/B/cC	C_2 = bB/cC	1, 6	−6	−2	−6
	mixture of C_1+C_2 aA/cC+bB/cC	42	9	30	25

An indirect proof of the intervention of HLA-D genes in the generation of cytotoxic effectors is again found in the inhibitory effect of treatment of stimulant cells by heat or UV light which suppresses the MLR while leaving intact the serologically detectable antigens HLA-A and B. Under these conditions no effector cells appear in the culture.[47]

The kinetics of the *in vitro* primary CML in man shows that it begins on the fourth day, ends around the twelfth, with a peak around the sixth to the tenth day (Figures 5 and 7).

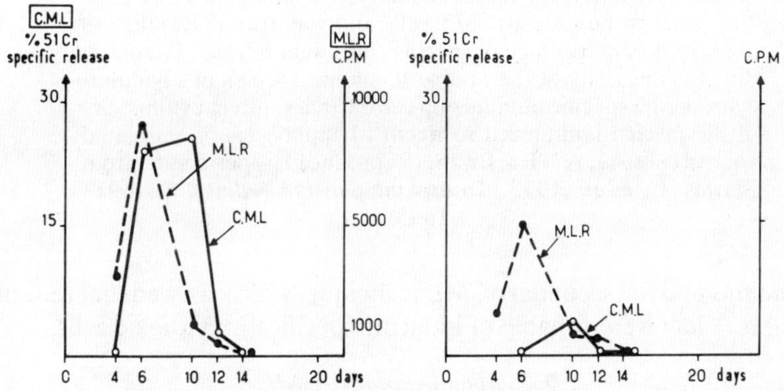

Figure 5. Kinetics and strength of the primary MLR and CML towards *four* HLA incompatibilities (left) versus only one (right)

In vivo, after immunization of a volunteer by one injection of 150×10^6 leukocytes from his pheno-identical brother, killer lymphocytes directly cytotoxic on the target cell were found in the peripheral blood from the twelfth to the eighteenth day only, with a peak at the fourteenth day (Figure 6).[48] One possible explanation for their rapid disappearance is that cytotoxic lymphocytes spend little time in the blood circulation after *in vivo* immunization.

The lack of sensitivity of the primary CML has occasionally resulted in mistaken conclusions. In the family studied by Eijsvoogel *et al.*[7] the effector cells apparently only appeared against the antigens of the HLA-B locus. This result was not confirmed using

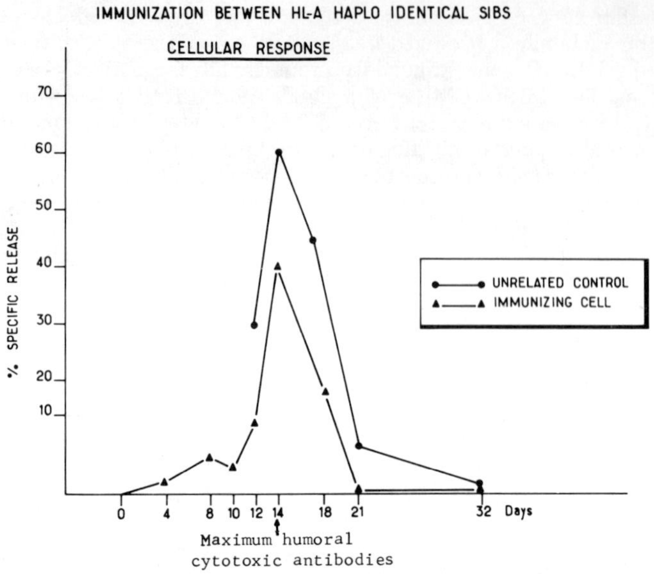

Figure 6. Cellular cytotoxicity of peripheral lymphocytes collected from an *in vivo* immunized volunteer. A volunteer Aw24, Bw18/Aw26, B5 received one injection of 150×10^6 lymphocytes from his haploidentical brother Aw32, Bw27/Aw26, B5. Blood samples were drawn from the recipient on day 0 and on various days after immunization. Lymphocytes were isolated from each sample and tested for direct cellular cytotoxicity on PHA blast ^{51}Cr labelled target cells. E/T cell ratio was 100/1. Results are expressed as percentage of specific chromium release. Triangles: direct cytotoxicity of the *in vivo* immunized recipient's lymphocytes on the specific immunizing cells. Circles: direct cytotoxicity of the *in vivo* immunized recipient's lymphocytes on unrelated control cells A1, A3; B8, Bw18. (Reproduced by permission from Mawas, C. *et al.*, 1973, *Transplantation Proceedings*, **5**, 1691–1695)

other combinations and the secondary CML technique, which showed that the antigens of two principal HLA loci were capable of inducing specifically cytotoxic cells.

2.2.2 *The secondary CML*

2.2.2.1 *Restimulation*

This process has been extensively studied in our laboratory by Mawas, Charmot and coworkers.[37,47,49] After 14 days of co-culture, no cells capable of directly producing cytolysis are left in the culture. However, restimulation will allow the rapid appearance of a cytotoxic action which is clearly more intense than a primary reaction, although strictly maintaining the same specificity (Figure 7).

What are the characteristics of this restimulation? It is not a classical immunological secondary response such as that observed against environmental antigens: there is no restimulation and no appearance of cytotoxic cells if the responding primed lymphocytes are confronted with cells presenting only the HLA-A and B antigens of the priming cell while being identical with the responding cell for the HLA-D specificity (Table 7). There is, however, generation of killer cells if the lymphocytes used for restimulation do possess

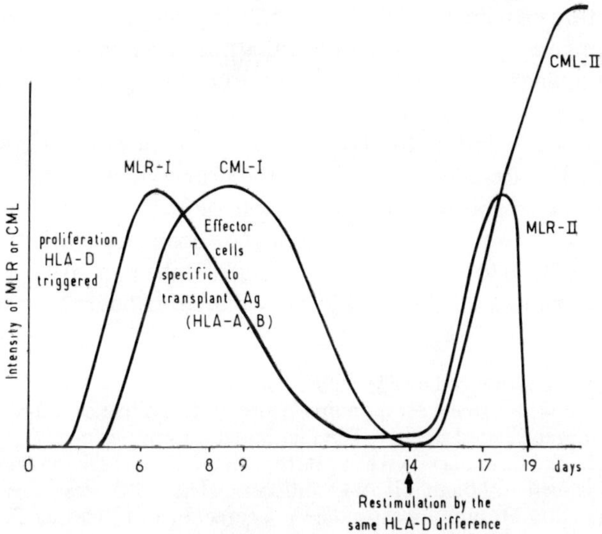

Figure 7. Kinetics of alloimmunization *in vitro* between cells with HLA complex differences. A schematic representation of the kinetics and usual level of proliferation or ^{51}Cr specific release in primary and secondary MLR and CML (arbitrary scale)

Table 7. No secondary CML in the absence of MLR (in a family with a recombination between HLA-B and D. Cells from family LE were used. D cells (A3, B7, Dw2/A3, B7, Dw2) were used as responding cells. B cells (Aw24, −, −/A3, B7, Dw2) were used as the priming cells. Restimulation for secondary CML was performed using the abovementioned cells and with the sib F, which is a recombinant between HLA-B/D (Aw24, −, Dw2/A3, B7, Dw2) and therefore HLA-A and B identical with the priming cell B although HLA-D different. An unrelated cell X was also used. The CML test was done 5 days after restimulation. Results are expressed as a percentage of ^{51}Cr specific release. Restimulation by F cells induces neither secondary MLR nor production of effector cytotoxic cells

Priming Combination	Restimulation	Secondary MLR	Target: D 3, 7, DW2 / 3, 7, DW2	B 3, 7, DW2 / W24, −, −	F recombinant 3, 7, DW2 / W24, −/ DW2	X 1, 3, 8, W40
	None	no	−	2	0	0
	D	no	−	10	4	−
D + B_{14}	B	yes	−	18	10	−
	F	no	−	5	0	−
	X	yes	−	18	13	0

the HLA-D structure specificity of the cell used for priming while being different for the HLA-A and B antigens. It seems, therefore, that the secondary CML is dependent on a reactivation phenomenon governed by HLA-D structure differences while HLA-A and B antigens play no role.

The secondary reactivation by the HLA-D structure, however, does not appear to be specific. Any HLA-D difference presented to the primed lymphocytes by the stimulating cells will lead to the generation of secondary cytotoxic lymphocytes. Furthermore, there is very little difference between the intensity of the secondary CML triggered off by the specific HLA-D (identical to that used for priming) and the non-specific HLA-D (different from that used for priming) (Table 8). Moreover, the secondary reactivation can also be

Table 8. Secondary CML is induced by a specific or non-specific HLA-D difference but effector cells remain specific for the priming cell. Responding lymphocyte aA/bB cells and priming lymphocyte cC/dD cells were from unrelated donors. The restimulation was performed using either these two abovementioned cells or a third party cell aE/bB from an unrelated individual HLA pheno-identical with the responding cell although HLA-D different. The CML test was done 4 days after restimulation. Results are expressed as a percentage of ^{51}Cr specific release

Priming combination	Restimulation	Targets aA/bB	cC/dD	aE/bB
aA/bB 1, 8 W33, W17 Cw3 + cC/dD 2, 5 W30, 27	None (day 4)	—	0	0
	aA/bB	—	5	0
	cC/dD	0	71	11
	aE/bB 1, 8 W33, W17, Cw3	0	58	4

brought about by non-specific mitogens in sharp contrast to what is obtained in a primary CML[45] (Table 9). In both cases, the effector cells remain specific for the priming cells.

The treatment of stimulant cells by heat or UV light prevents the induction of the secondary CML, providing yet another argument in favour of unique action of HLA-D product in restimulation.

It seems clear, therefore, that a fourteenth day primary culture contains 'resting' killer cells which can be quickly turned into the state of active effector cells by specific or non-specific reactivation. However, there seems to be two kinds of resting killer cells: one kind, which is reactivated by non-specific mitogens, can be eliminated from the fourteenth day primary culture by means of successive adsorptions on a monolayer of target cells;[47] another kind, which is reactivated by the specific priming cell, is only slightly adsorbable on the target cell monolayer (Table 10). What is observed in the secondary *in vitro* CML is therefore probably reflecting the development of two cell populations mixed together and maybe interacting. On the fourteenth day there would be two kinds of memory cells in the culture—some containing specific receptors available and ready to attach themselves to the corresponding antigen, but devoid of cytotoxic strength for lack of 'activation', others with no receptors expressed at the cell surface but capable of rapid transformation under the double action of the antigen and activation.

Table 9. Secondary CML is also induced by mitogens but effector cells remain specific for the priming cell. Responding cells A (A1, Aw33; B8, Bw17) were primed *in vitro* with an unrelated irradiated stimulating cell B (A2, Aw30; B5, B27). Restimulation was performed with B cells, PHA and PW mitogens. Targets for the secondary CML were either from unrelated donors or from family members. The CML test was done 4 days after restimulation. Results are expressed as a percentage of ^{51}Cr specific release

Priming Combination	Restimulation	Targets: unrelated 1,8 W33, W17 A	unrelated 2,5 W30, 27 B	unrelated 1,8 W33, W17	sibs 1,8 3,7	sibs 1,8 3,7	sibs 1,8 3, W35
A + B	None	-	0	0	0	0	0
	B	0	71	11	-	-	-
	PHA	0	62	0	0	0	14
	PW	0	46	0	0	0	26

Table 10. Secondary CML after absorption on a monolayer of the immunizing cell B on the fourteenth day of primary culture. A cells (A1, A2; B7, B8) were used as responding cells; irradiated B cells (Aw25, Aw19; Bw14, Bw35) were used as priming cells. Fourteen-day-primed A cells were either absorbed or not absorbed on three successive B monolayers. Both groups were restimulated by the specific priming cell and the mitogens PHA and PW. Results are expressed as a percentage of ^{51}Cr specific release. PHA and PW restimulated cells seem to be absorbed on the specific monolayer, while specifically reactivated cells are not

Priming Combination	Restimulation	Targets: A non absorbed	A absorbed 3 times	B non absorbed	B absorbed 3 times
A + B	None	3	0	12	8
	B	5	1	63	52
	PHA	3	0	36	8
	PW	5	-	47	15

Restimulation can be applied to cells which were sensitized *in vivo*. Charmot *et al.*[37] have shown that the peripheral blood lymphocytes in a volunteer immunized by a single injection of blood behave in every way, from the eighth day to the fifteenth day after immunization, like cells stimulated *in vitro*. They can be restimulated by the specific cell, third party cell or by non-specific mitogen such as pokeweed (Figure 8). No restimulation can be induced by UV light treated cells (Table 11).

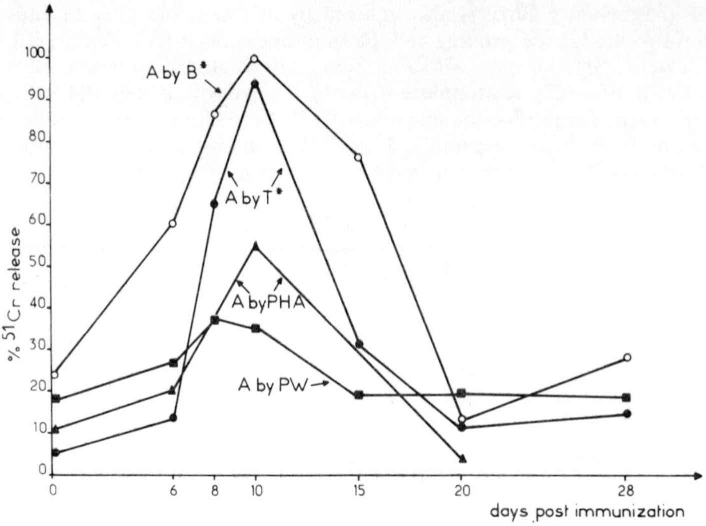

Figure 8. Specific cytotoxic activity of cells primed *in vivo* and reactivated *in vitro* on various days after immunization. *In vivo* immunization was performed between a volunteer recipient A (A2, A3; B7, Bw22) who received 150×10^6 lymphocytes subcutaneously once from donor B (A3, A11; B5, B12). Blood was drawn from the recipient before and on various days after immunization. Each sample of lymphocytes was frozen on the day of collection (same experiment as in Figure 3). One month later all frozen cell samples were thawed on the same day and restimulated for 5 days with either the specific priming cell B, an unrelated control T cell, PHA or PW, and with an A cell as control (not shown). The A cells collected on various days after *in vivo* immunization and reactivated *in vitro* by the abovementioned means were tested for cytotoxicity on the specific ^{51}Cr labelled B cells. Asterisks indicate 2000 R irradiated cells used for secondary *in vitro* restimulation

2.2.2.2 *Kinetics*

When *in vitro* primed lymphocytes are taken out of a fourteenth day culture and restimulated *in vitro* by the specific priming cell, the cellular cytotoxicity rises very fast and is detectable within 24 hours. The cytotoxicity strength then rises parallel to the MLR proliferation reaching a very high level of cytotoxicity, usually 40% to 60% of ^{51}Cr specific release on day 4–5. It is usually higher than in a primary CML.

When *in vivo* primed lymphocytes are used for secondary CML, an even higher level of cytotoxic activity (usually around 100% of ^{51}Cr specific release) is rapidly obtained by specific restimulation when the *in vivo* primed peripheral lymphocytes are collected from the volunteer around the eighth to the fifteenth day after one leukocyte injection. If collected later, around day 20 after immunization, the cytotoxic potency of the effector cells generated by restimulation *in vitro* will be identical to that observed in a primary CML. This suggests the absence of CML memory cells from the circulating blood at that time after immunization. It seems to parallel the published results on the modification of the kinetics and level of the MLR after *in vivo* human alloimmunization: at least when sensitization is weak.[36]

Table 11. Secondary CML using cells primed *in vivo*. Volunteer A (A2, A3; B7, Bw22) was immunized *in vivo* by one subcutaneous injection of 150×10^6 lymphocytes from donor B (A3, A11; B5, B12). Fourteen days after the injection, the A lymphocytes primed *in vivo* were restimulated *in vitro* either by irradiated B cells or B cells exposed to ultraviolet light, which inactivates the HLA-D stimulatory product but does not modify the HLA-A or B antigens. The CML test was performed 4 days after *in vitro* restimulation. Results are expressed as a percentage of ^{51}Cr specific release. Buv treated cells do not induce the reappearance of effector lymphocyte cells

Priming Combination (in vivo)	Restimulation (in vitro) by secondary MLR		Targets A 2, 3, 7, W22	Targets B 3, 11, 5, 12	Targets T 11, W19. 2, 13, W35
	A	no	0	0	0
A + B_{14}	B	yes	0	76	37
	B<u>uv</u>	no	0	7	0

Buv = cell treated by ultra-violet light which inactivates the HLA-D products but does not modify the HLA-A or B antigens

2.2.2.3 *Specificity*

In the future, the extreme sensitivity of this technique will allow a precise definition of the targets against which cytotoxic effectors develop *in vitro* or *in vivo*.

(a) *The main targets are the antigens of the HLA-A and B locus.* The CML technique in the mouse indicates that the main targets are the transplantation antigens governed by the D and K locus of the H-2 complex, but that there are many other possible targets.

In man, similar results are observed.[50] Nevertheless, there is still a need for statistical studies to show that lymphocytes sensitized by cells containing HLA-A2 antigen, for example, regularly kill all the targets containing this antigen. Such a study is now in progress. The intensity of the destruction depends moreover on the number of HLA antigens by which two cell populations brought together differ (Figure 5).

(b) *There are other targets governed by the HLA complex.* In experiments using families with chromosomal recombinations it has previously been found that CML was positive only if: (i) the MLR was positive and (ii) HLA incompatibilities were present in the A and/or B locus. However, numerous exceptions have been found in the primary[51,52] (Table 12) or secondary CML[40] where effectors have appeared while the HLA antigens were apparently identical (Table 13).

In all these examples, lytic activity followed an HLA haplotype indicating that the target was coded by the HLA complex while it was most certainly not the serologically recognized antigenic determinant in the two principal HLA-A and/or B loci. There are several possibilities to consider: namely, that these targets are (i) the products of the HLA-D genes, (ii) the antigenic determinants expressed on the HLA-A or B molecule but different from antibody recognized determinants—the existence of multiple determinants on these molecules, corresponding to the public specificities in the H-2 system, was

Table 12. Primary CML not directed against the serologically detectable HLA-B determinant. A (mother) and B (father) are unrelated parents sharing the serologically detectable B5, A2, B12 antigens. Note that in the three combinations A, Bm, C BM and D Fm, the only incompatible antigen was the father's B5. Cytotoxic effector cells appear to kill cells bearing this antigen but not those bearing the mother's B5. The effector cells were not directed against the serologically recognized B5 determinant

		Target cells					
		A	B	C	D	E	F
	Anti-	1, 5^M 2, 12^M	2, 5^F 2, 12^F	2, 12^M 2, 12^F	2, 12^M 2, 12^F	1, 5^M 2, 5^F	2, 12^M 2, 5^F
A Bm		-	42	0	0	28	21
C Bm	5^F	8	51	0	0	40	37
D Bm	5^F	3	44	0	0	31	39
D Fm	5^F	2	32	0	-	24	17
D Em	$5^{M,F}$	14	24	0	0	31	17
E Dm	$12^{M,F}$	21	15	10	5	0	12

Table 13. Secondary CML against a determinant present in the HLA complex (cC) but different from HLA-A, B private specificities and D products. a, b, c, d and y are the antigens of the HLA-A and B loci (a = A3, B7; b and c = Aw23, Bw21; d = Aw24, Bw35; y = Aw24, Bw40). A, B, C and D are the HLA-D products. The father's aA/bB cells were primed *in vitro* by the cC haplotype from a child bB/cC and restimulated with the same stimulating cells. The reactivated lymphocytes were tested for secondary CML on members of the family and a control cell. This control cell, although unrelated is mutually MLR negative with sib aA/cC and, therefore HLA-D identical with him.

In this family the killing of one of the sibs bB/dD was unexpected. Another observation is that the target cannot be the HLA-D product since the control cell possessing the same HLA-D product as that used for priming and restimulation is not killed

		Targets						
Priming combination	Restimulation	Father aA/bB	Mother cC/dD	Four sibs aA/dD	Two sibs bB/dD	One sib bB/cC (specific)	One sib aA/cC	Control aA/yC
aA + cC 3, 7, Dw2 W23, W21 W23, W21	cC	6	70	21 (mean)	64 37	78	73	11

a, b, c, d are the HLA-A and B antigens.
A, B, C, D are the HLA-D products.

postulated[53] and later proved by Legrand and Dausset[54]—(iii) the antigens of the HLA-C locus, (iv) the products of genes as yet unknown.

We can eliminate the first hypothesis: in the course of a secondary CML performed between the members of one family (Table 13), it was noted that cytotoxic effectors developed against one of the children but not against a control cell which was mutually MLR negative with this child. It seems therefore that the target is not the product of the HLA-D gene.[40]

There is no decisive argument enabling us to choose between the three latter hypotheses. However, the following experiments seem to support the existence of genes as yet unknown.

Immunization of an A2, A3, B7, B12 cell against a homozygous cell (A3, B7, Dw2/A3, B7, Dw2) generates cytotoxic effectors capable of lysing cells containing Dw2 independently of A3 or B7 antigens (Table 14). These results suggest the existence of a gene close to and in linkage disequilibrium with this Dw2 allele. This haplotype is already known for its strong linkage disequilibrium between the genes of the HLA complex.

Table 14. Secondary CML using cells primed by an homozygous 3, 7, Dw2 cell. Cells B and R are unrelated cells, HLA-A and B pheno-identical, but both HLA-Dw2 negative. Cells B and R were primed *in vitro* against the homozygous cell from family Le (A3, B7, Dw2/A3, B7, Dw2) and reactivated for secondary CML with the same homozygous cell. The highly cytotoxic effector cells produced were tested against different varieties of target cells. These effector cells recognize a specificity different from HLA-Dw2 but presenting a high degree of linkage disequilibrium with this HLA-Dw2 specificity

Priming Combination	Restimulation	Targets: 4 cells 3+7+DW2+	Targets: 5 cells 3−7−DW2+	Targets: 6 cells 3+7+DW2−	Targets: 6 cells 3−7−DW2−
B: 2,7,– / 2,12,– + L: 3,7,DW2 / 3,7,DW2	L	4	4	1	1
R: 2,7,– / 2,12,– + L: 3,7,DW2 / 3,7,DW2	L	4	4	1	1

DW2	Killed	non-Killed
+	8	1
−	2	10

$p < 0{,}001$

Similarly, when eight pairs of HLA cells phenotypically identical to the HLA-A and B loci were tested with reciprocal secondary CML, effectors appeared more frequently in the combinations where cells possessing haplotypes with no known linkage disequilibrium were co-cultured rather than the opposite. This would mean that this hypothetical gene governing the unknown target very often has the same allele on haplotypes with a high linkage disequilibrium.

The CML described here lyses targets having a correlation with HLA-D alleles, like anti-B lymphocyte antibodies mentioned previously.

(c) *There are targets independent of the HLA complex.* Finally, CML has shown that effectors can develop against targets completely independent of the HLA complex.

Lymphocytes from a father immunized 40 months previously with a skin graft and leukocyte injection from his pheno-identical HLA daughter, were used in a classical CML with stimulating cells taken from different members of the family. The cytotoxic lymphocytes generated in the father seemed to detect a specificity different from HLA-A and B antigens.[51] This father was later reimmunized *in vivo* with the same daughter, and also developed a cytotoxic antibody which specifically reacted with B lymphocytes, thus enabling the first lymphocyte system independent of the HLA system (Ly-Co) to be described[55] (Figure 9).

The fact that grafts between identical HLA siblings are constantly rejected after some time (20 to 25 days), clearly indicates that in man, as in the mouse, there are numerous histocompatibility systems apart from the HLA system. But their *in vivo* effect is usually masked by the stronger effect of the MHC. It is possible that the classical process of antigen competition constantly favours the MHC and that it is only when this latter is neutralized by matching, as in the above experiment, that the effect of these weaker histocompatibility systems becomes apparent. In this respect, Bevan[56] has shown that effector cells against minor histocompatibility antigens are only active on targets identical with them at the major histocompatibility complex.

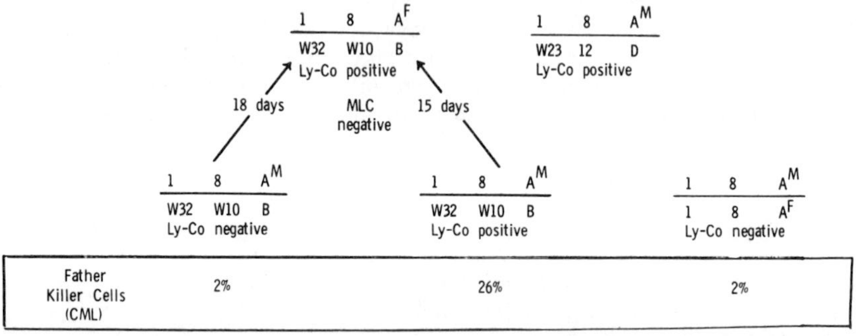

Figure 9. Lymphocyte system Ly-Co. A father simultaneously received two skin grafts from two HLA identical children. Father and children were mutually MLR negative. Forty months later the father was reimmunized by leukocyte injection from the child whose skin was rejected in 15 days. He developed an antilymphocyte B antibody active against the immunizing child but not against the other whose skin was rejected in 18 days. This antibody was also active against the mother. The father's lymphocytes immunized *in vitro* against the mother killed the immunizing child's lymphocytes but not those of the other child

2.3 Cytotoxic antibodies and their relationship with MLR and CML

In the course of the *in vivo* allogeneic response, not only are cytotoxic effectors (T lymphocytes) generated but also B lymphocytes secrete antibodies which may or may not be complement fixing. We know that the H-2 complex genes (especially the I region genes) are of considerable importance in the cooperation leading to the immunization against T dependent environmental antigens. Our knowledge of the mechanism of the production of antibodies in human alloimmunization is very limited. However, it is interesting to mention here the time and specificity relationships which can exist between lymphocyte proliferation (MLR), the appearance of cytotoxic effectors (CML), and the secretion of serologically detectable complement fixing antibodies.

Haplo-identical sibs immunized with injections of buffy-coat cells showed almost complete synchronization between the humoral and cellular immune response. The cytotoxic effectors, as well as the complement fixing antibodies produced, seemed to react with determinants other than the classical HLA private specificities.

In a series of other studies, both *in vivo* and *in vitro* primed lymphocytes were found to obey the same requirements for secondary *in vitro* restimulation. They were also shown to display the same secondary CML activity.

These experiments are the only ones, to our knowledge, in which the human allogeneic response has been systematically studied. These are the models most closely resembling transplantation immunization. The total conformity of facts observed *in vivo* and *in vitro* leads us to think that, in most cases, the application of laboratory experiments to clinical transplantation is partly justified.

2.4 Conclusion

From all this, we would like to try and construct a temporary general model for *in vivo* alloimmunization in man—which is, however, far from complete (Figure 10).

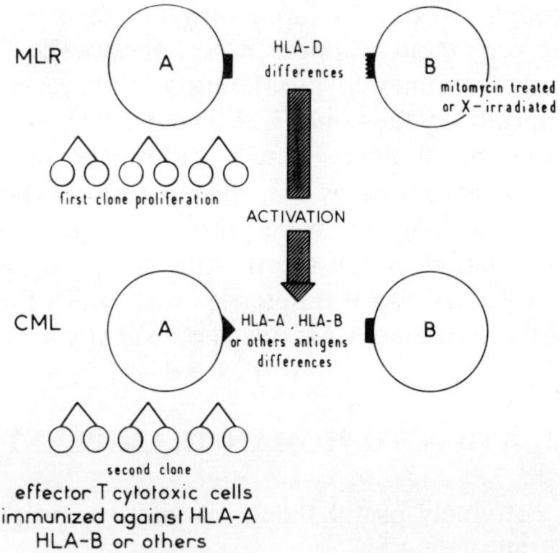

Figure 10. MLR-CML relationship. Schematic representation of the process of immunization against the transplantation antigens. A first clone of T lymphocytes recognizes the HLA-D difference and proliferates. This proliferation is accompanied by the 'activation' of a second clone of T lymphocytes which possesses the specific receptors for the foreign HLA-A or B difference. The clone differentiates in cytotoxic effector lymphocytes

It would seem that the first stage of the allogeneic recognition is triggered off by differences at the HLA-D locus. This recognition is clonal and leads to a proliferative reaction. The proliferation is accompanied by an 'activation' process of an unknown nature which appear to be a prerequisite to the immunization of another lymphocytic subpopulation, also belonging to T lymphocytes. The nature of the 'signal' sent from one population to another is unknown. Does it act by contact or through a diffusible substance? The effector cells make an early appearance in the culture at a time when there is a barely detectable proliferation, which may indicate that this activation might not necessitate cell division.

At the same time, T–B cell cooperation necessary for the development of circulating antibodies takes place *in vivo*. Here again we have to imagine a 'signal' sent by one cell population to the other. MHC compatibility is necessary for cell cooperation in certain murine systems but this is not obviously applicable to man. The antibodies which are secreted are mainly directed against transplantation antigens of the HLA complex (either private or public) and perhaps other minor histocompatibility systems. Antibodies can also develop against differentiation antigens of lymphocytes, granulocytes or platelets.

The circulating antibodies and cytotoxic cells peak at the same time (fourteenth day). Their specificities apparently also coincide, but more extensive studies must be made to confirm the identity of these specificities. It is possible that, as in the other immunological responses, the targets recognized by effector T lymphocytes are broader, less specific, and similar to 'carriers', while those recognized by antibodies are more exclusively specific, and similar to 'haptens'.[57]

With our present techniques, no directly cytotoxic cells can be found in the peripheral blood of immunized people on day 21 (after only one leukocyte injection), and yet restimulation will cause very rapid reappearance of these cells. This is not a classical immunological booster but a renewed proliferation with 'reactivation'. This can be triggered off either specifically, by the same HLA-D difference which was used in the first stage, or non-specifically, by another HLA-D difference, or even by non-specific mitogens. The effector cells generated by non-specific restimulation *strictly maintaining their specificity*. This indicates that they were present as resting, primed non-cytotoxic lymphocytes. With restimulation by the specific cell presenting both the HLA-D and serologically detectable HLA-A and B differences with which the responding cell was primed, the intensity of the secondary CML reaches a slightly higher level.

3. APPLICATION TO HUMAN TRANSPLANTATION

The preceding model, extremely useful though it may be, only partly applies to the complexities of human transplantation.

In our experiments antigen has usually only been administered once, whereas in real human transplantation the grafted organ remains permanently exposed to the immunological lymphoid system. Antigenic stimulation is therefore constant and, moreover, the simultaneous presence of circulating antibodies and of antigens favours the appearance of antigen–antibody complexes which can play an important role in the survival of the graft.

We must be extremely careful when applying experimentation to real organ transplantation. Also, immunological unresponsiveness, tolerance, suppressive cells and facilitating antibodies have not yet been studied on a practical level in man.

3.1 Role of HLA-D disparities

The role of HLA-D disparities between donor and recipient has been clearly defined *in vitro*. Its *in vivo* role is hypothetical.

3.1.1 *Skin grafts*

However, the results of skin grafts in man carried out by Koch *et al.*[58,59] and by Blussé *et al.* (to be published, 1976) are impressive (Table 15). The identical HLA-A, B and D grafts from unrelated donors survive the longest (19 days), almost as long as grafts carried out between HLA identical sibs (20–25 days). The grafts performed between people with

Table 15. MLC and unrelated skin grafts. Skin grafts were performed between unrclated people. The importance of serologically detectable identity (HLA-A and B) versus lymphocyte stimulatory product (HLA-D) indentity was studied comparing the skin survival time in the different donor–recipient compatibility situations. From data of Koch *et al.*[58,59] and of Blussé *et al.* (unpublished). (Reproduced by permission from Van Rood *et al.*, 1975, *Transplantation Proceedings*, **7**, 25–30)

HLA	HLA	No. of grafts	Survival (days)
A, B≠	D≠	24	10
A, B=	D≠	14	12
A, B≠	D=	6	13·8
A, B=	D=	7	17

differences at the three HLA loci are rejected the most rapidly. There are intermediate survivals for grafts between HLA-A and B identical but HLA-D different donors, and vice versa. On the whole, an HLA-D identity seems to allow an appreciable gain in time over an HLA-A and B identity. We must bear in mind, however, that one can never be sure of the perfect identity of all the antigens in the MHC complex between two unrelated individuals. HLA-C and other still unknown genes can differ, while the HLA-D identity can be deduced from a reciprocal negativity in MLR. These skin graft results seem to fit well into the general lymphocyte activation model presented here; without activation by an HLA-D difference, immunization against major histocompatibility antigens is either prevented or inhibited. The graft is rejected nevertheless, as in the case of grafts between identical HLA siblings. This would indicate that *in vivo* there is either (a) an accessory activation route or (b) the possibility of immunization occurring against minor histocompatibility antigens without detectable activation (at the threshold sensitivity of our techniques).

3.1.2 *Kidney grafts*

Despite the numerous non-immunological factors complicating analysis of the allogeneic response during the kidney grafts, these remain the only grafts currently practised that can guide us.

Several teams have tried to prove that the intensity of the MLR carried out between donor and recipient before a graft has a predictive value: a bad prognosis is observed with intense MLR and vice versa. In situations where donor and recipient only differed by a single haplotype, Hamburger *et al.*[60] found a significant correlation. Similarly, Cochrum *et al.*[61] confirmed this tendency in a prospective study of cadaver kidney grafts. These studies deserved to be followed by *in vitro* experimentation showing a relationship between MLR and CML intensity. This correlation is not particularly evident from our own results although it has not been specifically researched. It is nonetheless possible that activation is not an all-or-nothing phenomenon like a switch, and that the degree of activation allows a larger or smaller number of effector cells to appear.

Effector cells have been looked for in the peripheral blood of grafted patients, but so far without success, which is not surprising in view of the experiments mentioned above, in which cytotoxic effector lymphocytes were detected between day 12 and 18 and then disappeared very rapidly from the peripheral blood of an immunized volunteer. Resting memory cells with or without specific receptor might be detected at certain stages after kidney grafting, either in the peripheral blood for a short period, or rather in the efferent lymph nodes of the graft, or the spleen.

Another statistical argument in favour of intervention of HLA-D genes has been looked for in the longer survival time of kidney grafts exchanged between individuals both possessing haplotypes known to be in linkage disequilibrium (HLA-A3, B7, Dw2 or HLA-A1, B8, Dw3). Thus grafts carried out with such haplotypic identity have more chance than others of being identical for the HLA-D locus as well. Such a tendency has in fact been observed[62,63] but one must not forget that *all the* genes of these haplotypes are in disequilibrium and that, consequently, not only HLA-D genes but also many other genes of the HLA complex are likely to be identical.

Pretreatment of cadaver kidney donors, in order to eliminate as many lymphocytes as possible from the kidney, has been reported, with encouraging results.[64] One interpretation of the good results obtained may be that such treatment prevents, or at least minimizes the recognition of the lymphocyte stimulatory product of the donor's lymphocytes by the recipient's immunological system. This is very questionable, however, since endothelial cells also seem to express lymphocyte stimulating products possibly identical with the structure coded by the HLA-D genes.[65] Several reports have also pointed out that circulating MLC inhibitor in the recipient's blood appear at the time of rejection.[66,67]

None of the above arguments are decisive but together they may indicate that HLA-D disparities between donor and recipient play a role in the *in vivo* alloimmunization.

3.2 Role of incompatibilities in the HLA-A and/or HLA-B loci

Although this has long been a subject of much controversy it now seems clear that a graft's survival time is greatly influenced by incompatibilities in the HLA-A and B loci of the HLA complex. This has been shown in two stages. Firstly, it was discovered that skin graft and kidney graft survival depended on *the number* of HLA complexes genetically identical in the donor and recipient. Secondly, it was observed that graft survival depended on the number of incompatible HLA(A or B) antigens. This was initially determined by a skin graft on preimmunized recipients,[68,69] then on normal volunteers,[70] and finally on kidney graft recipients[62,71] (Figures 11 and 12).

These statistics, however, did not prove that HLA antigens have any role. The real structures involved could be coded by genes very near to and in linkage disequilibrium with HLA genes. Although this possibility is unlikely, it cannot be totally excluded.

It is known that incompatibilities in the H2-D region in the mouse are less crucial for graft survival than those in the H2-K region. The equivalent of the D region is thought to be the HLA-A locus. In fact, kidney grafts from unrelated donors differing from the recipient only by an HLA-A incompatibility are slightly more successful than those differing only by an HLA-B incompatibility.

One of the characteristics of HLA molecules is that there are frequent cross-reactions between them which could be due to (a) similarities of structure, (b) antigenic communities,[53] (c) a combination of both. Analysis of kidney graft survival shows that cross-reactions resemble identities more than incompatibilities. The same conclusions were reached when studying skin grafts.

If, however, the incompatibilities of the two A and B loci in the HLA system correlate well with the statistical results, they do not explain all the facts observed. Certain kidneys phenotypically identical to the recipient are rapidly rejected, and vice versa. It seems likely that the HLA complex has other structures under its control. The practical importance of

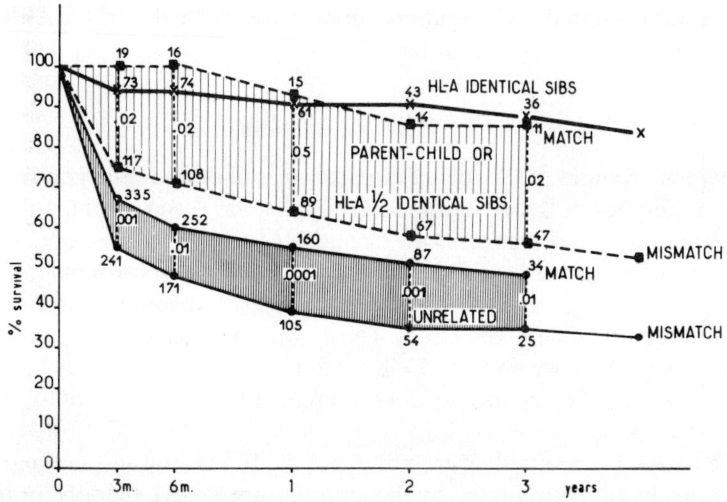

Figure 11. 1143 first kidney grafts performed since 1959 in the France-Transplant network. Actuarial curves of kidney survival. The figures on the lines mean the number of patients at risk during the corresponding interval. The best results are obtained when donor and recipients are identical HLA sibs. The upper shaded area illustrates the results obtained when donor and recipient shared one HLA complex genetically. The matched kidney (no incompatibilities) behaved almost as well as the identical HLA sib category. The lower shaded area represents the results obtained with unrelated kidney grafts (no HLA genetically identical complex shared). (For definition of match and mismatch in the unrelated kidney graft, see the caption to Figure 13)

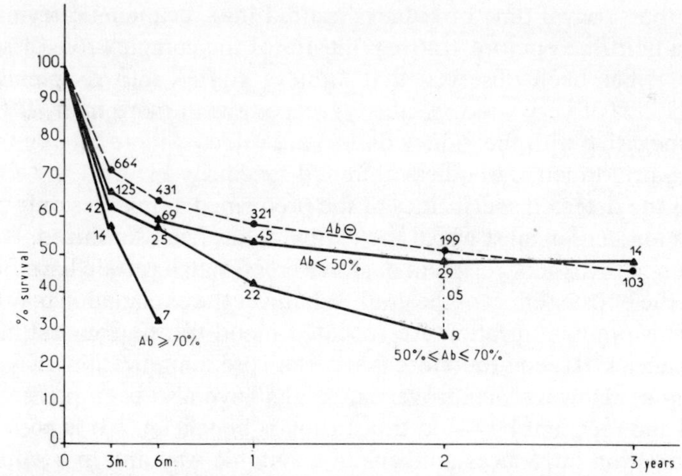

Figure 12. 845 first kidney unrelated grafts. Actuarial survival curves of first cadaver kidney grafts, according to the presence (Ab) or absence (Ab−) of preformed antibodies in the recipient's circulation. The survival is the same in the absence of antibody and when the antibodies react against less than 50% of the random panel cells tested. In contrast, the survival is shortened when the antibodies react with more than 50% and especially when they react with more than 70% of the cells tested

the latter is probably limited under immunosuppression, since the grafts of identical HLA siblings are usually remarkably well tolerated.

3.3 Role of antibodies

The most obvious example of the role of circulating antibodies in the graft is given by the action of the antibodies of the ABO blood group system. ABO incompatible skin grafts applied to ABO immunized recipients are accurately rejected in a few hours.[72] In normal unimmunized recipients, there is a 2 or 3 day survival difference for skin grafts performed between ABO compatible or incompatible individuals.[73] Incompatible kidney grafts in this system undergo a violent rejection. These antibodies act mainly on the vascular endothelium cells which they destroy. Cellular immunity, however, may also play a role. On the other hand, ABO compatibility does not seem to be so important for bone marrow grafting.

Circulating anti-HLA antibodies are not generally found in transplant patients, either because their production is inhibited by the immunosuppressive therapy, or because they are completely bound to the graft. This often seems to be the case since antibodies frequently appear in the circulation after graft removal.

Detection of antibodies, however, does not indicate clearly their influence on graft survival. According to their intensity, nature and relationships with antigens liberated by the graft they can be harmful (cytotoxic antibodies or lymphocyte dependent antibodies (LDA)) or beneficial. Anti-HLA antibodies blocking complement fixation have been found either after hyperimmunization, or in patients tolerating an incompatible graft.[74] We know of the existence of antibodies blocking the MLR.[20] the presence of the latter in transplant patients could constantly prevent or weaken the lymphocytic activation essential for the appearance of cytotoxic effectors. Finally, the formation of antigen–antibody complexes can by itself induce graft tolerance or facilitate the growth of a tumour.

The study of the survival time of kidneys grafted into recipients carrying anti-tissue antibodies, already formed before grafting, illustrates the complex role of antibodies in transplantation: it has been observed that kidneys grafted into recipients possessing preformed antibodies of very wide specificity (reacting with more than 70% of random individuals and negative with the kidney donor) are rejected more rapidly than grafts in recipients having preformed antibodies of limited specificity Figure 13).[62] One explanation may be that the detected specificities of the preformed antibodies only partly reflect the spectrum on antigens against which the recipient has been sensitized. It is therefore possible that even with a donor recipient negative cross-match certain harmful antibodies reappear due to the booster effect of the graft. Whatever the explanation may be, the main causes of recipient preimmunization are repeated blood transfusion and more seldom, numerous pregnancies. It seems therefore that certain preimmunizations may threaten the survival of the graft. However, controversial results have also been presented in which pretreatment of the recipient by blood transfusion is beneficial.[75] It is possible that this latter preimmunization introduces antigens in a suitable way and in a suitable dose to favour the appearance of facilitating antibodies.

Lastly, it is certain that individual capacity to produce circulating antibodies after immunization is very variable. It is possible to define 'responders' and 'non-responders', as with mice. However, our experience of planned immunization shows that so-called 'non-responders' are generally capable of responding but only after a large number of stimulations. Logically 'non-responders' should tolerate grafts longer if antibodies are harmful. These graft results have been observed by Opelz *et al.*[76]

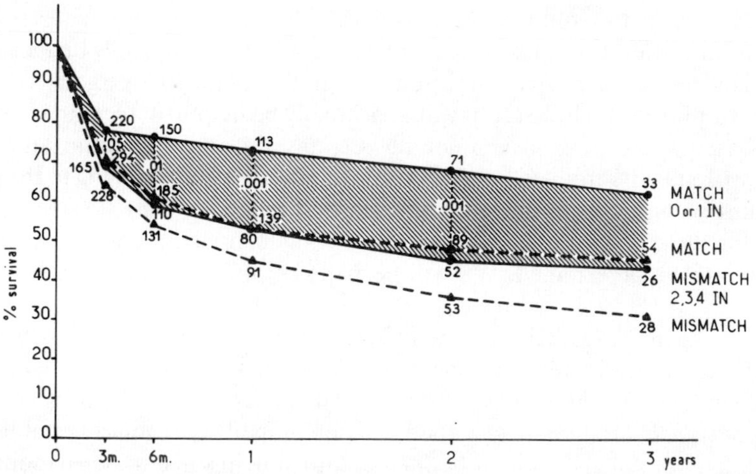

Figure 13. 907 first kidneys unrelated grafts. Actuarial survival curves of first cadaver kidney grafts carried out in the France-Transplant network. The shaded area shows the results obtained by teams where the influence of the HLA matching is most obvious. The difference between the most compatible grafts (match 0 or 1 incompatibility) and the less compatible (mismatch 2, 3 or 4 incompatibilities) is significant. The dotted lines are the corresponding most compatible and less compatible grafts' survival obtained by the remaining teams

3.4 Conclusion

The development of human transplantation obviously depends on a better knowledge of alloimmunization. Two directions of research may lead to better graft survival: one is *selection*, that is, the effort to find each recipient an identical donor, or one who is the most compatible for the greatest number of loci concerned with transplantation; the other is *specific tolerance*, that is, the development in the recipient of a refractory immunological state, before or immediately after transplantation, specific for the antigens of the future donor.

It is certainly still possible to improve donor selection. HLA-D grouping will very probably make rapid progress: present cellular immunological technique will be simplified. The use of *in vitro* presensitized cells may become effective. Serological techniques detecting either HLA-D gene products themselves, or products coded by nearby genes, might gradually replace cellular techniques. A very rapid MLR technique which could be applied to an unrelated cadaver kidney donor would be welcome. HLA grouping could be further refined and its quality improved. The choice of recipient should be made from an international list on the widest possible scale. This would make it easier to carry out identical grafts for the four alleles of the A and B loci, to avoid incompatibilities in the B locus and, lastly, in the case of incompatibility, to choose incompatible antigens showing cross-reactions with the recipient's antigens. Matching haplotypes should then be encouraged, especially where haplotypes in linkage disequilibrium are concerned.

Other antigenic systems, such as the P system, probably have a role in transplantation. With a long waiting list it might be possible to match them as well.

But there are obvious limits to selection: as the number of loci to be neutralized by selection increases, the choice of recipient becomes more difficult, if not impossible.

We must therefore turn our attention in another direction: towards the establishment of a state of specific tolerance. This specific tolerance would theoretically be more efficient if established by methods inhibiting the initial recognition of the foreign grafted tissue by the recipient's lymphocytes. However, it would probably be helpful in blocking the afferent or efferent phase of the allogeneic immune response. Planned immunization could be attempted with specific antigens before, during or after the graft, leading to the production of antibodies or antigen–antibody complexes that facilitate graft survival. Methods of administering these injections in man (nature of antigen, injection route, dosage, periodicity) have yet to be determined. We will perhaps be able to define these by successive approaches.

However, a better knowledge of the balance mechanism between destructive factors (cytotoxic antibodies, effector cells) and beneficial factors (facilitating antibodies, antigen–antibody complexes, suppressor cells) may lead us to discover other means of improving the results of transplantation. It may then become possible to regularly establish the state of partial tolerance which is often obtained empirically in some of our patients today.

4. REFERENCES

1. Bain, B., Vas, M. R. and Lowenstein, L. (1963). *Fed. Proc.*, **22**, 428 (abstract).
2. Bain, B., Vas, M. R. and Lowenstein, L. (1964). *Blood*, **23**, 108–116.
3. Bach, F. H. and Hirschhorn, K. (1964). *Science*, **142**, 813–814.
4. Amos, D. B. and Bach, F. H. (1968). *J. exp. Med.*, **128**, 623–637.
5. Bach, F. H. and Amos, D. B. (1967). *Science*, **156**, 1506–1508.
6. Yunis, F. J. and Amos, D. B. (1971). *Proc. Nat. Acad. Sci.*, **68**, 3031–3035.
7. Eijsvoogel, J. P., du Bois, M. J. G., Melief, C. I. M., de Groot-Kooy, M. L., Koning, C., Van Rood, J. J., Van Leeuwen, A., du Toit, E. and Schellekens, P. Th. A. (1972). In *Histocompatibility Testing 1972* (Dausset, J. and Colombani, J., Eds.), Munksgaard, Copenhagen, pp. 501–508.
8. Dupont, B., Nielsen, L. S. and Svejgaard, A. (1971). *Lancet*, **2**, 1336–1340.
9. Lebrun, A., Sasportes, M., Lebrun, D. and Dausset, J. (1971). *C.R. Acad. Sci. Ser. D*, **273**, 2130–2133.
10. Thorsby, E. and Piazza, A. (1975). In *Histocompatibility Testing 1975* (Kissmeyer-Nielsen, F., Ed.), Munksgaard, Copenhagen, pp. 414–458.
11. Thorsby, E., Hirschberg, H. and Helgesen, A. (1973). *Transplant. Proc.*, **5**, 1523–1528.
12. Sasportes, M., Mawas, C., Bernard, A., Christen, Y. and Dausset, J. (1973). *Transplant. Proc.*, **5**, 1517–1522.
13. Suciu-Foca, N. and Dausset, J. (1975). *Immunogenetics*, **2**, 389–391.
14. Festenstein, H., Sachs, J. A., Abbasi, K. and Oliver, R. T. (1972). *Transplant. Proc.*, **4**, 219–222. 219–222.
15. Bradley, B. A., Edwards, J. M., Dunn, D. C. and Calne, R. Y. (1972). *Nature* (*New Biol.*), **240**, 54–56.
16. Mempel, W., Grosse-Wilde, H., Baumann, P., Netzel, B., Steinbauer-Rosenthal, I., Scholz, S., Bertrams, J. and Albert, E. S. (1973). *Transplant. Proc.*, **5**, 1529–1534.
17. Dupont, B., Jersild, G., Hansen, G. S., Nielsen, L. S., Thomsen, M. and Svejgaard, A. (1973). *Transplant. Proc.*, **5**, 1543–1549.
18. Jørgensen, F., Lamm, L. N. and Kissmeyer-Nielsen, F. (1973). *Tissue Antigen*, **3**, 323–339.
19. Van den Tweel, J. G., Blussé van Oud Alblas, A., Keuning, J. J., Goulmy, E., Termijtelen, A., Bach, M. L. and Van Rood, J. J. (1973). *Transplant. Proc.*, **5**, 1535–1542.
20. Meo, T., David, C. S., Rijnbeek, A. M., Nabholz, M., Miggiano, V. C. and Shreffler, D. C. (1975). *Transplant. Proc.*, **7** (suppl. 1), 127–129.
21. Walford, R. L., Zeller, E., Combs, L. and Konrad, P. (191). *Transplant. Proc.*, **3**, 1297–1300.
22. Van Rood, J. J., Van Leeuwen, A., Keuning, J. J. and Blussé van Oud Alblas, A. (1975). *Tissue Antigen*, **5**, 73–79.
23. Wernet, P., Winchester, R., Kunkel, H. G., Wernet, D., Giphart, M., Van Leeuwen, A. and Van Rood, J. J. (1975). *Transplant. Proc.*, **7** (suppl. 1), 193–200.

24. Legrand, L. and Dausset, J. (1975). In *Histocompatibility Testing 1975* (Kissmeyer-Nielsen, F., Ed.), Munksgaard, Copenhagen, pp. 665–670.
25. Van Rood, J. J., Van Leeuwen, A., Termijtelen, A. and Keuning, J. J. (1976). In *The Role of Products of the Histocompatibility Complex in Immune Responses* (Katz, D. H. and Benacerraf, B., Eds.), Academic Press, New-York, pp. 31–51.
26. Sheehy, M. J., Sondel, P. M., Bach, M. L., Wank, R. and Bach, F. H. (1975). *Science*, **188**, 1308–1310.
27. Fradelizi, F., Mawas, C. E., Charmot, D. and Sasportes, M. (1975). In *Histocompatibility Testing 1975* (Kissmeyer-Nielsen, F., Ed.), Munksgaard, Copenhagen, pp. 584–587.
28. Dupont, B., Good, R. A., Hansen, G. S., Jersild, C., Staub-Nielsen, L., Park, B. H., Svejgaard, A., Thomsen, M. and Yunis, E. (1974). *Proc. Nat. Acad. Sci.*, **71**, 52–56.
29. Thorsby, E. (1974). *Transplant. Rev.*, **18**, 51–129.
30. Kennedy, J. L. and Expaha-Mensah, J. A. (1973). *J. Immunol.*, **111**, 1639-1652.
31. Dausset, J., Lebrun, A. and Sasportes, M. (1972). *C.R. Acad. Sci. Paris*, **275**, 2279–2282.
32. Zosche, D. C. and Bach, F. H. (1971). *Science*, **172**, 1350–1352.
33. Salmon, S. E., Krakauer, R. S. and Whitmore, W. F. (1971). *Science*, **172**, 490–492.
34. Hirschberg, H. and Thorsby, E. (1973). *J. Immunol. Method*, **3**, 251–264.
35. Dausset, J., Le Calvez, H., Sasportes, M. and Rapaport, F. T. (1967). *Nouv. Rev. Franç. Hém.*, **7**, 643–662.
36. Bondevick, H. and Thorsby, E. (1973). *Transplant. Proc.*, **4**, 1477–1480.
37. Charmot, D., Mawas, C. E., Legrand, L. and Sasportes, M. (1976). *Immunogenetics*, **3**, 157–165.
38. Fradelizi, D., Charmot, D., Mawas, C. E. and Sasportes, M. (1976). *Immunogenetics*, **3**, 29–40.
39. Fradelizi, D. and Dausset, J. (1975). *Eur. J. Immunol.*, **5**, 295–301.
40. Mawas, C. E., Charmot, D. and Sasportes, M. (1975). *Immunogenetics*, **2**, 449–463.
41. Haÿry, P. and Defendi, V. (1970). *Science*, **168**, 133–135.
42. Solliday, S. and Bach, F. H. (1970). *Science*, **170**, 1406–1409.
43. Lightbody, J., Bernoco, D., Miggiano, V. C. and Ceppellini, R. (1971). *G. Batt. Virol. Immunol.*, **64**, 243–260.
44. Eijsvoogel, V. P., du Bois, R., Melief, C. J. M., Zeylemaker, W. P., Koning, L. and de Groot-Kooy, L. (1973). *Transplant. Proc.*, **5**, 415–420.
45. Eijsvoogel, V. P., du Bois, M. L. G. S., Meinesz, A., Bierhorst-Eijlander, A., Zeylemaker, W. P. and Schellekens, P. Th. A. (1973). *Transplant. Proc.*, **5**, 1675–1678.
46. Mawas, C. E., Sasportes, M., Christen, Y., Bernard, A., Dausset, J., Alter, B. J. and Bach, M. L. (1973). *Transplant. Proc.*, **5**, 1683–1689.
47. Charmot, D., Mawas, C. E. and Sasportes, M. (1975). *Immunogenetics*, **2**, 465–483.
48. Mawas, C., Christen, Y., Legrand, L. and Dausset, J. (1973). *Transplant. Proc.*, **5**, 1691–1695.
49. Mawas, C. E., Charmot, D. and Sasportes, M. (1976). *Immunogenetics*, **3**, 41–51.
50. Grunnet, N., Kristensen, T. and Kissmeyer-Nielsen, F. (1975). *Tissue Antigen*, **6**, 205–220.
51. Mawas, C. E., Christen, Y. Legrand, L., Sasportes, M. and Dausset, J. (1974). *Transplantation*, **18**, 256–266.
52. Mawas, C., Sasportes, M. and Christen, Y. (1975). *Transplant. Proc.*, **7** (suppl. 1), 53-55.
53. Dausset, J., Ivanyi, P. and Ivanyi, D. (1965). In *Histocompatibility Testing 1965* (Balner, H., Cleton, F. L. and Eernisse, J. G., Eds.), Munksgaard, Copenhagen, pp. 51–62.
54. Legrand, L. and Dausset, J. (1974). *Tissue Antigen*, **4**, 329–345.
55. Legrand, L. and Dausset, J. (1975). *Transplant. Proc.*, **7**, 5–8.
56. Bevan, M. J. (1975). *J. exp. Med.*, **142**, 1349–1364.
57. Alkan, S. S. and El-Khateeb, M. (1975). *Eur. J. Immunol.*, **5**, 766–770.
58. Koch, C. T., Frederiks, E., Eijsvoogel, V. P. and Van Rood, J. J. (1971). *Lancet*, **2**, 1334–1336.
59. Koch, C. T., Van Hooff, J. P., Van Leeuwen, A., Van den Tweel, J. G., Frederiks, E., Van der Steen, G. J., Schippers, H. M. A. and Van Rood, J. J. (1973). In *Histocompatibility Testing 1972* (Dausset, J. and Colombani, J., Eds.), Munksgaard, Copenhagen, pp. 521–526.
60. Hamburger, J., Descamps, B. and Fermanian, J. (1971). *C.R. Acad. Sci.*, **272**, 2029–2031.
61. Cochrum, K. C., Perkins, H. A., Payne, R. O., Kountz, S. L. and Belzer, F. O. (1973). *Transplant. Proc.*, **5**, 391–396.
62. Van Hooff, J. P., Schippers, H. M. A., Hendriks, G. F. J. and Van Rood, J. J. (1974). *Lancet*, **1**, 1130–1132.

63. Dausset, J., Hors, J. and Busson, M. (1976). *Proceedings of VIth International Congress of Nephrology*, Florence 1975, Karger, Basel, pp. 728–735.
64. Guttman, R. D., Beaudoin, J. G., Morehouse, D. D., Klassen, J., Knaack, J., Jeffery, J., Chassot, P. G. and Abbou, C. C. (1975). In *Séminaire d'Uronéphrologie. Pitié Salpétrière 1975* (Küss, R. and Legrain, M., Eds.), Masson, Paris, pp. 110–116.
65. Hirschberg, H., Evensen, S. A., Henriksen, T. and Thorsby, E. (1975). *Transplantation*, **19**, 191–194.
66. Suciu-Foca, N., Buda, J. A., Thiem, T., Almajera, P. and Reemtsma, K. (1974). *Lab. Invest.*, **31**, 1–5.
67. Miller, J., Lifton, J., Rood, F. and Hattler, B. G., Jr. (1975). *Transplantation*, **20**, 53–62.
68. Dausset, J., Rapaport, F. T., Ivanyi, P. and Colombani, J. (1965). In *Histocompatibility Testing 1965* (Balner, H., Cleton, F. L. and Eernisse, J. G., Eds.), Munksgaard, Copenhagen, pp. 63–72.
69. Van Rood, J. J., Van Leeuwen, A., Schippers, A. M. J., Vooys, W. H., Frederiks, E., Balner, H. and Eernisse, J. G. (1965). In *Histocompatibility Testing 1965* (Balner, H., Cleton, F. L. and Eernisse, J. G., Eds.), Munksgaard, Copenhagen, pp. 37–50.
70. Dausset, J., Rapaport, F. T., Legrand, L., Colombani, J. and Marcelli-Barge, A. (1970). In *Histocompatibility Testing 1970* (Terasaki, P. I., Ed.), Munksgaard, Copenhagen, pp. 381–397.
71. Dausset, J., Hors, J., Busson, M., Festenstein, H., Oliver, R. T. D., Paris, A. M. I. and Sachs, J. A. (1974). *New Engl. J. Med.*, **290**, 979–983.
72. Dausset, J. and Rapaport, F. T. (1966). *Nature*, **209**, 209–210.
73. Ceppellini, R., Bigliari, S., Curtoni, E. S. and Leigheb, G. (1969). *Transplant. Proc.*, **1**, 390–394.
74. Colombani, J., Colombani, M. and Dausset, J. (1973). *Transplantation*, **16**, 257–260.
75. Opelz, G., Sengar, D. P. S. and Mickey, M. R. (1973). *Transplant. Proc.*, **5**, 253–259.
76. Opelz, G., Mickey, M. R. and Terasaki, P. I. (1973). *Transplantation*, **16**, 649–654.

Chapter 19

Histocompatibility Associated Diseases

L. P. RYDER

A. SVEJGAARD

1. INTRODUCTORY REMARKS

One of the most active areas of research within the field of immunology is the unravelling of major histocompatibility systems in terms both of structure and function. It has become increasingly clear that each vertebrate species has one such system which controls transplantation antigens and, more importantly, a variety of immune functions. As discussed elsewhere in this volume, the immune functions controlled by these systems mainly involves the activity of T lymphocytes which are responsible for cell-mediated immunity and which cooperate with B lymphocytes in the induction of humoral antibody responses towards most antigens.

In an attempt to understand the function of major histocompatibility systems it is, of course, important to consider all the characters which are known to be controlled or at least influenced by them. One of the most fascinating aspects in this relation is the observation that the major human histocompatibility system, HLA, seems to be deeply involved in the aetiology and/or pathogenesis of a variety of diseases. The purpose of this survey is to discuss these associations between HLA and diseases: (a) how they can be detected, (b) which have been found and (c) how they can be explained. It should be noted that we do not attempt to cover the entire literature, but refer mainly to a few key references, primarily reviews.

1.1 The HLA system

The genetic and molecular aspects of the HLA system have been dealt with in greater depth elsewhere (e.g. References 1–3), but we think that a few summary statements are in place here.

It is well known that the 'classical' serologically defined (SD) HLA antigens of the A, B and C series are glycoproteins floating rather freely in the plasma membrane of most—if not all—nucleated cells (and thrombocytes) in the human organism. These proteins display an allotypic variation in the human population, which is the basis for their (allo)-antigenic properties, for their importance in the matching of donor–recipient pairs for transplantation, and, of course, for their recognition by serological/immunological means. At the present time three series of such antigens are recognized, and they are coded for by genes at three distinct, although closely linked loci. The genes are sited at one of the autosomes, chromosome No. 6, and the three loci are, according to WHO recommendations,[4] designated locus A, locus B and locus C. Another locus is designated locus D (see below).

There is now general agreement about the occurrence at locus A of about 17 different alternative forms (alleles) in Caucasian populations. The HLA make-up of other ethnic groups is not so well investigated, but a few more alleles must certainly be allowed for. This comment also applies to the number (=20) of alleles recognized at the B locus as well as to

Chromosome 6

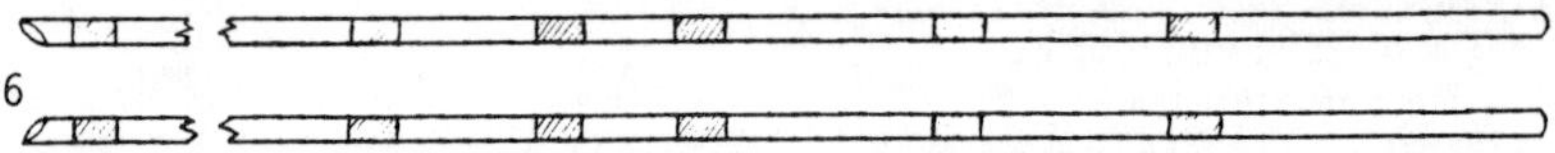

Locus	PGM_3	HLA-D	Bf	HLA-B	HLA-C	HLA-A
Synonyms	-	LD	GBG	2nd	3rd	1st
	-	MLC	-	FOUR	AJ	LA
No. alleles	2	>8	3	>20	>5	>17
Determinants		HLA-Dw1		HLA-B5	HLA-Cw1	HLA-A1
		- Dw2		- B7	- Cw2	- A2
		- Dw3		- B8	- Cw3	- A3
		- Dw4		- B12	- Cw4	- Aw23 } A9
		- Dw5		- B13	- Cw5	- Aw24 } A9
		- Dw6		- B14	- T7	- Aw25 } A10
		- LD107		- B27		- Aw26 } A10
		- LD108		- Bw15		- A11
				- Bw17		- A28
				- Bw21		- A29 } Aw19
				- Bw22		- Aw30 } Aw19
				- Bw35		- Aw31 } Aw19
				- Bw37		- Aw32 } Aw19
				- Bw38 } Bw16		- Aw33 } Aw19
				- Bw39 } Bw16		- Aw34 } Aw19
				- Bw40		- Aw36
				- Bw41		- Aw43
				- Bw42		

Figure 1. The human major histocompatibility complex. Genes for C2, C4 and for the blood group Bg, Chido and Rogers are also linked to HLA. For references see Reference 3

the recently accepted C locus (five alleles). The presence of 'null' alleles at any of the three loci would usually be explained simply as being due to lack of antisera recognizing the corresponding antigens.

The biochemical characterization of the HLA antigens of series A, B and C is advancing rapidly, and it suffices to state that all three series of antigens show great homologies—it would probably be reasonable to regard them as isotypes in the sense of Milstein[5] and Heremans.[6]

Besides the abovementioned three series of SD alloantigens several other properties are now recognized as being controlled by genes located in the same region of chromosome No. 6 (see Figure 1). First there is the stimulatory capacity in the mixed lymphocyte culture test. This phenomenon is due to a similar allotypic variation at the locus now called HLA-D, at which the existence of at least eight alleles have been agreed upon.[7] Furthermore, genes coding for (or controlling synthesis/expression of) properdin factor Bf (also called C3 proactivator), complement factor 2(C2), perhaps complement factor 4, and most probably a series of Ia-like antigens are located between or within a short distance of the four loci, HLA-A, B, C, D.[8–11]

This brings us to a question of definition: which of these characters are to be considered as belonging to the HLA system. The requirement has been proposed that there must be definite linkage disequilibrium between at least some alleles from the different loci, which thus constitute a supergene.[12] This is the case, and it is very pronounced indeed, for alleles at the HLA-A, B, C, and D loci; and it seems to hold true for the C2 deficiency[13] and for electrophoretically distinguishable variants of properdin factor Bf.[9,15] The concept of linkage disequilibrium is also a key concern in many of the proposed explanations for the disease–HLA associations and is briefly described hereafter.

Consider two genetic loci A and B, each with a number of different alleles $A_1, A_2, \ldots, A_I$; $B_1, B_2, \ldots, B_K$. It is well known that under ideal circumstances of (a) random mating in an infinite population, (b) absence of selection with respect to the characters in question, (c) no migration, etc. (see, for example, References 15 and 16), the Hardy–Weinberg equilibrium for each locus will be attained within one generation of breeding, if it was not present beforehand. The Hardy–Weinberg law states that the alleles at locus A (respectively B) in the formation of the diploid individuals combine by twos in a random fashion. The expected frequencies of the different genotypes at locus A (and of the corresponding phenotypes) are thus predictable by probabilistic arguments. In the simplest case,

$$p(A_i, A_j) = p(A_i)p(A_j)\delta_{ij} \tag{1}$$

where $p(A_i, A_j)$ is the frequency of genotype $(A_i A_j)$, $p(A_i)$ and $p(A_j)$ are the frequency of alleles A_i and A_j in the total pool of alleles in the population, and δ_{ij} is the number of ways in which the genotype (A_i, A_j) can arise:

$$\delta_{ij} = \begin{cases} 1 \text{ for } i = j \text{ (homozygous individuals)} \\ 2 \text{ for } i \neq j \text{ (heterozygous individuals)} \end{cases}$$

Having investigated a population sample, the gene frequencies can be estimated,[17] and the fit of the expected phenotype distribution to that observed can be tested, e.g. by a chi-square test. A failing fit may be due to violating the above premises or it may be the consequence of incomplete serology.

The simultaneous behaviour of alleles at two loci, A and B, depends somewhat on their linkage relationships. If A and B are not linked, their alleles will combine at random in the population and the frequency of a given genotype $(A_i, A_j; B_k, B_l)$ will be $\delta_{ij} \times p(A_i) \times p(A_j)$, $\delta_{kl} \times p(B_k) \times p(B_l)$, and this is usually also the phenotype frequency. If A and B are linked, i.e. sited close to each other on the same chromosome, it is appropriate to define a haplotype as a pair of alleles at A, respectively B, locus present on the same chromosome.[18] In family investigations such haplotypes usually are inherited as single units, but there is a low-frequency occurrence of recombinants which are due to chromosomal cross-over events. The recombination frequency is a measure of the distance between the two loci on the chromosome, and for the HLA-A and B loci it amounts to about 1%.

A priori one would expect all the different possible haplotypes to occur in the population with frequencies determined solely by the frequencies of their constituent A and B alleles:

$$p(A_i, B_k) = p(A_i)p(B_k) \tag{2}$$

In a hypothetical situation of random breeding in a population in which alleles A_1 and B_1 always occur together initially (absolute association) even a low recombination frequency of 1% will lead in the long run to complete breaking up of the haplotype which will be recombined at random.

The deviation from linkage equilibrium in a population can be measured in a variety of ways.[12,19] Most common is the delta value (Δ) which is defined as the difference between the observed haplotype frequency $h(A_i, B_j)$ and its equilibrium expectation $p(A_i)p(B_j)$:

$$\Delta(A_i, B_j) = h(A_i, B_j) - p(A_i)p(B_j) \tag{3}$$

In the hypothetical breeding experiment, this deviation from equilibrium declines each generation with a factor $(1-c)$, c being the frequency of crossing-over, $(1-c)$ the frequency of non-recombination. For example, it can be estimated that with the recombination frequency of 1% between the A and B loci, the 'half-life' of a Δ value is approximately 69 generations.

The treatment of haplotypes involving more loci becomes progressively more complicated as the number of loci increases;[20] the problem of the three HLA loci has been discussed by Piazza.[21]

In the human population, some of the haplotypes involving the HLA loci show very marked deviations from equilibrium. The situation, however, is complicated by the diploid organization of the human being: in population studies it is not possible to determine which of the two conceivable haplotype constellations $(A_1, B_1/A_2, B_2)$ or $(A_1, B_2/A_2, B_1)$ an individual of a given phenotype $(A_1, A_2; B_1, B_2)$ represents. This makes it difficult to estimate haplotype frequencies from a population sample, but approximate methods have been developed.[22] The statistical testing of the hypothesis of no linkage disequilibrium between characters A_i and B_j can be tested by confronting in a 2×2 table the number of individuals positive or negative for antigen A_i and B_j, respectively. This bears a more than mathematical resemblance to the study of disease–HLA associations, as will be seen later. The reason there are such marked linkage disequilibria between the HLA loci is unknown, but both migration between different populations and natural selection on haplotype level are possible explanations.[23,24]

1.2 Premises for the study of disease associations

There has been some controversy concerning the statistical as well as the biological significance of disease–ABO blood group associations.[25,26] The major objection was the absence of *a priori* reasons for such associations, in statistical terms somewhat related to be Bayesian philosophy of statistics, and biologically based on the lack of evidence concerning the function of the ABO blood groups. The biological arguments may be relevant in the sense that the statistically highly significant observations,[26,27] have not as yet given hints, either to the biological 'meaning' of the blood group or to the aetiologies of the diseases.

For the disease–HLA studies the situation is quite different. Animal studies have clearly indicated the existence of some disease-related biological function of the major histocompatibility loci in other species (Table 1). The initial observations by Lilly and coworkers of the influence of the mouse H-2 system on the susceptibility to development of leukaemia upon Gross-leukaemia virus infection was followed by the demonstration of analogous H-2 dependent susceptibility to several other viruses of different types causing different diseases. More recently, the relevance of these studies has become even more obvious with the discovery of Ir genes within the major histocompatibility complexes of a variety of animal species. Finally, the observation that in the mouse, the H-2 system influences the ability of the animal to develop choriomeningitis and experimental autoimmune thyroiditis has made human autoimmune diseases obvious candidates for HLA association studies.

Table 1. Histocompatibility related properties relevant to disease

Property	Species	Key references
Histocompatibility—graft rejection	All vertebrates	*legio*
Susceptibility to development of leukaemia upon inoculation with Gross leukaemia virus	Mouse	28
Susceptibility to development of leukaemia upon inoculation with Friend leukaemia virus	Mouse	28
Susceptibility to development of choriomeningitis upon inoculation with LCM-virus	Mouse	28
Immune response to restricted antigens	Mouse, guinea pig, rhesus monkey	28, 29, 30
Cell interaction between T and B lymphocytes	Mouse	31
T–B cooperation factors and acceptors	Mouse	32
Susceptibility to development of autoimmune thyroiditis upon inoculation with thyroglobulin	Mouse	33
Susceptibility to development of experimental autoallergic encephalomyelitis upon inoculation with basic protein	Rat	34
Susceptibility to development of spontaneous autoimmune thyroiditis	Chicken	35
Level of cyclic AMP in liver cells	Mouse	61
Variants of certain complement factors	Mouse, man	14

2. THE BASIC OBSERVATIONS AND DATA ANALYSIS

The observations of most publised disease–HLA relationships have been on the antigen profiles of a patient sample compared to a sample of normal control individuals originating from the same background population. The concept of association and relative risk is related to this situation (cf. next section). Family investigations have been carried out; they are suited for linkage studies but alone they do not tell us anything about associations at the population level. The analysis of population data is mainly hampered by the many possible comparisons within a system as polymorphic as the HLA, while the linkage studies take advantage of this, but suffer from difficulties such as obtaining informative families, the variability in age at disease-onset and statistical analysis. The two methods of investigation are more or less complementary for the understanding of disease relations to HLA; but, as most observations have been made using population studies, we will mainly be concerned with this aspect here.

2.1 Phenotype tables

Phenotype tables of patient and control samples constitute the raw material derived from population studies. The phenotype table is best presented as a triangular table showing the frequency of all the possible combinations of alleles within one locus ($n(n+1)/2$ combinations where n = number of alleles). From such a table the allele frequencies can easily be estimated, the fit to Hardy–Weinberg equilibrium can be tested and the total number of individuals expressing the different antigens counted (this is referred to as the antigen frequency of the antigen).

The comparisons between patients and controls can of course, in principle, be done by a classical $2\times(n(n+1)/2)$ contingency table comparing cell by cell the two phenotype tables; but this would give a very inaccurate test because of the many $(n(n+1)/2)$ comparisons and the consequently low expected numbers in most cells. A more attractive method of analysis is described in the next section.

However, the more complete information contained in the phenotype tables may be needed when studying possible interactions between various alleles.

2.2 Estimation and test of association

In the biomedical sciences, the application of statistics has three chief purposes, namely, once the observations have been made, (a) to provide the best estimate of parameters in proposed models, (b) to test the fit of alternative parameter values or models to the observations and (c) in some cases to give guidelines for practical decisions. All three aspects are relevant in the study of disease–HLA associations.

What is to be estimated, of course, is the effect of having a given HLA-type on the susceptibility to a disease. The HLA-type could mean the entire HLA-A, B, C, D type of the individual, but this would give rise to the statistical problem of many comparisons between small numbers. A reasonable simplification seems to be to consider in turn simply the effect of presence or absence of the alleles at the respective HLA loci. This amounts to comparing the antigen frequencies instead of phenotype tables (or even worse, individual HLA-A, B, C, D phenotypes).

In order to estimate the effect, we must define a measure of it, and an intuitive way of doing this is to use the *relative risk*.[36] The relative risk *of* having the disease *for* individuals carrying a given antigen is defined as the ratio of the probability for an antigen-positive individual of having the disease over that of an antigen-negative individual:

$$\text{relative risk} = p(\text{Disease}|\text{Antigen-positive})/p(\text{Disease}|\text{Antigen-negative}) \qquad (4)$$

i.e. the factor by which the presence of an antigen augments disease probability. As demonstrated below this quantity is well approximated by the simple cross-ratio obtained by comparing the antigen frequencies of patient and control sample in a 2×2 table, when the disease frequency is low in the background population. The cross-ratio itself can be regarded as a measure of the strength of association in the 2×2 tables. It has, however, often been pointed out that it is far more convenient to use instead the natural logarithm of the cross-ratio, because this quantity follows approximately the normal Gaussian distribution with a certain variance as given below.[37,38] This allows one to combine evidence from different studies of the same disease in order to obtain a closer estimate of the effect, if it is in common.

Another way of looking at the effect is analogous to the binomial response situation in biological assay.[39,40] For example, the disease state can be taken as a positive and the normal state as a negative response, and the antigen presence or absence can be considered as two treatments. It is called binomial because each individual has two alternative ways of responding, either by having the disease (with a certain probability π) or by being normal (with probability $(1-\pi)$). This would apply correctly only to studies of disease occurrence in individuals selected as antigen-positive and antigen-negative; but is approximately valid in the case-control study of individuals ascertained as patients or normal controls. In this case the measure of the effect is defined as

$$\ln\frac{\pi+}{1-\pi+}-\ln\frac{\pi-}{1-\pi-}$$

$\pi+$ and $\pi-$ representing probability of disease in antigen-positive and antigen-negative groups, respectively. The treatment is analogous to the method given below and will not be discussed further.

The alternative models to be tested are, in the simplest case, the compatibility of the association in the actual 2×2 tables (e.g. measured by the relative risk or the ln (cross-ratio)) with the hypothesis of no association (relative risk $=1$, ln (cross-ratio) $=0$). For a single study this is best done by Fisher's exact test, as it does not rely upon large sample approximations, and the p-value (significance) tells how often one would expect a measure of association as large or larger in 2×2 tables with the same marginal frequencies, if in fact there is no association.

In the situation where several studies have been combined, we are faced with two problems. First, we must test whether the data are compatible with the hypothesis that the effect of having the antigen is the same in all the investigations (i.e. ln (cross-ratio) common to them). This is a test of heterogeneity between the ln (cross-ratio)s and, as each of these approximately follows the Gaussian distribution, it can be carried out in a way analogous to the analysis of variance. If the existence of a common effect is accepted it remains to be tested whether it is really different from no effect. This is easily done, as in this case, the estimate also follows approximately the Gaussian distribution with a given variance.

In judging the significance of the associations based on the 2×2 comparisons of several (n) antigen frequencies in patient and control samples, the conventional levels of significance cannot be used.[41,42] Had the presence of an antigen no effect on disease susceptibility and were the n comparisons independent, one would expect that the number (i) of comparisons showing significance at a given level (p) follows a binomial distribution with parameters p and n:

$$p(i)=\binom{n}{i}p^{i}(1-p)^{n-i}$$

Thus the probability of finding one or more significances merely by chance would be the sum of terms:

$$p(1)+p(2)+\ldots+p(n)=np(1-p)^{n-1}+\frac{n(n-1)}{2}p^{2}(1-p)^{n-2}+\ldots+p^{n}$$

and neglecting terms of higher order in p one is left with $np(1-p)^{n-1}$ which, when p is small, is nearly equal to np, i.e. n times to the nominal significance level. A conservative correction would be to multiply the obtained significance with the number of comparisons, n, or in other words, to require that the reciprocal of the p-value exceeds the number n by, say two orders of magnitude, before significance is stated.

Relative risk and 2 × 2 tables

Consider a population of individuals who can be classified according to two different criteria, e.g. antigen-presence (Ag+) or -absence (Ag−) and disease (Dis) or normal (Nor) state, respectively.

Each individual in the population must thus fall in one of the four groups: (Ag+, Dis), (Ag−, Dis), (Ag+, Nor) or (Ag−, Nor). The relative risk of having the disease for antigen-positive individuals is defined by

$$\text{relative risk} = p(\text{Dis}|\text{Ag}+)/p(\text{Dis}|\text{Ag}-) \tag{4}$$

where $p(\text{Dis}|\text{Ag}+)$ and $p(\text{Dis}|\text{Ag}-)$ are the probabilities of the disease occurring in the antigen-positive and antigen-negative individuals, respectively. As there are two mutually exclusive ways of being antigen-positive, namely as diseased or normal, we have

$$p(\text{Dis}|\text{Ag}+) = p(\text{Dis, Ag}+)/\{p(\text{Dis, Ag}+) + p(\text{Nor, Ag}+)\} \tag{5}$$

and analogously

$$p(\text{Dis}|\text{Ag}-) = p(\text{Dis, Ag}-)/\{p(\text{Dis, Ag}-) + p(\text{Nor, Ag}-)\} \tag{6}$$

thus giving the

$$\text{relative risk} = \frac{p(\text{Dis, Ag}+)\{p(\text{Dis, Ag}-) + p(\text{Nor, Ag}-)\}}{\{p(\text{Dis, Ag}+) + p(\text{Nor, Ag}+)\}p(\text{Dis, Ag}-)} \tag{7}$$

If the disease is rare, then $p(\text{Dis, Ag}+)$ is usually negligible in relation to $p(\text{Nor, Ag}+)$ as is $p(\text{Dis, Ag}-)$ compared to $p(\text{Nor, Ag}-)$, and we obtain the approximation:

$$\text{relative risk} \simeq \frac{p(\text{Dis, Ag}+)p(\text{Nor, Ag}-)}{p(\text{Dis, Ag}-)p(\text{Nor, Ag}+)} \tag{8}$$

This fraction is called the (population) cross-ratio, or odds-ratio, and it can be proven that if disease and antigen occurrence are independent of each other, both the relative risk and the cross-ratio equal 1.

If we want an estimate of the relative risk, we can in a *cohort* study[40] investigate a representative sample of size N, taken at random from the population, and count the numbers of individuals falling in the four groups, obtaining $n(\text{Dis, Ag}+)$, $n(\text{Dis, Ag}-)$, $n(\text{Nor, Ag}+)$ and $n(\text{Nor, Ag}-)$. A direct estimate of the relative risk will be found simply by substituting n for p in expression (7) above and a direct estimate of the cross-ratio by doing so in expression (8).

When the disease is rare, it requires a very large sample size N in order to obtain a reasonably precise estimate of the relative risk, and a far more economic way is to investigate a sample of patients (N(Dis)) and a sample of normal controls (N(Nor)) in a *case control* study. The patients fall in two categories: $N(\text{Dis, Ag}+)$ are antigen-positive and $N(\text{Dis, Ag}-)$ are negative, and so do the controls in numbers $N(\text{Nor, Ag}+)$ and $N(\text{Nor, Ag}-)$. In this sampling situation it is impossible to find a direct estimate of the relative risk, but it is easily shown that the fraction

$$x = \frac{N(\text{Dis, Ag}+)N(\text{Nor, Ag}-)}{N(\text{Dis, Ag}-)N(\text{Nor, Ag}+)} \tag{9}$$

is a direct estimate of the population cross-ratio. As the latter approximately equals the relative risk, when the disease is rare, the estimated cross-ratio, x, will also do so under this condition.

In the sampling situation, the cross-ratio's obtained will fluctuate around the true (population) value, but as they can only attain values from zero to infinity, the distribution will be very unsymmetrical. It has been shown[37] that by taking the natural logarithm of the cross-ratio this skewness is eliminated and if further we add 1/2 to each of the four numbers before forming the fraction, we have approximately corrected a statistical bias and eliminated the risk of taking the logarithm of a zero.[38,39]

This quantity

$$y = \ln\left(\frac{\{N(\text{Dis, Ag}+) + \frac{1}{2}\}\{N(\text{Nor, Ag}-) + \frac{1}{2}\}}{\{N(\text{Dis, Ag}-) + \frac{1}{2}\}\{N(\text{Nor, Ag}+) + \frac{1}{2}\}}\right) \tag{10}$$

follows approximately the Gaussian distribution with a mean $y = \ln$ (population cross-ratio) and a sampling variance

$$\text{Var}(y) = \frac{1}{N(\text{Dis, Ag}+) + 1} + \frac{1}{N(\text{Dis, Ag}-) + 1} + \frac{1}{N(\text{Nor, Ag}+) + 1} + \frac{1}{N(\text{Nor, Ag}-) + 1} \tag{11}$$

which allows one to combine evidence from different studies. Each value of y obtained, y_i, has to be given a weight, which is taken to be the reciprocal of its sampling variance

$$w_i = 1/\text{Var}(y_i) \tag{12}$$

and a weighted mean can be found

$$\bar{y} = \sum y_i w_i / \sum w_i \tag{13}$$

which has a sampling variance

$$\text{Var}(\bar{y}) = \frac{1}{\sum w_i} \tag{14}$$

It can be tested whether there is significant heterogeneity between the obtained values of y_i by comparing

$$X^2_{(n-1)} = \sum y_i^2 w_i - \frac{(\bar{y})^2}{\text{Var}(\bar{y})} \tag{15}$$

to a table of chi-square with $n-1$ degrees of freedom, where n is the number of investigations. The significance of the main effect, i.e. the deviation of the mean y, $\bar{y}$, from its true value if there is no association (namely 0) can be found from a table of chi-square with one degree of freedom, comparing this criterion with

$$X^2_{(1)} = \frac{(\bar{y}-0)^2}{\text{Var}(\bar{y})} \tag{16}$$

Figure 2 illustrates this procedure for the available data on HLA-B27 in ankylosing spondylitis. For a review and extension of the theory see Reference 43.

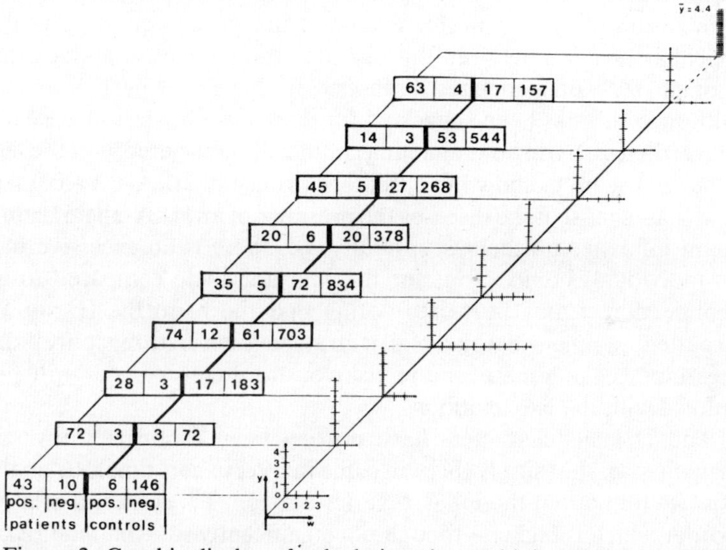

Figure 2. Graphic display of calculations in combining nine case control studies of HLA-B27 in ankylosing spondylitis. For each investigation, the numbers of patients and normal controls carrying and lacking HLA-B27 are given; the vertical bars in the right part of the figure indicate the corresponding ln (cross-ratio) = y_i (expression (10)) and the horizontal bars indicate the weight w_i of each y_i-value (expression (11)). The rearmost vertical bar indicates the combined weighted estimate of the ln (cross-ratio), $\bar{y} = 5{\cdot}5$. In this series of investigations there is no sign of heterogeneity (chi-square = 8·11 with eight degrees of freedom, $p = 0{\cdot}4$, *expression* (15)) while the significance of the main effect, i.e. the deviation of $\bar{y} = 4{\cdot}4$ from 0, is astronomic (chi-square = 730 with one degree of freedom, $p \ll 10^{-6}$ expression (16)). The combined estimate of the cross-ratio itself is $e^{\bar{y}} = e^{4{\cdot}4} = 81$, which is an approximation to the relative risk of having ankylosing spondylitis for HLA-B27 positive individuals

3. SURVEY OF DISEASES INVESTIGATED FOR HLA ASSOCIATION

Many aspects of human pathology have already been investigated with respect to HLA-associations; Table 2 surveys most of the disorders. A few notes of caution must be taken into account when looking at such tables. Firstly, in evaluating the observations, two types of statistical error can occur: (a) one can believe in an association which is not there or has nothing to do with the phenomenon one is investigating, or (b) a true association may escape attention, especially if the statistical tools have insufficient power or the investigation is too small. One or the other of these errors is likely to arise in studies where patients and normal controls are not carefully taken from the same background population. Another source of error is combining data from investigations which do not apply exactly the same criteria for selecting the patients; this is especially likely to occur in severe diseases where newly diagnosed cases may differ from long-term survivors.

From Table 2 it appears that most of the significant associations concern increases in antigen frequencies among patients (relative risks above 1). This may be due to low power in the statistical tests; it requires large numbers of patients (and controls) to demonstrate the significance of a decrease in an antigen frequency which is already low, as is the case for most HLA antigens except HLA-A2.

It appears that two HLA antigens occur very frequently in Table 2, namely HLA-B8 and HLA-B27. Table 3 shows a list of disorders associated with these two antigens. The HLA-B27 associated diseases seem to comprise a series of arthropathies involving mainly the spine. The inclusion of acute anterior uveitis in this group conforms well with the well-known clinical association between this disorder and ankylosing spondylitis.

In contrast, not all HLA-B8 associated diseases listed have previously been recognized to be related, although this has been suggested for diabetes, Graves's disease and Addison's disease. Nevertheless, pernicious anaemia is clinically associated with the latter three disorders, but does not seem to show increased occurrence of HLA-B8 among patients.

It may be noted that in most of the diseases the presence of an HLA-allele seems to exert a simple dominant influence on the susceptibility, i.e. there is no excessive increase of susceptibility in individuals homozygous for the relevant allele compared to heterozygotes. In this connection it may be worth noting that the fit to the Hardy–Weinberg equilibrium in a patient sample will usually fail if an association is strong; only if the risk for the homozygotes equals the square of the risk for the heterozygotes will phenotype frequencies conform with the expectations.

The HLA-C and D locus characters have not yet been extensively investigated in relation to human diseases, but surely this will come in a very short time. Re-analysing our data on psoriasis, we found that the HLA-C series antigen, T7, confers a relative risk of 14·9 of this disorder which is higher—though not significantly so—than the risk (9·9) for HLA-B13, Bw17 and/or Bw37 positives. Thus, the association between HLA and psoriasis may primarily concern the HLA-C series. The sparse evidence concerning the HLA-D determinants points in some cases to a stronger association with disease, than that of the HLA-A and B series, but in other disease this does not hold; e.g. the relative risk of multiple sclerosis for HLA-Dw2 positive individuals is significantly higher than for individuals carrying HLA-B7[44] and it also seems as if Dw3 is more strongly associated with the group of HLA-B8 associated disorders than B8 itself.

4. MODELS AND POSSIBLE EXPLANATIONS

Scientifically it would be most enticing to regard the various disease–HLA associations as being due basically to a common mechanism, but considering the wide range of diseases

Table 2. Association between HLA and disease

Disease	Antigen	Combined estimate of relative risk	Significance of relative risk deviating from 1 (p)	No. of studies	No. of patients investigated
Arthropathies					
Ankylosing spondylitis	HLA-B27	81·0	$\ll 10^{-6}$	9	445
Reiter's syndrome	HLA-B27	48·0	$\ll 10^{-6}$	6	
Reactive arthritis	HLA-B27	increased in Yersinia and salmonella arthritis, but not in meningococcal arthritis			
Psoriatic arthritis	HLA-B27	5·4	$\ll 10^{-6}$	5	220
Juvenile arthritis	HLA-B27	apparently increased when the spine is involved			
Rheumatoid arthritis		no definite associations		5	313
Gout		no deviations			66
Eye diseases					
Acute anterior uveitis	HLA-B27	16·9	$\ll 10^{-6}$	3	207
Chronic anterior uveitis		no deviations		1	13
Skin diseases					
Psoriasis vulgaris	HLA-B13	4·3	$< 10^{-6}$	8	820
	HLA-Bw17	4·8	$< 10^{-6}$	8	820
	HLA-Bw37	8·4	$< 10^{-6}$	1	220
Pustular psoriasis		no deviations		1	31
Pemphigus	HLA-A10	perhaps increased in Jewish patients		3	61
Dermatitis herpetiformis	HLA-B8	4·3	$< 10^{-6}$	3	89
Endocrine diseases					
Insulin-dependent diabetes mellitus	HLA-B8	1·9	10^{-4}	5	350
	HLA-Bw15	2·3	$< 10^{-6}$	5	350
Non-insulin-dependent diabetes mellitus		no deviations		>2	100
Graves's disease	HLA-B8	2·5	10^{-6}	4	282
Graves's disease among Japanese	HLA-Bw35	5·1	10^{-4}	1	44
Idiopathic Addison's disease	HLA-B8	6·4	10^{-5}	1	30
Neurologic diseases					
Multiple sclerosis	HLA-B7	1·5	10^{-4}	5	>1000
	HLA-Dw2	5·0	$< 10^{-6}$	3	95
Optic neuritis		probably as M.S.			
Paralytic polio		discordant results		2	136

Table 2—continued

Disease	Antigen	Combined estimate of relative risk	Significance of relative risk deviating from 1 (p)	No. of studies	No. of patients investigated
Allergies					
Childhood asthma		no definite deviations		2	71
Atopic dermatitis		no deviations		2	88
Ragweed hay-fever		no deviations, but perhaps linkage in family studies		3	
Infections					
Leprosy		no definite deviations		1	39
Tuberculosis		no definite deviations		1	119
Haemophilus influenzae		no definite deviations		1	65
Infectious mononucleosis		no definite deviations		1	40
Intestinal diseases					
Coeliac disease	HLA-B8	9·5	$<10^{-6}$	5	248
Ulcerative colitis		no deviations		4	111
Crohn's disease		no deviations		5	139
Liver diseases					
Chronic autoimmune hepatitis	HLA-B8	3·6	$<10^{-6}$	3	170
Acute hepatitis, chronic persistent hepatitis, liver cirrhosis, healthy Au-carriers		no deviations			
Systemic diseases and 'others'					
Myasthenia gravis	HLA-B8	5·0	$\ll 10^{-6}$	5	260
Sjögren's disease (including Sicca syndrome)	HLA-B8	3·2	10^{-6}	3	
Sarcoidosis		no deviations		3	262
S.L.E.		no deviations		4	64
Cystic fibrosis		no deviations		1	24
Polymyalgia rheumatica		no deviations			
Rheumatic fever		no definite deviations			
Chronic glomerulonephritis	HLA-A2	1·5	10^{-3}	>1	485
Behçet's disease	HLA-B5	4·6	10^{-4}	3	49
Essential hypertension		no deviations			
Asbestosis		no deviations			
Periodontitis		no deviations			
Gingivitis		no deviations			
Schizophrenia		no deviations			

Malignant diseases					
Hodgkin's disease	HLA-A1	1·4	$<10^{-6}$	17	1508
	HLA-B5	1·6	$<10^{-6}$	17	1508
	HLA-B8	1·3	10^{-4}	17	1508
	HLA-B18	1·9	$<10^{-6}$	12	1165
Acute lymphatic leukaemia	HLA-A2	1·3	$<10^{-2}$	10	527
	HLA-B8	1·3	$<0{\cdot}05$	10	527
	HLA-B12	1·2	0·05	10	527
Mammary carcinoma	no deviations			4	593
Malignant melanoma	no deviations			5	349

A full list of references will be available on request to the authors.

Table 3. The HLA-B27 and the HLA-B8 associated diseases

HLA-B8 Disease	Relative risk	HLA-B27 Disease	Relative risk
Dermatitis herpetiformis	4·3	Ankylosing spondylitis	81·0
Coeliac disease	9·5	Reiter's disease	48·0
Chronic autoimmune hepatitis	3·6	*Juvenile* rheumatoid arthritis	3·5
Sjögren's disease	3·2	Psoriatic arthritis	5·4
Myasthenia gravis	5·0	*Acute* anterior uveitis	16·9
Diabetes mellitus, insulin-dependent	1·9		
Graves's disease	2·5		
Addison's disease	7·0		
Hodgkin's disease	1·3		

showing strong HLA associations, it is an extremely difficult task to construct a testable hypothesis explaining most of the observations. We are hampered on two sides: the lack of knowledge of the biological function of the HLA itself (especially the classical SD markers) and the incomplete understanding of the diseases. From the survey of diseases showing HLA associations it appears that all of them have an unclear or unknown aetiology, and that many are suspected of having an element of autoimmunity. Furthermore, some of the diseases listed have a well recognized familial aggregation, but in no case is the inheritance clear-cut or known. It has often been pointed out that it is not the diseases that are inherited, but rather the liability to become affected, when exposed to the relevant pathogenic environment.

Retrospectively, most of the models proposed have resorted to the concept of genes strongly affecting susceptibility, located near or within the HLA region and with alleles in linkage disequilibrium with the recognizable HLA markers, rather than considering the known HLA-markers themselves. The term linkage disequilibrium between alleles at different loci is another word for association between them, and we could just as well take the cross-ratio instead of the conventional Δ-value as a measure of the degree of association.[19] To illustrate this relationship, we have analysed our normal material (2398 individuals) with respect to the well known strong association between the presence of HLA-A1 and B8. This gave an estimated cross-ratio = 20·9 with a significance chi-square = 618·2, DF = 1 (the cross-ratio is in this case no good approximation to the relative risk of having HLA-A1 for HLA-B8 positive individuals, because all four combinations occur in appreciable frequencies in the population). In comparison, the association between ankylosing spondylitis and HLA-B27 has a strength of estimated overall cross-ratio beyond 80 (combination of nine materials, Figure 1) which is an extremely high degree of association indeed. What this means is difficult to say, at least it does not necessarily imply a correspondingly close linkage between a putative major ankylosing spondylitis disposing gene and the HLA-B locus. Family investigations of this disease show in some cases independent segregation of HLA-B27 and manifest disease, with the disease occurring without B27 and vice versa, indicating strong influence also by genes other than HLA-B27.

4.1 Threshold models

A general class of models originally developed for the hereditary predisposition to diseases (e.g. to diabetes and various malformations) has its background in the quantitative genetics developed for domestic animal breeding. Common to these models is the concept that every individual possesses a certain underlying liability to become affected by a disease.[15,45] Assuming that the liability is mainly genetically determined and that the alleles at many different loci each exert a moderate influence on the liability in an additive fashion, it can be shown that the liabilities in the population constitute a Gaussian distributed continuum with a certain mean and variance. One of the models introduced by Falconer states that the disease is contracted when a certain threshold level of liability is reached.[46] This threshold model has mostly been applied to account for the familial aggregation of certain diseases, as the correlation between polygenically controlled quantitative characters in close relatives follows relatively well-fixed laws (first degree relatives—parent–child and sib–sib combinations—share approximately one half of their genome, more distantly related family members less of it). According to the model, the liabilities of the relatives to the propositi also follow a Gaussian distribution, but with a higher mean-liability than that of the background population due to their sharing of high-liability alleles with the propositi.

Following this line of thought let us assume that a gene (or a few genes) have alleles which exert a major influence on the liability, and further that they occur in linkage disequilibrium with (or that they are identical to) the alleles for the known HLA markers. The result would be that a high proportion of antigen-positive individuals would have a high liability, or, in other words, the mean-liability of antigen-positive individuals would be higher than that of antigen-negative individuals, giving a high relative risk of developing the disease to the antigen-possessing individuals.[47] The correspondence between antigen presence (or absence) and high (or low) liability need not be absolute; owing to the assumed polygenic control, the simultaneous action of other genes in the opposite direction could very well in some cases level out the resultant liability.

In fact, a recent analysis by Goodman and Chung[48] of a large series of family data on diabetes mellitus showed that: (a) in the early onset group the results were compatible with the presence of major genes, while (b) the middle and late onset groups seemingly did not fit well to such a model, suggesting that the juvenile and maturity onset diabetes may be two different diseases. Unfortunately, the families were not investigated for HLA markers, but the findings corroborate the totally independent results from HLA studies: the existence of major genes is clearly implied[49] by the mere association between HLA-B8 and Bw15 and juvenile diabetes. The absence of detectable associations between HLA markers and maturity onset diabetes suggest that this disease is distinct from the juvenile form.

The outlined threshold model and other equivalent models are very general, and not incompatible with most other models, e.g. the ultimate outcome of a physiological immune reaction could very well be a polygenic matter. They have the serious drawback of being very difficult to test. This is particularly due to the abstract nature of the liability as defined until a measurable quantity, such as the level of a metabolite or hormone, or a receptor function, can be held responsible.

4.2 Ir genes

For the time being, by far the most fashionable model is the hypothetical existence of Ir genes with alleles in linkage disequilibrium with the alleles at the other HLA-loci (primarily the B and D loci). However, the definition of Ir genes in man is rather vague, the strongest indication being in fact the disease associations themselves, and thus we may be dealing with a circular argument.

The reasons for the popularity of Ir genes are obvious. Firstly, the evidence of histocompatibility-linked Ir genes in animals were among the premises for these studies, and many HLA workers come from the field of immunology. Secondly, the array of different Ir genes in a given animal seems to control immune responsiveness to a wide variety of different antigens, which may be relevant in view of the many different diseases associated with HLA. Thirdly, the lack of concrete evidence about the function of Ir genes in man allows rather free speculation on their relevance for human disease. Different alleles at various Ir loci could control both the quantitative aspects of immune responsiveness and in a broader sense the mode of reaction in some cases leading to 'lacunar immunodeficiencies'[50] in other cases conferring an exaggerated immune reactivity, i.e. leading to autoimmune disease. Both mechanisms may act in concert: e.g., a virus could infect an organism lacking the relevant Ir allele, while the simultaneous action of another Ir gene could lead to a later substantial immune reaction against altered cell surface components in virus infected target organs and even to breakdown of self-tolerance.

Such models could fit most of the diseases listed in Table 2, even the endocrine disorders.

Perhaps the best available evidence of Ir genes playing a role in human disease is the observation that auto-antibodies against adrenal tissues occur significantly more often in the group of HLA-B8 positive patients with idiopathic Addison's disease than in the group of HLA-B8 negative.[51] In fact, this observation may be the best evidence that HLA contains Ir genes, as the interpretation of data on the familial co-occurrence of ragweed allergy and certain HLA types[52,53] have recently been subjected to re-evaluation.[54]

In Graves's disease, the association between HLA and autoimmunity is not so clear-cut, although this disease may be a parallel of the histocompatibility influenced thyroiditis in mice and chicken.

In juvenile diabetes mellitus, virus infection as well as autoimmunity have been suspected:[55,56] the target being in this disease the insulin-producing beta-cells of the islets of Langerhans. Moreover, juvenile diabetes is associated with two different HLA antigens of the same series, B8 and Bw15, which could indicate that the same (Ir(?)) allele occurred in linkage disequilibrium with both of these, a situation not uncommon for the HLA-A and B loci. However, the observation that the relative risk for heterozygotes of type B8, Bw15 seems to be higher than that for other B8 or Bw15 heterozygotes or homozygotes has led to the hypothesis of an additive effect of two different (Ir(?)) alleles (at the same or different loci) in linkage disequilibrium with B8 and Bw15, respectively.[51] As the effect is not additive in the homozygous individuals, this would imply that two different mechanisms of action are involved. This may be seen in relation to the recent finding by Munro and Taussig that two levels of control of the immune response are exerted by products of two distinct genes within the I region of the H-2 complex in the mouse.[32]

4.3 'Molecular mimicry'

Turning to the models involving the known HLA markers themselves, most of the theories are closely related to those proposed for the biological function and diversity of the HLA characters.

Firstly, a very straightforward model would be an immunological cross-reaction between some of the allelic variants of the HLA antigens and antigens of an infectious agent ('molecular mimicry'), causing cross-tolerance to this agent in individuals carrying the corresponding HLA-antigen.[57,58] This phenomenon, however, has never been convincingly demonstrated. A variation of this theme is the theory of 'passenger' proteins.[59] It is known that many viruses after intracellular multiplication and maturation are released by budding from the cell membrane, presumably incorporating some host-membrane components in the virus coat. If these components included HLA antigens, one would expect an increased susceptibility in (homozygote or heterozygote) individuals of frequent HLA types and especially a high contagiousness of viruses derived from frequent homozygotes. This could lead to a frequency-dependent selection favouring rareness of individual HLA types, thus explaining the extreme polymorphism within the HLA system, but it would hardly explain the association observed between disease and specific antigens.

4.4 Receptor functions

Other models, not necessarily involving immunological mechanisms, can be envisaged. It has been proposed that the HLA molecules on the cell surface may act as (or interact with) membrane receptors for either viruses or physiological ligands (hormones, metabolites, etc.). In order to explain the association between diseases and single antigen-specificities this model requires that the serologically recognizable alloantigenic variation reflects also a variation in thermodynamic binding constants.

The *virus* receptor theory has been directly investigated recently by Dausset and Hors in two ways.[60] They studied the susceptibility of fibroblasts derived from HLA-typed donors to the cytopathogenic effect of four different viruses: herpes hominis, adenovirus, echovirus and measles virus without finding any differences attributable to HLA. In another series of experiments, they studied the interdependence between the occurrence or absence of HLA antigens and virus receptors (for Echo 11 and Coxsackie B3) on somatic mouse–human hybrid cell-lines. They came to the conclusion that there was no relation between the presence of virus receptor and HLA antigen on the cells, suggesting that the genes controlling virus receptors are not even located on the same chromosome.

So far the only evidence pointing towards *hormone* receptor involvement comes from mouse studies. Recently Meruelo and Edidin[61] reported that in the mouse, the H-2 locus exerts a major influence on the level of cyclic AMP in liver cells. The mechanism of action has not been unravelled, but both the possibility of genes, within H-2, coding for variants of cAMP processing enzymes or adenylcyclase activating hormone receptors, and the possibility of interaction between the H-2 antigens and hormone receptors in the cell membrane, remains open.

It is tempting to speculate that the decreased concentration of cAMP in the epidermal cells of psoriatic plaques (Søndergaard, personal communication) may indicate that similar HLA influenced mechanism(s) operate in this disorder.

In view of the widespread physiological functions of cAMP, involving the cellular immune reactions as well as many other signal/response mechanisms, it appears, that eventually a great many diseases could be linked together by such a common factor. The differential expression of the various diseases may arise as a result of modulation by other unrelated genetic or environmental sources.

4.5 Differentiation antigens

It has been proposed that the HLA antigens may be *differentiation antigens*, controlling cellular interactions. Naturally, variant HLA antigens could lead to defective cellular interactions and thus to various diseases. This seems to be a rather speculative explanation and the only evidence favouring it is some similarity in the mouse between H-2 and

T/t-complex gene products.[62] The T/t complex affects mainly the development of the spine in the embryo and appears to involve genuine differentiation antigens.

5. CONCLUDING REMARKS

The mere observation and description of associations between diseases and HLA as listed in Table 2 is, of course, of limited value in itself. In a few cases the knowledge is likely to be useful diagnostically, but the widest implications are that the strong associations clearly point to major genetic components in the aetiologies of diseases.

In the search for the nature of the biochemical pathways or biophysical interactions involved, the investigation of genetic markers is only one, but clearly a very important, stage. Obviously, a further step would be to find out whether the observed HLA markers themselves play a direct role. It should be possible to gain some information on this point by studying the HLA associations of the same disease in several remote populations. Thus, if one disease showed a strong association with one HLA antigen in population A but with another antigen in population B, it would be fair to conclude that it most probably is not the HLA alleles themselves which are responsible, but rather a causative allele at another locus, in linkage disequilibria, with the two different HLA markers in population A and B, respectively. This may be the case for Graves's disease which is associated with HLA-B8 in Caucasians[63] and has been reported in association with HLA-Bw35 in Japanese.[64]

The reverse observation of an association between one disease and a constant antigen in several populations is not as conclusive, because we may still be dealing with linkage disequilibria between HLA markers and causative alleles at other loci. Nevertheless, most linkage disequilibria between HLA-A and B alleles seem to be completely different in different races, and the finding that ankylosing spondylitis is associated with the HLA-B27 antigen both in Caucasians, Japanese and American Indians (P. Terasaki, personal communication) might indicate that HLA-B27 itself is the causative factor.

Finally, it is clear that most of the diseases have to be re-investigated in terms of Ia antigens when these become more accurately defined in humans. If Ir genes are truly responsible for the associations, and if Ia and Ir genes are as close in humans as in mice, a stronger association with Ia than with HLA-B and perhaps D antigens may be expected. On the other hand, the other models discussed above may just as well operate for the HLA-A, B and C determinants, and *a priori*, it may perhaps be suggested that a primary association with one of these determinants would argue against Ir genes being involved.

Acknowledgements

This study was supported by grants from the Danish Medical Research Council and the Danish Blood-Donor Foundation. For the excellent secretarial assistance we are grateful to Mrs Elly Andersen.

6. REFERENCES

1. Kissmeyer-Nielsen, F. and Thorsby, E. (1970). *Transplant. Rev.*, **4**, 1–176.
2. Thorsby, E. (1974). *Transplant. Rev.*, **18**, 51–129.
3. Svejgaard, A. (1975). *Monographs in Human Genetics*, Vol. 7, S. Karger, Basel.
4. WHO–IUIS Terminology Committee (1975). In *Histocompatibility Testing 1975* (Kissmeyer-Nielsen, F., Ed.), Munksgaard, Copenhagen, pp. 5–20.
5. Milstein, C. (1968). *Biochem. J.*, **110**, 26P–27P.

6. Heremans, J. F. (1969). Protides of Biological Fluids. *Proc. 17th Colloquium, Bruges* (Peeters, H., Ed.), pp. 3–23.
7. Joint Report (MLC) (1975). *Histocompatibility Testing 1975* (Kissmeyer-Nielsen, F., Ed.), Munksgaard, Copenhagen, pp. 414–458.
8. Allen, F. H. Jr. (1974). *Vox Sang.*, **27**, 382–384.
9. Fu, S. M., Kunkel, H. G., Brusman, H. P., Allen, F. M. Jr. and Fotino, M. (1974). *J. exp. Med.*, **140**, 1108–1110.
10. Rittner, C. H., Hauptmann, G., Grosse-Wilde, H., Grosshans, E., Tongio, M. M. and Mayr, S. (1975). *Histocompatibility Testing 1975* (Kissmeyer-Nielsen, F., Ed.), Munksgaard, Copenhagen, pp. 945–954.
11. Rood, J. J. van, Leeuwen, A. van., Keuning, J. J., Blussé van Oud Alblas, A. (1975). *Tissue Antigens*, **5**, 73–79.
12. Turner, J. R. G. (1968). *Genetica*, **39**, 82–93.
13. Fu, S. M., Stern, R., Kunkel, H. G., Dupont, B., Hansen, J. A., Day, N. K., Good, R. A., Jersild, C. and Fotino, M. (1975). *J. exp. Med.*, **142**, 495–506.
14. Jersild, C., Rubinstein, P. and Day, N. K. (1975). In *Biological Amplification Systems* (Good, R. A. and Day, S., Eds.), Plenum Press, New York (in press).
15. Cavalli-Sforza, L. L. and Bodmer, W. F. (1971). *The Genetics of Human Populations*, W. H. Freeman and Co., San Francisco.
16. Li, C. C. (1961). *Human Genetics*, McGraw-Hill, New York.
17. Yasuda, N. and Kimura, M. (1968). *Ann. Hum. Genet.*, **31**, 409–420.
18. Ceppellini, R., Curtoni, E. S., Mattiuz, P. L., Miggiano, V., Scudeller, G. and Serra, A. (1967). *Histocompatibility Testing 1967*, Munksgaard, Copenhagen, pp. 149–185.
19. Svejgaard, A., Jersild, C., Nielsen, L. Staub and Bodmer, W. F. (1974). *Tissue Antigens*, **4**, 95–105.
20. Hill, W. G. (1974). *Theoretical Population Biology*, **6**, 184–198.
21. Piazza, A. (1975). *Histocompatibility Testing 1975* (Kissmeyer-Nielsen, F., Ed.), Munksgaard, Copenhagen, pp. 923–927.
22. Mattiuz, P. L., Ihde, D., Piazza, A., Ceppellini, R. and Bodmer, W. F. (1970). *Histocompatibility Testing 1970*, Munksgaard, Copenhagen, pp. 193–205.
23. Li, W.-H. and Nei, M. (1974). *Theoretical Population Biology*, **6**, 173–184.
24. Degos, L. and Dausset, J. (1973). *C.R. Acad. Sc. Paris*, **277D**, 2433–2436.
25. Wiener, A. S. (1970). *Amer. J. Hum. Genet.*, **22**, 476–483.
26. Vogel, F. (1970). *Amer. J. Hum. Genet.*, **22**, 464–475.
27. Vogel, F. and Helmold, W. (1972). *Humangenetik*, Vol. 1 (Becker, P. E., Ed.), Georg Thieme Verlag, Stuttgart, pp. 129–557.
28. Klein, J. (1975). *Biology of the Mouse Histocompatibility-2 Complex*. Springer-Verlag, New York.
29. Ellman, L., Green, I., Martin, W. J. and Benacerraf, B. (1970). *Proc. Nat. Acad. Sci.*, **66**, 322–328.
30. Ziegler, J. B., Alper, C. A. and Balner, H. (1975). *Nature*, **254**, 609–611.
31. Katz, D. H., Graves, M., Dorf, M. E., Dimuzio, H. and Benacerraf, B. (1975). *J. exp. Med.*, **141**, 263–267.
32. Munro, A. J. and Taussig, M. J. (1975). *Nature*, **256**, 103–106.
33. Rose, N. R., Kite, J. H. Jr., Vladutiu, A. O., Tomazie, V. and Bacon, L. D. (1973). *Int. Arch. Allergy Appl. Immunol.*, **45**, 138–149.
34. McFarlin, D. E., Hsu, S. C.-L., Slemenda, S. B., Chou, F. C.-H. and Kibler, R. F. (1975). *J. exp. Med.*, **141**, 72–81.
35. Bacon, L. D., Kite, J. H. Jr. and Rose, N. R. (1973). *Transplantation*, **16**, 591–598.
36. Cornfield, J. (1956). *Proc. Third Berkeley Symposium on Mathematical Statistics and Probability*, Vol. IV (Neyman, J., Ed.), University of California Press, Berkeley, Los Angeles, pp. 135–148.
37. Woolf, B. (1955). *Ann. Hum. Genet.*, **19**, 251–253.
38. Haldane, J. B. S. (1956). *Ann. Hum. Genet.*, **20**, 309–311.
39. Anscombe, F. J. (1956). *Biometrika*, **43**, 461–464.
40. Armitage, P. (1971). *Statistical Methods in Medical Research*, Blackwell, London.
41. Edwards, J. H. (1974). *J. Immunogenetics*, **1**, 249–257.
42. Miller, R. G. (1966). *Simultaneous Statistical Inference*, McGraw-Hill, New York.
43. Gart, J. (1971). *Rev. Internat. Stat. Inst.*, **39**, 148–169.

44. Platz, P., Dupont, B., Fog, T., Ryder, L. P., Thomsen, M., Svejgaard, A. and Jersild, C. (1974). *Proc. Roy. Soc. Med.*, **67**, 1133–1136.
45. Curnow, R. N. and Smith, C. (1975). *J. R. Statist. Soc. A*, **138**, 131–169.
46. Falconer, D. S. (1965). *Ann. Hum. Genet.*, **29**, 51–76.
47. Edwards, J. H. (1965). *Ann. Hum. Genet.*, **29**, 77–83.
48. Goodman, M. J. and Chung, C. S. (1975). *Clin. Genetics*, **8**, 66–74.
49. Smith, C. (1974). *Lancet*, **1**, 450.
50. Jersild, C., Ciongoli, A. K., Fog, T., Good, R. A., Platz, P. J., Svejgaard, A., Thomsen, M. and Dupont, B. (1975). In *Infection and Immunity in Rheumatic Diseases* (Dumonde, D. C., Ed.), Blackwell Scientific, Oxford (in press).
51. Thomsen, M., Platz, P., Andersen, O. Ortved, Christy, M., Lyngsøe, J., Nerup, J., Rasmussen, K., Ryder, L. P., Nielsen, L. Staub and Svejgaard, A. (1975). *Transplant. Rev.*, **22**, 125–147.
52. Levine, N. B., Stember, R. H. and Fotino, M. (1972). *Science*, **178**, 1201–1203.
53. Blumenthal, M. N., Amos, D. B., Noreen, H., Mendell, N. R. and Yunis, E. J. (1974). *Science*, **184**, 1301–1303.
54. Bias, W. and Marsh, D. G. (1975). *Science*, **188**, 375–377.
55. Nerup, J., Andersen, O. Ortved, Bendixen, G., Egebjerg, J., Gunnarsson, R., Kromann, H. and Poulsen, J. E. (1974). *Proc. Roy. Soc. Med.*, **67**, 506–513.
56. Gamble, D. R., Taylor, K. W. and Cumming, H. (1973). *Brit. Med. J.*, **4**, 260–262.
57. Snell, G. D. (1968). *Folia Biologica (Prague)*, **14**, 335–358.
58. Bodmer, W. F. (1972). *Nature*, **237**, 139–183.
59. Nandi, S. (1967). *Proc. Nat. Acad. Sci.*, **58**, 485–492.
60. Dausset, J. and Hors, J. (1975). *Transplant. Rev.*, **22**, 44–74.
61. Meruelo, D. and Edidin, M. (1975). *Proc. Nat. Acad. Sci.*, **72**, 2644–2648.
62. Artzt, K. and Bennett, D. (1975). *Nature*, **256**, 545–547.
63. Seignalet, J., Jaffiol, C., Baldet, L., Robin, M. and Lapinski, H. (1974). *Rev. Franc. Transp.*, **XVII**, 305–321.
64. Grumet, F. C., Payne, R. O., Konishi, J., Mori, T. and Kriss, J. P. (1975). *Tissue Antigen.*, **6**, 347–352.

Chapter 20

An Integration of B and T Lymphocytes in Immune Activation

P. A. BRETSCHER

Abbreviations

(A): concentration of specific associative antibody
B cells: precursors of the plasma cells
BSA: bovine serum albumin
CMI: cell-mediated immunity
(d)FGG: (de-aggregated) fowl gammaglobulin
F, f: foreign antigen, determinant
GAT: (Glu, Ala, Tyr)
HGG: human gammaglobulin
LPS: lipopolysaccharide
P_{CMI} cells: precursors of the cells involved in CMI
PLL: poly-L-lysine
PVP: polyvinyl pyrrolidone
S, s: self component, determinant
S III: pneumococcal type III polysaccharide
SRBC: sheep red blood cells
T cells: cells involved in the synthesis of A
t cells: precursors of T cells
TGAL: (Tyr, Glu)-Ala–Lys
TNPMRBC: trinitrophenylated mouse red blood cells

1. INTRODUCTION

The wish of the editors is that this chapter should provide an integrated view of the immune system. I shall discuss some ideas on the mechanisms by which cells interact and lay emphasis on the possible physiological significance of such interactions. Many of the ideas I discuss are at least controversial, and the evidence in any case is not sufficient to regard them as established. Even so it seems to me that there are two good reasons for attempting to provide integrated descriptions of the immune system at this time. Firstly, it is only by attempting to provide and test such descriptions that we shall discover their inadequacies, and so be able to construct a better integrated view which is our long-term aim. Secondly, such integrated descriptions, if carefully analysed and examined, often bring out the importance of novel questions even if the description itself is subsequently found to be unsatisfactory.

In this introduction I shall describe briefly a *basis* for constructing an integrated view of the immune system. It seems to me that the most powerful restrictions on such views come from a recognition that they must be consistent with certain physiological considerations. In order to make both this rather abstract statement clearer and more specific, and the subsequent discussion easier to follow, I shall initially outline some grounds for believing that certain physiological problems are particularly important.

One approach in trying to assess the nature of and the necessity for the more complex properties of the immune system is to imagine simpler defence mechanisms, and to analyse in what ways such systems differ from the immune system. One can imagine, for example, a primitive defence mechanism which consists of a few, say twenty, recognition molecules that have specificity for determinants present on foreign but not on self antigens. These molecules could be induced by the simple binding of 'antigen' to receptors on the cell. The antigen could be analogous to a hormone, such as insulin, in influencing the metabolism of the cell in the required way on interacting with appropriate surface receptors. There is no reason why sophisticated effector mechanisms could not be part of this primitive system. Conceptually, there appears to be no need for cell interactions.

Such a primitive system differs from the immune system in one obvious respect. The immune system can respond to almost everything macromolecular that is foreign to the host. If we imagine a process of greatly expanding the library of recognition units of the primitive system, such that it is more similar to that of the immune system, it is inevitable that some of the new members will have specificity for self components. There are both strong experimental and theoretical grounds for the belief that an immunologically capable individual has the *potentiality* for making anti-self antibodies, as described below; it thus seems that a mechanism for preventing the induction of anti-self immunity is necessary before the library of recognition units can be enormously expanded.

The ability to respond predominantly against foreign as opposed to self antigens, that is to discriminate between self and non-self, could in principle be due to one of two classes of mechanism. It could be due to the fact that the individual does not have the genetic capacity to synthesize recognition molecules able to bind self, or the individual has such a capacity but a regulatory mechanism normally prevents it from being expressed. I shall refer to the former as a genetic, and to the latter as a regulatory mechanism of self–non-self discrimination. One type of regulatory mechanism is that in which self–non-self discrimination is learnt. Both experimental evidence and theoretical considerations show that there must be a learning mechanism of self–non-self discrimination in the immune system. Certain procedures, for example, are capable of regularly inducing auto-antibodies, as discussed later, demonstrating that antibodies able to bind and damage self can be synthesized. Theoretical considerations also provide three arguments in favour of a learning mechanism and suggest some advantages of such a mechanism for the immune system. Firstly, the immune system of an individual can respond to non-self antigens that belong to a sibling; it seems impossible to imagine a way by which the correct genes could in general be handed down to an individual such that he can respond to those of his sibling's antigens that are foreign but not to his own. Secondly, a defence mechanism that could respond to almost all but self antigens, and in which self–non-self discrimination was genetically determined, would interfere with the evolution of many antigens; many such antigens would be recognized as foreign, and consequently attacked. Thirdly, a learning mechanism has the advantage that the generation of diversity of recognition units can produce some having anti-self activity without ill effects. It appears that if no learning mechanism existed the generation of recognition units able to bind a wide range of antigens would have to occur very slowly and painfully during a considerable period of evolutionary time in order to establish a large library without anti-self activity.

All these arguments for a learning mechanism of immunological self–non-self discrimination depend on the wide diversity of the immune response. It thus seems that a primitive defence mechanism, limited in its repertoire of recognition units, could be different from the immune system in that the self–non-self discrimination could be genetically determined by the specificity of its recognition units. Thus the generation of *diversity* of the immune response appears to be dependent on a *learning* mechanism of self–non-self discrimination.

In the first part of this chapter I shall discuss possible mechanisms by which self–non-self discrimination can be learnt. It appears, however, that the actual mechanism is not perfect as small amounts of auto-antibody are frequently found and severe cases of autoimmunity occasionally occur. The observation that mild autoimmunity is common poses the question of whether there are ways of minimizing its *effects* as opposed to minimizing its induction. I shall discuss a means by which this might be accomplished. Many features of the immune system are consistent with the view that much of its complexity is due to mechanisms that minimize the disadvantageous consequences of autoimmunity while allowing an effective response against foreign antigens.

A prerequisite to a discussion of ways by which the consequences of autoimmunity could be minimized is a knowledge of the circumstances under which it is likely to be induced. Although the theory of self–non-self discrimination I favour accounts for these circumstances, as discussed later, it is sufficient for the present discussion to know what these circumstances are. Autoimmunity is often induced when a foreign antigen that bears some chemical structures similar to those of self-components provokes a response. The autoimmunity is specific for these similar structures.

It would seem to be only advantageous for the effector mechanism of a primitive system to be very efficient in causing damage to and clearance of foreign matter. This is not necessarily true for the immune system, however, as autoimmunity is sometimes induced; the more efficient the effector mechanism is the more serious are the consequences of autoimmunity.

This argument can be most easily developed and examined by considering a particular example. It is known that IgG-mediated complement-dependent lysis requires the binding of two IgG molecules close together on the surface of a cell.[1] This requirement makes such lysis inefficient or even ineffective against cells with a low density of foreign sites (as examples, to be discussed later, show). Such a requirement is not disadvantageous, however, if the IgG response is only induced by a cell with a high density of foreign sites and if sufficient IgG antibody has been synthesized; this type of requirement of antigen recognition in effector function, moreover, has the advantage of minimizing the consequences of IgG autoimmunity. Consider a foreign cell that induces IgG immunity and that bears a determinant s that is similar or idential to one on a self cell. The anti-s IgG auto-antibody that may be induced when this foreign cell provokes a response would not damage the self cell via complement-dependent lysis unless the determinant s were unusually heavily represented on the self cell (in analogy with the fact that 'direct' IgG plaque-forming cells are rare).

Consider now a cell with a very low density of foreign sites. Such a cell cannot in general be damaged by an effector mechanism that requires the recognition of the two sites that are close together on the cell surface. An effector mechanism against such a cell can only be efficient if the recognition of only one site, or two sites that can be far apart, is necessary to trigger effector function.

I shall describe later the experimental grounds for suggesting that antigens with few foreign sites induce cell-mediated immunity, and that the antigen-recognition requirements for effector function in cell-mediated immunity may be less stringent than for humoral, particularly IgG, immunity. In addition, there are reasons for believing that IgG antibody is induced by antigens with many, or a high density of, foreign sites (in the absence of complicating factors such as adjuvants). The suggestion that different classes of immunity are induced by different antigens, dependent at least in part on their degree of foreignness, such that the consequences of autoimmunity are minimized while allowing an effective response against the antigens, is thus attractive.

If one accepts these suggestions one can see teleological reasons for a further aspect of the regulation between different classes of immunity. It has been known for some time that the induction of humoral immunity to an antigen inhibits the induction of cell-mediated immunity, and vice versa. (The experimental basis for these statements will be discussed later.) These observations also make teleological sense in the context of the ideas discussed above. Consider again the case of a cell with a high density of foreign sites and that bears a determinant s that is similar to one on a self cell. The IgG antibody induced, when made in sufficient amounts, will be efficient in damaging and clearing the cell; the induction of cell-mediated immunity would not be necessary to clear the antigen and would have the disadvantage that any cell-mediated immunity induced against s would be more damaging to the self cell than would IgG immunity. Hence one can see a teleological reason for the inhibition of cell-mediated immunity when a strong IgG response is mounted.

Consider again a cell with a low density of foreign sites. Such a cell will induce cell-mediated immunity. IgG antibodies would in general be ineffective in damaging it and, if induced, might reduce the efficiency of the effector function of cell-mediated immunity by binding to, and thereby masking, the foreign sites of the antigen. If there were any tendency for IgG antibody to be induced it would be advantageous for the induction of such immunity to be inhibited.

These considerations suggest the following means by which the consequences of autoimmunity can be minimized. There exists a hierarchy of classes of immune responses. The class highest in the hierarchy is induced by antigens with few foreign sites, and the antigen-recognition requirements for effector function are minimal. Any autoimmunity induced in this class would be particularly damaging to self. The lower classes of the hierarchy would normally be induced by antigens with a greater number or higher density of foreign sites, and the antigen-recognition requirements for effector function would be more stringent. Autoimmunity of a class low in the hierarchy would generally be less harmful than that of a class higher in the hierarchy due to the stringent antigen-recognition requirements of effector function. In order to make this scheme more tangible I shall discuss a teleological explanation for certain aspects of the humoral response.

Consider a cell that bears a high density of foreign sites. Such a cell normally induces IgM antibody before it induces IgG antibody. The induction of IgG antibody would not initially be effective in eliminating the cell if all IgG-mediated effector functions have similar antigen-recognition requirements as does complement-mediated lysis; early after infection, before there has been time to induce the synthesis of much antibody, it is unlikely that the low amounts of IgG that could be synthesized would result in two IgG molecules binding close together on the surface of a cell even if the cell displayed a high density of foreign sites. It seems likely that this is why IgM antibody is initially induced. It is certainly more efficient as far as complement-mediated lysis is concerned. Its higher valency, and wider span, make it more efficient than IgG antibody particularly at low concentrations. The price paid for initially inducing the more efficient IgM immunity is that any auto-antibody that arises of IgM type is more likely to cause damage to self-antigens than would IgG antibody of the same specificity. Once sufficient time has elapsed that IgG antibody can be made in large enough amounts, a switch to IgG production occurs. This IgG antibody can be efficient in damaging cells that bear a high density of foreign sites. The advantage of this switch is that any IgG auto-antibodies induced will be less damaging to self components than would IgM auto-antibodies.

In this introduction I have attempted to develop some ideas that, if accepted, put restrictions on integrated descriptions of the cellular interactions of the immune system. I shall assume that any integrated description of how cells interact must:

(a) provide a learning mechanism of self–non-self discrimination;

(b) account for the circumstances under which autoimmunity is induced;
(c) provide a mechanism that explains how different antigens induce different classes of immunity, dependent at least in part on their degree of foreignness; and
(d) describe how the induction of one class of immunity affects the induction of other classes; in particular, provide an explanation for the observed tendency for mutual exclusiveness between the induction of cell-mediated and IgG immunity.

These assumptions allow the development of a view of cellular interactions that is at best incomplete and may be incorrect; but it is one that is at least sufficiently defined to be experimentally tested in several ways.

2. SELF–NON-SELF DISCRIMINATION

The 'theory of self–non-self discrimination' I shall outline here has been discussed elsewhere in some detail.[2,3] I shall therefore try to make the arguments in its favour as concise as possible.

An analysis can be made based on the four assumptions.

(1) The 'clonal selection theory' is correct.
(2) Precursor cells begin to be generated sometime during foetal development, and are thereafter generated throughout life.
(3) Every precursor cell that is inducible is also paralysable and vice versa.
(4) An understanding of the antigen-specific steps in the induction and paralysis of precursor cells must lead to an explanation of self–non-self discrimination. This assumption is a statement of the belief that paralysis is relevant to an understanding of why strong immunity to self is not commonly found.

These assumptions, with the possible exception of (3), appear to be generally accepted. Assumption (3) will be discussed later (see Section 2.1).

It is worthwhile, before discussing any possible mechanisms of self–non-self discrimination, to consider what property of self components distinguishes them from foreign antigens. The immune system must rely on such a property in order to be able to distinguish self from non-self. The only plausible such property of self antigens that has been proposed is their continuous presence. Evidence in favour of this view was obtained some years ago by Triplett.[4] He showed that an animal, from which a self component is removed for some time, rejects this self component when it is grafted back.

Consider an adult animal that generates two B cells, one specific for a self component S and the other specific for a foreign antigen F. Both F and S are present in the animal at concentrations at which they can interact with their respective precursor cells. If there is to be self–non-self discrimination S must paralyse the anti-S B cell whereas F may, in general, induce the anti-F B cell. This requires that there be molecules, specific for S and/or F, that allow S and F to interact differently with their respective precursor cells. (For the present I ignore the possibility that these molecules could in principle be specific for idiotypic determinants of the receptors on the precursor cells.) It is natural in view of the observations on the carrier effect and the phenomenon of specific B–specific T cell collaboration in the induction of humoral responses to suppose it is the induction of B cells that requires the recognition of two sites on an antigen, and that paralysis of B cells can occur as a result of the binding of antigen alone to their receptors (see Figure 1(a)).

These arguments suggest the following model. Antigen binding to the receptors results in the generation of signal ①, and signal ① alone leads to paralysis. Induction requires the

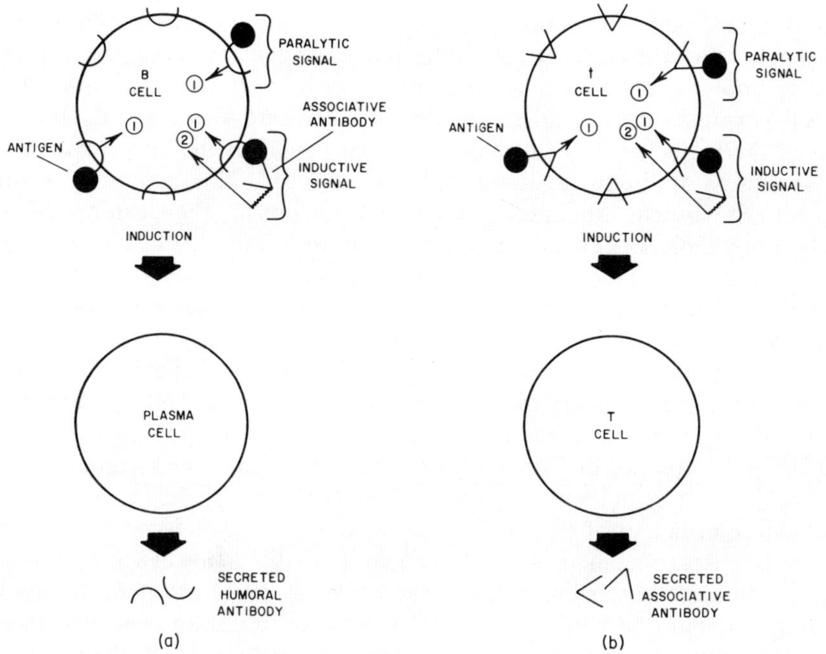

Figure 1

B cell to receive both signal ① and signal ②, the latter being generated as a consequence of the binding of a second antigen-specific molecule, called associative antibody. Associative antibody is believed to be derived from a helper T cell.

According to this model there is a lack of associative antibody specific for self components, and thus anti-self B cells are paralysed, whereas there is associative antibody, or the machinery to produce it, specific for foreign antigens. Associative antibody must be an antigen-specific molecule whose synthesis is paralysable as anti-self associative antibody is not induced.

The above type of analysis, leading to the suggestion that the induction of B cells requires the recognition of two sites on an antigen molecule and that the binding of naked antigen paralyses B cells, can be applied to the induction and paralysis of the synthesis of associative antibody. Just as the presence or absence of associative antibody determines whether antigen induces or paralyses a B cell, so antigen-specific molecules could determine whether antigen induces or paralyses those precursor cells that give rise to helper T cells. Evidence suggests that the induction of associative antibody requires the recognition of two sites on an antigen.[5] Since associative antibody has the capacity, on binding antigen, to initiate the delivery of signal ② to B cells, it is attractive to suppose that it can also initiate the delivery of signal ② to those precursor cells committed to the synthesis of associative antibody (see Figure 1(b)). This suggestion has two appealing features. Firstly, it means that the mechanism of induction and paralysis of B cells and those precursor cells committed to the synthesis of associative antibody, called t cells, are qualitatively the same. Secondly it provides, at the cellular level, a complete qualitative description of the antigen-specific events involved in the induction and paralysis of B and t cells. If another class of specific molecule, say molecule X, regulated the induction/paralysis decision of t cells, one would have to consider how the induction and paralysis of the synthesis of molecule X was regulated. Suppose an antigen-specific molecule *Y* regulated the induction/paralysis decision of the synthesis of molecule *X*; we would again have to consider how the synthesis of molecule *Y* was regulated. This process is unending until we postulate a system where a regulatory molecule, already defined, regulates the expression of a regulatory molecule, e.g. where *X* regulates the induction and paralysis of the molecule *Y*. The simplest of such models is the one favoured above in which associative antibody regulates its own synthesis. This argument is somewhat analogous to St Thomas Aquinas's first cause argument for the existence of God: the existence of anything depends on the existence of other things, and at some stage there must be something, i.e. God, which contains the reason for its own existence.

This theory of induction and paralysis of precursor cells accounts for self–non-self discrimination. Induction of immunity by an antigen requires the presence of at least two cells specific for the antigen, whereas a single precursor cell will be paralysed on contact with antigen. Precursor cells specific for a self component S will thus be paralysed by S one at a time as they arise, whereas precursor cells specific for a foreign antigen F can accumulate in the absence of F; when F impinges on the immune system these cells can collaborate to result in an immune response against F.

The evidence in favour of this model is now rather strong as far as the induction and paralysis of B cells is concerned.[6] I shall limit myself to a discussion of a few observations. B cells can be paralysed by giving rather large doses of antigen to nude mice which are relatively deficient in T cells including those of the helper class, whereas such doses do not paralyse B cells of normal mice.[7] This observation is consistent with the hypothesis that helper T cells determine whether B cells are induced or paralysed in the manner described by the theory. According to the theory an antigen with one foreign site can only paralyse B cells. Haptens can be so coupled to self components that the are weakly, if at all, immunogenic for a humoral response and such antigens can paralyse anti-hapten B cells.[8,9] Antigens similar to self, such as human gammaglobulin (HGG), paralyse murine B cells when administered in a de-aggregated form and are immunogenic when aggregated. De-aggregated HGG paralyses t cells, and the aggregated form induces them.[10] This provides indirect evidence that the qualitative rules for inducing and paralysing t cells are the same as for B cells.

All the general evidence for specific B–specific T cell collaboration is consistent with the theory, but sheds little light on the mechanism of interaction between these cells. The strongest evidence for the two-signal model of induction comes from systems in which signals ① and ② can be generated separately. For example, if the antigen is paralytic when

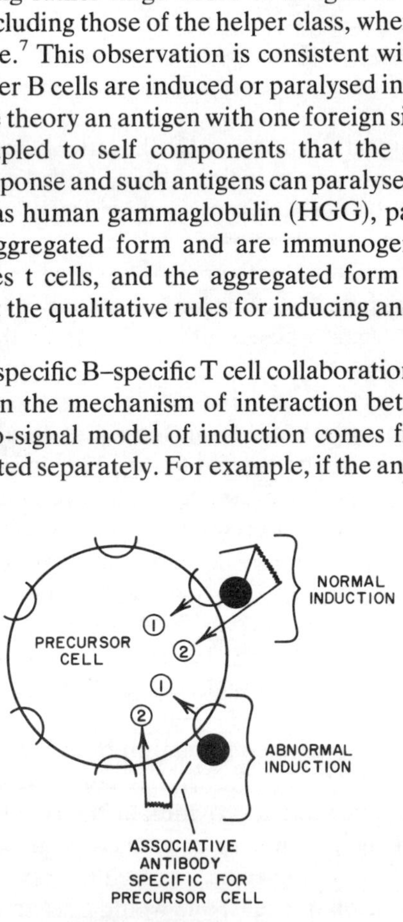

Figure 2

administered under certain conditions it should, according to the theory, be delivering only signal ①. Various agents, such as lipopolysaccharide LPS or allogeneic cells, can prevent this paralysis and, when given with the antigen, provoke an immune response.[11,12] These results are consistent with induction requiring two signals, the one delivered by the binding of antigen to the receptor causing paralysis when generated alone. An interpretation of the action of allogeneic cells in this type of reaction fits naturally into the framework of the two-signal model, and is shown in Figure 2. Consistent with this interpretation is the observation that allogeneic cells exert their effect by virtue of being able to recognize the host cells as foreign.[13] There would not be T cells specific for host B cells in a normal animal, and so this kind of induction would not occur; it is referred to as abnormal induction.

2.1 Autoimmunity

The induction of autoimmunity is sometimes observed when antigens that share some structures with self are administered to an animal. When, for example, antisera to various thymus antigens are raised by the injection of allogeneic thymocytes auto-antibodies against the host's thymocytes are often induced.[14] Consider an antigen that shares some determinants with self components. We can represent such an antigen as having determinants f_1–f_{10} s_1–s_3, say, where the f determinants are foreign and the s determinants are shared with self components. Such an antigen has the possibility of inducing an anti-s_2 B cell, for example, by virtue of the T helper cells specific for the f determinants. Another example of such breaking of tolerance appears to occur in certain streptococcal infections. Streptococcal infections can induce rheumatoid heart disease, and streptococci have antigens similar to some present on heart tissue. The disease is associated with auto-antibodies to heart tissue.[15]

It was assumed above that every precursor cell that is inducible is paralysable, and vice versa. This assumption is particularly crucial as the only other hypothesis proposed for the paralysis and induction of precursor cells, that accounts for self–non-self discrimination, violates this assumption. According to Lederberg's theory precursor cells pass through a stage where antigen can only paralyse them; a cell that is paralysed does not give rise to one that is inducible, and so inducible anti-self cells are not generated, owing to the continuous presence of self components. The main argument against Lederberg's hypothesis is that auto-antibodies are often induced by immunizing with antigens that cross-react with self. This observation is accounted for by the two-signal model, as discussed above, and demonstrates that inducible anti-self B cells are generated. There is also suggestive evidence that some anti-self precursor cells involved in cell-mediated immunity can be similarly induced.[16] This suggests that Lederberg's hypothesis does not describe the maturation sequence of precursor cells involved in cell-mediated immunity.

2.2 Quantitative aspects of the two-signal model

The 'self–non-self discrimination theory' has been given a semi-quantitative description.[3] Here I shall only describe and try to make plausible the main features of the quantitative formulation.

In the absence of antigen neither signal ① nor signal ② can be generated. Under conditions of excess antigen, i.e. when the free antigen concentration is well above the binding constants of both the receptor and associative antibody, these antigen binding molecules will be almost completely saturated. Consequently only a few, if any, inductive complexes of receptor, antigen and associative antibody form, and signal ① is predominantly generated (see Figure 3). The maximum number of inducing complexes occurs at a concentration of antigen around the binding constants of the receptor and the associative antibody. At such concentrations half the receptor and associative antibody molecules will be binding antigen, and it is intuitively plausible that this provides conditions for maximal formation of complexes of receptor, antigen and associative antibody.

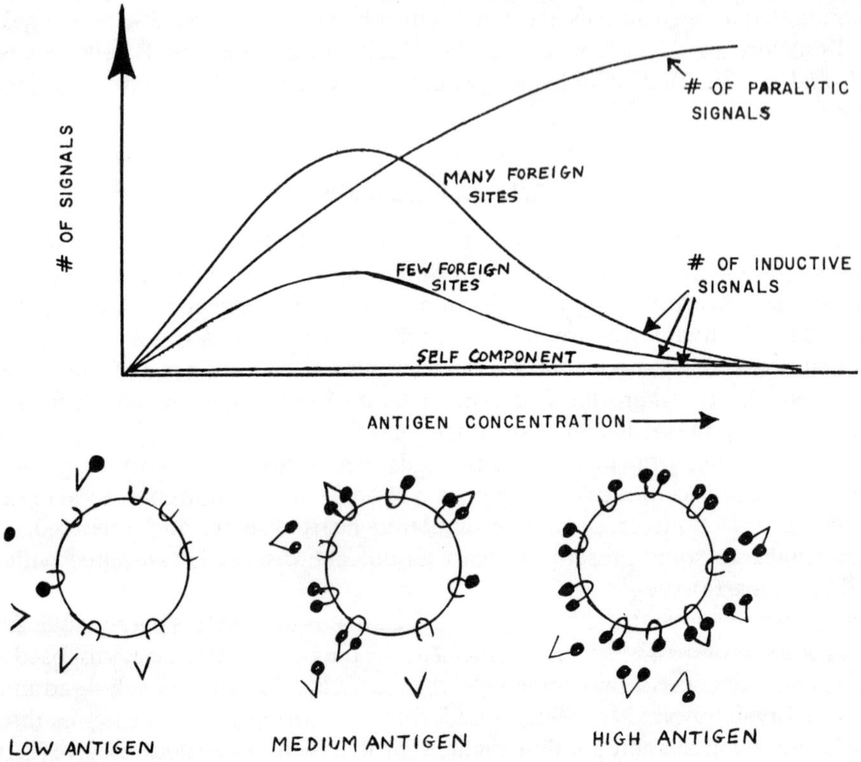

Figure 3

Figure 3 also illustrates another feature. No signals ② will be generated at any concentration of a self-antigen as there are no helper T cells specific for self components. There will, in general, be more T cells specific for an antigen with many different foreign sites on one physical entity compared to an antigen with few foreign sites. The size of the curves, describing how the number of signals ② varies with antigen concentration, will be different for different antigens as shown in Figure 3.

3. A THEORY OF IMMUNE CLASS REGULATION

I shall discuss here a theory that attempts to describe some of the conditions that determine the class of immune response an antigen induces, and to explain how the induction of one class of immunity affects the induction of other classes.[17] The theory will be limited to a consideration of cell-mediated, IgM and IgG immunity.

In order to make an assessment of the theory it is necessary to look both at how it accounts for various observations and at alternative explanations for these observations. I shall first state the theory, then examine those observations it accounts for, and subsequently look at other explanations for these observations. The theory will be referred to as a 'theory of immune class regulation'.

3.1 Assumptions

Certain assumptions made are ones of convenience. The validity of the theory does not depend upon them, as shown elsewhere[17] but they allow a simpler exposition of the theory. These assumptions are:

(1) All classes of precursor cell have the same number of receptors.
(2) There are separate IgM and IgG B cells. A discussion of how the theory can be slightly modified if this assumption is incorrect is discussed later.

The substantial assumptions of the theory are:

(3) The induction and paralysis of all precursor cells is correctly described in a qualitative sense by the 'self–non-self discrimination theory' (i.e. signal ① alone leads to paralysis, whereas signals ① and ② are required for induction). The quality of evidence in favour of this assumption depends on the class of precursor cell considered. For B cells it is strong,[6] whereas the evidence is either indirect or only suggestive when precursor cells for helper T cells or cell-mediated immunity are considered.[5,18,19] The main reasons for accepting this assumption as a working theoretical hypothesis is that it appears to apply to B cells and by analogy it is likely to apply to other precursor cells; furthermore it is one of very few descriptions of the induction and paralysis of precursor cells that accounts for self–non-self discrimination for any class of immunity to which it applies.
(4) The binding constants of all classes of antibody (i.e. all antigen recognition elements) have the same broad range of affinity for antigen. The number of inductive complexes of receptor, antigen and associative antibody, and hence the number of signals ② generated, is governed solely by the number of receptors, the concentrations of associative antibody and antigen, and the binding constants of the receptor and associative antibody for antigen. This assumption, together with the assumption (1) that the number of receptors on all classes of precursor cell are the same, means that for any given concentration of antigen and associative antibody the number of signals ② generated for precursor cells belonging to different classes is in the same range.
(5) There are precursor cells, called t cells, that can be induced to give rise to T cells that produce associative antibody; these t cells are different from those precursor cells that can be induced to allow cell-mediated immunity (CMI) to be detected by the delayed reaction test. Such cells are referred to as P_{CMI} cells. This assumption, for which there is rather strong though indirect evidence, will be discussed later (see Section 6).

3.2 The scope of the theory

The detailed experimental basis of the theory will be described later. It is convenient, however, in order to provide some general context in which postulates of the theory can be understood, to describe the type of evidence on which it is based. The theory is designed to account for the following kinds of observation:

(1) Some antigens induce a predominantly cell-mediated response, whereas others can induce a humoral and/or a cell-mediated response.

(2) The class of response induced, by those antigens that can induce both a cell-mediated and a humoral response, depends on the dose of antigen administered. Low doses of such antigens, for example, commonly favour the induction of cell-mediated immunity.
(3) The induction of one class of immunity can affect the induction of other classes. It is known, for example, that the induction of a humoral response can prevent the appearance of cell-mediated immunity on challenge with an antigen dose that produces a delayed reaction in a naive animal.

3.3 The theory

The theory is based on the semi-quantitative formulation of the 'self–non-self discrimination theory'. The most important idea of the theory is that different classes of precursor cell require the generation of different numbers of signals ② to be induced. The theory is represented in Figures 4 and 5. Figure 4 applies only to the particular concentration of

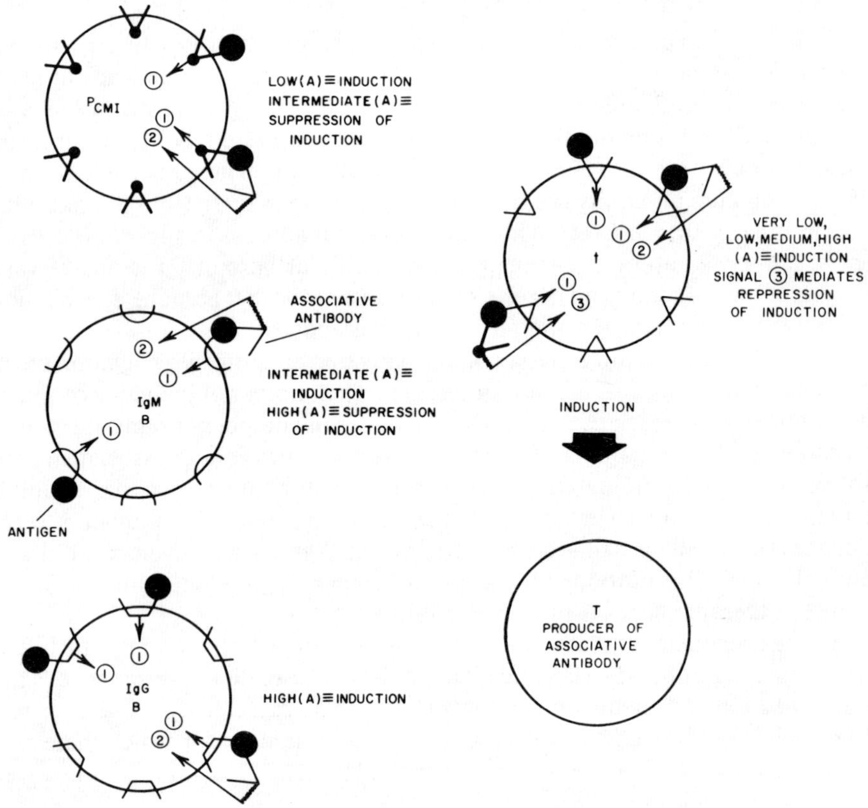

Figure 4

antigen that allows the maximum formation of inductive complexes of receptor, antigen and associative antibody, i.e. at an antigen concentration around the binding constants of the receptor and associative antibody for antigen. At such an antigen concentration the class of immunity induced depends on the concentration of specific associative antibody, referred to as (A). If it is low, few signals ② are generated, and only P_{CMI} cells are induced.

If it is medium, medium levels of signal ② are generated and IgM B cells are induced whereas the induction of P_{CMI} cells is suppressed. If it is high, high levels of signal ② are generated and IgG B cells are induced, whereas P_{CMI} and IgM B cells are suppressed.

The precursor cell for the cell producing associative antibody, the t cell, is induced at all values of (A), unless it receives sufficient amount of another signal: signal ③. Signal ③ represses t cell induction, and is generated as a result of cell-mediated effector antibody, or an antigen specific molecule co-ordinately expressed with cell-mediated effector antibody, binding to an antigen molecule that is also bound by a t cell receptor.

Figure 5 shows what happens when the concentration of associative antibody is held constant and the antigen concentration is varied. If (A) is so low that only P_{CMI} cells are induced when the antigen is at a concentration that results in the maximum formation of inducing complexes, i.e. curve Y, increasing or decreasing the antigen concentration will result in the generation of fewer signals ②, and may lead to paralysis of precursor cells. When (A) is high, such that IgG B cells are induced when the antigen is at a concentration that results in the maximum formation of inducing complexes, i.e. curve X, medium doses of antigen result in the induction of IgG B cells and the suppression of IgM B and P_{CMI} cells. Increasing or decreasing the antigen concentration results in fewer signals ② being generated. If the antigen concentration is increased or decreased moderately, IgM B cells are induced, whereas IgG B cells are not; the P_{CMI} cell is still suppressed. If the antigen concentration is drastically increased only the P_{CMI} cells are induced.[17]

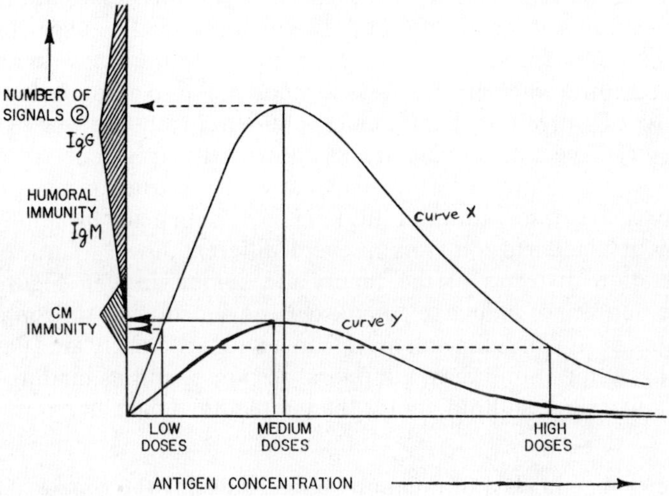

Figure 5

4. EVIDENCE

I shall outline here the evidence that is most obviously pertinent to the theory. Later some observations will be described that can be interpreted in terms of the theory but which are not sufficiently complete to be used as strong evidence in favour of it.

4.1 Characteristics and doses of antigen that determine the class of immunity induced

It has been known for some years that certain antigens induce a good delayed response but induce little, if any, humoral immunity. As emphasized by Pearson and Raffel[20] such antigens are either small in size or bear few foreign sites. Examples of the former are small synthetic copolymers of tyrosine and glutamic acid, and arsenobenzoate-*N*-acetyltyrosine; of the latter, slightly altered homologous serum proteins, a peptide derived from adrenocoricotropin hormone (ACTH) and haptenated self components obtained by skin painting.

Antigens have also been deliberately modified so as to reduce their ability to bind to antibody and hence, presumably, their ability to form inductive complexes. Such treatment is believed to either destroy determinants on the antigen, or greatly reduce the affinity of the antigen for antibody. Antigens can be so modified that they only induce a delayed response. Examples are acetoacetylated flagellin[21] and periodate-treated acetoacetylated sheep red blood cells.[22] All these observations, taken together, provide strong evidence that antigens with few foreign sites induce only cell-mediated immunity.

Some antigens are known to be able to induce both a delayed and a humoral response; the induction of a delayed response is favoured by administering very low doses of antigen which may be given either in complete Freund's adjuvant or saline.[21,23] It has also been observed that the administration of high doses of antigen can give rise to a delayed response, the best example being the fragment A of flagellin.[21] These observations show that low and high doses of an antigen, that induce a humoral response at medium doses, can induce a delayed response.

The attraction of the hypothesis that the induction of a cell-mediated response requires the generation of fewer signals ② than does the induction of a humoral response is that this one hypothesis accounts for two classes of observation, namely: antigens with few foreign sites induce a predominantly cell-mediated response, and antigens with more foreign sites induce such a response when given at very low doses. It also accounts for the observation that a high dose of antigen, that can induce a humoral response when given at medium doses, induces cell-mediated immunity as shown with fragment A of flagellin. This observation, however, is not experimentally known to be true for many antigens.

The most extensive observations of this kind have been made by administering various modified forms of flagellin to adult rats of several different doses.[21] These observations and their interpretation in terms of the theory are summarized in Figure 5. The curve describing the number of inductive complexes as a function of antigen concentration is higher for fragment A of flagellin (curve X) than it is for acetoacetylated flagellin (curve Y). The latter antigen does not induce a humoral response, and its binding to anti-flagellin antibody is about a thousandfold poorer than unmodified flagellin.

4.2 The change in class of immunity induced during the course of a response

The number of t and T cells, and consequently the amount of associative antibody specific for an antigen, depends in general on the number of foreign sites on the antigen. There will in general be fewer cells specific for an antigen with few foreign sites than for an antigen with many foreign sites. Antigens with few foreign sites will tend to induce cell-mediated immunity. Consider what happens if strong cell-mediated immunity is established before t cells have been sufficiently induced to raise (A) to a level where, at the concentration of antigen present, the induction of P_{CMI} cells is suppressed. The cell-mediated effector antibody will, in the presence of antigen, allow signal ③ to be generated, and the induction of t cells will be repressed. Stable cell-mediated immunity will be established.

For an antigen with more foreign sites the initial concentration of (A) will tend to be higher, and the likelihood of stable cell-mediated immunity being established will be low unless the antigen is at either very high or very low concentrations (see Figure 5). For moderate doses of such antigens cell-mediated immunity, even if initially induced, is likely to be transitory and suppressed as the concentration of (A) increases. IgM B cells will be induced, and if the concentration of (A) is allowed to rise sufficiently IgM B cells will be suppressed and IgG B cells will be induced. If, as some suggest, IgM B cells can give rise to IgG producing cells, the switch from IgM to IgG production would occur when IgM B cells receive high levels of signal ②.

Consistent with the theory is the evidence that the induction of IgG antibody synthesis is more thymus-dependent than is IgM.[24–26] Neonatal thymectomy, for example, affects the IgG response more than the IgM response. The T cell helper activity present in a primed animal is greater than that of an unprimed animal, and the ratio of IgG to IgM antibody induced is much greater in a secondary response than it is in a primary. These observations suggest that more specific T cells may be required to induce an IgG response than an IgM response. Further evidence is in accord with this suggestion and provides evidence that the function of the T cells required is to deliver signal ②. For example, allogeneic spleen cells can cause TGAL-non-responder mice, which under standard conditions of challenge only synthesize IgM antibody, to switch to IgG antibody production.[27] The anti-hapten response to a hapten-polysaccharide conjugate, which is usually of the IgM class, can similarly be caused to switch to the production of anti-hapten IgG antibody.[28] Thus the abnormal delivery of signals ② to precursor cells can allow IgG B cells to be induced under conditions where they, as opposed to IgM B cells, would otherwise not be.

The change in class of immunity induced during the course of a response can be accounted for in a rough manner as described above. Such a description, however, is not completely satisfactory as the concentration of antigen in the environment of precursor cells will change during the course of the response; the way in which the number of signals ② changes with time will not therefore depend solely on how the amount of associative antibody changes. Indeed it is possible within the context of the theory that cell-mediated immunity can be induced after an IgG response. The removal of antigen by IgG-mediated mechanisms could leave significant but very low concentrations of antigen that induce a delayed response.

4.3 The exclusiveness between different classes of immunity

The evidence for the exclusiveness between IgM and IgG antibody synthesis is so well known that no evidence will be cited in its favour. The only point to note is that this exclusiveness is not complete.

Many old observations demonstrate that the induction of a humoral response tends to preclude the induction of a delayed response. Pretreatment of adults with antigen in saline, for example, can inhibit the induction of a delayed response by a subsequent challenge of the antigen in complete Freund's adjuvant, and this inhibition is often accompanied by the induction of humoral antibody.[29–34] The tendency for exclusiveness between the delayed and humoral responses can also be seen during the course of the response to one antigen.[23] These observations appear somewhat analogous to the observation that less IgM antibody is sometimes produced during a secondary response compared to a primary response, and the switch to IgG synthesis is more rapid. The explanation for both phenomena is that the pretreatment of the animal with antigen resulted in increased numbers of specific T and t cells such that the induction of P_{CMI} and IgM B cells was more rapidly suppressed on subsequent challenge.

Fewer observations have been deliberately made to see whether the induction of a cell-mediated response can inhibit a humoral response. It seems likely, however, that several unresponsive states at the humoral level are due to the induction of cell-mediated immunity as discussed later. The best examples of cell-mediated immunity inhibiting the induction of a humoral response are provided by fragment A of flagellin and acetoacetylated flagellin. The former, if given at low or high doses to adult rats, induces a cell-mediated response that renders the rats partially unresponsive, at the humoral level, to a challenge with flagellin.[21] Acetoacetylated flagellin has a similar effect. The 'theory of immune class regulation' suggests reasons why it may often be difficult to observe a dramatic inhibitory effect of cell-mediated immunity on a humoral response if it is looked for in the customary fashion and suggests a procedure that should reveal more clearly this inhibitory activity (i.e. repression) as described later (see Section 7.3).

5. FURTHER ASPECTS OF REGULATION

The theory, as described above, is a minimal one in that only those regulatory elements required to provide a coherent description of what is currently known about the control between cell-mediated, IgM and IgG immunity have been discussed. It is worth while to discuss certain detailed aspects of the theory that are not essential to account for these rather general observations.

5.1 The generation of signal ③

It is quite likely that, just as there are different classes of IgG antibody, there are different classes of antibody involved in phenomena regarded as manifestations of cell-mediated immunity. The generation of signal ③ could be initiated by the binding of any antigen-specific molecule to antigen with one provision. Such a molecule should be induced under the same conditions as cell-mediated effector antibody for only then can the repression it mediates account for the control of delayed hypersensitivity. It therefore may be cell-mediated effector antibody and I shall for convenience refer to it as such.

The location of the antibody that initiates the generation of signal ③, during the repression of t cell induction, is not crucial to an understanding of the general observations discussed above. It seems likely, however, that it is a cytophilic molecule. This would allow it to interact more easily with antigen that is bound to a t cell receptor than if it were associated with a unispecific cell.

5.2 Signal ③ and other precursor cells

The theory has been made as conservatively as possible and the existence of controlling signals has been postulated only where necessary. It is quite possible that signal ③ also acts on B and P_{CMI} cells though there is no evidence at present to suggest this.

5.3 Suppression of IgG B and t cells

It is not clear whether the excessive generation of signals ② can suppress IgG B and t cell induction.

5.4 The timing of the generation of regulatory signals to precursor cells

During the normal course of a humoral response, in which IgM antibody synthesis is followed by IgG synthesis, the number of signals ② that each precursor cell receives will initially increase smoothly with time. When one considers what might happen at the biochemical level as a result of the generation of signals ① and ② it is difficult to guess whether the way in which the number of signals ② changes with time is crucial in determining whether a precursor cell is induced or whether only the average number of signals ② generated is important. One could imagine, for example, that a precursor cell that receives a number of signals ② that increases linearly from none to twenty over a two-day period is induced, but that a precursor cell which receives the same average number of signals ②, namely ten, but constantly over this time, is not. The point I wish to make is that the way in which the number of signals ② is generated over a time comparable to the induction period may well be important in determining whether induction takes place.

It is also plausible that a cell that has already received signal ③ for some time may require more signals ② to be induced than a virgin cell that is confronted with a certain ratio of signals ② and ③.

5.5 Other classes of immunity

The theory, as described above, does not take account of the existence of subclasses of IgG antibody, nor has the regulation of IgE and IgA antibody been considered. Where such classes fit into the hierarchy is at present unclear. There are reasons for the existence of different classes of immunity other than the teleological ones proposed in the introduction. Different classes are required to perform different effector functions (e.g. some classes fix complement whereas others do not), and their induction is known to depend on the site in the body where immunity is needed, e.g. IgA immunity being preferentially induced by antigens in the gut.

5.6 Cell-mediated autoimmunity

On the basis of the theory one would expect cell-mediated autoimmunity to occur and to be damaging to self. The frequency with which such autoimmunity occurs must depend on the rate at which P_{CMI} cells are generated and on the frequency and duration of cell-mediated responses against antigens that cross-react with self. The reputation of cell-mediated immunity as being the most damaging type of autoimmunity is in accord with the teleological ideas of the theory.

The ease with which the theory can explain the breaking of the unresponsive state to self at the cell-mediated level provides some grounds for uneasiness and may make one question the reasonableness of the theory. It is possible, however, to think of means by which the occurrence of such autoimmunity could be considerably lessened. I shall give one example of such a mechanism. My purpose in doing so is mainly illustrative; if such mechanisms are conceivable then the above grounds against the theory are not so serious.

Suppose virgin P_{CMI} cells require the generation of 10 signals ② to be induced. Consider an antigen, that induces a cell-mediated response and cross-reacts with self, arising as the result of a chronic infection. Such an antigen will induce t cells until there is a level of associative antibody at least sufficient to generate ten signals ② at the concentration of antigen present. Once P_{CMI} cells are induced the induction of t cells will be repressed. A stable (or oscillating) state should arise in which the level of associative antibody is sufficient to generate, let us say, 15 signals ②. This figure must obviously be above ten.

Suppose memory P_{CMI} cells require the generation of fewer signals ② to be induced than do virgin P_{CMI} cells. As memory cells are generated the level of associative antibody would fall to another stable level. If, for example, memory cells only require the generation of one signal ② to be induced the level of associative antibody might fall to a stable level at which 1·5 signals ② were generated. Newly arising virgin P_{CMI} cells would not now be induced as this would require the generation of ten signals ②. Thus an ongoing cell-mediated response would not result in the induction of a newly generated virgin P_{CMI} cell that has specificity for a cross-reacting self site.

One means by which a memory cell could require less associative antibody than a virgin cell to be induced is by having a larger number of receptors. This possibility is reasonable as some evidence suggests that memory B cells have a higher density of receptors than virgin B cells.[28,35]

6. ARGUMENTS FOR AND ALTERNATIVES TO THE THEORY

I shall describe here some of the alternative explanations that have been, or might be, put forward to account for the observations on which the theory is based, and the reasons why I favour the interpretations that the theory provides.

6.1 Cell-mediated effector cells and T helper cells

It has often been suggested that helper T cells and the effector cells involved in cell-mediated immunity are one and the same cell. This seems unlikely at face value as the expression of help is not co-ordinately expressed with cell-mediated immunity. Under conditions where a secondary humoral response is under way, for example, delayed hypersensitivity usually cannot be elicited. In order to account for this fact it could be argued that under such circumstances the cell-mediated effector function, as opposed to helper function, was being masked by either antigen or humoral antibody. This is unlikely as peritoneal cells from animals expressing no delayed hypersensitivity but a good humoral response are unable to confer a delayed hypersensitive state on a normal animal, whereas peritoneal cells from an animal expressing delayed hypersensitivity can.[21] It thus seems likely that the cell-mediated effector cell and the T helper cell are separate entities, and have different precursor cells as assumed in the formulation of the theory.

6.2 The induction of cell-mediated immunity

It has been suggested that the induction of cell-mediated immunity does not require the recognition of two sites on an antigen but that direct interaction of antigen with a precursor cell can result in its induction. Good evidence, recently obtained, shows that precursor cells for cytotoxic T cells, regarded as a manifestation of cell-mediated immunity, are different from t cells.[35,36] There appears, furthermore, to be synergy between such precursor cells and a help-like T cell in the induction of cytotoxic T cells.[18] No evidence that I know of bears directly on whether cellular cooperation is required for a delayed response.

6.3 Suppression of cell-mediated immunity and the IgM to IgG switch

It could be suggested that a humoral response prevents the induction of cell-mediated immunity by causing the removal of antigen. This possibility is unlikely to be correct on both theoretical and experimental grounds. As the induction of cell-mediated immunity is favoured by administering lower doses of antigen than are required for the induction of humoral immunity it seems unlikely that humoral antibody can prevent the induction of cell-mediated immunity, by removing antigen, *under conditions where it is itself induced.* In agreement with this argument is the observation that passively administered antibody, given in varying but small amounts, reduces the humoral response and increases the delayed response to an injection of antigen that would normally give a predominantly humoral response.[21] The finding that specific T cells inhibit the induction and elicitation of cell-mediated immunity, as assayed by skin painting,[37] is consistent with the idea that helper T cells can suppress the induction of cell-mediated immunity. According to the theory the cessation of IgM synthesis, during the IgM to IgG switch, is due to IgM B cells being suppressed on receiving many signals ②. Other alternatives could be imagined. It could be suggested that the initial IgG synthesized aids the removal of antigen and thereby preferentially exerts a feedback effect on IgM B cell induction. This is, however, unlikely if the induction of IgM B cells requires the formation of fewer inductive complexes than the induction of IgG B cells. Lowering the antigen concentration to low levels by IgG-mediated feedback would be expected to affect the induction of IgG B cells before it affects the induction of IgM B cells. Indeed, low doses of antigen appear to favour the induction of IgM rather than IgG antibody.[38] It therefore appears unlikely that the removal of antigen by IgG antibody can cause the shut-off of IgM induction *under conditions where IgG antibody itself is induced.*

6.4 The inhibitory effect of cell-mediated immunity on a humoral response

There is some evidence that low doses of antigen or modified antigen, that induce an almost pure delayed response, can also induce helper function, i.e. induce t cells.[39,40] This evidence supports the proposition that t cells can be induced under conditions where few signals ② are generated. This proposition is required by the theory as helper T cells are postulated to be necessary for the induction of P_{CMI} cells, and so t cells should be inducible on receiving at least as few signals ② as P_{CMI} cells require to be induced.

It is essential, given the proposal that the generation of too many signals ② can suppress the induction of P_{CMI} cells, that cell-mediated effector antibody can inhibit, i.e. repress, the induction of help; otherwise cell-mediated immunity could not be stable. As far as I am aware no other coherent explanation for the inhibitory effect of cell-mediated immunity on the induction of a humoral response has been suggested.

7. THE RELEVANCE OF THE THEORY TO OTHER OBSERVATIONS

7.1 Unresponsive states

According to the theory there can be different reasons for an animal being unresponsive at the humoral level. Animals in which B and t cells have been paralysed will be unresponsive, as will those animals that have strong cell-mediated immunity such that t cell induction is fully repressed. I shall discuss three cases of unresponsive animals.

7.1.1 *Fowl gammaglobulin in mice*

De-aggregated fowl gammaglobulin (dFGG) induces a clean unresponsive state for IgG antibody induction if given to CBA mice at the massive dose of 15 mg. Unresponsive mice can be shown to contain anti-FGG B cells.[41] The same dose of dFGG paralyses B cells in (t and T cell deficient) nude mice.[7] It seems highly plausible that the differences between the effects of dFGG on B cells in CBA and nude mice is due to the differences in the numbers of t and T cells present in these two strains. B cells in nude mice are paralysed as they receive very few, if any, signals ② and the observation that B cells in CBA mice are not paralysed strongly suggests that they receive a sufficient number of signals ② to prevent paralysis. It is therefore expected on the theory that the administration of dFGG to CBA mice allows some signals ② to be generated, and that a class of immunity will be induced that is high in the hierarchy, probably cell-mediated immunity. Such dFGG-treated mice would be expected to contain repressors, and it has been shown that spleen cells from such mice can specifically inhibit the humoral response of a naive animal to FGG. Furthermore, these inhibitory cells preferentially affect the induction of IgG as opposed to IgM antibody. This is expected on the theory as repressors will inhibit the induction of T cell help and the induction of IgG synthesis is postulated to be more help-dependent than is IgM synthesis.

7.1.2 Human gammaglobulin in mice

Human gammaglobulin (HGG), de-aggregated in a swinging bucket centrifuge, renders A/J mice unresponsive at the humoral level to a dose of aggregated HGG that induces a good response in naive animals. Mixing studies have shown that dHGG-treated A/J mice have a deficiency of both B and t cells specific for HGG[10] and parabiosis between normal and unresponsive mice shows the unresponsive state to be recessive.[42] The unresponsive state of these mice thus appears to be principally due to clonal deletion and is analogous, according to the theory, to the normal state of the immune system to self components.

7.1.3 *Sheep red blood cells in rats and mice*

Unresponsive states to sheep red blood cells (SRBC) can be induced in rats and mice.[43,44] Rats neonatally treated with enormous doses of the antigen display a dominant unresponsive state as shown by transfer experiments. Cells from an unresponsive rat, when given in sufficient numbers, can inhibit the response of normal cells in an irradiated animal.[45] Such rats can be induced to make a large and very rapid response to SRBC when allogeneic spleen cells are given with the challenge of SRBC, though the rats are unresponsive if challenged with SRBC alone. The peak of this large IgM response occurs around 48 hours after challenge and has been shown to be due to host cells.[13] This response shows that IgM B cells exist in the unresponsive animals, and suggests that they are in some sense primed. This is consistent with the view that the IgM B cells in the unresponsive animal receive some signals ② but insufficient to give rise to IgM secreting cells. The characteristics of these animals, and the conditions required to induce the unresponsive state, are consistent with the unresponsive state being due to the induction of cell-mediated immunity and repressors.

7.2 'Low-zone paralysis'

It has been observed that some antigens, if given at a dose lower than that required to induce a humoral response, render animals partially unresponsive at the humoral level. This phenomenon is commonly referred to as 'low-zone paralysis'. It seems very likely that such a state is not due to paralysis but rather to the induction of cell-mediated immunity. The classic example is bovine serum albumin (BSA) in mice. Mitchison has estimated that a molarity of BSA of 10^{-7} M in the environment of precursor cells gives rise to the induction of antibody synthesis, whereas a concentration of 10^{-8} M results in 'low-zone paralysis'.[46] Another example of 'low-zone paralysis' is known to be dominant. The dominant state can be detected by mixing unresponsive and normal cells and challenging with antigen. At least some of the inhibitory cells are θ bearing cells.[47,48] The theory predicts that such mice should have cell-mediated immunity to the antigen.

7.3 Jumping out of the cell-mediated state

Cell-mediated immunity can be stably induced by the repeated injection of low doses of antigen as exemplified by fragment A of flagellin in adults rats. When rats so treated are challenged with an immunogenic dose of flagellin the humoral response observed is less than that in an untreated rat, but is by no means negligible. Similar observations have been made on rats given acetoacetylated flagellin. The induction of a humoral response is quite consistent with the theory. The induction of t cells is repressed in rats that have stable cell-mediated immunity to low doses of fragment A. On challenge with flagellin, given at a much higher dose than the fragment A, the associative antibody present will allow the generation of many more signals ② than were generated by the low dose of fragment A. Consequently the induction of cell-mediated immunity may well be partially *suppressed.* The *repression* of t cell induction may be thus partially lifted, and hence t cells may be induced to allow the induction of a humoral response. The phenomenon of obtaining a humoral response to an antigen, against which stable cell-mediated immunity has been induced, on dramatically increasing the antigen dose will be referred to as 'jumping out of the cell-mediated state'.

The phenomenon of jumping is relevant to the conditions under which one might observe 'low-zone paralysis'. For BSA the concentration required for 'low-zone paralysis' only differs from the immunogenic concentration by a factor of ten, and so jumping would be expected to be undramatic. It is expected to be generally easier to see the phenomenon of 'low-zone paralysis' if the lowest dose of antigen that can induce a measurable humoral response is employed for challenge.

Some unresponsive states at the humoral level, established by giving massive doses of antigens such as dFGG, are probably due to the induction of stable cell-mediated immunity. It may be that 'high dose establishment' of cell-mediated immunity is often more resistant than 'low-dose establishment' to challenge with a medium dose of the antigen that induces a humoral response in a normal animal. As the antigen is cleared after the administration of massive doses its concentration decreases in a continuous manner, and may therefore avoid the jumping that could occur by suddenly moving from high to medium concentrations of antigen.

7.4 Class switching

Methods for converting a cell-mediated response to a humoral one, and vice versa, are suggested by the theory.[17] Consider an antigen Q against which a pure cell-mediated response has been induced. Immunizing such an animal with an antigen R, that does not cross-react with Q and induces a strong humoral response, will result in the induction of t cells specific for R. On administering the conjugate QR to the animal the induction of P_{CMI} cells specific for Q will be suppressed by the large amounts of associative antibody specific for R, and consequently the repression of the induction of t cells specific for Q should be lifted. A subsequent challenge with Q should result in a humoral response to Q.

It should also be possible to strongly favour the induction of cell-mediated immunity to an antigen D against which the animal normally makes or is already making a mild humoral response. If the animal is induced by various means to have strong cell-mediated immunity to an antigen E, the administration of the conjugate DE will favour the conversion of the anti-D response to one of pure cell-mediated immunity. Cell-mediated effector antibody specific for E should repress the induction of t cells specific for D in the presence of DE. T cells most probably have a fairly short half-life, and so the level of associative antibody specific for D will decay. Once it falls below a certain level cell-mediated immunity will be induced by D itself. This type of procedure is relevant to an interpretation of some phenomena observed in studies on the genetics of the response (discussed under Ir-1 phenomena).

7.5 Suppressors

The administration of an immunogenic dose of an antigen that induces a good humoral response should result in the induction of T helper cells specific for the antigen. Such cells, if given to naive animals in sufficient numbers, are expected to suppress the induction of those classes of immunity that are high in the hierarchy and may also enhance others lower in the hierarchy.

It has been observed that thymocytes or spleen cells from mice, primed with sheep erythrocytes to provoke a strong humoral response, contain cells that inhibit certain responses in normal mice. Such cells, when given to recipients, inhibit the IgM anti-hapten response that is normally induced on challenge with the hapten coupled to sheep erythrocytes. Particularly dramatic inhibition is observed seven days after challenge if the cells are given on day five. The inhibition appears to be specific and is abolished by treating the cells with anti-θ serum and complement.[49] It seems likely that this inhibition is an example of suppression.

It has also been shown that T cells specific for the trinitrophenyl group (TNP) can specifically inhibit the delayed reaction to this group in a system where sensitization is achieved by skin painting. It is likely that such T cells are producers of associative antibody that suppresses the induction of P_{CMI} cells.[37]

7.6 Ir-1 phenomena

The Ir-1 genetic region, best characterized in mice and guinea pigs, controls the magnitude of the humoral response in an antigen-specific manner. The simplest interpretation of Ir-1 gene function is that it controls the amount of help available either by coding for associative antibody itself or for an antigen-specific molecule that regulates the induction of associative antibody. Another interpretation is that it affects the efficiency of B–T cell collaboration;[50] such an interpretation is difficult to reconcile with the observation that mice that are low responders for one antigen are high responders for another. The lesion in low responders appears operationally to be due to a lack of sufficient specific T cells. The theory of immune class regulation may be relevant to an understanding of the immunological characteristics of responder and non-responder mice. I shall discuss four cases of Ir-1 controlled responses.

7.6.1 *PLL in guinea pigs*

Non-responders do not mount a humoral response to haptens coupled to poly-L-lysine, whereas responders do. Non-responders respond to haptens conjugated to immunogenic carriers and cannot, as contrasted to responders, be induced to make a delayed response to PLL.[51] These observations suggest that very few helper T cells specific for PLL are induced in non-responder guinea pigs.

7.6.2 *TGAL in mice*

The Ir-1 region controls the response to an antigen called TGAL, made from tyrosine, glutamic acid, alanine and lysine. 'Non-responders' make a small IgM response when given the standard challenge of TGAL; responders make both an IgM and IgG response. Non-responders, once challenged, become unresponsive in the sense that they do not make another IgM response on re-challenge.[52]

'Non-responders' can be induced to make an IgG response when TGAL is coupled to an immunogenic carrier as well as by repeated immunization with a lower dose of TGAL than is employed in the standard challenge.[26] This latter observation, as well as the fact that non-responders make an IgM response, suggests that non-responders have some t and T cells specific for TGAL, but not as many as responders. It seems likely that non-responders, on challenge with the standard dose of TGAL, lapse into a purely cell-mediated state and are hence resistant to re-challenge as far as an IgM response is concerned. The most striking prediction of this interpretation is that non-responders to TGAL should display greater delayed hypersensitivity to TGAL than responders do after challenge with a *standard* dose of TGAL.

7.6.3 *GAT in mice*

Another Ir-1 gene controls the response to GAT, a polymer made from glutamic acid, alanine and tyrosine. Gershon[53] was the first to suggest that 'non-responders' were not completely deficient of thymus-derived cells specific for GAT on the grounds that GAT stimulated the incorporation of thymidine by non-responder spleen cells. GAT does not induce a humoral response when incubated *in vitro* with 'non-responder' spleen cells, though a GAT humoral response does occur when GAT is conjugated to an immunogenic carrier. The presence of spleen cells, from 'non-responder' mice that have been given GAT, strikingly inhibits the response of normal 'non-responder' spleen cells to the GAT conjugate.[54] If 'non-responders' have a level of associative antibody that allows them to make a cell-mediated response on challenge with the standard dose of GAT, as is suggested here, these results are expected. 'Non-responder' cells, primed with GAT, will have cell-mediated immunity to GAT, and consequently will tend to repress the induction of any t cell specific for an epitope that is on or linked to GAT. Thus the GAT conjugate is less able to induce specific associative antibody, in the presence of GAT-primed 'non-responder' cells, and hence the presence of these cells inhibits the humoral response.

7.6.4 *Ir-1 region and the control of other responses*

When mice are challenged with varying but low doses of a weak antigen it is found that the lowest dose able to induce a humoral response is different for different strains.

A strain of mouse able to respond to a rather low dose of one antigen is sometimes found to require a relatively high 'low dose' of a different antigen in order for a humoral response to be induced. The ability to respond to the lower dose is H-2 associated, and appears to be governed by the Ir-1 complex.[55–57]

These observations are expected on the theory if different strains of mice have significantly different levels, or ability to produce, associative antibody specific for such antigens. Mice of a strain that has a comparatively high level of associative antibody for a particular antigen are expected to be able to mount a humoral response to lower doses of the antigen than mice of a strain with a lower level. It is inevitable, in terms of the theory, that the administration of antigen at doses slightly subimmunogenic for a humoral response should result in the induction of delayed hypersensitivity.

7.7 Stable IgM synthesis and 'thymus-independency'

Some antigens induce a predominantly IgM response that appears by various criteria to be 'thymus-independent'. Examples of such antigens are pneumococcal type III polysaccharide (SIII),[58] polyvinyl pyrrolidons (PVP) and lipopolysaccharide (LPS),[59] and trinitrophenylated mouse red blood cells (TNPMRBC).[60] It may be that such antigens have some intrinsic property that allows them to generate the equivalent of signal ② in the absence of associative antibody; in this case they really would be thymus-independent.[61] Here I shall consider another possibility.

Most of these antigens are polymeric and are not readily metabolized in mice. Their polymeric structure probably allows them to be cleared very efficiently and rapidly by IgM antibody due to the high valency of this molecule. It is known that SIII, for example, can continuously aid the removal of antibody, and is itself kept at low concentrations by this process.[62] The concentration of such antigens in the environment of precursor cells is expected to fall rapidly and reach a semi-steady state soon after IgM antibody begins to be synthesized. This in itself could provide a partial explanation for the lack of induction of IgG synthesis. The concentration of antigen could be rapidly decreased to a low level and a steady state be established before helper T cells are induced sufficiently to allow the induction of IgG B cells.

The IgM response to SIII,[63] PVP[64] and TNPMRBC[60] have all been shown to be partially inhibited by thymus-dependent cells. Irradiated mice reconstituted with bone marrow, for example, make a better anti-TNP response on challenge with TNPMRBC than if reconstituted with bone marrow and thymus cells. Nude mice make a better response to TNPMRBC than their wild-type littermates, and anti-lymphocyte serum, when given to normal mice, enhances their anti-TNP response.[60] It seems likely, as the exclusiveness between the induction of IgG and cell-mediated immunity is not complete, that these antigens induce cell-mediated immunity as well as IgM antibody. A classical way of obtaining delayed type hypersensitivity is to couple haptens to self-components. It would not be surprising if TNPMRBC induced cell-mediated immunity. Suppose these antigens induce a cell-mediated response. An explanation for the observed enhancing effect on the IgM response on decreasing the T cell population, as well as the apparent thymus-independency of the response, appears then to be possible. According to this view nude mice, for example, have a deficiency in the ability to provide helper T cells and an even greater deficiency in the ability to induce cell-mediated immunity and repressors. The anti-TNP response induced in nudes by TNPMRBC is thus higher than in wild-type mice. In addition the repressors associated with cell-mediated immunity could contribute to an inhibition of the induction of more helper T cells, and thus be partly responsible for the lack of a switch to the synthesis of IgG antibody.

These views of the regulation of the response to such antigens are rather similar to those suggested by Baker *et al.*[63] based on his studies with SIII.

8. 'DOMINANT' TOLERANCE AND ANTI-IDIOTYPE NETWORKS

Two views concerning rather general aspects of the immune system have recently been much discussed. These views are: (a) maintenance of tolerance to self components occurs by an inhibitory mechanism active on anti-self precursor cells and (b) lymphocytes form an interacting network of cells due to the reactivities of some lymphocytes for idiotypes present on receptors of other lymphocytes.[65,66] Sometimes these views are combined to suggest that maintenance of tolerance to self could be due, at least in part, to immune reactivity against the idiotypes of those receptors with anti-self activity. These views, if correct, would require the theories discussed here to be modified drastically or, more probably, abandoned.

It appears attractive to postulate mechanisms positively inhibiting those precursor cells with anti-self reactivity when considering means of minimizing the induction of autoimmunity. In trying to assess the plausibility of such mechanisms it is worth while to attempt to see what consequences they might have.

I shall initially ignore the possibility that dominant forms of tolerance are due to anti-idiotypic reactivity against those cells with anti-self activity. Consider two precursor cells generated in an adult, one with specificity for a self-component S and the other specific for a foreign antigen F. If there is a dominant form of tolerance there must be a certain class of molecules specific for S, as opposed to F, that inhibits the anti-S precursor cell. Consider the consequences of such an inhibitory mechanism on the induction of a response against a slightly modified self-antigen. If the inhibitory mechanism has significant strength the response against the foreign site of such an antigen would be inhibited. In order to avoid this the inhibitory mechanism would have to be very weak indeed. The plausibility of such a mechanism thus appears to depend on one's assessment of the importance of responses against slightly altered self. Viruses often cause small changes to the surface of cells and such altered cells induce immune responses. For this reason such an inhibitory mechanism seems unappealing to me.

A main impetus for the idea of a dominant mechanism of self-tolerance has undoubtedly been the discovery of dominant unresponsive states. Most of the observations on such states can be accounted for by the 'theory of immune class regulation' discussed here. It is clear, however, that completely compelling evidence on whether tolerance to self is due to an inhibitory mechanism or is due purely to deletion is lacking.

The type of analysis discussed above depends on distinguishing the two precursor cells specific for S and F by virtue of the receptor's specificity for antigen. Consider how one might try to construct a detailed form of a network theory that accounts for a reliable inhibitory mechanism acting on anti-self precursor cells. Suppose the anti-S precursor cell has idiotype i_s and the anti-F cell idiotype i_f. The induction of a reaction against i_s but not i_f must involve some way of distinguishing i_s from i_f. It would seem that this requires S and F to bind to their respective precursor cells and for a class of molecules specific for S or F to bind to the appropriate antigen. In this way the anti-S and anti-F cells can be distinguished and a reaction against the anti-S but not against the anti-F cell can be induced. This inhibitory mechanism has similar features to the one discussed above in that the response to the foreign sites on an antigen that was slightly altered self would be inhibited.

A network theory is not necessary *a priori*. It is possible that as new idiotypes arise they paralyse precursor cells able to recognize them. In an adult, for example, a new idiotype may consist of only one new 'foreign' site and thus lead to paralysis of the anti-idiotype precursor cells.[67] It seems that a network in the immune system would cause havoc unless some precise rules governed its behaviour. If some rules that are molecularly plausible can be suggested or found for controlling the network the idea of such a system would be more appealing.

9. CONCLUDING REMARKS

The 'theory of immune class regulation' might appear at first sight to be rather extravagant in its complexity and assumptions. There are, however, only three new elements in the theory beyond what has already been postulated in the 'self–non-self discrimination theory'. The evidence obtained in the last three years makes the latter theory more reasonable as a basis for considering other properties of the immune system.

The three new elements are as follows. (1) Cell-mediated immunity is induced to antigens that have few foreign sites. This generalization, based on considerable evidence, has been stressed by Pearson and Raffel.[20] It leads to the idea that different classes of precursor cell require the generation of different numbers of signals ② to be induced. (2) The generation of too much help, i.e. of signals ②, can inhibit a response. This idea has been suggested by several people (e.g. Reference 49). (3) The induction of help can be repressed by a molecule whose induction is coordinately expressed with the induction of cell-mediated effector antibody. This idea was necessary to explain how stable cell-mediated immunity could occur. It may be relevant to an understanding of the dominant unresponsive states that have been recently discovered in increasing numbers.

The feature of the theory that is most attractive to me is the way in which diverse observations are unified in a manner that makes teleological sense. The predictions the theory makes as regards effector function are particularly important. The recent demonstration that IgG antibody-dependent cell-dependent killing requires a high number of sites on the target cell[68] is consistent with the theory, as is the evidence suggesting that cells with low amounts of H-2 can be killed by cytotoxic T cells, whose presence is assumed to be a manifestation of cell-mediated immunity, whereas they appear not to be lysed by IgG antibody and complement.[69] The theory postulates two types of specific inhibitory cell. Repressors, i.e. those cells able to mediate the generation of signal ③, can be preferentially induced when antigens that are able to provoke a humoral response are administered under relatively non-immunogenic conditions. Repressors inhibit the induction of IgG antibody more than IgM. Under some conditions helper T cells can inhibit the induction of some classes of immunity, and are then referred to as suppressors. The way in which helper T cells control the response is determined by the number of signals ② they generate. When few signals ② are generated cell-mediated immunity is induced; when a medium number is generated cell-mediated immunity is suppressed and IgM immunity is induced, and when a large number is generated both the induction of cell-mediated immunity and IgM antibody synthesis is suppressed and the induction of IgG synthesis occurs. With the exception of signal ① all specific signals, whether stimulatory or inhibitory, require the associative recognition of two sites on an antigen.

Should the teleological ideas behind the theory be substantiated by a deeper knowledge of the antigen recognition requirements in effector function they should be useful in understanding some pathological conditions and why the immune system is sometimes not effective. An example of such a speculation concerns the inefficiency of the immune system in rejecting some tumours. Some tumours can vary the extent to which they express their tumour specific antigens. A Moloney lymphoma line, for example, that is usually killed by Moloney-specific humoral antiserum in the presence of complement, was passaged through syngeneic mice immunized to Moloney-specific antigens. A stable subline was isolated from these mice that was not killed in the presence of complement by the Moloney-specific antiserum. This subline still expressed Moloney-specific surface antigens but at approximately one-tenth the level of the parental line.[70] The change in the amount of antigen on the surface of the cell could presumably lead to its being insensitive to IgG-dependent complement-mediated lysis as two molecules of IgG antibody, bound to sites close together on the surface of the cell, are required to obtain such lysis. These observations suggest, particularly within the context of the theory, that some tumour cells may escape effective immune surveillance if they can induce some humoral immunity and subsequently give rise to variants that are resistant to humoral effector mechanisms. It seems plausible that some tumours arise with a high density of tumour specific antigens and are able to induce t cells sufficiently so that when variants with a low density of antigen occur they can induce humoral antibody which is ineffective in damaging them. If the presence of humoral antibody can inhibit the effector function of cell-mediated immunity, as suggested by the Hellströms,[71] the tumour would progress. Such a view of the way by which some tumours might avoid rejection supports the idea that pre-immunization of an animal in a manner that favours the induction of cell-mediated immunity might aid in the efficient rejection of some tumours.

Acknowledgements

I am grateful to A. Cunningham and I. Ramshaw for their comments on the manuscript. Figures 1 to 5 of this chapter were reproduced from the author's paper in *Cellular Immunology*, **13**, 171 (1974), by permission of Academic Press Inc.

10. REFERENCES

1. Humphrey, J. and Dourmashkin, R. (1969). *Advan. Immunol.*, **11**, 75–116.
2. Bretscher, P. A. and Cohn, M. (1970). *Science*, **169**, 1042–1049.
3. Bretscher, P. A. (1972). *Transplant. Rev.*, **11**, 217–267.
4. Triplett, E. L. (1962). *J. Immunol.*, **89**, 505–510.
5. Feldmann, M., Kilburn, D. and Kevy, T. (1975). *Nature*, **256**, 751–743.
6. Bretscher, P. A. (1975). *Transplant. Rev.*, **23**, 37–48.
7. Schrader, J. W. (1974). *J. exp. Med.*, **139**, 1303–1316.
8. Hamilton, J. A. and Miller, J. F. A. P. (1973). *Eur. J. Immunol.*, **3**, 457–460.
9. Walters, C. S., Moorhead, J. W. and Claman, H. N. (1972). *J. exp. Med.*, **136**, 546–555.
10. Chiller, J. M., Havicht, G. S. and Weigle, W. O. (1970). *Proc. Nat. Acad. Sci.*, **65**, 551–556.
11. Katz, D. H., Davie, J. M., Paul, W. E. and Benacerraf, B. (1971). *J. exp. Med.*, **134**, 201–223.
12. Schrader, J. W. (1974). *Eur. J. Immunol.*, **4**, 20–24.
13. McCullagh, P. J. (1970). *J. exp. Med.*, **132**, 916–925.
14. Boyse, E. A., Bressler, E., Iritani, L. A. and Lardis, M. (1970). *Transplantation*, **9**, 339–341.
15. Kaplan, M. and Rakita, L. (1971). In *Immunological Diseases*, 2nd ed. (Samter, M., Ed.), Little, Brown and Co., Boston, pp. 1367–1384.
16. Seger, R., Rogers, K. and Catty, D. (1974). *Eur. J. Immunol.*, **4**, 524–526.
17. Bretscher, P. A. (1974). *Cell. Immunol.*, **13**, 171–195.
18. Cantor, H. and Boyse, E. A. (1975). *J. exp. Med.*, **141**, 1390–1399.
19. Lafferty, K. J., Misko, I. S. and Cooley, M. A. (1974). *Nature*, **249**, 275–276.
20. Pearson, M. N. and Raffel, S. (1971). *J. exp. Med.*, **133**, 494–505.
21. Parish, C. R. (1972). *Transplant. Rev.*, **13**, 35–66.
22. Parish, C. R. (1972). *Eur. J. Immunol.*, **2**, 143–151.
23. Salvin, S. B. (1958). *J. exp. Med.*, **107**, 109–124.
24. Taylor, R. B. and Wortis, H. H. (1968). *Nature*, **220**, 927–928.
25. Cheers, C. and Miller, J. F. A. P. (1972). *J. exp. Med.*, **136**, 1661–1665.
26. Grumet, F. C. (1972). *J. exp. Med.*, **135**, 110–125.

27. Ordal, J. C. and Grumet, F. C. (1972). *J. exp. Med.*, **136**, 1195–1206.
28. Klaus, G. G. B. and Humphrey, J. H. (1975). *Transplant. Rev.*, **23**, 105–118.
29. Asherson, G. L. and Stone, S. H. (1962). *Immunology*, **9**, 205–217.
30. Battisto, J. R. and Miller, J. (1962). *Proc. Soc. exp. Biol. Med.*, **111**, 111–115.
31. Borel, Y., Fanconnel, M. and Miescher, P. A. (1966). *J. exp. Med.*, **123**, 585–598.
32. Crowle, A. J. and Hu, C. C. (1966). *Clin. exp. Immunol.*, **1**, 322–335.
33. Dvorak, H. F., Billtoe, J. B., McCarthy, J. S. and Fax, H. M. (1965). *J. Immunol.*, **94**, 966–975.
34. Loewi, G. E., Holborow, E. J. and Temple, A. (1966). *Immunology*, **10**, 339–347.
35. Klinman, N. R. (1972). *J. exp. Med.*, **136**, 241–260.
36. Cantor, H. and Boyse, E. A. (1975). *J. exp. Med.*, **141**, 1376–1389.
37. Zembala, N. and Asherson, G. L. (1973). *Nature*, **244**, 227–228.
38. Wortis, H. H., Taylor, R. B. and Dresser, D. W. (1966). *Immunology*, **11**, 603–616.
39. Kettman, J. (1972). *Immunol. Commun.*, **1**, 289–295.
40. Liew, F. Y. and Parish, C. R. (1974). *J. exp. Med.*, **139**, 779–784.
41. Basten, A. (1975). In *Immunological Tolerance* (Katz, D. H. and Benacerraf, B., Eds.), Academic Press, New York, pp. 107–122.
42. Zolla, S. and Naor, D. (1974). *J. exp. Med.*, **140**, 648–654.
43. McCullagh, P. J. (1972). *Transplant. Rev.*, **12**, 180–197.
44. Gershon, R. K. and Kondo, K. (1970). *Immunology*, **18**, 723–737.
45. McCullagh, P. J. (1974). *Eur. J. Immunol.*, **4**, 540–545.
46. Mitchison, N. A. (1969). In *Regulation and the Antibody Response* (Cinader, B., Ed.), Thomas, Springfield, pp. 54–67, C.
47. Weber, G. and Kolsch, E. (1973). *Eur. J. Immunol.*, **3**, 767–772.
48. Kolsch, E., Mengersen, R. and Weber, C. (1974). In *The Immune System: Genes, Receptors, Signals* (Sercarz, E., Williamson, A. and Fox, C. F., Eds.), Academic Press, New York, pp. 447–453.
49. Okumura, K. and Tada, T. (1973). *Nature* (*New Biol.*), **245**, 180–182.
50. Munro, A. J. and Taussig, M. J. (1975). *Nature*, **256**, 103–106.
51. Green, I., Paul, W. E. and Benacerraf, B. (1966). *J. exp. Med.*, **123**, 859–879.
52. Mitchell, G. F., Grumet, F. C. and McDevitt, H. O. (1972). *J. exp. Med.*, **135**, 126–135.
53. Gershon, R. K., Maurer, P. H. and Merryman, C. F. (1973). *Proc. Nat. Acad. Sci.*, **70**, 250–254.
54. Kapp, J. A., Pierce, C. W., Scholssman, S. and Benacerraf, B. (1974). *J. exp. Med.*, **140**, 648–659.
55. Vaz, N. M., Phillips-Quagliata, J. M., Levine, B. B. and Vaz, E. M. (1971). *J. exp. Med.*, **134**, 1335–1348.
56. Melchers, I., Rajewsky, K. and Schreffler, D. C. (1973). *Eur. J. Immunol.*, **3**, 754–761.
57. Durnham, E. I., Dorf, M. E., Scheffler, D. C. and Benacerraf, B. (1973). *J. Immunol.*, **111**, 1621–1625.
58. Howard, J. G., Christie, G. H., Courtenay, B. M., Leuchars, B. and Davies, A. J. S. (1971). *Cell. Immunol.*, **2**, 614–626.
59. Andersson, B. and Blomgren, H. (1971). *Cell. Immunol.*, **2**, 411–424.
60. Naor, D., Saltoun, R. and Falkenberg, F. (1975). *Eur. J. Immunol.*, **5**, 220–223.
61. Coutinho, A. and Möller, G. (1974). *Scand. J. Immunol.*, **3**, 133–146.
62. Howard, J. G., Christie, G. H. and Courtenay, B. M. (1971). *Proc. Roy. Soc. London B*, **178**, 417–438.
63. Baker, P. K., Prescott, B., Stashak, P. W. and Amsbaugh, D. F. (1974). In *The Immune System: Genes, Receptors, Signals* (Sercarz, E., Williamson, A. and Fox, C. F., Eds.), Academic Press, New York, pp. 415–430.
64. Kerbel, R. S. and Eidinger, D. (1972). *Eur. J. Immunol.*, **2**, 114–118.
65. Jerne, N. K. (1974). *Ann. Immunol. Inst. Pasteur.*, **125**c, 373–389.
66. Lindenmann, J. (1974). *Ann. Immunol. Inst. Pasteur.*, **124**c, 171–184.
67. Iverson, G. M. and Dresser, D. W. (1970). *Nature*, **227**, 274–276.
68. Wiedermann, G., Denk, H., Stemberger, H., Eckersforfer, R. and Tappeiner, G. (1975). *Cell. Immunol.*, **17**, 440–446.
69. Lesley, Jayne, Hyman, H. and Dennert, G. (1974). *J. Nat. Cancer Inst.*, **53**, 1759–1765.
70. Fenyö, E. M., Klein, E., Klein, G. and Sweich, K. (1968). *J. Nat. Cancer Inst.*, **40**, 69–89.
71. Hellström, K. E. and Hellström, I. (1970). *Annu. Rev. Microbiol.*, **24**, 373–398.

Note added in proof

Since this chapter was written, some experiments to test crucial features of the theory of immune class regulation have been performed (Ramshaw, I. A., Bretscher, P. A. and Parish, C. R., *Eur. J. Immunol.* (in press); Ramshaw, I. A., McKenzie, I. F. C., Bretscher, P. A. and Parish, C. R. (submitted)). These experiments show that mice making a strong humoral response to horse erythrocytes (HRBC) have T cells able to inhibit the *in vivo* induction of delayed hypersensitivity (DTH) to HRBC. These cells are specific, sensitive to treatment with complement and anti-θ or anti-Ly1 serum, but not anti-Ly2 or anti-Ia serum. They require the linked associative recognition of antigen to exert their inhibitory effect. All these observations are consistent with helper T cells suppressing the induction of DTH in the manner described by the theory.

Mice were also made specifically responsive at the humoral level to HRBC by treating them with HRBC and cyclophosphamide. Such mice show no defect in their B cell population. They harbour T cells that inhibit the induction of the humoral response that is usually mounted by normal spleen cells in the environment of an irradiated host. This inhibition is specific and abolished by treating the cells from unresponsive donors with complement and anti-θ, anti-Ly2 or anti-Ia serum but not anti-Ly1 serum. These mice, unresponsive at the humoral level, display high levels of DTH to HRBC.

These experiments define two classes of inhibitory T cell. One, induced under conditions where a strong humoral response is provoked, suppresses the induction of DTH and has the markers present on helper T cells. The other, present under conditions where a strong DTH response is provoked, can repress the induction of a humoral response, and has different surface markers from the DTH effector, helper and suppressor T cells. The markers on our repressor T cell appear to be the same as those on the inhibitory T cell found in mice made unresponsive at the humoral level to heterologous γ-globulins (Vadas, M. A., Miller, J. F. A. P., McKenzie, I. F. C., Chism, S. E., Shen, F. W., Boyse, E. A., Gamble, J. R. and Whitelaw, A. M. (1976) *J. Exp. Med.*, **144**, 10).

Index

31-2
48-9
166
176

206-248
293
348-391

48-9 *